CAMBRIDGE STUDIES IN
ADVANCED MATHEMATICS: 55
EDITORIAL BOARD
W. FULTON, D.J.H. GARLING, K. RIBET,
P. WALTERS

AUTOMORPHIC FORMS AND REPRESENTATIONS

CAMBRIDGE STUDIES IN ADVANCED MATHEMATICS 55

Already published

2 K. Petersen *Ergodic theory*
3 P.T. Johnstone *Stone spaces*
4 W.H. Schikhof *Ultrametric calculus*
5 J.-P. Kahane *Some random series of functions, 2nd edition*
7 J. Lambek & P.J. Scott *Introduction to higher-order categorical logic*
8 H. Matsumura *Commutative ring theory*
9 C.B. Thomas *Characteristic classes and the cohomology of finite groups*
10 M. Aschbacher *Finite group theory*
11 J.L. Alperin *Local representation theory*
12 P. Koosis *The logarithmic integral I*
13 A. Pietsch *Eigenvalues and s-numbers*
14 S.J. Patterson *An introduction to the theory of the Riemann zeta-function*
15 H.J. Baues *Algebraic homotopy*
16 V.S. Varadarajan *Introduction to harmonic analysis on semisimple Lie groups*
17 W. Dicks & M. Dunwoody *Groups acting on graphs*
18 L.J. Corwin & F.P. Greenleaf *Representations of nilpotent Lie groups and their applications*
19 R. Fritsch & R. Piccinini *Cellular structures in topology*
20 H. Klingen *Introductory lectures on Siegel modular forms*
21 P. Koosis *The logarithmic integral II*
22 M.J. Collins *Representations and characters of finite groups*
24 H. Kunita *Stochastic flows and stochastic differential equations*
25 P. Wojtaszczyk *Banach spaces for analysts*
26 J.E. Gilbert & M.A.M. Murray *Clifford algebras and Dirac operators in harmonic analysis*
27 A. Frohlich & M.J. Taylor *Algebraic number theory*
28 K. Goebel & W.A. Kirk *Topics in metric fixed point theory*
29 J.F. Humphreys *Reflection groups and Coxeter groups*
30 D.J. Benson *Representations and cohomology I*
31 D.J. Benson *Representations and cohomology II*
32 C. Allday & V. Puppe *Cohomological methods in transformation groups*
33 C. Soule, *et al.* *Lectures on Arakelov geometry*
34 A. Ambrosetti & G. Prodi *A primer of nonlinear analysis*
35 J. Palis & F. Takens *Hyperbolicity, stability and chaos at homoclinic bifurcations*
37 Y. Meyer *Wavelets and operators 1*
38 C. Weibel *An Introduction to Homological Algebra*
39 W. Bruns & J. Herzog Cohen *Macaulay rings*
40 V. Snaith *Explicit Brauer induction*
41 G. Laumon *Cohomology of Drinfeld modular varieties: Part I*
42 E.B. Davies *Spectral theory of differential operators*
43 J. Diestel, H. Jarchow & A. Tonge *Absolutely summing operators*
44 P. Mattila *Geometry of sets and measures in Euclidean spaces*
45 R. Pinsky *Positive harmonic functions and diffusion*
46 G. Tenenbaum *Introduction to analytic and probabilistic number theory*
47 C. Peskine *Complex projective geometry*
48 Y. Meyer & R. Coifman *Wavelets and Operators II*
49 R. Stanley *Enumerative combinatorics*
50 I. Porteous *Clifford algebras and the classical groups*

AUTOMORPHIC FORMS AND REPRESENTATIONS

DANIEL BUMP

Stanford University

PUBLISHED BY THE PRESS SYNDICATE OF THE UNIVERSITY OF CAMBRIDGE
The Pitt Building, Trumpington Street, Cambridge CB2 1RP

CAMBRIDGE UNIVERSITY PRESS
The Edinburgh Building, Cambridge CB2 2RU, United Kingdom
40 West 20th Street, New York, NY 10011-4211, USA
10 Stamford Road, Oakleigh, Melbourne 3166, Australia

© Cambridge University Press 1998

This book is in copyright. Subject to statutory exception
and to the provisions of relevant collective licensing agreements,
no reproduction of any part may take place without
the written permission of Cambridge University Press.

First published 1996
First paperback edition 1998

Library of Congress Cataloging-in-Publication Data is available.

A catalog record for this book is available from the British Library.

ISBN 0 521 55098 X hardback
ISBN 0 521 65818 7 paperback

Transferred to digital printing 2004

Contents

Preface

The theory of automorphic forms, rightly or wrongly, has a reputation of being difficult for the student. I felt that there was a need for a book that would present the subject in a style that was accessible yet based on complete proofs and revealed clearly the uniqueness principles that underlie the basic constructions. I have been lecturing on automorphic forms and representation theory at Stanford and the Mathematical Sciences Research Institute since 1990, and this book is the end result.

The level of this book is intermediate between an advanced textbook and a monograph. I hope that it will be interesting to experts as well as graduate students. Its aim is to cover a substantial portion of the theory of automorphic forms on $GL(2)$. Both the "classical" and "representation theoretic" viewpoints are covered.

There are significant omissions from my treatment, most seriously the Selberg trace formula. It has not been my aim to achieve complete coverage of the topics treated or to write a reference book. I feel that the existing reference material is adequate, and that it was not feasible to cover any single topic with the thoroughness I would have liked. I hope that the reader will begin studying the reference material (such as the Corvallis volume (Borel and Casselman, 1979) and above all Jacquet and Langlands (1970)) in the course of reading this book. If I have done my job well, the task of approaching Jacquet and Langlands should be made easier by the current volume.

I can imagine a useful sequel to this book. A second volume is therefore a possibility, but not for several years.

I would like to thank William Banks, Antonia Bluher, Aleksandr Brener, David Cardon, Jim Cogdell, Anton Deitmar, David Feldman, Solomon Friedberg, Masaaki Furusawa, Steve Gelbart, Tom Goetze, David Goldberg, Jiandong Guo, Jeffrey Hoffstein, Özlem Imamoglu, David Joyner, Chris Judge, Par Kurlberg, Annette Klute, David Manderscheid, Greg Martin, Andrei Paraschivescu, Ralph Phillips, Freydoon Shahidi, Tom Shemanske, Trask Stalnaker, Steve Rallis, Ken Ribet, Dinakar Ramakrishnan, Julie Roskies, San Cao Vo, James Woodson – and probably others I've forgotten – for helpful comments, corrections, discussions, or other feedback. Thanks also to Lauren Cowles of Cambridge University Press for her interest in the manuscript and for her guidance, to Ellen Tirpak and the staff at TechBooks for their expert treatment of the manuscript, to Reid Augustin for helping me set up my Linux machine, and to the MSRI for their help and support during 1994–1995. And thank you, my wife Kathi, and my parents Kenneth and Ellen Bump, for your support, which was always there when I needed it most.

Parts of this book were written with the support of the American Mathematical Society Centennial Research Fellowship and grants from the National Science Foundation.

Advice to the Reader

It has not been my intention to write a reference book on automorphic forms. I have cut corners in many places. For example, although I treat the representation theory of $GL(2, \mathbb{R})$ from the viewpoint of $(\mathfrak{g}, K)$-modules with some degree of completeness, I ignore $GL(2, \mathbb{C})$ completely. In laying the foundations for automorphic representations, I have concentrated on the cuspidal representations and given only indications of what happens with the continuous spectrum. I do discuss the theory of Eisenstein series but only in so far as necessary for my limited goals – I want to discuss the Rankin–Selberg method, and I want to show how the intertwining integrals that pervade the local theory arise from the constant terms of the Eisenstein series. My proofs of some foundational results in Section 3.3 are complete only when the ground field is $\mathbb{Q}$. I feel that despite these omissions, I have been able to treat my subject matter with some degree of depth.

In some places I have left details to the reader in the form of exercises. If the result of an exercise is required in the text, I have usually provided enough in the way of hints that the reader will be able to fill the gaps. Hints are enclosed in square brackets, sometimes explicitly labeled as such, sometimes not. Some exercises are trivial and some (those not needed for the text) are genuinely difficult.

Each of the four chapters is itself a complete course. The material of the four chapters is complementary, but each can be studied on its own. However, Chapter 3 should be read after Sections 1–4 of Chapter 2. Also, Chapter 3 makes use of results on the representation theory of $GL(2, F)$, where F is a non-Archimedean local field, whose proofs are postponed until Chapter 4. The reader may take these on faith during a first reading of Chapter 3. Chapter 3 may be more difficult than Chapter 4. Some readers will want to start with Chapter 4.

I have tried to write in a style that encourages the reader to skip around or to start in the middle. Definitions are sometimes repeated, and there is a lot of cross referencing.

Chapter 1 is written in the classical language. It is based on the paper of Doi and Naganuma (1969) which exhibits a rich variety of ideas and techniques. The chapter contains an introduction to the Langlands conjectures, which are taken up from a more sophisticated point of view later in the book. The phenomenon of *base change*, which was discovered by Doi and Naganuma, is an example of a *lifting* of automorphic forms. Such liftings are systematically predicted by the conjectures of Langlands. The method by which Doi and Naganuma proved their result is also suggestive – it is based on the Rankin–Selberg method and the converse theorem, a pleasant combination of techniques that is at the heart of the present-day project of Piatetski–Shapiro and his collaborators to prove liftings of automorphic forms from classical groups to $GL(n)$.

The reader approaching this subject for the first time might want to read Sections 1.1–1.4, 1.6, and 1.8, skimming Sections 1.5 and 1.7 before attacking Section 1.8. Sections 1.7, 1.9, and 1.10 are more difficult, and the details of how the approach to base change laid out in Section 1.8 are carried out are only important if you care about them. Be aware that the rest of the book is largely independent of Chapter 1.

There is an apparent dichotomy in our field between the "classical" and "representation theoretic" approaches. But in fact this dichotomy is illusory, and it is important for the worker to understand both languages. Although Chapter 1 could be skipped by the reader, it is included for a good reason. The adele group is a large and complex object, and we derive our intuition to a large extent from the example of $GL(2, \mathbb{R})$. Moreover, adelic statements often reduce in the end to classical ones.

Chapter 2 marks the introduction of representation theory into the study of automorphic forms. There are two types of classical automorphic forms, namely, Maass forms and modular forms. This dichotomy reflects the fact that the irreducible representations of $SL(2, \mathbb{R})$ fall into two main classes – the principal series and the discrete series. In Chapter 2, we study the representation theory of $GL(2, \mathbb{R})$ and the spectral theory of compact quotients of the upper half plane and make clear the relationship between these two topics.

Chapter 2 ends with two special results that are needed for Chapter 3, namely, the uniqueness of Whittaker models and a theorem of Harish–Chandra. We do not discuss the representation theory of $GL(2, \mathbb{C})$, though it is in many respects simpler than the representation theory of $GL(2, \mathbb{R})$.

In Chapter 3, we introduce the adeles and the modern approach to automorphic forms. We cover Tate's thesis, the discreteness of the cuspidal spectrum, the tensor product theorem of Flath, and, following Jacquet and Langlands (1970), the implications of the uniqueness of Whittaker models–the strong multiplicity one theorem and the construction of L-functions. We consider the standard L-functions, the Rankin–Selberg L-functions, and (briefly) the triple L-functions. We discuss the Eisenstein series in enough detail for our study of the Rankin–Selberg method, but we do not prove much about the continuous spectrum.

Chapter 4 is independent of the first three chapters, and some readers will want to begin with it. In contrast to Chapter 3, I tried for some measure of completeness in Chapter 4, which is devoted to the representation theory of $GL(2)$

over a local field. I begin the chapter with a long section on the representation theory of $GL(2)$ over a finite field, where complete proofs may be given in just a few pages. The exercises for Section 4.1 emphasize the Weil representation and the philosophy of cusp forms, where representations are built up from cuspidal atoms by parabolic induction. Turning to $GL(2)$ over a non-Archimedean local field, I follow Bernstein and Zelevinsky (1976) in emphasizing sheaves and distributions in the proofs, among other things. I prove the uniqueness of Whittaker models and the fact that the character of a representation is invariant under transpose. As in Bernstein and Zelevinsky (1976), I treat the character as a distribution, without establishing its nature as a locally integrable function. I give fairly complete discussions of the principal series and spherical representations, including the Macdonald formula for the spherical function and the explicit formula for the spherical Whittaker function, which I prove by the method of Casselman and Shalika (1980). I show how the local functional equations follow from a uniqueness principle and how supercuspidal representations may be constructed by means of the Weil representation. I also prove the multiplicativity of the local ϵ-factors by means of the Weil representation. My discussion of the Weil representation is based on Jacquet and Langlands (1970). The chapter ends with a discussion of the local Langlands conjectures.

The reader may take the chapters in any order. Because Chapter 3 contains forward references to Chapter 4 (which is independent of Chapter 3), the *logical* sequence of the chapters is 1, 2, 4, then 3. However, there is a rationale for putting Chapter 3 ahead of Chapter 4. This is that the *motivation* for many of the topics in Chapter 4 comes from automorphic forms.

I would appreciate any comments that you may have. I may be able to take them into account in revising this text some day. Particularly, if you find mistakes – whether typographical errors or more serious mathematical or historical mistakes – I would like to be informed. My e-mail address is `bump@math.stanford.edu`. I intend to maintain a list of errata on my web page at `http://math.stanford.edu/~bump`.

Prerequisites

We assume a basic knowledge of algebraic number theory, the representation theory of finite groups, and of Fourier analysis on locally compact abelian groups (Pontriagin duality). Moreover, we will assume the existence and basic properties of Haar measure on locally compact groups. On the other hand, we develop all the Lie theory and much of the functional analysis that we need from scratch, and the prerequisites from Fourier analysis are reviewed as they arise.

In Chapter 2, Section 9, we use the fact that a solution to an elliptic differential equation is analytic. In Chapter 3, Section 4, we expect the reader to consult Knapp and Vogan (1995) for the properties of a certain Hecke algebra of distributions.

Notations

We have attempted to write in a style that allows the reader to start in the middle if desired. Definitions are sometimes repeated, and there is a lot of cross referencing. We attempt to avoid "global" notations that are defined throughout the book. We did compromise on this point in the matter of matrix transposes. If g is a matrix, ${}^{\top}g$ is its transpose.

We may denote the identity matrix as either I or 1. We will often omit zero entries from a matrix. Thus

$$\begin{pmatrix} a & b \\ & d \end{pmatrix} \quad \text{means} \quad \begin{pmatrix} a & b \\ 0 & d \end{pmatrix}.$$

Occasionally, we will denote an "arbitrary" matrix entry with an asterisk. Thus for example

$$f\begin{pmatrix} y & * \\ & y^{-1} \end{pmatrix} = |y|^s$$

means that this identity is true regardless of the value of $*$.

"Almost all" means for all but finitely many.

If G is an affine algebraic variety, particularly an affine algebraic group, defined over a field F, and if A is a commutative ring containing F, we will denote by $G(A)$ or G_A the points of G with coordinates in A. Some affine varieties, such as $GL(n)$, are really defined over $\mathbb{Z}$. Thus $GL(n, A)$ is defined if A is any commutative $\mathbb{Z}$-algebra, that is, any commutative ring. We will encounter mostly affine algebraic varieties. Our notations for affine algebraic varieties are discussed in more detail at the beginning of Section 3.3.

A *character* of a group is a continuous homomorphism into the group of complex numbers of absolute value 1. A continuous homomorphism into $\mathbb{C}^{\times}$ is called a *quasicharacter*.

A *global field* F is by definition an algebraic number field, or else a "function field," which is a finitely generated field of transcendence degree 1 over a finite field. If v is a place of F, we will denote by F_v the completion of F at v. If v is non-Archimedean, we will denote the ring of integers in F by $\mathfrak{o}_v$, its maximal ideal by $\mathfrak{p}_v$, and the cardinality of $\mathfrak{o}_v/\mathfrak{p}_v$ by q_v. By ϖ_v, we will denote an arbitrarily selected generator of $\mathfrak{p}_v$. We will denote by $\text{ord} : F^\times \to \mathbb{Z}$ the valuation, so that $\text{ord}(\epsilon\varpi_v^m) = m$ when $\epsilon \in \mathfrak{o}_v^\times$.

Let F be a global field and A its adele ring. We will use the notation $a = (a_v)$ for elements of A. This notation means that for each place v of F, a_v is the vth component of the adele a. Thus $a_v \in F_v$ and $a_v \in \mathfrak{o}_v$ for almost all v. Similarly, if G is an affine algebraic variety, we will denote elements of $G(A)$ by $g = (g_v)$, where $g_v \in G(F_v)$ and g_v has coordinates in $\mathfrak{o}_v$ for almost all v. This notation requires some explanation. Let us assume that G is the locus in affine n-space of some system of equations. Then *a priori* an element of g is an n-tuple $(g^1, \cdots, g^n)$ with $g^i = (g_v^i) \in A$. For each v, we find that $g_v = (g_v^1, \cdots, g_v^n) \in G(F_v)$, and we write $g = (g_v)$.

If R is a topological ring, then for $a \in R$, we will denote by $|a|$ the module of the endomorphism $x \mapsto ax$ of R, namely, the factor by which this transformation multiplies the additive Haar measure. Thus if $R = \mathbb{R}$, then $|\ |$ is the usual absolute value, while if $R = \mathbb{C}$, then $|\ |$ is the square of the usual absolute value. Also for a topological ring R, we will denote the additive Haar measure by dx and the multiplicative Haar measure by $d^\times x$. These are subject to normalizations that will be discussed when they arise. We will use these notations if R is a local field or the adele ring of a global field.

If F is a local field, we will often denote by $\psi : F \to \mathbb{C}$ a fixed nontrivial character of the additive group F. Alternatively, if A is the adele ring of a global field F, we will denote by ψ a fixed nontrivial character of A that is trivial on F. In either case, we will normalize the additive Haar measure on F or A to be self-dual for Fourier transform with respect to ψ. That is, if f is a compactly supported continuous function on F or A, define its Fourier transform

$$\hat{f}(x) = \int f(y)\,\psi(xy)\,dy.$$

The Fourier inversion formula asserts that

$$\hat{\hat{f}}(x) = f(-x),$$

and this is true for a unique choice of Haar measure. This is the Haar measure that we will usually use. With this normalization in the global case, the Haar volume of the compact quotient A/F is one, as is proved in Proposition 3.1.3.

If G is a topological group, we will always denote the *left* Haar integral by $\int_G dg$.

The end-of-proof symbol is ■, and we will also use □ to indicate the end of the proof of a lemma that is interpolated in the middle of another proof.

1
Modular Forms

In this chapter, we will introduce the study of modular forms through the paper of Doi and Naganuma (1969). This paper gave one of the first historical examples of a *functorial lifting* of automorphic forms, a phenomenon now codified in Langlands' important functoriality conjecture. The paper also uses (following a suggestion of Shimura) a beautiful L-function technique based on the Rankin–Selberg method and the so-called converse theorem for $GL(2)$. Though this method has been somewhat eclipsed by the trace formula, it is still being used to good effect by I. Piatetski-Shapiro and his coworkers in constructing liftings from classical groups to $GL(n)$.

The unifying theme of this chapter and those that follow will be *L-functions*. Briefly, an L-function is a *Dirichlet series*, that is, a series of the form

$$\phi(s) = \sum_{n=1}^{\infty} c(n)\, n^{-s},$$

which has an *Euler product* and a *functional equation*. We may illustrate this with two examples. The first is the Riemann zeta function

$$\zeta(s) = \sum_{n=1}^{\infty} n^{-s},$$

which is convergent for $\mathrm{re}(s) > 1$ and has meromorphic continuation to all s, with just a simple pole at $s = 1$. The *Euler product* expresses this as a product over all primes p:

$$\zeta(s) = \prod_{p} \left(1 - \frac{1}{p^s}\right)^{-1}.$$

The *functional equation* asserts that if

$$\xi(s) = \pi^{-s/2}\, \Gamma\left(\frac{s}{2}\right) \zeta(s),$$

then

$$\xi(s) = \xi(1-s).$$

The second example is due to Ramanujan (1916), who defined a certain function $\tau(n)$ by equating the coefficients in the series

$$q \prod (1-q^n)^{24} = \sum_{n=1}^{\infty} \tau(n)\, q^n. \tag{0.1}$$

Thus $\tau(1) = 1$, $\tau(2) = -24$, $\tau(3) = 252$, etc. Ramanujan's intuition had led him to a function with very remarkable properties. To begin with, the coefficients are multiplicative: if $(n, m) = 1$, then $\tau(nm) = \tau(n)\,\tau(m)$. Ramanujan considered the Dirichlet series

$$L(s, \Delta) = \sum_{n=1}^{\infty} \tau(n)\, n^{-s}.$$

He conjectured, and Mordell (1917) later proved, that

$$L(s, \Delta) = \prod_p \left(1 - \tau(p)\, p^{-s} + p^{11-2s}\right)^{-1}.$$

This is the *Euler product*. As for the *functional equation*, $L(s, \Delta)$ has analytic continuation to all s, with no poles, and if we define

$$\Lambda(s, \Delta) = (2\pi)^{-s}\, \Gamma(s)\, L(s, \Delta),$$

we have

$$\Lambda(s, \Delta) = \Lambda(12 - s, \Delta).$$

The explanation of these formulas is connected with the fact that the coefficients $\tau(n)$ are the Fourier coefficients of a *modular form*. Let $z = x + iy$ where $x, y \in \mathbb{R}$, $y > 0$. Then with $q = e^{2\pi i z}$, (0.1) becomes a function of z:

$$\Delta(z) = e^{2\pi i z} \prod_{n=1}^{\infty} (1 - e^{2\pi i n z})^{24}.$$

This function, *Ramanujan's discriminant function* is a *modular form of weight 12*, which means that

$$\Delta\left(\frac{az+b}{cz+d}\right) = (cz+d)^{12}\, \Delta(z)$$

for $a, b, c, d \in \mathbb{Z}$, $ad - bc = 1$. It turns out that there are lots of modular forms besides Δ; associated with such modular forms are L-functions having Euler products and functional equations. We will study the modular forms by studying their L-functions.

We will begin by studying a more basic type of L-function than those associated with modular forms; these are the L-functions associated with Dirichlet characters, of which the Riemann zeta function may be regarded as the prototype. Proving the Euler products and functional equations of these L-functions is the subject of Section 1.1.

Sections 1.2–1.4 form a basic course in modular forms, culminating with the Hecke theory. There are many exercises, particularly in Section 1.3, and this part of the book could be studied by an advanced undergraduate. Section 1.5 has as its goal Weil's *converse theorem*, which allows us to assert the existence of an automorphic form if sufficiently many functional equations can be proved, and though this particular theorem is not needed in the remainder of the book, variants of it are applied in Sections 1.9 and 1.10 to prove the existence of automorphic forms. Still, the proof of Weil's theorem could be skipped without loss of continuity. Sections 1.6 and 1.7 introduce tools we need, the Rankin–Selberg method and Hilbert modular forms, and at the end of Section 1.7, the result of Doi and Naganuma is formulated. This result asserts that given a modular form for $SL(2, \mathbb{Z})$, there exists a Hilbert modular form – the *base change lift* – whose L-function is described in terms of the L-function of the given modular form for $SL(2, \mathbb{Z})$. Section 1.8 explains how this result fits into the very general functoriality conjecture of Langlands. To apply the converse theorem, the functional equations of many L-functions must be proved. Sections 1.9 and 1.10, which are more technical than the rest of the chapter, carry out this program by constructing certain nonholomorphic automorphic forms (the first application of a converse theorem). They then form Rankin–Selberg convolutions of these auxiliary forms with the given one in order to prove the functional equations that are needed to conclude (in the second application of a converse theorem) that the base change lift exists.

1.1 Dirichlet L-Functions

The results of this section will be generalized in Section 1.7, and again in Section 3.1.

Let N be an integer. A *Dirichlet character* modulo N is a function $\chi : \mathbb{Z} \to \mathbb{C}$, which is periodic with period N, such that

$$|\chi(n)| = \begin{cases} 1 & \text{if } (n, N) = 1, \\ 0 & \text{otherwise,} \end{cases}$$

and such that $\chi(nm) = \chi(n)\,\chi(m)$. To obtain a Dirichlet character, start with a character of the finite Abelian group $(\mathbb{Z}/N\mathbb{Z})^\times$, and extend it to a function on all of $\mathbb{Z}/N\mathbb{Z}$ by making it zero on the residue classes not prime to N; then compose this function with the canonical map $\mathbb{Z} \to \mathbb{Z}/N\mathbb{Z}$.

If $N_1 | N$, there are canonical maps $\mathbb{Z}/N\mathbb{Z} \to \mathbb{Z}/N_1\mathbb{Z}$ and $(\mathbb{Z}/N\mathbb{Z})^\times \to (\mathbb{Z}/N_1\mathbb{Z})^\times$. If χ is a Dirichlet character modulo N_1, we may take the corresponding character of $(\mathbb{Z}/N_1\mathbb{Z})^\times$, pull it back to $(\mathbb{Z}/N\mathbb{Z})^\times$, and obtain a

Dirichlet character modulo N. If χ is obtained this way from a character modulo a proper divisor N_1 of N, then χ is called *imprimitive*. If χ is not imprimitive, it is called *primitive*. If χ is primitive modulo N, we say that N is the *conductor* of χ.

If χ is a Dirichlet character, whether primitive or not, we may define an L-function

$$L(s,\chi) = \sum_{n=1}^{\infty} \chi(n)\, n^{-s}.$$

By comparison with the Riemann zeta function, it is absolutely convergent for $re(s) > 1$. It has an *Euler product*, as we can see as follows. We will prove that for $\mathrm{re}(s) > 1$, we have

$$L(s,\chi) = \prod_p \left(1 - \frac{\chi(p)}{p^s}\right)^{-1}. \tag{1.1}$$

Indeed, expanding a geometric series, each individual Euler factor

$$\left(1 - \frac{\chi(p)}{p^s}\right)^{-1} = \sum_{k=0}^{\infty} \chi(p)^k\, p^{-ks}.$$

Now for any positive integer n, there is exactly one way to write n^{-s} as a product of factors p^{-ks}; so if $n = \prod p_i^{k_i}$, on expanding

$$\prod_p \sum_{k=0}^{\infty} \chi(p)^k\, p^{-ks},$$

the coefficient of n^{-s} is

$$\prod_i \chi(p_i)^{k_i} = \chi(n).$$

Hence the right side of Eq. (1.1) is equal to the left side, as required.

Dirichlet L-functions have another, deeper property, namely *analytic continuation* and a *functional equation*, which may be regarded as aspects of the interplay between additive and multiplicative Fourier analysis. The functional equation only works well for primitive characters. The remainder of this section will be devoted to developing the functional equations of Dirichlet L-functions.

It is a useful property of primitive characters that there is a convenient way to interpolate the character, originally defined on $\mathbb{Z}$, to a smooth function on $\mathbb{R}$. To this end, we introduce *Gauss sums*.

Let χ be a primitive character modulo N. The *Gauss sum* $\tau(\chi)$ or τ_χ is defined by the formula

$$\tau(\chi) = \sum_{n \bmod N} \chi(n)\, e^{2\pi i n/N}. \tag{1.2}$$

We will prove that

$$\sum_{n \bmod N} \chi(n)\, e^{2\pi i n m/N} = \overline{\chi(m)}\, \tau(\chi). \tag{1.3}$$

There are two cases. If $(m, N) = 1$, we have $|\chi(m)| = 1$. Making the change of variables $n \to nm$, we see that

$$\tau(\chi) = \sum_{n \bmod N} \chi(nm)\, e^{2\pi i n m/N} = \chi(m) \sum_{n \bmod N} \chi(n)\, e^{2\pi i n m/N},$$

and multiplying by $\chi(m)^{-1} = \overline{\chi(m)}$, we obtain Eq. (1.3).

On the other hand, if $(m, N) > 1$, we have $\chi(m) = 0$, so it is sufficient to show the left side of Eq. (1.3) vanishes. Suppose that $m = dM$ and $N = dN_1$ where $d > 1$. Let us show first that the primitivity of χ implies that there exists $c \equiv 1 \bmod N_1$ such that $(c, N) = 1$ and $\chi(c) \neq 1$. If not, $\chi(c) = 1$ for all c prime to N such that $c \equiv 1 \bmod N_1$, which implies that $\chi(n) = \chi(n')$ whenever n, n' are prime to N and $n \equiv n' \bmod N_1$. Hence χ is well defined modulo N_1, a proper divisor of N, and χ is the pullback of a character under the canonical map $(\mathbb{Z}/N\mathbb{Z})^\times \to (\mathbb{Z}/N_1\mathbb{Z})^\times$, contradicting the primitivity of χ. Now observe that the left side of Eq. (1.3) equals

$$\sum_{n \bmod N} \chi(n)\, e^{2\pi i n M/N_1} = \sum_{r \bmod N_1} \left\{ \sum_{\substack{n \bmod N \\ n \equiv r \bmod N_1}} \chi(n) \right\} e^{2\pi i r M/N_1}. \tag{1.4}$$

Now the substitution $n \to cn$ permutes the residue classes $n \bmod N$ such that $n \equiv r \bmod N_1$ amongst themselves, so

$$\sum_{\substack{n \bmod N \\ n \equiv r \bmod N_1}} \chi(n) = \sum_{\substack{n \bmod N \\ n \equiv r \bmod N_1}} \chi(cn) = \chi(c) \sum_{\substack{n \bmod N \\ n \equiv r \bmod N_1}} \chi(n);$$

as $\chi(c) \neq 1$, this expression must equal zero. Hence Eq. (1.4) vanishes, completing the proof of Eq. (1.3).

Now we need to know that $\tau(\chi) \neq 0$. In fact, we will prove that

$$|\tau(\chi)| = \sqrt{N}. \tag{1.5}$$

Because

$$\overline{\chi(n)\, e^{2\pi i n m/N}} = \overline{\chi(n)}\, e^{-2\pi i n m/N},$$

we have

$$\left| \sum_{n \bmod N} \chi(n)\, e^{2\pi i n m/N} \right|^2 = \sum_{\substack{n_1, n_2 \bmod N \\ (n_1 n_2, N) = 1}} \chi(n_1)\overline{\chi(n_2)} e^{2\pi i (n_1 - n_2) m/N}.$$

By Eq. (1.3), this equals $|\tau(\chi)|^2$ if $(m, N) = 1$ and zero if $(m, N) \neq 1$. Summing over all m modulo N, we obtain

$$\phi(N)\,|\tau(\chi)|^2 = \sum_{m \bmod N} \sum_{\substack{n_1, n_2 \bmod N \\ (n_1 n_2, N) = 1}} \chi(n_1)\overline{\chi(n_2)} e^{2\pi i (n_1 - n_2) m/N},$$

where ϕ is the Euler totient function: $\phi(N)$ is the number of residue classes modulo N prime to N, or in other words, the cardinality of $(\mathbb{Z}/N\mathbb{Z})^\times$. Because

$$\sum_{m \bmod N} e^{2\pi i a m/N} = \begin{cases} N & \text{if } N|a, \\ 0 & \text{otherwise,} \end{cases}$$

only terms with $n_1 \equiv n_2 \bmod N$ contribute; for these, $\chi(n_1)\overline{\chi(n_2)} = 1$, so we obtain

$$\phi(N)\,|\tau(\chi)|^2 = \sum_{\substack{n_1, n_2 \bmod N \\ (n_1 n_2, N) = 1 \\ n_1 \equiv n_2 \bmod N}} N = \phi(N)\,N.$$

Hence we have Eq. (1.5).

Now that we know the Gauss sums do not vanish, we may explain how to interpolate a primitive Dirichlet character between the integers. Replace χ by $\overline{\chi}$ and rewrite Eq. (1.3) as follows:

$$\chi(n) = \frac{1}{\tau(\overline{\chi})} \sum_{m \bmod N} \overline{\chi(m)}\, e^{2\pi i n m/N}, \tag{1.6}$$

so by Eq. (1.5) and Exercise 1.1.1, we have

$$\chi(n) = \frac{\chi(-1)\,\tau(\chi)}{N} \sum_{m \bmod N} \overline{\chi(m)}\, e^{2\pi i n m/N}. \tag{1.7}$$

Observe that the right-hand side is defined when n is an arbitrary real number. We have therefore obtained our goal of finding a natural way of interpolating a primitive Dirichlet character to an arbitrary real argument.

We now require the *Poisson summation formula.* Let f be a function on $\mathbb{R}$ that is sufficiently well behaved. For example, it is sufficient if f is piecewise continuous with only finitely many discontinuities, of bounded total variation, satisfies

$$f(a) = \tfrac{1}{2}\left[\lim_{x \to a-} f(x) + \lim_{x \to a+} f(x)\right]$$

for all a, and

$$|f(x)| < c_1 \min(1, x^{-c_2})$$

for some $c_1 > 0$, $c_2 > 1$. We define the *Fourier transform*

$$\hat{f}(x) = \int_{-\infty}^{\infty} f(y)\, e^{2\pi i x y}\, dy. \tag{1.8}$$

The Poisson summation formula asserts that

$$\sum_{n=-\infty}^{\infty} f(n) = \sum_{n=-\infty}^{\infty} \hat{f}(n). \tag{1.9}$$

To prove this, let

$$F(x) = \sum_{n=-\infty}^{\infty} f(x+n).$$

Then $F(x)$ is periodic with period 1, is of bounded variation, and satisfies

$$F(a) = \tfrac{1}{2}\left[\lim_{x \to a-} F(x) + \lim_{x \to a+} F(x)\right].$$

It is a standard theorem from Fourier analysis that $F(x)$ has a Fourier expansion that represents it for all values of x:

$$F(x) = \sum_{-\infty}^{\infty} a_m\, e^{2\pi i m x}.$$

(This follows, for example, from Whittaker and Watson (1927, 9.42 on p. 175)) The constants a_m are computed in the usual way, by orthogonality:

$$a_m = \int_0^1 F(x)\, e^{-2\pi i m x}\, dx = \int_0^1 \sum_{n=-\infty}^{\infty} f(x+n)\, e^{-2\pi i m x}\, dx.$$

Because $e^{2\pi i m x} = e^{2\pi i m(x+n)}$, this equals

$$\sum_{n=-\infty}^{\infty} \int_0^1 f(x+n)\, e^{-2\pi i m(x+n)}\, dx = \int_{-\infty}^{\infty} f(x)\, e^{-2\pi i m x}\, dx,$$

so $a_m = \hat{f}(-m)$. Thus

$$\sum_{n=-\infty}^{\infty} f(n) = F(0) = \sum_{n=-\infty}^{\infty} a_n = \sum_{n=-\infty}^{\infty} \hat{f}(n),$$

as required.

Actually, we require a slight generalization of the Poisson summation formula, which we call *twisted Poisson summation*; the formula is like the usual

Poisson formula, except that it is "twisted" by a Dirichlet character. Let χ be a primitive character modulo N. We will prove that

$$\sum_{n=-\infty}^{\infty} \chi(n)\, f(n) = \frac{\chi(-1)\,\tau(\chi)}{N} \sum_{n=-\infty}^{\infty} \overline{\chi(n)}\, \hat{f}(n/N). \tag{1.10}$$

To prove this, let us observe that by Eq. (1.7), the left side of Eq. (1.10) equals $\sum f_1(n)$, where

$$f_1(x) = \frac{\chi(-1)\,\tau(\chi)}{N} \sum_{m \bmod N} \overline{\chi(m)}\, e^{2\pi i x m/N}\, f(x).$$

We may thus apply the Poisson summation formula. It is easy to check that

$$\hat{f}_1(x) = \frac{\chi(-1)\,\tau(\chi)}{N} \sum_{m \bmod N} \overline{\chi(m)}\, \hat{f}\left(\frac{Nx+m}{N}\right).$$

Thus the left side of Eq. (1.10) equals

$$\frac{\chi(-1)\,\tau(\chi)}{N} \sum_{m \bmod N} \sum_{n=-\infty}^{\infty} \overline{\chi(m)}\, \hat{f}\left(\frac{Nn+m}{N}\right).$$

Because $\chi(m) = \chi(Nn+m)$, and because $Nn+m$ runs uniquely through $\mathbb{Z}$ when m runs through a set of residue classes modulo N and n runs through $\mathbb{Z}$, this equals the right side of Eq. (1.10). This completes the proof of the twisted Poisson summation formula.

Now we need to compute the Fourier transform of the *Gaussian distribution*. Let t have positive real part, and let

$$f_t(x) = e^{-\pi t x^2}. \tag{1.11}$$

Then f_t is of faster-than-polynomial decay as $x \to \pm\infty$. We will prove that

$$\hat{f}_t = \frac{1}{\sqrt{t}}\, f_{1/t}. \tag{1.12}$$

Because both sides are analytic functions of t defined when the real part of t is positive, it is sufficient to prove Eq. (1.12) when t is real, which we now assume. Completing the square

$$\hat{f}_t(x) = \int_{-\infty}^{\infty} e^{-\pi(ty^2-2ixy)}\, dy = e^{-\pi x^2/t} \int_{-\infty}^{\infty} e^{-\pi(\sqrt{t}\,y - ix/\sqrt{t})^2}\, dy.$$

Now we may use Cauchy's theorem to shift the path of integration with respect to y vertically by a constant distance depending on x, amounting to replacing

y by $y + ix/t$. Thus

$$\hat{f}_t(x) = e^{-\pi x^2/t} \int_{-\infty}^{\infty} e^{-\pi t y^2}\, dy = \frac{c}{\sqrt{t}} e^{-\pi x^2/t}, \tag{1.13}$$

where

$$c = \int_{-\infty}^{\infty} e^{-\pi y^2}\, dy.$$

It is well known that $c = 1$; in fact, we may prove that right now by applying Eq. (1.13) twice to obtain

$$\hat{\hat{f}}_t = \frac{c}{\sqrt{t}} \hat{f}_{1/t} = c^2 f_t.$$

We recall the *Fourier inversion formula*

$$\hat{\hat{f}}(x) = f(-x),$$

which is valid if f is any continuous function such that both f and $\hat{f}$ are in $L^1(\mathbb{R})$. (See Katznelson, 1976, VI.1.12.) Applying this to the even function f_t, we have $\hat{\hat{f}}_t = f_t$. Therefore $c^2 = 1$. Evidently, $c > 0$, and so $c = 1$, from which we get Eq. (1.13).

Now we may construct some *theta functions*. Let χ be a primitive character modulo N. First assume that $\chi(-1) = 1$, and define

$$\theta_\chi(t) = \tfrac{1}{2} \sum_{n=-\infty}^{\infty} \chi(n)\, e^{-\pi n^2 t} = \tfrac{1}{2}\chi(0) + \sum_{n=1}^{\infty} \chi(n)\, e^{-\pi n^2 t} \tag{1.14}$$

when t has positive real part. (Here $\chi(0) = 0$ unless $N = 1$ because χ is primitive.) Using Eq. (1.12) and the twisted Poisson summation formula, we obtain the functional equation:

$$\theta_\chi(t) = \frac{\tau(\chi)}{N\sqrt{t}} \theta_{\overline{\chi}}\left(\frac{1}{N^2 t}\right). \tag{1.15}$$

Now suppose that $\chi(-1) = -1$. In this case, we cannot use the definition Eq. (1.14); the terms for n and $-n$ cancel, so Eq. (1.14) is zero in this case. So we must do something else. Let

$$g_t(x) = x\, e^{-\pi t x^2} = -\frac{1}{2\pi t}\left(\frac{df_t}{dx}\right)(x). \tag{1.16}$$

We will prove that

$$\hat{g}_t(x) = \frac{i}{t^{3/2}}\, g_{1/t}. \tag{1.17}$$

We have, by definition of the Fourier transform

$$\hat{g}_t(x) = -\frac{1}{2\pi t} \int_{-\infty}^{\infty} \left(\frac{df_t}{dy}\right)(y)\, e^{2\pi ixy}\, dy.$$

Integrating by parts, this equals

$$\frac{ix}{t} \int_{-\infty}^{\infty} f_t(y)\, e^{2\pi ixy}\, dy,$$

which by Eq. (1.12) equals the right side of Eq. (1.17), which is now proved. Now if $\chi(-1) = -1$, we define

$$\theta_\chi(t) = \tfrac{1}{2} \sum_{n=-\infty}^{\infty} n\, \chi(n)\, e^{-\pi n^2 t} = \sum_{n=1}^{\infty} n\, \chi(n)\, e^{-\pi n^2 t}. \tag{1.18}$$

Applying twisted Poisson summation to g_t, we obtain

$$\theta_\chi(t) = \frac{-i\, \tau(\chi)}{N^2\, t^{3/2}}\, \theta_{\overline{\chi}}\left(\frac{1}{N^2 t}\right). \tag{1.19}$$

Now we may prove the functional equations of Dirichlet L-functions.

Theorem 1.1.1 *Let χ be a primitive Dirichlet character with conductor N, and let $\epsilon = 0$ or 1 be chosen so that $\chi(-1) = (-1)^\epsilon$. Let*

$$\Lambda(s, \chi) = \pi^{-(s+\epsilon)/2}\, \Gamma\left(\frac{s+\epsilon}{2}\right) L(s, \chi).$$

Then $\Lambda(s, \chi)$ has meromorphic continuation to all s; indeed, if $\chi \neq 1$, it is entire, while if $\chi = 1$, it is analytic for all s except $s = 1$ or $s = 0$, where it has simple poles. We have the functional equation

$$\Lambda(s, \chi) = (-i)^\epsilon\, \tau(\chi)\, N^{-s}\, \Lambda(1 - s, \overline{\chi}). \tag{1.20}$$

Proof We will consider the case where $\chi \neq 1$. If χ is the trivial character, then the primitivity of χ implies that $N = 1$, and $L(s, \chi)$ is just the Riemann zeta function; we leave this case to the reader (Exercise 1.7). Because $\chi(0) = 0$, the series $\theta_\chi(t)$ is a sum of terms of the form $n^\epsilon\, \chi(n)\, e^{-\pi n^2 t}$ with $|n| \geq 1$, each of which is of very rapid decay as $t \to \infty$; combining Eqs. (1.15) and (1.19), we have

$$\theta_\chi(t) = \frac{(-i)^\epsilon\, \tau(\chi)}{N^{1+\epsilon}\, t^{\epsilon+1/2}}\, \theta_{\overline{\chi}}\left(\frac{1}{N^2 t}\right);$$

this implies that $\theta_\chi(t)$ is also of very rapid decay as $t \to 0$. Therefore the Mellin transform

$$\int_0^\infty \theta_\chi(t)\, t^{(s+\epsilon)/2} \frac{dt}{t} \tag{1.21}$$

is convergent for all s. If the real part of $s > 1$, we may use the identity

$$\int_0^\infty e^{-\pi t n^2} t^{(s+\epsilon)/2} \frac{dt}{t} = \pi^{-(s+\epsilon)/2} \Gamma\left(\frac{s+\epsilon}{2}\right) n^{-s-\epsilon}$$

to see that Eq. (1.21) equals $\Lambda(s, \chi)$. Because the integral Eq. (1.21) is convergent for all s, and clearly defines an analytic function, this gives the analytic continuation of $\Lambda(s, \chi)$. Now substituting Eq. (1.15) or Eq. (1.19) into Eq. (1.21) and making the change of variables $t \mapsto 1/N^2 t$, Eq. (1.21) equals

$$(-i)^\epsilon\, \tau(\chi)\, N^{-s} \int_0^\infty \theta_{\overline{\chi}}(t)\, t^{(1-s+\epsilon)/2} \frac{dt}{t}.$$

Hence we obtain Eq. (1.20). ■

Exercises

Exercise 1.1.1 Let χ be a primitive character modulo N. Show that $\tau(\overline{\chi}) = \chi(-1)\,\overline{\tau(\chi)}$.

Exercise 1.1.2: Dirichlet (a) Show that the identity

$$\sum_{n=1}^\infty \frac{x^n}{n} = -\log(1-x),$$

valid if $|x| < 1$, remains true if $|x| = 1$ and $x \neq 1$, in which case the series is conditionally convergent.

(b) Let χ be a nontrivial primitive character modulo N. Assume $N > 1$, so that χ is nontrivial. Use Eq. (1.7) to prove that

$$L(1, \chi) = \sum_{n=1}^\infty \frac{\chi(n)}{n} = -\frac{\chi(-1)\,\tau(\chi)}{N} \sum_{m \bmod N} \overline{\chi(m)} \log(1 - e^{2\pi i m/N}). \tag{1.22}$$

From this, deduce that

$$L(1,\chi) = \begin{cases} -\dfrac{\tau(\chi)}{N} \displaystyle\sum_{m \bmod N} \overline{\chi(m)} \log|1 - e^{2\pi i m/N}| & \text{if } \chi(-1) = 1; \\ \dfrac{i\pi\tau(\chi)}{N^2} \displaystyle\sum_{m=1}^{N} \overline{\chi(m)}\, m & \text{if } \chi(-1) = -1. \end{cases}$$

Recall that for the character χ modulo N to be *quadratic* means that $\chi(n) = \pm 1$ for all $(n, N) = 1$, but that χ is not identically one.

Exercise 1.1.3 Let p be an odd prime.

(a) Prove that there is a unique quadratic character χ modulo p.
(b) Prove that the number of solutions to $x^2 \equiv a \bmod p$ equals $1 + \chi(a)$.
(c) Show that

$$\tau(\chi) = \sum_{n=0}^{p-1} e^{2\pi i n^2/p}.$$

Exercise 1.1.4 Let $\tau = x + iy$, where $x, y \in \mathbb{R}$ and $y > 0$. Let k be an integer greater than or equal to 2. Define

$$f(u) = (u - \tau)^{-k}.$$

Use the residue theorem to show that

$$\hat{f}(v) = \begin{cases} 2\pi i \ \mathrm{res}\big(e^{2\pi i u v}\,(u-\tau)^{-k}\big)\,|_{u=\tau} & \text{if } v > 0; \\ 0 & \text{if } v \le 0. \end{cases}$$

Hence

$$\hat{f}(v) = \begin{cases} \frac{(2\pi i)^k}{(k-1)!}\, v^{k-1}\, e^{2\pi i v \tau} & \text{if } v > 0; \\ 0 & \text{if } v \le 0. \end{cases}$$

Conclude that

$$\sum_{n=-\infty}^{\infty} (n - \tau)^{-k} = \frac{(2\pi i)^k}{(k-1)!} \sum_{n=1}^{\infty} n^{k-1}\, e^{2\pi i n \tau}.$$

The Quadratic Reciprocity Law I will state without proof this fundamental theorem of Gauss. If p is an odd prime, the *Legendre symbol*

$$\left(\frac{a}{p}\right) = \begin{cases} 1 & \text{if } x^2 \equiv a \bmod p \text{ has two solutions;} \\ -1 & \text{if } x^2 \equiv a \bmod p \text{ has no solutions;} \\ 0 & \text{if } a \equiv 0 \bmod p. \end{cases}$$

The definition is extended so that $\left(\frac{a}{b}\right)$ is defined whenever b is an odd positive number by the rule

$$\left(\frac{a}{\prod_{i=1}^n p_i}\right) = \prod_{i=1}^n \left(\frac{a}{p_i}\right)$$

where $p_1, \cdots, p_n$ are primes. Extended in this way, the symbol is called the *Jacobi symbol*. The basic properties of the Jacobi symbol are as follows:

(i) $$\left(\frac{a}{b}\right) = \left(\frac{a'}{b}\right) \qquad \text{if } a \equiv a' \bmod b.$$

(ii) $$\left(\frac{aa'}{b}\right) = \left(\frac{a}{b}\right)\left(\frac{a'}{b}\right).$$

(iii) $$\left(\frac{a}{bb'}\right) = \left(\frac{a}{b}\right)\left(\frac{a}{b'}\right).$$

(iv) $$\left(\frac{-1}{b}\right) = (-1)^{(b-1)/2} = \begin{cases} 1 & \text{if } b \equiv 1 \bmod 4; \\ -1 & \text{if } b \equiv -1 \bmod 4. \end{cases}$$

(v) $$\left(\frac{2}{b}\right) = (-1)^{(b^2-1)/8} = \begin{cases} 1 & \text{if } b \equiv \pm 1 \bmod 8; \\ -1 & \text{if } b \equiv \pm 3 \bmod 8. \end{cases}$$

If a and b are *both* odd positive integers, we have

(vi) $$\left(\frac{a}{b}\right) = (-1)^{\frac{1}{2}(a-1)\cdot\frac{1}{2}(b-1)}\left(\frac{b}{a}\right) = \begin{cases} -\left(\frac{b}{a}\right) & \text{if } a \equiv b \equiv -1 \bmod 4; \\ \left(\frac{b}{a}\right) & \text{otherwise.} \end{cases}$$

Part (vi) is the *quadratic reciprocity law.*

Exercise 1.1.5: Quadratic Fields If K is a quadratic extension of $\mathbb{Q}$, let $\mathfrak{o}_K$ denote the ring of integers in K. Then $\mathfrak{o}_K \cong \mathbb{Z} \oplus \mathbb{Z}$ as an Abelian group. Let α, β be a $\mathbb{Z}$-basis of $\mathfrak{o}_K$. The *discriminant* of K is by definition

$$D_K = \begin{vmatrix} \alpha & \beta \\ \alpha' & \beta' \end{vmatrix}^2,$$

where $x \mapsto x'$ denotes conjugation, that is, the nontrivial Galois automorphism of K over $\mathbb{Q}$. Show that this definition is independent of the choice of basis and that $D_K \in \mathbb{Z}$.

Exercise 1.1.6: Fundamental Discriminants Part (c) of this exercise assumes the quadratic reciprocity law, and part (d) assumes the definition of a discriminant of a quadratic field.

(a) Prove that if q is a prime power, then there exists a primitive quadratic character modulo q if and only if q equals 4, 8, or is an odd prime; in each of these cases there is precisely one primitive quadratic character, except that if $q = 8$, there are two, one satisfying $\chi(-1) = 1$ and one satisfying $\chi(-1) = -1$.

(b) Show that if $(m, n) = 1$, then

$$(\mathbb{Z}/mn\mathbb{Z})^\times \cong (\mathbb{Z}/m\mathbb{Z})^\times \times (\mathbb{Z}/n\mathbb{Z})^\times,$$

and deduce that there exists a primitive quadratic Dirichlet character modulo a positive integer d if and only if d is the product of relatively prime factors, each of which is an odd prime, or else equals 4 or 8.
(c) Show that if D is an integer, positive or negative, and if there exists a quadratic character χ that is primitive modulo $|D|$ such that the sign of D is equal to $\chi(-1)$, then

$$\chi(n) = \left(\frac{D}{n}\right)$$

for odd positive integers n.

The integers D satisfying the condition of (c) are called *fundamental discriminants*. They are in one-to-one correspondence with the primitive quadratic characters.

The restriction to odd n in (c) is undesirable; it is sometimes removed by employing Kronecker's modification of the Jacobi symbol, in which $\left(\frac{a}{b}\right)$ is sometimes defined even when b is even. Using the Kronecker symbol, one may say that the unique quadratic character modulo $|D|$, where D is a fundamental discriminant, is $n \mapsto \left(\frac{D}{n}\right)$. This has some disadvantages; for example, (i) above is no longer true for the Kronecker symbol. (Shimura (1973) has proposed yet another modification of the quadratic symbol.) We will avoid using the Kronecker symbol. If D is a fundamental discriminant, we will denote the unique primitive quadratic character modulo $|D|$ such that $\chi(-1)$ has the same sign as D by χ_D.
(d) Prove if K is a quadratic extension of $\mathbb{Q}$, there exists a unique fundamental discriminant D such that $K = \mathbb{Q}(\sqrt{D})$; thus the fundamental discriminants are in one-to-one correspondence with quadratic fields.
(e) Prove that if D is a fundamental discriminant, then $D \equiv 0$ or $1 \bmod 4$, and that the ring of integers in $K = \mathbb{Q}(\sqrt{D})$ is $\mathbb{Z} \oplus \mathbb{Z}\tau$ where

$$\tau = \begin{cases} \frac{1}{2}\sqrt{D} & \text{if } D \equiv 0 \bmod 4; \\ \frac{1}{2}(\sqrt{D}+1) & \text{if } D \equiv 1 \bmod 4. \end{cases}$$

Conclude that D is the discriminant of K. Hence the fundamental discriminants are precisely the discriminants of quadratic fields.
(f) Let D be a fundamental discriminant, p a prime, and $K = \mathbb{Q}(\sqrt{D})$. Show that

$$\begin{cases} p \text{ splits in } K & \text{if and only if } \chi_D(p) = 1; \\ p \text{ remains prime in } K & \text{if and only if } \chi_D(p) = -1; \\ p \text{ ramifies in } K & \text{if and only if } \chi_D(p) = 0. \end{cases}$$

(Use the quadratic reciprocity law. The case $p = 2$ must be handled separately.)

Exercise 1.1.7 Explain how to modify the proof of Theorem 1.1 to handle the case where $N = 1$, so that $\chi = 1$ and $L(s, \chi)$ is the Riemann zeta function.

Exercise 1.1.8: Riemann (1892) Riemann gave two proofs of the functional equation of ζ, each important in its own way. The proof based on taking the Mellin transform of a theta function as in the proof of Theorem 1.1 is Riemann's second proof. (It was extended to L-functions by Hecke (1918) and (1920)). This exercise, based on Riemann's first proof of the functional equation, leads to a determination of the values of $\zeta(s)$ at the negative odd integers or equivalently, at the positive even integers. Riemann's paper is discussed at length in Edwards (1974). For the extension to L-functions, see Chapter 4 of Washington (1982).
(a) The *Hankel Contour C* begins and ends at ∞, circling the origin counter-clockwise:

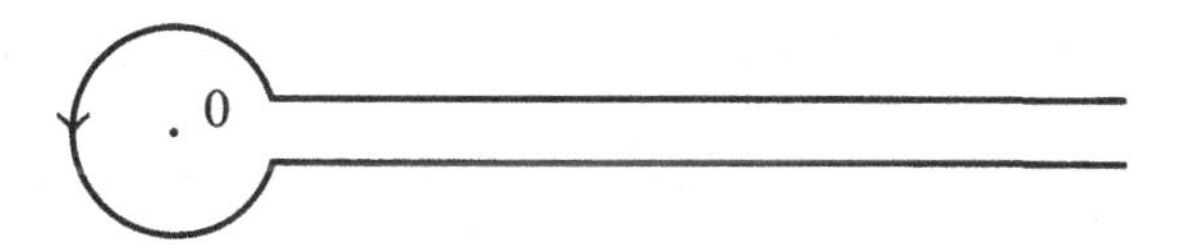

Prove that if $\mathrm{re}(s)$ is large

$$\int_C (-x)^{s-1} e^{-x}\, dx = -2i \sin(\pi s) \int_0^\infty t^{s-1} e^{-t}\, dt$$
$$= -2i \sin(\pi s)\, \Gamma(s).$$

In this integration, we define $(-x)^{s-1}$ to be $e^{(s-1)\log(-x)}$, where we choose the branch of log that is real when $(-x)$ is real and positive. In view of the well-known identity

$$\Gamma(s)\, \Gamma(1-s) = \frac{\pi}{\sin(\pi s)},$$

this may be rewritten

$$\frac{1}{\Gamma(1-s)} = \frac{i}{2\pi} \int_C (-x)^{s-1} e^{-x}\, dx.$$

Although we proved this only for $\mathrm{re}(s)$ large, observe that the integral is convergent for all s, so by analytic continuation, this formula is valid for all s.
(b) Use the geometric series identity

$$\frac{1}{e^x - 1} = \sum_{n=1}^{\infty} e^{-nx},$$

valid if $\mathrm{re}(x) > 0$, and adapt the calculation of (a) to show that

$$\zeta(s) = -\frac{\Gamma(1-s)}{2\pi i} \int_C \frac{(-x)^{s-1}}{e^x - 1}\, dx.$$

This formula is valid for all s.

The *Bernoulli numbers* are defined by the identity

$$\frac{t}{e^t - 1} = \sum_{n=0}^{\infty} B_n \frac{t^n}{n!}.$$

We have $B_0 = 1$, $B_1 = -1/2$, $B_2 = 1/6$, and $B_4 = -1/30$. It is not hard to see that $B_n = 0$ if n is odd and greater than 1; if n is even, it is clear that B_n is rational, and it will follow from (d) below that the sign of B_n is $-(-1)^{n/2}$.
(c) Use the functional equation to show that ζ vanishes at the negative even integers. Use the residue theorem to show that if n is a positive even integer, then $\zeta(1-n) = -B_n/n$.
(d) Use the functional equation to deduce that if n is a positive even integer

$$\zeta(n) = -\frac{2^{n-1}\,\pi^n\,(-1)^{n/2}\,B_n}{n!}.$$

Exercise 1.1.9 We return to the setting of Exercise 1.2(b). Assume that χ is quadratic, so its conductor $N = |D|$, where D is a fundamental discriminant. Then χ is the quadratic character attached to the quadratic extension $K = \mathbb{Q}(\sqrt{D})$. We recall the factorization of the Dedekind zeta function $\zeta_K(s) = \zeta(s)\, L(s, \chi)$ (see Lang (1970, Theorem XII.1, p. 230)). Thus $L(1, \chi)$ is the residue at $s = 1$ of ζ_K, which is computed classically as in Lang (1970, Theorem XIII.2, p. 259). Suppose that $\chi(-1) = -1$ so that K is imaginary quadratic. Then

$$L(1, \chi) = \frac{2\pi\, h}{w\sqrt{|D|}},$$

where D is the discriminant of K, h is its class number, and w is the number of roots of unity in K (two unless $D = -4$ or -3.) Thus by Exercise 1.2(b),

$$h = i\tau(\chi)\, w\, |D|^{-3/2}\, 2^{-1} \sum_{m=1}^{D} \chi(m)\, m.$$

But $\tau(\chi) = i\sqrt{D}$. (See Washington (1982, Corollary 4.6, p. 35) for the evaluation of quadratic Gauss sums. Also, compare Eq. (9.15) in section 1.9.) We obtain *Dirichlet's class number formula*

$$h = -\frac{w}{2|D|} \sum_{m=1}^{D} \chi(m)\, m. \tag{1.23}$$

Exercise 1.1.10 (a) Let χ be a primitive character modulo N. Prove, using the functional equation, that $L(s,\chi)$ has a simple zero at $s=0$ if $\chi(-1)=1$ and is nonzero at $s=0$ if $\chi(-1)=-1$.
(b) Stark (1971), (1975), (1976) and (1980) has conjectured that if ρ : $\mathrm{Gal}(\overline{\mathbb{Q}}/\mathbb{Q}) \to GL(n,\mathbb{C})$ is a Galois representation such that the Artin L-function $L(s,\rho)$ has a zero of order r at $s=0$, the leading coefficient in its Taylor expansion is essentially an $r\times r$ "Stark regulator" of units in some number field. The simplest open cases of the conjecture are when $r=1$. Artin's reciprocity law allows us to consider χ to be a Galois character, namely, it gives a character of $\mathrm{Gal}\big(\mathbb{Q}(\zeta)/\mathbb{Q}\big)\cong(\mathbb{Z}/N\mathbb{Z})^\times$, $\zeta=e^{2\pi i/N}$, where $a\in(\mathbb{Z}/N\mathbb{Z})^\times$ corresponds to $\sigma_a\in\mathrm{Gal}\big(\mathbb{Q}(\zeta)/\mathbb{Q}\big)$, $\sigma_a(\zeta)=\zeta^a$. With this identification, reinterpret (a) to show that $r=1$ if $\chi(-1)=1$, and $r=0$ if $\chi(-1)=-1$.
(c) Assume that $\chi(-1)=1$. In this case Exercise 1.2(b) verifies the Stark conjecture because it shows that

$$L(1,\chi)=-\frac{\tau(\chi)}{N}\sum_{m \bmod N}\overline{\chi(m)}\,\log|\epsilon_m|, \tag{1.24}$$

where $\epsilon_m=(1-e^{2\pi im/N})/(1-e^{2\pi i/N})$. Note that if m and N are coprime, ϵ_m is a unit in $\mathbb{Z}[\zeta]$.

1.2 The Modular Group

In this section, let $G=SL(2,\mathbb{R})$ and let $\mathcal{H}$ be the *Poincaré upper half plane* consisting of $z=x+iy$ where $x,y\in\mathbb{R}$, and $y>0$. G acts on $\mathcal{H}$ via linear fractional transformations:

$$g=\begin{pmatrix}a&b\\c&d\end{pmatrix}:z\to g(z)=\frac{az+b}{cz+d}. \tag{2.1}$$

It is easy to check that this is a bona fide group action, that is, $(g_1g_2)(z)=g_1\big(g_2(z)\big)$. This action is not quite faithful because $-I$ acts trivially, I being the identity matrix. If we wish to work with a group having a faithful action, we may pass to the group $PSL(2,\mathbb{R})=G/\{\pm I\}$, which may be identified with a group of transformations of $\mathcal{H}$. If $\Gamma\subset G$ is any group, we will denote by $\overline{\Gamma}$ its image in $\overline{G}=PSL(2,\mathbb{R})$. Thus $\overline{\Gamma}=\Gamma/\{\pm I\}$ if $-I\in\Gamma$, or $\overline{\Gamma}\cong\Gamma$ if $-I\notin\Gamma$. We may sometimes extend the action of $SL(2,\mathbb{R})$ to the group $GL(2,\mathbb{R})^+$ of 2×2 nonsingular matrices with positive determinant by the formula (2.1). Of course, the scalar matrices act trivially.

More generally, we allow $SL(2,\mathbb{C})$ (or $GL(2,\mathbb{C})$) to act on the Riemann sphere $\mathbb{P}^1(\mathbb{C})=\mathbb{C}\cup\{\infty\}$ by linear fractional transformations using the same formula (2.1). The subgroup of $SL(2,\mathbb{C})$ that maps the subspace $\mathcal{H}\subset\mathbb{P}^1(\mathbb{C})$ onto itself is just $SL(2,\mathbb{R})$.

The action of G on $\mathcal{H}$ is transitive because in fact the subgroup B of upper triangular matrices acts transitively. Indeed,

$$\begin{pmatrix} y^{1/2} & xy^{-1/2} \\ & y^{-1/2} \end{pmatrix} : i \to x + iy,$$

so every element of $\mathcal{H}$ is in the orbit of i. The stabilizer of i is the subgroup

$$SO(2) = \left\{ \begin{pmatrix} a & b \\ -b & a \end{pmatrix} \middle| a^2 + b^2 = 1 \right\}.$$

Recall that if X is any set, and if G is a group acting transitively on X, we may identify X with the set of cosets G/G_x, where G_x is the stabilizer of some fixed point $x \in X$. In the particular case where $G = SL(2, \mathbb{R})$, $X = \mathcal{H}$, and $x = i$, we see that we may identify $\mathcal{H}$ with the space of cosets $G/SO(2)$. Because, as we have just seen, B acts transitively on $\mathcal{H}$, it acts transitively on $G/SO(2)$, and so we obtain a geometric proof that $G = B \cdot SO(2)$. This relation is known as the *Iwasawa decomposition* for $SL(2, \mathbb{R})$.

We will be particularly interested in the subgroup $\Gamma(1) = SL(2, \mathbb{Z})$ of $SL(2, \mathbb{R})$ and certain subgroups that are called *congruence subgroups.* Let

$$\Gamma(N) = \left\{ \begin{pmatrix} a & b \\ c & d \end{pmatrix} \in \Gamma(1) \middle| a \equiv d \equiv 1 \bmod N, b \equiv c \equiv 0 \bmod N \right\}$$

Note that $\Gamma(N)$ is the kernel of the canonical map $\Gamma(1) \to SL(2, \mathbb{Z}/N\mathbb{Z})$. Because this is a finite group, we see that $\Gamma(N)$ is normal in $\Gamma(1)$ and of finite index. A subgroup of $SL(2, \mathbb{Z})$ is called a *congruence subgroup* if it contains $\Gamma(N)$ for some N.

The identity

$$\operatorname{im}\left(g(z)\right) = |cz + d|^{-2}\, y \tag{2.2}$$

$$\text{for } g = \begin{pmatrix} a & b \\ c & d \end{pmatrix} \in SL(2, \mathbb{R}), \quad z = x + iy \in \mathcal{H}$$

is readily confirmed (Exercise 1.2.1). If Γ is a subgroup of G, we say that the action of Γ on $\mathcal{H}$ is *discontinuous* if for any two compact subsets $K_1, K_2 \subset \mathcal{H}$, the set

$$\{\gamma \in \Gamma | K_2 \cap \gamma(K_1) \neq \emptyset\}$$

is finite. As a first application of Eq. (2.2), let us prove the following:

Proposition 1.2.1 *The group* $\Gamma(1)$ *acts discontinuously on* $\mathcal{H}$.

It may be shown more generally that a subgroup $\Gamma \subset SL(2, \mathbb{R})$ acts discontinuously on $\mathcal{H}$ if Γ is discrete in the topology that it inherits from $SL(2, \mathbb{R})$. For our purposes, however, Proposition 2.1 is sufficient.

Proof Let K_1 and K_2 be compact subsets of $\mathcal{H}$. There exists a constant $\epsilon > 0$ such that $\mathrm{im}(w) > \epsilon$ for all $w \in K_2$. Now for fixed $z = x + iy \in K_1$, note that $(c, d) \mapsto |cz + d|^2$ is a positive definite quadratic form. Applying Eq. (2.2), there is a constant $R(z)$ such that $\mathrm{im}\left(g(z)\right) = |cz+d|^{-2}\, y < \epsilon$ unless $|c|, |d| < R(z)$. Because K_1 is compact, $R = \max\{R(z) | z \in K_1\} < \infty$, and $K_2 \cap \gamma(K_1) = \emptyset$ unless $|c|, |d| < R$. This proves that there are only a finite number of possible bottom rows of $\gamma \in \Gamma(1)$ such that $K_2 \cap \gamma(K_1) \neq \emptyset$. We must therefore show that given c, d, there are only a finite number of possible γ with given bottom row (c, d) with $K_2 \cap \gamma(K_1) \neq \emptyset$. If γ_1 and γ_2 have the same bottom row, then $\gamma_2 = \gamma_0 \gamma_1$ where γ_0 has the form $\left(\begin{smallmatrix} 1 & n \\ & 1 \end{smallmatrix}\right)$. The effect of the matrix γ_0 is therefore translation by an integer distance n: $\gamma_0(z) = z + n$. For fixed γ_1, there can clearly be only finitely many γ_0 such that $\gamma_0\left(\gamma_1(K_1)\right) \cap K_2 \neq \emptyset$.

To summarize, there are only finitely many possible bottom rows (c, d) of γ such that $\gamma(K_1) \cap K_2 \neq \emptyset$, and for each (c, d) there are only finitely many possible γ with the prescribed bottom row such that $\gamma(K_1) \cap K_2 \neq \emptyset$. Hence the action of $\Gamma(1)$ is discontinuous. ■

As a second application of Eq. (2.2), we will determine a fundamental domain for Γ. If $\Gamma \subset SL(2, \mathbb{R})$ is a subgroup acting discontinuously on $\mathcal{H}$, a *fundamental domain* for Γ will be an open subset $F \subset \mathcal{H}$ such that (i) for every $z \in \mathcal{H}$, there exists $\gamma \in \Gamma$ such that $\gamma(z)$ is in the closure $\overline{F}$; and (ii) if $z_1, z_2 \in F$, and $\gamma(z_1) = z_2$ for some $\gamma \in \Gamma$, then $z_1 = z_2$, and $\gamma = \pm I$.

Let $F = \{z = x + iy \in \mathcal{H} | -\frac{1}{2} < x < \frac{1}{2},\ |z| > 1\}$.

Proposition 1.2.2 *The set F is a fundamental domain for $SL(2, \mathbb{Z})$.*

Proof Let $z \in \mathcal{H}$. Because $(c, d) \mapsto |cz + d|^2$ is a positive definite quadratic form, it has a minimum value as (c, d) runs through the pairs of relatively prime integers. It follows from Eq. (2.2) that $\mathrm{im}\left(\gamma(z)\right)$ has a maximum with $\gamma \in SL(2, \mathbb{Z})$, so let $\gamma \in \Gamma(1)$ maximize $\mathrm{im}\left(\gamma(z)\right)$. Now we can find $n \in \mathbb{Z}$ so that $\gamma(z) + n$ has a real part with absolute value $\leq 1/2$; replacing γ by $\left(\begin{smallmatrix} 1 & n \\ & 1 \end{smallmatrix}\right)\gamma$, we see that there exists γ such that $|\mathrm{re}\left(\gamma(z)\right)| \leq 1/2$ and $\mathrm{im}\left(\gamma(z)\right)$ is maximal for $\gamma \in SL(2, \mathbb{Z})$. This implies that $|\gamma(z)| \geq 1$, because otherwise the imaginary part of $\gamma_1(z)$ would be larger, where

$$\gamma_1 = \begin{pmatrix} & -1 \\ 1 & \end{pmatrix} \gamma, \qquad \mathrm{im}\left(\gamma_1(z)\right) = \frac{\mathrm{im}\ \gamma(z)}{|\gamma(z)|^2}.$$

This shows that every $\Gamma(1)$ orbit intersects the closure of F establishing property (i) in the definition of a fundamental domain.

Now suppose that $z = x + iy \in F$ and that $\gamma = \left(\begin{smallmatrix} a & b \\ c & d \end{smallmatrix}\right) \in \Gamma(1)$ such that $w = \gamma(z) \in F$. If $c = 0$, then $\gamma = \pm\left(\begin{smallmatrix} 1 & n \\ & 1 \end{smallmatrix}\right)$ for some $n \in \mathbb{Z}$, and $z, \gamma(z) \in F$ implies that $n = 0$, so $z = \gamma(z)$, in accordance with (ii) in the definition of a fundamental domain. Thus we may assume that $c \neq 0$. Observe that every element of F has imaginary part greater than $\sqrt{3}/2$. Also, clearly $|cz+d| \geq cy$.

We therefore have the inequalities

$$\frac{\sqrt{3}}{2} < \operatorname{im}(\gamma(z)) = \frac{y}{|cz+d|^2} \leq \frac{1}{c^2 y} < \frac{2}{c^2\sqrt{3}}.$$

Hence $c^2 < 4/3$, implying that $c = \pm 1$. Suppose that $c = \pm 1$. Because γ and $-\gamma$ have the same action on $\mathcal{H}$, we may assume without loss of generality that $c = 1$. Then $ad - bc = 1$ implies that

$$\gamma = \begin{pmatrix} a & b \\ 1 & d \end{pmatrix} = \begin{pmatrix} 1 & a \\ & 1 \end{pmatrix} \begin{pmatrix} & -1 \\ 1 & \end{pmatrix} \begin{pmatrix} 1 & d \\ & 1 \end{pmatrix}.$$

Now let $z_1 = z + d$ and $w_1 = w - a$. Because $|\operatorname{re}(z)| < 1/2$, we have $|z_1| \geq |z| > 1$, and similarly $|w_1| > 1$, yet $w_1 = \left(\begin{smallmatrix} & -1 \\ 1 & \end{smallmatrix}\right)(z_1)$. This is a contradiction. This proves that F satisfies (ii) in the definition of a fundamental domain. ∎

It is often convenient to have generators for the group $SL(2, \mathbb{Z})$. Let

$$T = \begin{pmatrix} 1 & 1 \\ & 1 \end{pmatrix}, \qquad S = \begin{pmatrix} & -1 \\ 1 & \end{pmatrix}. \tag{2.3}$$

Proposition 1.2.3 *$SL(2, \mathbb{Z})$ is generated by S and by T.*

The method of proof generalizes easily to give generators for other discontinuous groups. See Exercise 1.2.3(b) for an example.

Proof Let Γ be the subgroup of $\Gamma(1)$ generated by S and T. We will show that $\Gamma = \Gamma(1)$. Because $-I = S^2 \in \Gamma$, it is sufficient to show that the images $\overline{\Gamma} = \overline{\Gamma(1)}$ in $PSL(2, \mathbb{R})$. Let $\gamma \in \Gamma(1)$; we will describe a process by which γ may be reduced to a product of elements of the form S, T, and T^{-1}. We do not distinguish now between matrices A and $-A$ because these are equal in $PSL(2, \mathbb{R})$ and have the same effect on $\mathcal{H}$.

Because F is a fundamental domain for $\Gamma(1)$, $\mathcal{H}$ is the union of the closure $\overline{\gamma(F)}$ with $\gamma \in \Gamma(1)$, and these sets have disjoint interiors. We may therefore find a sequence $\gamma_1, \cdots, \gamma_n \in \Gamma(1)$ such that $\gamma_1(F) = F$ and $\gamma_n(F) = \gamma(F)$, and each $\gamma_k(F)$ is adjacent to $\gamma_{k+1}(F)$. Of course, this implies that $\gamma_1 = I$ and $\gamma_n = \gamma$. Observe that the domains γF that are adjacent to F are precisely $T(F)$, $T^{-1}(F)$, and $S(F)$ (cf. Figure 1).

Because $\gamma_k(F)$ is adjacent to $\gamma_{k+1}(F)$, we must have $\gamma_k^{-1}\gamma_{k+1}(F)$ adjacent to F, and so $\gamma_k^{-1}\gamma_{k+1}$ equals S, T, or T^{-1}. Thus $\gamma = \gamma_1^{-1}\gamma_n = \prod \gamma_k^{-1}\gamma_{k+1} \in \overline{\Gamma}$, as required. ∎

What is the "boundary" of the Poincaré upper half plane? If we embed $\mathcal{H}$ in the Riemann sphere $\mathbb{P}^1(\mathbb{C}) = \mathbb{C} \cup \{\infty\}$, the topological boundary is $\mathbb{P}^1(\mathbb{R}) = \mathbb{R} \cup \{\infty\}$. The point ∞ should be regarded as no different from the

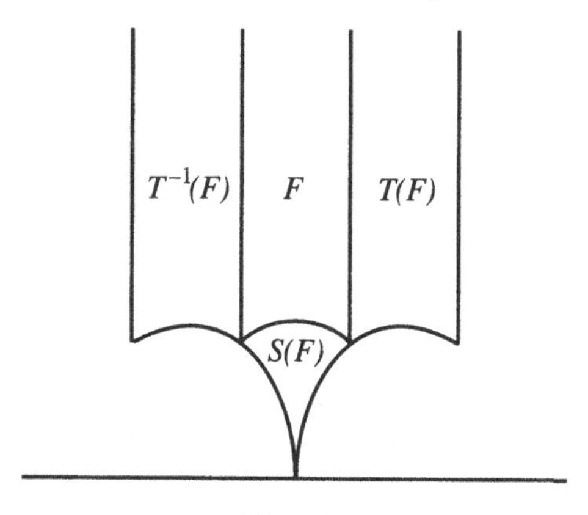

Figure 1

other boundary points. For example, the group $SL(2, \mathbb{R})$ acts transitively on $\mathbb{R} \cup \{\infty\}$. Another way of seeing this is to use the *Cayley transform* $\mathcal{C}$, which maps $\mathcal{H}$ to $\mathcal{D}$, the unit disk, defined by

$$\mathcal{C}(z) = \frac{z - i}{z + i}. \tag{2.4}$$

(Clearly $z \in \mathcal{H}$ if and only if z is closer to i than to $-i$, i.e., if and only if $\mathcal{C}(z) \in \mathcal{D}$.) Then $\mathcal{C}$ maps $\mathbb{R} \cup \{\infty\}$ onto the unit circle. This shows again that the points of $\mathbb{R} \cup \{\infty\}$ should be considered equivalent to each other. To give an example, the image under the Cayley transform of the fundamental domain F for $SL(2, \mathbb{Z})$ described above looks like that shown in Figure 2.

If Γ is a discontinuous group acting on $\mathcal{H}$, let $\Gamma\backslash\mathcal{H}$ be the quotient space consisting of the orbits of elements of $\mathcal{H}$ under the action of Γ. We topologize $\Gamma\backslash\mathcal{H}$ as a quotient: This means that a subset of $\Gamma\backslash\mathcal{H}$ is open if and only if its preimage in $\mathcal{H}$ under the canonical map $\mathcal{H} \to \Gamma\backslash\mathcal{H}$ is open.

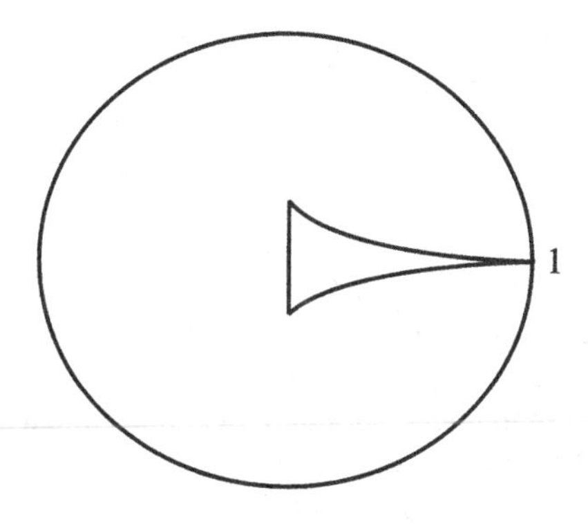

Figure 2

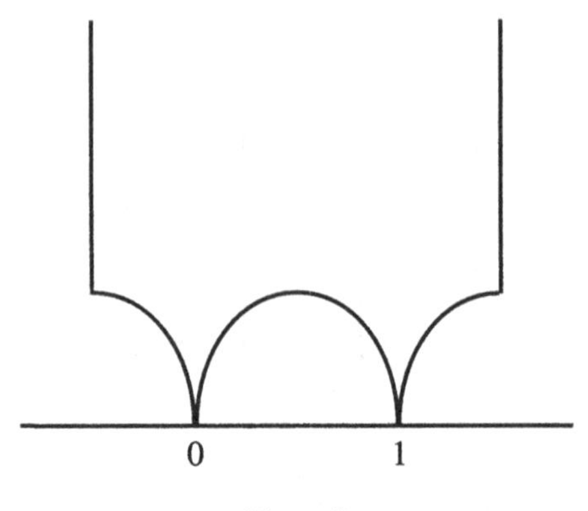

Figure 3

We now consider the *cusps* of a congruence group Γ. (We will extend this discussion shortly to the case of a discontinuous group Γ such that $\Gamma\backslash\mathcal{H}$ has finite volume with respect to the measure introduced in Exercise 1.2.6.) Intuitively, the cusps are the places where a fundamental domain for Γ touches the boundary of $\mathcal{H}$. For example, in Figure 2, we see that $SL(2,\mathbb{Z})$ has one cusp. On the other hand, $\Gamma(2)$ has three. Indeed, its fundamental domain looks like that shown in Figure 3 (Exercise 1.2.3).

Its image under the Cayley transform looks like the region shown in Figure 4. Evidently, there should be three cusps, if we can give the correct definition.

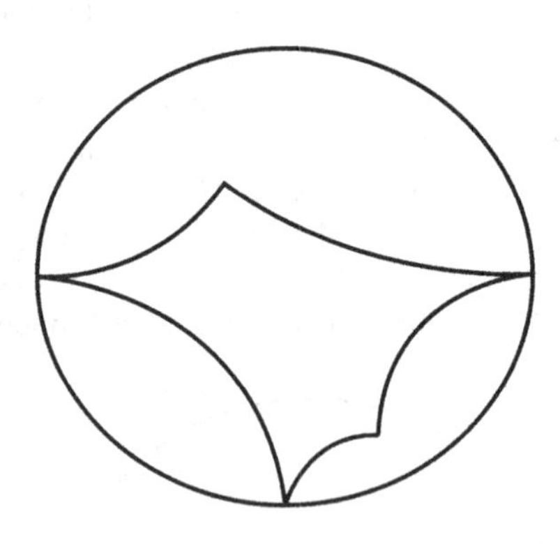

Figure 4

Let $\mathbb{P}^1(\mathbb{Q}) = \mathbb{Q}\cup\{\infty\}$ be the projective line over $\mathbb{Q}$; $SL(2,\mathbb{Z})$ acts transitively on $\mathbb{P}^1(\mathbb{Q})$, so a subgroup of finite index can have only finitely many orbits on this set. An orbit of Γ in $\mathbb{P}^1(\mathbb{Q})$ is called a *cusp* of Γ. In practical terms, these are the points where a fundamental domain for Γ must touch the boundary of $\mathcal{H}$.

More generally, if Γ is not assumed to be a congruence subgroup, but only a discontinuous group acting on $\mathcal{H}$ with $\Gamma\backslash\mathcal{H}$ having finite volume, the term *cusp* refers to either (i) a point of $a \in \mathbb{P}^1(\mathbb{R}) = \mathbb{R}\cup\{\infty\}$ such that Γ contains a parabolic element $\gamma \neq I$ with $\gamma(a) = a$, or (ii) an orbit of such points under the action of Γ. (We recall from Exercise 1.2.7(c) that $\gamma \neq I$ is called *parabolic* if $|\operatorname{tr}(\gamma)| = 2$.) See Exercise 1.2.10 for the relationship of this definition to the one previously given for congruence subgroups.

We now show how $\Gamma\backslash\mathcal{H}$, for Γ a congruence group, may be compactified to give a compact Riemann surface by adjoining a finite number of points; indeed, by adjoining precisely one point for each cusp. Again, the discussion generalizes easily to a discontinuous group Γ where $\Gamma\backslash\mathcal{H}$ has finite volume. We start with the topological space $\mathcal{H}^* = \mathcal{H} \cup \mathbb{Q} \cup \{\infty\}$. (For a general discontinuous group Γ, we would take $\mathcal{H}^*$ to be the union of $\mathcal{H}$ and the cusps of Γ in $\mathbb{P}^1(\mathbb{R})$.) We topologize $\mathcal{H}^*$ as follows. The set $\mathcal{H}$ is to be an open set with its usual topology. We must describe the topology in the neighborhood of a point $a \in \mathbb{Q} \cup \{\infty\}$. If $a = \infty$, we take as a neighborhood base at a the sets of the form $\{\infty\} \cup \{z | \operatorname{im}(z) > C\}$ for $0 \leq C \in R$. On the other hand, if $a \in \mathbb{Q}$, we take as a neighborhood base at a the sets $\{a\} \cup U$, where U is the interior of a circle contained in $\mathcal{H}$, tangent to the real line at the point a. With $\mathcal{H}^*$ topologized in this way, we give $\Gamma\backslash\mathcal{H}^*$ the quotient topology. This is a manifold. We will now specify charts around each point, which will give it a complex structure.

Around "most" points $a \in \Gamma\backslash\mathcal{H}$, we may simply take a neighborhood of a in $\mathcal{H}$ to be a chart. Certain points must be treated carefully: These are the elliptic points. We call a point $a \in \mathcal{H}$ an *elliptic point* if there exists a nontrivial subgroup $\overline{\Gamma_a}$ of the image $\overline{\Gamma}$ of Γ in $SL(2,\mathbb{R})/\{\pm I\}$ that stabilizes a. Such a group is necessarily cyclic (Exercise 1.2.4). Its order is called the *order* of a.

For example, if $\Gamma = \Gamma(1)$, the Γ-orbits of elliptic points are represented by i, with $\overline{\Gamma_i} = \left\langle \begin{pmatrix} & -1 \\ 1 & \end{pmatrix} \right\rangle$, and by $\rho = e^{2\pi i/3}$, with $\overline{\Gamma_\rho} = \left\langle \begin{pmatrix} & -1 \\ 1 & 1 \end{pmatrix} \right\rangle$. The orders of these points are 2 and 3, respectively. Most congruence subgroups do not have elliptic points.

Suppose that a is an elliptic point. How are we to construct a chart in a neighborhood of a? We use the modified Cayley transform

$$z \mapsto \frac{z-a}{z-\overline{a}}.$$

This maps $\mathcal{H}$ to the unit disk $\mathcal{D}$ but maps a to zero. Conjugation by this modified Cayley transform maps $\overline{\Gamma_a}$ to the group of rotations of the unit disk in angles that are a multiple of $2\pi/n$, where n is the order of a (Exercise 1.2.5(c)). If w is the coordinate function on $\mathcal{D}$, it is easy to see that $z \mapsto w^n$ maps a neighborhood of a in $\Gamma\backslash\mathcal{H}^*$ homeomorphically onto a neighborhood of the origin in $\mathbb{C}$, and we take this map to be a coordinate chart near a. This takes care of the remaining points in $\mathcal{H}$.

As for the cusps, if $a \in \mathbb{Q} \cup \{\infty\}$, let $\rho \in SL(2,\mathbb{Z})$ such that $\rho(a) = \infty$. Let $\overline{\Gamma_a}$ be the stabilizer of a in $\overline{\Gamma}$. Now $\overline{\rho\Gamma\rho^{-1}}$ is a subgroup of finite index in $\overline{\Gamma(1)}$, and the stabilizer of ∞ in this group is $\overline{\rho\Gamma_a\rho^{-1}}$. Hence this is a subgroup of finite index in the stabilizer of infinity in $\Gamma(1)$. The image of this stabilizer in

$PSL(2,\mathbb{R})$ is the infinite cyclic group generated by $z \mapsto z+1$. Thus $\overline{\rho\Gamma_a\rho^{-1}}$ is an infinite cyclic group generated by $z \mapsto z+n$ for some n. It is not hard to see then that $z \mapsto e^{2\pi i\rho(z)/n}$ maps a neighborhood of a in $\Gamma\backslash\mathcal{H}^*$ homeomorphically onto a neighborhood of the origin in $\mathbb{C}$, and we take this map to be a coordinate chart near a.

We have specified a coordinate chart near each point of $\Gamma\backslash\mathcal{H}^*$, which thus becomes a compact Riemann surface.

Exercises

Exercise 1.2.1 Prove Eq. (2.2).

Exercise 1.2.2 Let Γ be a discontinuous subgroup of $SL(2,\mathbb{R})$, and let Γ' be a subgroup. Let F be a fundamental domain for Γ, and let $\gamma_1, \cdots, \gamma_n$ be a set of coset representatives for $\Gamma'\backslash\Gamma$; that is, $\Gamma = \bigcup \Gamma'\gamma_i$ disjointly. Prove that $\bigcup \gamma_i(F)$ is a fundamental domain for Γ'.

Exercise 1.2.3 (a) Prove that a fundamental domain for $\Gamma(2)$ consists of $x+iy$ such that $-1/2 < x < 3/2$, $|z+1/2| > 1/2$, $|z-1/2| > 1/2$ and $|z-3/2| > 1/2$ (cf. Figure 3). [HINT: construct first a fundamental domain by means of Exercise 2.2, then modify it to obtain the domain in question.]
(b) Use the method of Proposition 2.3 to prove that $\Gamma(2)$ is generated by $\begin{pmatrix} 1 & 2 \\ & 1 \end{pmatrix}$ and $\begin{pmatrix} 1 & \\ 2 & 1 \end{pmatrix}$.

Exercise 1.2.4 Prove that the stabilizer of an elliptic point is cyclic.

Exercise 1.2.5 (a) Let $SL(2,\mathbb{C})$ act on $\mathbb{P}^1(\mathbb{C})$ by linear fractional transformations as in Eq. (2.1). Prove that the subgroup that maps the unit disk $\mathcal{D}$ onto itself is

$$SU(1,1) = \left\{ \begin{pmatrix} a & b \\ \overline{b} & \overline{a} \end{pmatrix} \middle| \; |a|^2 - |b|^2 = 1 \right\}.$$

(b) Prove that the group $SU(1,1)$ is conjugate to $SL(2,\mathbb{R})$ in $SL(2,\mathbb{C})$. [HINT: use the Cayley transform.]
(c) Prove that the subgroup of $SU(1,1)$ fixing $0 \in \mathcal{D}$ is the group of rotations

$$\begin{pmatrix} e^{i\theta/2} & \\ & e^{-i\theta/2} \end{pmatrix}.$$

Exercise 1.2.6 (a) *Bruhat decomposition:* Prove that if B is the *Borel subgroup* of $SL(2,\mathbb{R})$ consisting of upper triangular matrices and $S = \begin{pmatrix} & -1 \\ 1 & \end{pmatrix}$, then

$$SL(2,\mathbb{R}) = B \cup BSB,$$

and the union is disjoint. Thus $SL(2,\mathbb{R})$ is generated by matrices of the following types:

$$\begin{pmatrix} a & \\ & a^{-1} \end{pmatrix}, \quad \begin{pmatrix} 1 & b \\ & 1 \end{pmatrix}, \quad \begin{pmatrix} & -1 \\ 1 & \end{pmatrix}.$$

(Making use of the identity (IV.1.16) below, it is possible to dispense with the diagonal matrices here.)
(b) Show that the measure $|y|^{-2}\, dx\, dy$ is invariant under the action of $SL(2, \mathbb{R})$ by checking that it is invariant under generators in part (a).
(c) Show that the volume of $\Gamma(1)\backslash\mathcal{H}$ is finite with respect to this invariant measure.

Exercise 1.2.7 Let $\pm I \neq \gamma \in SL(2, \mathbb{R})$ acting on the Riemann sphere $\mathbb{P}^1(\mathbb{C}) = \mathbb{C} \cup \{\infty\}$.
(a) If $|\operatorname{tr}(\gamma)| < 2$, show that γ has two fixed points in $\mathbb{P}^1(\mathbb{C})$: one in $\mathcal{H}$ and its complex conjugate. Such an element is called *elliptic*.
(b) If $|\operatorname{tr}(\gamma)| > 2$, show that γ has two fixed points in $\mathbb{P}^1(\mathbb{R})$ and no other fixed points in $\mathbb{P}^1(\mathbb{C})$. Such an element is called *hyperbolic*.
(c) If $|\operatorname{tr}(\gamma)| = 2$, show that γ has a single fixed point in $\mathbb{P}^1(\mathbb{R})$ and no other fixed points in $\mathbb{P}^1(\mathbb{C})$. Such an element is called *parabolic*. If $\operatorname{tr}(\gamma) = 2$, then both eigenvalues of γ are one, in which case the matrix γ is called *unipotent*. If γ is parabolic, then either γ or $-\gamma$ is unipotent.

Exercise 1.2.8 (a) Let $\Gamma \subset SL(2, \mathbb{R})$ be a discontinuous group, and let $\pm I \neq \gamma \in \Gamma$. Show that γ is elliptic if and only if it has finite order. In this case, we call the fixed point of γ in $\mathcal{H}$ an *elliptic fixed point* for Γ.
(b) Show that there are only two orbits of elliptic fixed points for $SL(2, \mathbb{Z})$, represented by i and $e^{2\pi i/3}$, respectively. [Prove this by examining the fundamental domain.]

Exercise 1.2.9 Let $\Gamma \subset SL(2, \mathbb{R})$ be a discontinuous group such that the quotient $\Gamma\backslash\mathcal{H}$ has finite volume (cf. Exercise 2.6(c)). Show that $\Gamma\backslash\mathcal{H}$ is compact if and only if Γ contains no parabolic elements.

Exercise 1.2.10 Let Γ be a congruence subgroup of $SL(2, \mathbb{Z})$. Prove that if $a \in \mathbb{P}^1(\mathbb{R}) = \mathbb{R} \cup \{\infty\}$, then there exists a parabolic element $\gamma \in \Gamma$ such that $\gamma(a) = a$ if and only if $a \in \mathbb{P}^1(\mathbb{Q}) = \mathbb{Q} \cup \{\infty\}$.

Exercise 1.2.11 Let $\gamma \in \Gamma(1)$ be a hyperbolic element. Show that there exists a real quadratic field $K = \mathbb{Q}(\sqrt{D})$ with $D > 0$ such that the fixed points and eigenvalues of γ lie in K. Show that the eigenvalues of γ are a conjugate pair of units of norm one in K. *Make the following assumption about K: assume that the ring generated by the units of norm one in K is the full ring of integers.* This may or may not be true. Let ϵ, ϵ' be the eigenvalues of γ. If $\mathfrak{a}$ is a fractional ideal of K, then $\mathfrak{a}$ is a free $\mathbb{Z}$-module of rank 2; let $\{a_1, a_2\}$ be a basis. Then there exists an element $\gamma \in SL(2, \mathbb{Z})$ such that $\epsilon(a_1, a_2) = (a_2, a_2)\gamma$. Show that the $GL(2, \mathbb{Z})$-conjugacy class of γ depends only on the ideal class of $\mathfrak{a}$, and that the $GL(2, \mathbb{Z})$-conjugacy classes of hyperbolic elements with eigenvalues

ϵ and ϵ' are thus in bijection with the ideal classes of K. For a fuller discussion of the hyperbolic conjugacy classes in $SL(2, \mathbb{Z})$, see the references in Volume I of Terras (1985, p. 273).

1.3 Modular Forms for $SL(2, \mathbb{Z})$

Modular forms are certain holomorphic functions on $\mathcal{H}$ that have in common with Dirichlet characters the remarkable property of being associated with Euler products having functional equations. We will consider first the case of a modular form for $\Gamma(1) = SL(2, \mathbb{Z})$.

Let k be an even nonnegative integer. A *modular form of weight* k for $SL(2, \mathbb{Z})$ is a holomorphic function f on $\mathcal{H}$, which satisfies the identity

$$f\left(\frac{az+b}{cz+d}\right) = (cz+d)^k f(z) \qquad \text{for} \begin{pmatrix} a & b \\ c & d \end{pmatrix} \in \Gamma(1) \tag{3.1}$$

and which is *holomorphic at the cusp* ∞. The latter condition requires some discussion. Recall from the end of the previous section that we may choose the quantity $q = e^{2\pi i z}$ as a coordinate function near ∞ in $\Gamma(1)\backslash\mathcal{H}^*$. Because $\begin{pmatrix} 1 & 1 \\ & 1 \end{pmatrix} \in \Gamma(1)$, Eq. (3.1) implies that $f(z+1) = f(z)$, and thus any function satisfying Eq. (3.1) has a Fourier expansion

$$f(z) = \sum_{n=-\infty}^{\infty} a_n\, e^{2\pi i n z} = \sum_{n=-\infty}^{\infty} a_n q^n. \tag{3.2}$$

If for some sufficiently large N the coefficients a_n are zero for $n < -N$, we say the function f is *meromorphic* at ∞. If $a_n = 0$ for $n < 0$, we say that f is *holomorphic* at ∞. If f is holomorphic at ∞ and furthermore $a_0 = 0$, we say f *vanishes* or is *cuspidal at* ∞.

A modular form for $SL(2, \mathbb{Z})$ that vanishes at ∞ is called a *cusp form*. We denote the space of modular forms of weight k for $\Gamma(1) = SL(2, \mathbb{Z})$ as $M_k(\Gamma(1))$, and the space of cusp forms as $S_k(\Gamma(1))$. Our first objective is to prove that these spaces are finite dimensional.

There is a related notion of an *automorphic function*. We call f an *automorphic function* for Γ if

$$f\left(\frac{az+b}{cz+d}\right) = f(z)$$

and f is meromorphic on $\mathcal{H}$ and at ∞. Hence f may be regarded as a meromorphic function on the compact Riemann surface $\Gamma(1)\backslash\mathcal{H}^*$. Note that an automorphic function is allowed to have poles, while a modular form is not. It is a consequence of the maximum modulus principle that an automorphic function with no poles is constant. (An automorphic function with no poles is the same as a modular form of weight zero, so we may equally well state that a modular form of weight zero is constant.) This simple fact, together with the observation that if $f_1, f_2 \in M_k(\Gamma(1))$ then f_1/f_2 is an automorphic function, is quite a powerful tool that we will use systematically in determining the spaces $M_k(\Gamma(1))$. A first application of this principle is Proposition 1.3.2 below.

Proposition 1.3.1 *Let X be a compact Riemann surface, $P_1, \cdots, P_n \in X$, and let $r_1, \cdots, r_n$ be positive integers. Let V be the vector space of meromorphic functions on X, which are holomorphic except possibly at the points P_m, and which are holomorphic or else have poles of order at most r_m at P_m. Then the space V has dimension at most $r_1 + \ldots + r_m + 1$.*

More precise information about the dimension of this space is contained in the Riemann–Roch theorem.

Proof We will denote $r = r_1 + \ldots + r_m$. Let us choose a coordinate function $t = t_j$ in a neighborhood of P_j with respect to which P_j is the origin. If $\phi \in V$, it has a Laurent expansion:

$$\phi(t) = a_{j,-r_j}\, t^{-r_j} + a_{j,-r_j+1}\, t^{-r_j+1} + \ldots$$

We associate with ϕ the vector $A(\phi) \in \mathbb{C}^r$ whose coordinates are the r Taylor coefficients $a_{j,-h}$, $1 \le h \le r_j$. If $\phi_1, \cdots, \phi_N \in V$, and if $N > r$, we may find coefficients $c_1, \cdots, c_N$, not all zero, such that $\sum c_j\, A(\phi_j) = 0$. This means that $\sum c_j \phi_j$ has no poles. It is a consequence of the maximum modulus principle that any meromorphic function on a compact Riemann surface having no poles is automatically constant. Hence any vector subspace of V having dimension greater than r contains a nonzero constant function. This implies that dim $V \le r + 1$. ■

Proposition 1.3.2 *The space $M_k(\Gamma(1))$ is finite dimensional.*

Proof Let f_0 be a nonzero element of $M_k(\Gamma(1))$. Let X be the compactification of $\Gamma(1)\backslash\mathcal{H}$ described in the previous section. Let $P_1, \cdots, P_m$ be the zeros of f_0, and let $r_1, \cdots, r_m$ be the orders of vanishing of f_0 at these points. (Actually, we must count the order of vanishing at an elliptic fixed point carefully. The order of vanishing of a function on $\mathcal{H}$ at an elliptic point a whose stabilizer $\overline{\Gamma}_a$ has order e will be e times the order of vanishing of the corresponding function on X.) If $f \in M_k(\Gamma(1))$, then f/f_0 is an automorphic function, and indeed, $f \mapsto f/f_0$ is an isomorphism of $M_k(\Gamma(1))$ with the vector space V in Proposition 3.1. Thus $M_k(\Gamma(1))$ has dimension at most $r + 1$. ■

We would like to know, on the other hand, that modular forms do exist. A convenient construction is by means of *Eisenstein series*. Let us assume that k is an even integer ≥ 4. Define

$$E_k(z) = \frac{1}{2} \sum_{\substack{m,n \in \mathbb{Z} \\ (m,n) \neq (0,0)}} (mz+n)^{-k}. \tag{3.3}$$

The series is absolutely convergent (Exercise 1.3.1). Let us show that $E_k(z)$ is a modular form of weight k. We have

$$E_k\left(\frac{az+b}{cz+d}\right) = (cz+d)^k \, \tfrac{1}{2} \sum_{\substack{m,n \in \mathbb{Z} \\ (m,n) \neq (0,0)}} (m(az+b)+n(cz+d))^{-k}$$

$$= (cz+d)^k \, \tfrac{1}{2} \sum_{\substack{m,n \in \mathbb{Z} \\ (m,n) \neq (0,0)}} ((ma+nc)z+(mb+nd))^{-k}.$$

Observe that because c and d are coprime, $(m,n) \mapsto (ma+nc, mb+nd)$ permutes the nonzero elements of $\mathbb{Z} \times \mathbb{Z}$ amongst themselves, so we see that E_k satisfies Eq. (3.1). To show that it is analytic at ∞, let us compute its Fourier expansion. Firstly, the sum of the terms with $m = 0$ is clearly just $\zeta(k)$. For the terms with $m \neq 0$, because k is even, the terms ± 1 contribute equally, and we may consider only $m > 0$. By Exercise 1.1.4, these contribute

$$\frac{(2\pi i)^k}{(k-1)!} \sum_{n=1}^{\infty} n^{k-1} \, e^{2\pi i n m z}.$$

If r is a complex number, let us define the *divisor sum*

$$\sigma_r(n) = \sum_{d|n} d^r.$$

We see that the Fourier expansion of E_k has the form

$$E_k(z) = \zeta(k) + \frac{(2\pi)^k \, (-1)^{k/2}}{(k-1)!} \sum_{n=1}^{\infty} \sigma_{k-1}(n) \, q^n, \qquad q = e^{2\pi i z}. \tag{3.4}$$

Note that by Exercise 1.1.8, the Fourier coefficients of $G_k(z) = \zeta(k)^{-1} \, E_k(z)$ are rational numbers.

For given k, either $S_k(\Gamma(1)) = M_k(\Gamma(1))$ or else $\dim \; M_k(\Gamma(1)) = \dim S_k(\Gamma(1)) + 1$, because if there exists a modular form of weight k with nonvanishing constant Fourier coefficient, we may subtract a suitable multiple of that from any given modular form to obtain a cusp form. Because there exist Eisenstein series with nonvanishing constant coefficient for $k \geq 4$, we see that

$$\dim \; M_k(\Gamma(1)) = \dim \; S_k(\Gamma(1)) + 1 \qquad \text{for } k \geq 4. \tag{3.5}$$

Although we have now constructed some modular forms, we still have not constructed *cusp forms*. This may be accomplished as follows. The modular forms form a graded ring, because if $f \in M_k(\Gamma(1))$, $g \in M_l(\Gamma(1))$, then $fg \in M_{k+l}(\Gamma(1))$. One may thus construct a large number of modular forms by ring operations. For example, we see that

$$G_4(z) = 1 + 240 \sum_{n=1}^{\infty} \sigma_3(n) q^n, \qquad G_6(z) = 1 - 504 \sum_{n=1}^{\infty} \sigma_5(n) q^n.$$

From these we may compute the Fourier coefficients of $G_4^3 - G_6^2$, a modular form of weight 12. We find that

$$\tfrac{1}{1728}(G_4^3 - G_6^2) = q - 24\,q^2 + 252\,q^3 - 1472\,q^4 + 4830\,q^5 + \dots.$$

We have constructed a nontrivial cusp form of weight 12. This modular form is denoted $\Delta(z)$.

We will obtain another formula for $\Delta(z)$, due to Ramanujan (1916). It will be useful to have at our disposal a famous formula from the theory of elliptic functions, *Jacobi's triple product formula*:

$$\sum_{n=-\infty}^{\infty} q^{n^2} x^n = \prod_{n=1}^{\infty} (1 - q^{2n})(1 + q^{2n-1} x)(1 + q^{2n-1} x^{-1}), \tag{3.6}$$

valid if $0 < |q| < 1$ and $x \neq 0$. A proof of this is sketched in Exercise 1.3.2.

Now in Eq. (3.6), substitute $q^{3/2}$ for q and $-q^{-1/2}$ for x. (Because $q = e^{2\pi i z}$, a fractional power q^r is naturally interpreted as $e^{2\pi i r z}$.) We see that

$$\sum_{n=-\infty}^{\infty} (-1)^n q^{(3n^2+n)/2} = \prod_{n=1}^{\infty} (1 - q^{3n})(1 - q^{3n-1})(1 - q^{3n-2}) = \prod_{n=1}^{\infty} (1 - q^n).$$

Now completing the square in this identity, we see that

$$\sum_{n=-\infty}^{\infty} (-1)^n q^{(6n+1)^2/24} = q^{1/24} \prod_{n=1}^{\infty} (1 - q^n). \tag{3.7}$$

The function $\eta(z) = q^{1/24} \prod (1 - q^n)$ is known as the *Dedekind eta function.* Grouping the terms with n positive and negative together, we may rewrite this formula:

$$\eta(z) = \sum_{n=1}^{\infty} \chi(n)\, q^{n^2/24}, \tag{3.8}$$

where χ is the primitive quadratic character with conductor 12 (Exercise 1.1.6). We have

$$\chi(n) = \begin{cases} 1 & \text{if } n \equiv \pm 1 \bmod 12; \\ -1 & \text{if } n \equiv \pm 5 \bmod 12; \\ 0 & \text{otherwise.} \end{cases}$$

Now we show that if $\gamma = \begin{pmatrix} a & b \\ c & d \end{pmatrix} \in \Gamma(1)$, there exists a 24th root of unity $\epsilon(\gamma)$ such that

$$\eta\left(\frac{az+b}{cz+d}\right) = \epsilon(\gamma)\,(cz+d)^{1/2}\,\eta(z). \tag{3.9}$$

(There is an ambiguity in sign in the choice of square root $(cz+d)^{1/2}$, but because we are only asserting that $\epsilon(\gamma)$ lies in the group of 24th roots of unity, this is not a problem.)

From Proposition 1.2.3, it is sufficient to check this when $\gamma = T = \begin{pmatrix} 1 & 1 \\ & 1 \end{pmatrix}$ and $\gamma = S = \begin{pmatrix} & -1 \\ 1 & \end{pmatrix}$. The transformation property for $\gamma = T$ is clear from the right-hand side of Eq. (3.7), because under this transformation $q^{1/24} \mapsto e^{2\pi i/24}\, q^{1/24}$. On the other hand, if $\gamma = S$, Eq. (3.8) gives $\eta(z) = \theta_\chi(-iz/12)$ with θ_χ as in Eq. (1.14). In Eq. (1.15), we have $\tau(\chi) = 2\sqrt{3}$ and $N = 12$, so

$$\sqrt{-iz}\,\eta(z) = \eta\left(-\frac{1}{z}\right), \tag{3.10}$$

so we have Eq. (3.9) in this case also. This completes the proof of Eq. (3.9) for all $\gamma \in \Gamma(1)$.

Next, we raise Eq. (3.9) to the 24th power to get rid $\epsilon(\gamma)$. We see that if

$$\Delta(z) = \eta(z)^{24} = q\prod_{n=1}^{\infty}(1-q^n)^{24},$$

then Δ is a cusp form of weight 12. Observe that Δ is defined by a convergent infinite product, each of whose factors has no zero on $\mathcal{H}$. Consequently,

$$\Delta(z) \neq 0 \qquad \text{for } z \in \mathcal{H}.$$

Proposition 1.3.3 *The space $S_{12}(\Gamma(1))$ is one dimensional and spanned by Δ. In particular*

$$\Delta = \tfrac{1}{1728}(G_4^3 - G_6^2). \tag{3.11}$$

Proof If $f \in S_{12}(\Gamma(1))$, then f/Δ is an automorphic function. It clearly has no poles in $\mathcal{H}$. It is also holomorphic at the cusp because Δ has only a first-order zero there, while f also vanishes. Because an automorphic function without poles is constant, f/Δ is constant. In particular, $\frac{1}{1728}(G_4^3 - G_6^2) = c\Delta$ for some c; examining the Fourier coefficients, we see that $c = 1$. ∎

In general, one may give formulas for the dimensions of spaces of modular forms starting with either the Riemann–Roch theorem or the Selberg trace formula. For the group $SL(2, \mathbb{Z})$, however, we will obtain complete information using *ad hoc* tools.

Proposition 1.3.4 *Suppose that k is an even nonnegative integer. Let $k = 12j + r$ where $0 \le r \le 10$. Then*

$$\dim\, M_{12j+r}\big(\Gamma(1)\big) = \begin{cases} j+1 & \text{if } r = 0, 4, 6, 8 \text{ or } 10; \\ j & \text{otherwise.} \end{cases} \tag{3.12}$$

The ring $\bigoplus_{k=0}^{\infty} M_k\big(\Gamma(1)\big)$ of modular forms is generated by G_4 and G_6.

Proof First let us show that $M_k\big(\Gamma(1)\big)$ is one dimensional and generated by E_k if $k = 4, 6, 8$, or 10. Let $h = 6(12-k)$. If $f \in M_k\big(\Gamma(1)\big)$ is not in the one-dimensional space spanned by E_k, we may subtract a multiple of E_k to cancel the constant Fourier coefficient, and so we may assume that f is a nonzero element of $S_k\big(\Gamma(1)\big)$. We consider $E_h(f/\Delta)^6$. This is an automorphic function with no poles, and hence is constant. Therefore $E_h = c\Delta^6/f^6$ for some c. We see that E_h can have no zeros on $\mathcal{H}$. Now $h = 12H$ where $H = 1, 2, 3$, or 4. We consider Δ^H/E_h. This is a nonzero automorphic function with no poles but with a zero of order H at ∞, which is a contradiction. This shows that each of the spaces $M_k\big(\Gamma(1)\big)$ is one dimensional, spanned by E_k if $k = 4, 6, 8$, or 10.

Now let us show that $M_2\big(\Gamma(1)\big)$ is zero. Suppose that f is a nonzero element of this space. Then $fE_4 \in M_6\big(\Gamma(1)\big)$, so $fE_4 = cE_6$ for some nonzero constant c. Because $E_4(\rho) = 0$ with $\rho = e^{2\pi i/3}$ (Exercise 1.3.3), we see that $E_6(\rho) = 0$. Now Eq. (3.11) implies that $\Delta(\rho) = 0$, a contradiction. Hence $M_2\big(\Gamma(1)\big) = 0$. Of course $M_0\big(\Gamma(1)\big)$ is one dimensional, comprising the constant functions. Thus Eq. (3.12) is proved if $k < 12$.

If $k \ge 12$, we show that multiplication by Δ is an isomorphism of $M_{k-12}(\Gamma(1))$ with $S_k(\Gamma(1))$. Indeed, it is an injection of $M_{k-12}\big(\Gamma(1)\big)$ into $S_k\big(\Gamma(1)\big)$, and if $f \in S_k\big(\Gamma(1)\big)$, then f/Δ has no poles, and hence lies in $M_{k-12}\big(\Gamma(1)\big)$. Formula (3.12) now follows from Eq. (3.5).

As for the fact that G_4 and G_6 generate the ring of modular forms, let R be the subring generated by these. It follows from the one-dimensionality of M_8 and M_{10} that E_8 and E_{10} are constant multiples of E_4^2 and E_4E_6; M_k lies in R for $k \le 10$. Also, Eq. (3.11) implies that $\Delta \in R$. Let k be the first even positive integer such that $M_k\big(\Gamma(1)\big)$ is not contained in R; we see that $k \ge 12$. Now $R \supset \Delta M_{k-12}\big(\Gamma(1)\big) = S_k\big(\Gamma(1)\big)$. Moreover, R contains a noncuspidal modular form in $M_k\big(\Gamma(1)\big)$, namely, $E_4^rE_6^s$, where r and s are chosen so that $4r + 6s = k$. Hence by Eq. (3.5), R contains M_k. ■

There exists a natural inner product on $S_k\big(\Gamma(1)\big)$, known as the *Petersson inner product*. If $f(z)$, $g(z) \in S_k\big(\Gamma(1)\big)$, then it is a consequence of Eq. (2.2) that

$$f(z)\,\overline{g(z)}\,y^k$$

is invariant under $z \mapsto \frac{az+b}{cz+d}$. Hence by Exercise 1.2.6(b), the integral

$$\langle f, g \rangle = \int\limits_{\Gamma(1)\backslash\mathcal{H}} f(z)\,\overline{g(z)}\,y^k\,\frac{dx\,dy}{y^2} \tag{3.13}$$

is well defined. Because if $n > 0$, $q^n \to 0$ very rapidly as $z \to \infty$, and because a cusp form has a Fourier expansion $\sum a_n\,q^n$ with $a_n \neq 0$ only for $n > 0$, a cusp form $f(z)$ decays very rapidly as $y \to \infty$. Hence the integrand in Eq. (3.13) is very small near the cusp, and the integral is very rapidly convergent. Evidently, $\langle f, g \rangle$ is a positive definite Hermitian inner product.

Let $f(z) = \sum_{n=0}^{\infty} a_n\,q^n$ be an element of $M_k(\Gamma(1))$. Let

$$L(s, f) = \sum_{n=1}^{\infty} a_n\,n^{-s}.$$

This is known as the *L-function* of f. We need to know that this series is convergent for s sufficiently large. For this, the following estimate is sufficient.

Proposition 1.3.5 *If f is cuspidal, its Fourier coefficients satisfy $a_n \le C\,n^{k/2}$ for some constant C independent of n.*

This estimate, called the *trivial estimate*, is due to Hardy (1927) and (more simply) Hecke (1937). The correct estimate $a_n \le C\,n^{(k-1)/2+\epsilon}$ for any $\epsilon > 0$, was conjectured (for $f = \Delta$) by Ramanujan (1916); this famous statement, the *Ramanujan conjecture*, was finally proved around 1970 by Deligne (1971) using difficult techniques from algebraic geometry. See Section 3.5 for further discussion of this conjecture.

Proof It follows from Eqs. (2.2) and (3.1) that $|f(z)\,y^{k/2}|$ is $\Gamma(1)$ invariant. Because f is cuspidal, this function decays rapidly as z approaches the cusp, and so it is bounded on the fundamental domain; consequently, there exists a constant C_1 such that $|f(z)\,y^{k/2}| < C_1$ for all $z \in \mathcal{H}$. Now for fixed y,

$$|a_n|\,e^{-2\pi ny} = \left|\int_0^1 f(x+iy)\,e^{-2\pi inx}\,dx\right| \le \int_0^1 |f(x+iy)|\,dx < C_1\,y^{-k/2}.$$

This estimate is independent of n. We choose $y = 1/n$ and obtain

$$a_n < e^{2\pi}\,C_1\,n^{k/2}$$

as required. ■

If f is not a cusp form, this estimate is no longer valid. If f is the Eisenstein series E_k, the nth Fourier coefficient of f is $\sigma_{k-1}(n)$, which is bounded by a constant times $\log(n)\, n^{k-1}$. In any case, the L-series $L(s, f)$ is convergent for $\mathrm{re}(s)$ sufficiently large.

Proposition 1.3.6 *The L-function $L(s, f)$ has meromorphic continuation to all s and satisfies a functional equation. In fact, if*

$$\Lambda(s, f) = (2\pi)^{-s}\, \Gamma(s)\, L(s, f) \tag{3.14}$$

then $\Lambda(s, f)$ extends to an analytic function of s if f is a cusp form; if f is not cuspidal, then it has simple poles at $s = 0$ and $s = k$. It satisfies

$$\Lambda(s, f) = (-1)^{k/2}\, \Lambda(k - s, f). \tag{3.15}$$

Proof In this proof we will assume that f is a cusp form, leaving the remaining case to the reader (Exercise 1.3.5). Because f is cuspidal, $f(iy) \to 0$ very rapidly as $y \to \infty$. When $\gamma = S$, Eq. (3.1) implies that

$$f(iy) = (-1)^{k/2}\, y^{-k}\, f(i/y), \tag{3.16}$$

so $f(iy) \to 0$ very rapidly as $y \to 0$ also. Hence the integral

$$\int_0^\infty f(iy)\, y^s\, \frac{dy}{y} \tag{3.17}$$

is convergent for all s and clearly defines an analytic function of s. If $\mathrm{re}(s)$ is large, we may substitute the Fourier expansion for f. Noting that

$$\int_0^\infty e^{-2\pi n y}\, y^s\, \frac{dy}{y} = (2\pi)^{-s}\, \Gamma(s)\, n^{-s},$$

we see that Eq. (3.17) equals $\Lambda(s, f)$. Now substituting Eq. (3.16) into Eq. (3.17) and substituting $1/y$ for y, we see that Eq. (3.17) equals

$$(-1)^{k/2} \int_0^\infty f(iy)\, y^{k-s}\, \frac{dy}{y},$$

from which we get Eq. (3.15). ∎

The first historical hint that a Euler product should be associated with the L-series of a modular form came from Ramanujan's investigation of Δ. The Fourier coefficients of Δ comprise Ramanujan's tau function: $\Delta(z) = \sum \tau(n)\, q^n$. Ramanujan (1916) conjectured, and Mordell (1917) proved shortly afterward, that

$$\sum_{n=1}^\infty \tau(n)\, n^{-s} = \prod_p \left(1 - \tau(p)\, p^{-s} + p^{11-2s}\right)^{-1}. \tag{3.18}$$

The true explanation of this identity requires the theory of *Hecke operators*, which is our next topic. We will give the proof of Eq. (3.18) at the end of Section 4.

Exercises

Exercise 1.3.1 Verify that the Eisenstein series Eq. (3.3) is absolutely convergent if $k \geq 4$.

Exercise 1.3.2 This exercise outlines a proof of Jacobi's triple product formula (3.7). Let z and w be complex parameters such that $z \in \mathcal{H}$. Let $\Lambda \subset \mathbb{C}$ be the lattice $\{2mz + n | m, n \in \mathbb{Z}\}$. We will also let $q = e^{2\pi i z}$ and $x = e^{2\pi i w}$.
(a) An *elliptic function* with respect to the lattice Λ, is meant a meromorphic function f such that $f(u + \lambda) = f(u)$ for $\lambda \in \Lambda$. Use the maximum modulus principle to show that if f is an elliptic function that has no poles, then f is constant.
(b) Define

$$\vartheta(z, w) = \sum_{n=-\infty}^{\infty} q^{n^2} x^n,$$

and let

$$P(z, w) = \prod_{n=1}^{\infty} (1 + q^{2n-1} x)(1 + q^{2n-1} x^{-1}).$$

Prove that

$$\vartheta(z, w + 2z) = (qx)^{-1} \vartheta(z, w)$$

and that

$$P(z, w + 2z) = (qx)^{-1} P(z, w).$$

Hence for fixed z, $f(w) = \vartheta(z, w)/P(z, w)$ is an elliptic function.
(c) Prove for fixed z that if $P(z, w) = 0$ then either $w = \frac{1}{2} + z + \lambda$ or else $w = \frac{1}{2} - z + \lambda$ for some $\lambda \in \Lambda$. Show that these values of w are also zeros of $\vartheta(z, w)$, and conclude that $f(w)$ has no poles, and hence by (a) is constant. This shows that

$$\vartheta(z, w) = \phi(q) P(z, w)$$

where $\phi(q)$ is independent of w.
(d) The Jacobi triple-sum formula will follow if we know that

$$\phi(q) = \prod_{n=1}^{\infty} (1 - q^{2n}).$$

To this end, show that

$$\vartheta(4z, 1/2) = \vartheta(z, 1/4),$$

whereas

$$P(4z, 1/2)/P(z, 1/4) = \prod_{n=1}^{\infty}(1 - q^{4n-2})(1 - q^{8n-4}).$$

Then

$$\phi(q) = \frac{P(4z, 1/2)}{P(z, 1/4)}\,\phi(q^4).$$

Now show that $\phi(q) \to 1$ as $q \to 0$, and thus evaluate $\phi(q)$.

Exercise 1.3.3 Show that if $\rho = e^{2\pi i/3} \in \mathcal{H}$, and if $3 \nmid k$, then $f(\rho) = 0$ for any modular form of weight k. [HINT: Observe that $\gamma(\rho) = \rho$ where $\gamma = \begin{pmatrix} & 1 & 1 \\ -1 & \end{pmatrix}$, and apply Eq. (3.1).]

Exercise 1.3.4 Show that G_4 and G_6 are algebraically independent.

Exercise 1.3.5 Prove Proposition 1.3.6 in the case where f is not necessarily cuspidal.

Exercise 1.3.6 Show that the inner product Eq. (3.13) is defined if only one of f and g is cuspidal and the other is an arbitrary modular form. Prove that the Eisenstein series E_k is orthogonal to the cusp forms (cf. Exercise 1.6.4).

Exercise 1.3.7 (a) Let M be a compact Riemann surface, and let $f : M \to \mathbb{C}$ be a meromorphic function. Assume that f has only one pole, at $m \in M$, which is simple. Extend f to a mapping $M \to \mathbb{P}^1(\mathbb{C})$ by $f(m) = \infty$. Prove that f is an isomorphism of Riemann surfaces.
(b) Define a function $j : \mathcal{H} \to \mathbb{C}$ by $j(z) = G_4^3/\Delta$. Show that j is an automorphic function for $SL(2, \mathbb{Z})$ with a Fourier expansion

$$j(z) = \frac{1}{q} + 744 + 196884\,q + 21493760\,q^2 + 864299970\,q^3 + \dots.$$

Prove that $j(i) = 1728$ and $j(e^{2\pi i/3}) = 0$. Use part (a) to conclude that j is a bijection of the compactified space

$$SL(2, Z)\backslash\mathcal{H} \cup \{\infty\} \cong \mathbb{P}^1(\mathbb{C}).$$

We now recall some elementary facts from the topology of surfaces and the theory of compact Riemann surfaces, particularly the notions of genus and ramification. For further information, see Siegel (1969, 1971, and 1973), Lang (1982), and Gunning (1966).

If X is a (connected) compact Riemann surface, then as a topological space, X is a compact orientable surface, which is homeomorphic to a sphere with g handles attached, where the *genus* g of X is half the rank of the first homology group $H_1(X, \mathbb{Z}) \cong \mathbb{Z}^{2g}$. The second homology group $H_2(X, \mathbb{Z}) \cong \mathbb{Z}$. If $f : X \to Y$ is a holomorphic mapping of compact Riemann surfaces, then the *topological degree* of f is defined to be the positive integer n such that the map

$$\mathbb{Z} \cong H_2(X, \mathbb{Z}) \to H_2(Y, \mathbb{Z}) \cong \mathbb{Z}$$

induced by f is multiplication by n. Equivalently, f induces an injection of the field F_Y of meromorphic functions on Y into the field F_X of meromorphic functions on F_X, and n is the field degree $[F_X : F_Y]$. If y is a point of Y in general position, then the cardinality of the fiber $f^{-1}(y)$ is usually n. However, there can be a finite number of points y such that the cardinality of $f^{-1}(y)$ is strictly less than n; we say that these points *ramify*. Intuitively, we think of the mapping f as a covering of Y by X; however, if ramification occurs, it is not strictly a covering in the topological sense but a *ramified covering*. Here *ramification* means that some of the points of the fiber can coalesce when a is specialized to a point that ramifies.

Suppose that $y \in Y$ and that $x_1, \cdots, x_r$ are the points of $f^{-1}(y)$. Let e_i be the number of points of $f^{-1}(\eta)$ that are near x_i when $\eta \in Y$ is a nonramified point of Y that is near y. Then e_i is called the *ramification index* of x_i, and if $e_i > 1$, we say that x_i is *ramified*. It is clear that $\sum e_i = n$. Note that with this definition, $e_i = 1$ if x_i is not ramified. We also denote $e_i = e(x_i|y)$. There is a strong analogy between ramification in this geometric setting and the ramification of primes in a number field.

We have the *Hurwitz genus formula* (Lang (1982)). Let g_X and g_Y be the genera of X and Y, respectively, let n be the topological degree of f, let $P_1, \cdots, P_u$ be the finite number of ramified points in X, and let $Q_i = f(P_i)$, their images in Y. Then the genus formula asserts that

$$(2g_X - 2) = n(2g_Y - 2) + \sum \big(e(P_i|Q_i) - 1\big). \tag{3.19}$$

Exercise 1.3.8 Apply this now in the case of the canonical map $f : \Gamma(N)\backslash\mathcal{H}^* \to SL(2, \mathbb{Z})\backslash\mathcal{H}^*$, with $N \geq 2$. Show that the degree n of f is 6 if $N = 2$, and $\frac{1}{2}N^3 \prod_{p|N}(1-p^{-2})$ if $N > 2$. Show that the points of $SL(2, \mathbb{Z})\backslash\mathcal{H}^*$ that ramify are i, $e^{2\pi i/3}$, and ∞. Show that there are $n/2$ points in the fiber over i, each with ramification index 2, $n/3$ points in the fiber over $e^{2\pi i/3}$, each with ramification index 3, and n/N points in the fiber over ∞, each with ramification index N. Hence show that when $N = 2$ or 3, $\Gamma(N)\backslash\mathcal{H}^*$ has genus zero. Confirm this when $N = 2$ by examining the fundamental domain in Exercise 1.2.3.
(Note: The fact that all the points in the fiber over the three points that ramify all have the same ramification index is due to the fact that $\Gamma(N)$ is a normal subgroup of $SL(2, \mathbb{Z})$. This phenomenon does not occur for subgroups that are not normal.)

Exercise 1.3.9 Show that $\Gamma_0(11)\backslash\mathcal{H}^*$ has genus one.

Exercise 1.3.10: Picard's theorem Prove that if ϕ is an entire function on $\mathbb{C}$ such there are two complex numbers a and b such that $a, b \notin \phi(\mathbb{C})$, then ϕ is constant.

[HINT: By Exercise 1.3.8, $\Gamma(2)$ has three cusps and $\Gamma(2)\backslash\mathcal{H}^*$ has genus zero. Consequently, $\Gamma(2)\backslash\mathcal{H}$ is equivalent to the Riemann sphere minus three points, or $\mathbb{C} - \{a, b\}$. Making this identification, we may regard f as taking values in $\Gamma(2)\backslash\mathcal{H}$. Now mapping $\mathcal{H}$ onto the unit disk by the Cayley transform, we obtain a bounded entire function, which is therefore constant.]

Now we require some basic facts about (nonramified) covering spaces. This well-known and important theory has applications to ramified coverings because if $f : X \to Y$ is a ramified covering, and $P_1, \cdots, P_r$ are the points of Y that ramify, and if Y' is the (noncompact) space $Y - \{P_i\}$ and $X' = f^{-1}(Y')$, then X' is a *bona fide* covering space of Y'. For the theory of covering spaces, see Spanier (1966) and Hilton and Wylie (1960), or other standard references on topology.

Let U be a topological space, $x, y \in U$. A *path* from x to y is a continuous map t of the unit interval $[0, 1]$ to U with $t(0) = x$ and $t(1) = y$. x is called the *left endpoint* and y is called the *right endpoint*. U is called *path connected* if any two points may be joined by a path. The space U is called *contractible* if the identity map $U \to U$ is homotopic to a constant map.

We will consider topological spaces U satisfying the following axiom:

Axiom 1.3.1 *U is path connected and every point of U has a contractable neighborhood.*

For example, (connected) manifolds have this property. Let U and V satisfy Axiom 1.3.1, and let $p : V \to U$ be a continuous map. We say that p is a *covering* or *covering map* if the fibers $p^{-1}(u)$ are discrete and if every point $u \in U$ has a neighborhood N such that $p^{-1}(N)$ is homeomorphic to a direct product $N \times p^{-1}(u)$ in such a way that the composition

$$p^{-1}(N) \cong N \times p^{-1}(u) \to N,$$

where the second map is projection, coincides with p. Covering maps have the following important *path-lifting property*:

Property 1.3.1 *If $u : [0, 1] \to U$ is a path, and $v \in p^{-1}\big(u(0)\big)$, then there exists a unique path $\tilde{u} : [0, 1] \to V$ such that $u = p \circ \tilde{u}$ and $\tilde{u}(0) = v$.*

This property is crucial in supplying proofs in the theory that we now describe. One easy consequence of the path-lifting property is that the fibers $p^{-1}(u)$ all have the same cardinality. Indeed, if u' is another point, we choose a path from u to u'. Now for every $v \in p^{-1}(u)$, by lifting this path to V with v as the left endpoint, the right endpoint of the lifted path is an element of $p^{-1}(u')$, defining a bijection between the fibers. The cardinality of the fibers is called the *degree* of the covering.

We say a topological space V is *simply connected* if it is path connected and if every homeomorphism of the circle into V is homotopic to a constant map. If U satisfies Axiom 1.3.1, then U admits a simply connected cover $\widetilde{U}$, called the *universal covering space*. To construct it, we fix a base point $u_0 \in U$; let $\widetilde{U}$ be the space of all paths $h : [0, 1] \to U$ such that the left endpoint

$h(0) = u_0$, modulo the identification of two paths when they can be deformed one into the other by a homotopy fixing both endpoints. The projection map $\widetilde{p} = \widetilde{p}_U : \widetilde{U} \to U$ is given by $\widetilde{p}(h) = h(1)$. We may topologize $\widetilde{U}$ in a natural way (we leave this to the reader), and $\widetilde{p}$ is a local homeomorphism.

Defining the *fundamental group* $\pi_1(U)$ also requires fixing a base point $u_0 \in U$. Then $\pi_1(U)$ may be defined to be the set of all paths $h : [0, 1] \to U$ with *both* endpoints equal to u_0, modulo the identification of paths that can be deformed one into the other by a homotopy fixing both endpoints. Thus $\pi_1(U)$ is precisely the fiber $\widetilde{p}^{-1}(u_0)$ in the map $\widetilde{p} : \widetilde{U} \to U$. To make $\pi_1(U)$ a group, if $\gamma_1, \gamma_2 \in \pi_1(U)$, we define the product to be the homotopy class of paths obtained by gluing the right endpoint of γ_1 to the left endpoint of γ_2. Similarly, if $\gamma \in \pi_1(U)$ and $h \in \widetilde{U}$, we define γh by gluing the right endpoint of γ to the left endpoint of h, and we obtain an action of $\pi_1(U)$ on $\widetilde{U}$, and we may identify the quotient space $\Gamma\backslash\widetilde{U} = U$.

We say that two covering maps $p : V \to U$ and $p' : V' \to U$ are *equivalent* if there exists a homeomorphism $\phi : V \to V'$ such that $p = p' \circ \phi$.

Exercise 1.3.11 Prove that there is a bijection between equivalence classes of coverings of U and conjugacy classes of subgroups of the fundamental group $\pi_1(U)$, which associates with the subgroup $\Gamma \subset \pi_1(U)$ the covering map $\Gamma\backslash\widetilde{U} \to U$ induced by projection $\widetilde{p}_U : \widetilde{U} \to U$. Show that $\pi_1(\Gamma\backslash\widetilde{U}) \cong \Gamma$.

[HINT: using the the path-lifting property, show that any covering map $p : V \to U$ lifts to an isomorphism $\widetilde{V} \to \widetilde{U}$; that is, the existence of the covering p implies that $\widetilde{V}$ may be identified with $\widetilde{U}$ in such a way that $\widetilde{p}_U = p \circ \widetilde{p}_V$. We assume the base points $u_0 \in U$ and $v_0 \in V$ are chosen so that $u_0 = p(v_0)$; then the fundamental group $\pi_1(V) = \widetilde{p}_V^{-1}(v_0)$ is a subgroup of $\pi_1(U) = \widetilde{p}_U^{-1}(u_0)$. The flexibility in this construction is that we may change the base point v_0 to another element of the fiber $p^{-1}(u_0)$; this has the effect of replacing $p_V^{-1}(v_0)$ by another conjugate subgroup. Conversely, given a subgroup Γ of $\pi_1(U)$, we define a covering space of U as the quotient space $\Gamma\backslash\widetilde{U}$, where the action of Γ is inherited from the natural action of $\pi_1(U)$ on $\widetilde{U}$. These two constructions are inverses of each other.]

Let $p : V \to U$ and $p' : V' \to U$ be covering maps. We say that p *dominates* p' if there exists a covering map $q : V \to V'$ such that $p = p' \circ q$.

Exercise 1.3.12 Let $p : V \to U$ and $p' : V' \to U$ be covering maps, and let $\Gamma, \Gamma' \subset \pi_1(U)$ be the subgroups associated with these covering maps by Exercise 1.3.11. Show that p dominates p' if and only if Γ is conjugate in $\pi_1(U)$ to a subgroup of Γ'.

A covering $p : V \to U$ is called *regular* if the group $\Gamma \subset \pi_1(U)$ associated with the covering by Exercise 3.11 is normal. In this case, the quotient group $\pi_1(U)/\Gamma$ acts on V. Indeed, identifying V with $\Gamma\backslash\widetilde{U}$, if $\gamma \in \Gamma$, $\widetilde{u} \in \widetilde{U}$, let the coset $\overline{\gamma}$ of γ in $\pi_1(U)/\Gamma$ act by $\overline{\gamma}\Gamma\widetilde{u} = \Gamma\gamma\widetilde{u}$. The action of $\pi_1(U)/\Gamma$ commutes with the map p, and hence preserves the fiber $p^{-1}(u_0)$ and is transitive on the fiber.

Exercise 1.3.13 Conversely, show that if $p : V \to U$ is a covering, and if there exists a group G of automorphisms of V that commute with p such that G is transitive on the fiber $p^{-1}(u_0)$, then the covering p is regular, and if Γ is the subgroup of $\pi_1(U)$ associated with p by Exercise 1.3.11, then $G \cong \pi_1(U)/\Gamma$.

Exercise 1.3.14 Show that every covering is dominated by a regular covering.

A regular covering should be thought of as analogous to a Galois field extension, and the covering group $\pi_1(U)/\Gamma$ should be thought of as analogous to the Galois group.

Exercise 1.3.15 shows that there is a close connection between the topology of covering spaces and holomorphic mappings of compact Riemann surfaces. The covering map $p : V \to U$ is called *finite* if the fibers $p^{-1}(u)$ are finite for $u \in U$.

Exercise 1.3.15 Let X and Y be compact Riemann surfaces and let $f : X \to Y$ be a holomorphic mapping. Let $P_1, \cdots, P_r \in Y$ be the points that ramify. Let $U = Y - \{P_1, \cdots, P_r\}$, and let $V = f^{-1}(U)$. Then the restriction of f to V is a finite covering of U. Conversely, show that if $f' : V' \to U$ is any finite covering of U, then V' may be identified with an open subset of a compact Riemann surface X', and f' may be extended to a holomorphic mapping $X' \to Y$.

[HINT: V' inherits a complex structure from U by the requirement that f' be a holomorphic mapping. The problem is how to compactify V' by adjoining points to make up the fiber $f'^{-1}(P)$ when P is one of the exceptional points P_i. This is a purely local question. First solve the topological problem of constructing the fiber; what remains then is the analytic problem of imposing a complex structure in the neighborhood of a point $Q \in f'^{-1}(P)$. Let $e = e(Q|P)$ be the ramification index of Q. Let (U, y) be a chart near P so U is a small neighborhood of P and $y : U \to \mathbb{C}$ is a holomorphic equivalence of U with a domain in $\mathbb{C}$; assume that $y(P) = 0$. Show on purely topological grounds that $y \circ f' = x^e$ for a function x defined in the connected component of $f'^{-1}(U - P)$ whose closure contains Q. Now use a theorem on removable singularities, such as Rudin's Theorem 10.20 (Rudin, 1974), to extend x to a chart near Q, making X' a complex manifold.]

Exercise 1.3.16 In the setting of Exercise 1.3.15, the holomorphic mapping $f : X \to Y$ induces an inclusion $F_Y \to F_X$ of the fields of meromorphic functions. Show that the field degree $[F_X : F_Y]$ equals the degree of the cover $V \to U$ and that the cover is regular if and only if F_X/F_Y is a Galois extension, in which case the group $\pi_1(\Gamma\backslash\tilde{U}) \cong \Gamma$ of Exercise 1.3.11 is isomorphic to the Galois group $\mathrm{Gal}(F_X/F_Y)$.

Exercise 1.3.17 Let $Y = \mathbb{P}^1(\mathbb{C})$, let y_0, y_1, and y_∞ be three distinct points of Y, and let $U = Y - \{y_0, y_1, y_\infty\}$. Prove that there exists a regular cover of degree six of U, which can be extended to a holomorphic mapping $f : X \to Y$ of compact Riemann surfaces, and such that $f^{-1}(y_0)$ and $f^{-1}(y_1)$ each consist of three points, with ramification index two, and $f^{-1}(y_\infty)$ consists of two points, each with ramification index three. Use the genus formula (3.19) to show that X has genus zero. Let $p : Z \to Y$ be any holomorphic map from another Riemann surface to Y. Assume that only y_0, y_1, and y_∞ ramify, and that the ramification index of any point in the fiber over y_0 or y_1 is either 1 or 2 and that the ramification index of any point in the fiber over y_∞ is either 1 or 3. Prove that there exists a holomorphic mapping $q : X \to Z$ such that $f = p \circ q$.

[HINTS: For the first part, note that $\pi_1(U)$ is a free group with two generators γ_0 and γ_1. Here γ_i is a loop issuing out of the base point and circling the point y_i once counterclockwise before returning to the base point. Let Γ be the smallest normal subgroup of $\pi_1(U)$ containing by γ_0^2, γ_1^2, and $(\gamma_0\gamma_1)^3$, and let $V = \Gamma\backslash\widetilde{U}$. Note that $\pi_1(U)/\Gamma$ is the group with two generators g_0 and g_1 subject to the relations

$$g_0^2 = g_1^2 = (g_0 g_1)^3 = 1.$$

This group is isomorphic to the symmetric group S_3. Extend the cover $V \to U$ to a ramified cover of Y by a Riemann surface X by using Exercise 1.3.15. For the other part, first construct q over $f^{-1}(U)$ by Exercise 1.3.12, then extend it by Exercise 1.3.15.]

We now show how these ideas can be applied to the construction of automorphic functions for various groups.

Exercise 1.3.18 Prove that there exists an automorphic function z on $\Gamma(2)\backslash\mathcal{H}$ that satisfies the polynomial

$$z^3 - zj - 16j = 0.$$

[HINTS: By identifying $SL(2,\mathbb{Z})\backslash\mathcal{H}$ with $\mathbb{P}^1(\mathbb{C})$ by means of the map j as in Exercise 1.3.7 (b), we may take $y_0 = 0$, $y_1 = 1728$, and $y_\infty = \infty$ in Exercise 1.3.17. If $X = \Gamma(2)\backslash\mathcal{H}$, the projection

$$f : \Gamma(2)\backslash\mathcal{H} \to SL(2,\mathbb{Z})\backslash\mathcal{H}$$

has the ramification described for the map $X \to \mathbb{P}^1(\mathbb{C})$ in Exercise 1.3.17, so we may identify the covering space X with $\Gamma(2)\backslash\mathcal{H}$. Check that if $j_0 \in \mathbb{C}$, the polynomial $z^3 - zj_0 - 16j_0$ has no multiple roots unless $j_0 = 0$ or $j_0 = 1728$. (Compute the discriminant of this cubic polynomial.) Define, therefore, a threefold covering $Z \to SL(2,\mathbb{Z})\backslash\mathcal{H}$ by taking

$$Z = \{(z_0, \tau) \in \mathbb{C} \times SL(2,\mathbb{Z})\backslash\mathcal{H} | z_0^3 - z_0 j(\tau) - 16j(\tau) = 0\},$$

with the covering map $p : Z \to SL(2,\mathbb{Z})\backslash\mathcal{H}$ being the projection on the second component. Check that the hypotheses of Exercise 1.3.17 are satisfied, and conclude that there exists a holomorphic mapping $q : X \to Z$ such that $f = p \circ q$. Composing q with the projection on the first component gives the required holomorphic mapping.]

Exercise 1.3.19 Prove that there exists an automorphic function on $\Gamma(3)\backslash\mathcal{H}$ whose cube equals j.

1.4 Hecke Operators

Hecke (1937) introduced a certain ring of operators acting on modular forms. The commutativity of this ring leads to Euler products associated with modular forms. In the modern viewpoint, the Hecke ring is seen as a convolution ring of functions on $GL(2, A_f)$, where A_f is the ring of "finite adeles," which we introduce in Section 3.1. We will encounter Hecke operators in various forms throughout the book.

We are influenced in our treatment of this subject by the discussion in Shimura (1971). Let us fix a weight k, which is a positive integer. It may be *even* or *odd*.

If f is a holomorphic function on $\mathcal{H}$, and $\gamma = \left(\begin{smallmatrix} a & b \\ c & d \end{smallmatrix}\right) \in GL(2,\mathbb{R})^+$, we will denote by $f|\gamma$ the function

$$(f|\gamma)(z) = (\det\ \gamma)^{k/2}\,(cz+d)^{-k}\, f\left(\frac{az+b}{cz+d}\right). \tag{4.1}$$

That $(f|\gamma)|\gamma' = f|(\gamma\gamma')$ may be checked, so this is a bona fide right action on holomorphic functions on $\mathcal{H}$. Note that if k is even, scalar matrices act trivially; on the other hand, if k is odd,

$$f\left|\begin{pmatrix} \lambda & \\ & \lambda \end{pmatrix}\right. = \begin{cases} f & \text{if } \lambda > 0; \\ -f & \text{if } \lambda < 0. \end{cases}$$

It will be convenient to immediately generalize the definition of modular forms. Let Γ be a discontinuous subgroup of $SL(2,\mathbb{R})$ such that $\Gamma\backslash\mathcal{H}$ has finite volume with respect to the measure defined in Exercise 1.2.6, for example, a congruence subgroup of $\Gamma(1) = SL(2,\mathbb{Z})$. We say that a holomorphic function f on $\mathcal{H}$ is a *modular form* with respect to Γ if it satisfies Eq. (3.1) for all $\gamma \in \Gamma$ and is *holomorphic* at the cusps of $\Gamma\backslash\mathcal{H}$. Furthermore, it *vanishes* at all the cusps, we say it is a *cusp form.* The notion of holomorphicity or vanishing at the cusp $c \in \mathbb{R} \cup \{\infty\}$ is made precise as follows. Choose $\rho \in SL(2,\mathbb{R})$ such that $\rho(c) = \infty$. Then $f|\rho^{-1}$ is modular with respect to the group $\rho\Gamma\rho^{-1}$, which contains a translation $z \mapsto z + t$ for some real $t > 0$. Hence $f|\rho^{-1}$ has a Fourier expansion $\sum a_{\rho,n}\, e^{2\pi i n z/t}$. We say f is *meromorphic* at c if the coefficients $a_{\rho,n} = 0$ for $n < -C$ for some constant C; we say it is *holomorphic* at c if $a_{\rho,n} = 0$ for $n < 0$, and that it *vanishes* at c if the coefficients $a_{\rho,n} = 0$ for $n \le 0$.

We note that if $-I \in \Gamma$, then Eq. (3.1) is impossible unless k is even.

Lemma 1.4.1 *Let Γ be a congruence subgroup of $SL(2,\mathbb{Z})$, and let $\alpha \in GL(2,\mathbb{Q})^+$. Then there exists an integer M such that $\alpha^{-1}\Gamma\alpha \supseteq \Gamma(M)$. Consequently, $\alpha^{-1}\Gamma\alpha \cap \Gamma(1)$ is a congruence group.*

Proof Let N be such that $\Gamma(N) \subseteq \Gamma$. We may find positive integers M_1, M_2 such that $M_1\alpha$, $M_2\alpha^{-1} \in \mathrm{Mat}_2(\mathbb{Z})$. Let $M = M_1M_2N$. If $\gamma \in \Gamma(M)$, write

$\gamma = I + Mg$, where I is the 2×2 identity matrix, and $g \in \mathrm{Mat}_2(\mathbb{Z})$. Then $\alpha\gamma\alpha^{-1} = I + N(M_1\alpha)\, g\, (M_2\alpha^{-1})$. This is clearly an element of $\Gamma(N)$. ■

Now if f is a modular form for a congruence subgroup Γ, and $\alpha \in GL(2, \mathbb{Q})^+$, then $f|\alpha$ is modular with respect to $\alpha^{-1}\Gamma\alpha \cap \Gamma(1)$, which, we see, is a congruence group. Let us say that f is a *congruence modular form* or a *congruence cusp form* if it is a modular or cusp form, respectively, for some congruence subgroup of $\Gamma(1)$. We see that the action of $GL(2, \mathbb{Q})^+$ preserves the property of being a congruence modular form or cusp form.

If H is a group acting on the left on a set X, we will denote by $H\backslash X$ the set of orbits of X under this action. If X is a topological space, then $H\backslash X$ is given the *quotient topology* in which a subset is open if and only if its preimage under the natural map $X \to H\backslash X$ is open. The set $H\backslash X$ is variously known as a *quotient space, homogeneous space,* or *orbit space* – these terminologies are especially appropriate if X is a topological space but may be used in any case.

For example, if $G \supset H$ is a bigger group, H acts on G by left translation and $H\backslash G$ is the set of right cosets Hg for $g \in G$. Similarly, if H acts on X by right translation, we denote the set of orbits by X/H, so if $G \supset H$ is a bigger group, G/H is the set of left cosets gH. Of course, this is a group if H is normal, but otherwise it is just a set. If H_1 and H_2 are groups acting on X on the left and right, respectively, such that the actions are compatible

$$(h_1 x)h_2 = h_1(xh_2) \qquad \text{for } h_1 \in H_1, x \in X, h_2 \in H_2,$$

we again have a set of orbits; x and y will lie in the same orbit if $x = h_1 y h_2$ for some $h_1 \in H_1$ and $h_2 \in H_2$. The set of orbits in this situation is denoted $H_1\backslash X/H_2$. As a special case, if H_1 and H_2 are subgroups of a group G, $H_1\backslash G/H_2$ is the set of double cosets $H_1 g H_2$. One way to think of this is that H_1 acts on G/H_2 by left translation, and $H_1\backslash G/H_2$ is simply the set of orbits under this action; equivalently, H_2 acts on $H_1\backslash G$ by right translations, and $H_1\backslash G/H_2$ may be equally regarded as the set of orbits under this right action.

We will describe Hecke operators for $\Gamma(1) = SL(2, \mathbb{Z})$, leaving Hecke operators for congruence subgroups to the exercises. Because $-I \in SL(2, \mathbb{Z})$, as we have already noted, Eq. (3.1) requires that the weight k must be even, and we assume this for the remainder of this section, excluding the exercises, where we consider Hecke operators for congruence subgroups.

Proposition 1.4.1 *Let $\alpha \in GL(2, \mathbb{Q})^+$. Then the double coset $\Gamma(1)\alpha\Gamma(1)$ is a finite union of right cosets:*

$$\Gamma(1)\alpha\Gamma(1) = \bigcup_{i=1}^{N} \Gamma(1)\alpha_i, \qquad \alpha_i \in GL(2, \mathbb{Q})^+. \tag{4.2}$$

Indeed, the number of right cosets in this decomposition equals $[\Gamma(1) : \alpha^{-1}\Gamma(1)\alpha \cap \Gamma(1)]$, which is finite.

Proof We will show the cardinality of $\Gamma(1)\backslash\Gamma(1)\alpha\Gamma(1)$ is equal to $[\Gamma(1) : \alpha^{-1}\Gamma(1)\alpha \cap \Gamma(1)]$. (This cardinality is finite by Lemma 1.4.1 because $\alpha^{-1}\Gamma(1)\alpha \cap \Gamma(1)$ is a congruence subgroup.) Right translation by α^{-1} is a bijection of $GL(2, \mathbb{Q})^+$ onto itself, which induces a bijection of this set with

$$\Gamma(1)\backslash\Gamma(1)\alpha\Gamma(1)\alpha^{-1} \cong \left(\Gamma(1) \cap \alpha\Gamma(1)\alpha^{-1}\right)\backslash\alpha\Gamma(1)\alpha^{-1}.$$

Conjugating by α, this quotient has the same cardinality as $\left(\alpha^{-1}\Gamma(1)\alpha \cap \Gamma(1)\right)\backslash\Gamma(1)$. ■

If $\alpha \in GL(2, \mathbb{Q})^+$, we define the *Hecke operator* $T_\alpha = T(\alpha)$ on $M_k\left(\Gamma(1)\right)$ by

$$f|T_\alpha = \sum f|\alpha_i, \tag{4.3}$$

with the α_i as in Eq. (4.2). Observe that $f|T_\alpha$ is independent of the choice of representatives α_i because f is modular. Moreover $f|T_\alpha$ is modular, because if $\gamma \in \Gamma(1)$, it follows from Eq. (4.2) that the cosets $\Gamma(1)\,\alpha_i\gamma$ are the same as the $\Gamma(1)\,\alpha_i$ permuted, so there exist $\gamma_i \in \Gamma(1)$ such that the $\alpha_i\gamma$ are the $\gamma_i\alpha_i$ permuted, and then

$$(f|T_\alpha)|\gamma = \sum f|\alpha_i\gamma = \sum f|\gamma_i\alpha_i = \sum f|\alpha_i = f|T_\alpha.$$

Thus $f|T_\alpha$ is a modular form for $\Gamma(1)$, and T_α is a linear transformation of $M_k\left(\Gamma(1)\right)$. The space $S_k\left(\Gamma(1)\right)$ is clearly an invariant subspace.

If α, $\beta \in GL(2, \mathbb{Q})^+$, let α_i be as in Eq. (4.2) and also $\Gamma(1)\beta\Gamma(1) = \bigcup \Gamma(1)\beta_i$ (disjoint). We have

$$f|T_\alpha|T_\beta = \sum f|\alpha_i\beta_j = \sum_{\sigma\in\Gamma(1)\backslash GL(2,\mathbb{Q})^+} m(\alpha, \beta; \sigma)\, f|\sigma,$$

where σ runs through a set of representatives for $\sigma \in \Gamma(1)\backslash GL(2, \mathbb{Q})^+$, and $m(\alpha, \beta; \sigma)$ is the cardinality of the set of indices (i, j) such $\sigma \in \Gamma(1)\alpha_i\beta_j$. We see easily that $m(\alpha, \beta; \sigma)$ depends only on the double coset $\Gamma(1)\sigma\Gamma(1)$, so we may rewrite this

$$f|T_\alpha|T_\beta = \sum f|\alpha_i\beta_j = \sum_{\sigma\in\Gamma(1)\backslash GL(2,\mathbb{Q})^+/\Gamma(1)} m(\alpha, \beta; \sigma)\, f|T_\sigma, \tag{4.4}$$

where σ runs through a set of representatives for $\Gamma(1)\backslash GL(2, \mathbb{Q})^+/\Gamma(1)$. This prompts us to introduce a certain ring $\mathcal{R}$. Let $\mathcal{R}$ be the free Abelian group generated by the symbols $T_\alpha = T(\alpha)$ as α runs through a complete set for $\Gamma(1)\backslash GL(2, \mathbb{Q})^+/\Gamma(1)$. We define a multiplication in $\mathcal{R}$ by

$$T_\alpha \cdot T_\beta = \sum_{\sigma\in\Gamma(1)\backslash GL(2,\mathbb{Q})^+/\Gamma(1)} m(\alpha, \beta; \sigma)\, T_\sigma. \tag{4.5}$$

To check that this is associative, we note that if in addition to $\alpha, \alpha_i, \beta, \beta_i$ as above, γ and $\gamma_i \in GL(2,\mathbb{Q})^+$ with $\Gamma(1)\gamma\Gamma(1) = \bigcup \Gamma(1)\gamma_i$, we have

$$(T_\alpha \cdot T_\beta) \cdot T_\gamma = \sum_{\sigma \in \Gamma(1)\backslash GL(2,\mathbb{Q})^+/\Gamma(1)} m(\alpha, \beta, \gamma; \sigma)\, T_\sigma = T_\alpha \cdot (T_\beta \cdot T_\gamma),$$

where $m(\alpha, \beta, \gamma; \sigma)$ is the number of (i, j, k) such that $\sigma \in \Gamma(1)\alpha_i\beta_j\gamma_k$. With $\mathcal{R}$ given a ring structure in this manner, we see that $f \mapsto f|T_\alpha$ gives a right action of this ring on $M_k(\Gamma(1))$ or $S_k(\Gamma(1))$. The ring $\mathcal{R}$ is known as the *Hecke algebra* for $SL(2,\mathbb{Z})$.

Our objective is to determine the structure of the ring of Hecke operators. In particular, it is very important for us to show that this ring is commutative.

We recall the following theorem:

Theorem 1.4.1: Elementary divisor theorem *Let R be a principal ideal domain, let Λ_1 be a free R-module of rank n, and let Λ_2 be a free R-submodule, also of rank n. There exist a basis $\xi_1, \cdots, \xi_n$ of Λ_1 and nonzero elements $D_1, \cdots, D_n$ of R such that $D_{i+1}|D_i$ for $1 \le i < n$ and such that $D_1\xi_1, \cdots, D_n\xi_n$ is a basis of Λ_2.*

Proof This is a special case of Theorem III.7.8 of Lang (1993, p. 153). Note that in Lang's theorem, if the free module F is of finite rank (dimension) then the assumption that M be finitely generated is automatic by his Theorem III.7.1. ■

As an application we provide the following propositions.

Proposition 1.4.2 *A complete set of coset representatives for*

$$\Gamma(1)\backslash GL(2,\mathbb{Q})^+/\Gamma(1)$$

consists of the diagonal matrices

$$\operatorname{diag}(d_1, d_2) = \begin{pmatrix} d_1 & \\ & d_2 \end{pmatrix},$$

where $d_1, d_2 \in \mathbb{Q}$, and where d_1/d_2 is a positive integer.

Proof Let $\alpha \in GL(2,\mathbb{Q})^+$, and let N be a positive integer such that $N\alpha$ has integer coefficients. In the Elementary Divisor Theorem, we take $R = \mathbb{Z}$, we take $\Lambda_1 = \mathbb{Z}^2$, and we take Λ_2 to be the sublattice spanned by the rows of $N\alpha$. We see that there exists a basis ξ_1, ξ_2 of $\mathbb{Z}^2$ and positive integers D_1 and D_2 such that $D_2|D_1$, and such that $D_1\xi_1$ $D_2\xi_2$ is a basis of Λ_2. Replacing ξ_1 by $-\xi_1$ if necessary, we may assume that the unimodular matrix ξ with rows ξ_1 and ξ_2 has determinant one; this matrix is then an element of $SL(2,\mathbb{Z})$, and the rows of $\operatorname{diag}(D_1, D_2)\,\xi$ span the same lattice as $N\,\alpha$, so $\gamma\, N\alpha = \operatorname{diag}(D_1, D_2)\,\xi$ for some $\gamma \in GL(2,\mathbb{Z})$. The determinant of γ is positive because all the other matrices in this equation have a positive determinant, so actually $\gamma \in SL(2, \mathbb{Z})$. We see that if $d_i = D_i/N$, the matrices α and $\mathrm{diag}(d_1, d_2)$ generate the same double coset, as required.

To show that the double cosets generated by distinct diagonal matrices $\mathrm{diag}(d_1, d_2)$ with $d_1/d_2 \in \mathbb{Z}$ are disjoint, observe that d_1 and d_2 can be reconstructed from any representative α of the double coset: for d_2 is the greatest common divisor of the entries of α (a positive rational number), and $d_1 d_2$ is the determinant. ■

By an *antiautomorphism* of a group or ring, we mean a map $x \to x^*$ such that $(xy)^* = y^* x^*$.

Proposition 1.4.3 *We have $\Gamma(1)\alpha\Gamma(1) = \Gamma(1)^{\top}\alpha\Gamma(1)$.*

Proof This is a consequence of Proposition 1.4.2; each double coset contains a representative that is diagonal and hence symmetric. Hence each double coset is stable under transposition. ■

Lemma 1.4.2 *We may choose the coset representatives α_i in Eq. (4.2) so that also*

$$\Gamma(1)\alpha\Gamma(1) = \bigcup_{i=1}^{N} \alpha_i \Gamma(1).$$

Proof Because by Proposition 1.4.2, the double coset $\Gamma(1)\alpha\Gamma(1)$ contains a diagonal representative, it equals $\Gamma(1)^{\top}\alpha\Gamma(1) = \bigcup {}^{\top}\alpha_i\Gamma(1)$. Because α_i and ${}^{\top}\alpha_i$ generate the same double coset

$$\Gamma(1)\alpha_i \cap {}^{\top}\alpha_i\Gamma(1) \neq \emptyset.$$

Let β_i be an element of this intersection. Replacing α_i with β_i, we get a set of representatives with the required property. ■

Theorem 1.4.2 *The Hecke algebra $\mathcal{R}$ is commutative.*

The idea of the proof is that transposition $g \to {}^{\top}g$ is an antiautomorphism of $GL(2, \mathbb{Q})^+$, which thus induces an antiautomorphism of $\mathcal{R}$; by Proposition 1.4.3, this antiautomorphism of $\mathcal{R}$ is actually the identity map. But if the identity map is an antiautomorphism, the ring must be commutative! This idea is repeated in Theorems 2.2.3, 4.1.2, and 4.6.1.

Proof We give a definition of the structure constants $m(\alpha, \beta; \sigma)$ for the ring $\mathcal{R}$, which is more symmetrical than the one given before. Indeed, we will show that $m(\alpha, \beta; \sigma)$ is equal to $1/\deg(\sigma)$ times the cardinality of the set of indices (i, j) such that $\sigma \in \Gamma(1)\alpha_i\beta_j\Gamma(1)$, where $\deg(\sigma)$ is the cardinality of

$\Gamma(1)\backslash\Gamma(1)\sigma\Gamma(1)$. (This equals the cardinality of $\Gamma(1)\sigma\Gamma(1)/\Gamma(1)$, because, again by Lemma 4.5, the transpose mapping induces a bijection between these two sets.) To see this, write $\Gamma(1)\sigma\Gamma(1) = \bigcup \Gamma(1)\sigma_k$; then $\sigma \in \Gamma(1)\alpha_i\beta_j\Gamma(1)$ if and only if some $\sigma_k \in \Gamma(1)\alpha_i\beta_j$ and the number of σ_k is equal to $\deg(\sigma)$; consequently, the number of solutions to $\sigma \in \Gamma(1)\alpha_i\beta_j\Gamma(1)$ equals

$$\sum_k m(\alpha, \beta; \sigma_k) = \sum_k m(\alpha, \beta; \sigma) = \deg(\sigma)\, m(\alpha, \beta; \sigma).$$

We choose representatives α_i and β_i as in Lemma 1.4.2. We then have

$$\Gamma(1)\alpha\Gamma(1) = \bigcup \Gamma(1)\alpha_i = \bigcup \Gamma(1)\,{}^{\top}\alpha_i,$$

$$\Gamma(1)\beta\Gamma(1) = \bigcup \Gamma(1)\beta_i = \bigcup \Gamma(1)\,{}^{\top}\beta_i,$$

so

$$\begin{aligned} m(\alpha, \beta; \sigma) &= \frac{1}{\deg(\sigma)} \mathrm{Card}\{(i, j) | \sigma \in \Gamma(1)\,{}^{\top}\alpha_i\,{}^{\top}\beta_j\Gamma(1)\} \\ &= \frac{1}{\deg(\sigma)} \mathrm{Card}\{(i, j) | {}^{\top}\sigma \in \Gamma(1)\beta_j\alpha_i\Gamma(1)\} \\ &= m(\beta, \alpha; {}^{\top}\sigma). \end{aligned}$$

Because $T(\sigma) = T({}^{\top}\sigma)$, this proves that the ring $\mathcal{R}$ is commutative. ■

Because the Hecke algebra is commutative, there is no distinction between a right action and a left action. Hence if f is a modular form, we may write $T_\alpha f$ in place of $f|T_\alpha$.

We extend the Petersson inner product to all congruence cusp forms by

$$\langle f, g\rangle = \frac{1}{[\Gamma(1) : \Gamma(N)]} \int_{\Gamma(N)\backslash\mathcal{H}} f(z)\,\overline{g(z)}\, y^k\, \frac{dx\, dy}{y^2},$$

where N is chosen so that f and g are both modular forms for $\Gamma(N)$.

Theorem 1.4.3 *The T_α on $S_k(\Gamma(1))$ are self-adjoint with respect to the Petersson inner product.*

Proof By means of a simple change of variables $z \to \alpha^{-1}(z)$ in the definition of the inner product, it may be checked (Exercise 4.3) that

$$\langle f|\alpha, g\rangle = \langle f, g|\alpha^{-1}\rangle. \tag{4.6}$$

From the left expression, it is clear that this inner product is invariant under $\alpha \to \gamma\alpha$ where $\gamma \in \Gamma(1)$; from the right, it is invariant under $\alpha \to \alpha\gamma$. Hence Eq. (4.6) depends only on the coset of α in $\Gamma(1)\backslash GL(2, \mathbb{Q})^+/\Gamma(1)$.

Because the α_i in Eq. (4.2) all lie in this coset, we have, using Eq. (4.6)

$$\langle f|T_\alpha, g\rangle = \sum \langle f|\alpha_i, g\rangle = \deg(\alpha)\ \langle f|\alpha, g\rangle = \deg(\alpha)\ \left\langle f, g|\alpha^{-1}\right\rangle .$$

Using Proposition 1.4.3, this equals $\deg(\alpha)\left\langle f, g|{}^{\top}\alpha^{-1}\right\rangle$, or because scalar matrices act trivially, we have

$$\langle f|T_\alpha, g\rangle = \deg(\alpha)\,\langle f, g|\beta\rangle\,,$$

where $\beta = \det(\alpha)\,{}^{\top}\alpha^{-1}$. Now $\beta = S\alpha S^{-1}$, so α and β lie in the same double coset, and this equals

$$\deg(\alpha)\,\langle f, g|\alpha\rangle = \langle f, g|T_\alpha\rangle\,.$$

Hence T_α is self-adjoint with respect to the Petersson inner product. ∎

The Hecke algebra is a commutative family of self-adjoint operators on the finite-dimensional vector space $S_k\big(\Gamma(1)\big)$. As a consequence, there exists a basis of $S_k\big(\Gamma(1)\big)$ consisting of functions, each of which is an eigenfunction of all the Hecke operators (Exercise 1.4.1). If f is such a *Hecke eigenform*, let $f(z) = \sum A(n)\,q^n$ be its Fourier expansion. We will prove that $A(1) \neq 0$ and that if it is normalized so that $A(1) = 1$, then $L(s, f) = \sum A(n)\,n^{-s}$ has an Euler product.

Let n be a positive integer. Let $T(n)$ be the sum of the $T_{\alpha(d_1,d_2)}$ where d_1 and d_2 are integers such that $d_1 d_2 = n$, $d_2|d_1$, and

$$\alpha(d_1, d_2) = \begin{pmatrix} d_1 & \\ & d_2 \end{pmatrix}.$$

It is a consequence of Proposition 1.4.2 that if Δ_n is the subset of $\mathrm{Mat}(2, \mathbb{Z})$ consisting of elements of determinant n, and if

$$\Delta_n = \bigcup_j \Gamma(1)\,\delta_{n,j},$$

then $T(n)f = \sum f|\delta_{n,j}$. We may give an explicit decomposition of Δ_n into right cosets as follows:

Proposition 1.4.4 *We have*

$$\Delta_n = \bigcup_{\substack{a,d>0,\,ad=n\\ b \bmod d}} \Gamma(1)\begin{pmatrix} a & b \\ & d \end{pmatrix} \qquad \text{(disjoint)}.$$

Proof We leave the proof to the reader (Exercise 1.4.4). ∎

Let $f \in S_k\big(\Gamma(1)\big)$ have a Fourier expansion $\sum A(m)\,q^m$, and let $T(n)\,f$ have a Fourier expansion $\sum B(m)\,q^m$. We wish to express the coefficients $B(m)$ in terms of $A(m)$. It will be useful to extend the definition of $A(m)$ and $B(m)$ to all rational numbers by stipulating that $A(m) = B(m) = 0$ if $m \in \mathbb{Q} - \mathbb{Z}$. We have (denoting for convenience $e(x) = e^{2\pi i x}$)

$$
\begin{aligned}
(T(n)f)(z) &= \sum_{ad=n} \sum_{b \bmod d} \left(\frac{a}{d}\right)^{k/2} f\left(\frac{az+b}{d}\right) \\
&= \sum_{ad=n} \sum_{b \bmod d} \left(\frac{a}{d}\right)^{k/2} \sum_{m=1}^{\infty} A(m)\, e\left(\frac{amz}{d}\right) e\left(\frac{mb}{d}\right).
\end{aligned}
$$

Because $\sum_b e(mb/d) = d$ if $d|m$, or 0 otherwise, this may be written as

$$
\sum_{m=1}^{\infty} \sum_{\substack{ad=n \\ d|m}} \left(\frac{a}{d}\right)^{k/2} d\, e\left(\frac{amz}{d}\right) A(m).
$$

Consequently, we have

$$
B(m) = \sum_{\substack{ad=n \\ a|m}} \left(\frac{a}{d}\right)^{k/2} d\, A\left(\frac{md}{a}\right). \tag{4.7}
$$

Now suppose that f is a Hecke eigenform. Let us normalize the eigenvalues of the Hecke operators so that

$$
T(n)\, f = n^{1-k/2} \lambda(n)\, f. \tag{4.8}
$$

Proposition 1.4.5 *Let f be a Hecke eigenform with eigenvalues $\lambda(n)$ normalized as in Eq. (4.8), and Fourier coefficients $A(n)$. Then the Fourier coefficients $A(n)$ of f satisfy the following:*

(i) $A(1) \neq 0$.
(ii) If $A(1) = 1$, then $\lambda(n) = A(n)$ for all n.
(iii) If $A(1) = 1$, then the coefficients $A(n)$ are multiplicative; that is, if m and n are coprime, we have $A(nm) = A(n)\, A(m)$.

Proof By Eq. (4.7)

$$
n^{1-k/2} \lambda(n)\, A(m) = \sum_{\substack{ad=n \\ a|m}} \left(\frac{a}{d}\right)^{k/2} d\, A\left(\frac{md}{a}\right). \tag{4.9}
$$

Suppose that $(m, n) = 1$. Then the only a that divides both m and n is $a = 1$, so $d = n$ in this sum; it reduces to

$$
\lambda(n)\, A(m) = A(nm). \tag{4.10}
$$

Taking $m = 1$, this implies that $A(n) = \lambda(n)\,A(1)$. From this, we see that $A(1)$ cannot equal zero, because otherwise $A(n)$ would equal zero for all n. Parts (ii) and (iii) also follow immediately from Eq. (4.10). ■

By Proposition 1.4.5(i), we may adjust any Hecke eigenform by a constant so that $A(1) = 1$. Such a Hecke eigenform will be called *normalized*. We see that $S_k\big(\Gamma(1)\big)$ has a basis of normalized Hecke eigenforms. We now show that the L-function of a Hecke eigenform has a Euler product.

Theorem 1.4.4 *If f is a normalized Hecke eigenform, then*

$$L(s, f) = \sum A(n)\,n^{-s} = \prod_p \left(1 - A(p)\,p^{-s} + p^{k-1-2s}\right)^{-1}. \tag{4.11}$$

Proof It follows from the multiplicativity of the coefficients that

$$L(s, f) = \sum A(n)\,n^{-s} = \prod_p \left(\sum_{r=0}^{\infty} A(p^r)\,p^{-rs}\right).$$

We must evaluate the series on the right side. If $k \geq 1$, Eq. (4.9) and the equality $\lambda(p) = A(p)$ imply that

$$A(p^{r+1}) - A(p)\,A(p^r) + p^{k-1}\,A(p^{r-1}) = 0. \tag{4.12}$$

Equation (4.12) means that

$$\left(1 - A(p)\,X + p^{k-1}\,X^2\right)\left(\sum_{r=0}^{\infty} A(p^r)\,X^r\right) = 1. \tag{4.13}$$

Taking $X = p^{-s}$ in this identity, we obtain Eq. (4.11). ■

For example, $S_{12}\big(\Gamma(1)\big)$ is one dimensional, so Δ must be an eigenform. It is normalized because $\tau(1) = 1$. Hence we obtain the Euler product formula (3.18) of Ramanujan and Mordell.

We will consider the generalization of Hecke operator theory to congruence subgroups in the exercises.

Exercises

Let H be a Hilbert space, and let $T : H \to H$ be a bounded operator, that is, a continuous linear transformation. (The assumption of continuity is unnecessary if H is finite dimensional, the case of concern to us here.) The *adjoint* T^* of T is defined by the law $\langle Tx, y\rangle = \langle x, T^*y\rangle$. T is called *self-adjoint* if $T = T^*$ and *normal* if T and T^* commute. A self-adjoint operator is clearly normal.

Exercise 1.4.1 (a) Assume that H is finite dimensional. Let S be a family of normal operators on H such that if $T_1,\ T_2 \in S$, then T_1 and T_2 commute. Prove that H has a basis of simultaneous eigenvectors for the operators in S.
(b) Suppose that H is infinite dimensional, with an orthonormal basis x_n ($n \in \mathbb{Z}$). Let T_m be the "shift operator" $T_m x_n = x_{m+n}$. Show that this is a commutative family of normal operators, yet there are no simultaneous eigenvectors in H.

Exercise 1.4.2 Verify the statement that was made in connection with Eq. (4.4); that is, that $m(\alpha, \beta; \sigma)$, defined as in the text, depends only on the double coset $\Gamma(1)\sigma\Gamma(1)$.

Exercise 1.4.3 Verify Eq. (4.6).

Exercise 1.4.4 (a) Prove that if d_1, d_2 are positive integers, $d_2|d_1$, then $\Gamma(1)\begin{pmatrix} d_1 & \\ & d_2 \end{pmatrix}\Gamma(1)$ is the set of all matrices $\left(\begin{smallmatrix} a & b \\ c & d \end{smallmatrix}\right) \in \mathrm{Mat}_2(\mathbb{Z})$ such that $ad - bc = d_1 d_2$ and such that the greatest common divisor of a, b, c, d is d_2. (Note that this statement remains true if d_1 and d_2 are positive rational numbers such that d_1/d_2 is an integer if the "greatest common divisor of a, b, c and d" is interpreted to be the positive rational number that generates the same fractional ideal as is generated by a, b, c, and d.)
(b) Prove that if d_1, d_2 are positive integers, $d_2|d_1$, then

$$\Gamma(1)\begin{pmatrix} d_1 & \\ & d_2 \end{pmatrix}\Gamma(1) = \bigcup_{\substack{a,d>0,\ ad=d_1d_2 \\ b \bmod d \\ \gcd(a,b,d)=d_2}} \Gamma(1)\begin{pmatrix} a & b \\ & d \end{pmatrix} \qquad \text{(disjoint)}.$$

(c) Prove Proposition 1.4.4.

Exercise 1.4.5 Prove that the canonical map $\Gamma(1) \to SL(2, \mathbb{Z}/N\mathbb{Z})$ is surjective, and hence determine the index of $\Gamma(N)$ in $\Gamma(1)$.

Let N be a positive integer. The subgroups $\Gamma_0(N)$ and $\Gamma_1(N)$ are particularly important congruence subgroups of $SL(2, \mathbb{Z})$. $\Gamma_0(N)$ is the group of $\left(\begin{smallmatrix} a & b \\ c & d \end{smallmatrix}\right)$ such that $N|c$; $\Gamma_1(N)$ is the subgroup of $\Gamma_0(N)$ defined by the further conditions $a \equiv d \equiv 1 \bmod N$. Let χ be a Dirichlet character modulo N. We do *not* require χ to be primitive! We will assume that the weight k is positive and satisfies $\chi(-1) = (-1)^k$. Let $M_k\big(\Gamma_1(N)\big)$ and $S_k\big(\Gamma_1(N)\big)$ be defined as in the text, and let $M_k\big(\Gamma_0(N), \chi\big)$ and $S_k\big(\Gamma_0(N), \chi\big)$ be the subspaces of $S_k\big(\Gamma_1(N)\big)$ defined by the condition that

$$f|\gamma = \chi(d)\, f \qquad \text{for } \gamma = \left(\begin{smallmatrix} a & b \\ c & d \end{smallmatrix}\right) \in \Gamma_0(N).$$

Exercise 1.4.6 Prove that

$$S_k\big(\Gamma_1(N)\big) = \bigoplus_{\chi} S_k\big(\Gamma_0(N), \chi\big),$$

where the summation is over all Dirichlet characters χ modulo N. Moreover, show that the summands are orthogonal with respect to the Petersson inner product.

Exercise 1.4.7 Prove that if f is a modular form for a congruence group, there exists $\alpha \in GL(2, \mathbb{Q})^+$ such that $f|\alpha \in M_k(\Gamma_1(N))$ for some N.

Combining Exercises 1.4.6 and 1.4.7, we see that any congruence modular form may be expressed in terms of modular forms in the spaces $M_k(\Gamma_0(N), \chi)$. Thus there is no loss of generality in studying these modular forms.

Let Σ be the finite set of primes dividing N, and let $\mathbb{Z}_\Sigma$ be the localization of $\mathbb{Z}$ at the primes in Σ. Thus $\mathbb{Z}_\Sigma \subseteq \mathbb{Q}$ consists of fractions r/s with $r, s \in \mathbb{Z}$, and s coprime to N. Then $N\mathbb{Z}_\Sigma$ is an ideal of $\mathbb{Z}_\Sigma$, and $\mathbb{Z}_\Sigma/N\mathbb{Z}_\Sigma \cong \mathbb{Z}/N\mathbb{Z}$. We may therefore extend the character χ to a function on $\mathbb{Z}_\Sigma$ by $\chi(r/s) = \chi(r)/\chi(s)$.

Let $G_0(N)$ be the subgroup of $GL(2, \mathbb{Z}_\Sigma)$ consisting of elements $\alpha = \left(\begin{smallmatrix} a & b \\ c & d \end{smallmatrix}\right)$ with positive determinant such that $c \in N\mathbb{Z}_\Sigma$. If α is such a matrix, by abuse of notation, we will write $\chi(\alpha) = \chi(d)$, and this is a character of $G_0(N)$. We define, for a holomorphic function f on $\mathcal{H}$

$$f|_\chi \alpha = \chi(\alpha)^{-1} f|\alpha,$$

and this is a right action of $G_0(N)$.

Exercise 1.4.8 Prove that a complete set of coset representatives for $\Gamma_0(N)\backslash G_0(N)/\Gamma_0(N)$ consists of the matrices

$$\begin{pmatrix} d_1 & \\ & d_2 \end{pmatrix},$$

where d_2 and d_1 are positive elements of $\mathbb{Z}_\Sigma^\times$ and d_1/d_2 is a positive rational integer prime to N.

[HINT: If $\alpha \in G_0(N)$, first use Proposition 4.3 to write $\alpha = \gamma_1\delta\gamma_2$ with $\gamma_1, \gamma_2 \in SL(2, \mathbb{Z})$ where δ is the above diagonal matrix with eigenvalues d_1 and d_2; use Exercise 1.4.4(a) to determine the possible values of d_1 and d_2. Let $d = d_1/d_2$. Then d is a positive integer prime to N. Show that we have the freedom to adjust γ_2 on the left by an arbitrary element of $\Gamma_0(d)$, making a corresponding adjustment to γ_1. Using this freedom of choice, show without lose of generality that we may assume that $\gamma_2 \in \Gamma_0(N)$ and that this forces $\gamma_1 \in \Gamma_0(N)$ also.]

Exercise 1.4.9 Prove that if $d_2|d_1$, and d_1, d_2 are integers prime to N, then there are the same number of right cosets in

$$\Gamma_0(N)\backslash\Gamma_0(N)\alpha\Gamma_0(N), \qquad \alpha = \begin{pmatrix} d_1 & \\ & d_2 \end{pmatrix}$$

as in $\Gamma(1)\backslash\Gamma(1)\alpha\Gamma(1)$ and that we may use the same representatives in both cases.

Exercise 1.4.10 Define an abstract ring $\mathcal{R}_N$ of operators T_α using the double cosets $\Gamma_0(N)\alpha\Gamma_0(N)$ for $\alpha \in G_0(N)$. Prove that this abstract ring is isomorphic to a subring of the Hecke algebra $\mathcal{R}$ and hence is commutative.

Exercise 1.4.11 Define an action of $\mathcal{R}_N$ on $S_k(\Gamma_0(N), \chi)$ by

$$f|T_\alpha = \sum f|_\chi \alpha_i,$$

where $\Gamma_0(N)\alpha\Gamma_0(N) = \bigcup \Gamma_0(N)\alpha_i$. Prove that T_α is a normal operator with respect to the Petersson inner product.

Exercise 1.4.12 Prove that $S_k(\Gamma_0(N), \chi)$ has a basis consisting of elements that are eigenfunctions of the Hecke operators T_α for all $\alpha \in G_0(N)$.

Exercise 1.4.13 Determine the effect of the Hecke operators on the Fourier coefficients of $f \in S_k(\Gamma_0(N), \chi)$. Show that if $f = \sum A(n)\, q^n$ is an eigenfunction of all the Hecke operators, then

$$\begin{aligned} L(s, f) &= \sum A(n)\, n^{-s} \\ &= \left(\sum{}' A(n)\, n^{-s}\right) \prod_{p \nmid N} \left(1 - A(p)\, p^{-s} + \chi(p)\, p^{k-1-2s}\right)^{-1}, \end{aligned}$$

where $\sum' A(n)\, n^{-s}$ is the sum over all n whose only prime divisors are those primes dividing N.

Exercise 1.4.14 Consider $f(z) = \Delta(z) + \Delta(6z)$. Prove that $f \in S_{12}(\Gamma_0(6), 1)$ is an eigenfunction of the Hecke operators T_α with $\alpha \in G_0(6)$ but that $L(s, f)$ is not an Euler product because its coefficients $A(n)$ do not satisfy $A(2)\, A(3) = A(6)$.

This exercise illustrates that the theory we have presented for Hecke operators on congruence subgroups is incomplete. We will now describe without proof the solution to this dilemma.

If $dM|N$, and if $f \in S_k(\Gamma_0(M), \chi)$, then $z \mapsto f(dz)$ is in $S_k(\Gamma_0(N), \chi)$. We denote by $S_k(\Gamma_0(N), \chi)_{\text{old}}$ the space of "oldforms," which is the sum of the images of these spaces for proper divisors M of N. The orthogonal complement to $S_k(\Gamma_0(N), \chi)_{\text{old}}$ is denoted $S_k(\Gamma_0(N), \chi)_{\text{new}}$; this is the space of *newforms*. This space is stable under the action of the Hecke operators T_p for $p \nmid N$ (defined in the exercises for Section I.4) and has a basis of eigenforms of these Hecke operators. If f is such an eigenform, its first Fourier coefficient is automatically nonzero; we say that f is *normalized* if that coefficient is one. Thus the space of newforms has a basis consisting of normalized eigenforms.

Theorem 1.4.5: "Multiplicity one theorem" *$S_k(\Gamma_0(N), \chi)_{\text{new}}$ has a basis consisting of normalized newforms T_p for $p \nmid N$. If $f \in S_k(\Gamma_0(N), \chi)_{\text{new}}$, and $f' \in S_k(\Gamma_0(N'), \chi)_{\text{new}}$ are normalized newform eigenforms having the same eigenvalues for all but finitely many T_p, then $N = N'$ and $f = f'$.*

See Atkin and Lehner (1970), Miyake (1971), Casselman (1973), Deligne (1973b) and Li (1975). Accounts of Atkin–Lehner theory (including this result) may be found in the books of Lang (1976) and Miyake (1989). This result (whose proof we omit) is closely related to the adelic Theorem 3.3.11 (whose proof we give in Section 3.5). The relationship between the two theorems is discussed in Casselman (1973). Deligne (1973b, Théorème 2.2.6) gives a better proof of one of Casselman's main results. See also Gelbart (1975).

By restricting ourselves to normalized newforms, we obtain a completely satisfactory Hecke theory, without the pathology of Exercise 1.4.14. One of the goals of the Jacquet–Langlands theory is the reinterpretation of the multiplicity one theorem in representation theoretic terms.

An important aspect in the theory of modular forms is its interaction with algebraic geometry. We will not be able to discuss this subject in any detail but instead refer the reader to Shimura (1971) for further details. As an introduction to this circle of ideas, we consider only the very simplest case. For modular forms of weight two, the connection with algebraic geometry is less difficult than when the weight exceeds two, though even the weight-two case, known as *Eichler–Shimura theory* is already deep. If N is a positive integer, let $X = X_0(N)$ be the *modular curve*, that is, the Riemann surface obtained by adjoining the cusps to $\Gamma_0(N)\backslash\mathcal{H}$. There is a bijection between holomorphic modular forms of weight two and holomorphic differentials on $X_0(N)$; namely, if $f \in S_2\big(\Gamma_0(N)\big)$, then $f(z)\,dz$ is such a holomorphic one-form. By the Riemann–Roch theorem, it follows that

$$\dim S_2\big(\Gamma_0(N)\big) = \operatorname{genus}\big(X_0(N)\big). \tag{4.14}$$

Shimura has shown how to give X a natural structure of an algebraic variety defined over a cyclotomic number field. We can avoid introducing Abelian varieties if we specialize to the case where the genus of $X_0(N)$ is one. This is the case if $N = 11, 14, 15, 17, 19, 20, 21, 24, 27, 32, 36$, or 49. In this case, $X_0(N)$ is an elliptic curve defined over $\mathbb{Q}$, and if one is careful, it may even be realized as a scheme defined over $\mathbb{Z}$. (If the genus exceeds 1, we consider instead the Jacobian of $X_0(N)$.) By Eq. (4.14), $S_2\big(\Gamma_0(N)\big)$ is one dimensional and contains a unique normalized newform $f(z) = \sum a_n\, q^n$. The Euler product $L(s, f) = \sum a(n)\, n^{-1} = \prod_p Z_{f,p}(p^{-s})^{-1}$, where $Z_{f,p}(x) = 1 - a(p)\, x + p\, x^2$. The Fourier coefficients $a(p)$ have a remarkable concrete interpretation: $1 - a(p) + p$ is the cardinality of the group of $\mathbb{F}_p$-rational points of the reduction modulo p of X. If l is a prime distinct from p and also not dividing N, we may consider the action of $\mathrm{Gal}(\overline{\mathbb{Q}}/\mathbb{Q})$ on the points $X[l^n]$ of order l^n. Because $\varprojlim X[l^n] \cong \mathbb{Z}_l \times \mathbb{Z}_l$, we obtain a representation $\rho : \mathrm{Gal}(\overline{\mathbb{Q}}/\mathbb{Q}) \to GL(2, \mathbb{Z}_l)$, and we call $\varprojlim X[l^n]$ the *Tate module* of X. We find that if $\sigma_p \in \mathrm{Gal}(\overline{\mathbb{Q}}/\mathbb{Q})$ is a Frobenius element corresponding to p, then $\mathrm{tr}\,\big(\rho(\sigma_p)\big) = a(p)$ whereas $\mathrm{tr}\,\big(\rho(\sigma_p)\big) = p$.

More generally, if f is a normalized newform of arbitrary weight, Deligne and

Serre (1974) (see also Serre, (1977)) have associated with f a two-dimensional l-adic representation ρ whose character on Frobenius elements gives the Fourier coefficients of f. If the weight $k \neq 2$, higher dimensional algebraic varieties are required than just the modular curves X – instead, one looks for the l-adic representation in the étale cohomology of symmetric products of X. The case of weight one is interesting, for in this case Deligne and Serre (1974) showed that one may find a *complex* representation $\rho : \mathrm{Gal}(\overline{\mathbb{Q}}/\mathbb{Q}) \to GL(2, \mathbb{C})$ whose Artin L-function coincides with the L-function of f. The form of the Gamma factors in the functional equation show that a complex Galois representation can be associated with an automorphic form in this way only if the automorphic form is a modular form of weight one or a Maass form of weight zero with a Laplacian eigenvalue of $1/4$.

1.5 Twisting

We now reconsider the L-function of a modular form f. We will generalize Proposition 1.3.6, proving a functional equation for not one but an infinite family of *twisted L-functions* $L(s, f, \chi)$ associated with f and indexed by the primitive Dirichlet characters χ. Moreover, we will prove a converse to this theorem, due to Weil (1967), which asserts that the existence of sufficiently many such functional equations implies the existence of the modular form f. A theorem of this type is called a *converse theorem*. The particular theorem of Weil proved here is of historical importance, but we will not use it in the sequel. Rather, in Sections 1.9 and 1.10, we will use similar but different converse theorems to construct automorphic forms that we will use in the study of base change.

If $f \in S_k(\Gamma(1))$, a single functional equation is sufficient for the converse theorem. This case is atypical so we will later consider a modular form $f \in S_k(\Gamma_0(N), \psi)$, where k is a positive integer, and ψ is a character modulo N such that $\psi(-1) = (-1)^k$. To begin with the simplest case, let $N = 1$, so $\psi = 1$, $\Gamma_0(N) = \Gamma(1) = SL(2, \mathbb{Z})$, and k is even. As always, the variable $z \in \mathcal{H}$, and $q = e^{2\pi i z}$. In Sections 1.3 and 1.4, we saw that if $f \in S_k(\Gamma(1))$ is a normalized Hecke eigenform, then the L-function

$$L(s, f) = \sum A(n)\, n^{-s}, \qquad \text{where } f(z) = \sum A(n)\, q^n, \tag{5.1}$$

has analytic continuation, a functional equation

$$\Lambda(s, f) \equiv (2\pi)^{-s}\, \Gamma(s)\, L(s, f) = (-1)^{k/2}\, \Lambda(k - s, f), \tag{5.2}$$

and a Euler product

$$L(s, f) = \prod_p \left(1 - A(p)\, p^{-s} + p^{k-1-2s}\right)^{-1}. \tag{5.3}$$

There is a sense in which the converse is true.

Proposition 1.5.1 *Let $A(n)$ be a sequence of complex numbers such that $|A(n)| = O(n^K)$ for some sufficiently large real number K. Let $L(s, f)$ be defined by the series in Eq. (5.1), convergent for sufficiently large s. Assume that $\Lambda(s, f)$, defined in Eq. (5.2), has analytic continuation to all s, is bounded in every vertical strip $\sigma_1 \leq \mathrm{re}(s) \leq \sigma_2$, and satisfies the functional equation (5.2). Then $f(z)$, defined by the series in Eq. (5.1), is an element of $S_k(\Gamma(1))$. If the Euler product Eq. (5.3) is valid, it is a Hecke eigenform.*

Although Proposition 1.5.1 is too simple to be useful, we give its proof as an illustration. Before proving it, we need some preliminaries having to do with the Mellin inversion formula and the Phragmén–Lindelöf principle. In dealing with Dirichlet series, writing the complex variable s as $s = \sigma + it$, where σ and t are real is a tradition going back to Riemann (1859); as we shall soon see, σ is fixed in many arguments, and it is the behavior in t that is important. We will follow this convention.

Suppose that ϕ is a continuous function on the open interval $(0, \infty)$. We define the *Mellin transform*

$$\Phi(s) = \int_0^\infty \phi(y)\, y^s \,\frac{dy}{y} \tag{5.4}$$

wherever this integral is absolutely convergent. We observe that if

$$\int_0^1 \phi(y)\, y^s \,\frac{dy}{y}$$

is absolutely convergent for some value of s, it is absolutely convergent for any larger value, and if

$$\int_1^\infty \phi(y)\, y^s \,\frac{dy}{y}$$

is absolutely convergent for some s, it is absolutely convergent for any *smaller* value. Consequently, there exist $\sigma_1, \sigma_2 \in [-\infty, \infty]$ such that the integral defining $\Phi(s)$ is absolutely convergent for $\mathrm{re}(s) \in (\sigma_1, \sigma_2)$, whereas the integral is divergent for $\mathrm{re}(s) < \sigma_1$ or $\mathrm{re}(s) > \sigma_2$. Thus the domain of Φ, if nonempty, is a vertical strip. (If $\sigma_1 = -\infty$ or $\sigma_2 = \infty$, the domain may also be a half plane or all of $\mathbb{C}$. We can say nothing about convergence on $\mathrm{re}(s) = \sigma_1$ or σ_2.) Then with $\sigma \in (\sigma_1, \sigma_2)$ and $0 < y < \infty$, we will prove that

$$\phi(y) = \frac{1}{2\pi i} \int_{\sigma - i\infty}^{\sigma + i\infty} \Phi(s)\, y^{-s}\, ds. \tag{5.5}$$

This is known as the *Mellin inversion formula*.

We may deduce this from the Fourier inversion formula as follows. Let $\phi_\sigma(v) = \phi(e^v)\, e^{\sigma v}$. With our hypotheses, ϕ_σ is continuous and in $L^1(\mathbb{R}) \cap L^2(\mathbb{R})$, and putting $y = e^v$ in Eq. (5.4)

$$\Phi(\sigma + it) = \int_{-\infty}^{\infty} \phi_\sigma(v)\, e^{ivt}\, dv,$$

so by the Fourier inversion formula, we have

$$\phi_\sigma(v) = \frac{1}{2\pi} \int_{-\infty}^{\infty} \Phi(\sigma + it)\, e^{-ivt}\, dt.$$

Multiplying this formula by $e^{-\sigma v}$ and putting $y = e^v$, we obtain Eq. (5.5).

As an example, we have

$$e^{-y} = \frac{1}{2\pi i} \int_{\sigma - i\infty}^{\sigma + i\infty} \Gamma(s)\, y^{-s}\, ds, \tag{5.6}$$

if $\sigma,\ y > 0$. Note that by moving the path of integration to the left and summing the residues at $s = 0, -1, -2, \cdots$, we obtain the power series expansion for e^{-y}.

We also need the following proposition.

Proposition 1.5.2: (The Phragmén–Lindelöf principle) *Let $f(s)$ be a function that is holomorphic in the upper part of a strip*

$$\sigma_1 \le \mathrm{re}(s) \le \sigma_2, \qquad \mathrm{im}(s) > c,$$

such that $f(\sigma + it) = O\left(e^{t^\alpha}\right)$ for some real number $\alpha > 0$ when $\sigma_1 \le \sigma \le \sigma_2$. Suppose that $f(\sigma + it) = O(t^M)$ for $\sigma = \sigma_1, \sigma = \sigma_2$. Then $f(\sigma + it) = O(t^M)$ uniformly in σ for all $\sigma \in [\sigma_1, \sigma_2]$.

There are various forms of this result; another statement with an even weaker assumption on the growth in the strip may be found in Rudin (1974). It is not possible to dispense with the condition entirely, however (Exercise 1.5.2).

Proof See Lang (1970, p. 262). ∎

Finally, we need to know that the gamma function decays rapidly along vertical lines. For fixed σ, it may be deduced from Stirling's formula (see Whittaker and Watson, (1927), either 12.33 or 13.6) that

$$|\Gamma(\sigma + it)| \sim \sqrt{2\pi}\, |t|^{\sigma - 1/2}\, e^{-\pi |t|/2}, \tag{5.7}$$

as $t \to \pm\infty$.

We may now give the following proof.

Proof of Proposition 1.5.1 We observe first that the assumption that $|A(n)| = O(n^K)$ implies that $f(z)$ defined in Eq. (5.1) converges for $z \in \mathcal{H}$ and that $L(s, f)$, also defined in Eq. (5.1), converges in a right half plane. Because f is defined by a Fourier expansion, $f(z+1) = f(z)$, so in the notation of Section 2, f is invariant under the matrix $T = \begin{pmatrix} 1 & 1 \\ 0 & 1 \end{pmatrix}$. We will prove that f is also modular with respect to the matrix $S = \begin{pmatrix} 0 & -1 \\ 1 & 0 \end{pmatrix}$. Because S and T generate $\Gamma(1)$ by Proposition 1.2.3, it will follow that f is a modular form. It is cuspidal because by assumption the nonzero Fourier coefficients $A(n)$ have $n > 0$.

We must therefore prove that our hypotheses imply Eq. (3.16) when $z = iy \in \mathcal{H}$. It is sufficient to check this when y is a positive real number because if this is the case, the difference between the right and left sides of Eq. (3.16) is a holomorphic function on $\mathcal{H}$ vanishing on the positive imaginary axis. Because the zeros of a nonzero holomorphic function on $\mathcal{H}$ cannot have an accumulation point in the interior, we see that Eq. (3.16) for y real implies that $f(z) - z^{-k} f(-1/z)$ must vanish for all $z \in \mathcal{H}$, which is the required modularity of f for the second generator S of $\Gamma(1)$.

We will deduce Eq. (3.16) from the Mellin inversion formula. If σ is sufficiently large, we have, as in the proof of Proposition 1.3.6, the identity

$$\int_0^\infty f(iy)\, y^s \frac{dy}{y} = \Lambda(s, f),$$

so by the Mellin inversion formula

$$f(iy) = \frac{1}{2\pi i} \int_{\sigma - i\infty}^{\sigma + i\infty} \Lambda(s, f)\, y^{-s}\, ds = i^k \frac{1}{2\pi i} \int_{\sigma - i\infty}^{\sigma + i\infty} \Lambda(k - s, f)\, y^{-s}\, ds,$$

where we have used Eq. (5.2). Observe that $\Lambda(\sigma + it, f)$ is of rapid decay as $t \to \infty$ when $\sigma \gg 0$ because it is the product of an absolutely convergent Dirichlet series, itself bounded, with a gamma function that decays by Stirling's formula (5.7). Also, if $\sigma \ll 0$, we may use the functional equation to relate $\Lambda(\sigma + it, f)$ to $\Lambda(k - \sigma - it, f)$, which is of rapid decay as $t \to \infty$. We are assuming that $\Lambda(s, f)$ is bounded in vertical strips; by the Phragmén–Lindelöf principle, it follows that actually $\Lambda(\sigma + it, f) \to 0$ as $t \to \infty$ uniformly for σ in any compact set. This allows us to move the line of integration to the left by Cauchy's theorem; replacing $k - \sigma$ by σ and s by $k - s$, we obtain

$$f(iy) = i^k\, y^{-k} \frac{1}{2\pi i} \int_{\sigma - i\infty}^{\sigma + i\infty} \Lambda(s, f)\, y^s\, ds = i^k\, y^{-k} f\left(\frac{i}{y}\right).$$

Thus we obtain Eq. (3.16), as required.

The assertion that the Euler product Eq. (5.3) is equivalent to f being a Hecke eigenform follows from Eq. (4.7); the reasoning in Theorem 1.4.4 is easily seen to be reversible. ■

Next, we will generalize the functional equation (5.2) to obtain a functional equation for the *twisted L-function*:

$$L(s, f, \chi) = \sum \chi(n) A(n) n^{-s},$$

where χ is a primitive Dirichlet character.

We need to work in somewhat greater generality. Let $f \in S_k(\Gamma_0(N), \psi)$, where ψ is a Dirichlet character modulo N; this space was introduced before Exercise 1.4.6. The weight k is even if $\psi(-1) = 1$ and odd if $\psi(-1) = -1$. Let

$$w_N = \begin{pmatrix} & -1 \\ N & \end{pmatrix}.$$

Then w_N normalizes the group $\Gamma_0(N)$; in fact, we have

$$w_N \begin{pmatrix} a & b \\ c & d \end{pmatrix} w_N^{-1} = \begin{pmatrix} d & -c/N \\ -Nb & a \end{pmatrix},$$

so if $\gamma \in \Gamma_0(N)$, then $w_N \gamma w_N^{-1} \in \Gamma_0(N)$. Moreover, if $N|c$ and $ad - bc = 1$, then $\psi(d) = \overline{\psi(a)}$. It follows that if $f \in S_k(\Gamma_0(N), \psi)$, then for $\gamma = \left(\begin{smallmatrix} a & b \\ c & d \end{smallmatrix}\right) \in \Gamma_0(N)$ we have

$$f|w_N|\gamma = f|w_N \gamma w_N^{-1}|w_N = \overline{\psi(d)}\, f|w_N,$$

so $g = f|w_N \in S_k(\Gamma_0(N), \overline{\psi})$. Let $A(n)$ and $B(n)$ be the Fourier coefficients of f and g. Let

$$L(s, f) = \sum_{n=1}^{\infty} A(n) n^{-s}, \qquad L(s, g) = \sum_{n=1}^{\infty} B(n) n^{-s},$$

$$\Lambda(s, f) = (2\pi)^{-s} \Gamma(s) L(s, f), \qquad \Lambda(s, g) = (2\pi)^{-s} \Gamma(s) L(s, g).$$

Then $f = g|w_N^{-1}$ – which, we caution, is not the same as $f = g|w_N$ if k is odd – implies that

$$f(iy) = i^k N^{-k/2} y^{-k} g\left(\frac{i}{Ny}\right),$$

so arguing as at the end of Section 1.3, we obtain the analytic continuation of $\Lambda(s, f)$ and $\Lambda(s, g)$, and the functional equation

$$\Lambda(s, f) = i^k N^{-s+k/2} \Lambda(k - s, g). \tag{5.8}$$

More generally, let χ be a *primitive* character modulo D. (We do not assume that ψ is primitive modulo N, however.) We assume that N and D are coprime integers. Let

$$L(s, f, \chi) = \sum_{n=1}^{\infty} \chi(n)\, A(n)\, n^{-s}, \qquad L(s, g, \overline{\chi}) = \sum_{n=1}^{\infty} \overline{\chi(n)}\, B(n)\, n^{-s},$$

$$\Lambda(s, f, \chi) = (2\pi)^{-s}\, \Gamma(s)\, L(s, f, \chi), \quad \Lambda(s, g, \overline{\chi}) = (2\pi)^{-s}\, \Gamma(s)\, L(s, g, \overline{\chi}).$$

We will prove the functional equation

$$\Lambda(s, f, \chi) = i^k\, \chi(N)\, \psi(D)\, \frac{\tau(\chi)^2}{D}\, (D^2 N)^{-s+k/2}\, \Lambda(k - s, g, \overline{\chi}), \tag{5.9}$$

where the Gauss sum $\tau(\chi)$ is defined in Eq. (1.2). This reduces to Eq. (5.8) in the special case where $D = 1$ (so $\chi = 1$).

Let

$$f_\chi(z) = \sum_{n=1}^{\infty} \chi(n)\, A(n)\, q^n, \qquad g_{\overline{\chi}}(z) = \sum_{n=1}^{\infty} \overline{\chi(n)}\, B(n)\, q^n.$$

By Eq. (1.7), we have

$$\chi(n)\, q^n = \frac{\chi(-1)\, \tau(\chi)}{D} \sum_{\substack{m \bmod D \\ (m, D) = 1}} \overline{\chi(m)}\, e^{2\pi i n(z + m/D)}.$$

Therefore, we have

$$f_\chi = \frac{\chi(-1)\, \tau(\chi)}{D} \sum_{\substack{m \bmod D \\ (m, D) = 1}} \overline{\chi(m)}\, f \Big| \begin{pmatrix} D & m \\ & D \end{pmatrix}.$$

This, in view of the discussion following Lemma 1.4.1, shows that f_χ is a congruence cusp form of weight k. (See also Exercise 1.5.1.) We have

$$\begin{aligned} & f_\chi \Big| \begin{pmatrix} & -1 \\ D^2 N & \end{pmatrix} \\ &= f_\chi \Big| \begin{pmatrix} D & \\ & -1/ND \end{pmatrix} = \frac{\chi(-1)\, \tau(\chi)}{D} \sum_{\substack{m \bmod D \\ (m, D) = 1}} \overline{\chi(m)} \\ &\quad \times g \Big| \begin{pmatrix} & 1 \\ -N & \end{pmatrix} \begin{pmatrix} D & m \\ & D \end{pmatrix} \begin{pmatrix} D & \\ & -1/ND \end{pmatrix} \\ &= \frac{\chi(-1)\, \tau(\chi)}{D} \sum_{\substack{m \bmod D \\ (m, D) = 1}} \overline{\chi(m)}\, g \Big| \begin{pmatrix} D & -r \\ -Nm & s \end{pmatrix} \begin{pmatrix} D & r \\ & D \end{pmatrix}, \end{aligned}$$

where $r = r(m)$ and $s = s(m)$ are chosen so that $Ds - rNm = 1$. Observing that $\overline{\chi(m)} = \chi(-N)\,\chi(r)$, we obtain

$$f_\chi \Big| \begin{pmatrix} & -1 \\ D^2N & \end{pmatrix} = \chi(N)\,\frac{\tau(\chi)}{D} \sum_{\substack{r \bmod D \\ (r,D)=1}} \chi(r)\, g \Big| \begin{pmatrix} D & -r \\ -Nm & s \end{pmatrix} \begin{pmatrix} D & r \\ & D \end{pmatrix}. \tag{5.10}$$

Now because $g \in S_k\big(\Gamma_0(N), \overline{\psi}\big)$

$$g \Big| \begin{pmatrix} D & -r \\ -Nm & s \end{pmatrix} = \psi(D)\, g. \tag{5.11}$$

Moreover, we have

$$g_{\overline{\chi}} = \frac{\chi(-1)\,\tau(\overline{\chi})}{D} \sum_{\substack{r \bmod D \\ (r,D)=1}} \chi(r)\, g \Big| \begin{pmatrix} D & r \\ & D \end{pmatrix}. \tag{5.12}$$

Combining Eqs. (5.10), (5.11), and (5.12), we obtain

$$f_\chi \Big| \begin{pmatrix} & -1 \\ D^2N & \end{pmatrix} = \chi(-N)\,\psi(D)\,\frac{\tau(\chi)}{\tau(\overline{\chi})}\, g_{\overline{\chi}} = \chi(N)\,\psi(D)\,\frac{\tau(\chi)^2}{D}\, g_{\overline{\chi}}. \tag{5.13}$$

Now Eq. (5.9) follows from Eq. (5.13) by imitating the proof of Eq. (5.8).

Hecke conceived the idea that this reasoning might be reversible – that sufficiently many functional equations might imply the existence of a modular form – but he did not obtain a result assuming only primitive twists. This was done by Weil in (1967). A definitive converse theorem for automorphic representations of $GL(2)$ was included in Jacquet and Langlands (1970). The converse theorem for $GL(3)$ was proved by Jacquet, Piatetski–Shapiro, and Shalika (1979), and a result for $GL(n)$ was given by Cogdell and Piatetski–Shapiro (1994).

We note that for $f \in S_k(\Gamma_0(N), \psi)$, a single functional equation (as in Proposition 1.5.1) will definitely not suffice. In the proof of Proposition 1.5.1, we used a very special property of $SL(2, \mathbb{Z})$, namely, that it is generated by the matrices S and T in Proposition 1.2.3. To compensate for the fact that $\Gamma_0(N)$ requires more generators than $SL(2, \mathbb{Z})$, we must also assume functional equations for the twists $L(s, f, \chi)$.

Theorem 1.5.1: Weil *Suppose N is a positive integer and ψ is a Dirichlet character modulo N, not assumed to be primitive. Suppose that $A(n)$ and $B(n)$ are sequences of complex numbers such that $|A(n)|, |B(n)| = O(n^K)$ for some sufficiently large real number K. If D is relatively prime to N, and χ is a primitive Dirichlet character modulo D, let $L_1(s, \chi) = \sum \chi(n)\, A(n)\, n^{-s}$, $L_2(s, \overline{\chi}) = \sum \overline{\chi(n)}\, B(n)\, n^{-s}$, and let*

$$\Lambda_1(s, \chi) = (2\pi)^{-s}\,\Gamma(s)\,L_1(s, \chi), \qquad \Lambda_2(s, \overline{\chi}) = (2\pi)^{-s}\,\Gamma(s)\,L_2(s, \overline{\chi}).$$

Let S be a finite set of primes including those dividing N. Assume that whenever the conductor D of χ is either 1 or a prime $D \notin S$, $\Lambda_1(s,\chi)$ and $\Lambda_2(s,\overline{\chi})$, originally defined for $\mathrm{re}(s)$ *sufficiently large, have analytic continuation to all s, are bounded in every vertical strip $\sigma_1 \le \mathrm{re}(s) \le \sigma_2$, and satisfy the functional equation*

$$\Lambda_1(s,\chi) = i^k\,\chi(N)\,\psi(D)\,\frac{\tau(\chi)^2}{D}\,(D^2N)^{-s+k/2}\,\Lambda_2(k-s,\overline{\chi}). \qquad (5.14)$$

Then $f(z) = \sum A(n)\,q^n$ is a modular form in $M_k(\Gamma_0(N),\psi)$.

Remark As the proof will show, it is sufficient to assume the analytic continuation and functional equation (5.14) when D is a prime outside any given finite set of primes. Weil proved a slightly more general theorem in which Λ_1 and Λ_2 are allowed to have poles, which they may have if f is not cuspidal.

We will need the following lemma.

Lemma 1.5.1 *Let f be a holomorphic function on $\mathcal{H}$, and let $M \in SL(2,\mathbb{R})$ be an elliptic element of infinite order such that $f|M = f$. Then $f = 0$.*

Proof Let $a \in \mathcal{H}$ be the fixed point of the elliptic element M. We consider the Cayley transform

$$C = \begin{pmatrix} 1 & -a \\ 1 & -\overline{a} \end{pmatrix},$$

which maps $\mathcal{H}$ onto the unit disk $\mathcal{D}$, and takes a to the origin. Then

$$CMC^{-1} = \begin{pmatrix} \alpha & \\ & \alpha^{-1} \end{pmatrix},$$

where α and α^{-1} are the eigenvalues of M. They have absolute value one (because M is elliptic) but are not roots of unity (because M has infinite order). Let $f_1 = f|C^{-1}$. (We defined the $|$ operator in Eq. (4.1) for elements of $GL(2,\mathbb{R})^+$, but here define $|C^{-1}$ by the same formula. For the following argument, it is immaterial which square root is taken of the determinant in Eq. (4.1).) Then

$$f_1 = f_1\bigg|\begin{pmatrix} \alpha & \\ & \alpha^{-1} \end{pmatrix}, \qquad f_1(z) = \alpha^k\,f_1(\alpha^2 z).$$

Now f_1 has a power series expansion $f_1(z) = \sum \lambda_n\,z^n$. We have $\lambda_n = \alpha^{2n+k}\lambda_n$, and because α is not a root of unity, this implies that $\lambda_n = 0$ for all n. ■

Proof of Theorem 1.5.1 Let

$$f_\chi(z) = \sum_{n=1}^{\infty} A(n)\,\chi(n)\,q^n, \qquad g_{\overline{\chi}}(z) = \sum_{n=1}^{\infty} B(n)\,\overline{\chi(n)}\,q^n.$$

These series are convergent for $z \in \mathcal{H}$ because of the assumed growth conditions on $A(n)$ and $B(n)$. We will show that the Mellin inversion formula implies that if χ is a primitive character modulo D, where either $D = 1$ or D is a prime not in S, we have Eq. (5.13); this is similar to the proof of Proposition 1.5.1. It is sufficient that the left and right sides of Eq. (5.13) agree on the positive imaginary axis. If σ is sufficiently large, we have

$$\int_0^\infty f_\chi(iy)\, y^s \frac{dy}{y} = \Lambda_1(s, \chi),$$

so by the Mellin inversion formula

$$\begin{aligned} f_\chi(iy) &= \frac{1}{2\pi i} \int_{\sigma - i\infty}^{\sigma + i\infty} \Lambda_1(s, \chi)\, y^{-s}\, ds \\ &= i^k\, \chi(N)\, \psi(D)\, \frac{\tau(\chi)^2}{D} \frac{1}{2\pi i} \int_{\sigma - i\infty}^{\sigma + i\infty} (D^2 N)^{-s+k/2}\, \Lambda_2(k - s, \overline{\chi})\, y^{-s}\, ds, \end{aligned}$$

where we have used Eq. (5.14). Now, as in the proof of Proposition 1.5.1, the Phragmén–Lindelöf principle implies that $\Lambda_1(\sigma + it, \chi) \to 0$ as $t \to \infty$ for every value of σ, allowing us to move the line of integration to the left by Cauchy's theorem; replacing $k - \sigma$ by σ and s by $k - s$, we obtain

$$\begin{aligned} f_\chi(iy) &= i^k\, \chi(N)\, \psi(D)\, \frac{\tau(\chi)^2}{D}\, (D^2 N)^{-k/2}\, y^{-k} \\ &\quad \times \frac{1}{2\pi i} \int_{\sigma - i\infty}^{\sigma + i\infty} \Lambda_2(s, \overline{\chi})\, (D^2 N y)^s\, ds \\ &= i^k\, \chi(N)\, \psi(D)\, \frac{\tau(\chi)^2}{D}\, (D^2 N)^{-k/2}\, y^{-k}\, g_{\overline{\chi}}\left(\frac{i}{D^2 N y}\right), \end{aligned}$$

from which we get Eq. (5.13).

Now Eqs. (5.10) and (5.12), are known to be true for all D and primitive χ modulo D; these formulas do not require f and g to be automorphic. The derivation of Eq. (5.10) requires that $f|w_N = g$, but Eq. (5.13), which we have already proved when $D = 1$ (so $\chi = 1$), reduces to this fact.

Now let $D \notin S$ be prime. We would like to argue that Eqs. (5.10), (5.12), and (5.13) imply Eq. (5.11). This is not trivial, however, because we are only assuming functional equations for *primitive* characters modulo the prime $D \notin S$. What we can say immediately is that if $c(r)$ is any function of the nonzero residue classes modulo D such that $\sum c(r) = 0$, we have

$$\sum_{\substack{r \bmod D \\ (r,D)=1}} c(r)\, g\bigg|\begin{pmatrix} D & -r \\ -Nm & s \end{pmatrix}\begin{pmatrix} 1 & r/D \\ & 1 \end{pmatrix} = \sum_{\substack{r \bmod D \\ (r,D)=1}} c(r)\,\psi(D)\, g\bigg|\begin{pmatrix} 1 & r/D \\ & 1 \end{pmatrix}.$$

(Here, for given r, we choose $m = m(r)$ and $s = s(r)$ such that $Ds - Nmr = 1$.) To see this, note that by Eqs. (5.10), (5.12), and (5.13), this is true if $c(r) = \chi(r)$, where χ is a primitive Dirichlet character modulo D, and these characters form a basis of the $D-1$-dimensional space of functions $c(r)$. Let us express this in another way. Let us extend the right action of $GL(2,\mathbb{Q})^+$ on the holomorphic functions on $\mathcal{H}$ to a right action of the group algebra $\mathbb{C}[GL(2,\mathbb{Q})^+]$. Let Ω be the annihilator in this ring of g. It is a right ideal. Then we have

$$\sum_{\substack{r \bmod D \\ (r,D)=1}} c(r)\begin{pmatrix} D & -r \\ -Nm & s \end{pmatrix}\begin{pmatrix} 1 & r/D \\ & 1 \end{pmatrix} \equiv \sum_{\substack{r \bmod D \\ (r,D)=1}} c(r)\psi(D)\begin{pmatrix} 1 & r/D \\ & 1 \end{pmatrix}$$

mod Ω. In particular, if D is odd, we may take c to be the function that is 1 on the residue class of r, -1 on the class of $-r$, and zero elsewhere. We obtain

$$\left[\begin{pmatrix} D & -r \\ -Nm & s \end{pmatrix} - \psi(D)\right]\begin{pmatrix} 1 & r/D \\ & 1 \end{pmatrix} \equiv \left[\begin{pmatrix} D & r \\ Nm & s \end{pmatrix} - \psi(D)\right]\begin{pmatrix} 1 & -r/D \\ & 1 \end{pmatrix} \mod \Omega. \tag{5.15}$$

Now let us assume that D and s are distinct odd primes not in the set S such that $Ds \equiv 1 \bmod N$. Let r and m be chosen so that $Ds - Nmr = 1$. We have, in addition to Eq. (5.15), the equivalence

$$\left[\begin{pmatrix} s & -r \\ -Nm & D \end{pmatrix} - \psi(s)\right]\begin{pmatrix} 1 & r/s \\ & 1 \end{pmatrix} \equiv \left[\begin{pmatrix} s & r \\ Nm & D \end{pmatrix} - \psi(s)\right]\begin{pmatrix} 1 & -r/s \\ & 1 \end{pmatrix} \mod \Omega. \tag{5.16}$$

Now if we multiply Eq. (5.15) on the right by $\begin{pmatrix} 1 & r/D \\ & 1 \end{pmatrix}$ we obtain

$$\left[\begin{pmatrix} D & -r \\ -Nm & s \end{pmatrix} - \psi(D)\right]\begin{pmatrix} 1 & 2r/D \\ & 1 \end{pmatrix} \equiv \begin{pmatrix} D & r \\ Nm & s \end{pmatrix} - \psi(D) \tag{5.17}$$

mod Ω, and if we multiply Eq. (5.16) on the right by

$$-\psi(D)\begin{pmatrix} 1 & r/s \\ & 1 \end{pmatrix}\begin{pmatrix} D & -r \\ -Nm & s \end{pmatrix}\begin{pmatrix} 1 & 2r/D \\ & 1 \end{pmatrix},$$

we obtain

$$-\psi(D)\left[\begin{pmatrix} s & -r \\ -Nm & D \end{pmatrix} - \psi(s)\right]\begin{pmatrix} 1 & 2r/s \\ & 1 \end{pmatrix}\begin{pmatrix} D & -r \\ -Nm & s \end{pmatrix}\begin{pmatrix} 1 & 2r/D \\ & 1 \end{pmatrix}$$
$$\equiv \left[\begin{pmatrix} D & -r \\ -Nm & s \end{pmatrix} - \psi(D)\right]\begin{pmatrix} 1 & 2r/D \\ & 1 \end{pmatrix} \mod \Omega,$$

or

$$\left[\begin{pmatrix} D & r \\ Nm & s \end{pmatrix} - \psi(D)\right]\begin{pmatrix} s & -r \\ -Nm & D \end{pmatrix}\begin{pmatrix} 1 & 2r/s \\ & 1 \end{pmatrix}\begin{pmatrix} D & -r \\ -Nm & s \end{pmatrix}$$
$$\times \begin{pmatrix} 1 & 2r/D \\ & 1 \end{pmatrix} \equiv \left[\begin{pmatrix} D & -r \\ -Nm & s \end{pmatrix} - \psi(D)\right]\begin{pmatrix} 1 & 2r/D \\ & 1 \end{pmatrix} \mod \Omega. \tag{5.18}$$

Combining Eqs. (5.17) and (5.18), we see that

$$\left[\begin{pmatrix} D & r \\ Nm & s \end{pmatrix} - \psi(D)\right][1 - M] \in \Omega,$$

where

$$M = \begin{pmatrix} s & -r \\ -Nm & D \end{pmatrix}\begin{pmatrix} 1 & 2r/s \\ & 1 \end{pmatrix}\begin{pmatrix} D & -r \\ -Nm & s \end{pmatrix}\begin{pmatrix} 1 & 2r/D \\ & 1 \end{pmatrix}$$
$$= \begin{pmatrix} 1 & 2r/D \\ -2Nm/s & -3 + \frac{4}{Ds} \end{pmatrix}.$$

Let

$$g_1 = g\bigg|\begin{pmatrix} D & r \\ Nm & s \end{pmatrix} - \psi(D)g.$$

We see that g_1 is a holomorphic function on $\mathcal{H}$ such that $g_1 = g_1|M$.

Now, M is elliptic because $|\mathrm{tr}(M)| < 2$; its eigenvalues are not roots of unity, because their sum $\mathrm{tr}(M)$ is not an algebraic integer. Hence the order of M is infinite. It follows from Lemma 1.5.1 that $g_1 = 0$.

We may now prove that $g \in M_k(\Gamma_0(N), \overline{\psi})$. We wish to show that

$$g\bigg|\begin{pmatrix} a & b \\ Nc & d \end{pmatrix} = \psi(a)\, g.$$

We have proved this in the case where a and d are primes not in S – this is the assertion, just established, that $g_1 = 0$. In general, Dirichlet's theorem on primes in an arithmetic progression assures that we may find u and v such that $D = a - uNc$ and $s = d - vNc$ are such primes. Let $r = -av + uvNc + b - ud$. Then

$$g\bigg|\begin{pmatrix} a & b \\ Nc & d \end{pmatrix} = g\bigg|\begin{pmatrix} 1 & u \\ & 1 \end{pmatrix}\begin{pmatrix} D & r \\ Nc & s \end{pmatrix}\begin{pmatrix} 1 & v \\ & 1 \end{pmatrix} = \psi(D)\, g = \psi(a)\, g,$$

as required. We see that $g \in M_k(\Gamma_0(N), \overline{\psi})$; the corresponding property of f follows. ■

Exercises

Exercise 1.5.1 Prove that if $f \in S_k(\Gamma_0(N), \psi)$, then $f_\chi \in S_k(\Gamma_0(D^2N), \chi^2\psi)$.

Exercise 1.5.2 This example shows that the hypothesis of moderate growth in the Phragmén–Lindelöf principle (Proposition 1.5.2) is necessary. Let $f(s) = e^{e^{-is}}$. Show that f is bounded on the lines $\mathrm{re}(s) = \pm\pi/2$ but that f is not bounded in the strip $-\pi/2 < \mathrm{re}(s) < \pi/2$.

1.6 The Rankin–Selberg Method

Although earlier we reserved the term "automorphic" for automorphic functions in the sense of Section 1.3, modular forms are also examples of *automorphic forms*, a more general notion that includes *Maass forms* such as the Eisenstein series encountered in this section.

The converse theorem, of which Theorem 1.5.1 is the archetype, shows that a possible method of proving the existence of an automorphic form is to prove by any method the functional equations of sufficiently many of the the L-series attached to it. One of the most powerful methods of doing this is the *Rankin–Selberg method*, which we now study. The Rankin–Selberg method, which originated independently in the papers of Rankin (1939) and Selberg (1940), seeks to represent an L-function as an integral of one or more automorphic forms against an *Eisenstein series*, itself a type of automorphic form. The Eisenstein series itself has a functional equation, and so if the L-function can be represented as such an integral, it inherits this functional equation. We will return to the Eisenstein series, and to the Rankin–Selberg method, in Section 3.7. We will consider the Rankin–Selberg method from a more philosophical viewpoint in Section 1.8 of this chapter.

We will now define the Eisenstein series for $SL(2, \mathbb{Z})$. Let $z = x + iy \in \mathcal{H}$. We consider

$$E(z, s) = \pi^{-s}\,\Gamma(s)\;\tfrac{1}{2}\sum_{\substack{m,n \in \mathbb{Z}\\ (m,n) \neq (0,0)}} \frac{y^s}{|mz+n|^{2s}}. \tag{6.1}$$

The series is absolutely convergent if $\mathrm{re}(s) > 1$ (Exercise 1.6.1). It is similar to the holomorphic Eisenstein series that were introduced in Section 1.3, but there are two important differences: $E(z, s)$ is not holomorphic as a function of z, and it is strictly *automorphic* (not just modular); that is

$$E(\gamma(z), s) = E(z, s) \qquad \text{for } \gamma \in SL(2, \mathbb{Z}). \tag{6.2}$$

The proof is similar to the proof that the holomorphic Eisenstein series E_k is modular (Exercise 1.6.1).

The most significant property of $E(z, s)$ is its *analytic continuation and functional equation.*

Theorem 1.6.1 *$E(z, s)$, originally defined for* $\mathrm{re}(s) > 1$, *has meromorphic continuation to all s; it is analytic except at $s = 1$ and $s = 0$, where it has simple poles. The residue at $s = 1$ is the constant function $z = 1/2$. The Eisenstein series satisfies the functional equation*

$$E(z, s) = E(z, 1 - s). \tag{6.3}$$

We have

$$E(x + iy, s) = O(y^{\sigma}) \qquad \text{as } y \to \infty, \tag{6.4}$$

where $\sigma = \max\big(\mathrm{re}(s), 1 - \mathrm{re}(s)\big)$.

Proof One proof of most of these properties may be found in Exercise 1.6.2. We will give another proof, based on computing the Fourier coefficients of $E(z, s)$; this will have the additional benefit of giving us precise information about the behavior of $E(z, s)$ in the neighborhood of the cusp.

Before we prove Theorem 1.6.1, we will develop some properties of Bessel functions that we need. The *K-Bessel function,* also known as the *Macdonald Bessel function* is defined by

$$K_s(y) = \tfrac{1}{2} \int_0^{\infty} e^{-y(t+t^{-1})/2}\, t^s \,\frac{dt}{t}. \tag{6.5}$$

(We caution the reader that Whittaker and Watson (1927) regrettably do not follow the standard normalization of K_s.) If $y > 0$, the integrand in Eq. (6.5) decays rapidly as $t \to 0$ or ∞, so Eq. (6.5) is convergent for all values of s. We have the estimate

$$|K_s(y)| \le e^{-y/2}\, K_{\mathrm{re}(s)}(2) \qquad \text{if } y > 4. \tag{6.6}$$

To see this, note that if $a, b > 2$ we have $ab > a + b$, so $e^{-ab} < e^{-a}\, e^{-b}$. Applying this with $a = y/2$ and $b = t + t^{-1}$, and integrating with respect to t, we obtain Eq. (6.6). Because the integrand in Eq. (6.5) is invariant under $t \mapsto t^{-1}$, $s \mapsto -s$, we see that

$$K_s(y) = K_{-s}(y). \tag{6.7}$$

We will show that if $\mathrm{re}(s) > 1/2$, and r is real, we have

$$\left(\frac{y}{\pi}\right)^s \Gamma(s) \int_{-\infty}^{\infty} (x^2+y^2)^{-s}\, e^{2\pi i r x}\, dx$$
$$= \begin{cases} \pi^{-s+1/2}\,\Gamma\left(s-\frac{1}{2}\right) y^{1-s} & \text{if } r = 0; \\ 2\,|r|^{s-1/2}\,\sqrt{y}\,K_{s-1/2}(2\pi\,|r|\,y) & \text{if } r \neq 0. \end{cases} \tag{6.8}$$

Substituting the definition of the gamma function as a definite integral, the left side of Eq. (6.8) equals

$$\int_{-\infty}^{\infty}\int_{0}^{\infty} e^{-t}\left(\frac{ty}{\pi(x^2+y^2)}\right)^s e^{2\pi i r x}\,\frac{dt}{t}\,dx$$
$$= \int_{0}^{\infty}\int_{-\infty}^{\infty} e^{-\pi t(x^2+y^2)/y}\, t^s\, e^{2\pi i r x}\, dx\,\frac{dt}{t},$$

where we have interchanged the order of integration – the integral is absolutely convergent because $\mathrm{re}(s) > 1/2$ – and made a change of variables $t \mapsto \pi t(x^2+y^2)/y$. Now we have

$$\int_{-\infty}^{\infty} e^{-t\pi x^2/y}\, e^{2\pi i r x}\, dx = \begin{cases} \sqrt{\frac{y}{t}} & \text{if } r = 0; \\ \sqrt{\frac{y}{t}}\, e^{-y\pi r^2/t} & \text{if } r \neq 0 \end{cases}$$

(Cf. Eq. (1.12).) We substitute this into the previous integral. If $r = 0$, we obtain an Eulerian integral and we obtain Eq. (6.8); if $r \neq 0$, the substitution $t \to |r|t$ gives us the Macdonald integral, from which we get Eq. (6.8) in this case also.

We would like to compute the Fourier expansion of $E(z,s)$ – as a byproduct, we will obtain Theorem 1.6.1. We have $E(z,s) = E(z+1,s)$; this is Eq. (6.2) with $\gamma = T$ in the notation of Section 1.2. Consequently, $E(z,s)$ has a Fourier expansion in

$$E(z,s) = \sum_{r=-\infty}^{\infty} a_r(y,s)\, e^{2\pi i r x}. \tag{6.9}$$

We will compute the Fourier coefficients

$$a_r(y,s) = \int_0^1 E(x+iy,s)\, e^{-2\pi i r x}\, dx.$$

Firstly, there is the contribution of the terms with $m = 0$. Such a term is independent of x, and hence contributes only to a_0. Because n and $-n$

contribute equally, this contribution equals

$$\pi^{-s}\,\Gamma(s)\,y^s\sum_{n=1}^{\infty} n^{-2s} = \pi^{-s}\,\Gamma(s)\,\zeta(2s)\,y^s. \tag{6.10}$$

This is part, but not all, of a_0, as we shall see shortly.

Next we have the contribution of the terms with $m \neq 0$. Because (m, n) and $(-m, -n)$ contribute equally, we may simply sum the terms with $m > 0$. The contribution to a_r equals

$$\pi^{-s}\,\Gamma(s)\,y^s\sum_{m=1}^{\infty}\sum_{n=-\infty}^{\infty}\int_0^1 [(mx+n)^2+m^2y^2]^{-s}\,e^{2\pi irx}\,dx$$

$$= \pi^{-s}\,\Gamma(s)\,y^s\sum_{m=1}^{\infty}\sum_{n \bmod m}\int_{-\infty}^{\infty} [(mx+n)^2+m^2y^2]^{-s}\,e^{2\pi irx}\,dx.$$

Making the substitution $x \mapsto x - n/m$, this equals

$$\pi^{-s}\,\Gamma(s)\,y^s\sum_{m=1}^{\infty} m^{-2s}\sum_{n \bmod m} e^{2\pi irn/m}\int_{-\infty}^{\infty}(x^2+y^2)^{-s}\,e^{2\pi irx}\,dx.$$

Because

$$\sum_{n \bmod m} e^{2\pi irn/m} = \begin{cases} m & \text{if } m|r; \\ 0 & \text{otherwise,}\end{cases}$$

this equals

$$\pi^{-s}\,\Gamma(s)\,y^s\sum_{m|r} m^{1-2s}\int_{-\infty}^{\infty}(x^2+y^2)^{-s}\,e^{2\pi irx}\,dx. \tag{6.11}$$

There are now two cases. Firstly, if $r = 0$, the condition $m|r$ in Eq. (6.11) is vacuous, so using Eqs. (6.8), (6.11) equals

$$\pi^{-s+1/2}\,\Gamma\left(s-\tfrac{1}{2}\right)\,\zeta(2s-1)\,y^{1-s}.$$

Combining this with Eq. (6.10) and invoking the functional equation of $\zeta(s)$, we have computed the constant term

$$a_0(y,s) = \pi^{-s}\,\Gamma(s)\,\zeta(2s)\,y^s + \pi^{s-1}\,\Gamma(1-s)\,\zeta(2-2s)\,y^{1-s}. \tag{6.12}$$

If $r \neq 0$, then by Eqs. (6.8) and (6.11), we have

$$a_r(y,s) = 2\,|r|^{s-1/2}\,\sigma_{1-2s}(|r|)\,\sqrt{y}\,K_{s-1/2}(2\pi\,|r|\,y), \tag{6.13}$$

where as usual

$$\sigma_{1-2s}(r) = \sum_{m|r} m^{1-2s}.$$

If $\mathrm{re}(s) > 1$, we have proved the Fourier expansion Eq. (6.9) with coefficients given by Eqs. (6.12) and (6.13). Theorem 6.1 now follows from an examination of this expansion. Each individual term of the series has analytic continuation to all s, except that a_0 has simple poles at $s = 0$ and $s = 1$. (Each of the two terms in Eq. (6.12) has a pole at $s = 1/2$, but these cancel.) The convergence of the infinite series follows from the rapid decay of the Bessel function (cf. Eq. (6.6)). Thus we obtain the analytic continuation. To get the functional equation (6.3), we observe that

$$a_n(y, s) = a_n(y, 1-s).$$

This is clear if $n = 0$. If $n \neq 0$, this follows from Eq. (6.7) and the identity

$$r^s\,\sigma_{-2s}(r) = \prod_{d_1 d_2 = r} d_1^s\, d_2^{-s} = r^{-s}\,\sigma_{2s}(r).$$

Hence we have Eq. (6.3). Regarding the statement about the residues, the only term in Eq. (6.9) with a pole at $s = 1$ is a_0, and it is easy to check that the residue is the constant function $z = 1/2$. As for Eq. (6.4), it follows from Eq. (6.6) that the nonconstant terms (Eq. 6.13) decay rapidly as $y \to \infty$, so asymptotically the behavior of $E(z, s)$ is the same as its constant term Eq. (6.12). ∎

The Eisenstein series $E(z, s)$ has a certain group-theoretic interpretation. To sum over $(m, n) \in \mathbb{Z}^2$, not both zero, we may sum over all positive integers N and all pairs (c, d) of coprime integers, taking $(m, n) = (Nc, Nd)$. We associate a coset in $\Gamma_\infty\backslash\Gamma(1)$, with the ordered pair (c, d), where $\Gamma(1) = SL(2, \mathbb{Z})$, and

$$\Gamma_\infty = \left\{ \begin{pmatrix} 1 & n \\ & 1 \end{pmatrix} \middle| \, n \in \mathbb{Z} \right\}.$$

That is, we associate with (c, d) the set of all matrices in $\gamma \in \Gamma(1)$ with the same bottom row, which is a right coset of Γ_∞. We observe that by Eq. (2.2),

$$\frac{y^s}{|mz+n|^{2s}} = N^{-2s}\,\mathrm{im}\,(\gamma(z))^s.$$

Consequently, we have

$$\begin{aligned} E(z, s) &= \pi^{-s}\,\Gamma(s)\,\zeta(2s)\,\tfrac{1}{2} \sum_{\gamma \in \Gamma_\infty\backslash\Gamma(1)} \mathrm{im}\,(\gamma(z))^s \\ &= \pi^{-s}\,\Gamma(s)\,\zeta(2s) \sum_{\gamma \in \overline{\Gamma_\infty}\backslash\overline{\Gamma(1)}} \mathrm{im}\,(\gamma(z))^s, \end{aligned} \tag{6.14}$$

where, we recall, $\overline{\Gamma}$ denotes the image of a subgroup $\Gamma \subset SL(2,\mathbb{R})$ in $PSL(2,\mathbb{R}) = PGL(2,\mathbb{R})^+$. In the second expression, we dropped the $\frac{1}{2}$ because γ and $-\gamma$ have the same image in $\overline{\Gamma(1)}$.

We are now ready to consider the Rankin–Selberg method. Let ϕ be an automorphic function on $\mathcal{H}$, that is, a smooth function satisfying $\phi(\gamma(z)) = \phi(z)$ for $\gamma \in SL(2,\mathbb{Z})$. Let us suppose that

$$\phi(x+iy) = O(y^{-N}) \qquad \text{for all } N > 0 \text{ as } y \to \infty. \tag{6.15}$$

Because $\phi(z+1) = \phi(z)$, we have a Fourier expansion

$$\phi(z) = \sum_{-\infty}^{\infty} \phi_n(y)\, e^{2\pi i n x}, \qquad \phi_n(y) = \int_0^1 \phi(x+iy)\, e^{-2\pi i n x}\, dx \tag{6.16}$$

We naturally call ϕ_0 the *constant term* in the Fourier expansion of ϕ. Let

$$M(s,\phi_0) = \int_0^{\infty} \phi_0(y)\, y^s\, \frac{dy}{y} \tag{6.17}$$

be the Mellin transform of ϕ_0. With our hypotheses, ϕ is bounded on the fundamental domain; hence ϕ_0 is bounded as a function of y and, moreover, decays rapidly as $y \to \infty$. Hence Eq. (6.17) is absolutely convergent if $\mathrm{re}(s) > 0$. Let

$$\Lambda(s) = \pi^{-s}\, \Gamma(s)\, \zeta(2s)\, M(s-1,\phi_0). \tag{6.18}$$

Proposition 1.6.1 *With these hypotheses, we have*

$$\Lambda(s) = \int_{\Gamma(1)\backslash\mathcal{H}} E(z,s)\, \phi(z)\, \frac{dx\, dy}{y^2}. \tag{6.19}$$

Then $\Lambda(s)$ has meromorphic continuation to all s, with at most simple poles at $s=1$ and $s=0$. We have

$$\operatorname{res}\ \Lambda(s)|_{s=1} = \tfrac{1}{2} \int_{\Gamma(1)\backslash\mathcal{H}} \phi(z)\, \frac{dx\, dy}{y^2}. \tag{6.20}$$

This statement, with its characteristic "unfolding" proof, contains the essence of the Rankin–Selberg method.

Proof If we establish Eq. (6.19), the analytic properties of $\Lambda(s)$ evidently follow from the corresponding properties of $E(z,s)$ in Theorem 1.6.1. By Eq. (6.4), the right side of Eq. (6.19) is convergent for all s except $s=0$ and $s=1$, where $E(z,s)$ has poles. This gives the analytic continuation of the left side, originally defined only for $\text{re}(s) > 1$. Because the residue of the Eisenstein series is the constant function $1/2$, we obtain Eq. (6.20).

To prove Eq. (6.19), we assume that $\text{re}(s) > 1$. Then we may substitute Eq. (6.14). Using the fact that $\phi(\gamma(z)) = \phi(z)$ for $\gamma \in \Gamma(1)$, the right side of Eq. (6.19) equals

$$\pi^{-s}\,\Gamma(s)\,\zeta(2s) \sum_{\gamma\in\overline{\Gamma_\infty}\backslash\overline{\Gamma(1)}} \int_{\Gamma(1)\backslash\mathcal{H}} \text{im}\,(\gamma(z))^s\phi(\gamma(z))\,\frac{dx\,dy}{y^2}$$

$$= \pi^{-s}\,\Gamma(s)\,\zeta(2s) \int_{\Gamma_\infty\backslash\mathcal{H}} \text{im}(z)^s\,\phi(z)\,\frac{dx\,dy}{y^2}.$$

We may take the integration over any fundamental domain for Γ_∞. We choose the fundamental domain defined by $0 < x < 1$, $y > 0$. Our last integration equals

$$\pi^{-s}\,\Gamma(s)\,\zeta(2s) \int_0^\infty \int_0^1 y^s\,\phi(x+iy)\,y^{-1}\,dx\,\frac{dy}{y}.$$

This, of course, is just $\Lambda(s)$. ■

The original application of the Rankin–Selberg method was the result that if $f(z) = \sum A(n)\,q^n$ and $g(z) = \sum B(n)\,q^n$ are modular forms, then $\sum A(n)\,B(n)\,n^{-s}$ has analytic continuation and functional equation. This was discovered independently by Rankin (1939) and Selberg (1940). We will now explain this application.

We apply Proposition 1.6.1 with $\phi(z) = f(z)\,\overline{g(z)}\,y^k$. We assume that at least one of f or g is cuspidal, so that Eq. (6.15) is satisfied. We will assume that f and g are Hecke eigenforms. By Eq. (2.2), ϕ is automorphic with respect to $SL(2,\mathbb{Z})$. We have

$$\phi_0(y) = \int_0^1 f(x+iy)\,\overline{g(x+iy)}\,y^k\,dx$$

$$= \sum_{n=0}^\infty \sum_{m=0}^\infty \int_0^1 A(n)\,e^{2\pi inx}\,e^{-2\pi ny}\,\overline{B(m)}\,e^{-2\pi imx}\,e^{-2\pi my}\,y^k\,dx.$$

Because $\int_0^1 e^{2\pi i(n-m)x}\,dx = 0$ unless $n = m$,

$$\phi_0(y) = \sum_{n=0}^\infty A(n)\,\overline{B(n)}\,e^{-4\pi ny}\,y^k.$$

Consequently

$$M(s,\phi_0) = \sum_{n=0}^{\infty} A(n)\,\overline{B(n)} \int_0^{\infty} e^{-4\pi n y}\, y^{s+k}\, \frac{dy}{y}$$
$$= (4\pi)^{-s-k}\,\Gamma(s+k) \sum_{n=0}^{\infty} A(n)\,\overline{B(n)}\, n^{-s-k}.$$

Because the $B(n)$ are eigenvalues of the Hecke operators, which are self-adjoint, they are real, so $\overline{B(n)} = B(n)$. Consequently, in Proposition 1.6.1,

$$\Lambda(s) = 4^{-s-k+1}\,\pi^{-2s-k+1}\,\Gamma(s)\,\Gamma(s+k-1)\,\zeta(2s) \sum_{n=1}^{\infty} A(n)\,B(n)\,n^{-s-k+1}.$$

Now let us denote

$$L(s, f\times g) = \zeta(2s-2k+2) \sum_{n=1}^{\infty} A(n)\,B(n)\,n^{-s}, \tag{6.21}$$

$$\Lambda(s, f\times g) = (2\pi)^{-2s}\,\Gamma(s)\,\Gamma(s-k+1)\,L(s, f\times g). \tag{6.22}$$

We see that

$$\Lambda(s-k+1) = \pi^{k-1}\,\Lambda(s, f\times g). \tag{6.23}$$

Theorem 1.6.2 *With notation as above, $\Lambda(s, f\times g)$, originally defined for* $\mathrm{re}(s)$ *sufficiently large, has meromorphic continuation to all s, with holomorphic except for at most simple poles at $s = k$ and $s = k-1$. It satisfies a functional equation*

$$\Lambda(s, f\times g) = \Lambda(2k-1-s, f\times g). \tag{6.24}$$

The residue of Λ at $s = k$ is $\frac{1}{2}\pi^{1-k}\langle f, g\rangle$.

The inner product $\langle f, g\rangle$ is defined in Eq. (3.13).

Proof This is a simple consequence of Eq. (6.23) and Proposition 1.6.1. ■

We would like to rewrite the Euler product for $L(s, f\times g)$ in a more instructive form. To this end, we factor the Hecke polynomials

$$1 - A(p)\,X + p^{k-1}\,X^2 = \bigl(1-\alpha_1(p)\,X\bigr)\bigl(1-\alpha_2(p)\,X\bigr),$$
$$1 - B(p)\,X + p^{k-1}\,X^2 = \bigl(1-\beta_1(p)\,X\bigr)\bigl(1-\beta_2(p)\,X\bigr).$$

Theorem 1.6.3 *We have*

$$L(s, f \times g) = \prod_p \prod_{i=1}^{2} \prod_{j=1}^{2} \left(1 - \alpha_i(p)\, \beta_j(p)\, p^{-s}\right)^{-1}. \tag{6.25}$$

Proof This is a consequence of the following purely algebraic lemma.

Lemma 1.6.1 *If*

$$\sum_{r=0}^{\infty} A(r)\, x^r = (1 - \alpha_1 x)^{-1} (1 - \alpha_2 x)^{-1},$$

$$\sum_{r=0}^{\infty} B(r)\, x^r = (1 - \beta_1 x)^{-1} (1 - \beta_2 x)^{-1},$$

then

$$\sum_{r=0}^{\infty} A(r)\, B(r)\, x^r = (1 - \alpha_1\alpha_2\beta_1\beta_2\, x^2) \prod_{i=1}^{2} \prod_{j=1}^{2} (1 - \alpha_i\, \beta_j\, x)^{-1}.$$

Indeed, taking $x = p^{-s}$, the lemma implies that each Euler factor on the left side of Eq. (6.25) is equal to the corresponding factor on the right.

To prove the lemma, denote

$$\sum_{r=0}^{\infty} A(r)\, x^r = \phi(x), \qquad \sum_{r=0}^{\infty} B(r)\, x^r = \psi(x);$$

consider the integral

$$\frac{1}{2\pi i} \int \phi(xq)\, \psi(q^{-1})\, \frac{dq}{q},$$

where the path of integration circles the origin in the positive sense, so that the poles of $\phi(xq)$ are outside the path of integration, and the poles of $\psi(q^{-1})$ are inside. This is possible if x is sufficiently small. The integral equals

$$\sum_{r,r'} A(r)\, B(r')\, x^r \times \frac{1}{2\pi i} \int q^{r-r'-1}\, dq = \sum_r A(r)\, B(r)\, x^r.$$

Thus it is sufficient to show that

$$\frac{1}{2\pi i} \int (1 - \alpha_1 xq)^{-1} (1 - \alpha_2 xq)^{-1} (1 - \beta_1 q^{-1})^{-1} (1 - \beta_2 q^{-1})^{-1} \frac{dq}{q}$$

$$= \frac{1 - \alpha_1\alpha_2\beta_1\beta_2\, x^2}{(1 - \alpha_1\beta_1\, x)\, (1 - \alpha_2\beta_1\, x)(1 - \alpha_1\beta_2\, x)\, (1 - \alpha_2\beta_2\, x)}. \tag{6.26}$$

Indeed, the left side is equal to the sum of the residues at the poles at $q = \beta_1$ and $q = \beta_2$, and summing these gives the right-hand side. This completes the proof of the lemma and of Proposition 1.6.2. ■

Exercises

Exercise 1.6.1 Suppose that $\mathrm{re}(s) > 1$. Prove that the integral Eq. (6.1) is absolutely convergent and that $E(z, s)$ is automorphic, that is, prove Eq. (6.2).

Exercise 1.6.2 (a) Prove the Poisson summation formula for $\mathbb{R}^n$: Let

$$\langle x, y\rangle = x_1 y_1 + \ldots + x_n y_n$$

be the usual inner product; define the Fourier transform

$$\hat{f}(x) = \int_{\mathbb{R}^n} f(y)\, e^{2\pi i \langle x, y\rangle}\, dy.$$

Suppose that f is smooth and of rapid decay; that is, $f(x) = O\big((1 + |x|)^{-M}\big)$ for all $M > 0$. Prove that

$$\sum_{\xi \in \mathbb{Z}^n} f(\xi) = \sum_{\xi \in \mathbb{Z}^n} \hat{f}(\xi).$$

(b) If $z = x + iy \in \mathcal{H}$ and $t > 0$, let

$$\Theta(t) = \sum_{(m,n)\in\mathbb{Z}^2} e^{-\pi |mz+n|^2 t/y}.$$

Use the Poisson summation formula to show that $\Theta(t) = t^{-1}\,\Theta(t^{-1})$.
(c) Prove that

$$E(z, s) = \tfrac{1}{2} \int_0^\infty \big(\Theta(t) - 1\big)\, t^s\, \frac{dt}{t}.$$

Deduce the analytic continuation and functional equation of $E(z, s)$ from this identity and from (b).

Exercise 1.6.3 This exercise shows how to use a variant of the Rankin–Selberg method to compute the volume of a fundamental domain for $SL(2, \mathbb{Z})$.

The idea is that because $\mathrm{res}\ E(z, s)|_{s=1} = 1/2$, we should have

$$\mathrm{res}_{s=1} \int_{\Gamma(1)\backslash\mathcal{H}} E(z, s)\, \frac{dx\, dy}{y^2} = \tfrac{1}{2}\,\mathrm{vol}\big(\Gamma(1)\backslash\mathcal{H}\big).$$

Unfortunately, the integral on the left is only convergent for $\mathrm{re}(s) < 1$, whereas we would like to take $\mathrm{re}(s) > 1$ and unfold the integral. The solution to this dilemma is to take a modified version of the Eisenstein series. If $T > 0$ and $z = x + iy \in \mathcal{H}$, define a function

$$y_T(z) = \begin{cases} y & \text{if } y < T; \\ 0 & \text{if } y \geq T, \end{cases}$$

and let

$$E_T(z, s) = \pi^{-s}\,\Gamma(s) \sum_{\gamma \in \overline{\Gamma_\infty} \backslash \overline{\Gamma(1)}} y_T\big(\gamma(z)\big)^s.$$

(a) Prove that if $K \subset \mathcal{H}$ is compact, there exists a constant T_0 such that if $T > T_0$, then $E_T(z, s) = E(z, s)$ for $z \in K$.
(b) Prove that if $\mathrm{re}(s) > 1$

$$\int_{\Gamma(1)\backslash\mathcal{H}} E_T(z, s)\,\frac{dx\,dy}{y^2} = \pi^{-s}\,\Gamma(s)\,\zeta(2s)\,\frac{T^{s-1}}{s-1}.$$

(c) Assume that $T > 1$. Prove that if $z = x + iy$ is in the standard fundamental domain for $SL(2, \mathbb{Z})$, we have

$$E_T(z, s) = \begin{cases} E(z, s) & \text{if } y < T; \\ E(z, s) - \pi^{-s}\,\Gamma(s)\,\zeta(2s)\,y^s & \text{if } y \geq T. \end{cases}$$

Note that the difference is holomorphic at $s = 1$.
(d) Justify the following interchange of order:

$$\mathrm{res}_{s=1} \int_{\Gamma(1)\backslash\mathcal{H}} E_T(z, s)\,\frac{dx\,dy}{y^2} = \int_{\Gamma(1)\backslash\mathcal{H}} \mathrm{res}_{s=1}\,E_T(z, s)\,\frac{dx\,dy}{y^2},$$

and hence conclude that the volume of $\Gamma(1)\backslash\mathcal{H}$ is $\pi/3$.

Exercise 1.6.4 Now that you have the proper tools at your disposal, go back and do exercise 1.3.6.

Exercise 1.6.5 It follows from Proposition 1.3.5 that the series $L(s, f) = \sum a(n)\,n^{-s}$ is absolutely convergent if $\mathrm{re}(s) > \frac{k}{2} + 1$. The Ramanujan conjecture predicts that this can be improved to $\mathrm{re}(s) > \frac{k}{2} + \frac{1}{2}$. This exercise shows how the Rankin–Selberg method can be used to obtain this more precise estimate for the convergence of $L(s, f)$ *without* improving the estimate for each individual term. In other words, the Ramanujan conjecture is valid "on the average."

(a) Let $\sum c(n)\,n^{-s}$ be a Dirichlet series with positive coefficients which is convergent for $\mathrm{re}(s)$ sufficiently large. Let $\phi(s)$ be this function. Suppose that $\phi(s)$ has analytic continuation to a neighborhood of the half plane $\mathrm{re}(s) \geq \sigma$. Prove that $\sum c(n)\,n^{-s}$ is absolutely convergent in this half plane. [HINT: consider the values of the function on the real axis.]

(b) Let $a(n)$ be the Fourier coefficients of a cusp form f of weight k for $SL(2,\mathbb{Z})$. Prove that $\sum a(n)^2 n^{-s}$ is convergent for $re(s) > k$.

(c) Prove that $\sum a(n)\, n^{-s}$ is absolutely convergent if $s > (k+1)/2$. [HINT: Use the estimate $|a(n)| < \max(n^{\lambda}, n^{-\lambda}\,|a(n)|^2)$ to deduce a region of convergence for $\sum a(n)\, n^{-s}$ from one for $\sum a(n)^2 n^{-s}$. Choose λ to optimize this region.]

1.7 Hecke Characters and Hilbert Modular Forms

Let F be a totally real field of degree n over $\mathbb{Q}$. This means that every complex embedding of F lies in $\mathbb{R}$. Let $\mathfrak{o} = \mathfrak{o}_F$ be its ring of integers. We will describe analogs both of Dirichlet characters and of modular forms for F.

Let $\alpha \mapsto \alpha^{(j)}$, $j = 1, \cdots, n$ be the distinct real embeddings of F. We regard F as a subring of $\mathbb{R}^n$ by means of the embedding

$$\alpha \mapsto \left(\alpha^{(1)}, \cdots, \alpha^{(n)}\right). \tag{7.1}$$

We extend the norm and trace mappings $F \to \mathbb{R}$ to mappings $\mathbb{R}^n \to \mathbb{R}$ (or sometimes even $\mathbb{C}^n \to \mathbb{C}$) by the rules

$$\operatorname{tr}(x) = \sum_{j=1}^{n} x_j, \qquad N(x) = \prod_{j=1}^{n} x_j, \qquad x = (x_1, \cdots, x_n).$$

By a *place* v of F, we mean either a prime ideal of $\mathfrak{o}_F$ or one of the embeddings $\alpha \to \alpha^{(j)}$. In either case, associated with the place, there is an absolute value $|\ |_v$ on F such that we have the product formula $\prod_v |x|_v = 1$ for $x \in F^\times$. We recall the following theorem.

Theorem 1.7.1: The unit theorem *Let F be a totally real field of degree n over $\mathbb{Q}$. Then the units $\mathfrak{o}^\times$ form a finitely generated Abelian group. The torsion subgroup of $\mathfrak{o}^\times$ consists of the group $\{\pm 1\}$ of roots of unity in F. The mapping*

$$\alpha \to \left(\log|\alpha^{(1)}|, \cdots, \log|\alpha^{(n)}|\right)$$

induces an isomorphism of $\mathfrak{o}^\times/\{\pm 1\}$ onto a lattice of rank $n-1$ in the $n-1$-dimensional vector space $\{x \in \mathbb{R}^n \mid \sum x_j = 0\}$. Hence $\mathfrak{o}^\times/\{\pm 1\} \cong \mathbb{Z}^{n-1}$.

Proof For a proof, and for the statement when the number field F is not assumed to be totally real, see Lang (1970). ∎

An element a of F will be called *totally positive* if $a^{(j)} > 0$ for $j = 1, \cdots, n$. In this case, we write $a \gg 0$. If S is any subset of F, we will denote by S_+ the set of totally positive elements of S. If $\mathfrak{a}$ is an ideal of $\mathfrak{o}_F$, we will denote by $\mathrm{N}(\mathfrak{a})$ the absolute norm of $\mathfrak{a}$, that is, the cardinality of $\mathfrak{o}/\mathfrak{a}$.

The number field F has an ideal $\mathfrak{d}$, called the *different* of F, which is characterized by the property that for $x \in F$, we have $\operatorname{tr}(x\mathfrak{o}_F) \subset \mathbb{Z}$ if and only if

$x\mathfrak{d} \subset \mathfrak{o}_F$. For example, if $F = \mathbb{Q}(\sqrt{p})$, where p is prime, the different of F is the ideal generated by $\sqrt{p}$ if $p \equiv 1 \bmod 4$, by $2\sqrt{p}$ if $p \equiv 3 \bmod 4$, or if $p = 2$. The different has the property that $\mathbb{N}(\mathfrak{d}) = |D|$, where D is the discriminant of F. (Note that actually $D > 0$ because F is totally real.)

It will sometimes be useful to assume that the *narrow class number* of F is equal to one. This means that every ideal of F has a generator that is totally positive. However, we will not impose this condition until we need it.

Suppose that $F = \mathbb{Q}$ and that χ is a Dirichlet character modulo N. We regard χ as a character of *ideals* of $\mathbb{Z}$ as follows. If $\mathfrak{a} \subset \mathbb{Z}$ is an ideal, then $\mathfrak{a}$ has a unique generator a that is nonnegative; we let $\chi(\mathfrak{a}) = \chi(a)$. Thus $\chi(\mathfrak{a}) \neq 0$ if and only if $(\mathfrak{a}, N) = 1$, in which case $|\chi(\mathfrak{a})| = 1$.

Now if F is more generally a number field, there exists a class of characters of ideals of F, called *Hecke characters*, which generalize the Dirichlet characters. These were introduced by Hecke (1918) and (1920), who proved their functional equations. Hecke's theory was revisited in Tate's (1950) thesis. After studying Hecke's viewpoint in the present section (with simplifications resulting the assumption from that the ground field is totally real and of narrow class number one), we will return to study them from Tate's viewpoint in Section 3.1.

Because we are primarily interested in totally real fields, we will only define Hecke's L-functions in this case. Let $\mathfrak{f}$ be an ideal of $\mathfrak{o}$, and let $\chi_\mathfrak{f}$ be a character of $(\mathfrak{o}/\mathfrak{f})^\times$. As with Dirichlet characters, we say that $\chi_\mathfrak{f}$ is *primitive* if it does not factor through the canonical map $(\mathfrak{o}/\mathfrak{f})^\times \to (\mathfrak{o}/\mathfrak{f}_1)^\times$ for any proper divisor $\mathfrak{f}_1$ of $\mathfrak{f}$. We will assume this to be the case. We extend $\chi_\mathfrak{f}$ to a function on $\mathfrak{o}$ by

$$\chi_\mathfrak{f}(a) = \begin{cases} \chi_\mathfrak{f}(a \bmod \mathfrak{f}) & \text{if } (a, \mathfrak{f}) = 1; \\ 0 & \text{otherwise.} \end{cases}$$

We would like a character of ideals, or at least of *principal* ideals. The character $\chi_\mathfrak{f}$ may not be trivial on units, however. We remedy this situation as follows. Let ν_j $(j = 1, \cdots, n)$ be complex parameters. We will assume that they are purely imaginary, and we will also assume that $\sum_j \nu_j = 0$. (So if $n = 1$, $\nu_1 = 0$.) Let ϵ_j $(j = 1, \cdots, n)$ each equal either 0 or 1. We define a character χ_j of $\mathbb{R}^\times$ by $\chi_j(x) = \operatorname{sgn}(x)^{\epsilon_j} |x|^{\nu_j}$. We define a character χ_∞ of $\mathfrak{o}^\times$ by

$$\chi_\infty(x) = \prod_{j=1}^{n} \chi_j(x^{(j)}).$$

It is a consequence of the unit theorem that we may choose these data so that $\chi(x) = \chi_\infty(x)\,\chi_\mathfrak{f}(x)$ is trivial on $\mathfrak{o}^\times$. Indeed, by the unit theorem, the group $\mathfrak{o}_+^\times$ of totally positive units is free of rank $n - 1$, because $\mathfrak{o}_+^\times$ is a torsion-free subgroup of finite index in $\mathfrak{o}^\times$. (Indeed, it is easy to see that $[\mathfrak{o}^\times : \mathfrak{o}_+^\times] \leq 2^{n-1}$.) Let $u_1, \cdots, u_{n-1}$ be a basis of $\mathfrak{o}_+^\times$. Let $m_1, \cdots, m_{n-1}$ be integers. We choose

$\nu_1, \cdots, \nu_n$ to satisfy the n linear equations

$$\sum_{j=1}^{n} \nu_j = 0$$

$$\sum_{j=1}^{n} \nu_j \log\left(u_k^{(j)}\right) = 2\pi i m_k - \log \chi_{\mathfrak{f}}(u_k), \qquad k = 1, \cdots, n-1.$$

Note that the determinant of this system of linear equations is nonzero, being essentially the regulator. Here the logarithm $\log \chi_{\mathfrak{f}}(u_k)$ can be any branch, because changing the branch-of-log amounts to shifting the integers m_k. What is important is that $|\chi_{\mathfrak{f}}(u_k)| = 1$, so the right side is purely imaginary, and therefore the solutions ν_k to these equations are purely imaginary. We see that we may choose $\nu_1, \cdots, \nu_n$, subject to the single condition $\sum \nu_j = 0$ so that $\chi(x)$ is trivial on $\mathfrak{o}_+^\times$. Now if $\lambda \in \mathfrak{o}^\times$, $\chi(\lambda)$ depends only on the signs of the $\lambda^{(j)}$, so the ϵ_j may be chosen so that $\chi(\lambda) = 1$ for all units. In other words, because $\chi(x)$ is trivial on $\mathfrak{o}_+^\times$, any choice of the ϵ_j gives us a character of the finite group $\mathfrak{o}^\times/\mathfrak{o}_+^\times$, and by adjusting the ϵ_j, we may make this character trivial; hence we obtain a character that is trivial on units. We thus obtain a character of the principal ideals prime to $\mathfrak{f}$. A character of ideals whose restriction to the subgroup of principal ideals arises in this fashion is called a *Hecke character*. Note that once the character $\chi_{\mathfrak{f}}$ is chosen, the character χ still depends on the choice of the integers $m_1, \cdots, m_{n-1}$.

If χ is a Hecke character, we define

$$L(s, \chi) = \sum_{\mathfrak{a}} \chi(\mathfrak{a})\, \mathrm{N}(\mathfrak{a})^{-s} = \prod_{\mathfrak{p}} \left(1 - \tfrac{\chi(\mathfrak{p})}{\mathrm{N}(\mathfrak{p})^s}\right)^{-1}; \tag{7.2}$$

the sum is over all nonzero ideals $\mathfrak{a}$ of $\mathfrak{o}$, and the product is over all prime ideals. A special case (when $\mathfrak{f} = 1$, and χ is the trivial character) is the Dedekind zeta function

$$\zeta_F(s) = \sum_{\mathfrak{a}} \mathrm{N}(\mathfrak{a})^{-s} = \prod_{\mathfrak{p}} \left(1 - \tfrac{1}{\mathrm{N}(\mathfrak{p})^s}\right)^{-1}$$

of F.

We will now prove the analytic continuation and functional equations of the L-series $L(s, \chi)$ of a Hecke L-function, but under the special assumption that the field has narrow class number one. The arguments closely parallel those in Section 1.1. We define

$$\Lambda(s, \chi) = D^{s/2} \pi^{-ns/2} \prod_{j=1}^{n} \Gamma\left(\frac{s - \nu_j + \epsilon_j}{2}\right) L(s, \chi). \tag{7.3}$$

Because the narrow class number is one, every ideal has a totally positive generator. We select such a totally positive generator d for the different $\mathfrak{d}$, and a totally positive generator f for the ideal $\mathfrak{f}$. We define a *Gauss sum*

$$\tau_F(\chi_{\mathfrak{f}}) = \sum_{\alpha \in \mathfrak{o}/\mathfrak{f}} \chi_{\mathfrak{f}}(\alpha)\, e^{2\pi i\, \mathrm{tr}(\alpha/fd)}. \tag{7.4}$$

Note that this definition depends on the choices of generators d and f.

Theorem 1.7.2: Hecke (1918) and (1920) *Assume that the narrow class number of F is one. Then $\Lambda(s, \chi)$ has meromorphic continuation to all s; indeed, if $\chi \neq 1$, it is an entire function. If $\chi = 1$, $L(s, \chi)$ equals the Dedekind zeta function $\zeta_F(s)$ of F; in this case, $\Lambda(s, \chi)$ has simple poles at $s = 1$ and $s = 0$, with residue $2^{n-1} R$ at $s = 1$, where R is the regulator of the field F. $\Lambda(s, \chi)$ is bounded in vertical strips, except, of course, that if $\chi = 1$, it is obviously not bounded near the poles. We have the functional equation*

$$\Lambda(s, \chi) = \frac{\chi_{\mathfrak{f}}(-1)\, \chi_\infty(fd)\, \tau_F(\chi_{\mathfrak{f}})}{\sqrt{\mathbb{N}(\mathfrak{f})}}\, i^{\mathrm{tr}(\epsilon)}\, \mathbb{N}(\mathfrak{f})^{-s+1/2}\, \Lambda(1 - s, \chi^{-1}). \tag{7.5}$$

We observe that while $\chi_\infty(fd)$ and $\tau_F(\chi_{\mathfrak{f}})$ both depend on the choice of totally positive generators d and f of $\mathfrak{d}$ and $\mathfrak{f}$, their product does not, as may be easily checked.

Hecke (1918) and (1920) defined what we are calling Hecke characters over an arbitrary number field; he called them *Grössencharaktere* (a term often used in the literature). He proved the functional equations in complete generality. For simplicity, however, we are restricting our attention to the case of a totally real field of narrow class number one. Over an arbitrary number field, it is probably best to work adelically, as in Tate's (1950) thesis. We will discuss Tate's thesis in Section 3.1. The article of Heilbronn (1967) is also useful for its discussion of Hecke characters.

Proof We define the Fourier transform

$$\hat{f}(x) = \frac{1}{\sqrt{D}} \int_{\mathbb{R}^n} f(y)\, e^{2\pi i\, \mathrm{tr}(xy/d)}\, dy.$$

The definition of the Fourier transform thus depends not only on the field, but also on the choice of a totally positive generator d of the different. This particular normalization of the Fourier transform works rather well for our purposes.

We have the Poisson summation formula:

$$\sum_{\alpha \in \mathfrak{o}} F(\alpha) = \sum_{\alpha \in \mathfrak{o}} \hat{F}(\alpha). \tag{7.6}$$

This is valid, for example, when F lies in the *Schwartz space*. For example, the *Schwartz space* is the space of all smooth functions F, all of whose derivatives are of rapid decay; that is

$$\frac{\partial^k F}{\partial x_1^{k_1} \cdots \partial x_n^{k_n}}(x) = O\big((1 + |x|)^{-N}\big)$$

for $\sum k_j = k$ and all $N > 0$. The Schwartz space is invariant under the Fourier transform. Of course, the class of functions that satisfy the Poisson summation formula is actually much broader than this, but the functions in our applications will lie in the Schwartz space. The proof of Eq. (7.6) is similar to the one-dimensional formula (1.9) and is left to the reader (Exercise 1.7.2).

We have the following straightforward generalization of Eq. (1.6):

$$\tau_F(\overline{\chi}_\mathfrak{f})\, \chi_\mathfrak{f}(\alpha) = \sum_{\beta \in \mathfrak{o}/\mathfrak{f}} \overline{\chi_\mathfrak{f}(\beta)}\, e^{2\pi i \operatorname{tr}(\alpha\beta/fd)}. \tag{7.7}$$

This requires that $\chi_\mathfrak{f}$ be primitive, which we are assuming. The proof is a straightforward generalization of the proof of Eq. (1.6). Proceeding from this, we obtain, as in Section 1.1, the *twisted Poisson summation formula*, generalizing Eq. (1.10):

$$\tau_F(\overline{\chi}_\mathfrak{f}) \sum_{\alpha \in \mathfrak{o}} \chi_\mathfrak{f}(\alpha)\, F(\alpha) = \sum_{\alpha \in \mathfrak{o}} \overline{\chi_\mathfrak{f}(\alpha)}\, \hat{F}(\alpha/f). \tag{7.8}$$

Observe that for suitably chosen F, we may make the right side nonzero, so $\tau_F(\overline{\chi}_\mathfrak{f}) \neq 0$.

Let $t = (t_1, \cdots, t_n) \in \mathbb{R}_+^n$. We will calculate the Fourier transform of

$$F_t(x) = N(x^\epsilon)\; e^{-\pi \operatorname{tr}(tx^2/d)}, \qquad x \in \mathbb{R}^n.$$

Here of course, it is our notation that

$$N(x^\epsilon) = \prod_{j=1}^n x_j^{\epsilon_j}.$$

We have the one-dimensional integral

$$\frac{1}{\sqrt{d^{(j)}}} \int_{-\infty}^{\infty} e^{2\pi i x_j y_j/d^{(j)}}\; e^{-\pi t_j y_j^2/d^{(j)}}\; dy_j = \frac{1}{\sqrt{t_j}}\, e^{-\pi x_j^2/d^{(j)} t_j}, \tag{7.9}$$

which is a simple consequence of Eq. (1.12). Differentiating this equation under the integral sign with respect to x_j, we obtain

$$\frac{1}{\sqrt{d^{(j)}}} \int_{-\infty}^{\infty} e^{2\pi i x_j y_j/d^{(j)}}\; y_j\, e^{-\pi t_j y_j^2/d^{(j)}}\; dy_j = \frac{i x_j}{t_j^{3/2}}\, e^{-\pi x_j^2/d^{(j)} t_j}. \tag{7.10}$$

Combining these two identities, we obtain

$$\hat{F}_t(x) = N\left(\left(\frac{i}{t}\right)^{\epsilon}\right) \frac{1}{\sqrt{N(t)}} F_{1/t}(x). \tag{7.11}$$

Now let

$$\Theta_{\chi_{\mathfrak{f}}}(t) = \sum_{\alpha \in \mathfrak{o}} \chi_{\mathfrak{f}}(\alpha)\, N(\alpha^{\epsilon})\, e^{-\pi\, \mathrm{tr}(t\alpha^2/d)}. \tag{7.12}$$

We have, by Eqs. (7.8) and (7.11), the functional equation

$$\Theta_{\chi_{\mathfrak{f}}}(t) = \frac{1}{\tau_F(\overline{\chi}_{\mathfrak{f}})} \frac{1}{\sqrt{N(t)}}\, N\left(\frac{i}{ft}\right)^{\epsilon} \Theta_{\overline{\chi}_{\mathfrak{f}}}\left(\frac{1}{f^2 t}\right). \tag{7.13}$$

Applying this equation twice, we obtain

$$\tau_F(\chi_{\mathfrak{f}})\, \tau_F(\overline{\chi}_{\mathfrak{f}}) = (-1)^{\mathrm{tr}(\epsilon)}\, \mathbb{N}(\mathfrak{f}) = \chi_{\mathfrak{f}}(-1)\, \mathbb{N}(\mathfrak{f}).$$

Because it is easy to see that $\tau_F(\overline{\chi}_F) = \chi_{\mathfrak{f}}(-1)\, \overline{\tau_F(\chi_{\mathfrak{f}})}$, we obtain

$$|\tau_F(\chi_{\mathfrak{f}})| = \sqrt{\mathbb{N}(\mathfrak{f})}. \tag{7.14}$$

Of course, one may give a direct algebraic proof of this fact along the lines of Eq. (1.5), but it is also interesting to note that it drops out as a byproduct of our analytic considerations.

We will now prove the analytic continuation and functional equation of $\Lambda(s, \chi)$ by expressing it as an integral of $\Theta_{\chi_{\mathfrak{f}}}$. We assume first that $\mathfrak{f} \neq \mathfrak{o}$. Because $\chi_{\mathfrak{f}}$ is primitive, it is nontrivial, and so $\Theta_{\chi_{\mathfrak{f}}}(t)$ has no constant term; it is of rapid decay as the $y_j \to \infty$. We will prove (for sufficiently large s) that

$$\int_{\mathbb{R}^n_+/\mathfrak{o}^\times_+} \Theta_{\chi_{\mathfrak{f}}}(t)\, N\left(t^{(s+\epsilon-\nu)/2}\right) \prod_j \frac{dt_j}{t_j} = 2\pi^{-\mathrm{tr}(\epsilon)}\, N(d^{(\epsilon-\nu)/2})\, \Lambda(s, \chi). \tag{7.15}$$

That the integrand is invariant under the action of $\mathfrak{o}^\times_+$ can be checked; this will also be clear from the following discussion. Remembering that every ideal has a totally positive generator, the left side of Eq. (7.15) equals

$$\sum_{\alpha \in \mathfrak{o}} \chi_{\mathfrak{f}}(\alpha)\, N(\alpha^{\epsilon}) \int_{\mathbb{R}^n_+/\mathfrak{o}^\times_+} e^{-\pi\, \mathrm{tr}(t\alpha^2/d)} N\left(t^{(s+\epsilon-\nu)/2}\right) \prod_j \frac{dt_j}{t_j}$$

$$= \sum_{\alpha \in \mathfrak{o}_+/\mathfrak{o}^\times_+} \sum_{\eta \in \mathfrak{o}^\times} \chi_{\mathfrak{f}}(\eta\alpha)\, N\left((\eta\alpha)^{\epsilon}\right) \int_{\mathbb{R}^n_+/\mathfrak{o}^\times_+} e^{-\pi\, \mathrm{tr}(t\eta^2\alpha^2/d)} N\left(t^{(s+\epsilon-\nu)/2}\right) \prod_j \frac{dt_j}{t_j}.$$

Now we recall that $\chi(\eta) = 1$ when $\eta \in \mathfrak{o}^\times$, and from the definition of χ

$$\chi_{\mathfrak{f}}(\eta)\, N(\eta^{\epsilon}) = \chi_{\mathfrak{f}}(\eta)\, N\left(\mathrm{sgn}(\eta)^{\epsilon}\right) N(|\eta|^{\epsilon}) = N(|\eta|^{-\nu+\epsilon}).$$

Hence the last expression equals

$$\sum_{\alpha\in\mathfrak{o}_+/\mathfrak{o}_+^\times}\sum_{\eta\in\mathfrak{o}^\times}\chi_\mathfrak{f}(\alpha)\,N(\alpha^\epsilon)\int_{\mathbb{R}_+^n/\mathfrak{o}_+^\times} e^{-\pi\,\mathrm{tr}(t\eta^2\alpha^2/d)}\,N\big((\eta^2 t)^{(s+\epsilon-\nu)/2}\big)\prod_j\frac{dt_j}{t_j}$$

$$=2\sum_{\alpha\in\mathfrak{o}_+/\mathfrak{o}_+^\times}\chi_\mathfrak{f}(\alpha)\,N(\alpha^\epsilon)\int_{\mathbb{R}_+^n} e^{-\pi\,\mathrm{tr}(t\alpha^2/d)}\,N\big(t^{(s+\epsilon-\nu)/2}\big)\prod_j\frac{dt_j}{t_j},$$

because as η runs through $\mathfrak{o}^\times$, η^2 runs through $\mathfrak{o}_+^\times$ twice. The last integral splits as a product of gamma functions, and we obtain Eq. (7.15). This formula gives the analytic continuation of $\Lambda(s, f)$ because it is convergent for all s, and indeed, it is easy to see that the integral representation (Eq. 7.15) is bounded in vertical strips. Substituting Eq. (7.13) and replacing t by $1/f^2t$, we obtain

$$N\big(d^{(\epsilon-\nu)/2}\big)\,\Lambda(s,\chi)$$
$$=\tau_F(\overline{\chi}_\mathfrak{f})^{-1}\,N(i/f)^\epsilon\,N(f^{-s+1+\epsilon+\nu})\,N\big(d^{(\epsilon+\nu)/2}\big)\Lambda(1-s,\overline{\chi}).$$

Using Eq. (7.14), we obtain Eq. (7.5). This completes the proof if $\mathfrak{f}$ is nontrivial.

If $\mathfrak{f}=\mathfrak{o}$, we must modify the preceding argument. It is worth remarking that in this case, the ϵ_j are either all zero or all one, and (because $\chi(-1)=1$) the latter case can only occur if n is even. Assuming that $\mathfrak{f}=\mathfrak{o}$, we may take $f=1$. We will denote $\Theta_{\chi_\mathfrak{f}}$ as simply Θ. This has a constant term, corresponding to $\alpha=0$, that we must subtract. Arguing, as in the proof of Eq. (7.15), we have (for sufficiently large s)

$$2\,N\big(d^{(\epsilon-\nu)/2}\big)\,\Lambda(s,\chi)=\int_{\mathbb{R}_+^n/\mathfrak{o}_+^\times}\big(\Theta(t)-1\big)\,N\big(t^{(s+\epsilon-\nu)/2}\big)\prod_j\frac{dt_j}{t_j}. \tag{7.16}$$

We split this integral into two parts depending on $N(t)$. If $N(t)<1$, we substitute $1/f^2t$ for t as before and use Eq. (7.13), but now we must keep track of the part that has been subtracted off. We obtain

$$2\,N\big(d^{(\epsilon-\nu)/2}\big)\,\Lambda(s,\chi)$$
$$=\int_{\substack{\mathbb{R}_+^n/\mathfrak{o}_+^\times\\ N(t)>1}}\big(\Theta(t)-1\big)\,N\big(t^{(s+\epsilon-\nu)/2}\big)\prod_j\frac{dt_j}{t_j}$$
$$+\int_{\substack{\mathbb{R}_+^n/\mathfrak{o}_+^\times\\ N(t)>1}}\big(\Theta(t)-1\big)\,N\big(t^{(1-s+\epsilon+\nu)/2}\big)\prod_j\frac{dt_j}{t_j}$$
$$-\int_{\substack{\mathbb{R}_+^n/\mathfrak{o}_+^\times\\ N(t)<1}}N\big(t^{(s+\epsilon-\nu)/2}\big)\prod_j\frac{dt_j}{t_j}+\int_{\substack{\mathbb{R}_+^n/\mathfrak{o}_+^\times\\ N(t)>1}}N\big(t^{(1-s+\epsilon+\nu)/2}\big)\prod_j\frac{dt_j}{t_j}$$

Now if $\chi \neq 1$, then the ν_j are not all zero, and it is easy to see that the last two terms are zero (integrating first along $N(t)$ constant) and may be discarded. On the other hand, if $\chi = 1$, then the ν_j and the ϵ_j are both zero. In this case, which is identical to Section 3 of Chapter 13 of Lang (1970), the last two terms equal

$$-2^{n-1} R \left[\frac{1}{s} + \frac{1}{1-s} \right].$$

In any case, we obtain the analytic continuation and functional equations that we need. ■

Now suppose that $F/\mathbb{Q}$ is a quadratic extension. (We have chosen the ground field to be $\mathbb{Q}$, but actually the phenomenon that we will describe occurs for any quadratic extension of algebraic number fields, and indeed may be generalized without too much difficulty to arbitrary Abelian extensions.) Let D be the discriminant of F. We have seen (Exercise 1.1.6) that there exists a quadratic Dirichlet character χ_D with conductor $|D|$ such that if p is a prime number, then

$$\begin{cases} p \text{ splits in } F & \text{if and only if } \chi_D(p) = 1; \\ p \text{ remains prime in } F & \text{if and only if } \chi_D(p) = -1; \\ p \text{ ramifies in } F & \text{if and only if } \chi_D(p) = 0. \end{cases}$$

We have the identity

$$\zeta_F(s) = \zeta(s)\, L(s, \chi_D), \tag{7.17}$$

which we now prove. Both sides have Euler products. We will compare the individual Euler factors.

Suppose that p is a rational prime that splits in F. Then there are two primes of F with norm exactly p, so on the left we have $(1 - p^{-s})^{-2}$. If p is a rational prime that is inert in F, then there is a single prime of F with norm p^2, and so in this case, we have a contribution $(1 - p^{-2s})^{-1}$. Finally, if p ramifies, there exists a unique prime of F of norm p, and so we have a factor $(1 - p^{-s})^{-1}$. Thus

$$\begin{aligned} \zeta_F(s) &= \prod_{\chi_D(p)=1} (1-p^{-s})^{-2} \prod_{\chi_D(p)=-1} (1-p^{-2s})^{-1} \prod_{\chi_D(p)=0} (1-p^{-s})^{-1} \\ &= \prod_p \left(1-p^{-s}\right)^{-1}\left(1-\chi_D(p)\, p^{-s}\right)^{-1} = \zeta(s)\, L(s, \chi_D), \end{aligned}$$

from which we get Eq. (7.17).

Let us mention a very slight generalization of Eq. (7.17). Let ψ be a Dirichlet character of $\mathbb{Q}$. There is a Hecke character $\psi \circ N$ obtained by composing ψ with the norm map from ideals of $\mathfrak{o}$ to ideals of $\mathbb{Z}$. Then

$$L(s, \psi \circ N) = L(s, \psi)\, L(s, \psi\chi_D). \tag{7.18}$$

The proof is identical to Eq. (7.17) (cf. Exercise 1.7.1).

We regard the Hecke character $\psi \circ N$ as the *base change lift* of the Hecke character ψ relative to the quadratic extension $F/\mathbb{Q}$. This situation is characterized by the L-series factorization Eq. (7.18).

We wish to describe a similar "base change" phenomenon for modular forms. There is an important difference between the case of a Hecke character and that of a modular form. For Hecke characters, the proof of the existence of the base change amounts to a trivial construction, namely, composition with the norm map. For modular forms, even once the existence of a base change lift is suspected, its construction still requires a major amount of work.

We mention that in the adele language, Hecke characters may be interpreted as *automorphic forms on* $GL(1)$. Modular forms may be interpreted as *automorphic forms on* $GL(2)$ (cf. Section 1.8 and Chapter 3).

There are at least four approaches to base change: (i) the converse theorem; (ii) the theta correspondence; (iii) the twisted trace formula; and (iv) the relative trace formula. The most successful approach (yielding the best results) is the twisted trace formula. This program had its origins in the dissertation of Saito (1975), the work of Shintani (1979) and Langlands (1980), and culminated in the realization of cyclic base change for $GL(n)$ by Arthur and Clozel (1989). The trace formula approach has the advantage in that it gives a precise characterization of the image of the lift: using this method, it can be proved that a Hilbert modular form that is a Hecke eigenform (or more generally an automorphic representation) is a base change if and only if it is invariant under the action of the Galois group. This result is not obtainable by the converse theorem approach.

On the other hand, the converse theorem approach has been used by Jacquet, Piatetski-Shapiro, and Shalika (1981a) to prove base change for nonnormal cubic extensions, which has had applications to Artin's conjecture (Tunnell (1981)) and is one of the ingredients in the recent work of Wiles (1994). This result has not been proved by any other method. There is also the adjoint square lift from $GL(2)$ to $PGL(3)$ by Gelbart and Jacquet (1978), a result which has had many important applications. The converse theorem thus remains an important tool. Converse theorems for $GL(n)$ may be found in Cogdell and Piatetski-Shapiro (1994).

In order to describe the work of Doi and Naganuma (1973), we must introduce *Hilbert modular forms*. As references, there are Van der Geer (1987), Siegel (1961), Hirzebruch (1973), and Freitag (1990).

We return therefore, to the case of a totally real field F of degree n over $\mathbb{Q}$. Let $z = (z_1, \cdots, z_n) \in \mathcal{H}^n$, and let

$$\left(\begin{pmatrix} a_1 & b_1 \\ c_1 & d_1 \end{pmatrix}, \cdots, \begin{pmatrix} a_n & b_n \\ c_n & d_n \end{pmatrix} \right) \in SL(2, \mathbb{R})^n.$$

Actually, it will be convenient to denote this matrix by the simpler $\left(\begin{smallmatrix} a & b \\ c & d \end{smallmatrix}\right)$ where $a = (a_1, \cdots, a_n)$, etc. We denote the action of $SL(2, \mathbb{R})^n$ on $\mathcal{H}^n$

symbolically by

$$\begin{pmatrix} a & b \\ c & d \end{pmatrix} : z \to \frac{az+b}{cz+d}.$$

Now let us embed $SL(2, F)$ into $SL(2, \mathbb{R})^n$ via Eq. (7.1). The group $\Gamma = SL(2, \mathfrak{o})$ is called the *Hilbert modular group*. It may be shown that Γ acts discontinuously on $\mathcal{H}^n$ and that the quotient with respect to the invariant measure

$$\prod_i \frac{dx_i\, dy_i}{y_i^2}$$

has finite volume.

We consider the problem of compactifying $\Gamma\backslash\mathcal{H}^n$. Let $(\mathcal{H}^n)^* = \mathcal{H}^n \cup F \cup \infty$, where F is embedded into $\mathbb{R}^n$ via Eq. (7.1). The space $\Gamma\backslash(\mathcal{H}^n)^*$ is the correct compactification of $\Gamma\backslash\mathcal{H}^n$. The problem of compactification of symmetric quotients was considered by Satake (1956), Baily (1958), and by Baily and Borel (1966). This compactification is called the *Baily–Borel* compactification. It is an algebraic variety. It is definitely singular if $n > 1$ because the cusps are singular points.

The Baily–Borel compactification consists of adjoining to $\Gamma\backslash\mathcal{H}^n$ only a finite number of points. We consider the action of $SL(2, \mathfrak{o})$ on $\mathbb{P}^1(F) = F \cup \infty$. According to Exercise 1.7.3, the number of orbits in $SL(2, \mathfrak{o})\backslash\mathcal{H}^n$ is equal to the class number.

Let k be a nonnegative integer. We will assume, for simplicity, that k is even. A *Hilbert modular form* of weight k is a holomorphic function f on $\mathcal{H}^n$ that satisfies

$$f(z) = N(cz+d)^{-k} f\big(\gamma(z)\big), \qquad \gamma \in \Gamma, \tag{7.19}$$

where by definition

$$N(cz+d)^{-k} = \prod_{j=1}^{n} \big(c^{(j)} z_j + d^{(j)}\big)^{-k}.$$

(We could, more generally, take k to be a vector of weights $(k_1, \cdots, k_n)$.) If $n \geq 2$, it is not necessary to assume that f is holomorphic at the cusps – this is automatic, as is asserted by the *Koecher principle*, as we shall explain in a moment.

To discuss L-functions of Hilbert modular forms, we now impose the condition that the narrow class number of F is one. This implies that every ideal of F has a totally positive generator. In particular, let d be a totally positive generator of the different $\mathfrak{d}$.

Let f be a modular form of weight k. We have

$$f(z+a) = f(z) \qquad \text{if } a \in \mathfrak{o},$$

because $\left(\begin{smallmatrix} 1 & a \\ & 1 \end{smallmatrix}\right) \in \Gamma$, so f has a Fourier expansion. A complete set of characters of $\mathbb{R}^n/\mathfrak{o}$ is the set of functions

$$x \to e^{2\pi i\, \mathrm{tr}(\nu x/d)}, \qquad \nu \in \mathfrak{o}.$$

(This is essentially the definition of the different $\mathfrak{d} = (d)$.) The Fourier expansion therefore has the form

$$f(z) = \sum_{\nu \in \mathfrak{o}} A(\nu)\, e^{2\pi i\, \mathrm{tr}(\nu z/d)}. \tag{7.20}$$

The *Koecher principle* asserts that $A(\nu) = 0$ unless ν is totally positive or zero. For a proof, see Van der Geer (1987, p. 18).

If $A(0) = 0$, we say f is a *cusp form.* This must, of course, be replaced by a vanishing assumption at each cusp if the class number is greater than one.

We have

$$A(\nu) = A(\epsilon\nu) \tag{7.21}$$

if ϵ is a totally positive unit. Indeed, because we are assuming the narrow class number is one, every totally positive unit is the square of a unit (Exercise 1.7.5), and if $\epsilon = \epsilon_1^2$, where ϵ_1 is of course not necessarily totally positive, then, because f is modular with respect to

$$\begin{pmatrix} \epsilon_1 & \\ & \epsilon_1^{-1} \end{pmatrix},$$

we have

$$f(\epsilon z) = N(\epsilon_1)^{-k} f(z) = f(z) \tag{7.22}$$

for $z \in \mathcal{H}^n$, because $N(\epsilon_1) = \pm 1$ and k is even. Equation (7.22) implies Eq. (7.21). Because of this, we may regard A as a function of ideals. We have used the fact that the narrow class number of F is now assumed to be one.

We define the L-function

$$L(s, f) = \sum A(\mathfrak{a})\, \mathrm{N}(\mathfrak{a})^{-s}. \tag{7.23}$$

It is not hard to generalize Hecke's estimate in Proposition 1.3.5 to show that the Fourier coefficients of a cusp form satisfy

$$A(\mathfrak{a}) = O\big(\mathrm{N}(\mathfrak{a})^{k/2}\big). \tag{7.24}$$

The Fourier coefficients of Eisenstein series satisfy the weaker estimate $O\big(\mathrm{N}(\mathfrak{a})^{k-1}\big)$. In any case, the growth of the coefficients is such that the product and sum in Eq. (7.23) are absolutely convergent for s sufficiently large.

Hecke theory works as in the case of $SL(2, \mathbb{Z})$. One may define Hecke operators, which again form a commutative family of self-adjoint operators, and hence may be diagonalized. For an eigenfunction, we have

$$L(s, f) = \prod \big(1 - A(\mathfrak{p})\, \mathrm{N}(\mathfrak{p})^{-s} + \mathrm{N}(\mathfrak{p})^{k-1-2s}\big)^{-1}. \tag{7.25}$$

The proof is similar to the case of $SL(2, \mathbb{Z})$ (Exercise 1.7.4).

To prove a functional equation for $L(s, f)$, let

$$\Lambda(s, f) = (2\pi)^{-ns}\, \Gamma(s)^n\, D^s\, L(s, f). \tag{7.26}$$

We will prove that if f is a cusp form, $\Lambda(s, f)$ has analytic continuation to all s and that

$$\Lambda(s, f) = (-1)^{nk/2}\, \Lambda(k - s, f). \tag{7.27}$$

Indeed, we will prove that

$$\Lambda(s, f) = \int\limits_{\mathfrak{o}_+^\times \backslash \mathbb{R}_+^n} f(iy) \prod_{j=1}^{n} y_j^s \frac{dy_j}{y_j}. \tag{7.28}$$

We observe first that this integral is convergent for all s. Indeed, by the unit theorem, we may choose the fundamental domain for $\mathfrak{o}_+^\times \backslash \mathbb{R}_+^n$ so that the ratios y_i/y_j are all bounded. Consequently, the y_j all go to zero or infinity together. As the $y_j \to \infty$, the rapid decay of the cusp form assures convergence. As they go to zero, one uses the identity

$$f(i/y) = (-1)^{nk/2}\, N(y)^k\, f(iy) \tag{7.29}$$

to obtain the same result. As a consequence, the integral Eq. (7.28) is convergent for all s. We now show that it equals $\Lambda(s, f)$. Indeed, write the right side of Eq. (7.28) as

$$\sum_{\mathfrak{a}} A(\mathfrak{a}) \int\limits_{\mathfrak{o}_+^\times \backslash \mathbb{R}_+^n} \sum_{\substack{(\alpha) = \mathfrak{a} \\ \alpha \gg 0}} e^{-2\pi\, \mathrm{tr}(\alpha y/d)} \prod_{j=1}^{n} y_j^s \frac{dy_j}{y_j}.$$

Thus the summation is over all totally positive generators of the ideal $\mathfrak{a}$. Let us fix a totally positive generator α of $\mathfrak{a}$. Then every totally positive generator of $\mathfrak{a}$ has the form $\epsilon^2\alpha$ for some $\epsilon \in \mathfrak{o}^\times$. Of course, we are not allowed to assume $\epsilon \gg 0$. We have, by Eq. (7.21), that Eq. (7.28) equals

$$\begin{aligned}
&\sum_{\alpha \in \mathfrak{o}_+^\times \backslash \mathfrak{o}_+} \sum_{\epsilon \in \mathfrak{o}_+^\times} A(\alpha\mathfrak{o}) \int\limits_{\mathfrak{o}_+^\times \backslash \mathbb{R}_+^n} e^{-2\pi\, \mathrm{tr}(\alpha\epsilon y/d)} \prod_{j=1}^{n} y_j^s \frac{dy_j}{y_j} \\
&= \sum_{\alpha \in \mathfrak{o}_+^\times \backslash \mathfrak{o}_+} A(\alpha\mathfrak{o}) \int\limits_{\mathbb{R}_+^n} e^{-2\pi\, \mathrm{tr}(\alpha y/d)} \prod_{j=1}^{n} y_j^s \frac{dy_j}{y_j} \\
&= \sum_{\alpha \in \mathfrak{o}_+^\times \backslash \mathfrak{o}_+} A(\alpha\mathfrak{o}) \prod_{j=1}^{n} \left[(2\pi)^{-s} \left(d^{(j)}/\alpha^{(j)}\right)^s \Gamma(s)\right] \\
&= (2\pi)^{-ns}\, \Gamma(s)^n\, D^s \sum_{\mathfrak{a}} A(\mathfrak{a})\, \mathbb{N}(\mathfrak{a})^{-s}.
\end{aligned}$$

This proves Eq. (7.28); we see that $\Lambda(s, f)$ has analytic continuation to all s. We turn now to the proof of Eq. (7.27). The transformation $y \to y^{-1}$ preserves the coset space $\mathfrak{o}_+^\times\backslash\mathbb{R}_+^n$, and so

$$\Lambda(s, f) = \int\limits_{\mathfrak{o}_+^\times\backslash\mathbb{R}_+^n} f(i/y) \prod_{j=1}^{n} y_j^{-s} \frac{dy_j}{y_j}.$$

Now by Eq. (7.29), this equals

$$(-1)^{nk/2} \int\limits_{\mathfrak{o}_+^\times\backslash\mathbb{R}_+^n} f(iy) \prod_{j=1}^{n} y_j^{k-s} \frac{dy_j}{y_j},$$

from which we get Eq. (7.27).

We may now *state* without proof the fundamental theorem of quadratic base change. This was essentially proved by Doi and Naganuma (1973).

Theorem 1.7.3 *Let F be a real quadratic field of discriminant D whose narrow class number is one. Let $\phi(z) = \sum a(n)\, e^{2\pi i n z}$ be a Hecke eigenform of weight k for $SL(2, \mathbb{Z})$. Then there exists a Hilbert modular form f such that*

$$L(s, f) = L(s, \phi)\, L(s, \phi, \chi_D). \tag{7.30}$$

We discuss the proof of this result in Section 1.10, following a strategy that will be explained in Section 8. For the moment, we point out that the right side of Eq. (7.30) has at least the correct form and functional equation to be the L-function of a Hilbert modular form. A more sophisticated heuristic predicting the existence of the "base change lift" F will be presented in Section 1.8.

Let us describe the function f more specifically. It will be defined by a Fourier series (Eq. 7.20), where the function $A(n)$ is to be described in detail. The coefficients will satisfy Eq. (7.21), and hence may be described as a function of ideals. Let us show that Eq. (7.30) implies that we may reconstruct this function.

We have, by Proposition 1.3.6 and Eq. (5.9),

$$\Lambda(s, \phi) = (2\pi)^{-s}\, \Gamma(s)\, L(s, \phi) = (-1)^{k/2}\, \Lambda(k - s, \phi),$$

$$\Lambda(s, \phi, \chi_D) = (2\pi)^{-s}\, \Gamma(s)\, L(s, \phi, \chi_D) = D^{k-2s}\, (-1)^{k/2}\, \Lambda(k - s, \phi, \chi_D).$$

We have used the identity $\tau(\chi_D)^2 = D$. To see this, because $|\tau(\chi_D)| = \sqrt{D}$ by Eq. (1.5), it is sufficient to check that $\tau(\chi_D)$ is real, and this is easy to see. (What is still true, but more difficult, is that $\tau(\chi_D) = \sqrt{D}$. This will be shown in Section 1.9; cf. Eq. (9.14).) Multiplying these equations together, we see that the right side of Eq. (7.30) does satisfy the correct functional equation (7.27). To see that the Euler factors have the same form, let us compare the

Euler factors. We will denote by $a(n)$ the Fourier coefficients of ϕ and by $A(\mathfrak{a})$ the Fourier coefficients of the base change lift f. We factor the Hecke polynomials:

$$1 - a(p)\,x + p^{k-1}\,x^2 = \bigl(1 - \alpha(p)\,x\bigr)\,\bigl(1 - \beta(p)\,x\bigr),$$

$$1 - A(\mathfrak{p})\,x + \mathbb{N}(\mathfrak{p})^{k-1}\,x^2 = \bigl(1 - \alpha_f(\mathfrak{p})\,x\bigr)\,\bigl(1 - \beta_f(\mathfrak{p})\,x\bigr).$$

Suppose first that p splits in $\mathfrak{o}$, so that $p\,\mathfrak{o} = \mathfrak{p}_1\mathfrak{p}_2$. We have $\chi_D(p) = 1$, so the Euler factor on the right in Eq. (7.30) has the form

$$\bigl(1 - \alpha(p)\,p^{-s}\bigr)^{-2}\,\bigl(1 - \beta(p)\,p^{-s}\bigr)^{-2}.$$

On the other hand, we have $\mathbb{N}(\mathfrak{p}_1) = \mathbb{N}(\mathfrak{p}_2) = p$, so the Euler factor on the left has the form

$$\bigl(1-\alpha_f(\mathfrak{p}_1)\,p^{-s}\bigr)^{-1}\,\bigl(1-\beta_f(\mathfrak{p}_1)\,p^{-s}\bigr)^{-1}\,\bigl(1-\alpha_f(\mathfrak{p}_2)\,p^{-s}\bigr)^{-1}\,\bigl(1-\beta_f(\mathfrak{p}_2)\,p^{-s}\bigr)^{-1}.$$

These can be reconciled if we assume that $\alpha(p) = \alpha_f(\mathfrak{p}_1) = \alpha_f(\mathfrak{p}_2)$ and $\beta(p) = \beta_f(\mathfrak{p}_1) = \beta_f(\mathfrak{p}_2)$. In other words

$$A(\mathfrak{p}_1) = A(\mathfrak{p}_2) = a(p) \tag{7.31}$$

if $p = \mathfrak{p}_1\mathfrak{p}_2$ splits in $\mathfrak{o}$.

Next, suppose that p is inert, so $\mathfrak{p} = p\,\mathfrak{o}$ is a prime ideal of norm p^2. In this case, $\chi_D(p) = -1$, so the Euler factor on the right side of Eq. (7.30) is

$$\bigl(1 - \alpha(p)^2\,p^{-2s}\bigr)^{-1}\,\bigl(1 - \beta(p)^2\,p^{-2s}\bigr)^{-1}.$$

The Euler factor on the left is

$$\bigl(1 - \alpha_f(\mathfrak{p})\,p^{-2s}\bigr)^{-1}\,\bigl(1 - \beta_f(\mathfrak{p})\,p^{-2s}\bigr)^{-1},$$

and these may be reconciled if $A(\mathfrak{p}) = \alpha(p)^2 + \beta(p)^2$. Because $\alpha(p)\,\beta(p) = p^{k-1}$, we see that

$$A(\mathfrak{p}) = a(p)^2 - 2p^{k-1} \tag{7.32}$$

if p is inert, $\mathfrak{p} = p\,\mathfrak{o}$.

If p ramifies, $p\,\mathfrak{o} = \mathfrak{p}^2$ and it can be checked that we must have

$$A(\mathfrak{p}) = a(p). \tag{7.33}$$

Once the coefficients $A(\mathfrak{p})$ are known, the remaining coefficients are determined by the Hecke theory (Exercise 1.7.4). The prime power coefficients are determined by the formula

$$\sum_{k=0}^{\infty} A(\mathfrak{p}^k)\,x^k = \bigl(1 - A(\mathfrak{p})\,x + \mathbb{N}(\mathfrak{p})^{k-1}\,x^2\bigr)^{-1}, \tag{7.34}$$

and the general coefficients $A(\mathfrak{a})$ are then determined by the multiplicativity condition:

$$A(\mathfrak{a}\mathfrak{b}) = A(\mathfrak{a})\,A(\mathfrak{b}) \tag{7.35}$$

if $\mathfrak{a},\mathfrak{b}$ are coprime.

Exercises

Exercise 1.7.1 Prove Eq. (7.18).

Exercise 1.7.2 Prove the Poisson summation formula Eq. (7.6).

Exercise 1.7.3 Determine the orbits of Γ on the set $\mathbb{P}^1(F) = F \cup \infty$ as follows. Let A be an ideal class of $\mathfrak{o}_F$, and let $\mathfrak{a} \in A$. Find α and $\beta \in \mathfrak{o}$ such that $\mathfrak{a} = \alpha\mathfrak{o} + \beta\mathfrak{o}$. (This is always possible in a Dedekind domain.) Consider the cusp $\xi_A = \alpha/\beta$. Prove that the Γ-orbit of ξ_A depends only on the ideal class A and not on the choice of $\mathfrak{a}$, α, and β. Deduce that $A \to \xi_A$ is a bijection between the ideal class group and $\Gamma\backslash\mathbb{P}^1(F)$.

Exercise 1.7.4 Extend the theory of Hecke operators in Section 1.4 to Hilbert modular forms. Prove that the Hecke operators may be simultaneously diagonalized and that for an eigenform, we have Eq. (7.25).

Exercise 1.7.5 Prove that the totally real field F has narrow class number one if and only if F has class number one and every totally positive unit is the square of a unit, or equivalently, if and only if F has class number one, and F has units of every possible sign combination.

Exercise 1.7.6 Prove that if p is an odd prime, and if the narrow class number of $\mathbb{Q}(\sqrt{p})$ is one, then $p \equiv 1 \bmod 4$.

1.8 Artin L-Functions and Langlands Functoriality

The purpose of this section is to explain a highly important heuristic, a naive version of Langlands functoriality conjecture, which allows us to *predict* the existence of liftings, such as the Doi–Naganuma lift. We will also show how this heuristic predicts that we may use the Rankin–Selberg method and converse theorems to *prove* the existence of the Doi–Naganuma lift!

Langlands functoriality conjecture was first stated in Langlands (1970c). Evidence for the conjecture was given in Langlands (1971, 1980) and Jacquet and Langlands (1970). It is closely related to the local Langlands conjectures, which amount to a (conjectural) classification of the representations of reductive groups over local fields. We will discuss the Langlands functoriality conjecture naively in this section and more precisely in Section 3.9. We will consider the local Langlands conjectures in Section 4.9. Other references for the Langlands conjectures are Borel (1979) (which is the definitive account), Arthur (1981). Arthur and Gelbart (1991), Gelbart (1977, 1984), and (for the local conjectures) Kudla (1994) and Knapp (1994).

We will introduce the Artin L-functions. Let E/F be a finite Galois extension of number fields, and let $\rho : \mathrm{Gal}(E/F) \to GL(n, \mathbb{C})$ be a representation. Artin (1930) associated such a ρ with an L-function $L(s, \rho)$. He conjectured that it has analytic continuation and a functional equation. Although the *analytic* continuation has never been proved for all Artin L-functions, the *meromorphic* continuation and functional equation were reduced, by Artin, to a purely algebraic statement that was proved by Brauer (1947). The functional equation of the Artin L-series is always with respect to $s \to 1 - s$.

The definition of the Artin L-function is given as an Euler product. We will not attempt to describe the Euler factors at the finite number of primes that ramify in E, nor will we describe the precise functional equation. We will return to the Artin L-functions from a more sophisticated point of view in Section 4.9.

Let $\mathfrak{p}$ be a prime ideal of F that does not ramify in E. Let $\mathfrak{P}$ be a prime of E that divides $\mathfrak{p}$. There is an element $\sigma_{\mathfrak{P}|\mathfrak{p}}$ of $\mathrm{Gal}(E/F)$, called the Frobenius endomorphism, such that

$$\sigma_{\mathfrak{P}|\mathfrak{p}}(x) \equiv x^{\mathrm{N}(\mathfrak{p})} \bmod \mathfrak{P}, \qquad \text{for all } x \in \mathfrak{o}_E.$$

We would like it if $\sigma_{\mathfrak{P}|\mathfrak{p}}$ depended only on $\mathfrak{p}$. This is the case if E/F is Abelian. If the extension E/F is non-Abelian, $\sigma_{\mathfrak{P}|\mathfrak{p}}$ depends on the choice of the prime $\mathfrak{P}$, but *only up to conjugacy*. For our purposes, this is sufficient. We will let $\sigma_\mathfrak{p}$ denote any element of this conjugacy class. We define

$$L_\mathfrak{p}(s, \rho) = \det\left(I_n - \rho(\sigma_\mathfrak{p})\,\mathrm{N}(\mathfrak{p})^{-s}\right)^{-1},$$

where I_n is the $n \times n$ identity matrix. (Note that this depends only on the conjugacy class of $\sigma_\mathfrak{p}$ and hence is well defined!)

Thus, if $\alpha_1(\mathfrak{p}), \cdots, \alpha_n(\mathfrak{p})$ are the eigenvalues of $\rho(\sigma_\mathfrak{p})$

$$L_\mathfrak{p}(s, \rho) = \prod_{i=1}^{n} \left(1 - \alpha_i(\mathfrak{p})\,\mathrm{N}(\mathfrak{p})^{-s}\right)^{-1}. \tag{8.1}$$

This is independent of the choice of $\mathfrak{P}$. The *Artin L-function* is then defined by

$$L(s, \rho) = \prod_{\mathfrak{p}} L_\mathfrak{p}(s, \rho),$$

where the product is over all $\mathfrak{p}$ that do not ramify in E.

As a first example, we show that if χ is a Dirichlet character with conductor N, there is a one-dimensional representation $\rho : \mathrm{Gal}(E/\mathbb{Q}) \to GL(1, \mathbb{C}) = \mathbb{C}^\times$ such that $L(s, \rho) = L(s, \chi)$. Here $E = \mathbb{Q}(\zeta)$, where $\zeta = e^{2\pi i/N}$. The Galois group $\mathrm{Gal}(E/\mathbb{Q})$ may be identified with $(\mathbb{Z}/N\mathbb{Z})^\times$ in the following way: with $a \in \mathbb{Z}$ prime to N, let us associate the endomorphism $\sigma_a : \zeta \to \zeta^a$. This depends only on the image of a in $(\mathbb{Z}/N\mathbb{Z})^\times$, and so we obtain an isomorphism $(\mathbb{Z}/N\mathbb{Z})^\times \cong \mathrm{Gal}(E/\mathbb{Q})$. Thus χ, which is a character of $(\mathbb{Z}/N\mathbb{Z})^\times$, gives rise to a character of $\mathrm{Gal}(E/\mathbb{Q})$, which we call ρ. If p is a prime, then σ_p is the Frobenius element associated with the ideal p of $\mathbb{Q}$, and so $L(s, \rho) = L(s, \chi)$.

Now let us introduce the notion of an *automorphic cuspidal representation* of $GL(n)$ over a number field F. (This topic will be considered in detail in Chapter 3.) Let $A = A_F$ be the adele ring of F. The group $GL(n, F)$ is a discrete subgroup of $GL(n, A)$, and the quotient $Z_A\, GL(n, F)\backslash GL(n, A)$ has finite volume, where Z_A is the group of scalar matrices with entries in the idele group $A^\times$.

Let ω be a character of $A^\times/F^\times$. We assume that ω is unitary, that is, that $|\omega(z)| = 1$ for all $z \in A^\times$. Let $L^2\big(GL(n, F)\backslash GL(n, A), \omega\big)$ be the space of measurable functions f on $GL(n, A)$ such that

$$f\left(\begin{pmatrix} z & & \\ & \ddots & \\ & & z \end{pmatrix} g\right) = \omega(z)\, f(g) \qquad \text{for } z \in A^\times,$$

and such that

$$\int_{Z_A\, GL(n,F)\backslash GL(n,A)} |f(g)|^2\, dg < \infty,$$

where dg is the Haar measure. Let $L_0^2\big(GL(n, F)\backslash GL(n, A), \omega\big)$ be the closed subspace of functions that further satisfy

$$\int_{(A/F)^{r(n-r)}} f\left(\begin{pmatrix} I_r & x \\ & I_{n-r} \end{pmatrix} g\right) dx = 0$$

for almost all $g \in GL(n, A)$ and for every $1 \le r < n$. (Here x is an $r \times (n-r)$ block matrix.) This is the space of *cusp forms*.

We have the *right regular representation*

$$\rho : GL(n, A) \to \mathrm{End}\big(L^2(GL(n, F)\backslash GL(n, A), \omega)\big),$$

in which the group action is by right translation:

$$\big(\rho(g) f\big)(g') = f(g'g).$$

The space of cusp forms is invariant under this representation and has the advantage of decomposing into an infinite direct sum of irreducible invariant subspaces – if $n = 2$ we will prove this in Theorem 3.3.2. If π is a representation of $GL(n, A)$ that is isomorphic to the representation on one of these invariant subspaces (for some ω), we call π an *automorphic cuspidal representation*. We call ω the *central character* of π.

(More generally if π is an irreducible unitary representation of $GL(n, A)$ that occurs in $L^2\big(GL(n, F)\backslash GL(n, A), \omega\big)$, we call π an *automorphic representation*. However, it must be understood that π may not occur *discretely* – $L^2\big(GL(n, F)\backslash GL(n, A), \omega\big)$ does *not* decompose as an infinite direct sum of irreducible representations, for there is a continuous spectrum. We will avoid this nuance by only considering automorphic *cuspidal* representations.)

There is a close connection between classical automorphic forms and automorphic representations. Because we are only attempting to explain a heuristic, we do not need to be too precise about this. What we need to know is the following:

(1) If χ is a Hecke character of F, then χ may be canonically associated with an automorphic (cuspidal) representation of $GL(1)$ over F. We note that because $GL(1, A) \cong A^\times$ is Abelian, an automorphic cuspidal representation of $GL(1, A)$ is one dimensional – an automorphic cuspidal representation of $GL(1)$ over F is just a character of $A^\times/F^\times$. If $F = \mathbb{Q}$, the character of $A^\times/F^\times$ associated with a Hecke (i.e., Dirichlet) character will be constructed in Proposition 3.1.2.
(2) If F is totally real, and if f is a Hilbert modular cusp form over F (for the full modular group, or more generally, some congruence subgroup), and if f is an eigenform for the Hecke operators, then f may be canonically associated with an automorphic cuspidal representation of $GL(2)$ over F. If $F = \mathbb{Q}$, we will construct the automorphic representation associated with a modular form in Section 3.6.
(3) If $F = \mathbb{Q}$ and $n = 2$, the automorphic cuspidal representations fall cleanly into two disjoint classes: those arising from modular forms, as in point 2, and those arising from *Maass forms*, another class of automorphic forms that will be introduced in Section 9. This dichotomy follows from the classification of the irreducible representations of $GL(2, \mathbb{R})$ in Theorem 2.5.4, together with the results of Chapter 3.

If π is an automorphic representation of $GL(n)$ over F, Godement and Jacquet (1972) have associated with π an L-function, which has the form

$$L(s, \pi) = \prod_{\mathfrak{p}} \big(1 - \alpha_1(\mathfrak{p})\, \mathbb{N}(\mathfrak{p})^{-s}\big)^{-1} \cdots \big(1 - \alpha_n(\mathfrak{p})\, \mathbb{N}(\mathfrak{p})^{-s}\big)^{-1}.$$

Here, as with the Artin L-functions, we are ignoring a finite number of "bad" primes and the Archimedean places. Of course, a complete theory must deal with these eventually, but for the purposes of our heuristic considerations, this is not necessary.

If $n = 1$, this agrees with the L-function of the Hecke character that we have already considered. If $n = 2$, and if π is associated with a Hilbert cusp form f, it agrees with the L-function of f, except for one detail: the functional equation of the Godement–Jacquet L-functions are normalized so that the functional equation is with respect to $s \to 1 - s$. (The Artin L-functions are also usually normalized this way.) However, the functional equations of the L-functions of modular forms are normalized so that the functional equation is with respect to $s \to k - s$, where k is the weight. Suppose that f is a Hilbert modular form of weight k and that π is the automorphic representation that corresponds as in point 2 above. The L-functions are related by a shift:

$$L(s, \pi) = L\big(s + (k - 1)/2, f\big). \tag{8.2}$$

Thus if

$$L_{\mathfrak{p}}(s, f) = \prod_{i=1}^{n} \big(1 - \alpha_i(\mathfrak{p})\, \mathbb{N}(\mathfrak{p})^{-s}\big)^{-1}, \tag{8.3}$$

then $L(s, \pi)$ has the same form

$$L_{\mathfrak{p}}(s, \pi) = \prod_{i=1}^{n} \big(1 - \alpha_i'(\mathfrak{p})\, \mathbb{N}(\mathfrak{p})^{-s}\big)^{-1}, \tag{8.4}$$

where

$$\alpha_i(\mathfrak{p}) = \mathbb{N}(\mathfrak{p})^{(k-1)/2}\, \alpha_i'(\mathfrak{p}). \tag{8.5}$$

Observe that the form of the L-function of an automorphic representation of $GL(n)$ over F is the same as that of an Artin L-function corresponding to an n-dimensional representation of $\mathrm{Gal}(E/F)$.

Conjecture 1.8.1: Langlands *Let $\rho : \mathrm{Gal}(E/F) \to GL(n, \mathbb{C})$ be a representation. Then there exists an automorphic representation $\pi(\rho)$ of $GL(n)$ over F whose L-function agrees with the Artin L-function of ρ.*

The converse to Conjecture 1.8.1 is not true, for there are automorphic forms that do not come from Galois representations.

Conjecture 1.8.1 was made in the context of a far more sweeping conjecture, known as *Langlands functoriality conjecture*. We will attempt to give a naive and intuitive explanation of this conjecture. We will be unable to reveal its full implications, which have been revolutionary in the field of automorphic forms. Nor will we be able to be very precise in our description of the conjecture. For a precise statement of the conjectures, see Borel (1979).

Conjecture 1.8.2: Langlands functoriality, naive form *Operations on Artin L-functions correspond to operations on automorphic forms.*

We will illustrate this conjecture by giving examples. We consider three basic types of operations on Artin L-functions; by functoriality, these should correspond to operations on automorphic forms; we shall describe these operations as follows:

(a) composition with a homomorphism $GL(n, \mathbb{C}) \to GL(m, \mathbb{C})$, or more generally, of products $GL(n_1, \mathbb{C}) \times \cdots \times GL(n_r, \mathbb{C}) \to GL(m, \mathbb{C})$;
(b) restriction to a Galois subgroup; and
(c) induction from a Galois subgroup.

We will give examples of all these possibilities. Let us begin with operation (a). If r is a positive integer, $GL(2, \mathbb{C})$ has an irreducible representation $\vee^r : GL(2, \mathbb{C}) \to GL(r+1, \mathbb{C})$, called the rth symmetric power representation. (If V is a vector space, the rth symmetric power $\vee^r V$ of V is constructed in Lang (1993, Section XVI.8). Because it is a functor, there is a homomorphism $GL(V) \to GL(\vee^r V)$. We are concerned here with the case $V = \mathbb{C}^2$. The symmetric power representations are studied extensively in Fulton and Harris (1991).) It is also possible to twist $\vee^r$ by a power of the determinant, and $\det^{-1} \otimes \vee^2$ is sometimes called the *adjoint square* representation $GL(2, \mathbb{C}) \to GL(3, \mathbb{C})$. Suppose now that $\rho : \mathrm{Gal}(E/F) \to GL(2, \mathbb{C})$ is a Galois representation. The composition of ρ with $\vee^r$ is another Galois representation $\sigma : \mathrm{Gal}(E/F) \to GL(r+1, \mathbb{C})$. If α_1 and α_2 are the eigenvalues of $\rho(g)$, then

$$\alpha_1^r, \alpha_1^{r-1}\alpha_2, \cdots, \alpha_2^r$$

are the eigenvalues of $\sigma(g)$. Thus if

$$L(s, \rho) = \prod_p \left(1 - \alpha_1(\mathfrak{p})\, \mathbb{N}(\mathfrak{p})^{-s}\right)^{-1} \left(1 - \alpha_2(\mathfrak{p})\, \mathbb{N}(\mathfrak{p})^{-s}\right)^{-1},$$

then

$$L(s, \sigma) = \prod_p \left(1 - \alpha_1(\mathfrak{p})^r \mathbb{N}(\mathfrak{p})^{-s}\right)^{-1} \left(1 - \alpha_1(\mathfrak{p})^{r-1} \alpha_2(\mathfrak{p}) \mathbb{N}(\mathfrak{p})^{-s}\right)^{-1} \cdots \left(1 - \alpha_2(\mathfrak{p})^r \mathbb{N}(\mathfrak{p})^{-s}\right)^{-1}.$$

According to the functoriality conjecture, there should be a corresponding operation on automorphic forms. Thus we arrive at the following prediction.

Prediction 1.8.1 *If π is an automorphic representation of $GL(2)$ such that*

$$L(s, \pi) = \prod_p \left(1 - \alpha_1(\mathfrak{p})\, \mathbb{N}(\mathfrak{p})^{-s}\right)^{-1} \left(1 - \alpha_2(\mathfrak{p})\, \mathbb{N}(\mathfrak{p})^{-s}\right)^{-1}, \tag{8.6}$$

then there exists an automorphic cuspidal representation $\vee^r \pi$ such that

$$L(s, \vee^r \pi) = \prod_p \left(1 - \alpha_1(\mathfrak{p})^r \mathbb{N}(\mathfrak{p})^{-s}\right)^{-1} \left(1 - \alpha_1(\mathfrak{p})^{r-1} \alpha_2(\mathfrak{p}) \mathbb{N}(\mathfrak{p})^{-s}\right)^{-1} \cdots \left(1 - \alpha_2(\mathfrak{p})^r \mathbb{N}(\mathfrak{p})^{-s}\right)^{-1}. \tag{8.7}$$

If $r = 1$, this is trivial – $\vee^1 \pi = \pi$. If $r = 2$, this was proved by Gelbart and Jacquet (1978). They make use of the converse theorem for $GL(3)$, due to

Jacquet, Piatetski–Shapiro and Shalika (1979), and the fact that the functional equation of the L-function of Eq. (8.7) can be proved using the Rankin–Selberg method, as in Shimura (1975). If $r > 2$, the existence of the automorphic representation $\vee^r \pi$ is not known, but even though the automorphic representation $\vee^r \pi$ has not been constructed, its L-function can be studied when $r \leq 5$. There is considerable literature on this subject and we refer to the survey paper of Shahidi (1994) for a synopsis.

We consider next a homomorphism $GL(n, \mathbb{C}) \times GL(m, \mathbb{C}) \to GL(nm, \mathbb{C})$, namely, the tensor product. This gives rise to an operation on Artin L-functions. Suppose that

$$\rho_1 : \mathrm{Gal}(E_1/F) \to GL(n, \mathbb{C}), \qquad \rho_2 : \mathrm{Gal}(E_2/F) \to GL(m, \mathbb{C})$$

are Galois representations. Let E be the compositum of E_1 and E_2. We may compose ρ_i with the projection map $\mathrm{Gal}(E/F) \to \mathrm{Gal}(E_i/F)$, and this does not change the Artin L-function. We may therefore assume, without loss of generality, that $E_1 = E_2 = E$. We consider the Artin L-function associated with the nm-dimensional representation of $\mathrm{Gal}(E/F)$ defined by $\rho(g) = \rho_1(g) \otimes \rho_2(g)$. We see the following: if $L(s, \rho_1)$ and $L(s, \rho_2)$ are Artin L-functions, then there exists an Artin L-function $L(s, \rho)$ such that, for a prime $\mathfrak{p}$ of F, if

$$L_{\mathfrak{p}}(s, \rho_1) = \prod_{i=1}^{n} \left(1 - \alpha_i(\mathfrak{p})\, \mathbb{N}(\mathfrak{p})^{-s}\right)^{-1},$$

$$L_{\mathfrak{p}}(s, \rho_2) = \prod_{i=1}^{m} \left(1 - \beta_i(\mathfrak{p})\, \mathbb{N}(\mathfrak{p})^{-s}\right)^{-1},$$

then

$$L_{\mathfrak{p}}(s, \rho) = \prod_{i,j} \left(1 - \alpha_i(\mathfrak{p})\, \beta_j(\mathfrak{p})\, \mathbb{N}(\mathfrak{p})^{-s}\right)^{-1}.$$

According to the functoriality conjecture, there should be a corresponding operation on automorphic forms. To be precise, we have the following prediction.

Prediction 1.8.2 *If π_1 and π_2 are automorphic representations of $GL(n)$ and $GL(m)$, respectively, over F, then there exists an automorphic representation π of $GL(nm)$ such that for a prime $\mathfrak{p}$ of F, if*

$$L_{\mathfrak{p}}(s, \pi_1) = \prod_{i=1}^{n} \left(1 - \alpha_i(\mathfrak{p})\, \mathbb{N}(\mathfrak{p})^{-s}\right)^{-1}, \tag{8.8}$$

$$L_{\mathfrak{p}}(s, \pi_2) = \prod_{i=1}^{m} \left(1 - \beta_i(\mathfrak{p})\, \mathbb{N}(\mathfrak{p})^{-s}\right)^{-1}, \tag{8.9}$$

then

$$L_{\mathfrak{p}}(s,\pi)=\prod_{i,j}\left(1-\alpha_i(\mathfrak{p})\,\beta_j(\mathfrak{p})\,\mathbb{N}(\mathfrak{p})^{-s}\right)^{-1}. \tag{8.10}$$

This is not known unless n or m is equal to one, although a proof for $n=m=2$ should eventually follow from the converse theorem for $GL(4)$ and the Rankin–Selberg method. However, even though we cannot construct the automorphic representation π, we can at least construct the L-function $L(s,\pi)$. This was done for modular forms in Section 1.6 – an examination of Theorem 1.6.3 (taking into account the shift Eq. (8.5)) shows that the L-function $L(s, f\times g)$ constructed there, if shifted so that its functional equation is with respect to $s\to 1-s$, is precisely the L-function $L(s,\pi)$ of the conjectured lift to $GL(4)$. We will return to the study of these L-functions in Section 3.8. For arbitrary n and m, the construction of the L-functions $L(s,\pi)$ was carried out by Jacquet, Piatetski-Shapiro, and Shalika and (using a different method) by Shahidi. See Jacquet and Shalika (1981, 1990a), Jacquet, Piatetski–Shapiro and Shalika (1981b) and Cogdell and Piatetski–Shapiro (1994) for the Rankin–Selberg approach. for Shahidi's method, see the discussion on p. 418 of Shahidi (1990b). In the appendix to Moeglin and Waldspurger (1989), the poles of these L-functions are studied by a method related to Shahidi's.

Because the existence of π is unknown, but the functional equation of its L-function is proved, we will denote this L-function as $L(s,\pi_1\times\pi_2)$.

Now let us consider operation (b). Let K/F be an extension, and let $E\supset K$ be a bigger field, Galois over F. Then $\mathrm{Gal}(E/K)$ is a subgroup of $\mathrm{Gal}(E/F)$. Let $\rho:\mathrm{Gal}(E/F)\to GL(n,\mathbb{C})$ be a representation. Let $\rho':\mathrm{Gal}(E/K)\to GL(n,\mathbb{C})$ be the restriction of ρ. Corresponding to this operation on Galois representations, there should be a corresponding operation on automorphic forms. It should associate an automorphic representation π of $GL(n)$ over F with an automorphic representation π' of $GL(n)$ over K. This operation is called *base change*.

Let us make the operation more explicit in the case where K/F is quadratic. We return to the case of a Galois representation, and compute the Artin L-function $L(s,\rho')$. In this case, there exists a Hecke character χ of F such that

$$\chi(\mathfrak{p})=\begin{cases}-1 & \text{if } \mathfrak{p} \text{ splits in } K;\\ -1 & \text{if } \mathfrak{p} \text{ is inert in } K;\\ -0 & \text{if } \mathfrak{p} \text{ ramifies.}\end{cases}$$

Thus if $F=\mathbb{Q}$ and D is the discriminant of the quadratic field K, then $\chi=\chi_D$ in the notation of Exercise 1.1.6.

Let $\mathfrak{p}$ be a prime of F. Let

$$L_{\mathfrak{p}}(s,\rho)=\prod_{i=1}^{n}\left(1-\alpha_i(\mathfrak{p})\,\mathbb{N}(\mathfrak{p})^{-s}\right)^{-1}.$$

Suppose that $\mathfrak{p}$ splits in K and that $\mathfrak{p}\mathfrak{o}_K = \mathfrak{P}_1\mathfrak{P}_2$. Then $\sigma_{\mathfrak{P}_1} = \sigma_{\mathfrak{P}_2} = \sigma_{\mathfrak{p}}$ and $\mathrm{N}(\mathfrak{P}_1) = \mathrm{N}(\mathfrak{P}_2) = \mathrm{N}(\mathfrak{p})$, and so

$$L_{\mathfrak{P}_1}(s, \rho') = L_{\mathfrak{P}_2}(s, \rho') = L_{\mathfrak{p}}(s, \rho).$$

Thus we have a contribution to $L(s, \rho')$ of

$$L_{\mathfrak{P}_1}(s, \rho')\, L_{\mathfrak{P}_2}(s, \rho') = \prod_{i=1}^{n} \left(1 - \alpha_i(\mathfrak{p})\, \mathrm{N}(\mathfrak{p})^{-s}\right)^{-2}.$$

Next suppose that $\mathfrak{p}$ is inert, $\mathfrak{p}\mathfrak{o}_K = \mathfrak{P}$. Then $\sigma_{\mathfrak{P}} = \sigma_{\mathfrak{p}}^2$ and $\mathrm{N}(\mathfrak{P}) = \mathrm{N}(\mathfrak{p})^2$, so we have a contribution of

$$L_{\mathfrak{P}}(s, \rho') = \prod_{i=1}^{n} \left(1 - \alpha_i(\mathfrak{p})^2\, \mathrm{N}(\mathfrak{p})^{-2s}\right)^{-1}.$$

In either case, the contribution may be written

$$L_{\mathfrak{p}}(s, \rho)\, L_{\mathfrak{p}}(s, \chi \otimes \rho),$$

where

$$L_{\mathfrak{p}}(s, \chi \otimes \rho) = \prod_{i=1}^{n} \left(1 - \chi(\mathfrak{p})\, \alpha_i(\mathfrak{p})\, \mathrm{N}(\mathfrak{p})^{-s}\right)^{-1}.$$

We see that

$$L(s, \rho') = L(s, \rho)\, L(s, \chi \otimes \rho).$$

Let us translate our conclusions into a prediction about automorphic forms. If

$$L_{\mathfrak{p}}(s, \pi) = \prod_{i=1}^{n} \left(1 - \alpha_i(\mathfrak{p})\, \mathrm{N}(\mathfrak{p})^{-s}\right)^{-1},$$

let us denote

$$L_{\mathfrak{p}}(s, \pi, \chi) = \prod_{i=1}^{n} \left(1 - \chi(\mathfrak{p})\, \alpha_i(\mathfrak{p})\, \mathrm{N}(\mathfrak{p})^{-s}\right)^{-1},$$

and $L(s, \pi, \chi) = \prod_{\mathfrak{p}} L_{\mathfrak{p}}(s, \pi, \chi)$. Thus in the notation of Conjecture 1.8.1, if $\pi = \pi(\rho)$, then $L(s, \pi, \chi) = L\big(s, \pi(\chi \otimes \rho)\big)$. We have the following prediction.

Prediction 1.8.3 *Let K/F be an extension of number fields, and let π be an automorphic representation of $GL(n)$ over F. Then there should be an automorphic representation π' of $GL(n)$ over K canonically associated with*

π called the base change lift. *If K/F is quadratic, and χ is the quadratic Hecke character of F corresponding to this quadratic extension, we should have*

$$L(s,\pi') = L(s,\pi)\,L(s,\pi,\chi). \tag{8.11}$$

It is possible to give a similar factorization of $L(s,\pi')$ even if K/F is only Abelian. If K/F is cyclic, the existence of a lift has been proved for $GL(n)$ by Arthur and Clozel (1981), using the trace formula. (Note that a solvable extension may be built up by a series of cyclic extensions.) If K/F is a cubic extension, even possibly not normal, and if $n = 2$, the existence of π' was proved by Jacquet, Piatetski–Shapiro and Shalika (1981a) using the converse theorem. We remark that this result was one ingredient in Tunnell's (1981) proof of the octahedral case of Artin's conjecture, which was itself used by Wiles (1994) in his proof of the semistable case of the Taniyama–Shimura conjecture and Fermat's last theorem.

Now let us consider operation (c), induction. Let K/F and E be as in the previous discussion of operation (b), and let $\tau : \mathrm{Gal}(E/K) \to GL(n,\mathbb{C})$ be a representation. Let $r = [K : F]$. Then we may induce τ to a representation $\tau' : \mathrm{Gal}(E/F) \to GL(rn,\mathbb{C})$.

It may be shown that

$$L(s,\tau') = L(s,\tau). \tag{8.12}$$

(Actually, there is a slight nuance here in that we have only defined the Euler factors for $L(s,\tau)$ corresponding to primes that do not ramify in E. Thus there could be a few Euler factors missing from the right side of Eq. (8.12), corresponding to the finite number of primes of F that ramify in K. We will ignore this difficulty – its proper resolution is to define Euler factors for *all* primes, not just those that are nonramified.)

We will prove Eq. (8.12) under the simplifying assumption that K/F is quadratic. We leave the general case to the reader (Exercise 1.8.1). The method of proof is to compare the Euler factors on both sides. We will only consider Euler factors corresponding to nonramified primes.

Assuming K/F is quadratic, $\mathrm{Gal}(E/K)$ is of index two in $\mathrm{Gal}(E/F)$. Let γ be a representative for the nonidentity coset of $\mathrm{Gal}(E/K)$ in $\mathrm{Gal}(E/F)$. We take the following concrete realization of the induced representation τ' in matrices. If $\sigma \in \mathrm{Gal}(E/F)$, let

$$\tau'(\sigma) = \begin{cases} \begin{pmatrix} \tau(\sigma) & \\ & \tau(\gamma\sigma\gamma^{-1}) \end{pmatrix} & \text{if } \sigma \in \mathrm{Gal}(E/K); \\ \begin{pmatrix} & \tau(\sigma\gamma^{-1}) \\ \tau(\gamma\sigma) & \end{pmatrix} & \text{otherwise.} \end{cases} \tag{8.13}$$

Suppose now that $\mathfrak{p}$ is a prime of F that splits in K, $\mathfrak{p}\mathfrak{o}_K = \mathfrak{P}_1\mathfrak{P}_2$. Then $\sigma_{\mathfrak{p}}$, $\sigma_{\mathfrak{P}_1}$, and $\sigma_{\mathfrak{P}_2}$ are all conjugate in $\mathrm{Gal}(E/F)$. They are *not* necessarily conjugate in $\mathrm{Gal}(E/K)$, and it is important to be clear about this. Recall that

$\sigma_{\mathfrak{p}} = \sigma_{\mathfrak{Q}|\mathfrak{p}}$ actually depends on the choice of a prime $\mathfrak{Q}$ of E dividing $\mathfrak{p}$. Then $\mathfrak{Q}$ divides either $\mathfrak{P}_1$ or $\mathfrak{P}_2$. Let us assume that $\mathfrak{Q}$ divides $\mathfrak{P}_1$. Then because $\sigma_{\mathfrak{p}}$ is conjugate in $\mathrm{Gal}(E/K)$ to $\sigma_{\mathfrak{P}_1}$, and because γ interchanges $\mathfrak{P}_1$ and $\mathfrak{P}_2$, $\gamma\sigma_{\mathfrak{p}}\gamma^{-1}$ is conjugate in $\mathrm{Gal}(E/K)$ to $\sigma_{\mathfrak{P}_2}$. Suppose now that $\alpha_1, \cdots, \alpha_n$ are the eigenvalues of $\tau(\sigma_{\mathfrak{P}_1})$ and that $\beta_1, \cdots, \beta_n$ are the eigenvalues of $\tau(\sigma_{\mathfrak{P}_2})$. Because

$$\tau'(\sigma_{\mathfrak{p}}) = \begin{pmatrix} \tau(\sigma_{\mathfrak{p}}) & \\ & \tau(\gamma\sigma_{\mathfrak{p}}\gamma^{-1}) \end{pmatrix},$$

the eigenvalues of $\tau'(\sigma_{\mathfrak{p}})$ are $\alpha_1, \cdots, \alpha_n, \beta_1, \cdots, \beta_n$. Hence

$$L_{\mathfrak{p}}(s, \tau') = L_{\mathfrak{P}_1}(s, \tau)\, L_{\mathfrak{P}_2}(s, \tau).$$

On the other hand, suppose that $\mathfrak{p}$ is inert, $\mathfrak{p}\mathfrak{o}_K = \mathfrak{P}$. We have $\sigma_{\mathfrak{p}}^2 = \sigma_{\mathfrak{P}}$. Suppose that $\alpha_1, \cdots, \alpha_n$ are the eigenvalues of $\tau(\sigma_{\mathfrak{P}})$. We will show that

$$\sqrt{\alpha_1}, -\sqrt{\alpha_1}, \cdots, \sqrt{\alpha_n}, -\sqrt{\alpha_n} \tag{8.14}$$

are the eigenvalues of $\tau'(\sigma_{\mathfrak{p}})$. Indeed, let $\beta_1, \cdots, \beta_{2n}$ be the eigenvalues of this matrix, so $\sum \beta_i^r = \mathrm{tr}\ \tau'(\sigma_{\mathfrak{p}}^r)$. If r is odd, then $\tau'(\sigma_{\mathfrak{p}}^r)$ has no diagonal entries by Eq. (8.13), because $\sigma_{\mathfrak{p}}^r \notin \mathrm{Gal}(E/K)$. On the other hand, if r is even, it follows from Eq. (8.13) that

$$\tau'(\sigma_{\mathfrak{p}}^r) = \begin{pmatrix} \tau(\sigma_{\mathfrak{P}}^{r/2}) & \\ & \tau(\gamma\sigma_{\mathfrak{P}}^{r/2}\gamma^{-1}) \end{pmatrix},$$

so $\mathrm{tr}\ \tau'(\sigma_{\mathfrak{p}}^r) = 2\sum \alpha_i^{r/2}$. Because

$$\sum \beta_i^r = \begin{cases} 2\sum \alpha_i^{r/2} & \text{if } r \text{ is even;} \\ 0 & \text{if } r \text{ is odd,} \end{cases}$$

we see that the eigenvalues β_i are indeed given by Eq. (8.14). Now we have

$$\begin{aligned} L_{\mathfrak{p}}(s, \tau') &= \prod_{i=1}^{n} \left(1 - \sqrt{\alpha_i}\, \mathrm{N}(\mathfrak{p})^{-s}\right)^{-1} \left(1 + \sqrt{\alpha_i}\, \mathrm{N}(\mathfrak{p})^{-s}\right)^{-1} \\ &= \prod_{i=1}^{n} \left(1 - \alpha_i\, \mathrm{N}(\mathfrak{p})^{-2s}\right)^{-1} = L_{\mathfrak{P}}(s, \tau). \end{aligned}$$

Combining the two cases, we obtain Eq. (8.12). Now let us translate all this into a statement about automorphic representations.

Prediction 1.8.4 *Let K/F be an extension of number fields, and let ψ be an automorphic representation of $GL(n)$ over K. Then there exists an automorphic representation θ_ψ of $GL(nr)$ over F, where $r = [K : F]$, such that*

$$L(s, \psi) = L(s, \theta_\psi). \tag{8.15}$$

Suppose that $r = 2$. If $\mathfrak{p}$ and $\mathfrak{P}$ are primes of F and K, respectively, let

$$L_{\mathfrak{p}}(s, \theta_{\psi}) = \prod_{i=1}^{2n} \left(1 - \alpha_i'(\mathfrak{p})\, \mathrm{N}(\mathfrak{p})^{-s}\right)^{-1},$$

$$L_{\mathfrak{P}}(s, \psi) = \prod_{i=1}^{n} \left(1 - \alpha_i(\mathfrak{P})\, \mathrm{N}(\mathfrak{P})^{-s}\right)^{-1}.$$

If $\mathfrak{p}$ is a prime of F that splits in K, $\mathfrak{p}\mathfrak{o}_K = \mathfrak{P}_1 \mathfrak{P}_2$, then the $\alpha_i'(\mathfrak{p})$ are

$$\alpha_1(\mathfrak{P}_1), \cdots, \alpha_n(\mathfrak{P}_1), \alpha_1(\mathfrak{P}_2), \cdots, \alpha_n(\mathfrak{P}_2),$$

whereas if $\mathfrak{p}$ is inert, $\mathfrak{p}\mathfrak{o}_K = \mathfrak{P}$, then the $\alpha_i'(\mathfrak{p})$ are

$$\pm\sqrt{\alpha_1(\mathfrak{P})}, \cdots, \pm\sqrt{\alpha_n(\mathfrak{P})}.$$

When $r = 2$ and $n = 1$, examples of this phenomenon were discovered by Hecke (1926, 1927) and by Maass (1949); in this case, automorphic forms may be constructed by the converse theorem, the theta correspondence (Weil representation), or the trace formula. They were later investigated more systematically by Shalika and Tanaka (1969), and by Jacquet and Langlands (1970). If $n = 1$ and $r = 3$, these forms have been constructed by Jacquet, Piatetski–Shapiro and Shalika (1981a). If K/F is cyclic, $n = 1$, and r is arbitrary, they have been investigated from the point of view of the trace formula by Kazhdan (1983), and by Arthur and Clozel (1989). It is Maass' examples that were employed by Doi and Naganuma (1973).

Later in this book, we will devote considerable space to a *local* analog of the phenomenon described in Prediction 1.8.4. Namely, if F is a local field, and E/F is an extension of degree n, and if we are given an irreducible admissible representation of $GL(r, E)$, there should be an associated irreducible admissible representation of $GL(nr, F)$. If $n = 1$ and $r = 2$, these representations will figure in Chapter 4, Sections 4.1, 4.8, and 4.9.

Finally, we turn our attention to the question of proofs. First, a word about the converse theorem, which is the generalization of Theorem 1.5.1. Suppose that we wish to prove the existence of an automorphic representation π for $GL(2)$ over F, such that $L(s, \pi) = \prod_{\mathfrak{p}} L_{\mathfrak{p}}(s, \pi)$, and such that the local factors $L_{\mathfrak{p}}(s, \pi)$ are defined by specifying, for each prime, $\alpha_1(\mathfrak{p})$ and $\alpha_2(\mathfrak{p})$, and by definition

$$L_{\mathfrak{p}}(s, \pi) = \left(1 - \alpha_1(\mathfrak{p})\, \mathrm{N}(\mathfrak{p})^{-s}\right)^{-1} \left(1 - \alpha_2(\mathfrak{p})\, \mathrm{N}(\mathfrak{p})^{-s}\right)^{-1}.$$

In order to be able to infer the existence of π, we need to consider twists of this Dirichlet series by Hecke characters. Let ψ be a Hecke character of F. Define $L(s, \pi, \psi) = \prod_{\mathfrak{p}} L_{\mathfrak{p}}(s, \pi, \psi)$, where

$$L_{\mathfrak{p}}(s, \pi, \psi) = \left(1 - \psi(\mathfrak{p})\, \alpha_1(\mathfrak{p})\, \mathrm{N}(\mathfrak{p})^{-s}\right)^{-1} \left(1 - \psi(\mathfrak{p})\, \alpha_2(\mathfrak{p})\, \mathrm{N}(\mathfrak{p})^{-s}\right)^{-1}.$$

If we can prove the analytic continuation and functional equations of these L-functions, then we may conclude the existence of the automorphic representation π. In this generality, a precise converse theorem was proved by Jacquet and Langlands (1970).

A similar theorem is true for $GL(3)$, due to Jacquet, Piatetski-Shapiro and Shalika (1979). For $GL(n)$ (see Cogdell and Piatetski-Shapiro (1994)), however, the converse theorem will require twists not only by Hecke characters but by automorphic representations. On $GL(4)$, for example, one requires twists by automorphic representations on $GL(2)$.

If $n = 1$ and $r = 2$, the existence of θ_ψ in Prediction 1.8.4 may be established by the converse theorem: We may prove a functional equation for $L(s, \theta_\psi)$ and all its twists by Hecke characters λ of F, because $L(s, \theta_\psi, \lambda) = L\big(s, (\lambda \circ N)\psi\big)$ is the L-series of a Hecke character for K, and as such, its functional equation is amenable to proof. Here $\lambda \circ N$ is the composition of λ with the norm map from ideals of K to ideals of F; in other words, $\lambda \circ N$ is the base change lift of λ.

We now offer a strategy for proving quadratic base change for $GL(2)$. Let π be an automorphic representation of $GL(2)$ over F. Let K/F be a quadratic extension. We define a Dirichlet series that we will eventually see is the L-function of an automorphic representation π' on $GL(2)$ over K. Thus we will denote this Dirichlet series as $L(s, \pi')$. It is defined to be $L(s, \pi)\, L(s, \pi, \chi)$, where χ is the quadratic character of F corresponding to the extension K.

Because we can prove functional equations for $L(s, \pi)$ and $L(s, \pi, \chi)$, we may prove a functional equation for $L(s, \pi')$. This is not, however, sufficient because we also need functional equations for the twists of this Dirichlet series by Hecke characters. We will obtain these by the Rankin–Selberg method. Indeed, let ψ be a Hecke character of K. We will show that

$$L(s, \pi', \psi) = L(s, \pi \times \theta_\psi). \tag{8.16}$$

Because the existence of θ_ψ may be proved by means of the converse theorem, we may thus obtain enough functional equations to deduce the existence of the automorphic representation π'. Indeed, this is exactly how Doi and Naganuma proceeded, acting on a suggestion of Shimura, and this is the program that we will follow in the remaining sections of this chapter.

To prove Eq. (8.16), denote, for a prime $\mathfrak{p}$ of F

$$L_\mathfrak{p}(s, \pi) = \big(1 - \alpha_1(\mathfrak{p})\, \mathrm{N}(\mathfrak{p})^{-s}\big)^{-1} \big(1 - \alpha_2(\mathfrak{p})\, \mathrm{N}(\mathfrak{p})^{-s}\big)^{-1},$$

$$L_\mathfrak{p}(s, \theta_\psi) = \big(1 - \beta_1(\mathfrak{p})\, \mathrm{N}(\mathfrak{p})^{-s}\big)^{-1} \big(1 - \beta_2(\mathfrak{p})\, \mathrm{N}(\mathfrak{p})^{-s}\big)^{-1},$$

and for a prime $\mathfrak{P}$ of K, let

$$L_\mathfrak{P}(s, \pi') = \big(1 - \alpha_1'(\mathfrak{P})\, \mathrm{N}(\mathfrak{P})^{-s}\big)^{-1} \big(1 - \alpha_2'(\mathfrak{P})\, \mathrm{N}(\mathfrak{P})^{-s}\big)^{-1}.$$

Suppose that $\mathfrak{p}$ splits in K, $\mathfrak{p}o_K = \mathfrak{P}_1\mathfrak{P}_2$. According to Prediction 1.8.4, $\beta_1(\mathfrak{p}) = \psi(\mathfrak{P}_1)$, $\beta_2(\mathfrak{p}) = \psi(\mathfrak{P}_2)$. Thus by Prediction 1.8.2

$$\begin{aligned}
&L_{\mathfrak{p}}(s, \pi \times \theta_\psi)\\
&= \left(1 - \psi(\mathfrak{P}_1)\,\alpha_1(\mathfrak{p})\,\mathrm{N}(\mathfrak{p})^{-s}\right)^{-1}\left(1 - \psi(\mathfrak{P}_1)\,\alpha_2(\mathfrak{p})\,\mathrm{N}(\mathfrak{p})^{-s}\right)^{-1}\\
&\qquad\times\left(1 - \psi(\mathfrak{P}_2)\,\alpha_1(\mathfrak{p})\,\mathrm{N}(\mathfrak{p})^{-s}\right)^{-1}\left(1 - \psi(\mathfrak{P}_2)\,\alpha_2(\mathfrak{p})\,\mathrm{N}(\mathfrak{p})^{-s}\right)^{-1}\\
&= \left(1 - \psi(\mathfrak{P}_1)\,\alpha_1'(\mathfrak{P}_1)\,\mathrm{N}(\mathfrak{P}_1)^{-s}\right)^{-1}\left(1 - \psi(\mathfrak{P}_1)\,\alpha_2'(\mathfrak{P}_1)\,\mathrm{N}(\mathfrak{P}_2)^{-s}\right)^{-1}\\
&\qquad\times\left(1 - \psi(\mathfrak{P}_2)\,\alpha_1'(\mathfrak{P}_1)\,\mathrm{N}(\mathfrak{P}_1)^{-s}\right)^{-1}\left(1 - \psi(\mathfrak{P}_2)\,\alpha_2'(\mathfrak{P}_1)\,\mathrm{N}(\mathfrak{P}_2)^{-s}\right)^{-1}\\
&= L_{\mathfrak{P}_1}(s, \pi', \psi)\,L_{\mathfrak{P}_2}(s, \pi', \psi).
\end{aligned}$$

On the other hand, suppose that $\mathfrak{p}$ is inert, $\mathfrak{p}o_K = \mathfrak{P}$. Then according to Prediction 1.8.4, $\beta_1(\mathfrak{p}) = \sqrt{\psi(\mathfrak{P})}$, $\beta_2(\mathfrak{p}) = -\sqrt{\psi(\mathfrak{P})}$, so by Prediction 1.8.2

$$\begin{aligned}
&L_{\mathfrak{p}}(s, \pi \times \theta_\psi)\\
&= \left(1 - \sqrt{\psi(\mathfrak{P})}\,\alpha_1(\mathfrak{p})\,\mathrm{N}(\mathfrak{p})^{-s}\right)^{-1}\left(1 + \sqrt{\psi(\mathfrak{P})}\,\alpha_1(\mathfrak{p})\,\mathrm{N}(\mathfrak{p})^{-s}\right)^{-1}\\
&\qquad\times\left(1 - \sqrt{\psi(\mathfrak{P})}\,\alpha_2(\mathfrak{p})\,\mathrm{N}(\mathfrak{p})^{-s}\right)^{-1}\left(1 + \sqrt{\psi(\mathfrak{P})}\,\alpha_2(\mathfrak{p})\,\mathrm{N}(\mathfrak{p})^{-s}\right)^{-1}\\
&= \left(1 - \psi(\mathfrak{P})\,\alpha_1(\mathfrak{p})^2\,\mathrm{N}(\mathfrak{p})^{-2s}\right)^{-1}\left(1 - \psi(\mathfrak{P})\,\alpha_2(\mathfrak{p})^2\,\mathrm{N}(\mathfrak{p})^{-2s}\right)^{-1}\\
&= \left(1 - \psi(\mathfrak{P})\,\alpha_1'(\mathfrak{P})\,\mathrm{N}(\mathfrak{P})^{-2s}\right)^{-1}\left(1 - \psi(\mathfrak{P})\,\alpha_2'(\mathfrak{P})\,\mathrm{N}(\mathfrak{P})^{-2s}\right)^{-1}\\
&= L_{\mathfrak{P}}(s, \pi', \psi).
\end{aligned}$$

Combining the two cases, we obtain Eq. (8.16).

Exercises

Exercise 1.8.1 Prove Eq. (8.10) in general.

Exercise 1.8.2 (a) Let G be a finite group and H a subgroup, let ρ be a one-dimensional representation of G, and let τ be a finite-dimensional representation of H. We will denote by ρ_H the restriction of ρ to H, and by τ^G the representation of G induced by τ. Prove that $(\rho_H \otimes \tau)^G \cong \rho \otimes \tau^G$ as G-modules.
(b) Interpret this fact in terms of Artin L-functions.

1.9 Maass Forms

In this section, we will construct certain automorphic forms (in accordance with Prediction 1.8.4, with $n = 1$ and $r = 2$) that we will need. These are not modular forms, but rather nonholomorphic automorphic forms known as "Maass forms." We will study Maass forms in more detail in Chapter 2. The construction of these particular automorphic forms, which are associated with Hecke characters of real quadratic fields, is due to Maass (1949); similar modular forms of weight one, associated with Hecke characters of imaginary quadratic fields, were constructed by Hecke (1926).

The *non-Euclidean Laplacian* Δ is a second-order differential operator acting on functions on $\mathcal{H}$. It is given by

$$\Delta = -y^2 \left(\frac{\partial^2}{\partial^2 x} + \frac{\partial^2}{\partial^2 y} \right).$$

It is invariant under the action of $SL(2, \mathbb{R})$; that is, if $g \in SL(2, \mathbb{R})$, and if $\Delta(f) = F$, then $\Delta(f \circ g) = F \circ g$ for any smooth function f on $\mathcal{H}$ (Exercise 1.9.1).

We define a *Maass form* for $\Gamma(1) = SL(2, \mathbb{Z})$ to be a smooth function on $\mathcal{H}$ such that (i) $f\big(\gamma(z)\big) = f(z)$ for $\gamma \in \Gamma(1)$; (ii) f is an eigenfunction of Δ; and (iii) $f(x + iy) = O(y^N)$ as $y \to \infty$ for some N.

Furthermore, if

$$\int_0^1 f(z + x)\, dx = 0,$$

we call f a Maass *cusp form*.

As an example, let us show that the Eisenstein series $E(z, \nu + \frac{1}{2})$ introduced in Section 1.6 is a Maass form. Conditions (i) and (iii) are clear. It is necessary for us to check that f is an eigenfunction. If $\mathrm{re}(\nu) > 1/2$, we may check this directly from the series definition of $E(z, \nu + \frac{1}{2})$. We have

$$\Delta\, y^{\nu+1/2} = (\tfrac{1}{4} - \nu^2)\, y^{\nu+1/2}.$$

Because $y^{\nu+1/2}$ is an eigenfunction, and because Δ is invariant under the action of $SL(2, \mathbb{R})$, so is

$$\mathrm{im}\big(\gamma(z)\big)^{\nu+1/2} = \frac{y^{\nu+1/2}}{|cz + d|^{2\nu+1}}.$$

Because $E\left(z, \nu + \frac{1}{2}\right)$ is a sum of functions of this type, we have

$$\Delta\, E(z, \nu + \tfrac{1}{2}) = (\tfrac{1}{4} - \nu^2)\, E(z, \nu + \tfrac{1}{2}), \tag{9.1}$$

at least when $\mathrm{re}(\nu) > 1/2$.

We wish to extend the validity of Eq. (9.1) to all $\nu \neq \pm\frac{1}{2}$. There are, of course, different ways of doing this. We will do this by considering the Fourier expansion of $E(z, \nu + \frac{1}{2})$. Specifically, we will prove that each Fourier coefficient $a_r(y, \nu + \frac{1}{2})\, e^{2\pi i r x}$ in Eq. (6.9) is an eigenfunction of Δ, with eigenvalue $\frac{1}{4} - \nu^2$.

First of all, both $y^{\nu+1/2}$ and $y^{-\nu+1/2}$ are eigenfunctions with this eigenvalue, and hence, by Eq. (6.12), the constant term $a_0(y, \nu+\frac{1}{2})$ is also an eigenfunction. To deal with the nonconstant terms, we must show that

$$\Delta\ \sqrt{y}\, K_\nu(2\pi\,|r|\,y)\, e^{2\pi i r x} = (\tfrac{1}{4} - \nu^2)\, \sqrt{y}\, K_\nu(2\pi\,|r|\,y)\, e^{2\pi i r x}. \tag{9.2}$$

This reduces easily to Bessel's differential equation

$$\left\{ y^2 \frac{d^2}{dy^2} + y \frac{d}{dy} - (y^2 + \nu^2) \right\} K_\nu(y) = 0, \tag{9.3}$$

which follows easily from the definition Eq. (6.5) by differentiation under the integral sign. Thus we obtain Eq. (9.1) for all ν.

More generally, suppose that f is an arbitrary Maass form. It follows from (i) with $\gamma = \begin{pmatrix} 1 & x \\ & 1 \end{pmatrix}$ that $f(x+1) = f(x)$, and so f has a Fourier expansion

$$f(z) = \sum_{r=-\infty}^{\infty} a_r(y)\, e^{2\pi i r x}. \tag{9.4}$$

Let us choose $\nu \in \mathbb{C}$ so that the eigenvalue of Δ on f is $\frac{1}{4} - \nu^2$. We will show that if $r \neq 0$, the Fourier coefficient

$$a_r(y) = a_r\, \sqrt{y}\, K_\nu(2\pi\,|r|\,y) \tag{9.5}$$

for some constant a_r. Indeed, for fixed r, define $k : \mathbb{R}_+ \to \mathbb{C}$ by

$$a_r(y) = \sqrt{y}\, k(2\pi\,|r|\,y).$$

Then (ii) reduces to showing that k is a solution to Bessel's equation (9.3), and moreover, it follows from (iii) that $k(y)$ is of at most polynomial growth as $y \to \infty$. Now Eq. (9.3) admits two linearly independent solutions: the Bessel functions K_ν and I_ν; of these, K_ν is of rapid decay as $y \to \infty$, whereas I_ν grows exponentially. We refer to Whittaker and Watson (1927) for a proof of this, but offer here instead a plausible indication that Eq. (9.3) should have two solutions, one of exponential decay and one of exponential growth. For if y is large, Eq. (9.3) is a perturbation of the differential equation

$$\frac{d^2 k}{dy^2} - k = 0;$$

this has solutions $k = e^{-y}$ and $k = e^{y}$, so we expect Eq. (9.3) to have two solutions with similar behavior as $y \to \infty$. We see that the growth condition on $a_r(y)$ implies that it is a constant multiple of $\sqrt{y}\, K_\nu(2\pi r y)$.

Thus if f is a Maass cusp form, so $a_0(y) = 0$, we have the Fourier expansion

$$f(z) = \sum_{\substack{r=-\infty \\ r \neq 0}}^{\infty} a_r \sqrt{y}\, K_\nu(2\pi |r| y)\, e^{2\pi i r x}. \tag{9.6}$$

Let $\iota : \mathcal{H} \to \mathcal{H}$ be the antiholomorphic involution $\iota(x+iy) = -x+iy$. If f is an eigenfunction of Δ, then $f \circ \iota$ is an eigenfunction with the same eigenvalue. Because $\iota^2 = 1$, its eigenvalues are ± 1. We may therefore diagonalize the Maass cusp forms with respect to ι. If $f \circ \iota = f$, we call f *even*. In this case, $a_r = a_{-r}$. If $f \circ \iota = -f$, then we call f *odd*. Then $a_r = -a_{-r}$.

We define the L-series

$$L(s, f) = \sum_{n=1}^{\infty} a_n\, n^{-s}. \tag{9.7}$$

This has analytic continuation and a functional equation. We will assume that f is a cusp form. We will prove that $L(s, f)$ has analytic continuation and a functional equation.

First of all, the coefficients satisfy

$$a_n = O(n^{1/2}), \tag{9.8}$$

as may be proved by a generalization of Proposition 1.3.5 (Exercise 1.9.3). Hence the L-series $L(s, f)$ is convergent for $\mathrm{re}(s) > 3/2$. In fact, it is convergent for $\mathrm{re}(s) > 1$, as may be proved by use of the Rankin–Selberg method (Exercise 1.6.5).

Lemma 1.9.1 *The integral*

$$\int_0^\infty K_\nu(y)\, y^s \,\frac{dy}{y} = 2^{s-2}\, \Gamma\left(\frac{s+\nu}{2}\right) \Gamma\left(\frac{s-\nu}{2}\right) \tag{9.9}$$

is absolutely convergent if $\mathrm{re}(s) > |\mathrm{re}(\nu)|$.

Proof Substituting the definition Eq. (6.5) of K_ν, the left side of Eq. (9.9) equals

$$\frac{1}{2} \int_0^\infty \int_0^\infty e^{-(t+t^{-1})y/2}\, t^\nu\, y^s \,\frac{dy}{y}\,\frac{dt}{t}.$$

We make the change of variables

$$u = \tfrac{1}{2} t y, \qquad v = \tfrac{1}{2} t^{-1} y, \qquad \frac{du}{u} \wedge \frac{dv}{v} = 2 \frac{dt}{t} \wedge \frac{dy}{y}.$$

The integral becomes

$$2^{s-2} \int_0^\infty \int_0^\infty e^{-u-v}\, u^{(s+\nu)/2}\, v^{(s-\nu)/2}\, \frac{du}{u}\, \frac{dv}{v}.$$

This integral is absolutely convergent in the indicated region, and splits into two gamma integrals, from which we get Eq. (9.9). ■

Proposition 1.9.1 *Let f be a Maass cusp form with eigenvalue $\frac{1}{4} - \nu^2$. Let $\epsilon = 0$ if f is even, -1 if f is odd. Let*

$$\Lambda(s, f) = \pi^{-s}\, \Gamma\left(\frac{s+\epsilon+\nu}{2}\right) \Gamma\left(\frac{s+\epsilon-\nu}{2}\right) L(s, f). \tag{9.10}$$

Then $\Lambda(s, f)$ has analytic continuation to all s and satisfies the functional equation

$$\Lambda(s, f) = (-1)^{\epsilon}\Lambda(1-s, f). \tag{9.11}$$

Proof First consider the case where f is even. We consider the integral

$$\int_0^\infty f(iy)\, y^{s-1/2}\, \frac{dy}{y}. \tag{9.12}$$

As $y \to \infty$, the integrand is vanishingly small because of the rapid decay of the Bessel function; and as $y \to 0$, one uses the identity

$$f(iy) = f\left(\frac{i}{y}\right) \tag{9.13}$$

to see that the integrand is small. Hence Eq. (9.12) is convergent for all s. If $\mathrm{re}(s)$ is large, one substitutes the Fourier expansion of f and uses Eq. (9.9) to see that Eq. (9.12) equals $\frac{1}{2}\Lambda(s, f)$; the functional equation (9.11) now follows from Eq. (9.13).

If f is odd, we consider instead the integral

$$\int_0^\infty g(iy)\, y^{s+1/2}\, \frac{dy}{y},$$

where

$$\begin{aligned} g(z) &= \frac{1}{4\pi i} \frac{\partial f}{\partial x}(z) \\ &= \sum_{n=1}^\infty a(n)\, n\, \sqrt{y}\, K_\nu(2\pi i |n| y) \cos(2\pi n x). \end{aligned}$$

We have

$$g(iy) = -\frac{1}{y^2} g\left(\frac{i}{y}\right),$$

and the functional equation follows as before. ■

Maass forms were first studied by H. Maass (1949). The existence of Maass forms is best considered in the context of the Selberg trace formula. We will content ourselves here with a few impressionistic remarks. The reader in search of further information may consult Chapter 2, as well as Hejhal (1976, 1983), Iwaniec (1984, 1995), Maass (1949), Roelcke (1966), Selberg (1956), Terras (1985 and 1987) and Venkov (1990).

Let Γ be a discontinuous subgroup of $SL(2,\mathbb{R})$ with $\Gamma\backslash\mathcal{H}$ having finite volume. We may consider Maass forms on $\Gamma\backslash\mathcal{H}$, extending the above definition for $\Gamma = \Gamma(1)$ in the obvious way. That the non-Euclidean Laplacian Δ is a positive semidefinite, symmetric operator is a consequence of Stoke's theorem. If $\Gamma\backslash\mathcal{H}$ is compact, this will be proved in Exercise 2.1.8, and in Section 2.3 we show that the Laplacian is a self-adjoint operator. Because of the strong analogy with the eigenvalue problem for the vibrating membrane, Maass called his eigenforms *waveforms* (*Wellenformen.*)

Because the Laplacian is positive semidefinite, the eigenfunctions that occur in $L^2(\Gamma\backslash\mathcal{H})$ have eigenvalue $\lambda = \frac{1}{4} - \nu^2$ real and nonnegative. This implies that ν is either purely imaginary or real of absolute value less than $1/2$. If ν is real and nonzero, so $\lambda < 1/4$, then λ is called an *exceptional* eigenvalue. Selberg (1965) showed that there are no exceptional eigenvalues if $\Gamma = \Gamma(1)$ and conjectured that there are none for congruence subgroups of Γ. It may be seen that this conjecture of Selberg is precisely analogous to the Ramanujan conjecture, which is discussed in this volume in Sections 1.3 and 3.5. On the other hand, for an arbitrary discontinuous group Γ, exceptional eigenvalues may indeed exist.

If the fundamental domain $\Gamma\backslash\mathcal{H}$ is compact, then the spectrum of Δ in $L^2(\Gamma\backslash\mathcal{H})$ is discrete. There are no cusps, so every Maass form is trivially a cusp form. The eigenvalues form an ascending sequence $0 = \lambda_0 < \lambda_1 \leq \lambda_2 \leq \cdots$, where $\lambda_0 = 0$ corresponds to the constant function $f = 1$, and the other eigenvalues correspond to Maass forms. The Selberg trace formula is a sort of analog of the Poisson summation formula, in which on one side an arbitrary function is summed over the lengths of closed geodesics, and on the other side, an integral transform of that function is summed over the eigenvalues λ_j. One consequence that emerges is an estimate for the density of the eigenvalues. We have *Weyl's law*:

$$\mathrm{Card}\{\lambda_j \mid |\lambda_j| < N\} \sim \frac{\mathrm{vol}(\Gamma\backslash\mathcal{H})}{4\pi} N,$$

as $N \to \infty$. In particular, Maass forms exist.

If Γ has cusps, there is both a discrete spectrum and a continuous spectrum. The continuous spectrum consists of the Eisenstein series. Thus we have $E\left(z, \nu+\frac{1}{2}\right)$ in the case of $SL(2, \mathbb{Z})$. This is not in $L^2(\Gamma\backslash\mathcal{H})$, but if ν is purely imaginary, it is very close to being square integrable. The discrete spectrum consists, as before, of the cusp forms and the constant function and behaves exactly as in the case of the compact quotient, although the proofs are more difficult. But in particular, Weyl's law remains valid for the discrete spectrum.

Maass constructed certain eigenforms as liftings from $GL(1)$ of a quadratic extension, in accordance with Prediction 1.8.3. It is a consequence of the Selberg trace formula that there are other cuspidal eigenforms besides these; for example if, $\Gamma = SL(2, \mathbb{Z})$, none of the eigenforms are lifts – the lifts are only automorphic with respect to certain congruence subgroups. In general, the density of the lifts is less than the density of cusp forms predicted by Weyl's law, and so the lifts are only a small part of the spectrum. It is the construction of these special Maass forms that will occupy the remainder of this section.

Lemma 1.9.2 *Let f be an eigenfunction of Δ on $\mathcal{H}$. Suppose that f is real analytic and*

$$f(iy) = \frac{\partial f}{\partial x}(iy) = 0$$

on the imaginary axis. Then f is identically zero.

Proof Because f is real analytic, it is sufficient to show that its partial derivatives all vanish at some point, for then f will be zero in a neighborhood of that point. It is sufficient to show that $\partial^n f/\partial x^n$ vanishes on the imaginary axis for all n, because then

$$\frac{\partial^{n+m} f}{\partial x^n\, \partial y^m}(iy) = \frac{\partial^m}{\partial y^m} 0 = 0.$$

Our hypothesis is that $\partial^n f/\partial x^n = 0$ when $n = 0$ or 1. If now $n \geq 2$, we have (assuming that $\Delta f = \lambda f$)

$$\begin{aligned} \frac{\partial^n f}{\partial x^n}(iy) &= \frac{\partial^{n-2}}{\partial x^{n-2}}\left[-\frac{1}{y^2}\Delta f(iy) - \frac{\partial^2 f}{\partial y^2}(iy)\right] \\ &= -\frac{\lambda}{y^2}\frac{\partial^{n-2} f}{\partial x^{n-2}}(iy) - \frac{\partial^2}{\partial y^2}\frac{\partial^{n-2} f}{\partial x^{n-2}}(iy), \end{aligned}$$

and by induction, this is zero. ■

Let F be a real quadratic field with discriminant D and having narrow class number one. Let $\chi = \chi_D$ be, as usual, the quadratic Dirichlet character with conductor D such that $\chi(-1) = 1$. Let q be a positive integer prime to D. Let

σ be a primitive Dirichlet character modulo q. Then $\sigma \circ N$ is a Hecke character of F.

Proposition 1.9.2 *The conductor $\mathfrak{f}$ of $\sigma \circ N$ is the unique ideal $\mathfrak{f}$ with $\mathbb{N}(\mathfrak{f}) = q^2$ that is invariant under the Galois automorphism of $F/\mathbb{Q}$. We have*

$$\tau_F(\sigma \circ N) = D^{-1/2}\,\tau(\sigma)\,\tau(\sigma\chi_D) = \sigma(D)\,\chi_D(q)\,\tau(\sigma)^2. \tag{9.14}$$

This is a variant of the Hasse–Davenport relation. It is, of course possible to give a purely algebraic proof of this, as for example in Weil (1949). We will give an analytic proof.

In this particular instance, the Gauss sum $\tau_F(\sigma \circ N)$ is independent of the choice of f and d, in the notation of Section 1.7; we recall that f and d are totally positive generators of the ideal $\mathfrak{f}$ and the different, respectively.

Proof Let $\delta = 0$ or 1 be determined by the condition that $\sigma(-1) = (-1)^\delta$. With notation as in Theorem 1.1.1, let

$$\Lambda(s,\sigma) = \pi^{-(s+\delta)/2}\,\Gamma\left(\frac{s+\delta}{2}\right)\,L(s,\sigma),$$

$$\Lambda(s,\sigma\chi_D) = \pi^{-(s+\delta)/2}\,\Gamma\left(\frac{s+\delta}{2}\right)\,L(s,\sigma\chi_D).$$

Then by Theorem 1.1.1,

$$\Lambda(s,\sigma) = (-i)^\delta\,\tau(\sigma)\,q^{-s}\,\Lambda(1-s,\sigma^{-1}),$$

$$\Lambda(s,\sigma\chi_D) = (-i)^\delta\,\tau(\sigma\chi_D)\,(Dq)^{-s}\,\Lambda(1-s,\sigma^{-1}\chi_D).$$

Now with notation as in Eq. (7.3), let

$$\Lambda(s,\sigma\circ N) = D^{s/2}\,\pi^{-s}\,\Gamma\left(\frac{s+\delta}{2}\right)^2\,L(s,\sigma\circ N).$$

Let $\mathfrak{f}$ be the conductor of $\sigma \circ N$, and let f be a totally positive generator of $\mathfrak{f}$. By Eq. (7.18), we have

$$\Lambda(s,\sigma\circ N) = D^{s/2}\,\pi^\delta\,\Lambda(s,\sigma)\,\Lambda(s,\sigma\chi_D),$$

and so we may compare the two functional equations. On the one hand,

$$D^{s/2}\,\Lambda(s,\sigma)\,\Lambda(s,\sigma\chi_D)$$

$$= (-1)^\delta\,\frac{\tau(\sigma)\,\tau(\sigma\chi_D)}{q\sqrt{D}}\,q^{1-2s}\,D^{(1-s)/2}\,\Lambda(1-s,\sigma^{-1})\,\Lambda(1-s,\sigma^{-1}\chi_D).$$

On the other hand, by Theorem 1.7.2,

$$\Lambda(s, \sigma \circ N) = \frac{(\sigma \circ N)_\infty(fd)\, \tau_F(\sigma \circ N)}{\sqrt{\mathbb{N}(\mathfrak{f})}} (-1)^\delta\, \mathbb{N}(\mathfrak{f})^{-s+1/2}\, \Lambda(1-s, \sigma^{-1} \circ N).$$

Now $(\sigma \circ N)_\infty(x) = \operatorname{sgn}\big(N(x)\big)^\delta$, and because $fd > 0$, $(\sigma \circ N)_\infty(fd) = 1$. Thus

$$\frac{\tau(\sigma)\, \tau(\sigma\chi_D)}{q\sqrt{D}} q^{-2s+1} = \frac{\tau_F(\sigma \circ N)}{\sqrt{\mathbb{N}(\mathfrak{f})}} \mathbb{N}(\mathfrak{f})^{-s+1/2}.$$

We see first that $\mathbb{N}(\mathfrak{f}) = q^2$, and it is evident from the definition of $\sigma \circ N$ that $\mathfrak{f}$ is invariant under the Galois automorphism. Taking $s = 1/2$, we obtain $\tau_F(\sigma \circ N) = D^{-1/2}\, \tau(\sigma)\, \tau(\sigma\chi_D)$. Now we recall that because σ and χ_D have coprime conductors,

$$\tau(\sigma\chi_D) = \sigma(D)\, \chi_D(q)\, \tau(\sigma)\, \tau(\chi_D).$$

To prove this, we note that as N runs through the residue classes modulo q and M runs through the residue classes modulo D, $n = ND + Mq$ runs through the residue classes modulo Dq, which is the conductor of $\sigma\chi_D$, and we have $\sigma(n) = \sigma(D)\, \sigma(N)$, $\chi_D(n) = \chi_D(M)\, \chi_D(q)$. Consequently, $\tau(\sigma\chi_D)$ equals

$$\begin{aligned} &\sigma(D)\, \chi_D(q) \sum_{N \bmod q} \sum_{M \bmod D} \sigma(N)\, \chi_D(q)\, e^{2\pi i (ND+Mq)/Dq} \\ &= \sigma(D)\, \chi_D(q)\, \tau(\sigma)\, \tau(\chi_D). \end{aligned}$$

Thus we have

$$\tau_F(\sigma \circ N) = D^{-1/2}\, \sigma(N)\, \chi_D(q)\, \tau(\sigma)^2\, \tau(\chi_D).$$

We may evaluate $\tau(\chi_D)$ by taking $q = 1$ and σ to be the trivial character. Then $\tau_F(\sigma \circ N)$ and $\tau(\sigma)$ both equal one, so we obtain the well-known identity

$$\tau(\chi_D) = \sqrt{D}. \tag{9.15}$$

Substituting this into our previous identity gives Eq. (9.14). ■

We will associate a Maass form for $\Gamma_0(D)$ with a Hecke character ψ of F. In the notation of Section 1.7, such a character has the form $\psi = \psi_\mathfrak{f}\psi_\infty$ for some conductor $\mathfrak{f}$; we will consider only characters with $\mathfrak{f} = \mathfrak{o}_F$, so that $\psi(\mathfrak{a}) = \psi_\infty(\alpha)$, where α is a totally positive generator of $\mathfrak{a}$. Let ν_1, ν_2 and ϵ_1, ϵ_2 be as in Section 1.7. We have $\nu_1 = -\nu_2$, and we will denote this complex number by ν. (This differs from the notation of Section 1.7, where ν was the vector (ν_1, ν_2), but for a quadratic field, this is a more convenient notation.) Thus if $x \in \mathbb{R}^2$, we have

$$\psi_\infty(x) = \operatorname{sgn}(x_1)^{\epsilon_1}\, \operatorname{sgn}(x_2)^{\epsilon_2} \left|\frac{x_1}{x_2}\right|^\nu. \tag{9.16}$$

It is necessary that $\psi_\infty(\eta) = 1$ for $\eta \in \mathfrak{o}^\times$. This implies that

$$|\eta^{(1)}|^{2\nu} = 1 \tag{9.17}$$

for $\eta^{(1)} \in \mathfrak{o}_+^\times$. However, Eq. (9.17) may or may not be true for $\eta \in \mathfrak{o}^\times$. If Eq. (9.17) is valid for all units, then $\epsilon_1 = \epsilon_2 = 0$. If Eq. (9.17) is not valid for all units, then $\epsilon_1 = \epsilon_2 = 1$. In either case, we will denote $\epsilon = \epsilon_1 = \epsilon_2$.

Let η_0 be the *fundamental unit* of F, that is, the generator of $\mathfrak{o}^\times/\{\pm 1\}$ characterized by the assumption that $\eta_0^{(1)} > 1$. Because the narrow class number of F is one, $N(\eta_0) = -1$, and so $\eta_0^{(2)} < 0$. We choose as a totally positive generator of the different $d = \eta_0\sqrt{D}$. Applying Eq. (9.17) to η_0^2, with η_0 as before the fundamental unit, we see that

$$\nu = \frac{mi\pi}{2\log(\eta_0^{(1)})}, \qquad m \in \mathbb{Z}.$$

If m is even, then $\epsilon = 0$, whereas if m is odd, then $\epsilon = 1$.

We will assume that $\nu \neq 0$. Let

$$\theta_\psi(z) = \begin{cases} \sum_{\mathfrak{a}} \psi(\mathfrak{a})\sqrt{y}\,K_\nu\big(2\pi\,\mathrm{N}(\mathfrak{a})\,y\big)\,\cos\big(2\pi\,\mathrm{N}(\mathfrak{a})\,x\big) & \text{if } \epsilon = 0; \\ \sum_{\mathfrak{a}} \psi(\mathfrak{a})\sqrt{y}\,K_\nu\big(2\pi\,\mathrm{N}(\mathfrak{a})\,y\big)\,\sin\big(2\pi\,\mathrm{N}(\mathfrak{a})\,x\big) & \text{if } \epsilon = 1, \end{cases} \tag{9.18}$$

where the summation is over all ideals $\mathfrak{a}$ of $\mathfrak{o}_F$. It is useful to note that the particular Hecke characters ψ under consideration have the following property: if $\mathfrak{a}$ is an ideal of $\mathfrak{o}_F$, and if $\mathfrak{a}'$ is its conjugate under the action of the Galois group, then $\psi(\mathfrak{a}) = \psi(\mathfrak{a}')^{-1}$. Hence

$$\theta_\psi = \theta_{\psi^{-1}}, \qquad L(s,\psi) = L(s,\psi^{-1}). \tag{9.19}$$

Theorem 1.9.1: Maass (1949) *The function θ_ψ is a Maass cusp form for the group $\Gamma_0(D)$. More precisely, it satisfies*

$$\theta_\psi\left(\frac{az+b}{cz+d}\right) = \chi_D(d)\,\theta_\psi(z) \qquad \textit{for } \gamma = \begin{pmatrix} a & b \\ c & d \end{pmatrix} \in \Gamma_0(D) \tag{9.20}$$

and

$$\theta_\psi(z) = (-1)^\epsilon\,\theta_\psi\left(\frac{-1}{Dz}\right) \tag{9.21}$$

Hecke (1926), gave a similar construction of cusp forms of weight one associated with imaginary quadratic fields. We will prove, by means of a converse theorem analogous to Theorem 1.5.1, that θ_ψ is a Maass form. It is a cusp form. (If $\nu = 0$, we should add a constant term $\frac{1}{2}R\sqrt{y}$ where $R = \log(\eta_0^{(1)})$ is the regulator of F. In this case, f may be constructed alternatively by extending the converse theorem method that we will use, or more simply by observing that in this special case, it is an Eisenstein series.)

Proof Let us establish Eq. (9.21) first. In the notation of Eq. (7.3), let

$$\Lambda(s,\psi)=D^{s/2}\,\pi^{-s}\,\Gamma\left(\frac{s+\epsilon+\nu}{2}\right)\Gamma\left(\frac{s+\epsilon-\nu}{2}\right)L(s,\psi).$$

By Theorem 1.7.1 and Eq. (9.19), we have

$$\Lambda(s,\psi)=\psi_\infty(d)\,(-1)^\epsilon\,\Lambda(1-s,\psi).$$

Here d is a totally positive generator of the different, which we may take to be $\eta_0\sqrt{D}$. Thus $\psi_\infty(d)=\operatorname{sgn}\left(N(\eta_0^\epsilon)\right)=(-1)^\epsilon$, and we may write this last identity

$$\Lambda(s,\psi)=\Lambda(1-s,\psi). \tag{9.22}$$

We first consider the case $\epsilon=0$. For sufficiently large s, we have

$$\begin{aligned}\int_0^\infty \theta_\psi(iy)\,y^{s-1/2}\,\frac{dy}{y}&=(2\pi)^{-s}\sum_{\mathfrak{a}}\psi(\mathfrak{a})\,\mathrm{N}(\mathfrak{a})^{-s}\int_0^\infty K_\nu(y)\,y^s\,\frac{dy}{y}\\&=\tfrac14\pi^{-s}\,\Gamma\left(\frac{s+\nu}{2}\right)\Gamma\left(\frac{s-\nu}{2}\right)L(s,\psi)\\&=\tfrac14\,D^{-s/2}\,\Lambda(s,\psi),\end{aligned}$$

as may be seen by substituting the series definition Eq. (9.18) and applying Eq. (9.10). Thus by the Mellin inversion formula (5.5)

$$\theta_\psi(iy)=\frac{\sqrt{y}}{8\pi i}\int_{\sigma-i\infty}^{\sigma+i\infty}D^{-s/2}\,\Lambda(s,\psi)\,y^{-s}\,ds$$

for σ sufficiently large. Now moving the path of integration to the left using Cauchy's theorem (and the Phragmén–Lindelöf principle), applying Eqs. (9.22), (9.19), and making the substitutions $s\to 1-s$ and $y\to 1/Dy$, this equals

$$\frac{\sqrt{1/Dy}}{8\pi i}\int_{\sigma-i\infty}^{\sigma+i\infty}D^{-s/2}\,\Lambda(s,\psi)\,(Dy)^s\,ds=\theta_\psi\left(\frac{i}{Dy}\right),$$

so

$$\theta_\psi(iy)=\theta_\psi\left(\frac{i}{Dy}\right).$$

Now applying Lemma 1.9.2 to $f(z)=\theta_\psi(z)-\theta_\psi(-1/Dz)$ we obtain Eq. (9.21) if $\epsilon=0$. (The x partial derivatives of both $\theta_\psi(z)$ and $\theta_\psi(-1/Dz)$ vanish along the imaginary axis.)

Suppose next that $\epsilon = 1$. We argue similarly, starting with

$$\int_0^\infty \frac{\partial\theta_\psi}{\partial x}(iy)\, y^{s+1/2}\,\frac{dy}{y} = \tfrac{1}{4}\, D^{-s/2}\Lambda(s,\psi).$$

Applying the Mellin inversion formula and moving the path of integration as before, we obtain

$$\frac{\partial\theta_\psi}{\partial x}(iy) = \frac{1}{Dy^2}\,\frac{\partial\theta_\psi}{\partial x}\left(\frac{i}{y}\right).$$

Applying Lemma 1.9.2 to $f(z) = \theta_\psi(z) + \theta_\psi(-1/Dz)$, we again obtain Eq. (9.21).

We now turn to the more elaborate proof of Eq. (9.20). The proof of this will be similar to that of Theorem 1.5.1, except that we will make use of both primitive and imprimitive characters. If f is an eigenfunction of Δ on $\mathcal{H}$ with eigenvalue $\lambda = \frac{1}{4} - \nu^2$, and if $\gamma \in GL(2,\mathbb{R})^+$, we denote $(f|\gamma)(z) = f\big(\gamma(z)\big)$. Then $f|\gamma$ is also an eigenfunction.

Let p be a prime not dividing D. We will assume that p is inert, so that $\chi_D(p) = -1$. Let σ be a function of the nonzero residue classes modulo p. We define

$$f_\sigma = \sum_{\substack{m \bmod p\\ p\nmid m}} \overline{\sigma(m)}\,\theta_\psi\bigg|\begin{pmatrix} p & m\\ & p\end{pmatrix}.$$

Let $\rho(m) = \sigma(r)$, where $-rmD \equiv 1 \bmod p$. We will prove that

$$f_\sigma\bigg|\begin{pmatrix} & -1\\ p^2D & \end{pmatrix} = (-1)^{\epsilon+1}\, f_\rho. \tag{9.23}$$

It is sufficient to prove this when σ is a (not necessarily primitive) character of the nonzero residue classes, because these span the space of all such functions. If σ is a character, then $\rho = \overline{\sigma(-D)}\,\overline{\sigma}$, so it is sufficient to show in this case that

$$f_\sigma\bigg|\begin{pmatrix} & -1\\ p^2D & \end{pmatrix} = (-1)^{\epsilon+1}\,\sigma(-D)\, f_{\overline{\sigma}}. \tag{9.24}$$

There are two cases, depending on whether σ is a primitive character, or $\sigma(m) \equiv 1$.

We have

$$\theta_\psi(z) = \sum_{\substack{n=-\infty\\ n\neq 0}}^{\infty} a(n)\,\sqrt{y}\,K_\nu(2\pi|n|y)\,e^{2\pi i n x},$$

where

$$a(n) = \begin{cases} \dfrac{1}{2i^{\epsilon}} \displaystyle\sum_{\mathbb{N}(\mathfrak{a})=|n|} \psi(\mathfrak{a}) & \text{if } n > 0; \\ \dfrac{1}{2(-i)^{\epsilon}} \displaystyle\sum_{\mathbb{N}(\mathfrak{a})=|n|} \psi(\mathfrak{a}) & \text{if } n < 0. \end{cases}$$

Suppose first that σ is a primitive character. We first consider the case where $\sigma(-1) = (-1)^{\epsilon}$. In this case, f_{σ} is an even form. Because

$$\sum_{\substack{m \bmod p \\ p \nmid m}} \overline{\sigma(m)}\, e^{2\pi i n(x+m/p)} = \tau(\overline{\sigma})\, \sigma(n)\, e^{2\pi i n x},$$

we have

$$\begin{aligned} f_{\sigma}(z) &= \tau(\overline{\sigma}) \sum_{n} \sigma(n)\, a(n)\, \sqrt{y}\, K_{\nu}(2\pi |n| y)\, e^{2\pi i n x} \\ &= \frac{1}{i^{\epsilon}}\, \tau(\overline{\sigma}) \sum_{\mathfrak{a}} \sigma\big(\mathbb{N}(\mathfrak{a})\big)\, \psi(\mathfrak{a})\, \sqrt{y}\, K_{\nu}\big(2\pi \mathbb{N}(\mathfrak{a}) y\big)\, \cos\big(2\pi \mathbb{N}(\mathfrak{a}) x\big). \end{aligned}$$

Thus

$$\begin{aligned} \int_0^{\infty} f_{\sigma}(iy)\, y^{s-1/2} \frac{dy}{y} &= \frac{1}{4i^{\epsilon}}\, \tau(\overline{\sigma})\, \pi^{-s}\, \Gamma\left(\frac{s+\nu}{2}\right) \Gamma\left(\frac{s-\nu}{2}\right) L(s, (\sigma \circ N)\psi) \\ &= \frac{1}{4i^{\epsilon}}\, \tau(\overline{\sigma})\, D^{-s/2}\, \Lambda(s, (\sigma \circ N)\psi), \end{aligned}$$

so by the Mellin inversion formula

$$f_{\sigma}(iy) = \frac{\sqrt{y}}{8\pi i^{\epsilon+1}}\, \tau(\overline{\sigma}) \int_{\sigma-i\infty}^{\sigma+i\infty} D^{-s/2} \Lambda(s, (\sigma \circ N)\psi)\, y^{-s}\, ds.$$

We now need the functional equation of $L(s, (\sigma \circ N)\psi)$. We have

$$\big((\sigma \circ N)\psi\big)_{\infty}(x) = \left| \frac{x^{(1)}}{x^{(2)}} \right|^{\nu}, \qquad x \in \mathbb{R}^2,$$

and so we have $\big((\sigma \circ N)\psi\big)_{\infty}(pd) = (-1)^{\epsilon}$. The conductor of $\sigma \circ N$ is, by Proposition 1.9.2, the ideal $\mathfrak{f} = p\mathfrak{o}$. We have $\big((\sigma \circ N)\psi\big)_{\mathfrak{f}}(-1) = \sigma(N(-1)) = \sigma(1) = 1$. The Gauss sum $\tau_F(\sigma \circ N)$ is, by Proposition 1.9.2, equal to $-\sigma(D)\, \tau(\sigma)^2$, because $\chi_D(p) = -1$. The vector denoted ϵ in Theorem 1.7.1

for $(\sigma \circ N)\psi$ is not our ϵ, but equals $(0, 0)$. Consequently, the functional equation reads

$$\Lambda\big(s, (\sigma \circ N)\psi\big) = (-1)^{\epsilon+1}\,\sigma(D)\,\frac{\tau(\sigma)^2}{p}\,p^{1-2s}\,\Lambda\big(1-s, \overline{(\sigma \circ N)\psi}\,\big).$$

Now we have $\tau(\sigma)\tau(\overline{\sigma}) = \sigma(-1)\,p$, so substituting this into our previous identity, moving the path of integration, and replacing s by $1-s$ we obtain

$$\begin{aligned} f_\sigma(iy) &= (-1)^{\epsilon+1}\,\sigma(-D)\,\frac{\sqrt{1/p^2Dy}}{8\pi i^{\epsilon+1}}\,\tau(\sigma) \\ &\quad\times \int_{\sigma-i\infty}^{\sigma+i\infty} D^{-s/2}\,\Lambda\big(s, \overline{(\sigma \circ N)\psi}\big)\,(p^2Dy)^s\,ds \\ &= (-1)^{\epsilon+1}\,\sigma(-D)\,f_{\overline{\sigma}}\left(\frac{i}{p^2Dy}\right), \end{aligned}$$

so using Lemma 1.9.2, we obtain Eq. (9.24) in the case where $\sigma(-1) = (-1)^\epsilon$. We leave the case where $\sigma(-1) = (-1)^{\epsilon+1}$ to the reader (Exercise 1.9.5).

Next we consider the case where $\sigma = 1$. In this case, we will show

$$f_\sigma = \theta_\psi \bigg| \begin{pmatrix} p & \\ & p^{-1} \end{pmatrix} - \theta_\psi.$$

We assume for definiteness that $\epsilon = 1$; the other case is identical with sines replacing cosines in the Fourier expansion. Because p is inert, the only ideals whose norm is a multiple of p are those of the form $p\mathfrak{a}$. Also, we recall that $\psi(p) = 1$. Thus

$$\begin{aligned} (f_\sigma + \theta_\psi)(z) &= \sum_{b \bmod p} \theta_\psi \bigg| \begin{pmatrix} p & b \\ & p \end{pmatrix} \\ &= p\sum_n a(pn)\,\sqrt{y}\,K_\nu(2\pi|pn|y)\,e^{2\pi i pnx} \\ &= p\sum_{\mathfrak{a}} \psi(p\mathfrak{a})\,\sqrt{y}\,K_\nu\big(2\pi \mathrm{N}(p\mathfrak{a})y\big)\,\cos\big(2\pi \mathrm{N}(p\mathfrak{a})x\big) \\ &= \sum_{\mathfrak{a}} \psi(\mathfrak{a})\,\sqrt{p^2y}\,K_\nu\big(2\pi \mathrm{N}(\mathfrak{a})\,p^2y\big)\,\cos\big(2\pi \mathrm{N}(\mathfrak{a})\,p^2x\big) \\ &= \theta_\psi(p^2z). \end{aligned}$$

Now using Eq. (9.21), we have

$$f_\sigma \bigg| \begin{pmatrix} & -1 \\ p^2 D & \end{pmatrix} = \theta_\psi \bigg| \begin{pmatrix} p & \\ & p^{-1} \end{pmatrix} \begin{pmatrix} & -p^{-1} \\ pD & \end{pmatrix} - \theta_\psi \bigg| \begin{pmatrix} & -p^{-1} \\ pD & \end{pmatrix}$$

$$= \theta_\psi \bigg| \begin{pmatrix} & -1 \\ D & \end{pmatrix} - \theta_\psi \bigg| \begin{pmatrix} & -1 \\ D & \end{pmatrix} \begin{pmatrix} p & \\ & p^{-1} \end{pmatrix}$$

$$= (-1)^\epsilon \left[\theta_\psi - \theta_\psi \bigg| \begin{pmatrix} p & \\ & p^{-1} \end{pmatrix} \right] = (-1)^{1+\epsilon} f_\sigma.$$

This completes the proof of Eq. (9.24), and hence of Eq. (9.23).

Let us fix m and r coprime to p such that $-Dmr \equiv 1 \bmod p$. Let s be such that $ps - Dmr = 1$. We apply Eq. (9.23) in the case where

$$\sigma(n) = \begin{cases} 1 & \text{if } n \equiv m \bmod p; \\ 0 & \text{otherwise.} \end{cases}$$

Then

$$\rho(n) = \begin{cases} 1 & \text{if } n \equiv r \bmod p; \\ 0 & \text{otherwise.} \end{cases}$$

We have

$$f_\sigma = \theta_\psi \bigg| \begin{pmatrix} p & m \\ & p \end{pmatrix}, \qquad f_\rho = \theta_\psi \bigg| \begin{pmatrix} p & r \\ & p \end{pmatrix}.$$

By Eq. (9.23), we have

$$\theta_\psi \bigg| \begin{pmatrix} p & m \\ & p \end{pmatrix} \begin{pmatrix} & -1/pD \\ p & \end{pmatrix} = (-1)^{1+\epsilon} \theta_\psi \bigg| \begin{pmatrix} p & r \\ & p \end{pmatrix}.$$

On the other hand, using Eq. (9.21)

$$\theta_\psi \bigg| \begin{pmatrix} p & m \\ & p \end{pmatrix} \begin{pmatrix} & -1/pD \\ p & \end{pmatrix}$$

$$= (-1)^\epsilon \theta_\psi \bigg| \begin{pmatrix} & 1 \\ -D & \end{pmatrix} \begin{pmatrix} p & m \\ & p \end{pmatrix} \begin{pmatrix} & -1/pD \\ p & \end{pmatrix}$$

$$= (-1)^\epsilon \theta_\psi \bigg| \begin{pmatrix} p & -r \\ -Dm & s \end{pmatrix} \begin{pmatrix} p & r \\ & p \end{pmatrix}.$$

Equating these expressions and translating on the right by $\begin{pmatrix} p & -r \\ & p \end{pmatrix}$, we obtain

$$\theta_\psi \bigg| \begin{pmatrix} p & -r \\ -Dm & s \end{pmatrix} = -\theta_\psi.$$

This proves Eq. (9.20) in the special case where $a = p$ is an inert prime. Now in general, if $\chi_D(d) = \chi_D(a) = -1$, we may, by Dirichlet's theorem on primes in an arithmetic progression, find u such that

$$\begin{pmatrix} 1 & u \\ & 1 \end{pmatrix}\begin{pmatrix} a & b \\ c & d \end{pmatrix} = \begin{pmatrix} p & -r \\ -Dm & s \end{pmatrix}$$

with p prime, and so we obtain Eq. (9.20) whenever $\chi_D(d) = -1$. If on the other hand $\chi_D(d) = 1$, the matrix may be written as a product of two matrices of this type, and so we obtain Eq. (9.20) in general.

It remains to be shown that θ_ψ is cuspidal. It clearly vanishes at ∞, and by Eq. (9.20), it also vanishes at 0. Now the hypothesis that the narrow class number of F is one implies that the discriminant D is prime (cf. Borevich and Shafarevich (1966), Corollary on p. 247). It follows that 0 and ∞ are the only two cusps of $\Gamma_0(D)$ (Shimura (1971), Proposition 1.43 on p. 25). Hence θ_ψ is a cusp form. ■

Exercises

Exercise 1.9.1 Prove that the non-Euclidean Laplacian is invariant under the action of $SL(2,\mathbb{R})$. [HINT: It is sufficient to check this for a set of generators.]

Exercise 1.9.2 Suppose that f is a Maass form for $SL(2,\mathbb{Z})$, with eigenvalue $s(1-s)$. Prove that the constant term $a_0(y)$ in the Fourier expansion Eq. (9.4) is a linear combination of y^s and y^{1-s}.

Exercise 1.9.3 Prove the estimate Eq. (9.8) for the Fourier coefficients of a Maass cusp form.

Exercise 1.9.4: Hecke theory for Maass forms Define Hecke operators for Maass forms for $SL(2,\mathbb{Z})$. For example, if p is a prime, let

$$T_p f = f\Big|\begin{pmatrix} p & \\ & 1 \end{pmatrix} + \sum_{b \bmod p} f\Big|\begin{pmatrix} 1 & b \\ & p \end{pmatrix}.$$

Show that these commute with the Laplacian and with each other, and are self-adjoint with respect to the inner product

$$\langle f, g\rangle = \int_{SL(2,\mathbb{Z})\backslash\mathcal{H}} f(z)\,\overline{g(z)}\,\frac{dx\,dy}{y^2}$$

on the space of Maass cusp forms. Conclude that one may find a basis of Hecke eigenforms that are also eigenforms of the Hecke operators. Prove that if

$$f(z) = \sum a(n)\,\sqrt{y}\,K_\nu(2\pi|n|y)\,e^{2\pi inx}$$

is a Maass form that is an eigenfunction of the Hecke operators, then

$$L(s, f) = \sum_{n=1}^{\infty} a(n)\, n^{-s} = \prod_p (1 - a(p)\, p^{-s} + p^{-2s})^{-1}.$$

Exercise 1.9.5 Complete the proof of Theorem 1.9.1 by considering the case where $\sigma(-1) = (-1)^{\epsilon+1}$.

1.10 Base Change

In this section, we will partly prove Theorem 1.7.3. Specifically, let ϕ be a modular form of weight k for $SL(2, \mathbb{Z})$, and let F be a real quadratic field with narrow class number one and discriminant D. As in Sections 1.7 and 1.9, d will denote a fixed totally positive generator of the different, which may be taken to be $\eta_0 \sqrt{D}$, where η_0 is the fundamental unit. We associate with ϕ the function f on $\mathcal{H} \times \mathcal{H}$ with Fourier expansion Eq. (7.20), where the coefficients $A(n)$ are as follows. We require Eq. (7.21) so that the function A may be regarded as a function of ideals; this function is determined from the Fourier coefficients of ϕ by Eqs. (7.31–7.35). It is a consequence of Eq. (7.21) that Eq. (7.19) is satisfied when

$$\gamma \in \Gamma_\infty(\mathfrak{o}) = \left\{ \begin{pmatrix} \epsilon & \beta \\ & \epsilon^{-1} \end{pmatrix} \middle| \epsilon \in \mathfrak{o}^\times, \beta \in \mathfrak{o} \right\}.$$

We will prove in this section that Eq. (7.19) is satisfied when $\gamma = w = \begin{pmatrix} & -1 \\ 1 & \end{pmatrix}$, in other words, that

$$f\left(\frac{-1}{z}\right) = N(z)^k f(z) \tag{10.1}$$

Consequently, we know that f is modular for the group generated by Γ_∞ and by w. It is a deep theorem of Vaserstein (1972) that the Hilbert modular group is always generated by these elements, so this implies that f is a Hilbert modular form. Although we cannot prove Vaserstein's result in these notes, we will show that it is true for $K = \mathbb{Q}(\sqrt{5})$ (Exercise 1.10.1), so at least we will have completely proved Theorem 1.7.3 for this particular field.

It general, one might prove Theorem 1.7.3 by constructing automorphic forms θ_ψ as in Theorem 1.9.3 that correspond to more general Hecke characters ψ than the limited ones considered there and by applying the analog of Weil's Theorem 1.5.1. I hope that it will be clear that this program should work.

Throughout this section, ψ will be a Hecke character with conductor $\mathfrak{f} = \mathfrak{o}$, as described in Section 1.9 before the proof of Theorem 1.9.1; let ν and ϵ be as described there. Our proof will be based on the following lemma.

Lemma 1.10.1 *Let $A(n)$ be a function on $\mathfrak{o}_+$ satisfying Eq. (7.21); we will also denote by $A(\mathfrak{a})$ the corresponding function on ideals. Assume that $A(\mathfrak{a}) =$*

$O\left(\mathbb{N}(\mathfrak{a})^K\right)$ for some constant K. Assume that $A(0) = 0$. Let f be defined by Eq. (7.20). If ψ is a Hecke character of F with trivial conductor, and if ν is as above, define

$$L(s, f, \psi) = \sum_{\mathfrak{a}} \psi(\mathfrak{a})\, A(\mathfrak{a})\, \mathbb{N}(\mathfrak{a})^{-s}, \tag{10.2}$$

$$\Lambda(s, f, \psi) = D^s\, (2\pi)^{-s}\, \Gamma(s+\nu)\, \Gamma(s-\nu)\, L(s, f, \psi). \tag{10.3}$$

Then the following two conditions are equivalent:

(i) f satisfies Eq. (7.19) with $\gamma = w$; and
(ii) the functions $\Lambda(s, f, \psi)$ have analytic continuation to all s, are bounded in vertical strips, and satisfy

$$\Lambda(s, f, \psi) = \Lambda(k-s, f, \psi^{-1}). \tag{10.4}$$

It is simple to generalize this to a totally real field (Exercise 1.10.2).

Proof We generalize slightly the proof of Eq. (7.27). We have (for sufficiently large s) that

$$\Lambda(s, f, \psi) = (-1)^\epsilon \int_{\mathfrak{o}_+^\times \backslash \mathbb{R}_+^2} f(iy)\, y_1^{s-\nu}\, y_2^{s+\nu}\, \frac{dy_1}{y_1}\, \frac{dy_2}{y_2}. \tag{10.5}$$

Indeed, if one unfolds the right side of Eq. (10.5) as in the proof of Eq. (7.27), one obtains $|d^{(1)}/d^{(2)}|^{-\nu}\, \Lambda(s, f, \psi)$; as we observed in the proof of Eq. (9.22), $|d^{(1)}/d^{(2)}|^\nu = \psi_\infty(d) = (-1)^\epsilon$, from which we get Eq. (10.5).

Now assume condition (i), that is, Eq. (7.29). By comparison with Eq. (7.28), the integral Eq. (10.5) is convergent for all s. As in the proof of Eq. (7.27), the functional equation (10.4) is a consequence of Eqs. (7.29) and (10.5). (We have $(-1)^{nk/2}$=1 because $n = 2$ and k is even.) Thus we have (ii).

Now let us assume (ii). We reparametrize the integrals Eq. (10.5) using the variables y_0 and t, where $y_1 = y_0 e^t$, $y_2 = y_0 e^{-t}$. With these coordinates, a fundamental domain for $\mathfrak{o}_+^\times \backslash \mathbb{R}_+^2$ is given by the inequalities $y_0 > 0$, $-R < t < R$, where $R = \log|\eta_0|$ is the regulator. We have

$$\Lambda(s, f, \psi) = 2(-1)^\epsilon \int_0^\infty \int_{-R}^{R} f(y_0 e^t, y_0 e^{-t})\, e^{-2t\nu}\, dt\, y_0^{2s}\, \frac{dy_0}{y_0},$$

so by the Mellin inversion formula

$$\int_{-R}^{R} f(y_0 e^t, y_0 e^{-t})\, e^{-2t\nu}\, dt = \frac{(-1)^\epsilon}{4\pi i} \int_{\sigma - i\infty}^{\sigma + i\infty} \Lambda\left(\frac{s}{2}, f, \psi\right)\, y_0^{-s}\, ds.$$

Moving the path of integration as usual, we obtain

$$\int_{-R}^{R} f(y_0 e^t, y_0 e^{-t})\, e^{-2tv}\, dt - \int_{-R}^{R} f(y_0^{-1} e^{-t}, y_0^{-1} e^t)\, y_0^{-2k}\, e^{-2tv}\, dt = 0. \quad (10.6)$$

Now let us reinterpret Eq. (10.6) as follows. The function

$$g(t) = f(y_0 e^t, y_0 e^{-t}) - f(y_0^{-1} e^{-t}, y_0^{-1} e^t)\, y_0^{-2k}$$

is a periodic function of t, with period $2R$. Recall that the possible values of ν are $m\pi i/2R$. We see that Eq. (10.6) is simply the mth Fourier coefficient. Because all of its Fourier coefficients vanish, we have $g(t) = 0$ identically, from which we find Eq. (7.29). Thus we have (i). ■

We will require certain new Eisenstein series. Let

$$E(z, s) = \sum_{\substack{c,\, d \in \mathbb{Z} \\ d \neq 0}} \chi_D(d)\, \frac{(cDz + d)^k\, y^{s+1/2}}{|cDz + d|^{2s+1}}, \quad (10.7)$$

absolutely convergent if $s > (k+1)/2$. It is automorphic (of weight $-k$) for the group $\Gamma_0(D)$, as may be readily checked.

Proposition 1.10.1 *The function $E(z, s)$ has analytic continuation to all s. For fixed s, it is of at most polynomial growth as $y \to \infty$. It satisfies the functional equation*

$$E^*(z, s) = z^k\, D^{2k-3s}\, E^*\left(\frac{-1}{Dz}, k - s\right), \quad (10.8)$$

where

$$E^*(z, s) = \pi^{-s-1/2}\, \Gamma\left(s + \tfrac{1}{2}\right)\, E(z, s).$$

We will indicate the basic outlines of a proof, leaving the details to the reader.

Proof (*sketch*) If $k = 0$, the proposition may be proved along the lines of Theorem 1.6.1, making use of the twisted Poisson summation formula. We leave this to the reader. If $k \neq 0$, a similar proof may be given, making use of a confluent hypergeometric function rather than the K-Bessel function (cf. Whittaker and Watson (1927), Chapter 16.) Alternatively, one may avoid introducing new special functions by deducing the general case from the case $k = 0$ by successively applying the differential operators L_k introduced in Section 2.1 to shift the weight or by adapting the method of Exercise 1.6.2. Again, we leave the details to the reader. ■

We may extract from the expression

$$\chi_D(d)\,\frac{(cDz+d)^k\,y^{s+1/2}}{|cDz+d|^{2s+1}}$$

the greatest common divisor of c and d, and we obtain

$$\begin{aligned}E(z,s) &= L(2s+1-k,\chi_D)\sum_{\substack{(c,d)=1\\ d\neq 0}}\chi_D(d)\,\frac{(cDz+d)^k\,y^{s+1/2}}{|cDz+d|^{2s+1}}\\ &= L(2s+1-k,\chi_D)\sum_{\gamma\in\overline{\Gamma_\infty(\mathbb{Z})}\backslash\overline{\Gamma_0(D)}}\chi(\gamma)\,j(\gamma,z)^k\,\operatorname{im}\big(\gamma(z)\big)^{s+1/2},\end{aligned} \tag{10.9}$$

where $\Gamma_\infty(\mathbb{Z})=\left\{\left(\begin{smallmatrix}1&n\\&1\end{smallmatrix}\right)|n\in\mathbb{Z}\right\}$, $\overline{\Gamma}$ denotes the image of a discrete group Γ in $PSL(2,\mathbb{R})$, and

$$j(\gamma,z)=cz+d,\qquad \chi(\gamma)=\chi_D(d)\qquad \text{for } \gamma=\begin{pmatrix}a&b\\cD&d\end{pmatrix}\in\Gamma_0(D).$$

We will require the integral identity

$$\int_0^\infty K_\nu(y)\,e^{-y}\,y^s\,\frac{dy}{y}=\sqrt{\pi}\,2^{-s}\,\frac{\Gamma(s+\nu)\,\Gamma(s-\nu)}{\Gamma\left(s+\frac{1}{2}\right)}. \tag{10.10}$$

A derivation of this formula may be found in Exercise 10.4.

We turn now to the proof of Eq. (10.1). Let ϕ and f be as in the opening paragraph of this section. We must prove the functional equations (10.4) for the twists $\Lambda(s,f,\psi)$. If $\psi=1$, we already know the functional equation because of Eq. (7.30). Hence we may assume that $\nu\neq 0$. In this case, we will deduce the functional equation by expressing $\Lambda(s,f,\psi)$ as the Rankin–Selberg convolution of ϕ with θ_ψ, constructed in Theorem 1.9.1.

We have Fourier expansions

$$\phi(z)=\sum_{n=1}^\infty a(n)\,e^{2\pi inz},$$

$$\theta_\psi(z)=\sum_{n\neq 0} b(n)\,\sqrt{y}\,K_\nu(2\pi|n|y)\,e^{2\pi inx}.$$

We will prove that

$$\int_{\Gamma_0(D)\backslash\mathcal{H}}\phi(z)\,\theta_\psi(z)\,E^*(z,s)\,\frac{dx\,dy}{y^2}=\frac{i^\epsilon}{2}\,D^{-s}\,\Lambda(s,f,\psi). \tag{10.11}$$

First take $\mathrm{re}(s)$ to be large. Substitute Eq. (10.9) into the left side. This equals

$$\pi^{-s-1/2}\,\Gamma\left(s+\tfrac{1}{2}\right)\,L(2s+1-k,\chi_D)$$
$$\times \sum_{\gamma\in\overline{\Gamma_\infty(\mathbb{Z})\backslash\Gamma_0(D)}}\ \int_{\Gamma_0(D)\backslash\mathcal{H}} \phi(z)\,\theta_\psi(z)\,\chi(\gamma)\,j(\gamma,z)^k\ \mathrm{im}\left(\gamma(z)\right)^{s+1/2}\frac{dx\,dy}{y^2}.$$

Now because for $\gamma\in\Gamma_0(D)$, we have $\phi(z)\,j(g,z)^k=\phi\big(\gamma(z)\big)$ and $\theta_\psi(z)\,\chi(\gamma)=\theta_\psi\big(\gamma(z)\big)$, we may rewrite this as

$$\pi^{-s-1/2}\,\Gamma\left(s+\tfrac{1}{2}\right)\,L(2s+1-k,\chi_D)$$
$$\times \sum_{\gamma\in\overline{\Gamma_\infty(\mathbb{Z})\backslash\Gamma_0(D)}}\ \int_{\Gamma_0(D)\backslash\mathcal{H}} \phi\big(\gamma(z)\big)\,\theta_\psi\big(\gamma(z)\big)\ \mathrm{im}\left(\gamma(z)\right)^{s+1/2}\frac{dx\,dy}{y^2}$$

$$=\pi^{-s-1/2}\,\Gamma\left(s+\tfrac{1}{2}\right)\,L(2s+1-k,\chi_D)\int_{\Gamma_\infty(\mathbb{Z})\backslash\mathcal{H}}\phi(z)\,\theta_\psi(z)\,y^{s+1/2}\frac{dx\,dy}{y^2}$$

$$=\pi^{-s-1/2}\,\Gamma\left(s+\tfrac{1}{2}\right)\,L(2s+1-k,\chi_D)\int_0^\infty\int_0^1\phi(z)\,\theta_\psi(z)\,y^{s-1/2}\,dx\,\frac{dy}{y}.$$

Now making use of the Fourier expansions, we have

$$\int_0^1\phi(z)\,\theta_\psi(z)\,dx=\sum a(n)\,b(m)\,e^{-2\pi ny}\,\sqrt{y}K_\nu(2\pi|m|y)\int_0^1 e^{2\pi i(n+m)x}\,dx$$
$$=\sum_{n=1}^\infty a(n)\,b(-n)\,e^{-2\pi ny}\,\sqrt{y}K_\nu(2\pi ny).$$

We have, for $n>0$

$$b(-n)=\frac{i^\epsilon}{2}\sum_{\mathbb{N}(\mathfrak{a})=n}\psi(\mathfrak{a}).$$

Making use of Eq. (10.10), we see that the left side of Eq. (10.11) equals

$$\frac{i^\epsilon}{2}\,(2\pi)^{-2s}\,\Gamma(s+\nu)\,\Gamma(s-\nu)\,L(2s+1-k,\chi_D)\sum a\big(\mathbb{N}(\mathfrak{a})\big)\,\psi(\mathfrak{a})\,\mathbb{N}(\mathfrak{a})^{-s}.$$

Hence it is sufficient to prove that

$$L(2s+1-k,\chi_D)\sum a\big(\mathbb{N}(\mathfrak{a})\big)\,\psi(\mathfrak{a})\,\mathbb{N}(\mathfrak{a})^{-s}=L(s,f,\psi). \tag{10.12}$$

Both sides of Eq. (10.12) are Euler products. We compare the individual Euler factors by use of Lemma 1.6.1: We define the α_i in that lemma by factoring the Hecke polynomial:

$$1 - a(p)\,x + p^{k-1}\,x^2 = (1 - \alpha_1 x)\,(1 - \alpha_2 x),$$

and we define the β_i as follows. If p splits, $p\mathfrak{o} = \mathfrak{p}_1\mathfrak{p}_2$, let $\beta_i = \psi(\mathfrak{p}_i)$. If p is inert, we define $\beta_1 = 1$, $\beta_2 = -1$. In either case, the identity of the lemma asserts the equality of the corresponding Euler factors in Eq. (10.12). (If $\mathfrak{p}$ ramifies, evaluation of the Euler factor is simple, and does not require the lemma.)

We now obtain the analytic continuation of the $\Lambda(s, f, \psi)$ from Eq. (10.11) by means of the analytic continuation of the Eisenstein series. To obtain the functional equation, we use Eq. (10.7) to rewrite the left side of Eq. (10.11) as

$$D^{2k-3s} \int\limits_{\Gamma_0(D)\backslash\mathcal{H}} z^k \phi(z)\,\theta_\psi(z)\,E^*\left(-\frac{1}{Dz}, k-s\right)\frac{dx\,dy}{y^2}.$$

Now we observe that $\Gamma_0(D)$ is normalized by the involution $z \to -1/Dz$. Hence we may make the change of variable $z \to -1/Dz$, and this becomes

$$D^{2k-3s} \int\limits_{\Gamma_0(D)\backslash\mathcal{H}} (Dz)^{-k}\phi(-1/Dz)\,\theta_\psi(-1/Dz)\,E^*(z, k-s)\,\frac{dx\,dy}{y^2}.$$

We now use Eq. (9.21) and the modularity of ϕ to rewrite this as

$$(-1)^\epsilon\,D^{2k-3s} \int\limits_{\Gamma_0(D)\backslash\mathcal{H}} \phi(Dz)\,\theta_\psi(z)\,E^*(z, k-s)\,\frac{dx\,dy}{y^2}.$$

Now if $\mathrm{re}(k-s)$ is large, we may unfold this as before to obtain

$$\frac{i^\epsilon}{2}(-1)^\epsilon\,D^{2k-3s}\,\pi^{2s-2k}\,\Gamma(k-s+\nu)\,\Gamma(k-s-\nu)\,L(k-2s+1, \chi_D)$$
$$\times \sum a(n)\,b(-nD)\,(nD)^{s-k}.$$

We observe that $b(-Dn) = \psi(\mathfrak{d})\,b(-n) = (-1)^\epsilon\,b(-n)$ because the different $\mathfrak{d}$ is the unique prime of $\mathfrak{o}$ with norm D. Hence this equals

$$\frac{i^\epsilon}{2}\,D^{k-2s}\,\pi^{2s-2k}\,\Gamma(k-s+\nu)\,\Gamma(k-s-\nu)\,L(k-2s+1, \chi_D)$$
$$\times \sum a(n)\,b(-nD)\,n^{s-k} = \frac{i^\epsilon}{2}\,D^{-s}\,\Lambda(k-s, f, \psi).$$

Now because $\Lambda(s, f, \psi)$ and $\Lambda(s, f, \psi^{-1})$ are actually equal, we obtain the required functional equation (10.3).

Exercises

Exercise 1.10.1 Prove that if $K = \mathbb{Q}(\sqrt{5})$, then the Hilbert modular group $SL(2, \mathfrak{o})$ is generated by Γ_∞ and by $w = \begin{pmatrix} & -1 \\ 1 & \end{pmatrix}$. [HINT: By the method of Proposition 1.2.3, this may be proved by determining a fundamental domain for $SL(2, \mathfrak{o})$. For this particular field, such a proof was given by Götzky. Consult H. Cohn (1965).]

Exercise 1.10.2 Generalize Lemma 1.10.1 to a totally real field F.

Exercise 1.10.3 Supply details for the proof of Proposition 1.10.1.

Barnes' lemma asserts that

$$\int_{\sigma-i\infty}^{\sigma+i\infty} \Gamma(s-\alpha)\,\Gamma(s-\beta)\,\Gamma(s+\gamma)\,\Gamma(s+\delta)\,ds$$

$$= \frac{\Gamma(\alpha+\gamma)\,\Gamma(\alpha+\delta)\,\Gamma(\beta+\gamma)\,\Gamma(\beta+\delta)}{\Gamma(\alpha+\beta+\gamma+\delta)},$$

where the path of integration is taken to the right of the poles of $\Gamma(s-\alpha)$ and $\Gamma(s-\beta)$, and to the left of the poles of $\Gamma(s+\gamma)$ and $\Gamma(s+\delta)$. For a proof, see Whittaker and Watson (1927, Section 14.52). Barnes' lemma is a close analog of Lemma 1.6.1.

Exercise 1.10.4

(a) Suppose that $\phi(y)$ and $\psi(y)$ are functions that are analytic in a neighborhood of the real axis and of rapid decay as $y \to \infty$. Let

$$f(s) = \int_0^\infty \phi(y)\, y^s \,\frac{dy}{y}, \qquad g(s) = \int_0^\infty \psi(y)\, y^s \,\frac{dy}{y},$$

so by the Mellin inversion formula

$$\phi(y) = \frac{1}{2\pi i} \int_{\sigma-i\infty}^{\sigma+i\infty} f(s)\, y^{-s}\, ds,$$

$$\psi(y) = \frac{1}{2\pi i} \int_{\sigma-i\infty}^{\sigma+i\infty} g(s)\, y^{-s}\, ds,$$

valid for sufficiently large σ. Show that

$$h(s) = \int_0^\infty \phi(y)\,\psi(y)\; y^s \,\frac{dy}{y},$$

where

$$h(s) = \frac{1}{2\pi = i} \int_{\sigma-i\infty}^{\sigma+i\infty} f(u)\, g(s-u)\, du$$

for sufficiently large σ. (Use the Mellin inversion formula.)

(b) Starting with

$$\int_0^\infty e^{-y}\, y^s \frac{dy}{y} = \Gamma(s)$$

and Eq. (9.9), prove Eq. (10.10). (Use the duplication formula

$$2^{2s-1}\Gamma(s)\,\Gamma(s+\tfrac{1}{2}) = \sqrt{\pi}\,\Gamma(2s)$$

and Barnes' lemma.)

2

Automorphic Forms and Representations of $GL(2, \mathbb{R})$

The spectral theory of automorphic forms was developed to a large extent by Maass, Roelcke, and Selberg without the benefit of the insights of representation theory. The first work to recognize the connection between representation theory and automorphic forms was the paper of Gelfand and Fomin (1952), but it was not until the 1960s that the systematic introduction of representation theory into the study of automorphic forms commenced in earnest.

In this chapter, we will study the connection between the representation theory of $GL(2, \mathbb{R})$ and automorphic forms on the Poincaré upper half plane in a classical setting. We will concentrate on the spectral theory of compact quotients and return to the noncompact case in Chapter 3.

In Sections 2.1 and 2.7, we will discuss the relationship between the spectral problem for compact quotients of the upper half plane. Section 2.1 introduces the problem, and Section 2.7 summarizes the implications of the results obtained in the intervening sections.

Section 2.2 gives various foundational results from Lie theory such as the construction of the universal enveloping algebra $U(\mathfrak{g}_{\mathbb{C}})$, and describes its center $\mathcal{Z}$ when $\mathfrak{g}$ is the Lie algebra of $GL(2, \mathbb{R})$. We interpret these as rings of differential operators and realize the Laplace–Beltrami operator as an element of $\mathcal{Z}$.

In Section 2.3, we show that the spectrum of the Laplacian on a compact quotient $L^2(\Gamma\backslash PGL(2, \mathbb{R})^+)$ is discrete and that the Laplacian admits an extension to a self-adjoint operator. We will show that $L^2(\Gamma\backslash PGL(2, \mathbb{R})^+)$ decomposes into a direct sum of irreducible unitary representations of the group $PGL(2, \mathbb{R})^+$. The key to these results is to replace the study of *differential* operators with *integral* operators, the latter having the advantage of being compact.

In Section 2.4, we present certain foundational results from representation theory and introduce $(\mathfrak{g}, K)$-modules, which allow us to replace analytic considerations with algebraic ones. In Section 2.5, we classify the irreducible admissible $(\mathfrak{g}, K)$-modules for $GL(2, \mathbb{R})$, and in Section 2.6, we determine which of those correspond to unitary representations. Sections 2.8 and 2.9 (both of which may be read after Section 2.4) cover the uniqueness of

Whittaker models and a theorem of Harish–Chandra concerning the C^∞ functions that are both $\mathcal{Z}$-finite and K-finite.

2.1 Maass Forms and the Spectral Problem

The theory of automorphic forms ultimately involves the representation theory of both real Lie groups, such as $GL(2,\mathbb{R})$, and p-adic groups, such as $GL(2,\mathbb{Q}_p)$. In order to bring out the role of the p-adic groups, one passes to the adele group. Even though adelization is required in order to understand the role of the p-adic groups, it is nevertheless possible to explain the role of the representation theory of $GL(2,\mathbb{R})$ in a purely classical context. This is the aim of this chapter.

Let $G = GL(2,\mathbb{R})^+$ be the group of real 2×2 matrices with positive determinant. Then G acts on the Poincaré upper half plane $\mathcal{H}$ by linear fractional transformations. Let Γ be a discontinuous subgroup of G such that the volume of $\Gamma\backslash\mathcal{H}$ is finite. (We shall soon specialize to the case where $\Gamma\backslash\mathcal{H}$ is compact, for reasons that we will explain.) We may assume that $-I\in\Gamma$, because if $-I\notin\Gamma$, we simply adjoin it, replacing Γ by the group generated by Γ and $-I$. (I is the 2×2 identity matrix.) On the other hand, there is no loss of generality in assuming that $\Gamma\subset SL(2,\mathbb{R})$, and we assume this also. As in Chapter 1, we will usually denote by $z = x+iy$ an element of $\mathcal{H}$.

By a *quasicharacter* χ of a group H we mean a homomorphism $\chi : H\to\mathbb{C}^\times$. A *character* is a quasicharacter satisfying $|\chi(\gamma)| = 1$ for all γ. Some authors call a quasicharacter a character, and a character a *unitary character*. If we wish to remind the reader of the property that $|\chi(\gamma)| = 1$, we may also use this expression.

We recognize several types of automorphic forms on $\Gamma\backslash\mathcal{H}$.

(a) *Holomorphic modular forms*. Let $\chi : \Gamma\to\mathbb{C}^\times$ be a character, and let k be a positive integer; recall that we are assuming that $-I\in\Gamma$. We require that $\chi(-I) = (-1)^k$. We define $M_k(\Gamma,\chi)$ to be the space of all holomorphic functions $f : \mathcal{H}\to\mathbb{C}$ that satisfy the conditions that (i) $f(\gamma z) = \chi(\gamma)(cz+d)^k f(z)$ for $\gamma = \left(\begin{smallmatrix} a & b\\ c & d\end{smallmatrix}\right)\in\Gamma$; and (ii) f be holomorphic at the cusps of Γ (cf. Section 1.4 for the definition of this term).

We call the elements of $M_k(\Gamma,\chi)$ *holomorphic modular forms*. Furthermore if f vanishes at the cusps, we call f a *holomorphic cusp form*. The space of cusp forms is denoted $S_k(\Gamma,\chi)$.

(b) *Maass forms*. These were introduced in Section 1.9. However, the discussion below is independent of that in Section 1.9 and has a different emphasis. Maass forms, like holomorphic modular forms, are functions on the upper half plane; they are, however, nonholomorphic. Compensating for this defect, they satisfy a differential equation.

(c) *The constant function*. The constant function $f(z) = 1$ for all $z\in\mathcal{H}$, should be considered an automorphic form. It is invariant under (any!) discrete group Γ. More generally, for any $s \in \mathbb{C}$, consider the function $f(g) = \det(g)^s$ on G; it is right invariant by $K = SO(2) = \left\{ \left(\begin{smallmatrix} a & b \\ -b & a \end{smallmatrix} \right) \middle| a^2 + b^2 = 1 \right\}$, and hence may be regarded as a function on $\mathcal{H}$. Again, this function may be regarded as an automorphic form.

Question *Why are there precisely these types of automorphic forms on $\Gamma \backslash \mathcal{H}$ and no others?*

We will find that the answer comes from representation theory!

We will need a slightly more general notion of Maass forms here than was given in Section 1.9. The Maass forms of Section 1.9 are those of *weight zero.* Here we will encounter Maass forms of all weights, including odd weights.

We will see that there are two differential operators, called *Maass operators,* which shift the weight by two, up or down. Thus no real loss of generality would be entailed if we restricted our attention to Maass forms of weights zero and weight one. In addition to these differential operators, there is the Laplacian, which is involved in the very definition of Maass forms.

Question *Where do these differential operators come from, and why do they work?*

Again, we will find the answer comes from representation theory!

Maass forms of weight zero were introduced by H. Maass (1949). Maass operators were introduced in Maass (1953). See Hejhal (1976, 1983), Iwaniec (1984, 1995), Maass (1949), Roelcke (1966), Selberg (1956), Terras (1985 and 1987), and Venkov (1990) for further information about Maass forms.

Let k be a "weight," which may be a positive or negative integer. We will denote by $z = x + iy$ a complex variable. It will sometimes be useful to imagine z and $\overline{z} = x - iy$ as "independent variables" with corresponding partial derivatives

$$\frac{\partial}{\partial z} = \frac{1}{2}\left(\frac{\partial}{\partial x} - i\frac{\partial}{\partial y}\right), \qquad \frac{\partial}{\partial \overline{z}} = \frac{1}{2}\left(\frac{\partial}{\partial x} + i\frac{\partial}{\partial y}\right).$$

This is primarily a computational device. For example, if f is a holomorphic function of z, then (by the Cauchy–Riemann equations) $\partial f(z)/\partial \overline{z} = 0$, etc.; this, in conjunction with the chain rule, is helpful in carrying out certain computations.

We define the following *Maass differential operators* on $C^\infty(\mathcal{H})$, the space of smooth functions of $\mathcal{H}$:

$$R_k = iy\frac{\partial}{\partial x} + y\frac{\partial}{\partial y} + \frac{k}{2} = (z - \overline{z})\frac{\partial}{\partial z} + \frac{k}{2}, \tag{1.1}$$

$$L_k = -iy\frac{\partial}{\partial x} + y\frac{\partial}{\partial y} - \frac{k}{2} = -(z - \overline{z})\frac{\partial}{\partial \overline{z}} - \frac{k}{2}, \tag{1.2}$$

and the *(weight k) non-Euclidean Laplacian*

$$\Delta_k = -y^2\left(\frac{\partial^2}{\partial x^2}+\frac{\partial^2}{\partial y^2}\right)+iky\frac{\partial}{\partial x}. \tag{1.3}$$

It is easily checked that

$$\Delta_k = -L_{k+2}R_k - \frac{k}{2}\left(1+\frac{k}{2}\right) = -R_{k-2}L_k + \frac{k}{2}\left(1-\frac{k}{2}\right). \tag{1.4}$$

For each k, we define a right action of $G = GL(2,\mathbb{R})^+$ on $C^\infty(\mathcal{H})$ by

$$f|_k g = \left(\frac{c\overline{z}+d}{|cz+d|}\right)^k f\left(\frac{az+b}{cz+d}\right), \qquad g = \begin{pmatrix} a & b \\ c & d\end{pmatrix}. \tag{1.5}$$

Lemma 2.1.1 *If $f \in C^\infty(\mathcal{H})$, $g \in G$, then*

$$(R_k f)|_{k+2} g = R_k(f|_k g), \tag{1.6}$$

$$(L_k f)|_{k-2} g = L_k(f|_k g), \tag{1.7}$$

and

$$(\Delta_k f)|_k g = \Delta_k(f|_k g). \tag{1.8}$$

Proof We prove Eq. (1.6), which is easy once a few simple facts have been assembled. Let $g = \left(\begin{smallmatrix} a & b \\ c & d\end{smallmatrix}\right)$, and let $w = g(z) = \frac{az+b}{cz+d}$. The meaning of $f|_k g$ is unchanged if the matrix g is multiplied by a real scalar, so we may assume that $ad - bc = 1$. Then, by the chain rule, and by the fact that $\partial\overline{w}/\partial z = 0$ by the Cauchy–Riemann equations, we have

$$\frac{\partial}{\partial z} = \frac{\partial w}{\partial z}\frac{\partial}{\partial w} + \frac{\partial \overline{w}}{\partial z}\frac{\partial}{\partial \overline{w}} = (cz+d)^{-2}\frac{\partial}{\partial w}.$$

Thus by (1.2.2), we have

$$(w-\overline{w})\frac{\partial}{\partial w} = \left(\frac{cz+d}{c\overline{z}+d}\right)(z-\overline{z})\frac{\partial}{\partial z}. \tag{1.9}$$

Next we show that if ϕ is any smooth function on $\mathcal{H}$, we have

$$\begin{aligned}(z-\overline{z})&\frac{\partial}{\partial z}\left(\frac{c\overline{z}+d}{|cz+d|}\right)^k \phi \\ &= (z-\overline{z})\left(\frac{c\overline{z}+d}{|cz+d|}\right)^k\frac{\partial\phi}{\partial z} + \frac{k}{2}\left[\left(\frac{c\overline{z}+d}{|cz+d|}\right)^{k+2} - \left(\frac{c\overline{z}+d}{|cz+d|}\right)^k\right]\phi.\end{aligned} \tag{1.10}$$

Indeed, by the Leibnitz rule, and the fact that $(\partial/\partial z)(c\bar{z}+d) = 0$ whereas

$$\frac{\partial}{\partial z}|cz+d| = \frac{c}{2}\frac{|cz+d|}{cz+d},$$

the left side of Eq. (1.10) equals

$$(z-\bar{z})\left(\frac{c\bar{z}+d}{|cz+d|}\right)^k \frac{\partial \phi}{\partial z} - \frac{k}{2}c(z-\bar{z})\frac{(c\bar{z}+d)^k}{|cz+d|^k(cz+d)}\phi(z).$$

Now writing $c(z-\bar{z}) = (cz+d) - (c\bar{z}+d)$, we obtain Eq. (1.10).

The right side of Eq. (1.6), evaluated at z, equals

$$\left[(z-\bar{z})\frac{\partial}{\partial z} + \frac{k}{2}\right]\left(\frac{c\bar{z}+d}{|cz+d|}\right)^k f(w),$$

which by Eqs. (1.10) and (1.9) equals

$$\left[(z-\bar{z})\left(\frac{c\bar{z}+d}{|cz+d|}\right)^k \frac{\partial}{\partial z} + \frac{k}{2}\left(\frac{c\bar{z}+d}{|cz+d|}\right)^{k+2}\right] f(w)$$

$$= \left(\frac{c\bar{z}+d}{|cz+d|}\right)^{k+2}\left[(w-\bar{w})\frac{\partial}{\partial w} + \frac{k}{2}\right] f(w) = \big((R_k f)|_{k+2} g\big)(z),$$

from whence comes Eq. (1.6). The proof of Eq. (1.7) is similar, and Eq. (1.8) follows from Eqs. (1.6), (1.7), and (1.4). ■

Because obviously not every L^2 function is differentiable, differential operators are only defined on a dense subset of a Hilbert space. Thus any Hilbert space theory that is broad enough to include differential operators requires a generalization of the notion of an operator. So if $\mathfrak{h}$ is a Hilbert space, an *operator* on $\mathfrak{h}$ will be a densely defined linear transformation, that is, an ordered pair (T, D_T), where D_T is a dense linear subspace of $\mathfrak{h}$, called the *domain* of T, and $T : D_T \to \mathfrak{h}$ is a linear transformation. The operator is called *closed* if its graph $\{(f, Tf) | f \in D_T\}$ is a closed subspace of $\mathfrak{h} \times \mathfrak{h}$. T is called *unbounded* if it is not continuous when D_T is given the subspace topology of $\mathfrak{h}$.

Unbounded operators (such as the Laplacian) are important in mathematical physics and have therefore been extensively studied. We recommend the book of F. Riesz and B. Sz.-Nagy (1960), Chapter VIII for a lucid discussion of the spectral theory of self-adjoint unbounded operators; see also Dunford and Schwartz (1963), Chapters XII and XIII for a more extensive discussion.

An operator (T, D_T) is called *symmetric* if

$$\langle Tf, g\rangle = \langle f, Tg\rangle \tag{1.11}$$

for $f, g \in D_T$, where $\langle , \rangle$ is the Hilbert space inner product on $\mathfrak{h}$. Symmetric operators include the important class of *self-adjoint* operators, which we shall next define. Self-adjoint operators are blessed with a spectral theorem; the larger class of symmetric operators is not.

Let (T, D_T) be an operator on $\mathfrak{h}$. We define the adjoint T^* of T. Let D_{T^*} be the space of all $g \in \mathfrak{h}$ such that $f \mapsto \langle Tf, g\rangle$ is a bounded linear functional on D_T. For such a g, the functional may be extended by continuity to a bounded linear functional on $\mathfrak{h}$. A well-known theorem of Riesz asserts that every bounded linear functional on $\mathfrak{h}$ has the form $f \mapsto \langle f, h\rangle$ for some $h \in \mathfrak{h}$, and so if $g \in D_{T^*}$, there exists a unique element $T^*g \in \mathfrak{h}$ such that Eq. (1.11) is valid. The operator (T^*, D_{T^*}) is called the *adjoint* of T.

The operator T is called *self-adjoint* if $D_T = D_{T^*}$ and $T = T^*$. This clearly implies that T is symmetric; it also implies that T is closed. A symmetric operator always admits a closed extension (just take T^{**}), but it may not admit an extension that is self-adjoint. The following example, taken from Riesz and Sz.-Nagy (1960) shows why.

We recall that a function defined on a measurable subset U of $\mathbb{R}$ is called *absolutely continuous* if for every $\epsilon > 0$ there exists a $\delta > 0$ such that whenever V is a subset of U having measure less than δ, the total variation of the restriction of f to V is less than ϵ. It is proved in Riesz and Sz.-Nagy (1960) that this condition is equivalent to f being the indefinite integral of some integrable function f', *which is determined modulo functions that vanish off sets of measure zero.* (If f is differentiable, then f' is its derivative.) Let D_T be the space of absolutely continuous functions f on $[0, 1]$ such that $f' \in L^2[0, 1]$ and $f(0) = f(1) = 0$. Let T be the operator $(Tf)(x) = if'(x)$. Integration by parts (which is justified by our assumption of absolute continuity) shows that

$$\langle Tf, g\rangle = i\big(f(1)\overline{g(1)} - f(0)\overline{g(0)}\big) + \langle f, Tg\rangle,$$

so this operator is symmetric on D_T. It is closed. Its adjoint $U = T^*$, however, has a properly larger domain, consisting of all absolutely continuous functions with square integrable derivative – with no condition at all at the endpoints.

To obtain a self-adjoint function of T, we may extend its domain of definition to all f such that $f(0) = f(1)$. Thus T admits a self-adjoint extension. The function U, however, admits no self-adjoint extension – the problem with U is that its domain is too large.

The issue of self-adjointness is an important one because examples show that the spectral theorem only works with this hypothesis. Nevertheless, simply knowing that an operator is symmetric allows us to draw certain conclusions: if f is an L^2 eigenfunction of a symmetric operator, so that $Tf = \lambda f$, then λ is real, and eigenvectors corresponding to distinct eigenvalues are orthogonal. The proofs of these facts are elementary and familiar.

Now let $L^2(\mathcal{H})$ be the Hilbert space of measurable functions on $\mathcal{H}$ that are square-integrable with respect to the G-invariant measure $y^{-2}dx \wedge dy$ (cf. Exercise 1.2.6(b)). Then Δ_k is defined on the dense subspace $C_c^\infty(\mathcal{H})$ of $L^2(\mathcal{H})$. (If M is a differentiable manifold, we define $C^\infty(M)$ to be the space of smooth functions on M and $C_c^\infty(M)$ to be the subspace of compactly supported functions. If X is a topological space, $C_c(X)$ is the space of compactly supported, continuous functions on X.)

Δ_k is not continuous in the L^2 topology: it is an unbounded operator.

Let

$$\Delta^{\mathrm{e}} = \frac{\partial^2}{\partial x^2} + \frac{\partial^2}{\partial y^2}$$

be the Euclidean Laplacian. We will denote by d the exterior derivative, which takes 1-forms to 2-forms. Let f and g be smooth functions defined in a neighborhood of a bounded region $\Omega \subset \mathbb{C}$ whose boundary is a smooth curve (or union of smooth curves) $\partial\Omega$. One easily checks the identity

$$d\left(g\left(\frac{\partial f}{\partial x}dy - \frac{\partial f}{\partial y}dx\right) - f\left(\frac{\partial g}{\partial x}dy - \frac{\partial g}{\partial y}dx\right)\right)$$
$$= (g\Delta^{\mathrm{e}} f - f\Delta^{\mathrm{e}} g)dx \wedge dy.$$

Thus by Stokes' theorem, we have

$$\int_{\Omega} (g\Delta^{\mathrm{e}} f - f\Delta^{\mathrm{e}} g)dx \wedge dy$$
$$= \int_{\partial\Omega} \left(g\left(\frac{\partial f}{\partial x}dy - \frac{\partial f}{\partial y}dx\right) - f\left(\frac{\partial g}{\partial x}dy - \frac{\partial g}{\partial y}dx\right)\right). \tag{1.12}$$

The direction of integration around an exterior arc of $\partial\Omega$ is taken in the clockwise sense. This identity is known as *Green's formula.*

Proposition 2.1.1 *The Laplacian Δ_k is a symmetric operator on $L^2(\mathcal{H})$ with domain $C_c^\infty(\mathcal{H})$.*

Proof Let f and g be compactly supported smooth functions on $\mathcal{H}$. By Green's identity (1.12),

$$\int_{\mathcal{H}} (\overline{g}\Delta^{\mathrm{e}} f - f\Delta^{\mathrm{e}} \overline{g})dx \wedge dy$$
$$= \int_{C} \left(\overline{g}\left(\frac{\partial f}{\partial x}dy - \frac{\partial f}{\partial y}dx\right) - f\left(\frac{\partial \overline{g}}{\partial x}dy - \frac{\partial \overline{g}}{\partial y}dx\right)\right),$$

where C is any contour that completely encloses the support of f and g. In the last integral, the integrand obviously vanishes on the contour of integration, and so we obtain

$$\int_{\mathcal{H}} \overline{g}\Delta^{\mathrm{e}} f dx \wedge dy = \int_{\mathcal{H}} f\Delta^{\mathrm{e}} \overline{g} dx \wedge dy. \tag{1.13}$$

Now let T be the operator $iy^{-1}\partial/\partial x$ on $C_c^\infty(\mathcal{H})$. We have, similarly,

$$\int_{\mathcal{H}} \left((Tf)\overline{g} - f\overline{(Tg)}\right) dx \wedge dy = i \int_{\mathcal{H}} y^{-1} \left(\frac{\partial f}{\partial x}\overline{g} + f\frac{\partial \overline{g}}{\partial x}\right) dx \wedge dy$$

$$= i \int_{\mathcal{H}} d(y^{-1} f\overline{g} dy) = i \int_{C} y^{-1} f\overline{g} dy = 0,$$

where again C is a contour surrounding the supports of f and g, and we have again used Stokes' Theorem. Thus

$$\int_{\mathcal{H}} (Tf)\overline{g} dx \wedge dy = \int_{\mathcal{H}} f\overline{Tg} dx \wedge dy. \tag{1.14}$$

Now we have

$$\langle \Delta_k f, g\rangle = \int_{\mathcal{H}} (\Delta_k f)\overline{g}\frac{dx \wedge dy}{y^2} = \int_{\mathcal{H}} (-\Delta^e f + kTf)\overline{g} dx \wedge dy,$$

and so it follows from Eqs. (1.13) and (1.14) that Δ_k is symmetric. ■

Because $\mathcal{H}$ has infinite volume, the Laplacian has too many eigenfunctions for the spectral theory here to be of real interest. A more interesting situation is to consider the decomposition of $L^2(\Gamma\backslash\mathcal{H})$, where Γ is a discontinuous group, and $\Gamma\backslash\mathcal{H}$ is compact or at least has finite volume. Thus we commence the spectral theory of automorphic forms. This leads to deep analytic questions, in particular, the Selberg trace formula.

Thus let Γ be a discontinuous subgroup of $G = GL(2,\mathbb{R})^+$; as before, we make the assumption that $-I \in \Gamma$. If this is not the case, we simply replace Γ by the group generated by itself and $-I$. Again, there is no loss of generality in assuming that $\Gamma \subset SL(2,\mathbb{R})$, and we will make this assumption.

We shall assume for the next few sections that $\Gamma\backslash\mathcal{H}$ is compact. This simplifies the spectral theory by excluding the continuous spectrum. It excludes the important example of $\Gamma = SL(2,\mathbb{Z})$ and its congruence subgroups, so we make a few remarks about how examples with $\Gamma\backslash\mathcal{H}$ compact do arise. There are two principal constructions.

Firstly, let M be a compact Riemann surface of genus $g \geq 2$. Then it is well known that the universal covering surface of M is conformally equivalent to $\mathcal{H}$. The fundamental group $\Gamma = \pi_1(M)$ is realized as a group of linear fractional transformations of $\mathcal{H}$; all nonidentity transformations in this group are hyperbolic. The quotient $\Gamma\backslash\mathcal{H}$ is conformally equivalent to M and is compact. For proofs of these facts, see C. L. Siegel (1969, Chapter 2).

The second construction is more closely connected with number theory. In this example, the unit group in a quaternion algebra is realized as a discontinuous group acting on $\mathcal{H}$ with compact quotient. Even though the $\Gamma\backslash\mathcal{H}$ is compact, a modular or Maass form for such a group has an L-series similar to those associated with automorphic forms for $SL(2,\mathbb{Z})$. For details about these groups, see T. Miyake (1989) and G. Shimura (1971).

Let χ be a character of Γ. Let $C^\infty(\Gamma\backslash\mathcal{H},\chi,k)$ be the space of smooth functions on $\mathcal{H}$ such that

$$\chi(\gamma)f(z)=\left(\frac{c\bar{z}+d}{|cz+d|}\right)^k f\left(\frac{az+b}{cz+d}\right),\qquad \gamma=\begin{pmatrix} a & b\\ c & d\end{pmatrix}\in\Gamma. \tag{1.15}$$

In order for this condition to be possible with nonzero f, it is necessary that $\chi(-I)=(-1)^k$, and we assume this. If $f,g\in C^\infty(\Gamma\backslash\mathcal{H},\chi,k)$, then $f\overline{g}$ is invariant under Γ, and so we may define

$$\langle f,g\rangle=\int\limits_{\Gamma\backslash\mathcal{H}} f(z)\overline{g(z)}\frac{dx\,dy}{y^2}.$$

Let $L^2(\Gamma\backslash\mathcal{H},\chi,k)$ denote the Hilbert space completion of $C^\infty(\Gamma\backslash\mathcal{H},\chi,k)$ with respect to this inner product.

Lemma 2.1.2 *Let ω be a smooth differential 1-form such that $\gamma(\omega)=\omega$ for all $\gamma\in\Gamma$. Then*

$$\int\limits_{\Gamma\backslash\mathcal{H}} d\omega=0. \tag{1.16}$$

Proof If M is a compact oriented manifold of dimension n, then Stokes' theorem implies that whenever ω is an $(n-1)$-form, we have $\int_M d\omega=\int_{\partial M}\omega=0$, because ∂M is the empty set. In particular, let $M=\Gamma\backslash\mathcal{H}$. The periodicity property of ω implies that it may be regarded as a differential form on M, and Eq. (1.16) follows. ■

Proposition 2.1.2 *The operators R_k and L_k map $C^\infty(\Gamma\backslash\mathcal{H},\chi,k)$ into the spaces $C^\infty(\Gamma\backslash\mathcal{H},\chi,k+2)$ and $C^\infty(\Gamma\backslash\mathcal{H},\chi,k-2)$, respectively. The space $C^\infty(\Gamma\backslash\mathcal{H},\chi,k)$ is invariant under Δ_k.*

Proof This is an immediate consequence of Lemma 2.1.1. ■

Proposition 2.1.3 *If $f\in C^\infty(\Gamma\backslash\mathcal{H},\chi,k)$ and $g\in C^\infty(\Gamma\backslash\mathcal{H},\chi,k+2)$, then*

$$\langle R_k f,g\rangle=\langle f,-L_{k+2}g\rangle\,. \tag{1.17}$$

Proof Let $\omega=y^{-1}f(z)\overline{g(z)}d\bar{z}$, where $d\bar{z}=dx-i\,dy$. We will show that $\gamma(\omega)=\omega$ for $\gamma\in\Gamma$. Indeed, let $w=u+iv=\gamma(z)$. Let $\gamma=\left(\begin{smallmatrix} a & b\\ c & d\end{smallmatrix}\right)$, where (because we are assuming $\Gamma\subset SL(2,\mathbb{R})$) we have $ad-bc=1$. We have

$$f(z)\overline{g(z)}=\left(\frac{cz+d}{c\bar{z}+d}\right)f(w)\overline{g(w)},$$

$$v=|cz+d|^{-2}y,\qquad dw=(cz+d)^{-2}dz,\qquad \overline{dw}=(c\bar{z}+d)^{-2}d\bar{z}.$$

Thus

$$v^{-1} f(w)\overline{g(w)}d\overline{w} = y^{-1} f(z)\overline{g(z)}d\overline{z},$$

as required.

By Lemma 2.1.2, we have

$$\begin{aligned}
0 &= \int_{\Gamma\backslash\mathcal{H}} d\big(y^{-1} f(z)\overline{g(z)}d\overline{z}\big) \\
&= \int_{\Gamma\backslash\mathcal{H}} \left[-\frac{\partial}{\partial y}(y^{-1} f\overline{g}) - i\frac{\partial}{\partial x}(y^{-1} f\overline{g})\right] dx \wedge dy \\
&= -\int_{\Gamma\backslash\mathcal{H}} \left[\left(iy\frac{\partial f}{\partial x} + y\frac{\partial f}{\partial y}\right)\overline{g} - \overline{\left(iy\frac{\partial g}{\partial x} - y\frac{\partial g}{\partial y}\right)} f - f\overline{g}\right] \frac{dx \wedge dy}{y^2} \\
&= -\int_{\Gamma\backslash\mathcal{H}} \left[(R_k f)\overline{g} + f\overline{(L_{k+2} g)}\right] \frac{dx \wedge dy}{y^2},
\end{aligned}$$

from which we get Eq. (1.17). ■

Proposition 2.1.4 *Δ_k is a symmetric (unbounded) operator on the Hilbert space $L^2(\Gamma\backslash\mathcal{H}, \chi, k)$.*

Proof Δ_k is defined on the dense subspace $C^\infty(\Gamma\backslash\mathcal{H}, \chi, k)$. By Eq. (1.4), it is sufficient to show that $L_{k+2} R_k$ (which, by Proposition 2.1.2, leaves this space invariant) is symmetric; this follows from Proposition 2.1.3. ■

We eventually prove that the Laplacian has a self-adjoint extension. We do not require this fact, however, because we will deduce the facts that interest us concerning its spectrum (in Section 2.3) from the spectral theory of certain integral operators, which are compact. Thus we will dodge the issue of self-adjointness, establishing this property (at least when $k = 0$) at the very end of the proofs in Section 2.3.

Spectral problem (version I). *Determine the spectrum of the symmetric unbounded operator Δ_k on $L^2(\Gamma\backslash\mathcal{H}, \chi, k)$.*

We will prove some results on this in Section 2.3. In the present section, our objective is to reformulate it in representation–theoretic terms. To summarize the solution to the spectral problem, however, if $\Gamma\backslash\mathcal{H}$ is compact, we will show that there exists a countable sequence of eigenvalues $0 = \lambda_0, \lambda_1, \lambda_2, \ldots$ with corresponding eigenvectors $1 = \phi_0$ (the constant function), $\phi_1, \phi_2, \ldots$ comprising an orthonormal basis of $L^2(\Gamma\backslash\mathcal{H}, \chi, k)$. Moreover, the eigenvalues λ_i tend to infinity, and hence have no accumulation point in $\mathbb{C}$. This situation is summarized by saying that in the case of a compact quotient, the Laplacian has a *discrete spectrum.*

If $\Gamma\backslash\mathcal{H}$ is not compact but has cusps, then in addition to the discrete spectrum (consisting of the constant function and the cusp forms defined below) it has a *continuous spectrum*, which may be described in terms of Eisenstein series. It is useful to have in mind an example of a self-adjoint unbounded operator with a continuous spectrum. Such an example is afforded by the Laplacian d^2/dx^2 on $L^2(\mathbb{R})$. The eigenfunctions $f_a(x) = e^{2\pi iax}, a \in \mathbb{R}$, are actually *not square integrable* and form a continuous family. However, any square-integrable function ϕ has a "spectral expansion"

$$\phi = \int_{\mathbb{R}} \hat{\phi}(a) f_a da,$$

which is of course just the (L^2 version of the) Fourier inversion formula, with $\hat{\phi}$ the Fourier transform of ϕ (cf. Eq. (1.8 in chapter 1)). This is a typical example of a continuous spectrum. By contrast, consider the discrete spectrum of the same Laplacian d^2/dx^2 on the "compact quotient" $L^2(\mathbb{Z}\backslash\mathbb{R})$, where the eigenfunctions $f_n(x) = e^{2\pi inx}$, $(n \in \mathbb{Z})$ are square integrable, form a discrete family, and form an orthonormal basis of $L^2(\mathbb{Z}\backslash\mathbb{R})$.

If $\Gamma\backslash\mathcal{H}$ is noncompact but has finite volume, then there can be both a discrete spectrum (containing the cusp forms) and a continuous spectrum (coming from the Eisenstein series). It is partly to avoid the difficulties of the continuous spectrum that we are assuming that $\Gamma\backslash\mathcal{H}$ is compact.

Let us now define *Maass forms* that are solution to the spectral problem. A *Maass form of weight k* for Γ is a smooth complex-valued function on $\mathcal{H}$ that satisfies the automorphicity condition (Eq. 1.15) and is an eigenfunction of Δ_k. (If $\Gamma\backslash\mathcal{H}$ is not compact but has finite volume, this definition must be modified to include the hypothesis of *moderate growth at the cusps* of Γ – see Section 3.2.)

Next, we recall the basic facts about Haar measure. Basic references for this material are Hewitt and Ross (1979, Chapter 4), Halmos (1950), and Bourbaki (1969).

If G is a locally compact group, it is well known that there exists a positive regular Borel measure on G that is left invariant; that is, denoting this measure by $d_L g$, we have

$$\int_G f(xg) d_L g = \int_G f(g) d_L g.$$

This measure is unique up to scalar multiple. It is called a *left Haar measure.* Similarly, there exists a right invariant Haar measure $d_R g$. These measures may or may not coincide. If left and right Haar measures are equal, the group is called *unimodular.* For example, any Abelian, compact, reductive, or nilpotent Lie group is unimodular. An example of a group that is *not* unimodular is the group B of upper triangular matrices in $GL(2, \mathbb{R})^+$. If

$$b = \begin{pmatrix} u & \\ & u \end{pmatrix} \begin{pmatrix} y^{1/2} & xy^{-1/2} \\ & y^{-1/2} \end{pmatrix} \in B,$$

then

$$d_L(b) = \frac{dx\,dy}{y^2}\frac{du}{u}, \qquad d_R(b) = dx\frac{dy}{y}\frac{du}{u}. \tag{1.18}$$

(Exercise 2.1.2).

Proposition 2.1.5

(i) Let H be a locally compact group and M be a compact subgroup. Then there exists a positive regular Borel measure on the quotient space H/M that is invariant under the action of H by left translation. This measure is unique up to constant multiple.

(ii) Let G be a unimodular group, and let P and K be closed subgroups such that $P \cap K$ is compact and $G = PK$. Let $d_L p$ and $d_R k$ be left Haar measure on P and right Haar measure on K. Then the Haar measure on G is given by

$$\int_G f(g)dg = \int_P \int_K f(pk) d_L p\, d_R k.$$

Proof We first prove (i), which we deduce from the existence and uniqueness of the Haar measure on H. We recall the Riesz representation theorem (Rudin, 1974, Theorem 2.14; Hewitt and Ross, 1979, Chapter 3, Section 11).

Let X be a locally compact space. A functional on $C_c(X)$ is called *positive* if its value on a nonnegative measurable function is nonnegative. The integral with respect to a positive Borel measure that is finite on compact sets has these properties. The Riesz representation theorem asserts that specifying a positive Borel measure on X that is regular and finite on compact sets is equivalent to specifying a positive linear functional on $C_c(X)$. (The assumption of regularity on the measure is required to guarantee uniqueness of the measure.)

Haar measure is automatically regular and finite on compact sets. Thus the theorem on existence and uniqueness of Haar measure asserts more than just the uniqueness of an invariant measure: it also asserts the existence of a unique invariant positive linear functional on $C_c(G)$.

Because M is compact, we may normalize the Haar measure on M so the volume of M is equal to one. Let $X = H/M$.

We have mappings $\sigma : C_c(H) \to C_c(X)$ and $\tau : C_c(X) \to C_c(H)$ defined as follows. If $\phi \in C_c(H)$, let $\sigma(\phi) = \Phi \in C_c(X)$ where

$$\Phi(hM) = \int_M \phi(hm)dm.$$

Note that this is well defined independent of the representative h of the coset hM by the left invariance of the Haar measure dm on M. If $\Psi \in C_c(X)$, define $\tau(\Psi) = \psi \in C_c(H)$, where $\psi(h) = \Psi(hM)$. The composition $\sigma \circ \tau$ is the identity on $C_c(X)$, and so σ is surjective. With these preliminaries, we may now establish the existence and uniqueness of an invariant measure on X.

To prove existence, define a positive linear functional on $C_c(X)$ by

$$f \mapsto \int_H (\tau f)(h)dh \qquad (dh = \text{left Haar measure}).$$

By the Riesz representation theorem, this linear functional arises from a positive Borel measure on X, which is invariant under the action of H because of the left invariance of the Haar measure dh.

To prove uniqueness, let there be given an invariant positive Borel measure on X, and let $\Lambda : C_c(X) \to \mathbb{C}$ be the corresponding invariant linear functional. Let $\lambda = \Lambda \circ \sigma : C_c(H) \to \mathbb{C}$. It is easily checked that λ is invariant under left translation by H and hence agrees (up to constant multiple) with the linear functional $f \mapsto \int_H f(h)dh$. Because σ is surjective, we see that Λ is uniquely determined up to constant multiple. This concludes the proof of (i).

To prove (ii), we apply $H = P \times K$, and let $M = P \cap K$ be embedded diagonally in H. We observe that there is a homeomorphism $H/M \to G$ given by $(p, k)M \mapsto pk^{-1}$. This induces a linear isomorphism $C_c(G) \cong C_c(H/M)$. The linear functional on $C_c(G)$ corresponding to Haar measure thus gives rise to a linear functional on $C_c(H/M)$, which by (i) must coincide with any other H-invariant linear functional on this space. Another invariant linear functional is given by

$$f \mapsto \int_P \int_K f\big((p, k^{-1})M\big)dp_L d_R k, \qquad f \in C_c(H/M).$$

The agreement of these two linear functional proves (ii). ■

For example, let $G = GL(2, \mathbb{R})^+$. We recall that every element of G has a representation in the form

$$g = \begin{pmatrix} u & \\ & u \end{pmatrix} \begin{pmatrix} y^{1/2} & xy^{-1/2} \\ & y^{-1/2} \end{pmatrix} \kappa_\theta, \quad \kappa_\theta = \begin{pmatrix} \cos(\theta) & \sin(\theta) \\ -\sin(\theta) & \cos(\theta) \end{pmatrix}, \tag{1.19}$$

where $x, y, u, \theta \in \mathbb{R}$, $u, y > 0$. The representation (Eq. 1.19) is unique, except of course θ is only determined modulo 2π (Exercise 2.1.3). Let $SO(2) = \{\kappa_\theta | \theta \in \mathbb{R}\}$. The Haar measure on the compact unimodular group $SO(2)$ may be taken to be $d\theta$. Taking $P = B$ and $K = SO(2)$ in Proposition 2.1.5(ii), we see that Haar measure on G in the coordinates (Eq. 1.19) is

$$dg = \frac{du}{u}\frac{dxdy}{y^2}d\theta. \tag{1.20}$$

If U is a locally compact group and V a topological vector space, then by a *representation* of U on V, we mean a mapping $\pi : U \to \mathrm{End}(V)$ that is a representation of U in the algebraic sense and for which the map $U \times V \to V$ given by $(u, v) \mapsto \pi(u)v$ is continuous. (We may also refer to the ordered pair (π, V) as a *representation*.) If V is a Hilbert space, and if $\pi(u)$ is unitary for all $u \in U$, then the representation π is called *unitary*. If V admits a continuous Hermitian inner product that is U invariant, then the representation π is called *unitarizable*. In this case, we may extend the representation π to the Hilbert space completion of $\mathfrak{h}$ of V, obtaining a unitary representation. However, the topology of V may well differ from the subspace topology of $\mathfrak{h}$. A representation (π, V) is called *irreducible* if V has no closed π-invariant subspaces.

Let Z be the center of $G = GL(2, \mathbb{R})^+$ consisting of the group of scalar matrices, and let Z^+ be the subgroup of scalar matrices with positive diagonal entries. Let $G_1 = SL(2, \mathbb{R})$. We observe that there exists a unique element of Z^+ with any given positive determinant, and so $G_1 \cong G/Z^+$. Let $C^\infty(\Gamma\backslash G, \chi)$ be the space of smooth complex-valued functions F on G satisfying

$$F(\gamma g u) = \chi(\gamma)F(g), \qquad \gamma \in \Gamma, u \in Z^+, g \in G, \tag{1.21}$$

and let $C(\Gamma\backslash G, \chi)$ be the space of continuous functions having this property. Let $\mathfrak{h} = L^2(\Gamma\backslash G, \chi)$ be the space of measurable functions satisfying Eq. (1.21) that are square integrable with respect to Haar measure on $G/Z^+ \cong G_1$. (It might be more natural to denote this space as $L^2(\Gamma\backslash G/Z^+, \chi)$ or $L^2(\Gamma\backslash G_1, \chi)$, but we will avoid these notations.) Thus $\mathfrak{h}$ is a Hilbert space.

Proposition 2.1.6 *The space $C^\infty(\Gamma\backslash G, \chi)$ is dense in $L^2(\Gamma\backslash G, \chi)$.*

Proof First we show that $C(\Gamma\backslash G, \chi)$ is dense in $L^2(\Gamma\backslash G, \chi)$. Let $\mathcal{F}$ be an open fundamental domain for Γ in G/Z^+. Then we may extend any element of $C_c(\mathcal{F})$ to an element of $C^\infty(\Gamma\backslash G, \chi)$, and by the same token, we may identify $L^2(\Gamma\backslash G, \chi)$ with $L^2(\mathcal{F})$. Thus it is sufficient to show that $C_c(\mathcal{F})$ is dense in $L^2(\mathcal{F})$; this follows from Rudin (1974, Theorem 3.14).

Now that we know $C(\Gamma\backslash G, \chi)$ is dense in $L^2(\Gamma\backslash G, \chi)$, it is sufficient to show that an element of $C(\Gamma\backslash G, \chi)$ may be uniformly approximated by an element of $C^\infty(\Gamma\backslash G, \chi)$. Because $\Gamma\backslash G$ is compact, this will imply that it may be approximated in the L^2 norm. The method of approximation is a standard one: *convolution with a smooth function.* Let ϕ_1 and ϕ_2 be continuous functions on G, of which we assume that one is compactly supported: then we define the convolution

$$(\phi_1 * \phi_2)(g) = \int_G \phi_1(gh)\phi_2(h^{-1})dh = \int_G \phi_1(h)\phi_2(h^{-1}g)dh.$$

(The last equality follows from the change of variables $h \mapsto g^{-1}h$.) Suppose that ϕ_2 is smooth then $\phi_1 * \phi_2$ is automatically smooth, as is clear from the second expression here. Suppose that $\phi_1 \in C(\Gamma\backslash G, \chi)$. Then it is clear from the first expression that $\phi_1 * \phi_2 \in C(\Gamma\backslash G, \chi)$ also. Moreover, because $\Gamma\backslash G/Z^+$ is compact, ϕ_1 is automatically uniformly continuous, so if $\epsilon > 0$ is given, there exists a neighborhood U of the identity in G such that if $h^{-1} \in U$, then $|\phi_1(gh) - \phi_1(g)| < \epsilon$ for all $g \in G$. We may find a function ϕ_2 that is smooth, nonnegative, and has compact support contained in U and that satisfies $\int_G \phi_2(h)dh = 1$. Then

$$|(\phi_1 * \phi_2)(g) - \phi_1(g)| = |\int_G [\phi_1(gh) - \phi_1(g)]\,\phi_2(h)dh|$$

$$\leq \int_G |\phi_1(gh) - \phi_1(g)|\phi_2(h)dh \leq \epsilon.$$

Thus $\phi_1 * \phi_2$ is a smooth element of $C(\Gamma\backslash G, \chi)$ uniformly approximating ϕ_1. ■

We have a representation $\rho : G \to \mathrm{End}(\mathfrak{H})$, called the *right regular representation* defined by

$$\big(\rho(g)f\big)(x) = f(xg), \qquad g, x \in G. \tag{1.22}$$

Proposition 2.1.7 *The action (Eq. 1.22) is a representation of G.*

Proof We must verify continuity: That is, we must show that $f \in L^2(\Gamma\backslash G, \chi)$ and g is near the identity in G, then $\rho(g)f$ is near f (Exercise 2.1.9). Using the density of $C^\infty(\Gamma\backslash G, \chi)$ in $L^2(\Gamma\backslash G, \chi)$ (Proposition 2.1.6), it is sufficient to check this when f is smooth, in which case f is uniformly continuous because $\Gamma\backslash G/\mathbb{Z}^+$ is compact; hence if g is near the identity, $\pi(g)f$ is near f in the L^∞ norm, and hence also in the L^2 norm. ■

It is clear that ρ is a unitary representation.

Spectral problem (version II) *Determine the decomposition of the Hilbert space $L^2(\Gamma\backslash G, \chi)$ into irredicible subspaces.*

Let $K = SO(2)$. Then K is a maximal compact subgroup of G, and any maximal compact subgroup of G is conjugate to K. We may identify $\mathcal{H}$ with $G/Z^+K \cong G_1/K$, and because K is compact, $\Gamma\backslash\mathcal{H}$ is compact if and only if $\Gamma\backslash G/Z^+ \cong \Gamma\backslash G_1$ is compact. We caution the reader that if $\Gamma\backslash G_1$ is noncompact, then $L^2(\Gamma\backslash G, \chi)$ does not have a decomposition into a direct sum of irreducible subspaces, owing to the existence of a continuous spectrum in addition to the discrete spectrum. It is important to understand the situation when $\Gamma\backslash G_1$ has finite volume but is noncompact; however, we will only address the simpler situation where $\Gamma\backslash G_1$ is compact to avoid the technicalities of the continuous spectrum.

Our purpose here is not to solve either version of the spectral problem, but rather to explain the relationship between the two problems; this will only be completed in Section 2.2. However, we make a few remarks on the second version of the spectral problem.

Assuming $\Gamma\backslash G_1$ (or, equivalently, $\Gamma\backslash\mathcal{H}$) is compact, we will prove in Section 2.3 that the Hilbert space $L^2(\Gamma\backslash G,\chi)$ decomposes into a countable direct sum of irreducible representations. Except for the one-dimensional subspace spanned by the constant function, each of these subrepresentations is infinite dimensional.

The first step in relating the two versions of the spectral problem is to explain precisely the relationship between $L^2(\Gamma\backslash G,\chi)$ and the spaces $L^2(\Gamma\backslash\mathcal{H},\chi,k)$. We first restrict the action of the representation ρ to the maximal compact group $K=SO(2)$. It is *very typical* to study a representation of a noncompact group by restricting it to a maximal compact subgroup. The space $L^2(\Gamma\backslash G,\chi)$ then decomposes into a Hilbert space direct sum of the eigenspaces of the irreducible characters of K (Exercise 2.1.5). That is, we have a Hilbert-space direct sum decomposition

$$L^2(\Gamma\backslash G,\chi)=\bigoplus_{k\in\mathbb{Z}}L^2(\Gamma\backslash G,\chi,k), \tag{1.23}$$

where we define $L^2(\Gamma\backslash G,\chi,k)$ to be the subspace of $L^2(\Gamma\backslash G,\chi)$ consisting of functions such that with κ_θ as in Eq. (1.19), we have $\rho(\kappa_\theta)F=e^{ik\theta}F$; that is

$$F(g\kappa_\theta)=e^{ik\theta}F(g). \tag{1.24}$$

Also, let $C^\infty(\Gamma\backslash G,\chi,k)$ be the subspace of smooth functions in the Hilbert space $L^2(\Gamma\backslash G,\chi,k)$.

Proposition 2.1.8 *The Hilbert spaces $L^2(\Gamma\backslash G,\chi,k)$ and $L^2(\Gamma\backslash\mathcal{H},\chi,k)$ are isomorphic. Specifically, there is a Hilbert space isomorphism*

$$\sigma_k : L^2(\Gamma\backslash\mathcal{H},\chi,k)\to L^2(\Gamma\backslash G,\chi,k)$$

given by

$$(\sigma_k f)(g)=(f|_k g)(i), \qquad g\in G, \tag{1.25}$$

for $f\in L^2(\Gamma\backslash\mathcal{H},\chi,k)$.

Proof It is easy to check that if f satisfies Eq. (1.15), then the function $F=\sigma_k f$ satisfies Eqs. (1.21) and (1.24) (Exercise 2.1.6). Moreover, given a function F satisfying Eqs. (1.21) and (1.24), define

$$f(z)=F\left(\begin{pmatrix} y & x \\ & 1\end{pmatrix}\right);$$

it is not hard to check that this satisfies Eq. (1.15) (Exercise 2.1.6), and $F \mapsto f$ defined this way is an inverse map to σ.

The only remaining question is whether the map σ_k is compatible with the inner products on $L^2(\Gamma\backslash\mathcal{H}, \chi)$ and $L^2(\Gamma\backslash G, \chi)$. This follows immediately, however, from the measure computation Proposition 2.1.5(ii). ∎

Now that we understand the relationship between the Hilbert spaces $L^2(\Gamma\backslash G, \chi)$ and $L^2(\Gamma\backslash\mathcal{H}, \chi, k)$, let us define operators on $L^2(\Gamma\backslash G, \chi)$ to play the role of Δ_k, R_k and L_k.

We may use x, y, u, and θ in Eq. (1.19) as coordinates on G. If F is an element of $C^\infty(\Gamma, \chi, k)$, we have by Eq. (1.24)

$$\frac{\partial F}{\partial \theta} = ikF. \tag{1.26}$$

We define the following differential operators on G, in the coordinates Eq. (1.19):

$$R = e^{2i\theta}\left(iy\frac{\partial}{\partial x} + y\frac{\partial}{\partial y} + \frac{1}{2i}\frac{\partial}{\partial \theta}\right), \tag{1.27}$$

$$L = e^{-2i\theta}\left(-iy\frac{\partial}{\partial x} + y\frac{\partial}{\partial y} - \frac{1}{2i}\frac{\partial}{\partial \theta}\right), \tag{1.28}$$

and the *Laplace–Beltrami* operator

$$\Delta = -y^2\left(\frac{\partial^2}{\partial x^2} + \frac{\partial^2}{\partial y^2}\right) + y\frac{\partial^2}{\partial x \partial \theta}. \tag{1.29}$$

Combining Eq. (1.26) with Eqs. 1–1.3), we see that

$$\sigma_{k+2} \circ R_k = R \circ \sigma_k, \quad \sigma_{k-2} \circ L_k = L \circ \sigma_k, \quad \sigma_k \circ \Delta_k = \Delta \circ \sigma_k. \tag{1.30}$$

Consequently, in $L^2(\Gamma\backslash G, \chi)$, the operators R, L, and Δ play the roles that were played by R_k, L_k, and Δ_k in the previous upper half plane setting.

In Section 2.7, we will present a theorem that precisely describes the relationship between the two versions of the spectral problem. The proof will tie together much of the material in Sections 2.2–2.6. However, the statement of Theorem 2.7.1 can be understood without reading the intervening material, and we invite the reader to look ahead at it.

Exercises

Exercise 2.1.1 Prove that $R_k \circ \Delta_k = \Delta_{k+2} \circ R_k$ and $L_k \circ \Delta_k = \Delta_{k-2} \circ L_k$.

Exercise 2.1.2 Verify the formulas (1.18) for left and right Haar measures on B.

Exercise 2.1.3: Iwasawa Decomposition Prove that each element of $G = GL(2, \mathbb{R})^+$ has a unique representation (Eq. 1.19).

Exercise 2.1.4 Check that the right regular representation, defined by Eq. (1.22), satisfies $\rho(gg')f = \rho(g)\rho(g')f$. (This is a trivial verification, but it is also a possible source of confusion.)

Exercise 2.1.5 Let H be a Hilbert space, let

$$K = SO(2) = \left\{ \begin{pmatrix} a & b \\ -b & a \end{pmatrix} \middle| a, b \in \mathbb{R}, \quad a^2 + b^2 = 1 \right\},$$

and let $\rho : K \to \mathrm{End}(H)$ be a unitary representation of K. For $k \in \mathbb{Z}$, let

$$H_k = \left\{ v \in H \middle| \rho \begin{pmatrix} \cos(\theta) & \sin(\theta) \\ -\sin(\theta) & \cos(\theta) \end{pmatrix} v = e^{ik\theta} v \right\}.$$

Prove that the spaces H_k are orthogonal, and that

$$H = \bigoplus H_k \tag{1.31}$$

as a *Hilbert space direct sum.* This means that every vector $v \in H$ has a unique representation as the sum of a series $v = \sum_{k=-\infty}^{\infty} v_k$, with $v_k \in H_k$ (series convergent in H).

[HINTS: It is easy to see that the spaces H_k are orthogonal. To see that they span H, let v lie in their orthogonal complement. Let U be a neighborhood of the identity in $\mathbb{R}$ modulo 2π, and let $\epsilon > 0$. Then there exists a trigonometric polynomial

$$F(\theta) = \sum_{n=-N}^{N} a_n e^{in\theta}$$

such that $\frac{1}{2\pi} \int_0^{2\pi} F(\theta) d\theta = 1$, whereas $|F(\theta)| < \epsilon$ off of U; this follows from the Stone–Weierstrass theorem, or one may pick F from a specific sequence of trigonometric polynomials constructed by Féjer (see Katznelson, 1976, 1.2.5). If v is nonzero, then U and ϵ may be chosen so that

$$\frac{1}{2\pi} \int_0^{2\pi} F(\theta) \pi(\kappa_\theta) v d\theta$$

lies arbitrarily near v. But this vector is a finite linear combination of elements of H_k with $-N \leq k \leq N$, and hence is orthogonal to v, which is a contradiction if $v \neq 0$.]

Exercise 2.1.6 Verify the assertions in the proof of Proposition 2.1.6.

Exercise 2.1.7: Holomorphic modular forms as Maass forms Because the raising and lowering operators shift the weight by two, it might seem that by applying them repeatedly we can shift any element of $C^\infty(\Gamma\backslash\mathcal{H}, \chi, k)$ into $C^\infty(\Gamma\backslash\mathcal{H}, \chi, \epsilon)$ where $\epsilon = 0$ or 1 and reduce the study of Maass forms to weights zero or one. This is almost true; however, there is an exception to this.

(a) Show that if f is a modular form of weight $k > 0$, then $y^{k/2} f(z)$ is a Maass form of weight k (by our definition), with the eigenvalue $\lambda = \frac{k}{2}(1 - \frac{k}{2})$ of Δ_k, but that it is annihilated by the lowering operator L_k.
(b) Show that this is the only case where a Maass form with weight can be annihilated by a lowering operator. Describe the Maass forms that are annihilated by raising operators.

Exercise 2.1.8: Positivity of the Laplacian (a) Prove that Δ_0 is a *positive operator* on $C^\infty(\Gamma\backslash\mathcal{H}, \chi) = C^\infty(\Gamma\backslash\mathcal{H}, \chi, 0)$. That is, we have $\langle \Delta f, f \rangle \geq 0$, with equality only if f is equal to a constant function. Conclude that the eigenvalues of f are all nonnegative real numbers. [HINT: Use Eqs. (1.4) and (1.17).]
(b) Prove that if λ is an eigenvalue of Δ_1, then $\lambda \geq \frac{1}{4}$.
(c) Prove that if $k \geq 0$, and if λ is an eigenvalue of Δ_k, then either $\lambda = \frac{l}{2}(1 - \frac{l}{2})$, where l is an integer such that $1 \leq l \leq k, l \equiv k \bmod 2$; or else $\lambda \geq 0$, and in fact $\lambda \geq \frac{1}{4}$ if k is odd.

Exercise 2.1.9 Let H be a Hilbert space, let G be a group, and let $\pi : G \to \mathrm{End}(V)$ be a group action such that $\pi(g) : H \to H$ is unitary for all $g \in G$. Suppose that for all $v \in H$ the map $g \mapsto \pi(g)v$ is continuous. Prove that π is a representation. [HINT: If v_1 is near v_2 and g_1 is near g_2 then

$$|\pi(g_1)v_1 - \pi(g_2)v_2| \leq |\pi(g_1)v_1 - \pi(g_1)v_2| + |\pi(g_1)v_2 - \pi(g_2)v_2|.$$

Show that this is small, and conclude that $(g, v) \mapsto \pi(g)v$ is continuous.

2.2 Basic Lie Theory

In this section, we will reinterpret the raising and lowering operators R and L and the Laplace–Beltrami operator Δ, introduced in Eqs. (1.27–1.29), as elements in the *universal enveloping algebra* of the Lie algebra $\mathfrak{gl}(2, \mathbb{R})$ of $GL(2, \mathbb{R})$. We will begin, therefore, by discussing the Lie algebra. We will not assume familiarity with Lie algebras, giving *ad hoc* definitions valid for $GL(n, \mathbb{R})$.

By a (real or complex) *Lie algebra* we mean a (real or complex) vector space $\mathfrak{g}$ equipped with a bilinear operation, called the *Lie bracket*, satisfying certain axioms. The bracket operation is denoted $X, Y \mapsto [X, Y]$ for $X, Y \in \mathfrak{g}$ and is assumed to satisfy

$$[X, Y] = -[Y, X], \qquad [X, X] = 0 \tag{2.1}$$

and the "Jacobi identity"

$$[[X, Y], Z] + [[Y, Z], X] + [[Z, X], Y] = 0. \tag{2.2}$$

To give a first example, let A be an associative algebra over $\mathbb{R}$ or $\mathbb{C}$. Let $\mathrm{Lie}(A)$ be A as a vector space, and define the Lie bracket operation by

$[X,Y] = XY - YX$, where the multiplication on the right side is the multiplication in the algebra A. One may easily check that Eqs. (2.1) and (2.2) are satisfied.

This functor, which associates an associative algebra A with a Lie algebra $\mathrm{Lie}(A)$ has an adjoint, which associates a Lie algebra $\mathfrak{g}$ with an associative algebra $U(\mathfrak{g})$, called the *universal enveloping algebra*. Even if $\mathfrak{g}$ is finite dimensional, $U(\mathfrak{g})$ will be infinite dimensional. To construct $U(\mathfrak{g})$ we start with the tensor algebra $\bigotimes\mathfrak{g}$, which is by definition the direct sum

$$\bigoplus_{k=0}^{\infty} \otimes^k \mathfrak{g}, \qquad \otimes^k\mathfrak{g} = \mathfrak{g}\otimes\ldots\otimes\mathfrak{g}\ (k \text{ times}),$$

where the multiplication

$$\otimes^k\mathfrak{g}\times\otimes^l\mathfrak{g}\to\otimes^{k+l}\mathfrak{g}$$

is the tensor product ($\otimes = \otimes_{\mathbb{R}}$ or $\otimes_{\mathbb{C}}$ according as $\mathfrak{g}$ is a real or complex Lie algebra). Let I be the ideal generated by elements of the form

$$X\otimes Y - Y\otimes X - [X,Y], \qquad X, Y\in\mathfrak{g},$$

and let $U(\mathfrak{g})$ be the quotient $\bigotimes\mathfrak{g}/I$. If $\xi,\eta\in U(\mathfrak{g})$, we will denote their product by $\xi\circ\eta$, which is the image of $\xi\otimes\eta$. The injection $\mathfrak{g}\mapsto\bigotimes\mathfrak{g}$ induces a homomorphism of vector spaces $\mathfrak{g}\mapsto U(\mathfrak{g})$, which is known to be injective. This injectivity is part of the Poincaré–Birkoff–Witt theorem, which is not hard to prove. We thus identify $\mathfrak{g}$ with its image in $U(\mathfrak{g})$, and then we have the identity

$$[X,Y] = X\circ Y - Y\circ X \tag{2.3}$$

in $U(\mathfrak{g})$. If no confusion can result, we may denote the product in $U(\mathfrak{g})$ by $\xi\eta$ instead of $\xi\circ\eta$. (We also use the notation $\circ$ to denote composition of functions.)

A *Lie group* is a group that is also a differentiable manifold such that the group operations are smooth maps. If G is a Lie group, there is a canonical method of associating G with a Lie algebra $\mathrm{Lie}(G)$. The dimension of $\mathrm{Lie}(G)$ is equal to the dimension of G. We will not need this well-known construction in general, but will give an *ad hoc* substitute for the general construction in the case where $G = GL(n,\mathbb{R})$. (The Lie algebra of a Lie group is equal to the Lie algebra of its connected component, so $\mathrm{Lie}(G)$ is the same as the Lie algebra of $GL(n,\mathbb{R})^+$.) Later in this section, we shall introduce the Lie algebra of a Lie subgroup of $GL(n,\mathbb{R})$.

Thus let $G = GL(n,\mathbb{R})^+$. Let $\mathfrak{g} = \mathfrak{gl}(n,\mathbb{R})$ be the set of $n\times n$ real matrices, equipped with the following Lie bracket operation: If X, $Y\in\mathfrak{g}$, we define $[X,Y] = XY - YX$, where on the right side, XY and YX are computed using ordinary matrix multiplication. The Lie bracket operation is bilinear and satisfies Eqs. (2.1) and (2.2). We will also use the complex Lie algebra $\mathfrak{gl}(n,\mathbb{C})$, which is the set of $n \times n$ complex matrices, with the same definition for the Lie bracket.

By a *(Lie algebra) representation* of a Lie algebra $\mathfrak{g}$, we mean an ordered pair (π, V), where V is a vector space and $\pi : \mathfrak{g} \to \mathrm{End}(V)$ is a linear mapping such that

$$\pi(X) \circ \pi(Y) - \pi(Y) \circ \pi(X) = \pi([X, Y]). \tag{2.4}$$

If $\mathfrak{g}$ is the Lie algebra of a Lie group, in particular of $G = GL(n, \mathbb{R})^+$, then representations of $\mathfrak{g}$ arise by differentiating representations of G. To see this, let us start with a finite-dimensional representation (π, V) of G. We assume that for $v \in V$, the mapping $g \mapsto \pi(g)\, v$ is a smooth function $G \to V$. However, this hypothesis is not really necessary – it is well known that any finite dimensional representation of $GL(n, \mathbb{R})^+$ is not only smooth but analytic. Let $X \in \mathfrak{g}$. We define an endomorphism $d\pi X$ of V by the formula

$$(d\pi X)(v) = \frac{d}{dt}\pi\big(\exp(tX)\big)\, v|_{t=0}, \tag{2.5}$$

where the "exponential map" $\mathfrak{g} \to G$ is defined by

$$e^X = \exp(X) = I + X + \tfrac{1}{2}X^2 + \tfrac{1}{6}X^3 + \dots . \tag{2.6}$$

Our aim is to show that $d\pi : \mathfrak{g} \to \mathrm{End}(V)$ is a representation, that is, it satisfies Eq. (2.4). The Lie algebra representation $d\pi$ is called the *differential* of the Lie group representation π.

We recall that if W and U are finite-dimensional vector spaces over a field F, then a *quadratic map* $W \to U$ is a polynomial function whose components (when referred to any linear system of coordinates on W and U) are homogeneous polynomials of degree 2 with coefficients in F.

Lemma 2.2.1 *Let W and U be finite-dimensional vector spaces over a field F of characteristic not equal to 2, and let q be a quadratic map $W \to U$. Then there exists a symmetric bilinear map $B : W \times W \to U$ such that*

$$q(w) = B(w, w).$$

The symmetric bilinear map B is called the *polarization* of q.

Proof Without loss of generality, we may assume that $W = F^n$ and $U = F^m$, so elements of W are n-tuples $w = (w_1, \cdots, w_n)$. By definition, q is a homogeneous polynomial of degree two, so there exist coefficients $a_{ijk} \in F$ $(1 \le i \le j \le n, 1 \le k \le m)$ such that

$$q(w) = \left(\sum_{1 \le i \le j \le n} a_{ij1} w_i w_j, \cdots, \sum_{1 \le i \le j \le n} a_{ijm} w_i w_j \right).$$

Define

$$B_1(w, w') = \left(\sum_{1 \le i \le j \le n} a_{ij1} w_i w'_j, \cdots, \sum_{1 \le i \le j \le n} a_{ijm} w_i w'_j \right)$$

for $w = (w_1, \cdots, w_n)$ and $w' = (w'_1, \cdots, w'_n) \in F^n$. Then B_1 is a bilinear map such that $q(w) = B_1(w, w)$. However, B_1 is not symmetric, so we define

$$B(w, w') = \tfrac{1}{2}[B_1(w, w') + B_1(w', w)].$$

This is possible because the characteristic of F is not equal to 2. ■

Proposition 2.2.1 *Let $G = GL(n, \mathbb{R})^+$, and let $\mathfrak{g} = \mathfrak{gl}(n, \mathbb{R})$ be its Lie algebra. Let (π, V) be a finite-dimensional representation of G such that $g \mapsto \pi(g)\, v$ is a smooth function for all $v \in V$. Then $d\pi : \mathfrak{g} \to \mathrm{End}(V)$ is a Lie algebra representation of $\mathfrak{g}$.*

Proof We must verify that

$$d\pi(X) \circ d\pi(Y) - d\pi(Y) \circ d\pi(X) = d\pi([X, Y]). \tag{2.7}$$

We observe that the two paths $t \mapsto \exp(tX)$ and $t \mapsto I + tX$ are tangent in G. (For $X \in \mathfrak{g}$ and $t \in \mathbb{R}$ small, $I + tX$ lies in $G = GL(n, \mathbb{R})^+$.) Consequently, we may simplify Eq. (2.5) to read

$$(d\pi\, X)(v) = \frac{d}{dt} \pi(I + tX)\, v|_{t=0}. \tag{2.8}$$

(Nevertheless, Eq. (2.5) is the correct definition: The apparently simpler formula
Eq. (2.8) only works for $G = GL(n, \mathbb{R})^+$, whereas if G is an arbitrary Lie group and $\mathfrak{g}$ is its Lie algebra, there is an exponential map defined from $\mathfrak{g} \to G$, so the definition (Eq. 2.5) makes sense.) Let $v \in V$. Because $\pi(g)\, v$ is a smooth function, the map $X \mapsto \pi(I + X)\, v$ has a Taylor expansion near $X = 0$ in $\mathfrak{g}$. Thus there exist homogeneous polynomial maps $c_i : \mathfrak{g} \to V$ (with degree i) and a remainder term $r : \mathfrak{g} \to V$ such that

$$\pi(I + X)\, v = c_0(X) + c_1(X) + c_2(X) + r(X),$$

where $r(X)$ vanishes to order 3 at $X = 0$. Now $c_0(X) = c_0$ is a constant, and we let $B : \mathfrak{g} \times \mathfrak{g} \to V$ be the polarization of c_2 as in Lemma 2.2.1, so

$$\pi(I + X)\, v = c_0 + c_1(X) + B(X, X) + r(X).$$

We have

$$\begin{aligned}(d\pi\, X)(v) &= \frac{d}{dt} \pi(I + tX)\, v|_{t=0} \\ &= \frac{d}{dt} \big(c_0 + t c_1(X) + t^2 B(X, X) + r(tX)\big)|_{t=0} = c_1(X). \end{aligned} \tag{2.9}$$

Now let us compute

$$
\begin{aligned}
(d\pi X \circ d\pi Y)(v) &= (d\pi X)\left(\frac{d}{du}\pi(I+uY)\,v|_{u=0}\right)\\
&= \frac{\partial}{\partial t}\frac{\partial}{\partial u}\pi\big((1+tX)(1+uY)\big)\,v|_{t=u=0}\\
&= \frac{\partial}{\partial t}\frac{\partial}{\partial u}\big(c_0 + c_1(tX+uY+tuXY)\\
&\quad + B(tX+uY+tuXY, tX+uY+tuXY)\\
&\quad + r(tX+uY+tuXY)\big)|_{t=u=0}.
\end{aligned}
$$

We may omit r from this computation, because it vanishes to third order. Expanding the linear and bilinear maps c_1 and c_2, we obtain

$$
\begin{aligned}
(d\pi X \circ d\pi Y)(v) &= \frac{\partial}{\partial t}\frac{\partial}{\partial u}\big(c_0 + tc_1(X) + uc_1(Y)\\
&\quad + tuc_1(XY) + t^2B(X,X) + u^2B(Y,Y)\\
&\quad + t^2u^2B(XY,XY) + 2tuB(X,Y)\\
&\quad + 2t^2uB(X,XY) + 2tu^2B(Y,XY)\big)|_{t=u=0}\\
&= c_1(XY) + 2B(X,Y).
\end{aligned}
$$

Similarly

$$
(d\pi Y \circ d\pi X)(v) = c_1(YX) + 2B(X,Y),
$$

and subtracting (to cancel the unwanted term $2B(X,Y)$)

$$
(d\pi X \circ d\pi Y - d\pi Y \circ d\pi X)(v) = c_1(XY - YX) = d\pi([X,Y])(v)
$$

by Eq. (2.9). ∎

The assumption of finite dimensionality in the above proof is not really necessary – see Proposition 2.4.1 for a generalization to unitary representations. Proposition 2.4.1 will be deduced from a special case that we need right now. This is the case of the right regular representation ρ of $G = GL(n,\mathbb{R})^+$ on $C^\infty(G)$, which is defined by $\big(\rho(g)f\big)(x) = f(xg)$. If $X \in \mathfrak{g} = \mathfrak{gl}(n,\mathbb{R})$, we will denote $d\rho\, X$ by simply dX. It is defined by the following formula, for $g \in G$ and $f \in C^\infty(G)$:

$$
(dX\,f)(g) = \frac{d}{dt}\big(\rho(\exp(tX))\,f\big)(g)|_{t=0} = \frac{d}{dt}\,f\big(g\exp(tX)\big)|_{t=0}. \tag{2.10}
$$

As before, it is convenient to rewrite this as

$$
(dX\,f)(g) = \frac{d}{dt}\,f\big(g(I+tX)\big)|_{t=0}, \tag{2.11}
$$

which is more convenient for computation, although the formula (2.11) is special to $GL(n, \mathbb{R})^+$ and cannot be adapted to other Lie groups.

Proposition 2.2.2 *We have*

$$dX \circ dY - dY \circ dX = d([X, Y]). \tag{2.12}$$

Thus $X \to dX$ is a Lie algebra representation of $\mathfrak{g}$ on $C^\infty(G)$.

Proof The proof is similar to that of Proposition 2.2.1. We fix a function $f \in C^\infty(G)$ and an element $g \in G$. We write, for X near 0

$$f\big(g(I+X)\big) = c_0 + c_1(X) + B(X, X) + r(X),$$

where c_1 is linear, B symmetric and bilinear, and r vanishes to order 3 at $X = 0$. As in the proof of Proposition 2.2.1, we obtain

$$(dX\, f)(g) = c_1(X),$$

$$(dX \circ dY\, f)(g) = c_1(XY) + 2B(X, Y),$$

from whence we get

$$\big((dX \circ dY - dY \circ dX)\, f\big)(g) = c_1(XY - YX) = \big(d[X, Y]\, f\big)(g).$$

Because the details are similar, we leave them to the reader. ■

In addition to the right regular representation ρ of G on $C^\infty(G)$, there is the *left* regular representation $\lambda : G \to C^\infty(G)$, defined by

$$\big(\lambda(g)\, f\big)(x) = f(g^{-1}x). \tag{2.13}$$

Obviously, the left and right regular representations commute with each other:

$$\lambda(g_1) \circ \rho(g_2) = \rho(g_2) \circ \lambda(g_1). \tag{2.14}$$

It follows from Eqs. (2.14) and (2.10) that if $X \in \mathfrak{g}$ and $g \in G$, then

$$\lambda(g) \circ dX = dX \circ \lambda(g). \tag{2.15}$$

In other words, the elements of the Lie algebra are realized as *left-invariant differential operators* on G.

We return now to the case of a general Lie algebra $\mathfrak{g}$ and its enveloping algebra $U(\mathfrak{g})$.

Proposition 2.2.3 *Let $\mathfrak{g}$ be a Lie algebra, and let $U(\mathfrak{g})$ be its enveloping algebra. Let $\pi : \mathfrak{g} \to \mathrm{End}(V)$ be a representation. Then π may be extended to a representation of the associative algebra $U(\mathfrak{g})$.*

(We recall that if A is an associative algebra, a *representation* of A is an algebra homomorphism $A \to \mathrm{End}(V)$, where V is some vector space.)

Proof We recall the following property.

Universal property of the tensor algebra *Let V be a vector space, A an associative algebra, and $l : V \to A$ a linear map. Then there exists a unique algebra homomorphism $L : \bigotimes V \to A$ extending l.*

This is easy to verify: on the homogeneous elements of degree k, we have

$$L(v_1 \otimes \ldots \otimes v_k) = l(v_1) \cdots l(v_k).$$

To deduce Proposition 2.2.3, we use the universal property to extend π to a homomorphism $\bigotimes \mathfrak{g} \to \mathrm{End}(V)$. Recall that $U(\mathfrak{g})$ is defined as $\bigotimes \mathfrak{g}$ modulo an ideal I; this ideal is contained in the kernel of the homomorphism to $\mathrm{End}(V)$. Indeed, this is a consequence of the assumption that π is a Lie algebra representation. Hence an algebra homomorphism $U(\mathfrak{g}) \to \mathrm{End}(V)$ is induced, as required. ∎

In particular, let $G = GL(n, \mathbb{R})^+$, and let $\mathfrak{g} = \mathfrak{gl}(n, \mathbb{R})$ be its Lie algebra. We consider the representation of $\mathfrak{g}$ on $C^\infty(G)$ in Proposition 2.2.2. *There is therefore a representation of $U(\mathfrak{g})$ as a ring of differential operators on G.* These operators are *left-invariant* as a consequence of Eq. (2.15).

It may be shown that the representation of $U(\mathfrak{g})$ on $C^\infty(G)$ is *faithful*, that is, has trivial kernel, and in fact $U(\mathfrak{g})$ may be identified with the algebra of left-invariant differential operators on G. Although we will not prove that the representation is faithful, we will sometimes not distinguish between an element of $U(\mathfrak{g})$ and the corresponding differential operator. Thus if $D \in U(\mathfrak{g})$, we will sometimes use the notation $D(f)$ to denote the action of D on an element $f \in C^\infty(G)$.

Lemma 2.2.2 *Let $G = GL(n, \mathbb{R})^+$, and let X be an element of the Lie algebra $\mathfrak{g} = \mathfrak{gl}(n, \mathbb{R})$ of G. Suppose that $\phi \in C^\infty(G \times \mathbb{R})$ satisfies*

$$\frac{\partial}{\partial t}\phi(g, t) = dX\,\phi(g, t) \tag{2.16}$$

and the boundary condition $\phi(g, 0) = 0$. Then $\phi(g, t) = 0$ for all $t \in \mathbb{R}$.

Proof For $g \in G$, define $\phi_g \in C^\infty(\mathbb{R} \times \mathbb{R})$ by $\phi_g(u, t) = \phi\big(g \exp(uX), t\big)$. Recalling the definition (Eq. 2.10) of dX, Eq. (2.16) amounts to the assumption that

$$\frac{\partial}{\partial t}\phi_g(0, t) = \frac{\partial}{\partial u}\phi_g(0, t).$$

Using the identity $\phi_{g\exp(vX)}(u, t) = \phi_g(u + v, t)$, we have therefore

$$\frac{\partial}{\partial t}\phi_g(u, t) = \frac{\partial}{\partial u}\phi_g(u, t) \tag{2.17}$$

for all $u \in \mathbb{R}$. We now make the change of variables

$$t = v + w, \qquad u = v - w, \qquad \frac{\partial}{\partial w} = \frac{\partial}{\partial t} - \frac{\partial}{\partial u}.$$

In these new coordinates, Eq. (2.17) becomes

$$\frac{\partial}{\partial w}\phi_g(v - w, v + w) = 0,$$

so $\phi_g(v - w, v + w)$ is independent of w. There exists, therefore, a function $F_g \in C^\infty(\mathbb{R})$ such that $\phi_g(v - w, v + w) = F_g(v)$, or

$$\phi_g(u, t) = F_g\big(\tfrac{1}{2}(u + t)\big).$$

The boundary condition $\phi_g(u, 0) = 0$ now implies that $F_g = 0$, so $\phi_g = 0$, as required. ■

Proposition 2.2.4 *Let $G = GL(n,\mathbb{R})^+$, and let $\mathfrak{g} = \mathfrak{gl}(n,\mathbb{R})$ be its Lie algebra. Let D be an element of the center of $U(\mathfrak{g})$. Then the differential operator D is invariant under both the left and right regular representations of G.*

Proof The left invariance has already been proved for an arbitrary element of $U(\mathfrak{g})$. For the right invariance, the assumption that D lies in the center of $U(\mathfrak{g})$ is essential.

It is sufficient to prove that $\rho(h) \circ D = D \circ \rho(h)$ for h in a set of generators of G. Any neighborhood of the identity generates an open subgroup of G, and because G is connected, it is therefore generated by any neighborhood of the identity. Because the image of $\exp : \mathfrak{g} \to G$ contains a neighborhood of the identity, it is sufficient to show that

$$\rho(g_t) \circ D = D \circ \rho(g_t), \qquad g_t = \exp(tX)$$

for $X \in \mathfrak{g}$, $t \in R$. Let

$$\phi(g, t) = \big(D\,\rho\big(\exp(tX)\big)\,f - \rho(\exp(tX))\,D\,f\big)(g).$$

We will prove that $\phi(g, t)$ satisfies the hypotheses of Lemma 2.2.2. The boundary condition $\phi(g, 0) = 0$ is obvious. We verify Eq. (2.16). We have

$$\begin{aligned}\frac{\partial}{\partial t}\phi(g, t) = \frac{\partial}{\partial u}\phi(g, t + u)|_{u=0} &= \frac{\partial}{\partial u}\big((D\rho(\exp((t + u)X))f\big)(g)|_{u=0}\\ &\quad - \frac{\partial}{\partial u}\big(\rho(\exp((t + u)X)\,Df)\big)(g)|_{u=0}\\ &= \frac{\partial}{\partial u}\big((D\rho(\exp(uX))\rho(\exp(tX))f\big)(g)\\ &\quad - \frac{\partial}{\partial u}(\rho\big(\exp(uX)\big)\rho\big(\exp(tX)\big)\,Df))(g)|_{u=0}\\ &= \big(D\,dX\,\rho\big(\exp(tX)\big)f - dX\,\rho(\exp(tX))\,Df\big)(g).\end{aligned}$$

Because D lies in the center of $U(\mathfrak{g})$, it commutes with dX, and we may write the right side of this as

$$(dX\, D\, \rho\big(\exp(tX)\big) f - dX\, \rho\big(\exp(tX)\big)\, Df)(g) = dX\, \phi(g, t).$$

This shows that Eq. (2.16) is satisfied. It follows that $\phi(g, t) = 0$ for all t, which implies that D commutes with $\rho(tX)$. As we have already noted, elements of this type generate G, from whence we get Proposition 2.2.4. ■

Now let us construct an element of the center of $U(\mathfrak{g})$ when $\mathfrak{g} = \mathfrak{gl}(2, \mathbb{R})$. We require the following elements of $\mathfrak{g}$:

$$\hat{R} = \begin{pmatrix} 0 & 1 \\ 0 & 0 \end{pmatrix}, \quad \hat{L} = \begin{pmatrix} 0 & 0 \\ 1 & 0 \end{pmatrix}, \quad \hat{H} = \begin{pmatrix} 1 & 0 \\ 0 & -1 \end{pmatrix}, \quad Z = \begin{pmatrix} 1 & 0 \\ 0 & 1 \end{pmatrix}. \tag{2.18}$$

We easily compute

$$[\hat{H}, \hat{R}] = 2\hat{R}, \qquad [\hat{H}, \hat{L}] = -2\hat{L}, \qquad [\hat{R}, \hat{L}] = \hat{H}. \tag{2.19}$$

Now let

$$-4\Delta = \hat{H}^2 + 2\hat{R}\hat{L} + 2\hat{L}\hat{R} \tag{2.20}$$

(Multiplication here is the multiplication in $U(\mathfrak{g})$, not matrix multiplication!)

As an element of $U(\mathfrak{g})$, Δ is called the *Casimir element*; if $\mathfrak{g}$ is the Lie algebra of an arbitrary semisimple or reductive group, a Casimir element may be defined. Reinterpreted as a differential operator, Δ is called the *Laplace–Beltrami operator*, as we noted in connection with Eq. (1.29), which we shall presently identify with Eq. (2.20).

Theorem 2.2.1 *The element Δ of $U(\mathfrak{g})$, where $\mathfrak{g} = \mathfrak{gl}(2, \mathbb{R})$, lies in the center of $U(\mathfrak{g})$.*

Proof From Eq. (2.19), we obtain the following identities in $U(\mathfrak{g})$.

$$\hat{H}^2\hat{R} = \hat{R}\hat{H}^2 + 2\hat{R}\hat{H} + 2\hat{H}\hat{R}, \qquad \hat{H}\hat{R}\hat{L} = \hat{R}\hat{L}\hat{H}, \qquad \hat{H}\hat{L}\hat{R} = \hat{L}\hat{R}\hat{H},$$
$$\hat{R}\hat{L}\hat{R} = \hat{R}^2\hat{L} - \hat{R}\hat{H}, \qquad \hat{L}\hat{R}^2 = \hat{R}\hat{L}\hat{R} - \hat{H}\hat{R},$$
$$\hat{R}\hat{L}^2 = \hat{L}\hat{R}\hat{L} + \hat{H}\hat{L}, \qquad \hat{L}\hat{R}\hat{L} = \hat{L}^2\hat{R} + \hat{L}\hat{H},$$
$$\hat{H}^2\hat{L} = \hat{L}\hat{H}^2 - 2\hat{L}\hat{H} - 2\hat{H}\hat{L}.$$

From these identities, one checks that Δ commutes with $\hat{H}$, $\hat{R}$, and $\hat{L}$. It also commutes with Z, because Z lies in the center of $\mathfrak{g}$; that is, $[Z, X] = 0$ for all $X \in \mathfrak{g}$. Now $\hat{R}$, $\hat{L}$, $\hat{H}$, and Z generate $\mathfrak{g}$ as a vector space, and hence generate $U(\mathfrak{g})$ as an $\mathbb{R}$-algebra, and it follows that Δ is in the center of $U(\mathfrak{g})$. ■

Combining Proposition 2.2.4 and Theorem 2.2.1, we have thus constructed a differential operator Δ, which is invariant under both left and right translations.

Our final task in this section will be to show that Δ agrees with the differential operator on G defined by Eq. (1.29) and to introduce certain other differential operators, living in the complexification of $U(\mathfrak{g})$, which include the weight raising and lowering operators of Section 1. We must therefore begin with some remarks about the complexification of $U(\mathfrak{g})$ and its interpretation as an algebra of differential operators.

If V is a real vector space, we define its *complexification* to be $V_\mathbb{C} = \mathbb{C} \otimes_\mathbb{R} V$. This is a complex vector space, whose multiplication law $\mathbb{C} \times V_\mathbb{C} \to V_\mathbb{C}$ satisfies

$$a(b \otimes v) = (ab) \otimes v, \qquad a, b \in \mathbb{C},\ v \in V.$$

The complex dimension of $V_\mathbb{C}$ is equal to the real dimension of V; indeed, if $x_1, \cdots, x_d$ is a basis of V, then $1 \otimes x_1, \cdots, 1 \otimes x_d$ is a basis of $V_\mathbb{C}$. We may identify V with its image in $V_\mathbb{C}$ under the map $v \mapsto 1 \otimes v$; then, because $\mathbb{C} = \mathbb{R} \oplus i\mathbb{R}$ as a vector space, we have

$$V_\mathbb{C} = V \oplus iV. \tag{2.21}$$

If $\mathfrak{g}$ is a real Lie algebra, its *complexification* $\mathfrak{g}_\mathbb{C}$ is a complex Lie algebra; the bracket operation is extended by the requirement that it be complex bilinear and agrees with the Lie bracket on $\mathfrak{g}$. It is simple to verify that with this extension of the operation, the identities Eqs. (2.1) and (2.2) remain valid.

If $\rho : \mathfrak{g} \to \mathrm{End}(V)$ is a representation of a real Lie algebra, where V is a complex vector space, then we may extend ρ to a representation $\mathfrak{g}_\mathbb{C} \to \mathrm{End}(V)$ as follows: If $X \in \mathfrak{g}_\mathbb{C}$, write $X = X_1 + iX_2$. Then let

$$\rho(X) = \rho(X_1) + i\rho(X_2). \tag{2.22}$$

In particular, let $\mathfrak{g} = \mathfrak{gl}(n,\mathbb{R})$. We may identify its complexification $\mathfrak{g}_\mathbb{C}$ with $\mathfrak{gl}(n,\mathbb{C})$ in the obvious way. This is the Lie algebra of the complex Lie group $GL(n,\mathbb{C})$, and the exponential map $\mathfrak{gl}(n,\mathbb{C}) \to GL(n,\mathbb{C})$ is defined by the formula (2.6).

Now let $G = GL(2,\mathbb{R})^+$. We extend the action of $\mathfrak{gl}(2,\mathbb{R})$ on $C^\infty(G)$ to an action of $\mathfrak{gl}(2,\mathbb{C})$ by Eq. (2.22). Let

$$R = \frac{1}{2}\begin{pmatrix} 1 & i \\ i & -1 \end{pmatrix}, \qquad L = \frac{1}{2}\begin{pmatrix} 1 & -i \\ -i & -1 \end{pmatrix}, \qquad H = -i\begin{pmatrix} 0 & 1 \\ -1 & 0 \end{pmatrix}. \tag{2.23}$$

These elements satisfy the same relations as $\hat{H}$, $\hat{R}$, and $\hat{L}$:

$$[H, R] = 2R, \qquad [H, L] = -2L, \qquad [R, L] = H. \tag{2.24}$$

(Compare Eq. (2.19).) The explanation for this is that these three elements are conjugate inside of $SL(2,\mathbb{C})$ to $\hat{H}$, $\hat{R}$, and $\hat{L}$. Indeed, let

$$C = -\tfrac{1+i}{2}\begin{pmatrix} i & 1 \\ i & -1 \end{pmatrix}, \qquad C^{-1} = \tfrac{1+i}{2}\begin{pmatrix} 1 & 1 \\ i & -i \end{pmatrix} \in SL(2,\mathbb{C}). \tag{2.25}$$

We call C the *Cayley transform*. If we allow $SL(2,\mathbb{C})$ to act on the Riemann sphere by linear fractional transformations, C takes the upper half plane into the unit disk, and was introduced in Section 1.2. In the case at hand, we have

$$CHC^{-1} = \hat{H}, \qquad CRC^{-1} = \hat{R}, \qquad CLC^{-1} = \hat{L}. \tag{2.26}$$

Moreover, if

$$\kappa_\theta = \begin{pmatrix} \cos(\theta) & \sin(\theta) \\ -\sin(\theta) & \cos(\theta) \end{pmatrix},$$

then

$$\kappa_\theta = C^{-1} \begin{pmatrix} e^{i\theta} & \\ & e^{-i\theta} \end{pmatrix} C. \tag{2.27}$$

Proposition 2.2.5 *We have, using the coordinates Eq. (1.19)*

$$d\hat{R} = y\cos(2\theta)\frac{\partial}{\partial x} + y\sin(2\theta)\frac{\partial}{\partial y} + \sin^2(\theta)\frac{\partial}{\partial \theta}, \tag{2.28}$$

$$d\hat{L} = y\cos(2\theta)\frac{\partial}{\partial x} + y\sin(2\theta)\frac{\partial}{\partial y} - \cos^2(\theta)\frac{\partial}{\partial \theta}, \tag{2.29}$$

$$d\hat{H} = -2y\sin(2\theta)\frac{\partial}{\partial x} + 2y\cos(2\theta)\frac{\partial}{\partial y} + \sin(2\theta)\frac{\partial}{\partial \theta}, \tag{2.30}$$

$$dR = e^{2i\theta}\left(iy\frac{\partial}{\partial x} + y\frac{\partial}{\partial y} + \frac{1}{2i}\frac{\partial}{\partial \theta}\right), \tag{2.31}$$

$$dL = e^{-2i\theta}\left(-iy\frac{\partial}{\partial x} + y\frac{\partial}{\partial y} - \frac{1}{2i}\frac{\partial}{\partial \theta}\right), \tag{2.32}$$

and

$$dH = -i\frac{\partial}{\partial \theta}. \tag{2.33}$$

Thus the operators R and L are the same two that were introduced in Section 1 (Eqs. 1.27–1.28). Moreover, the differential operator Δ agrees with the Laplace–Beltrami operator Eq. (1.29).

Proof Of course, Eqs. (2.28–2.30) may be proved by direct computation, after which Eq. (1.29) may be deduced after some additional computation. However, we observe that Eqs. (2.31–2.33) have a simpler form than Eqs. (2.28–2.30), although they are obviously equivalent to Eqs. (2.28–2.30) because $\hat{H}$, $\hat{R}$, and $\hat{L}$ are related to H, R, and L by a linear change of variables. This suggests that we should proceed more indirectly by establishing Eqs. (2.31–2.33) first. It is first convenient to observe that

$$\text{(2.28) and (2.30) are true if } \theta = 0. \tag{2.34}$$

Indeed, suppose that $\theta = 0$ in Eq. (1.19). We find that

$$g\exp(t\hat{R}) = \begin{pmatrix} u & \\ & u \end{pmatrix}\begin{pmatrix} y^{1/2} & xy^{-1/2} \\ & y^{-1/2} \end{pmatrix}\begin{pmatrix} 1 & t \\ & 1 \end{pmatrix}$$

$$= \begin{pmatrix} u & \\ & u \end{pmatrix}\begin{pmatrix} y^{1/2} & (x+yt)y^{-1/2} \\ & y^{-1/2} \end{pmatrix},$$

so

$$(d\hat{R}\, f)(g) = \frac{d}{dt} f\left(g\exp(t\hat{R})\right)|_{t=0} = y\,\frac{\partial f}{\partial x}(g),$$

from which we get Eq. (2.28), the proof of Eq. (2.30) when $\theta = 0$ is similar.

Next we prove Eq. (2.33). We write

$$H = -iW, \qquad W = \begin{pmatrix} 0 & 1 \\ -1 & 0 \end{pmatrix}. \tag{2.35}$$

We have

$$\exp(tW) = k_t = \begin{pmatrix} \cos(t) & \sin(t) \\ -\sin(t) & \cos(t) \end{pmatrix}, \tag{2.36}$$

which one proves by simply multiplying out the matrices in the Taylor expansion for $\exp(tW)$ as given by Eq. (2.6) and observing that the individual coefficients are given by the well-known Taylor series for $\cos(t)$ and $\sin(t)$.

We now have

$$(dW\, f)(g) = \frac{\partial}{\partial t} f(gk_t)|_{t=0}.$$

Using the coordinates Eq. (1.19), we note that by Eq. (2.36)

$$g\,\exp(tW) = \begin{pmatrix} u & \\ & u \end{pmatrix}\begin{pmatrix} y^{1/2} & xy^{-1/2} \\ & y^{-1/2} \end{pmatrix} k_{t+\theta},$$

so

$$(dW\, f)(g) = \frac{d}{dt} f\left(g\,\exp(tW)\right)|_{t=0} = \frac{\partial}{\partial\theta} f(g), \tag{2.37}$$

which implies Eq. (2.33).

Next we prove Eq. (2.31). Let $x, y, u, \theta \in \mathbb{R}$, $u, y > 0$, and let

$$g = b\kappa_\theta,\ b = \begin{pmatrix} u & \\ & u \end{pmatrix}\begin{pmatrix} y^{1/2} & xy^{-1/2} \\ & y^{-1/2} \end{pmatrix},\ \kappa_\theta = \begin{pmatrix} \cos(\theta) & \sin(\theta) \\ -\sin(\theta) & \cos(\theta) \end{pmatrix}.$$

We will compute $dR\, f$ at g, where f is an analytic function defined in a neighborhood of g in $GL(2,\mathbb{R})^+$; that is, f is given by a convergent power series with complex coefficients. It is sufficient to compute $dR\, f$ under this assumption because a differential operator is determined by its effect on analytic functions. But because f has a power series expansion, we may extend it to an analytic function on a neighborhood of g in $GL(2, \mathbb{C})$. This has two consequences. Firstly, the formula (2.10) remains valid for $X \in \mathfrak{g}_{\mathbb{C}}$ as well as for $X \in \mathfrak{g}$, and secondly, it is not necessary that the variable t in Eq. (2.10) be real; specifically, we may rewrite Eq. (2.10) as

$$(dX\, f)(g) = \lim_{t \to 0} \frac{1}{t}\big(f(g\, \exp(tX)) - f(g)\big),$$

and because $f\big(g\, \exp(tX)\big)$ is an analytic function of t, it is not necessary for t to be a real variable in computing this limit.

With this in mind, we may compute $(dR\, f)(g)$ as follows. We have

$$\kappa_\theta\, \exp(tR) = \exp(e^{2i\theta} t R)\, \kappa_\theta. \tag{2.38}$$

Indeed, making use of Eqs. (2.27) and (2.26), we have

$$\begin{aligned}
\kappa_\theta\, \exp(tR)\, \kappa_\theta^{-1} &= C^{-1} \begin{pmatrix} e^{i\theta} & \\ & e^{-i\theta} \end{pmatrix} \exp(t\hat{R}) \begin{pmatrix} e^{i\theta} & \\ & e^{-i\theta} \end{pmatrix}^{-1} C \\
&= C^{-1} \begin{pmatrix} e^{i\theta} & \\ & e^{-i\theta} \end{pmatrix} \begin{pmatrix} 1 & t \\ & 1 \end{pmatrix} \begin{pmatrix} e^{i\theta} & \\ & e^{-i\theta} \end{pmatrix}^{-1} C \\
&= C^{-1} \begin{pmatrix} 1 & e^{2i\theta} t \\ & 1 \end{pmatrix} C = C^{-1}\, \exp(e^{2i\theta} t\hat{R})\, C = \exp(e^{2i\theta} t R).
\end{aligned}$$

Now let $\big(\rho(\kappa_\theta) f\big)(h) = f(h\kappa_\theta)$. This is an analytic function defined in a neighborhood of $h = b$. By Eq. (2.38), we have

$$\begin{aligned}
(dR\, f)(g) &= \frac{d}{dt} f\big(b\, \kappa_\theta\, \exp(tR)\big)|_{t=0} \\
&= \frac{d}{dt} f\big(b\, \exp(e^{2i\theta} t R)\, \kappa_\theta\big)|_{t=0} \\
&= e^{2i\theta} \frac{d}{dt} f\big(b\, \exp(tR)\, \kappa_\theta\big)|_{t=0} \\
&= e^{2i\theta} \frac{d}{dt} \big(\rho(\kappa_\theta) f\big)\big(b\, \exp(tR)\big)|_{t=0} \\
&= e^{2i\theta} \big(dR\, \rho(\kappa_\theta) f\big)(b).
\end{aligned}$$

Now because $R = \frac{1}{2}\hat{H} + \frac{1}{2}H + i\hat{R}$, this may be evaluated using Eqs. (2.34) and (2.33), which gives us Eq. (2.31). The proof of Eq. (2.32) is similar to the proof of Eq. (2.31), and as we have noted, Eqs. (2.28–2.30) follow from Eqs. (2.31–2.33).

Finally, we also have

$$-4\Delta = H^2 + 2RL + 2LR. \tag{2.39}$$

This may be deduced from Eq. (2.20) by direct computation in $U(\mathfrak{g}_\mathbb{C})$ or (better) by Exercise 2.2.5. The formula (1.29) may be deduced from either Eq. (2.39) or Eq. (2.20) by Eqs. (2.28–2.30) (Exercise 2.2.3). ■

We next discuss the Lie algebra of a Lie subgroup H of $GL(n,\mathbb{R})$. We recall that this means that H is a closed subgroup of G and also a differentiable submanifold.

Lemma 2.2.3 *Let H be a Lie subgroup of $G = GL(n,\mathbb{R})$. Let $g \in G - H$. Then there exists a function $\Phi \in C^\infty(G)$ such that $\Phi(xh) = \Phi(x)$ for $x \in G$, $h \in H$, and $\Phi|_H = 0$, yet $\Phi(g) \neq 0$.*

Proof First we find a nonnegative smooth function ϕ supported in a neighborhood of g such that $\phi(g) \neq 0$; if the neighborhood is small enough, it will be disjoint from H. Then we define $\Phi(x) = \int_H \phi(xh)\,dh$, where dh is left Haar measure on H. This function obviously has the required properties. ■

Proposition 2.2.6 *Let H be a Lie subgroup of $G = GL(n,\mathbb{R})^+$. Let $X \in \mathfrak{g} = \mathfrak{gl}(n,\mathbb{R})$. The following two conditions are equivalent:*

(i) The curve $t \mapsto \exp(tX)$ is tangent to H at $t = 0$.
(ii) We have $\exp(tX) \in H$ for all $t \in \mathbb{R}$.

Proof Obviously (ii) implies (i). We assume (i) and prove (ii). Suppose that $\exp(tX) \notin H$ for some t. Then by Lemma 2.2.3, we can find a smooth function Φ, constant on the cosets gH of H, such that if $\phi : \mathbb{R} \to \mathbb{C}$ is defined by $\phi(t) = \Phi\big(\exp(tX)\big)$, then $\phi(0) = 0$, but $\phi(t_0) \neq 0$ for some t_0. We will obtain a contradiction by showing that $d\phi/dt = 0$ for all t.

Indeed, let $u \in \mathbb{R}$, and let $g = \exp(uX)$. We observe that left translation by g is a diffeomorphism of G onto itself, taking the curve $t \mapsto \exp(tX)$ at $t = 0$ onto the curve $t \mapsto \exp\big((t+u)X\big)$, and taking H onto gH. Our hypothesis then shows that $t \mapsto \exp(tX)$ at $t = u$ is tangent to the coset gH. Because Φ is constant on this coset, $d\phi/dt = d\Phi\big(\exp(tX)\big)/dt$ is zero at $t = u$. This completes the proof that (i) implies (ii). ■

Theorem 2.2.2 *Let H be a Lie subgroup of $G = GL(n,\mathbb{R})$. Let $\mathfrak{H}$ be the set of all vectors $X \in \mathfrak{g} = \mathfrak{gl}(n,\mathbb{R})$ that satisfy the equivalent conditions of Proposition 2.2.6. Then $\mathfrak{H}$ is a Lie subalgebra of $\mathfrak{g}$. Its dimension equals that of H.*

The Lie algebra $\mathfrak{H}$ (sometimes denoted $\mathrm{Lie}(H)$) is called the *Lie algebra of H*.

Proof It is clear from condition (i) in Proposition 2.2.6 that $\mathfrak{H}$ is a vector space. We must show that if $X, Y \in \mathfrak{H}$, then $[X,Y] \in \mathfrak{H}$. First we observe that if $h \in H$, then $hYh^{-1} \in \mathfrak{H}$. (Here the product hYh^{-1} is obtained by ordinary matrix multiplication.) Indeed, we have $\exp(thYh^{-1}) = h\exp(tY)h^{-1} \in H$ for all t, so this follows by condition (ii) of Proposition 2.2.6.

Thus if $0 \neq t \in \mathbb{R}$, let $h = \exp(tX)$, and consider

$$t^{-1}(hYh^{-1} - Y) = XY - YX + \tfrac{t}{2}(X^2Y - 2XYX + YX^2) + \dots .$$

This is in $\mathfrak{H}$, and so letting $t \to 0$ to eliminate the higher-order terms, we obtain $XY - YX \in \mathfrak{H}$. This proves that $\mathfrak{H}$ is closed under commutation, and hence is a Lie subalgebra of $\mathfrak{g}$. Finally, it is clear from condition (i) of Proposition 2.2.6 that $\mathfrak{H}$ may be identified with the tangent space at the identity to H, so its dimension equals that of H. ■

The final topic of this section is the commutativity of certain convolution rings. These results will be needed in both Sections 2.3 and 2.4, and are included here so as not to interrupt the discussion later.

If ϕ_1 and $\phi_2 \in C_c^\infty(G)$, where as before $G = GL(n, \mathbb{R})$ or $GL(n, \mathbb{R})^+$, the convolution $\phi_1 * \phi_2 \in C_c^\infty(G)$ is defined by

$$(\phi_1 * \phi_2)(g) = \int_G \phi_1(gh)\, \phi_2(h^{-1})\, dh = \int_G \phi_1(h)\, \phi_2(h^{-1}g)\, dh. \tag{2.40}$$

The equality of these two expressions follows from the change of variables $h \to g^{-1}h$. This makes $C_c^\infty(G)$ a ring (without unit). We have already encountered convolutions in the proof of Proposition 2.1.6.

Proposition 2.2.7: The Cartan decomposition *(i) Let $\langle\, ,\rangle$ be a positive definite symmetric inner product on a finite-dimensional real vector space V, and let $T : V \to V$ be a linear transformation that satisfies $\langle Tx, y\rangle = \langle x, Ty\rangle$ for all $x, y \in V$. Then V has an orthonormal basis with respect to $\langle\, ,\rangle$ consisting of eigenvectors of T.*

(ii) Let S be a real symmetric matrix. Then there exists $\kappa \in SO(n)$ such that $\kappa^{-1}S\kappa$ is diagonal with decreasing entries.

(iii) Let $G = GL(n, \mathbb{R})$, and let $K = O(n)$, or else let $G = GL(n, \mathbb{R})^+$, and let $K = SO(n)$. In either case, every double coset in $K\backslash G/K$ has a unique representative of the form

$$\begin{pmatrix} d_1 & & \\ & \ddots & \\ & & d_n \end{pmatrix}, \qquad d_i \in \mathbb{R}, \qquad d_1 \geq d_2 \geq \dots \geq d_n > 0. \tag{2.41}$$

(iv) Let $\mathfrak{g} = \mathfrak{gl}(n, \mathbb{R})$, and let $\mathfrak{p}$ be the vector subspace of $\mathfrak{g}$ consisting of symmetric matrices. Let Σ be the set of positive definite symmetric matrices in $GL(n, \mathbb{R})$. Then $\exp : \mathfrak{p} \to \Sigma$ is a bijection.

(v) With G, K, and $\mathfrak{p}$ as in (ii) and (iii), every element of G can be written uniquely as $\kappa \exp(X)$ with $\kappa \in K$ and $X \in \mathfrak{p}$.

Either of the related decompositions in (iii) and (v) has been referred to as the *Cartan decomposition*. If $\mathfrak{k}$ is the Lie algebra of K, then $\mathfrak{k}$ consists of the skew-symmetric matrices in $\mathfrak{g}$ (Exercise 2.4.2 (a)), and $\mathfrak{p}$ consists of the symmetric matrices, so $\mathfrak{g} = \mathfrak{k} \oplus \mathfrak{p}$; this is also referred to as the *Cartan decomposition* of the Lie algebra $\mathfrak{g}$.

Proof Part (i), commonly known as the *principal axis theorem*, is well known. See Lang (1993, Theorem XV.7.1, p. 585). (Compare also Exercise 1.4.1.)

We now prove (ii). We give $\mathbb{R}^n$ the usual inner product $\langle x, y\rangle = \sum x_i y_i$. Let S be a real symmetric matrix. If T is the corresponding linear transformation of $\mathbb{R}^n$ (so if $x \in \mathbb{R}^n$ is regarded as a column vector, $Tx = Sx$), then the transformation T is symmetric, so by (i), $\mathbb{R}^n$ has an orthonormal basis $v_1, \cdots, v_n$ consisting of eigenvectors for T. We order them so that the eigenvalues λ_i are decreasing. Let κ be the orthogonal matrix with these columns v_i. We wish κ to have positive determinant: If its determinant is -1, we remedy this by replacing v_1 with $-v_1$. The identities $Sv_i = \lambda_i v_i$ imply that

$$S\kappa = \kappa \begin{pmatrix} \lambda_1 & & \\ & \ddots & \\ & & \lambda_n \end{pmatrix},$$

from whence we get (ii).

For part (iii), let $g \in GL(n, \mathbb{R})$. Consider $S = {}^\top g\, g$. This is a definite symmetric matrix, so by (ii), we may write

$$S = \kappa_1^{-1} \begin{pmatrix} \lambda_1 & & \\ & \ddots & \\ & & \lambda_n \end{pmatrix} \kappa_1,$$

where $\lambda_1 \geq \ldots \geq \lambda_n$ and $\kappa_1 \in SO(n)$. Because S is positive definite, the eigenvalues λ_i are all positive. Let $d_i = \sqrt{\lambda_i}$. Let

$$\kappa_2 = g\kappa_1^{-1} \begin{pmatrix} d_1^{-1} & & \\ & \ddots & \\ & & d_n^{-1} \end{pmatrix}.$$

We see that ${}^\top\kappa_2\kappa_2 = 1$, so κ_2 is orthogonal. Thus g and the matrix (Eq. 2.41) lie in the same $O(n)$ double coset. Furthermore, if $g \in GL(n, \mathbb{R})^+$, then because g, κ_1, and the matrix (Eq. 2.41) all have positive determinant, we also have $\kappa_2 \in SO(n)$. This completes the proof of (iii).

As for part (iv), we may write any positive definite symmetric matrix S as $\kappa d \kappa^{-1}$ where $\kappa \in SO(n)$ and d is diagonal with positive entries; if D is the diagonal matrix whose entries are the logarithms of those of d, then

$S = \kappa\, e^D\, \kappa^{-1} = \exp(\kappa D \kappa^{-1})$. This shows that $\exp : \mathfrak{p} \to \Sigma$ is surjective. We must show it is injective.

Let X and $X' \in \mathfrak{p}$ such that $\exp(X) = \exp(X')$. Let $d_1, \cdots, d_r$ be the distinct eigenvalues of X, arranged so $d_1 > \ldots > d_r$, and let n_i be the multiplicity of d_i. By (ii), there exists a matrix κ in K such that

$$X = \kappa \begin{pmatrix} d_1 I_{n_1} & & \\ & \ddots & \\ & & d_r I_{n_r} \end{pmatrix} \kappa^{-1}.$$

Now

$$e^X = \kappa \begin{pmatrix} D_1 I_{n_1} & & \\ & \ddots & \\ & & D_r I_{n_r} \end{pmatrix} \kappa^{-1},$$

where $D_i = e^{d_i}$. Because the exponential map $\mathbb{R} \to \mathbb{R}^+$ is injective, the D_i are all distinct. The injectivity of $\exp : \mathbb{R} \to \mathbb{R}^+$ also implies that the eigenvalues of e^X determine the eigenvalues of X, and because X and X' have the same exponential, the eigenvalues of X' are the same as the eigenvalues of X. We therefore have

$$X' = \kappa' \begin{pmatrix} d_1 I_{n_1} & & \\ & \ddots & \\ & & d_r I_{n_r} \end{pmatrix} \kappa'^{-1},$$

for some $\kappa' \in K$. Let $\tau = \kappa^{-1}\kappa'$. We then have

$$\begin{pmatrix} D_1 I_{n_1} & & \\ & \ddots & \\ & & D_r I_{n_r} \end{pmatrix} = \tau \begin{pmatrix} D_1 I_{n_1} & & \\ & \ddots & \\ & & D_r I_{n_r} \end{pmatrix} \tau^{-1}.$$

Consequently

$$\begin{pmatrix} D_1^k I_{n_1} & & \\ & \ddots & \\ & & D_r^k I_{n_r} \end{pmatrix} = \tau \begin{pmatrix} D_1^k I_{n_1} & & \\ & \ddots & \\ & & D_r^k I_{n_r} \end{pmatrix} \tau^{-1}$$

for $k = 0, 1, \cdots, r-1$. Now because the D_i are distinct, the Vandermonde determinant

$$\det \begin{pmatrix} D_1^{r-1} & \cdots & D_r^{r-1} \\ \vdots & & \vdots \\ 1 & \cdots & 1 \end{pmatrix} \neq 0,$$

and so there exist constants $c_0, \cdots, c_{r-1}$ such that $\sum_i c_i D_j^i = d_j$. Thus

$$\begin{pmatrix} d_1 I_{n_1} & & \\ & \ddots & \\ & & d_r I_{n_r} \end{pmatrix} = \sum c_i \begin{pmatrix} D_1^i I_{n_1} & & \\ & \ddots & \\ & & D_r^i I_{n_r} \end{pmatrix}$$

$$= \sum c_i \tau \begin{pmatrix} D_1^i I_{n_1} & & \\ & \ddots & \\ & & D_r^i I_{n_r} \end{pmatrix} \tau^{-1}$$

$$= \tau \begin{pmatrix} d_1 I_{n_1} & & \\ & \ddots & \\ & & d_r I_{n_r} \end{pmatrix} \tau^{-1}.$$

This implies that $X = X'$, completing the proof of (iv).

Finally, we prove part (v). Let $g \in G$. We write $g = \kappa_1 d \kappa_2$ as in (iii), with $\kappa_1, \kappa_2 \in K$ and d diagonal, with positive determinant; then $g = \kappa S$, where $S = \kappa_2^{-1} d \kappa_2$ is symmetric, and $\kappa = \kappa_1 \kappa_2 \in K$. By (iii), the matrix S may be written as $\exp(X)$ with $X \in \mathfrak{p}$. This proves that every matrix in G may be written as described in (v); we must show that the representation $\kappa \exp(X)$ is unique. If $g = \kappa' \exp(X')$ is another such representation, we have $\exp(2X) = {}^\top g g = \exp(2X')$, which by the injectivity assertion of (iv) implies that $2X = 2X'$, so $X = X'$ and $\kappa = \kappa'$. ■

The ring $C_c^\infty(G)$ is noncommutative. We will denote by $C_c^\infty(K\backslash G/K)$ the subring of functions that satisfy $\phi(\kappa_1 g \kappa_2) = \phi(g)$ for $\kappa_1, \kappa_2 \in K$. This algebra is very analogous to the Hecke algebra that was introduced in Section 1.4, and the commutativity of $C_c^\infty(K\backslash G/K)$ is analogous to the commutativity property Theorem 1.4.2. This analogy becomes clear in view of the fact that when automorphic forms are transferred to the adele group, the classical Hecke algebra of Section 1.4 is replaced by the spherical Hecke algebras $\mathcal{H}_K$ of Section 4.6, and the commutativity result there (Theorem 4.6.1) is clearly an exact analog of Theorem 2.2.3 below.

Theorem 2.2.3: Gelfand *Let $G = GL(n, \mathbb{R})$, $K = O(n)$, or else let $G = GL(n, \mathbb{R})^+$, $K = SO(n)$. Then $C_c^\infty(K\backslash G/K)$ is commutative.*

The idea of the proof is similar to that of Theorem 1.4.2: We show that transposition induces an antiinvolution of the ring, then observe that because $K\backslash G/K$ has representatives that are symmetric, this antiinvolution is actually the identity.

Proof We define a map $\phi \mapsto \hat{\phi}$ of $C_c^\infty(K\backslash G/K)$ to itself by

$$\hat{\phi}(g) = \phi({}^\top g).$$

We check that $\phi \mapsto \hat{\phi}$ is an antiinvolution of $C^\infty(K\backslash G/K)$. Indeed, we find that

$$\begin{aligned}
(\widehat{\phi_1 * \phi_2})(g) &= \int_G \phi_1({}^\top g h)\, \phi_2(h^{-1})\, dh \\
&= \int_G \hat{\phi}_2({}^\top h^{-1})\, \hat{\phi}_1({}^\top h g)\, dh \\
&= \int_G \hat{\phi}_2(h)\, \hat{\phi}_1(h^{-1} g)\, dh = (\hat{\phi}_2 * \hat{\phi}_1)(g),
\end{aligned}$$

as required.

Finally, we show that $\phi = \hat{\phi}$. Indeed, this is a consequence of Proposition 2.2.7(iii), because if $g = \kappa_1 d \kappa_2$ with $\kappa_1, \kappa_2 \in K$ and d diagonal, we have $\phi(g) = \phi(d) = \hat{\phi}(d) = \hat{\phi}(g)$. Because the identity map is thus an antiinvolution of $C_c^\infty(K\backslash G/K)$, this ring is commutative. ■

More generally, suppose that σ is a character of K, that is, a homomorphism $K \to \mathbb{C}^\times$. We define $C_c^\infty(K\backslash G/K, \sigma)$ to be the convolution subring of $C_c^\infty(G)$ consisting of (smooth, compactly supported) functions ϕ that satisfy $\phi(\kappa_1 g \kappa_2) = \sigma(\kappa_1)\, \phi(g)\, \sigma(\kappa_1)$ for $\kappa_1, \kappa_2 \in K$. This example is of interest when $G = GL(2, \mathbb{R})^+$, when the Abelian group $K = SO(2)$ has character $\kappa_\theta \mapsto e^{ik\theta}$, and where κ_θ is as in Eq. (1.19). If $n > 2$, the group $SO(n)$ has no nontrivial characters.

Proposition 2.2.8 *Let $G = GL(2, \mathbb{R})^+$, and let $K = SO(2)$. Let σ be a character of K. Then the algebra $C_c^\infty(K\backslash G/K, \sigma)$ is commutative.*

This result, by contrast with Theorem 2.2.3, is special to $GL(2, \mathbb{R})^+$. An important difference between $GL(2, \mathbb{R})^+$ and $GL(n, \mathbb{R})^+$ is that the maximal compact subgroup $SO(2)$ is commutative. If $n > 2$, the group $SO(n)$ is noncommutative and has no nontrivial characters except the trivial representation.

Proof We have $\sigma(\kappa_\theta) = e^{ik\theta}$ for some $k \in \mathbb{Z}$, where κ_θ is defined in Eq. (1.19). The proof is identical to that of Theorem 2.2.3, but we define instead

$$\hat{\phi}(g) = \phi\left(\begin{pmatrix} -1 & \\ & 1 \end{pmatrix} {}^\top g \begin{pmatrix} -1 & \\ & 1 \end{pmatrix}\right).$$

Everything works exactly as before with this modification. ■

Exercises

Exercise 2.2.1: The adjoint representations of G and $\mathfrak{g}$ Let $G = GL(n, \mathbb{R})$, and let $\mathfrak{g}$ be its Lie algebra. There exists an n^2-dimensional representation of G on the vector space $\mathfrak{g}$, called the *adjoint representation*, defined by

$$\mathrm{Ad}(g)(X) = gXg^{-1},$$

where the multiplications on the right-hand side are ordinary matrix multiplication. As we have shown, every finite-dimensional representation π of G has a differential $d\pi$. The differential of Ad is a representation of $\mathfrak{g}$ denoted $\mathrm{ad} : \mathfrak{g} \to \mathrm{End}(\mathfrak{g})$ (lower case "ad"). Prove the formula

$$(\mathrm{ad}\ X)(Y) = [X, Y].$$

Exercise 2.2.2 Let $G = GL(2, \mathbb{R})$. For each of the following Lie subgroups H of G, prove that the Lie algebra $\mathfrak{H}$ is as described.

(a) $H = SL(2, \mathbb{R})$; $\mathfrak{H} = \{X \in \mathfrak{gl}(2, \mathbb{R}) \mid \mathrm{tr}(X) = 0\}$.

(b) $H = \left\{ \begin{pmatrix} y^{1/2} & xy^{-1/2} \\ & y^{-1/2} \end{pmatrix} \middle| x, y \in \mathbb{R}, y > 0 \right\}$; $\mathfrak{H}$ is the linear span of $\hat{H}$ and $\hat{R}$.

(c) $H = \left\{ \begin{pmatrix} 1 & x \\ & 1 \end{pmatrix} \middle| x, y \in \mathbb{R}, y > 0 \right\}$; $\mathfrak{H}$ is the linear span of $\hat{R}$.

(d) $H = \left\{ \begin{pmatrix} y^{1/2} & \\ & y^{-1/2} \end{pmatrix} \middle| x, y \in \mathbb{R}, y > 0 \right\}$; $\mathfrak{H}$ is the linear span of $\hat{H}$.

(e) $H = SO(2)$; $\mathfrak{H}$ is the linear span of W.

Exercise 2.2.3 Finish the proof of Proposition 2.2.5 by verifying Eq. (1.29).

Exercise 2.2.4 (a) (i) Prove that if $X, Y \in \mathfrak{gl}(n, \mathbb{R})$, and if $[X, Y] = 0$, then

$$\exp(X + Y) = \exp(X) \exp(Y).$$

(b) Prove that if $X \in \mathfrak{gl}(n, \mathbb{R})$, then $t \mapsto \exp(tX)$ is a homomorphism $\mathbb{R} \to GL(n, \mathbb{R})$. Such a homomorphism is called a *one-parameter subgroup*.
(c) Prove that any continuous homomorphsim $\mathbb{R} \mapsto GL(n, \mathbb{R})$ is of the form $t \mapsto \exp(tX)$ for some $X \in \mathfrak{gl}(n, \mathbb{R})$.

Exercise 2.2.5 The adjoint representation defined in Exercise 2.2.1 extends to a an action $\mathrm{Ad} : G \to U(\mathfrak{g})$. Prove that if D is in the center of $U(\mathfrak{g})$ and $g \in G$ then $\mathrm{Ad}(g)\, D = D$.

Exercise 2.2.6 (a) Generalize Exercise 2.2.2(e) by proving that the Lie algebra $\mathfrak{k}$ of $K = SO(n)$ consists of the skew-symmetric matrices in $\mathrm{Mat}(n, \mathbb{R})$.
(b) Prove that the exponential map $\exp : \mathfrak{k} \to K$ is surjective.

[HINT: Prove this first when $n = 2$. For general n, if $\kappa \in SO(n)$, prove that there exists $\tau \in SO(n)$ such that $\tau\kappa\tau^{-1}$ is of the form

$$\begin{pmatrix} \kappa_1 & & & \\ & \ddots & & \\ & & \kappa_r & \\ & & & I_{n-2r} \end{pmatrix},$$

where each $\kappa_i \in SO(2)$.]

2.3 Discreteness of the Spectrum

We will prove in this section that the spectrum of the Laplace–Beltrami operator Δ is discrete when $\Gamma\backslash\mathcal{H}$ is compact.

The strategy is to deduce the spectral theorem for Δ from the spectral theorem for compact operators. The operator Δ is far from compact; however, it is possible to introduce certain *integral* operators, which are self-adjoined and compact and which commute with Δ. In the context of the spectral theory of the Laplacian, these integral operators were introduced by Selberg (1956) who gave the all-important *Selberg trace formula,* expressing the traces of these integral operators as sums over the conjugacy classes of Γ and allowing quantitative results for the distribution of the spectrum of the Laplacian to be obtained. The translation of Selberg's ideas into the context of representation theory was achieved by Gelfand, Graev, and Piatetski-Shapiro (1968).

We prove first a few facts about compact operators. We will work in the context of a Hilbert space $\mathfrak{H}$, which is assumed to be *separable,* that is, to have a countable orthonormal basis. For example, L^2 of a Riemannian manifold M is separable, assuming the very mild condition that there exists a countable family of disjoint open subsets $M_i \subseteq M$ such that M_i is diffeomorphic to $\mathbb{R}^n$ and $M - \bigcup M_i$ has measure zero. This example includes $\Gamma\backslash\mathcal{H}$ or $\Gamma\backslash G$ for Γ, a discontinuous subgroup.

A linear operator $T : \mathfrak{H} \to \mathfrak{H}$ is called *bounded* if its domain is all of $\mathfrak{H}$ and if there exists a constant C such that $|Tx| \leq C|x|$ for all $x \in \mathfrak{H}$. In this case, the smallest such C is called the *operator norm* of T, and is denoted $|T|$. The boundedness of the operator T is equivalent to its continuity. By contrast with unbounded operators, there is no need to distinguish between symmetric and self-adjoint bounded operators. A bounded operator is *symmetric* or *self-adjoint* if Eq. (1.11) holds for all $f, g \in \mathfrak{H}$. As usual, we call f an *eigenvector* with *eigenvalue* λ if $f \neq 0$ and $Tf = \lambda f$. Given λ, the set of eigenvectors with eigenvalue λ is called the λ-*eigenspace*. It follows from elementary and usual arguments that if T is a self-adjoint bounded operator, then its eigenvalues are real and the eigenspaces corresponding to distinct eigenvalues are orthogonal. Moreover, if $V \subset \mathfrak{H}$ is a subspace such that $T(V) \subset V$, it is easy to see that also $T(V^\perp) \subset V^\perp$.

An operator $T : \mathfrak{H} \to \mathfrak{H}$ is *compact* or *completely continuous* if T maps bounded sets into compact sets. If $\mathfrak{H}$ is separable, then a subset of $\mathfrak{H}$ is compact if and only if it is sequentially compact, so T is compact if and only if for every sequence $x_n \subset \mathfrak{H}$ of unit vectors, there is a subsequence y_n such that $T(y_n)$ is convergent. A compact operator is automatically bounded and hence continuous.

Theorem 2.3.1: Spectral theorem for compact operators *Let T be a compact self-adjoint operator on a separable Hilbert space $\mathfrak{H}$. Then $\mathfrak{H}$ has an orthonormal basis $\phi_i (i = 1, 2, 3, \cdots)$ of eigenvectors of T, so that $T\phi_i = \lambda_i \phi_i$. The eigenvalues $\lambda_i \to 0$ as $i \to \infty$.*

Because the eigenvalues $\lambda_i \to 0$, if λ is any nonzero eigenvalue, it follows from this statement that the λ-eigenspace is finite dimensional.

Proof This depends upon the equality

$$|T| = \sup_{0 \neq x \in \mathfrak{H}} \frac{|\langle Tx, x\rangle|}{\langle x, x\rangle}. \tag{3.1}$$

To prove this, let B denote the right-hand side. If $0 \neq x \in \mathfrak{H}$,

$$|\langle Tx, x\rangle| \leq |Tx| \cdot |x| \leq |T| \cdot |x|^2 = |T| \cdot \langle x, x\rangle,$$

so $B \leq |T|$. We must prove the converse. Let $\lambda > 0$ be a constant, to be determined later. Making use of the identity that $\langle T^2 x, x\rangle = \langle Tx, Tx\rangle$, we have

$$\begin{aligned}
\langle Tx, Tx\rangle &= \tfrac{1}{4}|\langle T(\lambda x + \lambda^{-1}Tx), \lambda x + \lambda^{-1}Tx\rangle \\
&\quad - \langle T(\lambda x - \lambda^{-1}Tx), \lambda x - \lambda^{-1}Tx\rangle| \\
&\leq \tfrac{1}{4}|\langle T(\lambda x + \lambda^{-1}Tx), \lambda x + \lambda^{-1}Tx\rangle| \\
&\quad + |\langle T(\lambda x - \lambda^{-1}Tx), \lambda x - \lambda^{-1}Tx\rangle| \\
&\leq \tfrac{1}{4}[B\langle \lambda x + \lambda^{-1}Tx, \lambda x + \lambda^{-1}Tx\rangle \\
&\quad + B\langle \lambda x - \lambda^{-1}Tx, \lambda x - \lambda^{-1}Tx\rangle] \\
&= \frac{B}{2}[\lambda^2 \langle x, x\rangle + \lambda^{-2}\langle Tx, Tx\rangle].
\end{aligned}$$

Now taking $\lambda = \sqrt{|Tx|/|x|}$, we obtain

$$|Tx|^2 = \langle Tx, Tx\rangle \leq B|x||Tx|,$$

so $|Tx| \leq B|x|$, which implies that $|T| \leq B$, from which we get Eq. (3.1).

We now prove the first assertion, that $\mathfrak{H}$ has an orthonormal basis consisting of eigenvectors of T. Let Σ be the set of all orthonormal subsets of $\mathfrak{H}$ whose elements are eigenvectors of T. Ordering Σ by inclusion, Zorn's lemma implies that it has a maximal element S. We must show that S is a basis of $\mathfrak{H}$. Let V be the closure of the linear span of S. We must prove that $V = \mathfrak{H}$. Let $\mathfrak{H}_0 = V^\perp$; we wish to show $\mathfrak{H}_0 = 0$. Observe that $\mathfrak{H}_0$ is stable under T, and T induces a compact self-adjoint operator on $\mathfrak{H}_0$. Suppose that $\mathfrak{H}_0 \neq 0$; if we show that T has an eigenvector in $\mathfrak{H}_0$, this will contradict the maximality of Σ. It is therefore sufficient to show that *a compact self-adjoint operator on a nonzero Hilbert space has an eigenvector.*

We are therefore reduced to the easier problem of showing that T has an eigenvector. By Eq. (3.1), there is a sequence $x_1, x_2, x_3, \cdots$ of unit vectors such that $|\langle Tx_i, x_i\rangle| \to |T|$. Observe that if $x \in \mathfrak{H}$, we have

$$\langle Tx, x\rangle = \langle x, Tx\rangle = \overline{\langle Tx, Tx\rangle},$$

so the $\langle Tx_i, x_i\rangle$ are real; we may therefore replace the sequence by a subsequence such that $\langle Tx_i, x_i\rangle \to \lambda$, where $\lambda = \pm|T|$. If $\lambda = 0$, then T is the zero operator, which clearly has eigenvectors, so we may assume that $\lambda \neq 0$. Because T is compact, there exists a further subsequence x_i such that Tx_i converges to a vector v. We will show that $x_i \to \lambda^{-1}v$.

Observe first that

$$|\langle Tx_i, x_i\rangle| \leq |Tx_i||x_i| = |Tx_i| \leq |T||x_i| = |\lambda|,$$

and because $\langle Tx_i, x_i\rangle \to \lambda$, we have $|Tx_i| \to |\lambda|$. Now

$$|\lambda x_i - Tx_i|^2 = \langle \lambda x_i - Tx_i, \lambda x_i - Tx_i\rangle = \lambda^2|x_i|^2 + |Tx_i|^2 - 2\lambda\langle Tx_i, x_i\rangle,$$

and because $|x_i| = 1$, $|Tx_i| \to |\lambda|$, $\langle Tx_i, x_i\rangle \to \lambda$, this converges to 0. Because $Tx_i \to v$, the sequence λx_i therefore also converges to v, and $x_i \to \lambda^{-1}v$. Now by continuity, $Tx_i \to \lambda^{-1}Tv$, so $v = \lambda^{-1}Tv$. This proves that v is an eigenvector, with eigenvalue λ. This completes the proof that $\mathfrak{H}$ has an orthonormal basis consisting of eigenvectors.

Now let $\phi_1, \phi_2, \cdots$ be this orthonormal basis, which is countable because we are assuming $\mathfrak{H}$ is separable, and let λ_i be the corresponding eigenvalues. (We recall that $\mathfrak{H}$ is assumed to be separable.) We wish to show that the $\lambda_i \to 0$. If not, there exists a constant $\epsilon > 0$ and a subsequence ϕ_i' of the ϕ_i such that $|T\phi_i'| > \epsilon$. Because the $T\phi_i'$ are orthogonal, this means that the sequence $T\phi_i'$ can have no convergent subsequence, even though the ϕ_i' are bounded. This is a contradiction, because T is a compact operator. ∎

Lemma 2.3.1 *Let T be a bounded operator on a Hilbert space $\mathfrak{H}$. Assume that for every $\epsilon > 0$ there exists a compact operator T_ϵ such that $|T - T_\epsilon| < \epsilon$. Then T is compact.*

Proof Let x_n be a sequence of unit vectors in $\mathfrak{H}$. We will inductively construct sequences $\{x_{r,1}, x_{r,2}, x_{r,3}, \cdots\}$ for $r = 0, 1, 2, 3, \cdots$ such that (i) $x_{0,i} = x_i$; (ii) $\{x_{r+1,1}, x_{r+1,2}, x_{r+1,3}, \cdots\}$ is a subsequence of $\{x_{r,1}, x_{r,2}, x_{r,3}, \cdots\}$; and (iii) $|Tx_{r,i} - Tx_{r,j}| < \frac{1}{r}$ for $1 \leq i, j \in \mathbb{Z}$.

Once this is a accomplished, we are done, for we choose the sequence $y_i = x_{i,i}$ of x_i; that is a Cauchy sequence by construction, and hence is convergent, proving that T is a compact operator.

Assume the sequence $\{x_{r-1,1}, x_{r-1,2}, x_{r-1,3}, \cdots\}$ has been constructed. Let $T_{1/3r}$ be a compact operator such that $|T - T_{1/3r}| < \frac{1}{3r}$. Because $T_{1/3r}$ is compact, the sequence $T(x_{r-1,i})$ has a convergent subsequence; because this is a Cauchy sequence, we may find a subsequence $\{x_{r,i}\}$ such that $|T_{1/3r}(x_{r,i}) - T_{1/3r}(x_{r,j})| < \frac{1}{3r}$. Now

$$|T(x_{r,i}) - T(x_{r,j})| \leq |T(x_{r,i}) - T_{1/3r}(x_{r,i})| + |T_{1/3r}(x_{r,i}) - T_{1/3r}(x_{r,j})|$$
$$+ |T_{1/3r}(x_{r,j}) - T(x_{r,j})| \leq \frac{1}{3r} + \frac{1}{3r} + \frac{1}{3r} = \frac{1}{r}.$$

This completes the proof of the lemma. ■

The following result is extremely well known.

Theorem 2.3.2: Hilbert–Schmidt *Let X be a locally compact space endowed with a positive Borel measure, and assume that $\mathfrak{H} = L^2(X)$ is a separable Hilbert space. Let $K \in L^2(X \times X)$. Then the operator*

$$(Tf)(x) = \int_X K(x, y) f(y)\, dy \tag{3.2}$$

is a compact operator on $\mathfrak{H}$.

Such a square-integrable kernel K is called a *Hilbert–Schmidt kernel*, and such an integral operator is called a *Hilbert–Schmidt operator.*

Proof Let $\phi_1, \phi_2, \phi_3, \cdots$ be an orthonormal basis of $\mathfrak{H}$. We expand the kernel K in terms of the ϕ_i:

$$K(x, y) = \sum f_i(x)\, \overline{\phi_i(y)}, \quad f_i(x) = \int_X K(x, y)\, \phi_i(y)\, dy,$$

where the convergence of the sum to $K(x, y)$ is in the sense of convergence in norm in $L^2(X \times X)$. One calculates easily

$$\sum_i |f_i|^2 = \int_X K(x, y)\, \overline{K(x, y)}\, dx\, dy < \infty \tag{3.3}$$

because $K \in L^2(X \times X)$.

We will use Lemma 2.3.1. If $0 < N \in \mathbb{Z}$, let

$$K_N(x, y) = \sum_{i=1}^{n} f_i(x)\, \overline{\phi_i(y)},$$

and let T_N be the corresponding integral operator

$$(T_N f)(x) = \int_X K_N(x, y) f(y)\, dy.$$

We assert that T_N is compact. Indeed, T_N has *finite rank*: its image is the finite-dimensional vector space spanned by $f_1, \cdots, f_N$, so clearly the image of the unit ball will be a compact set.

By Lemma 2.3.1, because T_N is compact, it is sufficient to show that $|T - T_N| \to 0$ as $N \to \infty$. Suppose that $\phi \in \mathfrak{H}$, $|\phi| = 1$. Then we can write

$$\phi = \sum a_n \phi_n, \quad \sum |a_n|^2 = 1.$$

Now $(T - T_N)\phi = \sum_{n>N} a_n f_n$, and by the Cauchy–Schwarz inequality,

$$|(T - T_N)\phi| \leq \sum_{n>N} |a_n||f_n| \leq \left(\sum_{n>N} |a_n|^2\right)^{1/2} \left(\sum_{n>N} |f_n|^2\right)^{1/2}$$
$$\leq \left(\sum_{n>N} |f_n|^2\right)^{1/2},$$

so $|T - T_N| \leq (\sum_{n>N} |f_n|^2)^{1/2} \to 0$ as $N \to \infty$, by Eq. (3.3). ■

Now let $G = GL(2, \mathbb{R})^+$, and let Γ be a discontinuous subgroup of G. We assume as always that $-I \in \Gamma$, that $\Gamma \subset G_1 = SL(2, \mathbb{R})$, and that $\Gamma\backslash\mathcal{H}$ is compact, or equivalently, that $\Gamma\backslash G_1 \cong \Gamma\backslash G/Z^+$ is compact, where Z^+ is the group of matrices of the form $\begin{pmatrix} z & \\ & z \end{pmatrix}$ with $z > 0$. Let χ be a unitary character of Γ, and let $\mathfrak{H} = L^2(\Gamma\backslash G, \chi)$ as in Section 2.1. Let $\phi \in C_c^\infty(G)$. We define an operator $\rho(\phi) : \mathfrak{H} \to \mathfrak{H}$ by

$$\big(\rho(\phi) f\big)(g) = \int_G f(gh)\, \phi(h)\, dh. \tag{3.4}$$

Let us observe that this is L^2. Indeed, we may write

$$\rho(\phi) f = \int_G \phi(h) \rho(h) f\, dh,$$

where ρ is the regular representation (Eq. 1.22). Then

$$\langle \rho(\phi)f, \rho(\phi)f\rangle = \int_G \int_G \phi(g)\overline{\phi(h)}\,\langle \rho(g)f, \rho(h)f\rangle \, dg\, dh.$$

But because ρ is unitary, $|\rho(g)f| = |\rho(h)f| = f$, and so by the Cauchy–Schwarz inequality, $|\langle \rho(g)f, \rho(h)f\rangle| \le \langle f, f\rangle$, and therefore

$$\langle \rho(\phi)f, \rho(\phi)f\rangle \le \left\{ \int_G |\phi(g)| dg \right\}^2 \langle f, f\rangle < \infty,$$

so indeed $\rho(\phi)f$ is L^2.

The unitaricity of ρ was not used in any essential way in this proof: The same argument would work for an arbitrary representation π of G on a Hilbert space because the norms of the operators $\pi(g)$ are bounded on the compact support of ϕ. Thus let (π, V) be any representation of G on the Hilbert space. If $\phi \in C_c^\infty(G)$, define $\pi(\phi) \in \mathrm{End}(V)$ by

$$\pi(\phi)v = \int_G \phi(g)\,\pi(g)v\,dg. \tag{3.5}$$

For example, if $\pi = \rho$ is the regular representation of G on $L^2(\Gamma\backslash G, \chi)$, then $\pi(\phi) = \rho(\phi)$ as defined by Eq. (3.4). The continuity of the representation is sufficient to guarantee that this integral converges, because the map $g \to \phi(g)\pi(g)v$ is continuous on the compact support K of ϕ. Hence it is Borel integrable, and Eq. (3.5) is defined. Here $\pi(\phi)$ is a bounded operator because if B is an upper bounded for the norms of the $\pi(g)$ on the support of ϕ, we have

$$|\pi(\phi)v| \le B \int_G |\phi(g)|\, dg\, |v|.$$

If $\pi = \rho$, then $\pi(\phi)$ is what we previously denoted $\rho(\phi)$.

Let Z^+ be the subgroup of scalar matrices in G with positive diagonal coefficient, and let $G_1 = SL(2,\mathbb{R})$. Let $K = SO(2)$. If $\theta \in \mathbb{R}$, let κ_θ be as in Eq. (1.19).

Proposition 2.3.1 *Let $\phi \in C_c^\infty(G)$.*

(i) The operator $\rho(\phi)$ is a Hilbert–Schmidt operator. In particular, the operator is compact. If $f \in L^2(\Gamma\backslash G, \chi)$ then $\rho(\phi)f \in C^\infty(\Gamma\backslash G, \chi)$.

(ii) If $\phi(g) = \overline{\phi(g^{-1})}$, then the operator $\rho(\phi)$ is self-adjoint. More generally, if $\pi : G \to \mathrm{End}(H)$ is a unitary representation of G on a Hilbert space H, and if $\phi(g) = \overline{\phi(g^{-1})}$, then $\pi(\phi)$ is self-adjoint.

(iii) If $\phi(\kappa_\theta g) = e^{-ik\theta}\phi(g)$, then $\rho(\phi)$ maps the Hilbert space $L^2(\Gamma\backslash G, \chi)$ into $C^\infty(\Gamma\backslash G, \chi, k)$.

Proof First making the substitution $h \mapsto g^{-1}h$ in Eq. (3.4), then replacing $h \in G$ by $\gamma h u$ with $\gamma \in \Gamma$, $h \in \Gamma\backslash G/Z^+$, and $u \in Z^+$, we have

$$(\rho(\phi)f)(g) = \int_G f(h)\,\phi(g^{-1}h)\,dh$$

$$= \int_{\Gamma\backslash G/Z^+} \int_{Z^+} \sum_{\gamma\in\Gamma} f(\gamma h)\,\phi(g^{-1}\gamma h u)\,du\,dh,$$

and because $f(\gamma h) = \chi(\gamma) f(h)$, this equals

$$\big(\rho(\phi)f\big)(g) = \int_{\mathcal{F}} f(h) K(g,h)\,dh, \tag{3.6}$$

where $\mathcal{F}$ is the closure of a fundamental domain for $\Gamma\backslash G/Z^+$ and

$$K(g,h) = \int_{Z^+} \sum_{\gamma\in\Gamma} \chi(\gamma)\,\phi(g^{-1}\gamma h u)\,du, \quad g, h \in G. \tag{3.7}$$

Because ϕ is smooth and compactly supported, it is not hard to see that this is a smooth function of g and h. In particular, it is a square integrable function on the compact set $\mathcal{F}\times\mathcal{F}$, hence $\rho(\phi)$ is Hilbert–Schmidt operator by Theorem 2.3.2. If $f \in L^2(\Gamma\backslash G, \chi)$, then $\int_{\Gamma\backslash G/Z^+} |f(g)|\,dg \le \|1\|_2 \cdot \|f\|_2 < \infty$ by the Cauchy–Schwarz inequality, so the integral on the right side of Eq. (3.6) converges uniformly in g; because $K(g,h)$ is smooth as a function of g, it follows that $\rho(\phi)f$ is smooth. This proves (i).

As for (ii), it is easy to see that $\phi(g) = \overline{\phi(g^{-1})}$ implies that

$$K(g,h) = \overline{K(h,g)},$$

which in turn implies that $\rho(\phi)$ is self-adjoint. Or, more generally, if π is a unitary representation, we have

$$\langle \pi(\phi)v, w\rangle = \int_G \phi(g)\,\langle \pi(g)v, w\rangle\,dg = \int_G \phi(g)\,\langle v, \pi(g^{-1})\rangle\,dg,$$

and making the change of variables $g \to g^{-1}$, this equals $\langle v, \pi(\phi)w\rangle$.

Finally, for (iii), we have $\big(\rho(\phi)f\big)(g\kappa_\theta) = \int_G f(g\kappa_\theta h)\,\phi(h)\,dh$. Making the change of variables $h \mapsto \kappa_\theta^{-1}h$ and using the assumption that $\phi(\kappa_\theta g) = e^{-ik\theta}\phi(g)$, this equals $e^{ik\theta}\big(\rho(\phi)f\big)(g)$, so $\rho(\phi)f \in C^\infty(\Gamma\backslash G, \chi, k)$. ■

Lemma 2.3.2 *Let $\pi : G \to \mathrm{End}(H)$ be a unitary representation of G on a Hilbert space H, and let $0 \ne f \in H$. Let $\epsilon > 0$ be given. Then there exists $\phi \in C_c^\infty(G)$ such that $\pi(\phi)$ (defined by Eq. (3.5)) is self-adjoint and $|\pi(\phi)f - f| < \epsilon$. In particular, if $\epsilon < |f|$, this implies that $\pi(\phi)f \ne 0$. Moreover, if $\pi(\kappa_\theta)f = e^{ik\theta}f$ for all $\kappa_\theta \in K = SO(2)$, then we may choose*

ϕ so that $\phi(\kappa_\theta g) = \theta(g\kappa_\theta) = e^{ik\theta}\phi(g)$. Furthermore, if the space $H(k)$ of $v \in H$ such that $\pi(\kappa_\theta)v = e^{ik\theta}v$ is finite dimensional, then we find ϕ such that $\pi(\phi)f = f$.

Proof Because the map $g \mapsto \pi(g)f$ is continuous from $G \to H$, we may find a neighborhood U of the identity in G such that $|\pi(g)f - f| < \epsilon$ for all $g \in U$. Let ϕ_0 be a positive real-valued function with compact support contained in U such that $\int_G \phi_0(g)dg = 1$. Then we have

$$|\pi(\phi_0)f - f| = \left|\int_G \phi_0(g)\big(\pi(g)f - f\big)dg\right| \le \int_G \phi_0(g)|\pi(g)f - f|dg < \epsilon.$$

If ϵ is less than the length of f, this implies that $\pi(\phi_0)f \neq 0$. We may also take $\phi_0(g) = \phi_0(g^{-1})$, which by Proposition 2.3.1(ii) implies that $\pi(\phi_0)$ is self-adjoint. For the first assertion, this is all we need, so we may take $\phi = \phi_0$.

For the second assertion, we must choose ϕ more carefully. We show first that we may assume that $\phi_0(\kappa g\kappa^{-1}) = \phi_0(g)$ for all $\kappa \in K$. To this end, we show that (with U as before) there exists a neighborhood V of the identity in G such that $\kappa V\kappa^{-1} \subset U$ for all $\kappa \in K$; indeed, the map $\sigma : G \times K \to G$ defined by $(g,\kappa) \mapsto \kappa g\kappa^{-1}$ is continuous, so $\sigma^{-1}(U)$ is open in $G \times K$. Clearly, $(1,\kappa) \in \sigma^{-1}(U)$ for all κ, and hence there exists an open neighborhood of this point of the form $V_\kappa \times W_\kappa$ contained in $\sigma^{-1}(U)$, where $V_\kappa \subset G$ and $W_\kappa \subset K$ are open. The open sets W_κ form an open cover of the compact set K, and hence admit a finite subcover $W_{\kappa_1}, \cdots, W_{\kappa_r}$. Let $V = V_{\kappa_1} \cap \cdots \cap V_{\kappa_r}$. Then V has the advertised property that $\kappa V\kappa^{-1} \subset U$ for all $\kappa \in K$. Now let ϕ_1 be a positive-valued function with support contained in V such that $\phi_1(g) = \phi_1(g^{-1})$, and let

$$\phi_0(g) = \int_K \phi_1(\kappa g\kappa^{-1})\,d\kappa.$$

Then ϕ_0 is a positive function with support contained in U that satisfies $\phi_0(g) = \phi_0(g^{-1})$, and furthermore has the property that $\phi_0(\kappa g\kappa^{-1}) = \phi_0(g)$ for all $\kappa \in K$.

Assume now that $\pi(\kappa_\theta)f = e^{ik\theta}f$. We use Proposition 2.1.5(ii) with $P = G$ and $K = SO(2)$ to write

$$\pi(\phi_0)f = \int_G \phi_0(h)\,\pi(h)f\,dh = \int_G\int_K \phi_0(h\kappa)\,\pi(h\kappa)f\,d\kappa\,dh$$

$$= \int_G \frac{1}{2\pi}\int_0^{2\pi} e^{ik\theta}\phi_0(h\kappa_\theta)\,d\theta\,\pi(h)f\,dh = \pi(\phi)f,$$

where

$$\phi(g) = \frac{1}{2\pi}\int_0^{2\pi} e^{ik\theta}\phi_0(g\kappa_\theta)\,d\theta = \frac{1}{2\pi}\int_0^{2\pi} e^{ik\theta}\phi_0(\kappa_\theta g)\,d\theta.$$

This ϕ satisfies $\phi(\kappa_\theta g) = \phi(g\kappa_\theta) = e^{-ik\theta}\phi(g)$ as well as $\phi(g) = \phi(g^{-1}) = \overline{\phi(g^{-1})}$, as required.

Finally, assume that $H(k)$ is finite dimensional. It is easy to see that the vectors of the fom $\pi(\phi)f$, where $\phi \in C_c^\infty(G)$ satisfies $\phi(\kappa_\theta g) = \phi(g\kappa_\theta) = e^{-ik\theta}\phi(g)$ and $\phi(g) = \overline{\phi(g^{-1})}$, are all in $H(k)$. These clearly form a real vector subspace of $H(k)$. We have proven that f lies in the closure of this space, and because a subspace of a finite-dimensional vector space is necessarily closed, f is actually in this space. This proves the final assertion. ■

A representation (π, H) of a group G on a Hilbert space H is called *irreducible* if H has no proper nonzero invariant closed subspaces. If π is unitary and is reducible, that is, if there is a proper nonzero invariant closed subspace $0 \subsetneq V \subsetneq H$, then the orthogonal complement U of V is also a nonzero invariant closed subspace, and $H = U \oplus V$. On the other hand, if π is not unitary, it is possible for there to be a closed invariant subspace V that is not complemented, that is, for which no invariant subspace U exists with $H = U \oplus V$. We will see examples of this in Theorem 2.5.2(ii) and 2.5.2(iii).

Proposition 2.3.2 *Let H be a nonzero closed subspace of $L^2(\Gamma\backslash G, \chi)$, which is closed under the action of G. Then H has a decomposition as a Hilbert space direct sum $\oplus_{k\in\mathbb{Z}} H_k$, where $\rho(\kappa_\theta)f = e^{2\pi ik\theta}f$ for $f \in H_k$. Let k be such that $H_k \neq 0$. Then Δ has a nonzero eigenvector in $H_k \cap C^\infty(\Gamma\backslash G, \chi)$.*

Proof The crucial property of $\rho(\phi)$ that we make use of is that it commutes with Δ; indeed, if $g \in G$, then $\Delta \circ \rho(g) = \rho(g) \circ \Delta$ by Proposition 2.2.4 and Theorem 2.2.1, and so

$$\big(\Delta \circ \rho(\phi)\big)f = \int_G \phi(g)\big(\Delta \circ \rho(g)\big)f\,dg$$

$$= \int_G \phi(g)\big(\rho(g) \circ \Delta\big)f\,dg = \big(\rho(\phi) \circ \Delta\big)f.$$

Let H be a closed G-invariant subspace of $L^2(\Gamma\backslash G, \chi)$ as in (i). By Exercise 2.1.5, we have a Hilbert space direct sum decomposition $H = \bigoplus H_k$, where $\rho(\kappa_\theta)f = e^{ik\theta}f$ for $f \in H_k$. Choose k so that H_k is nonzero. We will show that Δ has an eigenvector on H_k. Let f_0 be a nonzero vector in H_k. By Lemma 2.3.2 (with π the restriction of the regular representation ρ to H), there exists $\phi \in C_c^\infty(G)$ such that $\rho(\phi)f_0 \neq 0$. Moreover, by that lemma, we may assume that $\phi(\kappa_\theta g) = e^{-ik\theta}\phi(g)$, which by Proposition 2.3.1(iii) implies that $\rho(\phi)$ maps H into $H \cap C^\infty(\Gamma\backslash G, \chi, k) = H_k \cap C^\infty(\Gamma\backslash G, \chi)$. By Proposition 2.3.1, $\rho(\phi)$ induces a nonzero compact self-adjoint operator on the closed subspace $H_k \cap C^\infty(\Gamma\backslash G, \chi)$. By the spectral theorem for compact operators (Theorem 2.3.1), we may choose a nonzero eigenvector λ of $\rho(\phi)$ in H_k, and then the eigenspace of λ is finite dimensional. Because Δ commutes with $\rho(\phi)$, that eigenspace is invariant under Δ, and because any linear transformation of a finite-dimensional vector space has an eigenvector, some vector in the λ-eigenspace of $\rho(\phi)$ is also an eigenvector for Δ. ■

Theorem 2.3.3 *The space $L^2(\Gamma\backslash G, \chi)$ decomposes into a Hilbert space direct sum of subspaces that are invariant and irreducible under the right regular representation ρ.*

The proof is adapted from that of Lemma 3.2 in Langlands (1976).

Proof Let Σ be the set of all sets S of irreducible invariant subspaces of $L^2(\Gamma\backslash G, \chi)$ such that the elements of S are mutually orthogonal. By Zorn's lemma, Σ has a maximal element S. Let $\mathfrak{H}$ be the orthogonal complement of the closure of the direct sum of the elements of S. We will show that $\mathfrak{H} = 0$. If not, let $0 \neq f \in \mathfrak{H}$. We will construct an irreducible subspace of $\mathfrak{H}$, which will contradict the maximality of Σ. Choose ϕ so that $\rho(\phi)$ is self-adjoint, and $\rho(\phi)f \neq 0$. Again, $\rho(\phi)$ is a nonzero self-adjoint operator on $\mathfrak{H}$, and hence has a nonzero eigenvalue λ. Let $L \subset \mathfrak{H}$ be the eigenspace of $\rho(\phi)$. It is a finite-dimensional vector space. Let L_0 be a nonzero subspace of L that is minimal with respect to the property that L_0 may be expressed as the intersection of L with a nonzero invariant subspace. The existence of such an L_0 follows from the fact that L is finite dimensional.

Let V be the intersection of all closed invariant subspaces W of $\mathfrak{H}$ such that $L_0 = L \cap W$. We show that V is irreducible. If not, $V = V_1 \oplus V_2$, where V_1 and V_2 are smaller invariant subspaces. Now let $0 \neq f_0 \in L_0$. Write $f_0 = f_1 + f_2$, where $f_i \in V_i (i = 1, 2)$. We observe that from the definition of $\rho(\phi)$, any closed subspace of V that is invariant under the action of G is invariant under $\rho(\phi)$; in particular, V_1 and V_2 are invariant under $\rho(\phi)$. Thus $\rho(\phi) f_i - \lambda f_i \in V_i (i = 1, 2)$, and because

$$\big(\rho(\phi) f_1 - \lambda f_1\big) + \big(\rho(\phi) f_2 - \lambda f_2\big) = \rho(\phi) f_0 - \rho(\phi) f_0 = 0,$$

this means that the f_i are eigenfunctions of $\rho(\phi)$ with eigenvalue λ. Without loss of generality, assume that $f_1 \neq 0$. Then $f_1 \in L \cap V_1 \subseteq L_0$, and by the minimality of L_0, this means that $V_1 \cap L = L_0$. However, V was defined to be the intersection of all invariant subspaces W of $\mathfrak{H}$ such that $L_0 = L \cap W$, yet V_1 is a proper subspace of V. This is a contradiction. ■

Let k be a nonnegative real number, and let σ be the character $\sigma(\kappa_\theta) = e^{ik\theta}$ of $K = SO(2)$. We recall that in Proposition 2.2.8, we proved that the ring $C_c^\infty(K\backslash G/K,\sigma)$ of smooth compactly supported functions ϕ that satisfy $\phi(\kappa_1 g\kappa_2) = \sigma(\kappa_1)\sigma(\kappa_2)\phi(g)$ is commutative. A *character* of this ring is by definition a ring homomorphism into $\mathbb{C}$.

Theorem 2.3.4 *Let ξ be a character of $C_c^\infty(K\backslash G/K,\sigma)$, and let $H(\xi)$ be the space of $f \in L^2(\Gamma\backslash G,\chi,k)$ that satisfies $\pi(\phi)f = \xi(\phi)f$ for all $\phi \in C_c^\infty(K\backslash G/K,\sigma)$. The space $H(\xi)$ is a finite dimensional subspace of $C^\infty(\Gamma\backslash G,\chi,k)$. If ξ and η are distinct characters of $C_c^\infty(K\backslash G/K,\sigma)$, then $H(\xi)$ and $H(\eta)$ are orthogonal subspaces. $L^2(\Gamma\backslash G,\chi,k)$ is the Hilbert space direct sum of the spaces $H(\xi)$, where ξ ranges through all characters of $C_c^\infty(K\backslash G/K,\sigma)$ such that $H(\xi)$ is nonzero.*

Proof Suppose that $0 \neq f \in H(\xi)$. By Lemma 2.3.2, there exists $\phi \in C_c^\infty(K\backslash G/K,\sigma)$ such that $\rho(\phi)f \neq 0$. By hypothesis, $\rho(\phi)f = \xi(\phi)f$, so $\xi(\phi) \neq 0$. Let $\lambda = \xi(\phi)$. It follows from the spectral theorem 2.3.1 that $\lambda \neq 0$, and so the λ-eigenspace of $\xi(\phi)$ is finite dimensional. As $H(\xi)$ is by definition contained in this eigenspace, it is finite dimensional.

To show that the spaces $H(\xi)$ and $H(\eta)$ are orthogonal, where ξ and η are distinct characters of $C_c^\infty(K\backslash G/K,\sigma)$ choose $\phi \in C_c^\infty(K\backslash G/K,\sigma)$ such that $\xi(\phi) \neq \eta(\phi)$. We can write $\phi = \phi_1 + i\phi_2$, where

$$\phi_1(g) = \frac{1}{2}\left(\phi(g) + \overline{\phi(g^{-1})}\right), \quad \phi_2(g) = \frac{1}{2i}\left(\phi(g) - \overline{\phi(g^{-1})}\right).$$

The operators $\pi(\phi_1)$ and $\pi(\phi_2)$ are self-adjoint by Proposition 2.3.1(ii). Replacing ϕ by either ϕ_1 or ϕ_2, we may assume without loss of generality that $\pi(\phi)$ is self-adjoint. Now $H(\xi)$ and $H(\eta)$ are contained in distinct eigenspaces of the self-adjoint operator $\pi(\phi)$ and are therefore orthogonal.

Finally, we must show that the Hilbert space direct sum of the finite-dimensional orthogonal spaces $H(\xi)$ is all of H. It is sufficient to show that a vector f that is orthogonal to all the $H(\xi)$ is zero. If $f \neq 0$, then by Lemma 2.3.2, there exists $\phi_0 \in C_c^\infty(K\backslash G/K,\sigma)$ such that $\pi(\phi_0)f$ is arbitrarily near f, so we may arrange so that $\pi(\phi_0)f$ and f are not orthogonal.

Let λ_i $(i = 1, 2, 3, \cdots)$ be the nonzero eigenvalues of the self-adjoint compact operator $\pi(\phi_0)$ on $L^2(\Gamma\backslash G,\chi,k)$ as in the spectral theorem 2.3.1. We have a Fourier expansion $f = f_0 + f_1 + f_2 + \ldots$, where f_0 is in the zero eigenspace of $\pi(\phi_0)$ and f_i is the λ_i-eigenspace. Then $\pi(\phi_0)\, f = \lambda_1 f_1 + \lambda_2 f_2 + \ldots$, and because $\pi(\phi_0)f$ is not orthogonal to f, f_i is not orthogonal to f for some $i \geq 1$. Let V be the λ_i-eigenspace of $\pi(\phi_0)$. Then V is finite dimensional and invariant under all operators $\pi(\phi)$ with $\phi \in C_c^\infty(K\backslash G/K,\sigma)$, because this ring is commutative. It follows that V is the direct sum of the spaces $H(\xi)$, where ξ runs through the characters of $C_c^\infty(K\backslash G/K,\sigma)$ such that $\xi(\phi_0) = \lambda_i$. Because f is not orthogonal to $f_i \in V$, it follows that f is not orthogonal to all these spaces. This shows that the Hilbert space direct sum of the spaces $H(\xi)$ is all of H. ■

Corollary *The space $L^2(\Gamma\backslash\mathcal{H}, \chi, k)$ decomposes into a Hilbert space direct sum of eigenspaces for Δ_k.*

Proof This is equivalent to showing that $L^2(\Gamma\backslash G, \chi, k)$ decomposes into a direct sum of eigenspaces for Δ. Note that because Δ commutes with the operators in $C_c^\infty(K\backslash G/K, \sigma)$, the spaces $H(\xi)$ are Δ-invariant, and by Proposition 2.1.1, Δ induces a self-adjoint operator on each of the finite-dimensional vector spaces $H(\xi)$, so each of these decomposes as a direct sum of Δ-eigenspaces. Hence the corollary follows from Theorem 2.3.4. ■

We next turn to the problem of showing that the eigenvalues λ_i of Δ_k tend to infinity. In accordance with the corollary to Theorem 2.3.4, let ϕ_i be a basis of $L^2(\Gamma\backslash\mathcal{H}, \chi, k)$, which are eigenvectors for Δ_k, with $\Delta_k\phi_i = \lambda_i\phi_i$. We will prove in this section that

$$\sum \lambda_i^{-2} < \infty \tag{3.8}$$

provided $k = 0$; the general case may be proved using similar techniques, but we will content ourselves with this special case to give some flavor of one approach.

Let us tentatively assume that Eq. (3.8) is true, and consider how it might be proved. We will see later in Theorem 2.7.1 that each

$$\lambda_i > \tfrac{k}{2}\left(1 - \tfrac{k}{2}\right). \tag{3.9}$$

(This is provable using the methods of Section 2.1 only – it is Exercise 2.1.8.) Assuming this, Eq. (3.8) implies that there are at most finitely many negative eigenvalues (counting with multiplicity) and that

$$\lim_{i\to\infty} \lambda_i = \infty. \tag{3.10}$$

Assuming Eqs. (3.8) and (3.9), let I denote the identity operator on the Hilbert space $L^2(\Gamma\backslash\mathcal{H}, \chi, k)$. Then $\Delta_k + sI$ is a positive operator for real numbers $s > \frac{k}{2}(\frac{k}{2} - 1)$, with eigenvalues $s + \lambda_i > 0$. The inequality $\sum_i \lambda_i^2 < \infty$ is equivalent to $\sum(\lambda_i + s)^2 < \infty$. We will describe, therefore, a general strategy for proving that a positive self-adjoined operator L has eigenvalues μ_i that satisfy $\sum \mu_i^{-2} < \infty$; later, we will apply this strategy with $L = \Delta_k + s$.

Let L be a positive self-adjoint operator on $\mathfrak{H} = L^2(X)$, where X is some measure space. We assume that L has a discrete spectrum. Thus we have an orthogonal basis of eigenvectors ϕ_i of $\mathfrak{H}$, and $L\phi_i = \mu_i\phi_i$ with $\mu_i > 0$. We assume, actually, that the μ_i are bounded away from zero; that is, $\mu_i > C$ for some constant C. In this case, L has an inverse; that is, we may define a bounded operator on $\mathfrak{H}$ by

$$L^{-1}\left(\sum a_i\phi_i\right) = \sum \mu_i^{-1} a_i\phi_i.$$

Now suppose that $\sum \mu_i^{-2} < \infty$. Then L^{-1} is defined by an integral kernel; that is, let

$$G(z, \zeta) = \sum \mu_i^{-1} \phi_i(z) \overline{\phi_i(\zeta)}, \tag{3.11}$$

so that

$$L^{-1} f(z) = \int_X G(z, \zeta)\, f(\zeta)\, d\zeta, \tag{3.12}$$

or

$$f(z) = \int_X G(z, \zeta)\, Lf(\zeta)\, d\zeta. \tag{3.13}$$

Because the functions $\phi_i(z)\overline{\phi_j(\zeta)}$ are an orthonormal basis for $L^2(X \times X)$, our assumption that $\sum \mu_i^{-2} < \infty$. implies that Eq. (3.11) is convergent in $L^2(X \times X)$ and

$$\int_{X \times X} |G(z, \zeta)|^2\, dz\, d\zeta < \infty, \tag{3.14}$$

because this integral equals $\sum \mu_i^{-2}$ by Eq. (3.3). The function G is called a *Green's function.* Green's functions arise frequently in the analysis of differential and integral equations.

Now assume that X is a manifold and that L is a differential operator. Thus the domain of X will be the space of smooth functions satisfying some homogeneous boundary condition, such as vanishing in a certain direction. In this case, we can offer a heuristic that will lead to a description of the function G. Proceeding formally, let us write

$$L_z G(z, \zeta) = \sum_i \mu_i^{-1} (L\phi_i)(z) \overline{\phi_i(\zeta)} = \delta(z, \zeta), \tag{3.15}$$

where L_z is the operator L applied to the variable z, with ζ fixed, and

$$\delta(z, \zeta) = \sum_i \phi_i(z) \overline{\phi_i(\zeta)}. \tag{3.16}$$

The expression in the last identity is not convergent in $L^2(X \times X)$, but can be thought of as a distribution. To see this, let $f = \sum_i a_i \phi_i \in L^2(X)$, and let us formally compute the inner product

$$\int_X \delta(z, \zeta) f(\zeta)\, d\zeta = \sum a_i \phi_i(z) = f(z).$$

Thus $\delta(z, \zeta)$ should be thought of as the Dirac delta distribution concentrated on the diagonal of $X \times X$. Now the meaning of Eq. (3.15) is that $L_z G(z, \zeta)$

vanishes when $z \neq \zeta$ but is singular along the diagonal. Of course this is only a formal calculation, but it predicts the correct answer.

We many now predict that the Green's function $G(z,\zeta)$ should have the following properties:

(i) $L_z G(z,\zeta) = 0$ for $z \neq \zeta$;
(ii) $G(z,\zeta)$ blows up along the diagonal $z = \zeta$;
(iii) $G(z,\zeta) = \overline{G(\zeta,z)}$; and
(iv) as a function of z, $G(z,\zeta)$ satisfies the homogeneous boundary conditions associated with the differential operator L_z.

Indeed, (i) and (ii) are manifestations of our heuristic calculation that $L_z G(z,\zeta)$ is a Dirac distribution concentrated along the diagonal; (iii) is clear from Eq. (3.11) because the eigenvalues μ_i are real; and (iv) is a consequene of the fact that each of the eigenfunctions ϕ_i satisfy the homogeneous boundary conditions.

The plan is now roughly as follows. In many instances, conditions (i)–(iv) determine the function $G(z,\zeta)$ which can be directly checked to satisfy Eq. (3.12). Then if one verifies Eq. (3.14), it follows that L^{-1} is a Hilbert–Schmidt operator, from whence we get $\sum \mu_i^{-2} < \infty$. Actually, we will not follow this strategy exactly, but that is the basic idea.

We now construct Green's functions for the Laplacian. Our discussion is adapted from Hejhal (1976, Chapter 3, Section 2). See Iwaniec (1995) for further material on Green's functions and the spectral theory of automorphic forms.

We will work with the weight zero Laplacian Δ_0, leaving the weight k situation to the reader in Exercise 2.3.2. We recall that in Section 2.1, we studied the Laplacian in two distinct situations: first of all, as an operator on $L^2(\mathcal{H})$, as in Proposition 2.1.1, then as an operator on $L^2(\Gamma\backslash\mathcal{H},\chi)$ as in Proposition 2.1.4. (Note: Because we will be assuming $k = 0$ until lifting this restriction in the exercises, we will denote $L^2(\Gamma\backslash\mathcal{H},\chi,0)$ as simply $L^2(\Gamma\backslash\mathcal{H},\chi)$.) Correspondingly, there will be two different versions of the Green's function. First we construct a Green's function for the Laplacian on $L^2(\mathcal{H})$.

Let $\Delta = \Delta_0$, defined by Eq. (1.3), acting on $C_c^\infty(\mathcal{H})$. The operator Δ is positive because, using Eqs. (1.4) and (1.17), we have

$$\langle \Delta f, f\rangle = \langle -R_{-2}L_0 f, f\rangle = \langle L_0 f, L_0 f\rangle \geq 0.$$

(Note that Eq. (1.17) was proved for $f \in C^\infty(\Gamma\backslash\mathcal{H},\chi,k)$, but the proof is easily adapted to $C_c^\infty(\mathcal{H})$.) However, the spectrum of Δ is not bounded away from zero. (As an operator on $L^2(\mathcal{H})$, Δ has a continuous spectrum.) To remedy this defect, we add a constant times the identity operator. Thus we consider the operator L defined by

$$Lf = \Delta f + sf, \tag{3.17}$$

where s is an arbitrary positive constant. The spectrum of L is bounded away from zero.

Because L has a continuous spectrum, the heuristics leading up to conditions (i)–(iv) is on even shakier ground than before. Nevertheless, there does exist a function having the given properties, and this function does satisfy Eq. (3.14), as we shall prove directly. This is all that we require. We shall denote the Green's function for this L by $g(z, \zeta)$, for $z = x + iy$ and $\zeta = \xi + i\eta$ in $\mathcal{H}$. The function $g(z, \zeta)$ will turn out to be real valued, so we may omit the complex conjugation in (ii). Properties (i)–(iv) are thus explicitly, in this situation

$$\left[-y^2\left(\frac{\partial^2}{\partial x^2} + \frac{\partial^2}{\partial y^2}\right) + s\right] g(z, \zeta) = 0; \tag{3.18}$$

$$g(z, \zeta) \text{ is singular on the diagonal } z = \zeta; \tag{3.19}$$

$$g(z, \zeta) = g(\zeta, z); \tag{3.20}$$

and

$$g(z, \zeta) \to 0 \text{ as } y \to 0. \tag{3.21}$$

There is one other property that we expect. We require that

$$g\big(\gamma(z), \gamma(\zeta)\big) = g(z, \zeta), \quad \gamma \in GL(2, \mathbb{R})^+. \tag{3.22}$$

We will set out to construct a function with these properties. As in Section 2.1, we will denote by Δ^e the Euclidean Laplacian

$$\Delta^e = \frac{\partial^2}{\partial x^2} + \frac{\partial^2}{\partial y^2}.$$

Lemma 2.3.3 *Let Ω_1 and Ω_2 be two regions in the complex plane, and let $\phi : \Omega_1 \to \Omega_2$ be a holomorphic mapping that is bijective and such that ϕ' does not vanish on Ω_1. Let $z = x + iy \in \Omega_1$, and let $\phi(z) = w = u + iv$ be the corresponding point in Ω_2. Let f and g be functions on Ω_1, and let $F = f \circ \phi^{-1}$ and $G = g \circ \phi^{-1}$ be the corresponding function on Ω_2. Then*

$$\int_{\Omega_1} g(z)\, \Delta^e f(z)\, dx \wedge dy = \int_{\Omega_2} G(w)\, \Delta^e F(w)\, du \wedge dv.$$

In other words, the combination $\Delta^e dx \wedge dy$ is preserved under conformal mappings.

Proof This is easily checked. ∎

We will also require the theory of *regular singular points* in the theory of ordinary differential equations. This is a basic topic that is covered in every text on this subject.

We consider a second-order differential equation in a variable r. A point c is called a *regular singular point* for the equation if near $r = c$, it has the form

$$(r-c)^2 g''(r) + (r-c)P(r-c)g'(r) + Q(r-c)g(r) = 0, \tag{3.23}$$

where $P(r-c)$ and $Q(r-c)$ are analytic at $r = c$. Let $p = P(0)$ and $q = Q(0)$. The *indicial equation* is

$$\alpha^2 + (p-1)\alpha + q = 0. \tag{3.24}$$

The basic result is that Eq. (3.23) has two solutions that can be described in series expansions in a neighborhood of the regular singular point; and the behavior at $r = c$ is determined by the roots of the indicial equation. We state this more precisely in the following proposition.

Proposition 2.3.3 *Let Eq. (3.23) be a differential equation in the neighborhood of a regular singular point, and let α, β be the roots of the indicial equation (3.24).*

(i) If $\alpha - \beta$ is not an integer, then (3.23) has two solutions of the form $(r-c)^\alpha g_1(r-c)$ and $(r-c)^\beta g_2(r-c)$, where g_1 and g_2 are analytic at $r = c$ and are nonvanishing there.

(ii) Assume that $\alpha - \beta$ is a nonnegative integer. Then there exists a solution of the form $(r-c)^\alpha g_1(r-c)$, where $g_1(r-c)$ is analytic at $r = c$, and another of the form

$$(r-c)^\beta g_2(r-c) + C\log(r-c)(r-c)^\alpha g_1(r-c),$$

where $g_2(r-c)$ is analytic at $r = c$, and $g_2(0) = 0$ if and only if $\alpha = \beta$. If $\alpha = \beta$, the constant $C \neq 0$; if $\alpha \neq \beta$, the constant C may or may not vanish.

Proof We refer to Whittaker and Watson (1927, Sections 10.3–10.32). Proofs may be found in many other books as well. ■

Proposition 2.3.4 *Let s be a positive constant. There exists a function $g(z,\zeta)$, unique up to constant multiple, with the properties Eqs. (3.18–3.22). It may be normalized so that for $f \in C_c^\infty(\mathcal{H})$, we have*

$$\int_{\mathcal{H}} g(z,\zeta)\left[-\eta^2\left(\frac{\partial^2}{\partial\xi^2} + \frac{\partial^2}{\partial\eta^2}\right) + s\right] f(\zeta)\frac{d\xi\wedge d\eta}{\eta^2} = f(z). \tag{3.25}$$

There exists a function $g(r)$, analytic on the unit interval $(0,1)$ such that $g(z,\zeta) = g(r)$, where $r = |(z-\zeta)/(z-\bar{\zeta})|$. Near $r = 1$, $g(r) \sim (1-r)^\alpha$, where $\alpha = \frac{1}{2}(1+\sqrt{1+4s}) > 1$, while near $r = 0$, $g(r)$ has a logarithmic singularity; more precisely, $g(r) - \frac{1}{2\pi}\log(r)$ is a bounded function.

Proof Let us assume the existence of the function g. We will prove first that $g(z,\zeta)$ depends only on the value of

$$r = \left|\frac{z-\zeta}{z-\bar{\zeta}}\right|. \tag{3.26}$$

If $\zeta \in \mathcal{H}$, let C_ζ denote the Cayley transform $\mathcal{H} \to \mathcal{D}$ (the unit disk):

$$C_\zeta(z) = \frac{z-\zeta}{z-\bar{\zeta}}.$$

Note that C_ζ maps ζ to the origin. Now suppose that

$$\left|\frac{z-\zeta}{z-\bar{\zeta}}\right| = \left|\frac{z'-\zeta'}{z'-\bar{\zeta}'}\right|.$$

We will prove that there exists $\gamma \in GL(2,\mathbb{R})^+$ such that $\gamma(z) = z'$ and $\gamma(\zeta) = \zeta'$. Indeed, we see that $|C_\zeta(z)| = |C_{\zeta'}(z')|$, so there exists a constant u of absolute value one such that $uC_\zeta(z) = C_{\zeta'}(z')$. Let $R_u : \mathcal{D} \to \mathcal{D}$ be the map $w \mapsto uw$. Now $C_{\zeta'}^{-1} \circ R_u \circ C_\zeta$ is a linear fractional transformation of $\mathcal{H}$ that maps z to z' and ζ to ζ'. We have shown that if two pairs z, ζ and z', ζ' have the same value of r in Eq. (3.26), then they can be mapped to one another by an element of $GL(2,\mathbb{R})^+$, and it follows from Eq. (3.22) that the value of $g(z,\zeta)$ can depend only on r in Eq. (3.26).

Now let us fix ζ, and let $w = u + iv = C_\zeta(z)$, so $r = |w| = u^2 + v^2$. We postulate the existence of a function $g(r)$ such that $g(z,\zeta) = g(r)$. It is clear that if $g(z,\zeta)$ is derived from a one-variable function $g(r)$ in this way, then Eqs. (3.20) and (3.22) will be satisfied; we will see shortly that conditions Eqs. (3.18) and (3.21) are equivalent to the conditions that $g(r)$ satisfies a certain differential equation and boundary value condition, which determine the function uniquely, and then only checking that the function arrived at in this way satisfies Eq. (3.19) will remain.

Lemma 2.3.3 allows us to compute the Laplacian in the coordinates u, w. One easily shows that

$$\frac{dx \wedge dy}{y^2} = \frac{4du \wedge dv}{(1-r^2)^2}. \tag{3.27}$$

Thus the right side of Eq. (3.27) is the non-Euclidean volume element in the unit disk. Now by Lemma 2.3.3.

$$y^2\left(\frac{\partial^2}{\partial x^2} + \frac{\partial^2}{\partial y^2}\right) = \tfrac{1}{4}(1-r^2)^2\left(\frac{\partial^2}{\partial u^2} + \frac{\partial^2}{\partial v^2}\right). \tag{3.28}$$

Actually, we prefer to use polar coordinates because the function $g(|w|)$ is radially symmetric, so write $w = re^{i\theta}$. It may be checked that the Laplacian in polar coordinates is given by

$$\frac{\partial^2}{\partial u^2} + \frac{\partial^2}{\partial v^2} = \frac{\partial^2}{\partial r^2} + \frac{1}{r}\frac{\partial}{\partial r} + \frac{1}{r^2}\frac{\partial^2}{\partial \theta^2}. \tag{3.29}$$

Thus by Eqs. (3.28), (3.29), and (3.18), the function $g(r)$ must satisfy the differential equation

$$g''(r) + \frac{1}{r} g'(r) - \frac{4s}{(1-r^2)^2} g(r) = 0. \tag{3.30}$$

The condition Eq. (3.21) is equivalent to the boundary condition

$$\lim_{r \to 1} g(r) = 0. \tag{3.31}$$

At $r = 1$, the differential equation (3.30) has a regular singular point; the roots of the indicial equation (3.24) are $\frac{1}{2}(1 \pm \sqrt{1+4s})$. One root is negative; the other, $\alpha = \frac{1}{2}(1 + \sqrt{1+4s})$ is positive. Thus one of the solutions in Proposition 2.3.3(i) blows up, the other decays. We see that the differential equation (3.30) together with the boundary value condition Eq. (3.31) determine the function $g(r)$ uniquely up to a scalar multiple.

We next show that $g(r)$ has a logarithmic singularity at $r = 0$, or more precisely, that $g(r) - c\log(r)$ is bounded. We note that $r = 0$ is also a regular singular point and that the roots of the indicial equation there are $\alpha = \beta = 0$. Consequently, either $g(r)$ has a logarithmic singularity or else it is analytic at $r = 0$. If the latter is the case, $g(r)$ is analytic on the interval $[-1, 1]$. Consequently, it has either a maximum or minimum in the closed interval $[-1, 1]$. At such a point $g'(r) = 0$, and Eq. (3.30) implies that $g''(r)$ and $g(r)$ have the same sign, which is impossible at a maximum or minimum, because $g(1) = g(-1) = 0$. We have proved that $g(r)$ has a logarithmic singularity.

We have proved that the Green's function satisfying Eqs. (3.18–3.22) exists and is unique. To complete the proof, we must show that if it is normalized so that $g(r) - \frac{1}{2\pi}\log(r)$ is bounded, then it has the reproducing property Eq. (3.25). It follows from the analyticity of g_1 and g_2 in Proposition 2.3.3(ii) with $\alpha = \beta = 0$ that with this normalization, $g'(r) - \frac{1}{2\pi r}$ is analytic near zero.

Using Eq. (3.20), it is sufficient to prove that

$$\int_{\mathcal{H}} g(z, \zeta) \left[-y^2 \left(\frac{\partial^2}{\partial x^2} + \frac{\partial^2}{\partial y^2} \right) + s \right] f(z) \frac{dx \wedge dy}{y^2} = f(\zeta). \tag{3.32}$$

Let $w = C_\zeta(z)$ as before, and let F be the smooth and compactly supported function on D such that $F(w) = f(z)$, so that $F(0) = f(\zeta)$. Using Eqs. (3.27) and (3.28), the left side of Eq. (3.32) equals

$$\lim_{\epsilon \to 0} \int_{B_R - B_\epsilon} g(|w|) \left[-\left(\frac{\partial^2}{\partial u^2} + \frac{\partial^2}{\partial v^2} \right) + \frac{4s}{(1-|w|^2)^2} \right] F(w) du \wedge dv,$$

where B_r denotes the ball of radius r and where $0 < R < 1$ is large enough that the support of F is completely contained within B_R. Now we use Green's identity Eq. (1.12) (or Stokes' theorem) to write this as

$$\lim_{\epsilon\to 0}\int\limits_{B_R-B_\epsilon} F(w)\left[-\left(\frac{\partial^2}{\partial u^2}+\frac{\partial^2}{\partial v^2}\right)+\frac{4s}{(1-|w|^2)^2}\right]g(|w|)du\wedge dv$$

$$+\lim_{\epsilon\to 0}\int\limits_{C_\epsilon} F(w)\left(\frac{\partial g(|w|)}{\partial u}dv-\frac{\partial g(|w|)}{\partial v}du\right)$$

$$-\lim_{\epsilon\to 0}\int\limits_{C_\epsilon} g(|w|)\left(\frac{\partial F(w)}{\partial u}dv-\frac{\partial F(w)}{\partial v}du\right), \tag{3.33}$$

where C_ϵ is the path circling the origin counterclockwise around the circle with radius ϵ. (There should also be terms integrating around C_R, expect these are zero because they lie outside the support of F.).

The first term here is zero because of the differential equation satisfied by $g(r)$; we will show that the last term is also zero. It equals

$$\lim_{\epsilon\to 0} g(\epsilon)\int\limits_{C_\epsilon}\left(\frac{\partial F(w)}{\partial u}dv-\frac{\partial F(w)}{\partial v}du\right),$$

which by Green's identity Eq. (1.12) again equals

$$-\lim_{\epsilon\to 0} g(\epsilon)\int\limits_{B_\epsilon}\left(\frac{\partial^2 F(w)}{\partial u^2}+\frac{\partial^2 F(w)}{\partial v^2}\right)du\wedge dv.$$

If B is an upper bound for the Laplacian of F is $\mathcal{D}$, this is dominated by $B|g(\epsilon)|\pi\epsilon^2$; here ϵ^2 goes to zero faster than $g(\epsilon)$ blows up (i.e., logarithmically) as $\epsilon\to 0$. Thus the third term in Eq. (3.33) as well as the first is equal to zero, and we see that the left side in Eq. (3.32) equals

$$\lim_{\epsilon\to 0}\int\limits_{C_\epsilon} F(w)\left(\frac{\partial g(|w|)}{\partial u}dv-\frac{\partial g(|w|)}{\partial v}du\right).$$

Now we switch again to polar coordinates, letting $w=re^{i\theta}$. We have

$$\frac{\partial g(|w|)}{\partial u}=\frac{\partial r}{\partial u}g'(r)=r^{-1}ug'(r),\qquad \frac{\partial g(|w|)}{\partial v}=r^{-1}vg'(r),$$

and $d\theta=r^{-2}(udv-vdu)$. Hence Eq. (3.32) equals

$$\lim_{\epsilon\to 0}\epsilon^{-1}g'(\epsilon)\int\limits_{C_\epsilon}F(w)(udv-vdu)=\lim_{\epsilon\to 0}\epsilon g'(\epsilon)\int_0^{2\pi}F(re^{i\theta})d\theta$$

$$=F(0)=f(\zeta)$$

because $g'(\epsilon)\sim 1/(2\pi\epsilon)$ as $\epsilon\to 0$. ■

Now that we have the Green's function for $L^2(\mathcal{H})$, we can construct automorphic Green's functions as series analogous to Eq. (3.7). Let Γ be a discontinuous subgroup of $G = GL(2,\mathbb{R})^+$ as before, and let χ be a unitary character of Γ. As usual, we assume that $-I \in \Gamma$, that $\Gamma \subset G_1 = SL(2,\mathbb{R})$, and that $\Gamma\backslash\mathcal{H}$ is compact.

Proposition 2.3.5 *There exists a function G on $\mathcal{H} \times \mathcal{H}$, defined an analytic for all values of (z,ζ) except where $\zeta = \gamma(z)$ for some $\gamma \in \Gamma$, with the following properties. G satisfies the differential equation (3.18) and the periodicity property Eq. (3.22) as well as the following periodicity property:*

$$G(z,\zeta) = \chi(\gamma)^{-1}\, G(\gamma(z),\zeta) = \chi(\gamma)\, G(z,\gamma(\zeta)), \tag{3.34}$$

if $\gamma = \left(\begin{smallmatrix} a & b \\ c & d \end{smallmatrix}\right) \in \Gamma$. We have

$$G(z,\zeta) = \overline{G(\zeta,z),} \tag{3.35}$$

and $G(z,\zeta) \sim \frac{1}{2\pi}\log|z-\zeta|$ near $z = \zeta$. Finally, it has the property that for $f \in C^\infty(\Gamma\backslash\mathcal{H})$

$$\int\limits_{\Gamma\backslash\mathcal{H}} G(z,\zeta)\left[-\eta^2\left(\frac{\partial^2}{\partial\xi^2} + \frac{\partial^2}{\partial\eta^2}\right) + s\right] f(\zeta)\frac{d\xi\wedge d\eta}{\eta^2} = f(z). \tag{3.36}$$

Proof We make the definition

$$G(z,\zeta) = \sum_{\gamma\in\{\pm1\}\backslash\Gamma} \chi(\gamma)^{-1} g(z,\gamma(\zeta)) = \sum_{\gamma\in\{\pm1\}\backslash\Gamma} \chi(\gamma) g(\gamma(z),\zeta),$$

where the last equality is by Eq. (3.20). It is necessary to check convergence which is absolute and uniform in both variables. The absolute value of each term is just $g(z,\gamma(\zeta)) = g(\gamma^{-1}z,\zeta)$. Let B_R denote a ball of radius $R < 1$ with its center at the origin in $\mathcal{D}$. By Exercise 2.3.1, the member of γ such that $C_\zeta(\gamma^{-1}z) \in B_R$ is of the order $R^2/(1-R^2)$, and the function $g(r)$ of Proposition 2.3.4 is of the order $(1-\gamma)^\alpha$ near $r = 1$, where $\alpha > 1$; consequently, the convergence issue reduces to the convergence of the integral

$$\int\limits_0^1 (1-r)^\alpha \frac{d}{dr}\frac{r^2}{1-r^2}dr < \infty.$$

Nearly all the advertised properties of $G(z,\zeta)$ follow from the corresponding properties of $g(z,\zeta)$. The verfication of Eq. (3.36) requires some comment, however. One proceeds as in the proof of Eq. (3.25), except that there are two extra terms similar to the second and third terms of Eq. (3.33), but over the boundary of a fundamental domain. These terms vanish, however, bv lemma 2.1.2 ■

Theorem 2.3.5 *Let* Γ *be as in Theorem 3.6.*

(i) The eigenvalues λ_i *of* Δ_k *on* $L^2(\Gamma\backslash\mathcal{H}, \chi, k)$ *tend to* ∞*, and indeed satisfy* $\sum \lambda_i^2 < \infty$*.*
(ii) The Laplacian Δ_k *has an extension to a self-adjoint operator on the Hilbert space* $L^2(\Gamma\backslash\mathcal{H}, \chi, k)$*.*

Proof We will prove this for $k = 0$, leaving the general case to the reader. (Exercise 3.2.)

Let ϕ_i be a basis of $\mathfrak{H} = L^2(\Gamma\backslash\mathcal{H}, \chi, 0)$ consisting of eigenvectors of Δ_0, with corresponding eigenvalues λ_i. The existence of such a basis follows from the Corollary to Theorem 2.3.4.

We prove Theorem 2.3.5(i). Let $s > 0$ be a positive constant. The operator on $\mathfrak{H}$ whose integral kernel is the Green's function $G(z, \zeta)$ of Proposition 2.3.5. This is a Hilbert–Schmidt kernel, as is easily checked, because the logarithmic singularity along the diagonal is not sufficient to cause divergence of the integral

$$\int_{\Gamma\backslash\mathcal{H}} \int_{\Gamma\backslash\mathcal{H}} |G(z,\zeta)|^2 \frac{dx \wedge dy}{y^2} \frac{d\xi \wedge d\eta}{\eta^2} < \infty.$$

Now if ϕ is an eigenfunction of Δ_0 with eigenvalue λ, then it follows form Eq. (3.36) that it is also an eigenfunction of the integral operator with this kernel and eigenvalue $(\lambda + s)^{-1}$. Because this is a Hilbert–Schmidt kernel, we see that $\sum_i (\lambda_i + s)^{-2} < \infty$, from whence we get $\sum_i \lambda_i^{-2} < \infty$. This proves (i).

We now prove (ii). Let $\mathfrak{D}_\Delta$ be the linear subspace of $\mathfrak{H}$ consisting of elements of the form $\sum a_i\phi_i$ such that $\sum \lambda_i^2 |a_i|^2 < \infty$; on this space, define

$$\Delta\left(\sum a_i\phi_i\right) = \sum \lambda_i a_i \phi_i.$$

Because the λ_i tend to infinity, and in particular are bounded away from zero, it is not hard to check that this operator is closed and is in fact self-adjoint. This completes the proof of Theorem 2.3.5. ■

Exercises

Exercise 2.3.1 Let $\Gamma \subset SL(2, \mathbb{R})$ be a discontinuous group. Let $z, \zeta \in \mathcal{H}$. Prove that there exists a constant C such that for all $0 < r < 1$, the number of $\gamma \in \Gamma$ such that

$$\frac{|z - \zeta|}{|z - \bar{\zeta}|} < r$$

is less than $Cr^2/(1 - r^2)$.
[HINT: The volume of the ball B_r of radius r with center at the origin in the non-Euclidean metric (Eq. 3.27) is $4\pi r^2/(1 - r^2)$, so we expect the number

of γ s for which $C_\zeta(\gamma z)$ is in B_r should be approximately this number divided by the volume of a fundamental domain for the group $C_\zeta \Gamma C_\zeta^{-1}$ acting on $\mathcal{D}$. To make this rigorous, find a small ball B_0 containing $C_\zeta(z)$ that is completely contained within the given fundamental domain. If $C_\zeta(\gamma z)$ lies with in B_r, then at least half of $C_\zeta \gamma C_\zeta^{-1}(B_0)$ must lie within $B - r$.]

Exercise 2.3.2 Complete the proof of Theorem 2.3.5 by extending the theory of the Green's functions in Propositions 2.3.4 and 2.3.5 to the case of Maass forms of weight k.

2.4 Basic Representation Theory

We will develop in this section the standard notions *smooth vectors, K-finite vectors, admissible representations, $(\mathfrak{g}, K)$-modules,* and *infinitesimal equivalence.* Much of this foundational matter is due to Harish-Chandra. Two good references have appeared in recent years: Knapp (1986) and Wallach (1988). These, together with Warner (1972) treat the same material presented here but in greater depth and generality. Everything done here for $GL(n, \mathbb{R})$ may be readily generalized to an arbitrary reductive Lie group.

There are different notions of a representation of a Lie group G. Historically, the first notion was that of an irreducible unitary representation, which is an action on a Hilbert space. We recall from Section 2.1 that a *representation* $(\pi, \mathfrak{H})$ of G is an ordered pair in which $\mathfrak{H}$ is a topological vector space and $\pi : G \to \mathrm{End}(\mathfrak{H})$ is a homomorphism such that the map $G \times \mathfrak{H} \to \mathfrak{H}$ given by $(g, f) \mapsto \pi(g) f$ is continuous. If $\mathfrak{H}$ is a Hilbert space, the representation is called *unitary* if, furthermore, $\pi(g) : \mathfrak{H} \to \mathfrak{H}$ is a unitary operator for all $g \in G$. This is a simple notion, adequate for the study of $L^2(\Gamma\backslash G, \chi)$ when $\Gamma\backslash G/Z^+$ is compact. After all, the right regular representation on this space is a unitary representation, and we showed in Theorem 2.3.3 of the last section that it decomposes into a direct sum of irreducibles.

There are, however, good reasons not to consider only unitary representations. We offer two.

The first reason for considering nonunitary representations is that the irreducible unitary representations of an arbitrary Lie group are difficult to classify: This is an area of active research. The problem simplifies if one considers a larger class of *admissible representations.* Indeed, in 1973 Langlands (1989) proved the so-called *Langlands classification* of admissible representations of an arbitrary Lie group, which amounts to a satisfactory solution of the classification problem for admissible representations. The original classification problem for unitary representations now takes on a different nature: The problem is now to decide which of the admissible representations are unitary. In this section, we will define admissible representations of $GL(n, \mathbb{R})$. We will also introduce the correct notion of isomorphism in the category of admissible representations, called *infinitesimal equivalence*, and in the next section, we will prove a classification theorem for admissible representations of $GL(2, \mathbb{R})$.

The second reason for considering nonunitary representations is that even in the theory of automorphic forms, nonunitary representations arise in the theory of Eisenstein series. If $\Gamma\backslash G/Z^+$ is noncompact, the Eisenstein series are involved in the continuous part of the spectrum of $L^2(\Gamma\backslash G, \chi)$. We consider, for example, the Eisenstein series

$$E(z, s) = \pi^{-s}\,\Gamma(s)\,\tfrac{1}{2} \sum_{\substack{(m,n) \in \mathbb{Z}^2 \\ (m,n) \neq (0,0)}} \frac{y^s}{|mz+n|^{2s}}, \qquad z \in \mathcal{H},$$

previously introduced in Section 1.6. The series is convergent for $\mathrm{re}(s) > 1$, but as we proved there, it has analytic continuation to all s except $s = 0$ or $s = 1$. It is a Maass form of weight zero for the group $\Gamma = SL(2, \mathbb{Z})$. We pull this back to a function on $\Gamma\backslash G$ via the projection $G \mapsto G/KZ^+ \cong \mathcal{H}$; let $E_s : \Gamma\backslash G \to \mathbb{C}$ be the resulting function. Let V be the space of functions generated by right translates of E_s. The space V may be completed to a Fréchet space, and it may be shown that the resulting representation is irreducible, provided $s \notin \mathbb{Z}$. However, it is not unitary unless either $\mathrm{re}(s) = \frac{1}{2}$ or else s is a real number lying between 0 and 1. Only the Eisenstein series with $\mathrm{re}(s) = \frac{1}{2}$ are involved in the spectrum of the Laplacian. Nevertheless, Eisenstein series with $\mathrm{re}(s) \neq \frac{1}{2}$ must be studied, because the Eisenstein series is originally defined as a series convergent for $\mathrm{re}(s) > 1$ then analytically continued. This example shows that we really do not want to restrict ourselves to unitary representations alone.

The problem now is that of laying the foundations for a theory of representations on $GL(n, \mathbb{R})$ or $GL(n, \mathbb{R})^+$ that are not assumed to be unitary. There are several options. We could work with representations on a Banach space or a Fréchet space, or we could work with representations on a Hilbert space but relax the condition of unitaricity: Instead of assuming that $\langle \pi(g)v, \pi(g)w \rangle = \langle v, w \rangle$ for all $g \in G$, assume it merely for g in the standard maximal compact subgroup K of G. (Thus $K = O(n)$ if $G = GL(n, \mathbb{R})$ or $K = SO(n)$ if $G = GL(n, \mathbb{R})^+$.) There is a third option, the simplest of all, which is to consider Harish-Chandra's notion of a $(\mathfrak{g}, K)$-module, which has the advantage of being a purely algebraic concept. Thus analytic considerations will play little role in the proofs.

Let $(\pi, \mathfrak{H})$ be a representation of $G = GL(n, \mathbb{R})$ or $G = GL(n, \mathbb{R})^+$, where $\mathfrak{H}$ is assumed to be a Hilbert space, but the representation is not necessarily assumed to be unitary. From this starting point, we will gradually work our way to the concept of a $(\mathfrak{g}, K)$-module.

We would like to define a representation of the Lie algebra $\mathfrak{g} = \mathfrak{gl}(n, \mathbb{R})$ on $\mathfrak{H}$ by analogy with Proposition 2.2.1. However, this is not usually possible if $\mathfrak{H}$ is infinite dimensional. To obtain a representation of $\mathfrak{g}$, we will define a dense linear subspace of $\mathfrak{H}$, the space $\mathfrak{H}^\infty$ of *smooth vectors*. Here $\mathfrak{H}^\infty$ is not a closed subspace. (It can be given its own distinct topology as a Fréchet space, described in Section 2.8, but we will not need this.) The Lie algebra $\mathfrak{g}$ will then act on $\mathfrak{H}^\infty$.

We will define $\mathfrak{H}^\infty$ to be the largest subspace on which the definition (Eq. 2.5) makes sense. Specifically, we say that a vector $f \in \mathfrak{H}$ is C^1 if for all $X \in \mathfrak{g}$

$$\pi(X)\, f = Xf = \frac{d}{dt}\pi\big(\exp(tX)\big)\, f|_{t=0} := \lim_{t\to 0} \tfrac{1}{t}\big(\pi\big(\exp(tX)\big) f - f\big) \quad (4.1)$$

is defined. Then $g \mapsto \pi(g)\, Xf$ is a continuous function on G. (We will use the notations Xf and $\pi(X)\, f$ interchangeably, preferring the former if no confusion can arise.) We recursively define f to be C^k, $k > 1$ if f is C^1 and Xf is C^{k-1} for all $X \in \mathfrak{g}$. Finally, we say that f is C^∞ or *smooth* if it is C^k for all $k \geq 1$. The following proposition generalizes Propositions 2.2.1 and 2.3.3.

Lemma 2.4.1 *The space $\mathfrak{H}^\infty$ is invariant under the action of G.*

Proof Let $g \in G$, and let $X \in \mathfrak{g}$. In the notation of Exercise 2.2.1, let $\mathrm{Ad}(g)(X) = gXg^{-1}$, where the multiplication on the right side is matrix multiplication. We have

$$X\big(\pi(g) f\big) = \lim_{t\to 0} \tfrac{1}{t}\big(\pi(\exp(tX)g) f - \pi(g) f\big)$$

$$= \pi(g) \left(\lim_{t\to 0} \tfrac{1}{t}\big(\pi\big(\exp(t\,\mathrm{Ad}(g^{-1})X)\big)\big) f - f\right).$$

The limit exists and is continuous if f is C^1, so $\pi(g) f$ is C^1 if f is; iterating, we find that $\pi(g) f$ is C^k for all k if f is a smooth vector, and hence so is $\pi(g) f$. ■

Lemma 2.4.2 *Let f be a function defined on $G = GL(n,\mathbb{R})^+$. If $X \in \mathfrak{g}$, define $dX\, f$ by Eq. (2.10), provided the derivative on the right-hand side exists. A necessary and sufficient condition for f to be smooth is that f be continuous and that $dX_1 \circ \ldots \circ dX_r f$ exists and is continuous for all sequences $X_1, \cdots, X_r$ of elements of $\mathfrak{g}$.*

Proof This is simply a translation of the usual condition for a function to be smooth: it must have continuous partial derivatives of all orders. We will check first that f is smooth at the identity $g = 1$. Let $g = (g_{ij})$ be the usual coordinates on $GL(n,\mathbb{R})$ regarded as an open set in $\mathrm{Mat}(n,\mathbb{R}) = \mathbb{R}^{n^2}$. Let X_{ij} be the element of $\mathfrak{g}$ that, as a matrix, has a 1 in the i, j position and zeros elsewhere. It is easy to check that in the notation (Eq. 2.10), we have

$$(dX_{ij}\, f)(1) = \frac{\partial f}{\partial g_{ij}}(1)$$

for f a function on G defined in a neighborhood of the identity. It is clear, therefore, that there exist smooth functions m_{ijkl} defined in a neighborhood of the identity such that

$$\sum_{i,j} m_{ijkl}(g)\, dX_{ij} = \frac{\partial}{\partial g_{kl}}$$

near $g = 1$. Therefore our hypothesis implies that f has continuous partial derivatives of all orders at the identity and is smooth there.

Now at an arbitrary point $g \in G$, left translation $x \mapsto gx$ is a diffeomorphism $G \to G$ that commutes with the operators dX and maps a neighborhood of 1 to a neighborhood of g, so having proved f to be smooth near 1 is sufficient. ■

Proposition 2.4.1 *Let $(\pi, \mathfrak{H})$ be a Hilbert space representation of the group $G = GL(n, \mathbb{R})$ or $G = GL(n, \mathbb{R})^+$. Then the action of $\mathfrak{g}$ on the space $\mathfrak{H}^\infty$ of smooth vectors defined by Eq. (4.1) is a Lie algebra representation.*

Proof We may assume that $G = GL(n, \mathbb{R})^+$. Indeed, if $G = GL(n, \mathbb{R})$, then the restriction of π to $GL(n, \mathbb{R})^+$ is a representation. The two groups $GL(n, \mathbb{R})$ and $GL(n, \mathbb{R})^+$ both have the same Lie algebra, and it is clear that neither the definition of $\mathfrak{H}^\infty$ nor the action of $\mathfrak{g}$ depends on which group we use.

We must prove that if Xf is defined by Eq. (4.1), then

$$X(Yf) - Y(Xf) = [X, Y]f \tag{4.2}$$

for $X, Y \in \mathfrak{g} = \mathfrak{gl}(n, \mathbb{R})$, $f \in \mathfrak{H}^\infty$. We will deduce this from Proposition 2.2.2. It is sufficient to show that

$$\langle X(Yf) - Y(Xf), \phi \rangle = \langle [X, Y]f, \phi \rangle \tag{4.3}$$

for every $\phi \in \mathfrak{H}$. Let ϕ be fixed. We define a map $L : \mathfrak{H}^\infty \to C^\infty(G)$ by

$$(Lf)(g) = \langle \pi(g) f, \phi \rangle \, . \tag{4.4}$$

We must check that this is a smooth function. We have

$$\big((dX \circ L)f\big)(g) = \big((L \circ X)f\big)(g), \tag{4.5}$$

because by Eq. (2.10), the left side equals

$$\frac{d}{dt}(Lf)\big(ge^{tX}\big)|_{t=0} = \frac{d}{dt}\left\langle \pi(g)\, \pi\left(e^{tX}\right) f, \phi \right\rangle |_{t=0}$$

$$= \langle \pi(g) X\, f, \phi \rangle = \big((L \circ X)f\big)(g).$$

Our hypothesis that f is smooth implies that for $X_1, \cdots, X_r \in \mathfrak{g}$, $X_1 \circ \ldots \circ X_r\, f$ is defined, and so by Eq. (4.5), $dX_1 \circ \ldots \circ dX_r\, Lf$ is defined, and by Lemma 2.4.2, this means that Lf is smooth.

Now by Eqs. (4.5) and (2.12), we see that $L\big(X(Yf) - Y(Xf) - [X, Y]f\big) = 0$. Hence $X(Yf) - Y(Xf) - [X, Y]f$ is orthogonal to the vector ϕ. Thus we obtain Eq. (4.3), which as we have already noted implies Eq. (4.2). ■

We therefore have a representation of the Lie algebra $\mathfrak{g}$ not on the whole space $\mathfrak{H}$, but on the subspace $\mathfrak{H}^\infty$. We would like to know that this space is not too small. We will use the same device that we used in Section 2.3; namely, we will exploit the action of $C_c^\infty(G)$ on $\mathfrak{H}$. Thus, if $\phi \in C_c^\infty(G)$, let $\pi(\phi) : \mathfrak{H} \to \mathfrak{H}$ be defined by

$$\pi(\phi)\, f = \int_G \phi(g)\, \pi(g) f \, dg. \tag{4.6}$$

Proposition 2.4.2 *Let $(\pi, \mathfrak{H})$ be a Hilbert space representation of either $G = GL(n,\mathbb{R})$ or $G = GL(n,\mathbb{R})^+$.*
(i) If $\phi \in C_c^\infty(G)$, $f \in \mathfrak{H}$, then $\pi(\phi)\, f \in \mathfrak{H}^\infty$.
(ii) The space $\mathfrak{H}^\infty$ is dense in $\mathfrak{H}$.

Proof For (i), we have

$$X\,\pi(\phi)\, f = \frac{d}{dt}\pi\left(e^{tX}\right)\pi(\phi) f|_{t=0} = \frac{d}{dt}\int_G \phi(g)\,\pi\left(e^{tX}g\right) f\, dg|_{t=0}$$

$$= \frac{d}{dt}\int_G \phi\left(e^{-tX}g\right)\pi(g) f\, dg|_{t=0} = \int_G \phi_X(g)\,\pi(g) f\, dg,$$

where $\phi_X(g) = (d/dt)\,\phi(e^{-tX}g)|_{t=0}$. Thus $X\,\pi(\phi)\, f = \pi(\phi_X)\, f$ is defined. It is easily seen that $\pi(g)\,\pi(\phi_X)\, f$ is continuous. Thus $\pi(\phi)\, f$ is C^1, and iterating, it is C^k for all k. Hence $\pi(\phi)\, f \in \mathfrak{H}^\infty$.

For (ii), let $\epsilon > 0$ be given. Because the $(g, f) \mapsto \pi(g) f$ is a continuous map $G \times \mathfrak{H} \to \mathfrak{H}$, there exists a neighborhood U of the identity in G such that $|\pi(g) f - f| < \epsilon$ for all $g \in U$. We take $\phi \in C_c^\infty(G)$ to be positive valued, have support in U, and such that $\int_G \phi(g)\, dg = 1$. Then

$$|\pi(\phi)\, f - f| = \left|\int_G \phi(g)\left(\pi(g) f - f\right) dg\right| \le \int_G \phi(g)\, |\pi(g) f - f|\, dg \le \epsilon.$$

This proves that the smooth vectors are dense in $\mathfrak{H}$. ■

Next we begin to study how the representation $(\pi, \mathfrak{H})$ decomposes when restricted to the maximal compact subgroup K of G.

Lemma 2.4.3 *Let $(\pi, \mathfrak{H})$ be a representation of a compact group K on a Hilbert space $\mathfrak{H}$. There exists a Hermitian inner product $\langle\, ,\, \rangle$ on $\mathfrak{H}$ inducing the same topology as the original one and satisfying*

$$\langle \pi(\kappa)v, \pi(\kappa)w\rangle = \langle v, w\rangle \qquad \text{for all } \kappa \in K. \tag{4.7}$$

Proof Let $\langle\, ,\, \rangle_1$ be the given Hermitian inner product on $\mathfrak{H}$. Define the new inner product by

$$\langle v, w \rangle = \int_K \langle \pi(\kappa)v, \pi(\kappa)w \rangle_1 \, d\kappa.$$

With the new inner product, Eq. (4.7) will be valid. We must check that the two inner products induce the same topology on $\mathfrak{H}$. Because π is a representation, $\kappa \mapsto \langle \pi(\kappa)v, \pi(\kappa)v \rangle_1$ is a continuous, never-vanishing function on K. Hence for fixed v, the norms $|\pi(\kappa)\, v|_1$ ($\kappa \in K$) form a bounded family, and according to the *Principle of Uniform Boundedness* of Banach and Steinhaus (Rudin (1974), Theorem 5.8), this implies that the norms $|\pi(\kappa)|$ are themselves bounded. Thus there exists a constant C such that $|\pi(\kappa)\, v|_1 \le C|v|_1$ for $v \in \mathfrak{H}$, $\kappa \in K$, and because $\pi(\kappa^{-1})$ respects the same bound, we have $C^{-1}|v|_1 \le |\pi(\kappa)\, v|_1 \le C|v|_1$. It follows that

$$|v|^2 = \int_K \langle \pi(\kappa)v, \pi(\kappa)v \rangle_1 \, d\kappa$$

lies between $C^{-2}|v|_1^2$ and $C^2|v|_1^2$. Hence two vectors are close in the new topology if and only if they are close in the old topology: The topologies are the same. ∎

If (π_1, H_1) and (π_2, H_2) are representations of G, an *intertwining operator* is a continuous linear map $L : H_1 \to H_2$ such that

$$\pi_2(g) \circ L = L \circ \pi_1(g)$$

for all $g \in G$. An intertwining operator is just a homomorphism in the category of representations of G. We will also use the term *equivariant* or *G-equivariant* to indicate that a map $H_1 \to H_2$ is an intertwining operator.

If $(\pi, \mathfrak{H})$ is a representation of G on a Hilbert space, a *matrix coefficient* of the representation is a function of the form $g \mapsto \langle \pi(g)x, y \rangle$, where $x, y \in \mathfrak{H}$.

Proposition 2.4.3 *Let G be a compact group, and let (π_1, H_1) and (π_2, H_2) be representations. Assume that (π_2, H_2) is unitary. If there exist matrix coefficients f_1 and f_2 of π_1 and π_2 that are not orthogonal in $L^2(G)$, then there exists a nonzero intertwining operator $L : H_1 \to H_2$.*

This is, of course, a corollary to the Schur orthogonality relations, which generalize straightforwardly from finite groups to compact groups.

Proof Suppose that $x_1, y_1 \in H_1$ and $x_2, y_2 \in H_2$ such that

$$\int_G \overline{\langle \pi_1(g)x_1, y_1 \rangle} \, \langle \pi_2(g)x_2, y_2 \rangle \, dg \neq 0. \tag{4.8}$$

Define a bounded linear map $L : H_1 \to H_2$ by

$$L(v) = \int_G \langle \pi_1(g)v, y_1 \rangle \; \pi_2(g^{-1})y_2 \, dg.$$

This is an intertwining operator: Indeed

$$\big(\pi_2(h) \circ L\big)(v) = \int_G \langle \pi_1(g)v, y_1 \rangle \; \pi_2(hg^{-1})y_2 \, dg,$$

and the change of variables $g \mapsto gh$ shows that this equals $\big(L \circ \pi_1(h)\big)(v)$. We must show that $L \neq 0$. Indeed, using the fact that the inner product is conjugate linear in the second variable, we compute

$$\langle x_2, L(x_1) \rangle = \int_G \overline{\langle \pi_1(g)x_1, y_1 \rangle} \, \big\langle x_2, \pi_2(g^{-1})y_2 \big\rangle \; dg.$$

Because π_2 is unitary, this equals Eq. (4.8). Hence L is nonzero. ■

Theorem 2.4.1: Peter–Weyl *Let K be a compact subgroup of $GL(n,\mathbb{C})$.*

(i) The matrix coefficients of finite-dimensional unitary representations of K are dense in $C(K)$ and in $L^p(K)$ for all $1 \leq p < \infty$.
(ii) Any irreducible unitary representation of K is finite dimensional.
(iii) Let $(\pi, \mathfrak{H})$ be a unitary representation of K. Then $\mathfrak{H}$ decomposes into a (Hilbert space) direct sum of irreducible unitary subrepresentations.

Here the ring $C(K)$ of continuous functions on K is given the L^∞ norm. The hypothesis that K may be embedded in $GL(n,\mathbb{C})$ is unnecessary: Without the hypothesis, this is the *Peter–Weyl theorem*. The assumption that $K \subset GL(n,\mathbb{C})$ simplifies the proof.

Proof Because $GL(n,\mathbb{C})$ may be embedded in $GL(2n,\mathbb{R})$, we may assume that K is a subgroup of $GL(n,\mathbb{R})$ and change n to $2n$ if needed. We regard K as a subset of $\mathrm{Mat}(n,\mathbb{R}) = \mathbb{R}^{n^2}$. We call a function on K a *polynomial function* if it is the restriction to K of a polynomial with complex coefficients in n^2 variables, regarded as a function on $\mathrm{Mat}(n,\mathbb{R})$.

We first observe that *any polynomial function on K is a matrix coefficient of a finite-dimensional representation*. Indeed, let r be a positive integer, and let (ρ, R) be the representation of K on the space of polynomial functions f of degree $\leq r$ on $\mathrm{Mat}(n,\mathbb{C})$, where K acts by $\rho(g)f(x) = f(xg)$. By Lemma 2.4.3, we may find a Hermitian inner product on R so that the representation ρ is unitary. Every (bounded) linear functional on a Hilbert space is the inner product with some vector, so there exists an element $f_0 \in R$ such that $f(1) = \langle f, f_0 \rangle$.

Now let $f \in R$. We have

$$f(g) = \big(\rho(g)f\big)(1) = \langle \rho(g)f, f_0 \rangle,$$

so the function f is a matrix coefficient of R.

Part (i) now follows: by Rudin (1974) (Eq. 3.14) $C(K)$ is dense in $L^p(K)$ for any $1 \leq p < \infty$, so it is sufficient to show that the matrix coefficients are dense in $C(K)$. It is well known (and an immediate consequence of the Stone–Weierstrass theorem) that any function on a compact subset of (real) Euclidean space may be uniformly approximated by a polynomial, and we have just seen that the polynomial functions on $C(K)$ are matrix coefficients. This proves (i).

We prove next that *if* $(\pi, \mathfrak{H})$ *is a nonzero unitary representation of* K*, then* $\mathfrak{H}$ *admits a nonzero finite-dimensional invariant subspace.* Indeed, let ϕ be a nonzero matrix coefficient of $\mathfrak{H}$. For example, $\langle \pi(g)x, x \rangle$ does not vanish at the identity if x is any nonzero vector. We have just seen that we may approximate ϕ in $L^2(K)$ by a polynomial function, and in particular, we may find a polynomial function that is not orthogonal to ϕ. By Proposition 2.4.3, this implies the existence of a nonzero intertwining map $L : R \to \mathfrak{H}$ for the finite-dimensional representation R on the space of polynomials of degree $\leq r$ for some r. The image of L is a finite-dimensional invariant subspace of $\mathfrak{H}$.

Parts (ii) and (iii) now follow. If $(\pi, \mathfrak{H})$ is an irreducible unitary representation, it admits a nonzero finite-dimensional invariant subspace, which by irreducibility must be $\mathfrak{H}$ itself, from whence we get (ii). As for (iii), use Zorn's lemma to construct a maximal subspace that is a direct sum of irreducible subspaces; this must be all of $\mathfrak{H}$, because otherwise its orthocomplement contains a nonzero finite-dimensional invariant subspace, which must contain an irreducible subspace, and this is a contradiction. ∎

Now assume that either $G = GL(n, \mathbb{R})$ and $K = O(n)$ or else that $GL(n, \mathbb{R})^+$ and $K = SO(n)$. Let $(\pi, \mathfrak{H})$ be a representation of G on a Hilbert space. By Lemma 2.4.3, there is no loss of generality in assuming that the restriction of π to K is unitary. However, as we have mentioned, it may not be true that π is unitarizable as a representation of G.

By the Peter–Weyl Theorem 2.4.1(iii), $\mathfrak{H}$ decomposes into a direct sum of irreducible finite-dimensional representations of K.

A representation $(\pi, \mathfrak{H})$ of G on a Hilbert space $\mathfrak{H}$ is called *admissible* if each isomorphism class of finite-dimensional unitary representations of K occurs only finitely many times in some decomposition of $\mathfrak{H}$ into irreducible unitary representations of K. It is not hard to show that the multiplicity of a given finite-dimensional representation of K does not depend on the decomposition (Exercise 2.4.1).

Admissible representations are the right category to study: It is known that any irreducible unitary representation is admissible (cf. Knapp (1986, Theorem 8.1, p. 205) or Wallach (1988, Theorem 3.4.10, p. 85)). We will content ourselves with proving, in Corollary 2 to Theorem 2.4.3, that the irreducible unitary representations of $GL(2, \mathbb{R})^+$ that can occur in the spectral decomposition of $L^2(\Gamma\backslash G, \chi)$ (with Γ as in Theorem 2.3.3) are admissible. First we develop some further properties of the decomposition of $\mathfrak{H}$ over K.

The next result shows, with a technical hypothesis, that in the decomposition of $\mathfrak{H}$ over K, the multiplicity of the trivial representation of K is at most one. Indeed, let $\mathfrak{H}^K$ denote the space of K-fixed vectors in $\mathfrak{H}$. The space $\mathfrak{H}^K$ is invariant under $C_c^\infty(K\backslash G/K)$; in fact, if $f \in \mathfrak{H}$ and $\phi \in C_c^\infty(K\backslash G/K)$, then $\pi(\phi) f$ is easily seen to lie in $\mathfrak{H}^K$. Note that $\mathfrak{H}^K$ is closed, because it is the intersection of the closed subspaces $\{v|\pi(\kappa)v = v\}$ as κ runs through K.

Theorem 2.4.2: Uniqueness of the K-fixed vector *Let $G = GL(n, \mathbb{R})$ and $K = O(n)$, or else $G = GL(n, \mathbb{R})^+$ and $K = SO(n)$. Let $(\pi, \mathfrak{H})$ be an irreducible unitary representation of G. Assume that for $\phi \in C_c^\infty(K\backslash G/K)$, the operator $\pi(\phi)$ is a compact operator on $\mathfrak{H}^K$. Then the dimension of $\mathfrak{H}^K$ is at most one.*

The compactness assumption is unnecessary, because by the result quoted above, an irreducible unitary representation is automatically admissible, so $\mathfrak{H}^K$ is finite dimensional; thus any linear transformation of $\mathfrak{H}^K$ is automatically compact. However, we will content ourselves with a weaker result. We will prove an analog of this for $GL(2)$ of a p-adic field in Theorem 4.6.2.

Proof We will show that $\mathfrak{H}^K$ (if nonzero) is irreducible as a $C_c^\infty(K\backslash G/K)$ module. By this, we mean that the only closed subspaces of $\mathfrak{H}^K$ invariant under $\pi(\phi)$ for all $\phi \in C_c^\infty(K\backslash G/K)$ are $\mathfrak{H}^K$ and 0. We specify *closed* subspaces here, because we have not yet proved $\mathfrak{H}^K$ to be finite dimensional.

Indeed, suppose $V_0 \subset \mathfrak{H}^K$ is a closed nonzero subspace invariant under $C_c^\infty(K\backslash G/K)$. We will show that $V_0 = \mathfrak{H}^K$. If not, let $0 \neq x \in \mathfrak{H}^K$ be orthogonal to V_0. Let X be the closure of the space of all $\pi(\phi)x$ with $\phi \in C_c^\infty(G)$. We will show that X is orthogonal to V_0. Let $\phi \in C_c^\infty(G)$, $v \in V_0$. If κ_1 and $\kappa_2 \in K$, then $\pi(\kappa_1^{-1})x = x$ and $\pi(\kappa_2)v_0 = v_0$. Thus, normalizing the Haar measure on K so its volume equals one, we have

$$\langle \pi(\phi)x, v\rangle = \int_G \phi(g)\, \langle \pi(g)x, v\rangle \, dg$$

$$= \int_{K\times K}\int_G \phi(g) \left\langle \pi(g)\pi(\kappa_1^{-1})x, \pi(\kappa_2)v\right\rangle \, dg\, d\kappa_1\, d\kappa_2$$

$$= \int_{K\times K}\int_G \phi(g) \left\langle x, \pi(\kappa_1 g^{-1}\kappa_2)v\right\rangle \, dg\, d\kappa_1\, d\kappa_2.$$

Making the change of variables $g \mapsto \kappa_1 g^{-1} \kappa_2$, we obtain

$$\langle \pi(\phi)\, x, v\rangle = \int_G \phi_0(g)\, \langle x, \pi(g) v\rangle\, dg,$$

where

$$\phi_0(g) = \int_{K\times K} \phi(\kappa_1 g^{-1} \kappa_2)\, d\kappa_1\, d\kappa_2. \tag{4.9}$$

Obviously, $\phi_0 \in C_c^\infty(K\backslash G/K)$, so

$$\langle \pi(\phi)\, x, v\rangle = \langle x, \pi(\phi_0)\, v\rangle\,, \tag{4.10}$$

and because v lies in the space V_0, which is invariant under the action of $C_c^\infty(K\backslash G/K)$, we have $\pi(\phi_0)\, v \in V_0$, and thus $\pi(\phi)\, x$ is orthogonal to v. This proves that X is orthogonal to V_0.

It is not hard to see that X is G-invariant. It is a proper subspace because it is orthogonal to V_0, which is closed and nonzero; yet $0 \neq x \in X$, so X is nonzero. This contradicts the assumption that π is irreducible. We have proved that $\mathfrak{H}^K$ (if nonzero) is irreducible as a module over $C_c^\infty(K\backslash G/K)$.

Now we may prove that $\mathfrak{H}^K$ is finite dimensional. We assume that $\mathfrak{H}^K$ is nonzero. Let $0 \neq x \in \mathfrak{H}^K$. By Lemma 2.3.2, there exists $\phi \in C_c^\infty(G)$ such that $\pi(\phi)$ is self-adjoint, $\pi(\phi)\, x \neq 0$, and (taking $k = 0$ in the lemma), $\phi \in C_c^\infty(K\backslash G/K)$. By hypothesis, $\pi(\phi)$ is compact, so by the Spectral Theorem 2.3.1 for compact operators, $\pi(\phi)$ has a nonzero eigenvalue λ on $\mathfrak{H}^K$. Let $L \subseteq \mathfrak{H}^K$ be the corresponding eigenspace. It is nonzero and finite dimensional. It follows from the commutativity of $C^\infty(K\backslash G/K)$ (of which ϕ is an element) that L is invariant under this ring of operators. Because we have proved that the action of $C^\infty(K\backslash G/K)$ is irreducible, $L = \mathfrak{H}^K$. Thus $\mathfrak{H}^K$ is finite dimensional.

Now $C_c^\infty(K\backslash G/K)$ is realized as a commutative family of normal operators on a finite dimensional vector space, which may thus be simultaneously diagonalized (Exercise 1.4.1). We may therefore find a one-dimensional subspace V_0 of $\mathfrak{H}^K$ that is invariant under $C_c^\infty(K\backslash G/K)$, and we have proved above that $V_0 = \mathfrak{H}^K$. ∎

Corollary *Let $G = GL(n, \mathbb{R})$ and $K = O(n)$, or else $G = GL(n, \mathbb{R})^+$ and $K = SO(n)$. Let $(\pi, \mathfrak{H})$ be an irreducible admissible unitary representation of G. Then the dimension of $\mathfrak{H}^K$ is at most one.*

Proof Because π is admissible, the dimension of $\mathfrak{H}^K$ is finite, and hence any operator on $\mathfrak{H}^K$ is automatically compact. This follows, therefore, from the theorem. ∎

Theorem 2.4.3 *Let $G = GL(2, \mathbb{R})^+$ and $K = SO(2)$, and let $k \in \mathbb{Z}$. Let $(\pi, \mathfrak{H})$ be an irreducible unitary representation of G. Let $\mathfrak{H}_k$ be the space of vectors $v \in \mathfrak{H}$ satisfying $\pi(\kappa_\theta)\, v = \sigma_k(\kappa_\theta)\, v$, where $\sigma_k(\kappa_\theta) = e^{ik\theta}$. Assume* *that for $\phi \in C_c^\infty(K\backslash G/K, \sigma_k)$, the operator $\pi(\phi)$ is a compact operator on $\mathfrak{H}_k$. Then the dimension of $\mathfrak{H}_k$ is at most one.*

In contrast with Theorem 2.4.2 (which is valid for $GL(n,\mathbb{R})$), this is for $GL(2,\mathbb{R})$ only. For $GL(n,\mathbb{R})$, we can assert an upper bound for the multiplicity of a given finite-dimensional irreducible representation τ of $SO(n)$ in an irreducible unitary representation of $GL(n,\mathbb{R})$; that bound would not be one, but rather, the dimension of τ. See Knapp (1986, Theorem 8.1, p. 205), and the notes to Knapp's Chapter VIII for this theorem of Harish–Chandra.

Proof The proof is identical to that of Theorem 2.4.2, except that instead of Theorem 2.2.3, we use Proposition 2.2.8. ■

Corollary *Let $G = GL(2,\mathbb{R})^+$ and $K = SO(2)$, and let $k \in \mathbb{Z}$. Let $(\pi, \mathfrak{H})$ be an irreducible admissible unitary representation of G. Let $\mathfrak{H}_k$ be as in Theorem 2.4.3. Then the dimension of $\mathfrak{H}_k$ is at most one.*

In the following section, we will remove the assumption of unitaricity here. If unitaricity is assumed, the assumption of admissibility is not really needed.

Proof Because $\mathfrak{H}_k$ is finite dimensional, $\pi(\phi)$ induces a compact operator on $\mathfrak{H}_K$, so this follows from Theorem 2.4.3. ■

Corollary *Let $G = GL(2,\mathbb{R})^+$ and $K = SO(2)$, and let $k \in \mathbb{Z}$. Let Γ be a discontinuous subgroup of G containing $-I$ such that $\Gamma\backslash\mathcal{H}$ is compact, and let χ be a unitary character of Γ. Let $(\pi, \mathfrak{H})$ be an irreducible subrepresentation of $L^2(\Gamma\backslash G, \chi)$ as in Theorem 2.3.3. Let $\mathfrak{H}_k$ be as in Theorem 2.4.3. Then the dimension of $\mathfrak{H}_k$ is at most one. In particular, π is admissible.*

For the study of automorphic forms, the class of admissible representations is sufficiently general. This corollary shows this in a special case.

Proof The compactness of $\pi(\phi)$ is a consequence of Proposition 2.3.1. ■

Now let $(\pi, \mathfrak{H})$ be an admissible representation of $G = GL(n,\mathbb{R})$ or $GL(n,\mathbb{R})^+$; by Lemma 2.4.3, we may assume that the restriction of π to K is unitary. For each isomorphism class σ of irreducible unitary representations of the maximal compact subgroup K of G, let $\mathfrak{H}(\sigma)$ be the σ-isotypic part (Exercise 2.4.1). The admissibility of π means that the spaces $\mathfrak{H}(\sigma)$ are all finite dimensional. The space $\mathfrak{H}$ is a Hilbert space direct sum of the $\mathfrak{H}(\sigma)$. Let $\mathfrak{H}_{\text{fin}}$ be the algebraic direct sum of the $\mathfrak{H}(\sigma)$, that is, the space of all vectors f whose projection on only a finite number of $\mathfrak{H}(\sigma)$ is nonzero. Elements of $\mathfrak{H}_{\text{fin}}$ are called *K-finite vectors*. Let $\mathfrak{k}$ be the Lie algebra of K. It is not hard to see that $\mathfrak{k}$ consists of the skew-symmetric matrices in $\text{Mat}(n,\mathbb{R})$ (Exercise 2.2.6(a)).

Proposition 2.4.4 *In this situation, the following conditions are equivalent for $f \in \mathfrak{H}$.*

(i) f is a K-finite vector.
(ii) The vector space spanned by $\pi(\kappa) f$, for $\kappa \in K$, is finite dimensional.
(iii) The vector space spanned by $X f$, for $X \in \mathfrak{k}$, is finite dimensional.

Proof We will assume in the sequel that $G = GL(n, \mathbb{R})^+$ and $K = SO(n)$. This implies the same result for $GL(n, \mathbb{R})$ and $K = O(n)$, because if $(\pi, \mathfrak{H})$ is a representation of $GL(n, \mathbb{R})$, we may consider its restriction to $GL(n, \mathbb{R})^+$. It is easy to see that $f \in \mathfrak{H}$ is $O(n)$-finite if and only if it is $SO(n)$-finite, and also that the span W_0 of $\{\pi(\kappa)|\kappa \in SO(n)\}$ is finite dimensional if and only if the span W of $\{\pi(\kappa)|\kappa \in O(n)\}$ is, because if η is a representative of the nonidentity coset of $O(n)/SO(n)$, then $W = W_0 + \pi(\eta) W_0$.

That (i) is equivalent to (ii) is clear from Exercise 2.4.1. Indeed, if f is a K-finite vector, f lies in the direct sum of a finite number of the $\mathfrak{H}(\sigma)$, each of which is a finite-dimensional K-invariant subspace, and the space spanned by the K-translates of f is contained in this space. Conversely, if the K-translates of f span a finite-dimensional vector space V, this K-invariant space decomposes into a direct sum of irreducible subspaces, and if $\sigma_1, \cdots, \sigma_r$ are the isomorphism classes of irreducible representations of K occurring among these, it is clear that $V \subset \oplus \mathfrak{H}(\sigma_i)$, so f is a K-finite vector.

Next we show that (ii) implies (iii). If the K-translates of f lie in a finite-dimensional vector space V, and if $X \in \mathfrak{k}$, then because

$$Xf = \frac{d}{dt} \pi(e^{tX}) f|_{t=0},$$

and because the vectors $\pi(e^{tX}) f$ all lie in V, the vector Xf lies in V also. Thus (ii) implies (iii).

Finally, we show that (iii) implies (ii). Suppose that V is a finite-dimensional subspace of $\mathfrak{H}$ such that $Xf \in V$ for all $X \in \mathfrak{k}$. It follows from Exercise 2.4.3 that $\pi(\kappa) f \in V$ for all $\kappa \in K$ of the form $\exp(X)$ with $X \in \mathfrak{k}$. These elements exhaust K by Exercise 2.4.2, and so we are done. ■

Proposition 2.4.5 *Let $(\pi, \mathfrak{H})$ be an admissible Hilbert space representation of $G = GL(n, \mathbb{R})$ or $GL(n, \mathbb{R})^+$. The K-finite vectors are smooth. The space $\mathfrak{H}_{\text{fin}}$ is dense in $\mathfrak{H}$ and is invariant under the action of $\mathfrak{g} = \mathfrak{gl}(n, \mathbb{R})$ on $\mathfrak{H}^\infty$.*

Proof By Lemma 2.4.3, we may assume that the restriction of π to K is unitary.

Let us denote $\mathfrak{H}_0 = \mathfrak{H}^\infty \cap \mathfrak{H}_{\text{fin}}$. We show that $\mathfrak{H}_0$ is dense in $\mathfrak{H}$. Let $f \in \mathfrak{H}$. Let U be a small neighborhood of the identity in G, and let ϵ be a small positive constant. Suppose that ϕ is a smooth positive-valued function with support in KU such that $\int_G \phi(g)\, dg = 1$ and $\int_{G-U} |\phi(g)|\, dg < \epsilon$. If U and ϵ are sufficiently small, $\pi(\phi) f$ will be near f. Given such U and ϵ, we will show that ϕ may be chosen so that $\pi(\phi) f$ is K-finite.

Let U_1 be a neighborhood of the identity in G, and let V be a neighborhood of the identity in K such that $VU_1 \subset U$. Let ϕ_1 be a smooth-positive valued function supported in U_1 such that $\int_G \phi_1 = 1$. By the Peter–Weyl Theorem 2.4.1(i), there exists a matrix coefficient ϕ_0 of a finite-dimensional unitary representation of K such that $\int_K \phi_0 = 1$ and $\int_{K-V} |\phi_0| < \epsilon$. Let

$$\phi(g) = \int_K \phi_0(\kappa)\,\phi_1(\kappa^{-1}g)\,d\kappa.$$

It is clear that the support of ϕ is contained in $KU_1 \subset KU$. We show that $|\phi(g)| < \epsilon$ when $g \notin U$. Indeed, $g \notin U$ implies that $g \notin VU_1$. If $\phi_1(\kappa^{-1}g) \neq 0$ then $\kappa^{-1}g \in U_1$, which implies that $\kappa \notin V$. Thus

$$\begin{aligned}
\int_{G-U} |\phi(g)|\,dg &\leq \int_{G-U}\int_K |\phi_0(\kappa)|\,\phi_1(\kappa^{-1}g)\,d\kappa\,dg \\
&= \int_{G-U}\int_{K-V} |\phi_0(\kappa)|\,\phi_1(\kappa^{-1}g)\,d\kappa\,dg \\
&\leq \int_{K-V} |\phi_0(\kappa)| \int_G \phi_1(\kappa^{-1}g)\,dg\,d\kappa \\
&= \int_{K-V} |\phi_0(\kappa)|\,d\kappa < \epsilon.
\end{aligned}$$

Next, let us show that $\pi(\phi) f$ is K-finite. Let (ρ, R) be the finite-dimensional unitary representation of K of which ϕ_0 is a matrix coefficient. We can write

$$\phi_0(\kappa) = \langle \rho(\kappa)\xi, \eta\rangle$$

for ξ and $\eta \in R$. If $\kappa_1 \in K$, we have

$$\begin{aligned}
\phi(\kappa_1^{-1}g) &= \int_K \phi_0(\kappa)\,\phi_1(\kappa^{-1}\kappa_1^{-1}g)\,d\kappa \\
&= \int_K \langle \rho(\kappa)\xi, \eta\rangle\,\phi_1(\kappa^{-1}\kappa_1^{-1}g)\,d\kappa \\
&= \int_K \left\langle \rho(\kappa_1^{-1})\rho(\kappa)\xi, \eta\right\rangle\,\phi_1(\kappa^{-1}g)\,d\kappa \\
&= \int_K \langle \rho(\kappa)\xi, \rho(\kappa_1)\eta\rangle\,\phi_1(\kappa^{-1}g)\,d\kappa.
\end{aligned}$$

Thus the space of functions $\phi(\kappa_1^{-1}g)$ lies in the finite-dimensional space spanned by functions of the form

$$g \mapsto \int_K \langle \rho(\kappa)\xi, \zeta\rangle\, \phi_1(\kappa^{-1}g)\, d\kappa, \qquad \zeta \in R.$$

This is a finite-dimensional space of functions, so the space spanned by the vectors

$$\pi(\kappa_1)\,\pi(\phi)\, f = \int_G \phi(g)\,\pi(\kappa_1 g)\, f\, dg$$

$$= \int_G \phi(\kappa_1^{-1}g)\,\pi(g)\, f\, dg$$

is finite dimensional. It follows from Proposition 2.4.4 that $\pi(\phi)\, f$ is a K-finite vector. It is also in $\mathfrak{H}^\infty$ by Proposition 2.4.2(i). We have shown that $\mathfrak{H}_0$ is dense in $\mathfrak{H}$.

Now we may see that $\mathfrak{H}_{\text{fin}} \subseteq \mathfrak{H}^\infty$. Using Lemma 2.4.1, the space $\mathfrak{H}_0 = \mathfrak{H}_{\text{fin}} \cap \mathfrak{H}^\infty$ is K-invariant. If σ is an irreducible admissible representation of K, let $\mathfrak{H}_0(\sigma)$ denote the σ-isotypic part. Clearly, $\mathfrak{H}_0(\sigma) \subseteq \mathfrak{H}(\sigma)$, and because $\mathfrak{H}_{\text{fin}}$ is the algebraic direct sum of the $\mathfrak{H}(\sigma)$, it is sufficient to show that $\mathfrak{H}_0(\sigma) = \mathfrak{H}(\sigma)$. If not, let $0 \neq f \in H(\sigma)$ be orthogonal to $\mathfrak{H}_0(\sigma)$. This is clearly possible because $\mathfrak{H}(\sigma)$ is finite-dimensional by our assumption of admissibility, and we are assuming that $\mathfrak{H}_0(\sigma)$ is a proper subspace. Then f is orthogonal to all of $\mathfrak{H}_0$ because it is automatically orthogonal to all the other isotypic subspace $H_0(\tau)$ for $\tau \neq \sigma$ (Exercise 2.4.1(a)). This is a contradiction because we have shown that $\mathfrak{H}_0$ is dense in $\mathfrak{H}$.

Next, we show that the space of K-finite vectors is invariant under $\mathfrak{g}$. Let f be a K-finite vector, and let R be a finite-dimensional subspace of $\mathfrak{H}$ because is stable under $\mathfrak{k}$ containing X. Let R_1 be the space spanned by vectors $Y\phi$ with $Y \in \mathfrak{g}$ and $\phi \in R$; obviously, R_1 is a finite-dimensional vector space. We show that R_1 is invariant under $\mathfrak{k}$. If $X \in \mathfrak{k}$ and $Y\phi \in R_1$, we have $X(Y\phi) = [X, Y]\phi + Y(X\phi)$. Both terms on the right side are in R_1. This proves that R_1 is invariant under $\mathfrak{k}$. The elements of R_1 are thus K-finite vectors by Proposition 2.4.5, showing that Yf is a K-finite vector for any $Y \in \mathfrak{g}$. ■

Let $GL(n, \mathbb{R})$ or $G = GL(n, \mathbb{R})^+$, and let $(\pi, \mathfrak{H})$ be a Hilbert space representation of G. Let $X \in \mathfrak{g} = \mathfrak{gl}(n, \mathbb{R})$, let $g \in \mathfrak{g}$, and let $f \in \mathfrak{H}_\infty$. We make note of the identity

$$\pi(g)\,\pi(X)\,\pi(g^{-1}) f = \pi\big(\mathrm{Ad}(g)\, X\big)\, f, \tag{4.11}$$

where $\mathrm{Ad}(g)\,X$ was defined in Exercise 2.2.1. Indeed, using Eq. (4.1),

$$\begin{aligned}\pi(g)\,X\,\pi(g^{-1})f &= \frac{d}{dt}\pi(g)\,\pi(e^{tX})\,\pi(g^{-1})\,f|_{t=0}\\ &= \frac{d}{dt}\pi(g\,e^{tX}\,g^{-1})\,f|_{t=0}\\ &= \frac{d}{dt}\pi\left(\exp(t\,gXg^{-1})\right)f|_{t=0}\\ &= \left(\mathrm{Ad}(g)\,X\right)f.\end{aligned}$$

We now introduce the important notion of a $(\mathfrak{g}, K)$-*module*. In this definition, G will be a reductive Lie group, K a maximal compact subgroup of G, and $\mathfrak{g}$ its Lie algebra. We have not defined the notion of a reductive Lie group in this book (see, for example, Knapp (1986)), but the reader may assume for now that $G = GL(n,\mathbb{R})^+$ and $K = SO(n)$, or else that $G = GL(n,\mathbb{R})$ and $K = O(n)$.

Let G, K, and $\mathfrak{g}$ be as above. By a $(\mathfrak{g}, K)$-*module*, we mean a vector space V together with representations π of K and of $\mathfrak{g}$ subject to the following conditions.

(i) We require that V decompose into an algebraic direct sum of finite-dimensional invariant subspaces under the action of K.
(ii) We require that the representations of $\mathfrak{g}$ and K be compatible in that Eq. (4.1) be valid when $X \in \mathfrak{k}$ and $f \in V$.
(iii) We require that Eq. (4.11) be valid when $g \in K$ and $X \in \mathfrak{g}$.

The $(\mathfrak{g}, K)$-module is called *admissible* if in a decomposition of V into finite-dimensional irreducible representations, no isomorphism type of any irreducible representation of K occurs with infinite multiplicity. It is called *irreducible* if there are no proper nonzero subspaces that are invariant under both K and $\mathfrak{g}$.

If $G = GL(n,\mathbb{R})^+$ and $K = SO(n)$ (or more generally in any case where K is connected) then condition (iii) in the definition can be deduced from (i) and (ii) (Exercise 2.4.4). On the other hand, if K is not connected, as when $G = GL(n,\mathbb{R})$ and $K = O(n)$, condition (iii) is definitely needed. We will see this in Exercise 2.5.4 (b) below.

Now let $(\pi, \mathfrak{H})$ be an admissible representation of G on a Hilbert space. It follows from Proposition 2.4.5 that we have representations of both K and $\mathfrak{g}$ on $\mathfrak{H}_{\mathrm{fin}}$, and we therefore obtain examples of $(\mathfrak{g}, K)$-modules in this way. We say that two representations $(\pi, \mathfrak{H})$ and $(\pi', \mathfrak{H}')$ are *infinitesimally equivalent* if their $(\mathfrak{g}, K)$-modules are isomorphic. (The isomorphism must be a linear isomorphism $\mathfrak{H}_{\mathrm{fin}} \to \mathfrak{H}'_{\mathrm{fin}}$ that is simultaneously an isomorphism of K-modules and of $\mathfrak{g}$-modules.) There are examples of naturally occurring representations that are infinitesimally equivalent but that are not isomorphic – for this, see Exercise 2.6.1. Such representations should be considered essentially the same.

We emphasize that everything in this section generalizes without serious modification to the context of an arbitrary reductive group. We have made the assumption that $G = GL(n, \mathbb{R})$ only to achieve a minor shortening of the proofs. For the general theory, we refer to the books of Knapp, Wallach, and Warner mentioned at the beginning of the section.

Exercises

Exercise 2.4.1 Let K be a compact subgroup of $GL(n, \mathbb{C})$, and let (π, H) be a unitary representation, so that Theorem 2.4.2 applies.

(a) Let H_1 and H_2 be two irreducible invariant subspaces of H such that the representations of K on H_1 and H_2 are nonisomorphic. Prove that H_1 and H_2 are orthogonal.

(b) Let σ be an isomorphism class of irreducible representations of K. Let $H(\sigma)$ be the sum of all irreducible invariant subspaces of H that are isomorphic to σ. Prove that $H(\sigma)$ is an invariant subspace of H, and that

$$H = \bigoplus_{\sigma} H(\sigma),$$

where the sum runs over the isomorphism classes of irreducible representations of K. (Hilbert space direct sum.) $H(\sigma)$ is called the σ*-isotypic part* of H.

(c) In any decomposition of H into a direct sum of irreducible representations, prove that the (possibly infinite) number of terms isomorphic to σ is $\dim\big(H(\sigma)\big)/\dim(\sigma)$; in particular, the number of terms is independent of the particular direct sum decomposition.

Exercise 2.4.2 Let $G = GL(n, \mathbb{R})$, and let $X \in \mathfrak{g} = \mathfrak{gl}(n, \mathbb{R})$.

(a) Let $f \in C^\infty(G)$. Prove that

$$f(ge^X) = \sum_{n=0}^{\infty} \frac{1}{n!} (dX^n f)(g),$$

where $dX : C^\infty(G) \to C^\infty(G)$ is defined by Eq. (2.11).

[HINT: Use Lemma 2.2.2 with

$$\phi(g, t) = f(ge^{tX}) - \sum_{n=0}^{\infty} \frac{t^n}{n!} (dX^n f)(g).$$
]

(b) Let (π, H) be a representation of G on a Hilbert space, and let the representation of $\mathfrak{g}$ on the space H^∞ of smooth vectors be defined by Eq. (4.1). Prove that for $f \in H^\infty$, we have

$$\pi(e^X)f = \sum_{k=0}^{\infty} \frac{1}{n!} X^n f.$$

(Here of course $X^n f$ means X applied n times – or if you prefer, X^n is X raised to the nth power in the universal enveloping algebra $U(\mathfrak{g})$, which acts on H^∞ by Proposition 2.2.3.)

[HINT: Let $\phi \in H$. It is sufficient to show that

$$\left\langle \pi(e^{tX})f - \sum_{n=0}^{\infty} \frac{1}{n!} t^n X^n f, \phi \right\rangle = 0.$$

for all $t \in \mathbb{R}$. Make use of Eq. (4.5).]

Exercise 2.4.3 **Gelfand Pairs.** The purpose of this exercise is to shed some light on the relationship between Theorem 2.2.3 and Theorem 2.4.2. Let G be a topological group and K a closed subgroup. (We are making no assumptions of compactness – momentarily, we will assume for simplicity that G and K are both finite, but for the purpose of the present definition, we make no assumption.) We say (G, K) is a *Gelfand pair* if every irreducible unitary representation H of G has at most one linearly independent K-fixed vector. Assume that G and K are both finite. Let R be the algebra of K biinvariant functions $\phi : G \to \mathbb{C}$, with multiplication given by convolution:

$$(\phi_1 * \phi_2)(g) = \sum_{h \in G} \phi_1(gh)\, \phi_2(h^{-1}).$$

Prove that (G, K) is a Gelfand pair if and only if the convolution algebra R is commutative. [HINT: study the proof of Theorem 4.1.2.] See Gross (1991) for a survey of the important role of Gelfand pairs in representation theory and automorphic forms.

Exercise 2.4.4 Let $G = GL(n,\mathbb{R})^+$ and $K = SO(n,\mathbb{R})$. Prove that in the definition of a $(\mathfrak{g}, K)$-module, condition (iii) is automatically satisfied. [HINT: Let $X \in \mathfrak{g}$, $Y \in \mathfrak{k}$, let $v \in V$, and let Λ be any linear functional on $\mathfrak{g}$. Define $\phi : K \times \mathbb{R} \to \mathbb{C}$ by

$$\phi(g,t) = \Lambda\Big(\pi(ge^{tY})\, X\, \pi(e^{-tY}g^{-1})\, f - \pi(g)\, \big(\mathrm{Ad}(e^{tY})\, X\big)\, \pi(g^{-1})\, f\Big).$$

Verify that Lemma 2.2.2 remains valid when $G = SO(n)$, and apply it to this function ϕ. Now use Exercise 2.2.6 (b).]

Exercise 2.4.5 **Induction from a subgroup of finite index.** Let G be a reductive Lie group, K a maximal compact subgroup, and $\mathfrak{g}$ its Lie algebra. Let G_0 be a subgroup of finite index in G, and let $K_0 = G_0 \cap K$. Suppose that every connected component of G contains a unique connected component of K. Show that this implies that $[G : G_0] = [K : K_0]$. Let V_0 be a $(\mathfrak{g}, K_0)$-module. We will construct a $(\mathfrak{g}, K)$-module that is "induced" from V_0. We will denote the actions of both $\mathfrak{g}$ and K_0 on V_0 as π_0 to avoid confusion. Thus we will write $\pi_0(X)\, v$ instead of Xv for $X \in \mathfrak{g}$, $v \in V_0$.

The space V consists of all functions $F : K \to V_0$ that satisfy $F(k_0 k) = \pi_0(k_0)\, F(k)$ when $k_0 \in K_0$, $k \in K$. The action π of K on V is given by

$$\big(\pi(\kappa)\, F\big)(k) = F(k\,\kappa),$$

and the action π of $\mathfrak{g}$ is given by

$$\big(\pi(X)\, F\big)(k) = \pi_0\big(\operatorname{ad}(k)\, X\big)\, F(k).$$

Prove that with these actions of K and $\mathfrak{g}$, V is a $(\mathfrak{g}, K)$-module.

Exercise 2.4.6 SCHUR'S LEMMA FOR $(\mathfrak{g}, K)$-MODULES. Let $\mathfrak{g}$ and K be as in the definition of a $(\mathfrak{g}, K)$-module. Let V and W be irreducible admissible $(\mathfrak{g}, K)$-modules.

(a) Prove that the space of linear maps $\lambda : V \to W$ that commute with the actions of $\mathfrak{g}$ and K is at most one dimensional. [HINT: If V and W are not isomorphic, there are no such linear maps, so there is no loss of generality in assuming $V = W$. Let ρ be an irreducible finite-dimensional representation of K such that the isotypic space $V(\rho)$ is nonzero. Show that λ maps $V(\rho)$ to itself, and hence has an eigenvector on $V(\rho)$. Now show that the corresponding eigenspace on V is invariant under the actions of K and $\mathfrak{g}$, and deduce that λ acts by a scalar.]

(b) Denoting the $(\mathfrak{g}, K)$-actions of $\mathfrak{g}$ and K on V and on W by π_V and π_W, respectively, show that the space of bilinear maps $\langle\, ,\, \rangle : V \times W \to \mathbb{C}$ such that

$$\langle v, w \rangle = \langle \pi_V(k) v, \pi_W(k) w \rangle\,, \qquad k \in K \tag{4.12}$$

and

$$\langle \pi_V(X)\, v, w \rangle = -\, \langle v, \pi_W(X)\, w \rangle \tag{4.13}$$

is at most one dimensional.

(c) Recall that a pairing $V \times W \to \mathbb{C}$ is *sesquilinear* if it is linear in the first variable and conjugate linear in the second. Prove that the dimension of the space of sesquilinear maps $\langle\, ,\, \rangle : V \times W \to \mathbb{C}$ which satisfy Eqs. (4.12) and (4.13) is at most one dimensional.

2.5 Irreducible $(\mathfrak{g}, K)$-Modules for $GL(2, \mathbb{R})$

For most of this section, we let $G = GL(2, \mathbb{R})^+$ and $K = SO(2)$. At the end of the section, we will deduce the corresponding classification for $G = GL(2, \mathbb{R})$ from the results for $GL(2, \mathbb{R})^+$.

Let $\hat{H}$, $\hat{R}$, $\hat{L}$, Z, H, R, L, and W be elements of $\mathfrak{g}_{\mathbb{C}}$ defined by Eqs. (2.18), (2.23), and (2.35). Let $\Delta \in U(\mathfrak{g})$ be the Casimir element, defined by Eq. (2.20).

Let V be an irreducible admissible $(\mathfrak{g}, K)$-module. We will denote the action of K on V by $\pi : K \to \mathrm{End}(V)$. We decompose

$$V = \bigoplus V(\sigma), \qquad \text{(algebraic direct sum)} \tag{5.1}$$

where $V(\sigma)$ is the K-isotypic part. Because V is admissible, these spaces are finite dimensional.

We extend the action of $\mathfrak{g}$ to an action of $U(\mathfrak{g})$ by Proposition 2.2.3. Because V is a complex vector space, it is a module for the complexification for $\mathfrak{g}_{\mathbb{C}}$ and hence for the complex associative algebra $U(\mathfrak{g}_{\mathbb{C}})$; recall that H, R, and L are actually in $\mathfrak{g}_{\mathbb{C}}$. These elements of $\mathfrak{g}_{\mathbb{C}}$ are crucial to the classification, so our calculations will be done in $U(\mathfrak{g}_{\mathbb{C}})$.

Proposition 2.5.1 *Let V be an irreducible admissible $(\mathfrak{g}, K)$-module for $GL(2, \mathbb{R})^+$. Let D be an element of the center of $U(\mathfrak{g})$. Then D acts by a scalar on V. That is, there exists a constant λ such that $Dx = \lambda x$ for all $x \in V$.*

Thus both Z and Δ act as scalars on V. This is a form of Schur's Lemma.

Proof It follows from condition (iii) in the definition of a $(\mathfrak{g}, K)$-module that $D \circ \pi(\kappa) = \pi(\kappa) \circ D$ for $\kappa \in K$. Consequently, the isotypic subspaces $V(\sigma)$ in Eq. (5.1) are stable under D. Because these spaces are finite dimensional, D has a nonzero eigenvector in each nonzero $V(\sigma)$. Let $\lambda \in \mathbb{C}$ such that $Dx_0 = \lambda x_0$ for some $0 \neq x_0 \in V$, and let V_0 be the set of all x such that $Dx = \lambda x$. Because D is in the center of $U(\mathfrak{g})$, it commutes with the action of $\mathfrak{g}$ on V, and (by Eq. (4.11), which is assumed in the definition of a $(\mathfrak{g}, K)$-module, and by Exercise 2.2.5) it also commutes with the action of K. Therefore V_0 is a nonzero invariant subspace of V, and because V is assumed to be irreducible, $V = V_0$. ■

Because $K \cong \mathbb{R}/\mathbb{Z}$, the irreducible unitary representations of K are one dimensional and are parametrized by integers: They are the characters $\sigma_k(\kappa_\theta) = e^{ik\theta}$ with $k \in \mathbb{Z}$, where κ_θ was defined in Eq. (1.19). We will denote the isotypic component $V(\sigma_k)$ as simply $V(k)$. Thus Eq. (5.1) reads

$$V = \bigoplus_{k \in \mathbb{Z}} V(k). \qquad \text{(algebraic direct sum)} \tag{5.2}$$

Let Σ be the set of all integers such that $V(k) \neq 0$. We call Σ the *set of K-types of V*.

Proposition 2.5.2 *Let V be an irreducible admissible $(\mathfrak{g}, K)$-module for $GL(2, \mathbb{R})^+$.*

(i) *$V(k)$ is the space of all vectors $x \in V$ such that $Hx = kx$.*
(ii) *If $x \in V(k)$, then $Rx \in V(k+2)$ and $Lx \in V(k-2)$.*
(iii) *If $0 \neq x \in V(k)$, then $\mathbb{C}x = V(k)$, $\mathbb{C}R^n x = V(k+2n)$, $\mathbb{C}L^n x = V(k-2n)$, and*

$$V = \mathbb{C}x \oplus \bigoplus_{n>0} \mathbb{C}\,R^n x \oplus \bigoplus_{n>0} \mathbb{C}\,L^n x. \tag{5.3}$$

(iv) *Each space $V(k)$ is at most one dimensional. If $V(k)$ and $V(l)$ are both nonzero, then $k - l$ is an even integer.*
(v) *Suppose that λ is the eigenvalue of Δ on V. If $x \in V(k)$, then*

$$LRx = \left(-\lambda - \tfrac{k}{2}(1 + \tfrac{k}{2})\right) x, \qquad RLx = \left(-\lambda + \tfrac{k}{2}(1 - \tfrac{k}{2})\right) x. \tag{5.4}$$

(vi) *Suppose that λ is the eigenvalue of Δ on V. If $0 \neq x \in V(k)$ and $Rx = 0$, then $\lambda = -\frac{k}{2}(1 + \frac{k}{2})$, while if $Lx = 0$, then $\lambda = \frac{k}{2}(1 - \frac{k}{2})$.*
(vii) *Suppose that $\lambda = \frac{k}{2}(1 - \frac{k}{2})$ and $x \in V(l)$. If $Rx = 0$, then either $l = -k$ or $l = k - 2$, and if $Lx = 0$, then either $l = k$ or $l = 2 - k$.*

Proof If $x \in V(k)$, we have

$$Wx = \frac{d}{dt}\pi(e^{tW})x|_{t=0} = \frac{d}{dt}\pi(\kappa_t)x|_{t=0} = \frac{d}{dt}e^{ikt}x|_{t=0} = ikx.$$

Because $H = -iW$, we have

$$Hx = kx \qquad \text{for } x \in V(k) \tag{5.5}$$

By Eq. (5.2), V is the direct sum of the $V(k)$, so we may characterize $V(k)$ as the eigenspace of H with eigenvalue k. This proves (i).

As for (ii), let $x \in V(k)$, so $Hx = kx$. We recall from Eq. (2.24) that $[H, R] = 2R$, and so

$$HRx = [H, R]x + RHx = 2Rx + Rkx = (k+2)Rx,$$

so $Rx \in V(k+2)$; similarly, $Lx \in V(k-2)$. This proves (ii).

Next we prove (iii). By (ii) and by (i), we have $\mathbb{C}x \subseteq V(k)$, $\mathbb{C}R^n x \subseteq V(k+2n)$, $\mathbb{C}L^n x \subseteq V(k-2n)$, and so by Eq. (5.2), the sum on the right side of Eq. (5.3) is direct. We will prove that the right side of Eq. (5.3) is invariant under K and $\mathfrak{g}$; it is nonzero because it contains x, so because V is irreducible, it will follow that it is all of V. This in turn will imply that $\mathbb{C}x = V(k)$, and $\mathbb{C}R^n x = V(k+2n)$, and $\mathbb{C}L^n x = V(k-2n)$.

Because each summand in Eq. (5.3) is an eigenspace for K, the right side of Eq. (5.3) is invariant under K. To show that it is invariant under $\mathfrak{g}$, it is sufficient to check invariance under H, R, L, and Z, becuase these generate $\mathfrak{g}_{\mathbb{C}}$. Invariance under H follows from (i), becuase each summand is contained in $V(l)$ for some l. Invariance under Z follows from Proposition 2.5.1. We check invariance under R and L.

We will show that *if* $\xi \in V(l)$, *then* $RL\xi$, $LR\xi \in \mathbb{C}\xi$. Note that this will prove the invariance of Eq. (5.3) under R and L, becuase by (ii) each summand in Eq. (5.3) is contained in some $V(l)$. Indeed, by Proposition 2.5.1, there exists a constant λ such that $\Delta v = \lambda v$ for all $v \in V$. We have

$$\begin{aligned} -4\Delta &= H^2 + 2RL + 2LR \\ &= H^2 + 2H + 4LR \\ &= H^2 - 2H + 4RL. \end{aligned} \tag{5.6}$$

Thus

$$-4\lambda\xi = (l^2 + 2l)\xi + 4LR\xi = (l^2 - 2l)\xi + 4RL\xi. \tag{5.7}$$

This implies that $LR\xi$ and $RL\xi$ are multiples of ξ, as required.

Part (iv) follows from (iii).

Part (v) follows from Eq. (5.7) with $\xi = x$ and $l = k$.

Part (vi) follows from (v), becuase if $Rx = 0$, then $LRx = 0$ also.

Part (vii) is really a paraphrase of part (vi). If $Rx = 0$, then by (vi), we have $-\frac{l}{2}(1 + \frac{l}{2}) = \lambda = \frac{k}{2}(1 - \frac{k}{2})$, which implies that either $l = -k$ or $l = k - 2$. The proof is similar if $Lx = 0$. ■

It follows from Proposition 2.5.2(iv) that if V is an irreducible admissible $(\mathfrak{g}, K)$-module, then the set of K-types of V is a set of integers that are either all even or all odd. We say that the *parity* of V is either *even* or *odd* accordingly.

Theorem 2.5.1 *Let λ and μ be complex numbers. Assume that λ is not of the form $\frac{k}{2}(1 - \frac{k}{2})$ with k an even (resp. odd) integer. Then there exists at most one isomorphism class of even (resp. odd) $(\mathfrak{g}, K)$-modules V on which Δ and Z have respective eigenvalues λ and μ. For such a V, the set of K-types of V consists of all even (resp. odd) integers k.*

We will not prove until later that a $(\mathfrak{g}, K)$-module exists as described in the theorem, but at least we know that if it exists, it is unique up to isomorphism.

Proof The last assertion – that $V(k) \neq 0$ for all even (resp. odd) k – is proved as follows using Proposition 2.5.2(iii) and 2.5.2(vi). We choose one nonzero $x \in V(l)$ for some l, and then by our hypothesis that λ is not of the form $\frac{k}{2}(1 - \frac{k}{2})$ with k an even (resp. odd) integer, $L^n x \neq 0$ for all n; replacing k by $-k$ in our assumption shows that λ is also not of the form $-\frac{k}{2}(1 + \frac{k}{2})$ with k an even (resp. odd) integer, and so $R^n x \neq 0$ for all n also. This gives us nonzero vectors in $V(k + 2n)$ and $V(k - 2n)$ for all n.

Now let V' be another such $(\mathfrak{g}, K)$-module. We will construct an isomorphism $V \to V'$. Let k be an integer of the correct parity, and let $0 \neq x \in V(k)$

and $0 \neq x' \in V'(k)$. Then the x, $L^n x$, and $R^n x$, $(n > 0)$ form a basis of the vector space V, and x', $L^n x'$, and $R^n x'$ form a basis of V'. There exists, therefore, a linear isomorphism $\phi : V \to V'$ such that $\phi(x) = x'$, $\phi(L^n x) = L^n x'$, and $\phi(R^n x) = R^n x'$. Because this map takes each K-isotypic subspace of V into the corresponding isotypic part of V', it satisfies $\phi \circ \pi(\kappa) = \pi'(\kappa) \circ \phi$ for $\kappa \in K$. (We are denoting the representations of K on V and V' as π and π'.)

We must therefore show that ϕ is an isomorphism of modules for the Lie algebra $\mathfrak{g}$. It is sufficient to show that $\phi \circ X = X \circ \phi$ when $X = H, R, L$, and Z, because these are a basis of $\mathfrak{g}_{\mathbb{C}}$. This is clear for H and Z, which act as the same scalars k and μ on $V(k)$ as on $V'(k)$. As for R and L, all we really need to check is that $\phi(RL^n x) = RL^n x'$ and $\phi(LR^n x) = LR^n x'$, and this follows from Proposition 2.5.2(v), with $L^{n-1}x$ or $R^{n-1}x$ replacing x and with $k - 2n + 2$ or $k + 2n - 2$ replacing k. ■

Now if $\lambda = \frac{k}{2}(1 - \frac{k}{2})$, where k is an integer, we still have to classify the irreducible $(\mathfrak{g}, K)$-modules of parity K on which Δ has eigenvalue λ. We observe that replacing k by $2 - k$ does not change the value of λ, so we may assume that k is an integer ≥ 1.

Theorem 2.5.2 *Let $k \geq 1$ be an integer, and let $\lambda = \frac{k}{2}(1 - \frac{k}{2})$. Let V be an irreducible admissible $(\mathfrak{g}, K)$-module with parity the same as that of k. Let Σ be the set of K-types of V. Then Σ is one of the following sets:*

$$\Sigma^+(k) = \{l \in \mathbb{Z} | l \equiv k \bmod 2,\ l \geq k\};$$

$$\Sigma^-(k) = \{l \in \mathbb{Z} | l \equiv k \bmod 2,\ l \leq -k\};$$

$$\Sigma^0(k) = \{l \in \mathbb{Z} | l \equiv k \bmod 2,\ -k < l < k\}.$$

There exists at most one isomorphism class of $(\mathfrak{g}, K)$-module with each given Σ. (The case $\Sigma = \Sigma^0(k)$ clearly cannot occur if $k = 1$.)

Proof First suppose that $V(k) \neq 0$. Let $0 \neq x \in V(k)$. We claim that $Lx = 0$; indeed, if $Lx \neq 0$, let $y = Lx \in V(k-2)$. By Eq. (5.4), $Ry = 0$, and so

$$\bigoplus_{n \geq 0} L^n y$$

is closed under both R and L; it is also closed under H, Z, and the action of K, so it is an invariant submodule of V that is nonzero because it includes y, yet proper because it excludes x, contradicting the irreducibility of V. Thus $Lx = 0$. Now

$$\bigoplus_{n \geq 0} R^n x$$

is closed under R and L and hence is an invariant submodule of V and therefore equals V. None of the vectors $R^n x \in V(k+2n)$ is zero by Proposition 2.5.2(vii), so $\Sigma = \Sigma^+$ in this situation.

Next suppose that $V(k) = 0$ but $V(k-2) \neq 0$. Let $0 \neq x \in V(k-2)$. It follows from Proposition 2.5.2(vii) that $Lx, L^2x, \cdots L^{k-2}x$ are nonzero vectors in $V(k-2), \cdots, V(2-k)$. We claim that $L^{k-1}x = 0$. Let $L^{k-1}x = y$. If $y \neq 0$, then $Ry = RL(L^{k-2}x) = 0$, because $L^{k-2}x \in V(2-k)$, and RL annihilates this space by Eq. (5.4). Now

$$\bigoplus_{n\geq 0} L^n y$$

is closed under both R and L; it is also closed under H, Z, and the action of K, so it is an invariant submodule of V that excludes x, contradicting the irreducibility of V. Thus $L^{k-1}x = 0$. Now

$$\bigoplus_{n=0}^{k-2} L^n x$$

is closed under R and L and hence is an invariant submodule of V and therefore equals V. In this case, we see that $\Sigma = \Sigma^0$.

Finally, suppose that $V(k) = 0$ and $V(k-2) = 0$. Let l be any integer such that $V(l) \neq 0$. By hypothesis, the parity of V is the same as k, so $l \equiv k \bmod 2$. We will show that $l \leq -k$. We will obtain distinct contradictions in the cases $l \geq k$ and $2-k \leq l \leq k-2$.

If $l \geq k$, and if $V(l) \neq 0$, then by applying L multiple times to $V(l)$, one can obtain a nonzero element of $V(k)$; for $x \in V(l)$ with $l > k$, $Lx \neq 0$ by Proposition 2.5.2(vii). This contradicts the assumption that $V(k) = 0$.

If $2-k \leq l \leq k-2$, and if $V(l) \neq 0$, then we similarly apply R multiple times to get a nonzero vector in $V(k-2)$; again, we've used Proposition 2.5.2(vii) to see that $Rx \neq 0$ if $0 \neq x \in V(l)$ with $2-k \leq l < k-2$. This contradicts the assumption that $V(k-2) = 0$.

We see that $l \leq -k$ and $l \equiv k \bmod 2$. Applying R multiple times, we get a nonzero vector $x \in V(-k)$. Then $Rx = 0$ because $Rx \in V(2-k)$. However, $L^n x \neq 0$ for $n \geq 0$, and so $\Sigma = \Sigma^-$ in this case.

The proof that Σ determines the $(\mathfrak{g}, K)$-module up to isomorphism is similar to the isomorphism assertion in Theorem 2.5.1, and is left to the reader. ■

We now turn to the question of *existence* that was left open by Theorems 2.5.1 and 2.5.2. We ask firstly, do the $(\mathfrak{g}, K)$-modules described in these theorems actually exist? And can each such $(\mathfrak{g}, K)$-module be realized in the K-finite vectors in some irreducible admissible representation $(\pi, \mathfrak{H})$ of $GL(n,\mathbb{R})^+$?

To answer these questions, let there be given a parity $\epsilon = 0$ or 1 (i.e., $\epsilon = 0$ (resp. 1) if we are trying to construct an even (resp. odd) representation) and parameters s_1 and s_2. We will construct an admissible representation $\pi(s_1, s_2, \epsilon)$

of G on a Hilbert space. Let $\lambda = s(1-s)$, where $s = \frac{1}{2}(s_1 - s_2 + 1)$, and let $\mu = s_1 + s_2$. Then λ and μ will be the eigenvalues of Δ and Z on this representation. The set of K-types will be all integers congruent to ϵ modulo 2. We will see that if λ is not of the form $\frac{k}{2}(1-\frac{k}{2})$, where k is an integer congruent to ϵ modulo 2, then $\pi(s_1, s_2, \epsilon)$ is irreducible. This will afford examples of the representations in Theorem 2.5.1.

If on the other hand, $\lambda = \frac{k}{2}(1-\frac{k}{2})$, where k is an integer congruent to ϵ modulo 2, then $\pi(s_1, s_2, \epsilon)$ is reducible, although it does not split into a direct sum of irreducible representations. However, it does have a composition series and we will be able to find inside of it three irreducible subquotients affording us examples of the representations in Theorem 2.5.2. This slightly complex situation will be described in greater detail below.

Let $H^\infty(s_1, s_2, \epsilon)$ be the space of smooth functions $f \in C^\infty(G)$ that satisfy

$$f\left(\begin{pmatrix} y_1 & x \\ & y_2 \end{pmatrix} g\right) = y_1^{s_1+1/2}\, y_2^{s_2-1/2}\, f(g), \qquad y_1, y_2 > 0, \tag{5.8}$$

$$f\left(\begin{pmatrix} -1 & \\ & -1 \end{pmatrix} g\right) = (-1)^\epsilon f(g). \tag{5.9}$$

We let G act by right translation:

$$\big(\pi(g)f\big)(x) = f(xg). \tag{5.10}$$

Both properties Eqs. (5.8) and (5.9) are invariant under the action Eq. (5.10). We recall the *Iwasawa decomposition*: Every element of G may be written uniquely as

$$g = \begin{pmatrix} u & \\ & u \end{pmatrix} \begin{pmatrix} y^{1/2} & xy^{-1/2} \\ & y^{-1/2} \end{pmatrix} \kappa_\theta, \quad \kappa_\theta = \begin{pmatrix} \cos(\theta) & \sin(\theta) \\ -\sin(\theta) & \cos(\theta) \end{pmatrix}, \tag{5.11}$$

where $x, y, u, \theta \in \mathbb{R}$, $u, y > 0$ (Exercise 1.3). Thus $f \in H^\infty(s_1, s_2, \epsilon)$ is determined by its values on K. Moreover, the restriction of f to K can be any smooth function, subject to the condition

$$f(\kappa_{\theta+\pi}) = (-1)^\epsilon f(\kappa_\theta), \tag{5.12}$$

which is a consequence of Eq. (5.9), for given such an f, we can extend it to G by

$$f(g) = u^{s_1+s_2}\, y^s\, f(\kappa_\theta) \tag{5.13}$$

for g as in Eq. (5.11). We give $H^\infty(s_1, s_2, \epsilon)$ a Hermitian inner product by defining

$$\langle f_1, f_2 \rangle = \frac{1}{2\pi} \int_0^{2\pi} f_1(\kappa_\theta)\, \overline{f_2(\kappa_\theta)}\, d\theta. \tag{5.14}$$

Let $H(s_1, s_2, \epsilon)$ be the Hilbert space completion of this space. It can be identified with $L^2[-\pi/2, \pi/2]$ via the map $\theta \mapsto \kappa_\theta$.

We have defined the notion of smooth vectors in a representation of $GL(n,\mathbb{R})^+$, but this notion is valid for any Lie group. We consider smooth vectors in a representation $(\rho, \mathfrak{H})$ of $K = SO(2)$, where $\mathfrak{H}$ is a Hilbert space. We say a vector $f \in \mathfrak{H}$ is C^1 if the limit

$$f' = \lim_{t\to 0} \tfrac{1}{t}\left(\rho(\kappa_t) f - f\right)$$

exists in $\mathfrak{H}$ and if, moreover, $t \mapsto \rho(\kappa_t) f'$ is a continuous map $\mathbb{R} \to \mathfrak{H}$. If $k > 1$, we recursively define f to be C^k if it is C^1 and f' is C^{k-1}. We say f is C^∞ or *smooth* if it is C^k for all k.

Lemma 2.5.1 *Let ρ be the regular representation of K on the Hilbert space $L^2(K) = L^2[0, 2\pi]$. Then the smooth vectors in ρ are precisely the elements of $C^\infty(K)$.*

Proof Let $f \in L^2[0, 2\pi]$ have a Fourier expansion $f(x) = \sum a_n\, e^{2\pi i n x}$, where the convergence is in $L^2[0, 2\pi]$. If $t \in \mathbb{R}$, let f_t denote $\rho(\kappa_t) f$, so $f_t(x) = f(x+t)$ in the identification of $L^2(K)$ with $L^2[0, 2\pi]$. Suppose that f is a C^1 vector. Then there exists $g \in L^2[0, 2\pi]$ such that $\lim \frac{1}{t}(f_t - f) = g$, where again the convergence is with respect to the L^2 norm. Let $g(x) = \sum b_n\, e^{2\pi i n x}$. The Fourier expansion of $\frac{1}{t}(f_t - f)$ is

$$\sum_n \tfrac{1}{t}(e^{2\pi i n t} - 1)\, a_n\, e^{2\pi i n x},$$

so we have

$$\lim_{t\to 0} \sum_n \left|\tfrac{1}{t}(e^{2\pi i n t} - 1)\, a_n - b_n\right|^2 = 0.$$

Because the terms in this series are positive, we may truncate the series to conclude that

$$\begin{aligned} 0 &= \lim_{t\to 0} \sum_{-N\le n\le N} \left|\tfrac{1}{t}(e^{2\pi i n t} - 1)\, a_n - b_n\right|^2 \\ &= \sum_{-N\le n\le N} \lim_{t\to 0} \left|\tfrac{1}{t}(e^{2\pi i n t} - 1)\, a_n - b_n\right|^2 \\ &= \sum_{-N\le n\le N} |2\pi i n\, a_n - b_n|^2 . \end{aligned}$$

Consequently, $b_n = 2\pi\ i n\ a_n$ for all n. Because $\sum |b_n|^2 < \infty$, we see that if

f is a C^1 vector, then $\sum n^2 |a_n|^2 < \infty$. Iterating, if f is smooth, we have

$$\sum |p(n)\, a_n^2| < \infty \tag{5.15}$$

for any polynomial $p(n)$. This implies that

for any $N > 0$, there exists a constant C_N such that

$$|a_n| < C_N\, n^N \text{ for } n \neq 0. \tag{5.16}$$

(Indeed, we may assume without loss of generality that N is an integer, in which case $|a_n| < n^{-N}$ for all but finitely many n because otherwise Eq. (5.15) would fail with $p(n) = n^{2N}$; then C_N may be any number larger than 1 that is big enough to cover this finite number of exceptions.)

Now Eq. (5.16) implies that the series $f(x) = \sum a_n\, e^{2\pi i n x}$ converges absolutely and uniformly and that term-by-term differentiation is justified; moreover, the property Eq. (5.16) is preserved under differentiation, so f has continuous derivatives of all orders; hence f is a smooth function.

Conversely, it is well known and easy to show that the Fourier coefficients of a smooth function satisfy Eq. (5.16), which is enough to guarantee that f is a smooth vector in the regular representation – the reasoning is entirely reversible. ■

Proposition 2.5.3 *The action (Eq. 5.10) can be extended to $H(s_1, s_2, \epsilon)$. The subspace $H^\infty(s_1, s_2, \epsilon)$ is the space of smooth vectors for this representation.*

Proof To show that the action (Eq. 5.10) extends, we show that $\pi(g)$ defined by Eq. (5.10) is a bounded operator on $H^\infty(s_1, s_2, \epsilon)$; once we know it is a bounded operator on a dense subspace, it follows that it can be extended to the closure $\mathfrak{H} = H(s_1, s_2, \epsilon)$ by continuity.

It follows from the Cartan decomposition (Proposition 2.2.7) that G is generated by K together with the elements of the form

$$a(y_1, y_2) = \begin{pmatrix} y_1 & \\ & y_2 \end{pmatrix}, \qquad y_1, y_2 > 0, \tag{5.17}$$

and because $\pi(\kappa)$ clearly preserves the inner product $\langle\, , \,\rangle$, it is sufficient to prove that $\pi(g)$ is a bounded operator when $g = a(y_1, y_2)$.

We make use of the identity

$$\kappa_\theta\, a(y_1, y_2) = \begin{pmatrix} y_1 y_2\, D(\theta)^{-1} & \xi\, D(\theta)^{-1} \\ & D(\theta) \end{pmatrix} \kappa_{\theta'},$$

where

$$\theta' = \arctan\left(\frac{y_1}{y_2}\tan(\theta)\right), \qquad D(\theta) = \sqrt{y_1^2 \sin^2(\theta) + y_2^2 \cos^2(\theta)},$$

$$\xi = (y_2^2 - y_1^2)\sin(\theta)\cos(\theta).$$

Thus

$$\big(\pi(a(y_1, y_2))f\big)(\kappa_\theta) = (y_1 y_2)^{s_1+1/2}\, D(\theta)^{-s_1+s_2-1} f(\kappa_{\theta'}).$$

Because $d\theta' = y_1 y_2\, D(\theta)^{-2}\, d\theta$, we have

$$\int_0^{2\pi} \left|\big(\pi(a(y_1, y_2))f\big)(\kappa_\theta)\right|^2 d\theta = (y_1 y_2)^{s_1-1/2} \int_0^{2\pi} D(\theta)^{-s_1+s_2+1}\, |f(\kappa_{\theta'})|^2\, d\theta'. \tag{5.18}$$

Because $D(\theta)$ is bounded above and below, we see that $\pi\left(a(y_1, y_2)\right)$ is a bounded operator and hence extends to all of $H(s_1, s_2, \epsilon)$.

We thus obtain an action of G on a Hilbert space. To check that this is a representation, we must show that the map $(g, f) \mapsto \pi(g)f$ is continuous. Thus it is necessary to show that if g is near the identity and $|f|_2$ is small, then $|\pi(g)f|_2$ is small. If $\kappa \in K$, then $|\pi(\kappa)f|_2 = |f|_2$; we may, in general, write $g = \kappa_1 a(y_1, y_2)\kappa_2$ with $\kappa_1, \kappa_2 \in K$, and if g is near the identity, then so is $a(y_1, y_2)$, so it is sufficient to show that if g is near the identity and $|f|_2$ is small, then $|\pi(g)f|_2$ is small in the special case $g = a(y_1, y_2)$. This is again an easy consequence of Eq. (5.18). It follows that $(\pi, \mathfrak{H})$ is a representation.

Finally, we show that the space of smooth vectors in $\mathfrak{H}$ is precisely $H^\infty(s_1, s_2, \epsilon)$. It is clear that if f is a function on K, it is smooth if and only if its extension (Eq. 5.13) to a function on G is smooth. If this is the case, then $X f$ is smooth for all $X \in \mathfrak{g}$ and indeed is smooth for all iterates of such differential operators; it follows easily that the elements in $H^\infty(s_1, s_2, \epsilon)$ are all smooth vectors in the sense defined in Section 2.4. The point is that the limit Eq. (4.1) exists if convergence is taken in the sense of uniform convergence regarding f as a function defined on the compact set $[0, 2\pi]$, the L^2 norm is dominated by the L^∞ norm, so the limit Eq. (4.1) exists in L^2 as well.

Conversely, if f is a smooth vector, we consider Wf, with W as in Eq. (2.35). Considering just the restriction of f to K, identified with $[0, 2\pi]$ via $\theta \mapsto \kappa_\theta$, we assume that the limit Wf defined by Eq. (4.1) exists in L^2. We observe that by Eq. (4.1)

$$Wf = \lim_{t\to 0} \tfrac{1}{t}(f_t - f),$$

so f is a smooth vector in the sense of Lemma 2.5.1, so f is a smooth function on K, and its extension to G is therefore a smooth function also. ■

The representation $H(s_1, s_2, \epsilon)$ is an example of an *induced representation.* We recall the following construction for finite groups. (See Section 4.1 for further details.) Let G be a finite group, H a subgroup, and (ρ, V) a representation of H. There are different ways of defining an induced representation of G, but the definition that generalizes best is the following. We let V^G be the space of all functions $f : G \to V$ that satisfy $f(hg) = \rho(h) f(g)$ for $h \in H$ and let G act on this space by right translation.

If G is not a finite group, but rather a topological group, and H a closed subgroup, we make almost the same definition, but we impose some conditions on the functions: we ask that they be smooth, or (in some sense) square integrable, or (if we are only interested in a $(\mathfrak{g}, K)$-module) K-finite. The precise condition that we impose to limit the size of the space is not so important, because the infinitesimal equivalence class of the representation will not depend on it. There is another modification that we make, however, and that is that we incorporate into the definition of the space the *modular quasicharacters* δ_H and δ_G of the groups H and G. The modular quasicharacter of G is defined by the condition that $d_R(g) = \delta_G(g)\, d_L(g)$, where d_L and d_R are left and right Haar measures, respectively.

Now let (ρ, V) be a representation of H. The space V^G of the induced representation ρ^G is to consist of functions that satisfy

$$f(hg) = \sqrt{\frac{\delta_H(h)}{\delta_G(g)}}\, \rho(h)\, f(g), \qquad h \in H \tag{5.19}$$

subject, as we have stated, to some conditions such as smoothness or square-integrability, which may be chosen for convenience. The action of G is, again, by right translation. As we shall show in Theorem 2.6.1, the inclusion of the factors δ_H and δ_G has the important consequence that the property of unitaricity is preserved under induction. (It has other advantages also.)

For example, let $G = GL(2, \mathbb{R})^+$, and let $H = B(\mathbb{R})^+$ be the subgroup consisting of upper triangular matrices in G. We define a character of $B(\mathbb{R})^+$ by

$$\chi\begin{pmatrix} y_1 & x \\ & y_2 \end{pmatrix} = \operatorname{sgn}(y_1)^{\epsilon}\, |y_1|^{s_1}\, |y_2|^{s_2}. \tag{5.20}$$

(Note that y_1 and y_2 have the same sign if this matrix is in H.) We have $\delta_G(g) = 1$ for all g because G is unimodular (cf. Section 2.1), but by Eq. (1.18),

$$\delta_H\begin{pmatrix} y_1 & x \\ & y_2 \end{pmatrix}^{1/2} = y_1^{1/2}\, y_2^{-1/2}.$$

Consequently, the condition Eq. (5.19) is equivalent to Eqs. (5.8) and (5.9). We see that *if* $\rho : B(\mathbb{R})^+ \to \operatorname{End}(\mathbb{C})$ *is the one-dimensional representation corresponding to the character* χ*, then the representation* $H(s_1, s_2, \epsilon)$ *is obtained by inducing* ρ *from* $B(\mathbb{R})^+$ *to* $GL(2, \mathbb{R})^+$.

We shall develop the theory of induced representations in somewhat more detail in Section 2.6.

We now wish to study the $(\mathfrak{g}, K)$-module of K-finite vectors in $\mathfrak{H} = H(s_1, s_2, \epsilon)$. If $l \equiv \epsilon \bmod 2$, there exists a unique element f_l of $\mathfrak{H}$ whose restriction to K satisfies

$$f_l(\kappa_\theta) = e^{il\theta}; \tag{5.21}$$

the condition $l \equiv \epsilon \bmod 2$ is necessary for Eq. (5.12) to be satisfied. By Eq. (5.13), we have

$$f_l\left(\begin{pmatrix} u & \\ & u \end{pmatrix}\begin{pmatrix} y^{1/2} & xy^{-1/2} \\ & y^{-1/2} \end{pmatrix}\kappa_\theta\right) = u^{s_1+s_2}\, y^s\, e^{il\theta}. \tag{5.22}$$

We wish to compute the effect of R, L, Δ, and Z on these vectors.

Proposition 2.5.4 *Let $f_l \in H(s_1, s_2, \epsilon)$ be the vector (Eq. 5.22). We have*

$$Hf_l = lf_l, \qquad Rf_l = \left(s + \tfrac{l}{2}\right) f_{l+2}, \qquad Lf_l = \left(s - \tfrac{l}{2}\right) f_{l-2}, \tag{5.23}$$

and

$$\Delta f_l = \lambda f_l, \qquad Zf_l = \mu f_l, \tag{5.24}$$

where

$$\lambda = s(1-s), \qquad \mu = s_1 + s_2, \qquad s = \tfrac{1}{2}(s_1 - s_2 + 1). \tag{5.25}$$

Proof The formulas (5.23) are easily deduced from Eqs. (2.31–2.33). We also have

$$dZ = u\frac{\partial}{\partial u}, \tag{5.26}$$

which was not proved in Section 2.2, but is easily verified and which implies the second equality in Eq. (5.24). It follows from Eqs. (5.23) and (2.39) that

$$-4\Delta f_l = \left(l^2 + 2\left(s + \tfrac{l}{2} - 1\right)\left(s - \tfrac{l}{2}\right) + 2\left(s - \tfrac{l}{2} - 1\right)\left(s + \tfrac{l}{2}\right)\right) f_l,$$

which simplifies to give the first equality in Eq. (5.24). ■

Theorem 2.5.3 *Let $\mathfrak{H}$ be the $(\mathfrak{g}, K)$-module of K-finite vectors in the Hilbert space $H(s_1, s_2, \epsilon)$. Let $s = \frac{1}{2}(s_1 - s_2 + 1)$, $\lambda = s(1-s)$, and let $\mu = (s_1 + s_2)$. Then Δ and Z act as scalars on $\mathfrak{H}$, with eigenvalues λ and μ, respectively. The set of K-types of $\mathfrak{H}$ consists of all integers congruent to ϵ modulo 2.*

(i) Suppose that s is not of the form $\frac{k}{2}$, where k is an integer congruent to ϵ modulo 2. Then $\mathfrak{H}$ is irreducible.

(ii) Suppose that $s \geq \frac{1}{2}$ and that $s = \frac{k}{2}$, where $k \geq 1$ is an integer congruent to ϵ modulo 2. Then $\mathfrak{H}$ has two irreducible invariant subspaces $\mathfrak{H}_+$ and $\mathfrak{H}_-$. In the notation of Theorem 2.5.2, the set of K-types of $\mathfrak{H}_\pm$ is $\Sigma^\pm(k)$.

The quotient $\mathfrak{H}/(\mathfrak{H}_+ \oplus \mathfrak{H}_-)$ is irreducible, unless $k = 1$, in which case it is zero; its set of K-types is $\Sigma^0(k)$.

(iii) *Suppose that $s \leq \frac{1}{2}$ and that $s = 1 - \frac{k}{2}$, where $k \geq 1$ is an integer congruent to ϵ modulo 2. Then $\mathfrak{H}$ has an invariant subspace $\mathfrak{H}^0$ whose set of K-types is $\Sigma^0(k)$. Here $\mathfrak{H}_0$ is irreducible, unless $k = 1$, in which case it is zero. The quotient $\mathfrak{H}/\mathfrak{H}_0$ decomposes into two irreducible invariant subspaces $\mathfrak{H}^+$ and $\mathfrak{H}^-$; the set of K-types of $\mathfrak{H}^\pm$ is $\Sigma^\pm(k)$.*

Note that the cases (ii) and (iii) overlap when $k = \epsilon = 1$, $s = \frac{1}{2}$ but that their descriptions are consistent in this special case because $\mathfrak{H}_0 = 0$.

Proof The computation of the eigenvalues of Δ and Z is a consequence of Proposition 2.5.4, and the fact that $\mathfrak{H}$ is the vector space spanned by the f_k.

As for (i), the irreducibility of $\mathfrak{H}$ follows from Theorem 2.5.1.

For (ii), we observe that by Eq. (5.23), we have $Lf_k = Rf_{-k} = 0$. Consequently,

$$H_+ = \oplus_{n=0}^{\infty} R^n f_k, \qquad H_- = \oplus_{n=0}^{\infty} L^n f_{-k}$$

are invariant submodules whose sets of K-types are $\Sigma^+(k)$ and $\Sigma^-(k)$. They are irreducible by Theorem 2.5.2. The set of K-types of the quotient is then $\Sigma^0(k)$, and it too is irreducible by Theorem 2.5.2 (unless $k = 1$, in which case it is zero.)

The proof of (iii) is similar. By Eq. (5.23), $Rf_{k-2} = Lf_{2-k} = 0$, so the vectors $f_{k-2}, f_{k-4}, \cdots, f_{2-k}$ span an invariant subspace whose set of K-types is $\Sigma^0(k)$. By Theorem 2.5.2, it is either zero or irreducible. The quotient splits into two representations with K-types $\Sigma^\pm(k)$. Again, these are irreducible by Theorem 2.5.2. ■

Theorem 2.5.4: Classification of $(\mathfrak{g}, K)$-modules for $GL(2,\mathbb{R})^+$ *The following is a complete list of the irreducible admissible $(\mathfrak{g}, K)$-modules when $\mathfrak{g} = \mathfrak{gl}(2,\mathbb{R})$ and $K = SO(2)$. Every irreducible admissible $(\mathfrak{g}, K)$-module may be realized as the space of K-finite vectors in an admissible representation of $G = GL(2,\mathbb{R})^+$ on a Hilbert space. Let λ and μ be complex numbers, and let $\epsilon = 0$ or 1.*

(i) *If λ is not of the form $\frac{k}{2}(1 - \frac{k}{2})$, where k is an integer congruent to ϵ modulo 2, there exists a unique irreducible admissible $(\mathfrak{g}, K)$-module of parity ϵ on which Δ and Z act by scalars λ and μ. The set of K-types in this $(\mathfrak{g}, K)$-module is the set of all integers congruent to ϵ modulo 2.*

(ii) *If λ is of the form $\frac{k}{2}(1 - \frac{k}{2})$, where $k \geq 1$ is an integer congruent to ϵ modulo 2, there exist three irreducible admissible $(\mathfrak{g}, K)$-modules of parity ϵ on which Δ and Z act by scalars λ and μ, except that if $k = 1$, there are only two. In the notation of Theorem 2.5.2, the sets of K-types of these representations are $\Sigma^\pm(k)$ and (if $k > 1$) $\Sigma^0(k)$.*

Proof This classification theorem is completely contained in results that we have already proved. The existence of the representations, and their realization as spaces of K-finite vectors in representations on Hilbert spaces, is contained in Theorem 2.5.3. The uniqueness of the representations is a consequence of Theorems 2.5.1 and 2.5.2. ■

We introduce the following notation for these representations: If λ is not of the form $\frac{k}{2}(1-\frac{k}{2})$, where k is an integer congruent to ϵ modulo 2, let $\mathcal{P}_\mu(\lambda, \epsilon)$ denote the infinitesimal equivalence class of admissible representations of $GL(2, \mathbb{R})^+$ of parity ϵ on which Δ and Z have eigenvalues λ and μ, respectively. If $\mu = 0$, we denote this infinitesimal equivalence class of representations by $\mathcal{P}(\lambda, \epsilon)$. These representations are called *principal series* representations, although this term is sometimes reserved for the special case where λ is real and $\lambda \geq \frac{1}{4}$. The reason for this terminology will be clear in Section 2.6, where we discuss the unitaricity of the representations.

The finite-dimensional representations may all be realized in spaces of polynomials. Thus, let us consider the action of $G = GL(2, \mathbb{R})^+$ on homogeneous polynomials f of degree $k - 2$ in two variables x_1, x_2 by

$$\pi(g) f(x_1, x_2) = \det(g)^{(\mu-k-2)/2} f\big((x_1, x_2)g\big).$$

This is an irreducible admissible representation of degree $k - 1$ on which Z acts as the scalar μ, realizing the finite-dimensional representation occurring in Theorem 2.5.4(ii). These representations are less important for us, because (as we will show in Proposition 2.6.4) they are nonunitary. See Exercise 2.5.7 for more discussion of these representations.

If $k > 1$, an admissible representation of $GL(2, \mathbb{R})^+$ whose set of K-finite vectors is isomorphic to one of the two infinite-dimensional $(\mathfrak{g}, K)$-modules in Theorem 2.5.4(ii) is called a representation of the *discrete series*. We will denote the infinitesimal equivalence class of a discrete series representation whose set of K-types is $\Sigma^\pm(k)$ as $\mathcal{D}^\pm_\mu(k)$, or, if $\mu = 0$, as $\mathcal{D}^\pm(k)$. If $k = 1$, the representations $\mathcal{D}^\pm_\mu(1)$ are called *limit of discrete series*. They are not quite as nice as the discrete series representations.

We observe that our notation for $\mathcal{D}^+(k)$ and $\mathcal{D}^-(k)$ is the opposite of that in some of the literature. We insist on our notation, however, on the grounds that the representation of G associated with a holomorphic modular form of weight k should be $\mathcal{D}^+(k)$, not $\mathcal{D}^-(k)$.

* * *

Let us finally consider the representation theory of $GL(2, \mathbb{R})$. It is useful consider first the case of a group G and a subgroup H of index two. For example, we could have $G = GL(2, \mathbb{R})$ and $H = GL(2, \mathbb{R})^+$ or $G = O(2)$ and $H = SO(2)$. We consider the relationship between representations of G and H.

We say that irreducible representations (π, V) of G and (σ, W) of H are *related* if the restriction of π to H contains a submodule isomorphic to V. We say that the representation π of G is of Type I if its restriction to H is irreducible, and of Type II otherwise; and we will say that the representation σ of H is of Type I if it extends to G, and of Type II otherwise.

Let us consider the case where G and H are compact Lie groups.

Proposition 2.5.5 *Let G and H be compact Lie groups. Every (finite-dimensional) representation of either G or H is related to either one or two representations of the other group. More precisely, every Type I representation of H is related to exactly two Type I representations of G, and every Type I representation of G is related to exactly one Type I representation of H, and every Type II representation of H is related to exactly one Type II representation of G, and every Type II representation of G is related to exactly two Type II representations of H.*

Proof If (σ, W) is a (finite-dimensional) irreducible representation of H, let σ^G be the induced representation of G. Because H is of finite index, there are no subtleties to the notion of induction, and Frobenius reciprocity is valid in this context; it may be proved along the lines of Proposition 4.5.1.

If π is an irreducible admissible representation of G, then according to Frobenius reciprocity, the multiplicity of π in σ^G is the same as the multiplicity of σ in the restriction of π to H. What must be checked is that σ^G, if not irreducible, splits into two distinct irreducible representations of G. Once this fact is established, the proposition is simple to verify.

This may be seen as follows. By Frobenius reciprocity, the inner product in the character rings

$$\langle \sigma^G, \sigma^G \rangle_G = \langle \sigma^G, \sigma \rangle_H .$$

Because σ is irreducible, and the dimension of σ^G is twice that of σ, σ^G can contain at most two copies of σ, so this inner product is at most two. If it is one, σ^G is irreducible; and if it is two, σ^G is a direct sum of two distinct irreducibles. ■

Thus suppose that $H = SO(2)$ and $G = O(2)$. Because H is Abelian, its irreducible representations are all one dimensional; they are the characters $\kappa_\theta \mapsto e^{ik\theta}$, $k \in \mathbb{Z}$. If $k = 0$, this representation is Type I and admits two extensions to $O(2)$, namely, the trivial representation and det. If $k \neq 0$, it cannot be extended, and the two-dimensional induced representation of $O(2)$ is irreducible. The characters $\kappa_\theta \mapsto e^{ik\theta}$ and $e^{-ik\theta}$ both correspond to this same two-dimensional representation of $O(2)$. Thus we obtain a complete list of the irreducible admissible representations of $O(2)$.

We expect the same basic picture of Proposition 2.5.5 to extend to other categories of representations, such as $(\mathfrak{g}, K)$-modules. Instead of proving a general theorem of this type, we will be content to supply *ad hoc* proofs on a case-by-case basis that show, in the special case of $(\mathfrak{g}, K)$-modules for $G = GL(2, \mathbb{R})$ and $H = GL(2, \mathbb{R})^+$, the various cases in Theorem 2.5.4 all fit the pattern of either Type I or Type II. More precisely, the irreducible principal series and finite-dimensional representations are Type I (meaning that each has two distinct extensions to $GL(2, \mathbb{R})$), whereas the discrete series are Type II (meaning that $\mathcal{D}^+_\mu(k)$ and $\mathcal{D}^-_\mu(k)$ cannot be extended to $GL(2, \mathbb{R})$, but that $\mathcal{D}^+_\mu(k) \oplus \mathcal{D}^-_\mu(k)$ can be so extended).

We therefore consider the case when an irreducible $\big(\mathfrak{g}, SO(2)\big)$-module V can be extended to a $\big(\mathfrak{g}, O(2)\big)$-module. One obvious necessary condition is that the set Σ of K-types of V must be symmetrical under $k \mapsto -k$. This is because, as we have seen, every irreducible representation of $O(2)$ has this same symmetry property. Thus the discrete series representations $\mathcal{D}^{\pm}(k)$ cannot be extended to $O(2)$. We expect them to be Type II. On the other hand, the irreducible principal series representations and the finite-dimensional representations are candidates for extension, and indeed, we may see directly that the Hilbert space representations $H(s_1, s_2, \epsilon)$ may be extended.

The representation $H(s_1, s_2, \epsilon)$ was obtained by inducing the character (Eq. 5.20) from $B(\mathbb{R})^+$ to $GL(2, \mathbb{R})^+$. We can extend the character (5.20) to $B(\mathbb{R})$, which is the Borel subgroup of 2×2 upper triangular real matrices, in two different ways. Namely, let ϵ_1, $\epsilon_2 = 0$, or 1 be such that $\epsilon_1 + \epsilon_2 \equiv \epsilon$ modulo 2; of those there are two possible choices. Having fixed such a choice, let $\chi_i : \mathbb{R}^\times \to \mathbb{C}$ be $\chi_i(y) = \operatorname{sgn}(y)\, |y|^{s_i}$ $(i = 1, 2)$ and extend Eq. (5.20) to the character

$$\chi\begin{pmatrix} y_1 & x \\ & y_2 \end{pmatrix} = \chi_1(y_1)\, \chi_2(y_2).$$

We will change notation slightly and by $H(\chi_1, \chi_2)$ denote the representation of $GL(2, \mathbb{R})$, which is induced from χ, or by $H^\infty(\chi_1, \chi_2)$ denote the subspace of smooth vectors. We will denote by $\pi(\chi_1, \chi_2)$ the underlying $\big(\mathfrak{gl}(2, \mathbb{R}), O(2)\big)$-module of K-finite vectors. $H(\chi_1, \chi_2)$ has the same space as $H(s_1, s_2, \epsilon)$, because each function in $H(\chi_1, \chi_2)$ is easily seen to be determined by its restriction to $GL(2, \mathbb{R})^+$. We see thus that there are two extensions of the $GL(2, \mathbb{R})^+$-module structure on $H(s_1, s_2, \epsilon)$ to a $GL(2, \mathbb{R})$-module structure, and the same is true for the corresponding $(\mathfrak{g}, K)$-modules. According to Exercises 2.5.3 and 2.5.4, there are at most two such extensions (for the $(\mathfrak{g}, K)$-modules), at least in the case where $H(s_1, s_2, \epsilon)$ is irreducible.

Clearly, the representation $H(\chi_1, \chi_2)$, or its underlying $(\mathfrak{g}, K)$-module, will be irreducible if the representation $H(s_1, s_2, \epsilon)$ that it extends is irreducible. Thus by Theorem 2.5.3, we see that $H(\chi_1, \chi_2)$ and $\pi(\chi_1, \chi_2)$ are irreducible unless

$$\chi_1\chi_2^{-1}(y) = \operatorname{sgn}(y)^\epsilon\, |y|^{k-1},$$

where $\epsilon = 0$ or 1 and k is an integer of the same parity of ϵ.

We return now to the discrete series. These may be constructed as $(\mathfrak{g}, K)$-modules as follows. They may be realized as the quotients $\mathfrak{H}/\mathfrak{H}_0$ described in Theorem 2.5.3(iii) after the $\big(\mathfrak{gl}(2, \mathbb{R}), SO(2)\big)$-module structure on $\mathfrak{H}$ is extended to a $\big(\mathfrak{gl}(2, \mathbb{R}), O(2)\big)$-module, as described above. This realization not only provides us with a $(\mathfrak{gl}(2, \mathbb{R}), O(2))$-module structure on $\mathcal{D}_\mu^+(k) \oplus \mathcal{D}_\mu^-(k)$, but also shows that the $(\mathfrak{g}, K)$-module may be realized as the space of K-finite vectors in a representation of $GL(2, \mathbb{R})$ on a Hilbert space. We will denote this representation as $\mathcal{D}_\mu(k)$.

Lastly, the finite-dimensional representations are discussed in detail in Exercise 2.5.7, where they are realized as symmetric power representations, twisted by characters of $\mathbb{R}^\times$.

We summarize the results of this discussion in the following theorem.

***Theorem 2.5.5: Irreducible** $(\mathfrak{g}, K)$**-modules for** $GL(2, \mathbb{R})$* *The following is a complete list of the irreducible admissible $(\mathfrak{g}, K)$-modules when $\mathfrak{g} = \mathfrak{gl}(2, \mathbb{R})$ and $K = O(2)$. Every irreducible admissible $(\mathfrak{g}, K)$-module may be realized as the space of K-finite vectors in an admissible representation of $G = GL(2, \mathbb{R})^+$ on a Hilbert space.*

(i) The finite-dimensional representations are obtained by tensoring the symmetric powers of the standard representation with the one-dimensional representations of the form $\chi \circ \det$, where χ is a character of $\mathbb{R}^\times$.

(ii) If χ_1 and χ_2 are characters of $\mathbb{R}^\times$ such that $\chi_1\chi_2^{-1}$ is not a character of the form $y \mapsto sgn(y)^\epsilon\, |y|^{k-1}$, where $\epsilon = 0$ or 1 and k is an integer of the same parity as ϵ, then we have the irreducible $\big(\mathfrak{g}, O(2)\big)$-module $\pi(\chi_1, \chi_2)$.

(iii) If μ is a real number and k is an integer ≥ 1, then there are the representations $\mathcal{D}_\mu(k)$, called discrete series if $k \geq 2$ and limits of discrete series if $k = 1$.

Exercises

Exercise 2.5.1 Let (π, H) be an irreducible admissible representation of $GL(2, \mathbb{R})^+$. Let t be a complex number. Let (π', H) be the representation on the same space with

$$\pi'(g)\, v = \det(g)^t\, \pi(g)\, v.$$

This representation is sometimes denoted $\pi \otimes \det^t$. Prove that every irreducible admissible representation has a unique representation in the form $\pi \otimes \det^t$, where

$$\pi\begin{pmatrix} u & \\ & u \end{pmatrix} v = v.$$

Prove that this condition is equivalent to the assumption that the eigenvalue μ of Z in π is equal to zero.

Exercise 2.5.2 Describe the classification theorem for irreducible admissible $(\mathfrak{g}, K)$-modules and for infinitesimal equivalence classes of admissible representations of $SL(2, \mathbb{R})$. [HINT: use the corresponding theory for $GL(2, \mathbb{R})^+$.]

In the following exercises, we adapt the results of this section to $G = GL(2, \mathbb{R})$, $K = O(2)$.

Exercise 2.5.3 If V is a $(\mathfrak{g}, K)$-module for $GL(2, \mathbb{R})$, it is automatically also a $(\mathfrak{g}, K)$-module for $GL(2, \mathbb{R})^+$. Let $\pi : O(2) \to \mathrm{End}(V)$ denote the given representation of K. Let

$$\eta = \pi\begin{pmatrix} -1 & \\ & 1 \end{pmatrix} \in \mathrm{End}(V). \tag{5.27}$$

(a) Prove that the $(\mathfrak{g}, O(2))$-module structure of V is uniquely determined by specifying the $(\mathfrak{g}, SO(2))$-module structure, together with the action of η, which (by the definition of the $(\mathfrak{g}, k)$-module, part (iii)) must satisfy

$$\eta^2 = 1, \quad \eta H \eta = -H, \quad \eta R \eta = L, \quad \eta L \eta = R, \quad \eta Z \eta = Z. \tag{5.28}$$

(b) Prove, conversely, that given a $(\mathfrak{g}, SO(2))$-module V together with an endomorphism η satisfying Eq. (5.28), V may be given a unique $(\mathfrak{g}, O(2))$-module structure such that Eq. (5.27) is true.

Exercise 2.5.4 Let λ and μ be complex numbers and $\epsilon = 0$ or 1 such that λ is not of the form $\frac{k}{2}(1 - \frac{k}{2})$, where k is an integer congruent to ϵ modulo 2. Let V be the irreducible admissible $(\mathfrak{g}, SO(2))$-module $\mathcal{P}_{\lambda,\mu}(\epsilon)$ of Theorem 2.5.4(ii).

(a) Prove that there exist at most two $(\mathfrak{g}, O(2))$-module structures on this space extending the given $(\mathfrak{g}, SO(2))$-module structure. (Indeed, there are exactly two such structures.)

(b) Show that if condition (iii) were omitted from the definition of the $(\mathfrak{g}, K)$ module, there would be an infinite number of $(\mathfrak{g}, O(2))$-module structures on V.

[HINT: By Exercise 2.5.3, it is sufficient to show that there are at most two possible endomorphisms η satisfying Eq. (5.27). Handle the cases $\epsilon = 0$ or 1 separately. First show that η is determined up to at most two possibilities on $V(\epsilon)$, then work outward.]

Exercise 2.5.5 Prove that if V is a finite-dimensional $(\mathfrak{g}, SO(2))$-module, there exist at most two extensions of V to a $(\mathfrak{g}, O(2))$-module. (Again, there are exactly two.)

Exercise 2.5.6 Prove that any two extensions of $\mathcal{D}^+_\mu(k) \oplus \mathcal{D}^-_\mu(k)$ to a $(\mathfrak{g}, O(2))$-module give rise to isomorphic $(\mathfrak{g}, O(2))$-modules.

Exercise 2.5.7: Finite-dimensional representations of GL(2, $\mathbb{R}$) and GL(2, $\mathbb{R}$) If U is a complex vector space and k a nonnegative integer, let $\vee^h U$ denote the hth symmetric power of U; that is, $\vee^h U$ is quotient of the h-fold tensor product $\otimes^h U$ of U with itself by the subspace generated by elements of the form

$u_1 \otimes \cdots \otimes u_h - u_{\sigma(1)} \otimes \cdots \otimes u_{\sigma(h)}$, where σ is a permutation of $\{1, \cdots, h\}$. We will denote by $u_1 \vee \cdots \vee u_h$ the image of $u_1 \otimes \cdots \otimes u_h$ in $\vee^h U$. This construction is functorial so if $\phi : U \to U'$ is any linear transformation, there is induced a linear transformation $\vee^h \phi : \vee^h U \to \vee^h U'$. In particular, if $U = U', \phi \mapsto \vee^h \phi$ is a representation $GL(U) \to GL(\vee^h U)$.

Suppose that $U = \mathbb{C}^2$, and let $\{e_1, e_2\}$ be the standard basis. Then if $0 \leq r \leq h$, let

$$\hat{\xi}_r = e_1 \vee \cdots \vee e_1 \vee e_2 \vee \cdots \vee e_2, \qquad (r \text{ copies of } e_2 \text{ and } h - r \text{ copies of } e_2).$$

Let us denote (π, V) as the representation of $GL(2, \mathbb{C})$ on $V = \vee^h U$. Actually, we are most concerned with the representation of $GL(2, \mathbb{R}) \subset GL(2, \mathbb{C})$ on V, but it will be convenient to consider $\pi(g)$ for arbitrary elements of $GL(2, \mathbb{C})$. In particular, we will need $\pi(C)$ where C is the Cayley transform Eq. (2.25). Verify that with $\hat{H}$, $\hat{R}$, and $\hat{L}$ as in Eq. (2.18), we have

$$d\hat{H}\,\hat{\xi}_r = (2r - h)\,\hat{\xi}_r, \qquad d\hat{R}\,\hat{\xi}_r = (h - r)\,\hat{\xi}_{r+1}, \qquad d\hat{L}\,\hat{\xi}_r = r\,\hat{\xi}_{r-1}.$$

Thus if $\xi_i = \pi(C)^{-1}\,\hat{\xi}_i$, then by Eq. (2.26) we have

$$dH\,\xi_r = (2r - h)\,\xi_r, \qquad dR\,\xi_r = (h - r)\,\xi_{r+1}, \qquad dL\,\xi_r = r\,\xi_{r-1}.$$

Now let $k = h + 2$ and $s = 1 - \frac{k}{2}$. Let $\mu = h$, and let s_1 and s_2 be determined by Eq. (5.25). Let $\epsilon = 0$ or 1 be determined by $\epsilon \equiv k$ modulo 2. Show that mapping

$$\xi_r \to (-1)^r\, f_{2r-k+2}$$

gives a homomorphism of $\big(\mathfrak{gl}(2, \mathbb{R}), SO(2)\big)$-modules from V into $H(s_1, s_2, \epsilon)$. Because the representation V is finite dimensional, this is enough to show that the map is actually a homomorphism of $GL(2, \mathbb{R})$-modules.

We would like to modify the preceeding construction so that μ can be taken to be arbitrary, so as to obtain all the finite-dimensional representations of $GL(2, \mathbb{R})^+$ or $GL(2, \mathbb{R})$. Although V itself can be regarded as a module for $GL(2, \mathbb{C})$, these modifications only work for the real subgroups. Thus let $G = GL(2, \mathbb{R})^+$ or $GL(2, \mathbb{R})$, and let χ be a quasicharacter of $\mathbb{R}_+^\times$ or of $\mathbb{R}^\times$, respectively. Then we can tensor (π, V) with the character $g \mapsto \chi\big(\det(g)\big)$ of G. Show that (if $G = GL(2, \mathbb{R})^+$) this effects the value of μ but not s or λ with notation as in Eq. (5.25), as in Exercise 2.5.1. Thus we are able to obtain all the finite-dimensional representations whose sets of K-types are $\Sigma^0(k)$ in the notation of Theorem 2.5.4. Because each character χ of $\mathbb{R}_+^\times$ admits two distinct extensions to $\mathbb{R}^\times$, we see in this way that each finite-dimensional $GL(2, \mathbb{R})^+$-module admits two extensions to a $GL(2, \mathbb{R})$-module, and so these representations follow the pattern of the Type I representations in Proposition 2.5.5.

Exercise 2.5.8 Let $G = GL(2, \mathbb{R})^+$, $K = SO(2)$, and let V be an admissible $(\mathfrak{g}, K)$=module. Define the *contragredient module* to be the space $\hat{V}$ of linear functionals Λ on V such that the kernel of V contains $V(k)$ for almost all integers k, with the following action. Denoting the action of K on V by π, let $\hat{\pi} : K \to \mathrm{End}(\hat{V})$ be the action

$$\langle v, \hat{\pi}(g)\,\Lambda\rangle = \left\langle \pi(g^{-1})\,v, \Lambda\right\rangle,$$

and, if $X \in \mathfrak{g}$

$$\langle v, X\,\Lambda\rangle = -\,\langle X\,v, \Lambda\rangle\,.$$

Verify that this is a $(\mathfrak{g}, K)$-module. Show that if V is irreducible, then $\hat{\pi}$ is isomorphic to the $(\mathfrak{g}, K)$-module obtained by composing the action of π with the automorphism $g \mapsto {}^\top g^{-1}$ of $GL(2, \mathbb{R})^+$.

[HINT: Deduce this from the explicit classification of $(\mathfrak{g}, K)$-modules in Theorem 2.5.4.]

2.6 Unitaricity and Intertwining Integrals

We next consider the qurrestion of which of the admissible representations of $GL(2, \mathbb{R})^+$ constructed in the previous section are unitarizable. This is an important question, because, for example, only unitary representations can occur in $L^2(\Gamma\backslash G, \chi)$. We will first show that some representations certainly are unitarizable and some of them certainly are not; then, to obtain definitive results, we will introduce certain intertwining operators that play an essential role in this field.

We first prove that a representation induced from a unitary representation is unitary. To simplify the proofs slightly, we do this in less than complete generality; however, the result we state is sufficiently general to cover the case of parabolic induction in a Lie group or a p-adic group.

Let G be a topological group that is unimodular (left Haar measure equals right Haar measure), and let P and K be closed subgroups such that K is compact and $PK = G$. Let $\delta = \delta_P$ be the modular quasicharacter of P, defined by $d_R(p) = \delta_G(p)\,d_L(p)$, where d_L and d_R are left and right Haar measures on P, respectively. (We will sometimes denote the left Haar measure on P as simply dp.) Let $C(P\backslash G, \delta)$ denote the space of continuous functions on G that satisfy

$$f(pg) = \delta(p)\,f(g), \qquad p \in P. \tag{6.1}$$

Note that this space is invariant under the right regular representation ρ of G defined by $\big(\rho(g)f\big)(x) = f(xg)$. We define a linear functional $I : C(P\backslash G, \delta) \to \mathbb{C}$ by

$$I(f) = \int_K f(\kappa)\,d\kappa.$$

Lemma 2.6.1 *The linear functional I is invariant under the action ρ of G.*

Proof If $\phi \in C_c(G)$ (the space of compactly supported continuous functions on G), define $\Lambda\phi : G \to \mathbb{C}$ by

$$(\Lambda\phi)(g) = \int_P \phi(pg)\,dp,$$

where $dp = d_L(p)$ is the left Haar measure. We claim that $\Lambda\phi \in C(P\backslash G, \delta)$. Indeed, we have

$$(\Lambda\phi)(pg) = \int_P \phi(p_1pg)\,dp_1 = \int_P \phi(p_1pg)\,\delta^{-1}(p_1)\,d_Rp_1$$

$$= \int_P \phi(p_1g)\,\delta^{-1}(p_1p^{-1})\,d_Rp_1 = \delta(p)\,(\Lambda\phi)(g).$$

We will show that *the map* $\Lambda : C_c(G) \to C(P\backslash G, \delta)$ *is surjective.* First observe that the restriction of δ to $K \cap P$ is trivial, because $\delta(K \cap P)$ is a compact subgroup of $\mathbb{R}^\times$, which must be just $\{1\}$. Thus if $f \in C(P\backslash G, \delta)$, its restriction to K is constant on the cosets of $K \cap P$ and hence is a continuous function on the coset space $(K \cap P)\backslash K$. We thus obtain a linear map $C(P\backslash G, \delta) \to C(K \cap P\backslash K)$. Because $PK = G$, an element of $C(P\backslash G, \delta)$ is determined by its restriction to K, and this map is therefore injective. The surjectivity of Λ will follow, then, if we prove that the composition

$$C_c(G) \xrightarrow{\Lambda} C(P\backslash G, \delta) \longrightarrow C(K \cap P\backslash K) \tag{6.2}$$

is surjective. Let $f \in C(K \cap P\backslash K)$, and let $\phi_0 \in C_c(P)$ be chosen whose integral is nonzero. Replacing $\phi_0(g)$ by $\int_{K\cap P} \phi_0(gk)\,dk$, we may assume that $\phi_0(gk) = \phi_0(g)$ for $k \in K \cap P$; and multiplying ϕ_0 by a constant, we may assume that $\int_G \phi_0 = 1$. Now let $\phi \in C_c(G)$ be defined by $\phi(pk) = \phi_0(p)\,f(k)$. Note that this is well defined. We have $(\Lambda\phi)(k) = f(k)$, proving the surjectivity of Eq. (6.2) and hence of the map Λ.

Now to prove that the functional I on $C(P\backslash G, \delta)$ is G-invariant, let $f \in C(P\backslash G, \delta)$. We observe that if $f \in C_c(G)$, we have, by Proposition 2.1.5(ii) that $I(\Lambda f) = \int_G f(g)\,dg$. Thus $I \circ \Lambda$ is invariant under right translation by elements of G and, clearly, Λ commutes with right translation. Because Λ is surjective, it follows that I is invariant under right translation, as required. ∎

Lemma 2.6.2 *Let $\mathfrak{H}_0$ be a Hermitian inner product space, and let $\mathfrak{H}$ be its completion to a Hilbert space. Let G be a topological group, and let $\pi : G \to$* End$(\mathfrak{H})$ *be a group action by unitary operators such that $\mathfrak{H}_0$ is invariant under $\pi(g)$ for all $g \in G$ and such that for each $v_0 \in \mathfrak{H}_0$, $g \mapsto \pi(g)v_0$ is a continuous map $G \to \mathfrak{H}_0$. Then π is a representation; that is, the map $(g, v) \mapsto \pi(g)\,v$ is a continuous map from $G \times \mathfrak{H} \to \mathfrak{H}$.*

Proof Let $v \in \mathfrak{H}$ and $g \in G$. We must show that for all $\epsilon > 0$ there exist neighborhoods U and W of v and g in $\mathfrak{H}$ and G, respectively, such that if $v_1 \in U$ and $g_1 \in W$, then $|\pi(g_1)\, v_1 - \pi(g)\, v| < \epsilon$. Let v_0 be an element of $\mathfrak{H}_0$ such that $|v - v_0| < \epsilon/6$. We have the inequality

$$\begin{aligned}
&|\pi(g_1)\, v_1 - \pi(g)\, v| \\
&\quad = |\pi(g_1)\, v_1 - \pi(g_1)\, v_0 + \pi(g_1)\, v_0 - \pi(g)\, v_0 + \pi(g)\, v_0 - \pi(g)\, v| \\
&\quad \leq |\pi(g_1)\, v_1 - \pi(g_1)\, v_0| + |\pi(g_1)\, v_0 - \pi(g)\, v_0| + |\pi(g)\, v_0 - \pi(g)\, v| \\
&\quad = |v_1 - v_0| + |\pi(g_1)\, v_0 - \pi(g)\, v_0| + |v_0 - v|
\end{aligned}$$

Now let U be the ball of radius $\epsilon/6$ with center v. If $v_1 \in U$, then $|v_1 - v_0|$ and $|v_0 - v|$ are both less that $\epsilon/3$. We use our hypothesis that $g \mapsto \pi(g)v_0$ is a continuous map to choose a neighborhood W of g such that if $g \in W$, then $|\pi(g_1)\, v_0 - \pi(g)\, v_0| < \epsilon/3$. It follows that if $v_1 \in U$ and $g_1 \in W$, then $|\pi(g_1)\, v_1 - \pi(g)\, v| < \epsilon$, so π is indeed a representation. ■

Now let P and K be as above, and let (σ, V) be a unitary representation of P. Let $\mathfrak{H}_0$ be the space of all continuous functions $f : G \to V$ such that

$$f(pg) = \delta(p)^{1/2}\, \sigma(p)\, f(g). \tag{6.3}$$

We define an inner product $\langle\langle\ ,\ \rangle\rangle$ on $\mathfrak{H}_0$ in terms of the inner product $\langle\ ,\ \rangle$ on V by

$$\langle\langle f_1, f_2 \rangle\rangle = \int_K \langle f_1(\kappa), f_2(\kappa)\rangle\, d\kappa. \tag{6.4}$$

We let π be the action of G on $\mathfrak{H}_0$ induced by the right regular representation, so $\big(\pi(g)f\big)(x) = f(xg)$. Note that the property (Eq. 6.3) is preserved under right translation, and so $\mathfrak{H}_0$ is stable under this action.

Theorem 2.6.1 *Let G be a unimodular locally compact group, and let P be a closed subgroup. Assume that there exists a compact subgroup K such that $KP = G$. Let (σ, V) be a unitary representation of P on a Hilbert space V. Let $\mathfrak{H}_0$ be the space of continuous functions satisfying Eq. (6.3). Then the inner product (Eq. 6.4) on $\mathfrak{H}_0$ is preserved under the action of G by right translation, and the Hilbert space completion $\mathfrak{H}$ of $\mathfrak{H}_0$ affords a unitary representation of G.*

The use of the compact subgroup K to construct an inner product on G is unnecessary, as is the assumption of unimodularity – it is true in complete generality that a representation induced from a unitary representation is unitary. We have chosen to prove a weaker result that is sufficient for the important application of parabolic induction in the context of a Lie group or p-adic group.

Proof The function $\langle f_1(g), f_2(g)\rangle$ is in $C(P\backslash G, \delta)$, so the invariance of the inner product follows from Lemma 2.6.1. It follows that $\pi(g)$ defined on $\mathfrak{H}_0$ by $(\pi(g)f)(x) = f(xg)$ extends to a unitary transformation of $\mathfrak{H}$, and we need only check that the action $G \times \mathfrak{H} \to \mathfrak{H}$ is continuous. By Lemma 2.6.2, it is sufficient to show that for $f_0 \in \mathfrak{H}_0$, the map $g \mapsto \pi(g)f_0$ is continuous. Indeed, it is clear that it is continuous when $\mathfrak{H}_0$ is given the topology induced from the L^∞ norm on K, and because K is compact, the L^2 norm is dominated by the L^∞ norm. Hence Lemma 2.6.2 applies, and π is a representation. ■

Theorem 2.6.2 *Suppose that μ and λ are complex numbers such that λ is not of the form $\frac{k}{2}(1-\frac{k}{2})$ for an integer k congruent to ϵ modulo 2. In this case, we have defined an infinitesimal equivalence class $\mathcal{P}_\mu(\lambda, \epsilon)$ of admissible representations of $GL(2,\mathbb{R})^+$. There exists a unitary representation in this class if μ is pure imaginary and λ is real and $\lambda \geq \frac{1}{4}$.*

Proof Let $s - 1/2$ be a square root of $\frac{1}{4} - \lambda$. Because $\frac{1}{4} - \lambda$ is negative, $s - 1/2$ is purely imaginary. Let s_1 and s_2 be determined by the equations $\mu = s_1 + s_2$ and $s = \frac{1}{2}(s_1 - s_2 + 1)$.

Let B be the Borel subgroup of $G = GL(2,\mathbb{R})^+$ consisting of

$$b = \begin{pmatrix} y_1 & x \\ & y_2 \end{pmatrix}, \qquad y_1, y_2 \in \mathbb{R}^\times,\ x \in \mathbb{R},$$

and let σ be the one-dimensional representation

$$\sigma(b) = \operatorname{sgn}(y_1)^\epsilon\, |y_1|^{s_1}\, |y_2|^{s_2}.$$

This representation is unitary because s_1 and s_2 are pure imaginary. The hypotheses of Theorem 2.6.1 are satisfied with $K = SO(2)$, and so the representation $\mathfrak{H}(s_1, s_2, \epsilon)$ induced from the character σ is unitary. We proved in Theorems 2.5.3(i) and 2.5.4(i) that the infinitesimal equivalence class of this representation is $\mathcal{P}_\mu(\lambda, \epsilon)$. ■

We have proved that certain choices of λ, μ, and ϵ give rise to unitarizable $\mathcal{P}_\mu(\lambda, \epsilon)$. This does not exclude the possibility that there are other choices of data for which this is true. Our next task will be to prove that some of the irreducible $(\mathfrak{g}, K)$-modules in the classification of Section 2.5 are *not* unitarizable.

Proposition 2.6.1 *Let $(\pi, \mathfrak{H})$ be a unitary representation of the group $G = GL(n,\mathbb{R})^+$, and let $X \in \mathfrak{g} = \mathfrak{gl}(n,\mathbb{R})$. Then the action of X on the space of smooth vectors $\mathfrak{H}^\infty$ is skew-symmetric in the following sense: If $v, w \in \mathfrak{H}^\infty$, we have*

$$\langle Xv, w\rangle = -\langle v, Xw\rangle. \tag{6.5}$$

Proof We have

$$\langle Xv, w\rangle = \frac{d}{dt}\left\langle \pi(e^{tX})\,v, w\right\rangle|_{t=0} = \frac{d}{dt}\left\langle v, \pi(e^{-tX})\,w\right\rangle|_{t=0}$$
$$= -\frac{d}{dt}\left\langle v, \pi(e^{tX})\,w\right\rangle|_{t=0} = -\langle v, Xw\rangle\,,$$

because π is unitary. ■

As a particular case, let R and L be as in Eq. (2.23). We have

$$R = \tfrac{1}{2}(\hat{H} + i\,X),\;\; Y = \tfrac{1}{2}(\hat{H} - iX),\;\; \hat{H} = \begin{pmatrix} 1 & \\ & -1 \end{pmatrix},\;\; X = \begin{pmatrix} 0 & 1 \\ 1 & 0 \end{pmatrix},$$

so Eq. (6.5) implies that

$$\langle Rv, w\rangle = -\langle v, Lw\rangle\,. \tag{6.6}$$

Thus we recover the adjointness property (Eq. 1.17).

Theorem 2.6.3 *Let $(\pi, \mathfrak{H})$ be a unitary representation of $G = GL(2,\mathbb{R})^+$.*

(i) Assume that the elements Z and Δ of the center of $U\big(\mathfrak{gl}(2,\mathbb{R})\big)$ act by scalars μ and λ, respectively. Then μ is purely imaginary and λ is real.
(ii) Assume in the notation of Section 2.5 that the admissible equivalence class of $(\pi, \mathfrak{H})$ is $\mathcal{P}_\mu(\lambda,\epsilon)$, where λ does not equal $\frac{k}{2}(1-\frac{k}{2})$ for any integer $k \equiv \epsilon \bmod 2$. If $\epsilon = 0$, then $\lambda > 0$, and if $\epsilon = 1$, then $\lambda > \frac{1}{4}$.

Proof By Proposition 2.6.1, Z is skew-symmetric on the smooth vectors, and $-4\Delta = H^2 + 2RL + 2LR$ is symmetric; this implies that the eigenvalues of Z and Δ are purely imaginary and real, respectively, from whence we get (i).

We next prove (ii). By the results of Section 2.5, $V(k)$ is nonzero if $k \equiv \epsilon$ modulo 2. Let f_k denote a nonzero vector of $V(k)$. Because $-4\Delta - H^2 + 2H = 4RL$ and $Hf_k = kf_k$, we have, taking $k = \epsilon$

$$(-4\lambda - \epsilon^2 + 2\epsilon)\,f_\epsilon = 4RL\,f_\epsilon.$$

By Eq. (6.6)

$$\langle 4RL\,f_\epsilon, f_\epsilon\rangle = -4\,\langle Lf_\epsilon, Lf_\epsilon\rangle < 0,$$

so

$$-4\lambda - \epsilon^2 + 2\epsilon < 0.$$

This implies that $\lambda > 0$ if $\epsilon = 0$, and $\lambda > \frac{1}{4}$ if $\epsilon = 1$. ■

Theorems 2.6.2 and 2.6.3 settle the question of whether the infinitesimal equivalence classes $\mathcal{P}_\mu(\lambda,\epsilon)$ are unitarizable for most values of λ, μ, and ϵ.

However, they leave the question open if $\epsilon = 0$, μ is purely imaginary, and λ is a real number with $0 < \lambda < \frac{1}{4}$. It turns out that these representations *are* unitary, but this is surprising, because they are induced from nonunitary representations of the Borel subgroup – if an explanation for their unitaricity is to be found, it will *not* come from Theorem 2.6.2. These representations are called the *complementary series*, and the values $0 < \lambda < \frac{1}{4}$ of the non-Euclidean Laplacian are called *exceptional eigenvalues* when they occur.

In order to investigate the unitaricity of the complementary series, we will introduce certain *intertwining integrals* that are related to the constant terms of Eisenstein series. These integrals are extremely important in representation theory. They reappear in Chapter 4, Section 5 and are also related to the calculation of the constant term in the Eisenstein series in Eq. (6.12) of Chapter 1.

We begin with the observation that if s_1 and s_2 are interchanged, and if $\mu = s_1 + s_2$, $\lambda = s(1-s)$ with $s = \frac{1}{2}(s_1 - s_2 + 1)$, then neither μ nor λ are changed. If λ is not of the form $\frac{k}{2}(1 - \frac{k}{2})$, where k is an integer congruent to ϵ modulo 2, it follows from Theorem 2.5.4(i) that the representations $H(s_1, s_2, \epsilon)$ and $H(s_2, s_1, \epsilon)$ are infinitesimally isomorphic. There is no obvious reason for this isomorphism. Moreover, if $\lambda = \frac{k}{2}(1 - \frac{k}{2})$, where $k \equiv \epsilon \bmod 2$, the two representations $H(s_1, s_2, \epsilon)$ and $H(s_2, s_1, \epsilon)$ represent the two contrasting situations of Theorem 2.5.3(ii) and 2.5.3(iii): These representations are *not* isomorphic, although they have isomorphic composition factors.

We will construct by means of integrals certain (densely defined) intertwining operators $M(s)$ from $H(s_1, s_2, \epsilon)$ to $H(s_2, s_1, \epsilon)$. (It might be more natural to use the notation $M(s_1, s_2, \epsilon)$, but because interchanging s_1 and s_2 affects $s = \frac{1}{2}(s_1 - s_2 + 1)$ but not $\mu = s_1 + s_2$ or ϵ, no confusion can result from this notation.)

The operators $M(s)$ induce the infinitesimal equivalences between the representations when λ is not of the form $\frac{k}{2}(1 - \frac{k}{2})$, where k is an integer congruent to ϵ modulo 2. When $\lambda = \frac{k}{2}(1 - \frac{k}{2})$, they are not isomorphisms but are still very intimately connected with the more complex situation of Theorem 2.5.3(ii) and 2.5.3(iii).

The operators $M(s)$ are defined by

$$\big(M(s)\,f\big)(g) = \int_{-\infty}^{\infty} f\left(\begin{pmatrix} 0 & -1 \\ 1 & 0 \end{pmatrix}\begin{pmatrix} 1 & x \\ & 1 \end{pmatrix} g\right) dx. \tag{6.7}$$

We will see that if f is a smooth vector, this integral is absolutely convergent if $\mathrm{re}(s) > \frac{1}{2}$; if $\mathrm{re}(s) \leq \frac{1}{2}$, the operator $M(s)$ must be reinterpreted by some process of analytic continuation.

Proposition 2.6.2 *Suppose that $f \in H(s_1, s_2, \epsilon)$ is a smooth vector, and that $\mathrm{re}(s) > \frac{1}{2}$ where $s = \frac{1}{2}(s_1 - s_2 + 1)$. Then the integral (Eq. 6.7) is convergent, and $M(s)f$ is a smooth vector in $H(s_2, s_1, \epsilon)$. Denoting by $\rho(g)$ the action of*

$g \in GL(2,\mathbb{R})^+$ on smooth functions by right translation

$$\rho(g) \circ M(s) = M(s) \circ \rho(g), \tag{6.8}$$

so $M(s)$ is an intertwining operator on the smooth vectors. The K-finite vectors are invariant under $M(s)$, which therefore induces a homomorphism of $(\mathfrak{g}, K)$-modules

$$H(s_1, s_2, \epsilon)_{\mathrm{fin}} \to H(s_2, s_1, \epsilon)_{\mathrm{fin}}.$$

Proof Disregarding for the moment matters of convergence, we check Eq. (6.8). This is a simple consequence of the fact that left and right translations commute with each other: We have

$$\begin{aligned}\big(M(s) \circ \rho(g)\, f\big)(h) &= \int_{-\infty}^{\infty} \big(\rho(g) f\big)\left(\begin{pmatrix} 0 & -1 \\ 1 & 0 \end{pmatrix}\begin{pmatrix} 1 & x \\ & 1 \end{pmatrix} h\right) dx \\ &= \int_{-\infty}^{\infty} f\left(\begin{pmatrix} 0 & -1 \\ 1 & 0 \end{pmatrix}\begin{pmatrix} 1 & x \\ & 1 \end{pmatrix} hg\right) dx \\ &= \big(M(s)\, f\big)(hg) = \big(\rho(g) \circ M(s)\, f\big)(h).\end{aligned}$$

This proves Eq. (6.8) provided the integrals are convergent. Moreover, taking h to equal the identity element $1 \in GL(2,\mathbb{R})^+$, this calculation shows that checking the convergence of $M(s)\big(\rho(g) f\big)$ at the identity is equivalent to checking the convergence of $M(s) f$ at g. In other words, by replacing f by $\rho(g) f$, we are reduced to checking the convergence of Eq. (6.7) when $g = 1$. (Recall that the smooth vectors are invariant under the group action ρ.)

We now prove the convergence of Eq. (6.7) when $g = 1$. We use the identity

$$\begin{pmatrix} & -1 \\ 1 & \end{pmatrix}\begin{pmatrix} 1 & x \\ & 1 \end{pmatrix} = \begin{pmatrix} \Delta_x^{-1} & -x\Delta_x^{-1} \\ & \Delta_x \end{pmatrix} \kappa_{\theta(x)},$$

$$\Delta_x = \sqrt{1 + x^2}, \qquad \theta(x) = \arctan\left(-\frac{1}{x}\right). \tag{6.9}$$

Because $f \in H(s_1, s_2, \epsilon)$, we have, therefore

$$\big(M(s) f\big)(1) = \int_{-\infty}^{\infty} (1 + x^2)^{-s}\, f\big(\kappa_{\theta(x)}\big)\, dx. \tag{6.10}$$

Assuming f to be a smooth vector, $\kappa \mapsto f(\kappa)$ is a smooth function on K by Proposition 2.5.3 and hence is bounded, so for convergence purposes, Eq. (6.10) converges, when

$$\int |(1+x^2)^{-s}|\,dx < \infty,$$

that is, when $\mathrm{re}(s) > \frac{1}{2}$.

We now check that $M(s)f \in H^\infty(s_2, s_1, \epsilon)$. First, we check that it satisfies Eq. (5.8) with s_1 and s_2 reversed. It is sufficient to show separately that

$$\big(M(s)f\big)\left(\begin{pmatrix} 1 & \xi \\ & 1 \end{pmatrix} g\right) = \big(M(s)f\big)(g) \tag{6.11}$$

and

$$\big(M(s)f\big)\left(\begin{pmatrix} y_1 & \\ & y_2 \end{pmatrix} g\right) = |y_1|^{s_2+\frac{1}{2}}\,|y_2|^{s_1-\frac{1}{2}}\big(M(s)f\big)(g), \tag{6.12}$$

with $y_1, y_2 > 0$. The proof of Eq. (6.11) is a simple matter of replacing x by $x - \xi$ in the definition of $M(s)$ as an integral. The proof of Eq. (6.12) is more interesting. The left side equals

$$\int_{-\infty}^{\infty} f\left(\begin{pmatrix} 0 & -1 \\ 1 & 0 \end{pmatrix}\begin{pmatrix} 1 & x \\ & 1 \end{pmatrix}\begin{pmatrix} y_1 & \\ & y_2 \end{pmatrix} g\right) dx$$

$$= \int_{-\infty}^{\infty} f\left(\begin{pmatrix} 0 & -1 \\ 1 & 0 \end{pmatrix}\begin{pmatrix} y_1 & \\ & y_2 \end{pmatrix}\begin{pmatrix} 1 & y_1^{-1}y_2x \\ & 1 \end{pmatrix} g\right) dx.$$

Replacing x by $y_1y_2^{-1}x$ and dx by $y_1y_2^{-1}\,dx$, this equals

$$\frac{y_1}{y_2}\int_{-\infty}^{\infty} f\left(\begin{pmatrix} 0 & -1 \\ 1 & 0 \end{pmatrix}\begin{pmatrix} y_1 & \\ & y_2 \end{pmatrix}\begin{pmatrix} 1 & x \\ & 1 \end{pmatrix} g\right) dx$$

$$= \frac{y_1}{y_2}\int_{-\infty}^{\infty} f\left(\begin{pmatrix} y_2 & \\ & y_1 \end{pmatrix}\begin{pmatrix} 0 & -1 \\ 1 & 0 \end{pmatrix}\begin{pmatrix} 1 & x \\ & 1 \end{pmatrix} g\right) dx$$

$$= \frac{y_1}{y_2}\,y_2^{s_1+\frac{1}{2}}\,y_1^{s_2-\frac{1}{2}}\int_{-\infty}^{\infty} f\left(\begin{pmatrix} 0 & -1 \\ 1 & 0 \end{pmatrix}\begin{pmatrix} 1 & x \\ & 1 \end{pmatrix} g\right) dx.$$

This equals the right side of Eq. (6.12), completing the verification of Eq. (5.8). The verification of Eq. (5.9) is fairly trivial. If we check that the restriction of $M(s)f$ to K is smooth, this will complete the proof that it is a smooth vector in $H^\infty(s_2, s_1, \epsilon)$. Using Eq. (6.8),

$$\big(M(s)f\big)(\kappa_t) = \big(M(s)\rho(\kappa_t)f\big)(1),$$

which by Eq. (6.10) equals

$$\int_{-\infty}^{\infty} (1+x^2)^{-s}\, f\left(\kappa_{\theta(x)+t}\right) dx. \tag{6.13}$$

If $\mathrm{re}(s) > \frac{1}{2}$, the convergence is uniform in t, and this is a smooth function.

Finally, if f is a K-finite vector, then the functions $t \mapsto f\left(\kappa_{\theta(x)+t}\right)$ span a finite-dimensional vector space, and so do the functions (Eq. 6.13); this proves that $M(s)f$ is a K-finite vector. ■

We now obtain a formula for the effect of $M(s)$ on a K-finite vector. Because we will be considering elements of two distinct spaces $H(s_1, s_2, \epsilon)$ and $H(s_2, s_1, \epsilon)$, we will modify the notation (Eq. 5.21) slightly and denote

$$f_{k,s}\left(\begin{pmatrix} u & \\ & u \end{pmatrix}\begin{pmatrix} y^{1/2} & xy^{-1/2} \\ & y^{-1/2} \end{pmatrix}\kappa_\theta\right) = u^{s_1+s_2}\, y^s\, e^{ik\theta}, \tag{6.14}$$

so that $f_{k,s} \in H(s_1, s_2, \epsilon)$. (We note that interchanging s_1 and s_2 does not change either $\mu = s_1 + s_2$ or ϵ, so we suppress these data from the notation.)

Proposition 2.6.3 *Assume that the real part of $s = \frac{1}{2}(s_1 - s_2 + 1)$ is greater than $1/2$. Then*

$$M(s)\, f_{k,s} = (-i)^k \sqrt{\pi}\, \frac{\Gamma(s)\,\Gamma\left(s-\frac{1}{2}\right)}{\Gamma\left(s+\frac{k}{2}\right)\Gamma\left(s-\frac{k}{2}\right)}\, f_{k,1-s}. \tag{6.15}$$

Proof Because $M(s)\, f_{k,s}$ is a vector in $H(s_2, s_1, \epsilon)$ satisfying

$$\left(M(s)\, f_{k,s}\right)(g\kappa_\theta) = e^{ik\theta}\left(M(s)\, f_{k,s}\right)(g),$$

it equals $f_{k,1-s}$ times a constant, which must be $\left(M(s)\, f_{k,s}\right)(1)$, so we must prove that

$$\left(M(s)\, f_{k,s}\right)(1) = (-i)^k \sqrt{\pi}\, \frac{\Gamma(s)\,\Gamma\left(s-\frac{1}{2}\right)}{\Gamma\left(s+\frac{k}{2}\right)\Gamma\left(s-\frac{k}{2}\right)}. \tag{6.16}$$

By Eq. (6.10)

$$\left(M(s) f_{k,s}\right)(1) = \int_{-\infty}^{\infty} (1+x^2)^{-s} \exp\left(ik\theta(x)\right) dx. \tag{6.17}$$

We will use the auxiliary variable

$$y = \frac{x-i}{x+i}.$$

As x runs from $-\infty$ to ∞, y moves on a contour C consisting of the unit circle beginning and ending at $y = 1$ and circling the origin counterclockwise. If $u \in \mathbb{C}$, we will denote

$$(-y)^u = e^{u \log(-y)},$$

where the branch of logarithm is chosen so that $\log(-y)$ is real when y is a negative real number. Then

$$e^{ik\theta(x)} = \left(\frac{x-i}{\sqrt{x^2+1}}\right)^k = (-i)^k\,(-y)^{k/2}.$$

We have also

$$x^2 + 1 = \frac{4(-y)}{(1-y)^2}, \qquad dx = 2i(1-y)^{-2}\,dy.$$

Hence Eq. (6.17) equals

$$2i\,(-i)^k\,4^{-s}\int\limits_C (1-y)^{2s-2}\,(-y)^{\frac{k}{2}-s}\,dy.$$

This integral is convergent if $\mathrm{re}(s) > \frac{1}{2}$; in order to evaluate it, however, we may assume that $2s-1$ and $\frac{k}{2}-s$ both have positive real part. The evaluation for other values will then follow by analytic continuation, treating k as a variable, for although k was originally assumed to be an integer, our last integral is an analytic function of k.

We deform the contour C so that it proceeds directly from 1 to 0 along the real axis, circles the origin in the counterclockwise direction, then returns to 1 along the real line. After circling the origin, the value of $(-y)^{\frac{k}{2}-s}$ increases from $e^{i\pi(s-k/2)}\,y^{\frac{k}{2}-s}$ to $e^{-i\pi(s-k/2)}\,y^{\frac{k}{2}-s}$, so the integral equals

$$\begin{aligned}
&2i\,(-i)^k\,4^{-s}\left[e^{-i\pi(s-k/2)} - e^{i\pi(s-k/2)}\right]\int\limits_0^1 (1-y)^{2s-2}\,y^{\frac{k}{2}-s}\,dy \\
&= (-i)^k\,4^{1-s}\,\sin\left(\pi\left(s-\tfrac{k}{2}\right)\right)\,B\left(2s-1,\tfrac{k}{2}-s+1\right),
\end{aligned}$$

where the initial $-$ sign on the right side comes from replacing $\int_1^0$ with $\int_0^1$, and where the Euler beta function

$$B\left(2s-1,\tfrac{k}{2}-s+1\right) = \frac{\Gamma(2s-1)\,\Gamma\left(\frac{k}{2}-s+1\right)}{\Gamma\left(\frac{k}{2}+s\right)}.$$

Making use of the duplication formula

$$\Gamma(2s-1) = \frac{4^{s-1}}{\sqrt{\pi}}\,\Gamma\left(s-\tfrac{1}{2}\right)\,\Gamma(s)$$

and the relation

$$\Gamma(u)\,\Gamma(1-u) = \frac{\pi}{\sin(\pi u)},$$

our integral simplifies to Eq. (6.16). ■

We now are ready to prove the unitaricity of the complementary series.

Theorem 2.6.4 *Let μ be purely imaginary, and let $0 < \lambda < \frac{1}{4}$. Then the infinitesimal equivalence class $\mathcal{P}_\mu(\lambda, 0)$ of admissible representations of $GL(2, \mathbb{R})^+$ contains a representative that is a unitary representation.*

Proof Let s_1 and s_2 be complex numbers. We begin by constructing a Hermitian pairing

$$H^\infty(s_1, s_2, \epsilon) \times H^\infty(-\overline{s_1}, -\overline{s_2}, \epsilon) \to \mathbb{C}$$

by

$$(f, f') \mapsto \int_K f(\kappa)\,\overline{f'(\kappa)}\,d\kappa. \tag{6.18}$$

We observe that if $f \in H^\infty(s_1, s_2, \epsilon)$ and $f' \in H^\infty(-\overline{s_1}, -\overline{s_2}, \epsilon)$, then in the notation of Lemma 2.6.1, $g \mapsto f(g)\,\overline{f'(g)}$ is in $C(P\backslash G, \delta)$, where P is the Borel subgroup of $GL(2, \mathbb{R})^+$. Consequently, by Lemma 2.6.1 this pairing is invariant under $(f, f') \mapsto \big(\rho(g)f, \rho(g)f'\big)$, where as usual $\rho(g)$ is the right translation.

Now let us suppose that $-\overline{s_1} = s_2$ and $-\overline{s_2} = s_1$. We note that this assumption is equivalent to assuming that $s = \frac{1}{2}(s_1 - s_2 + 1)$ is real and that $\mu = s_1 + s_2$ is purely imaginary. We have $H^\infty(s_2, s_1, \epsilon) = H^\infty(-\overline{s_1}, -\overline{s_2}, \epsilon)$, so we may compose the pairing (Eq. 6.18) with the intertwining map $i^\epsilon M(s)$: $H^\infty(s_1, s_2, \epsilon) \to H^\infty(-\overline{s_1}, -\overline{s_2}, \epsilon)$ and obtain a Hermitian pairing of $H^\infty(s_1, s_2, \epsilon)$ with itself and defined by

$$\langle f, f' \rangle = \int_K f(\kappa)\,\overline{i^\epsilon\,M(s)\,f'(\kappa)}\,d\kappa; \tag{6.19}$$

we have

$$\langle \rho(g)\,f, \rho(g)\,f' \rangle = \langle f, f' \rangle.$$

If this pairing is positive definite, then it is a Hermitian inner product. We shall show that it is positive definite if $\epsilon = 0$ and $\frac{1}{2} < s < 1$.

Clearly an orthogonal basis for $H^\infty(s_1, s_2, 0)$ consists of the elements $f_{k,s}$ with k even, so we have to check the signs of $\langle f_{k,s}, f_{k,s} \rangle$. By Eq. (6.15)

$$\langle f_{k,s}, f_{k,s} \rangle = (-1)^{\frac{k}{2}}\,\sqrt{\pi}\,\frac{\Gamma(s)\,\Gamma\big(s-\frac{1}{2}\big)}{\Gamma\big(s+\frac{k}{2}\big)\,\Gamma\big(s-\frac{k}{2}\big)}.$$

We observe that if $\frac{1}{2} < s < 1$, then for even k, the sign of $\Gamma(s - \frac{k}{2})$ is equal to $(-1)^{\frac{k}{2}}$, and so we have constructed a positive definite Hermitian inner product on $H^\infty(s_1, s_2, 0)$. We now obtain a unitary representation on the Hilbert space completion of this inner product space.

We have constructed a unitary representation in the infinitesimal equivalence class $\mathcal{P}_\mu(\lambda, 0)$, with $\lambda = s(1-s)$. If μ is purely complex and $0 < \lambda < \frac{1}{4}$, we can find $\frac{1}{2} < s < 1$ such that $s(1-s) = \lambda$. This concludes the proof that the complementary series representations are unitarizable. ■

It remains for us to decide the irreducibility of the finite-dimensional representations and the discrete series. In contrast with compact groups, finite-dimensional representations of noncompact groups are usually *not* unitary. For example, we have the following proposition.

Proposition 2.6.4 *The only irreducible unitary representations of the group $GL(n, \mathbb{R})^+$ that are finite dimensional are the one-dimensional characters $g \mapsto \det(g)^r$, where r is purely imaginary.*

Proof Let $(\pi, \mathfrak{H})$ be an irreducible unitary representation of $GL(n, \mathbb{R})$ on a finite-dimensional Hilbert space $\mathfrak{H}$ of degree n. Choosing an orthonormal basis for $\mathfrak{H}$, we may identify it with $\mathbb{C}^n$, in which case π is a continuous map into the compact unitary group $U(n)$. Hence the image of π is compact.

The only normal subgroups of $SL(n, \mathbb{R})$ are itself, the trivial subgroup, and (when n is even) $\{\pm I\}$. Thus the only compact homomorphic image of $SL(n, \mathbb{R})$ is the trivial group. Therefore π is trivial on $SL(n, \mathbb{R})$. The representation π therefore factors through the determinant map $\det : GL(n, \mathbb{R})^+ \to \mathbb{R}_+^\times$, and because the only unitary representations of $\mathbb{R}_+^\times$ are of the form $t \mapsto t^r$ where r is purely imaginary, we are done. ■

Proposition 2.6.5 *The infinitesimal equivalence class of representations $\mathcal{P}_\mu(\lambda, \epsilon)$ or $\mathcal{D}_\mu^\pm(k)$ contains a unitary representation if and only if μ is purely imaginary, and the corresponding class $\mathcal{P}(\lambda, \epsilon)$ or $\mathcal{D}^\pm(k)$ (with μ adjusted to zero) also contains a unitary representation.*

Proof It was proved in Theorem 2.6.3(ii) that the eigenvalue μ of Z on a unitary representation must be purely imaginary. We may therefore assume that μ is purely imaginary.

If (π, V) is a representation of $G = GL(n, \mathbb{R})^+$, and r is a complex number, we let $\det^r \otimes \pi$ denote the representation whose underlying space is still V, but whose group action is "twisted" by $\det^r$. Thus if $g \in G$, g acts by

$$g : v \mapsto \det(g)^r \, \pi(g) \, v.$$

This operation does not affect the parameters λ or ϵ that we have used to describe the representation. However, it shifts μ by a factor of $2r$.

Now if r is pure imaginary, $\det(g)^r$ has absolute value one, so $\pi(g)$ is unitary if and only if $\det(g)^r$ is. Consequently, we may shift the value of μ by a purely imaginary number without changing the unitaricity of the representation. ■

We turn now to the unitaricity of the discrete series $\mathcal{D}_\mu^\pm(k)$. It is possible to adapt the argument of Theorem 2.6.4, making use of the intertwining integral (Exercise 2.6.2). However, this is far from straightforward, and we will give here an alternative procedure, which also yields more information. We will show that the discrete series representations are *square integrable*, a notion that we shall now define.

Let G be a unimodular locally compact group, and let Z be the center of G. Let χ be any unitary character of Z. Let $L^2(G, \chi)$ denote the space of all measurable functions $f : G \to \mathbb{C}$ such that $f(zg) = \chi(z) f(g)$ for $z \in Z$, $g \in G$, and such that

$$\int_{Z\backslash G} |f(g)|^2 \, dg < \infty.$$

If $\chi = 1$, this space is naturally identified with $L^2(G/Z)$, where G/Z is the quotient group. We have a unitary representation ρ on $L^2(G, \chi)$, the right regular representation. An irreducible unitary representation $(\pi, \mathfrak{H})$ of G is called *square integrable* if it is embeddable in $L^2(G, \chi)$ for some unitary character χ of Z.

Now let $G = GL(2, \mathbb{R})^+$. By Proposition 2.6.5, it is sufficient to study $\mathcal{D}^\pm(k)$ with $\mu = 0$. We shall realize these representations in certain spaces of holomorphic functions. We shall also realize them as invariant subspaces of $L^2(G/Z)$. Hence we shall show that these representations are square integrable.

We will be transferring functions between the group $GL(2, \mathbb{R})^+$ and the upper half plane by the same procedure used in Section 1, for example, Eq. (1.25). However, the operator $|_k$ defined in Eq. (1.5) does not preserve holomorphicity, so we will instead define the operator $|_k$ as in Section 1.4. Specifically, the group $GL(2, \mathbb{C})$ acts on the Riemann sphere, which contains the Poincaré upper half plane $\mathcal{H}$ as well as the unit disk $\mathcal{D}$. We fix an integer $k \geq 1$. (In Theorem 2.6.5, we will actually assume that $k > 1$.) Let f be a smooth (typically holomorphic) function on some region Ω of the Riemann sphere. Let $\gamma \in GL(2, \mathbb{C})$, and assume that $\det(\gamma)$ is a positive real number. Define a function $f|\gamma \equiv f|_k\gamma$ on $\gamma^{-1}(\Omega)$ by

$$(f|\gamma)(z) = \det(g)^{k/2} \, (cz + d)^{-k} \, f\left(\frac{az + d}{cz + d}\right), \quad \gamma = \begin{pmatrix} a & b \\ c & d \end{pmatrix}. \tag{6.20}$$

(This agrees with the definition of $|_k$ Section 1.4, but differs from that used in Section 2.1.)

Lemma 2.6.3 *There exists a bijection between holomorphic functions ϕ on $\mathcal{H}$ and smooth functions Φ on $G = GL(2, \mathbb{R})^+$ that satisfy*

$$\Phi(g\kappa_\theta) = e^{ik\theta} \, \Phi(g) \tag{6.21}$$

and

$$L\Phi = 0, \tag{6.22}$$

where L is the differential operator defined by Eq. (1.28) on $C^\infty(G)$. The correspondence is given by

$$\phi(z) = y^{-k/2}\,\Phi\begin{pmatrix} y^{1/2} & y^{-1/2}x \\ & y^{-1/2} \end{pmatrix}, \tag{6.23}$$

$$\Phi(g) = (\phi|_k\, g)(i). \tag{6.24}$$

Note the close relationship between this lemma, Proposition 2.1.6 and Exercise 2.1.7(a). The only significant difference is that we are normalizing the $|_k$ operator differently, so that it preserves the property of holomorphicity.

Proof We leave it to the reader to check that Eqs. (6.23) and (6.24) give a bijection between $C^\infty(\mathcal{H})$ and the subspace of functions in $C^\infty(G)$ that satisfy Eq. (6.21). As usual, let

$$\frac{\partial}{\partial \bar{z}} = \frac{1}{2}\left(\frac{\partial}{\partial x} + i\frac{\partial}{\partial y}\right),$$

so that (in the coordinates Eq. (1.19))

$$L = e^{-2i\theta}\left(-2iy\,\frac{\partial}{\partial \bar{z}} - \frac{1}{2i}\,\frac{\partial}{\partial \theta}\right).$$

Thus if Φ satisfies Eq. (6.21), the equation (6.22) is equivalent to

$$-2iy\,\frac{\partial \Phi}{\partial \bar{z}} = \frac{k}{2}\,\Phi,$$

and then if ϕ is given by Eq. (6.23), this is equivalent to $\partial\phi/\partial\bar{z} = 0$. Thus Φ satisfies Eq. (6.22) if and only if ϕ satisfies the Cauchy–Riemann equations, that is, is a holomorphic function. ■

Theorem 2.6.5 *Let $1 < k \in \mathbb{Z}$. Then there exists a unitary representation in the infinitesimal equivalence class $\mathcal{D}^\pm(k)$. More precisely,*

(i) let $\mathfrak{H}$ be the space of holomorphic functions f on the upper half plane $\mathcal{H}$ that satisfy

$$\int_{\mathcal{H}} |f(z)|^2\, y^k\, \frac{dx\,dy}{y^2} < \infty. \tag{6.25}$$

Define group actions π^+ and π^- of $G = GL(2,\mathbb{R})^+$ on $\mathfrak{H}$ by

$$\left(\pi^\pm(g)\,f\right)(z) = (ad - bc)^{k/2}\,(\mp bz + d)^{-k}\, f\left(\frac{az \mp c}{\mp bz + d}\right),$$

$$g = \begin{pmatrix} a & b \\ c & d \end{pmatrix}. \tag{6.26}$$

Then $(\pi^+, \mathfrak{H})$ and $(\pi^-, \mathfrak{H})$ are admissible unitary representations in the infinitesimal equivalence classes $\mathcal{D}^+(k)$ and $\mathcal{D}^-(k)$, respectively.

(ii) Let Z be the center of $G = GL(2,\mathbb{R})^+$. The right regular representation of G on $L^2(G/Z)$ contains an invariant subspace that affords an irreducible admissible representation of G in the infinitesimal equivalence class $\mathcal{D}^\pm(k)$.

Proof Let ι be the automorphism of G given by

$$^{\iota}g = \begin{pmatrix} a & b \\ c & d \end{pmatrix}, \qquad g = \begin{pmatrix} a & -b \\ -c & d \end{pmatrix}.$$

The representations π^+ and π^- are related by $\pi^+(g) = \pi^-({}^{\iota}g)$, and it is easily checked that the involution ι interchanges the infinitesimal equivalence classes $\mathcal{D}^+(k)$ and $\mathcal{D}^-(k)$. Thus for (i), it is sufficient to show that $(\pi^-, \mathfrak{H})$ is an irreducible admissible representation in the infinitesimal equivalence class $\mathcal{D}^-(k)$.

Actually, we will work with a slight modification of π^-. We will study instead the representation $(\pi, \mathfrak{H})$ defined by

$$\pi(g)\, f = f|_k\, g^{-1}. \tag{6.27}$$

We observe that the inner product

$$\langle f_1, f_2\rangle = \int_{\mathcal{H}} f_1(z)\,\overline{f_2(z)}\, y^k\, \frac{dx\,dy}{y^2} \tag{6.28}$$

is preserved under the action $f \mapsto f|_k\, g^{-1}$ of $G = GL(2,\mathbb{R})^+$; this is a simple consequence of Eq. (2.2) in Chapter 1. Consequently, π is a *bona fide* action of G on $\mathfrak{H}$. It is evidently unitary.

We now observe that

$$\pi^-(g)\, f = f|{}^{\top}g = f|\big(\det(g)^{-1}\,{}^{\top}g\big) = f|wg^{-1}w_0^{-1} = \pi(w_0 g w_0^{-1}) f,$$

$$w_0 = \begin{pmatrix} & -1 \\ 1 & \end{pmatrix}.$$

The representations $(\pi, \mathfrak{H})$ and $(\pi^-, \mathfrak{H})$ are therefore equivalent, and it is sufficient to show that $(\pi, \mathfrak{H})$ is an irreducible admissible representation in the infinitesimal equivalence class $\mathcal{D}^-(k)$.

Let $s = k/2$, and let s_1 and s_2 be real numbers such that $s_1 + s_2 \equiv \mu = 0$ and $\frac{1}{2}(s_1 - s_2 + 1) = s$. (Thus $s_1 = -s_2 = s - \frac{1}{2}$.) Let $\epsilon = 0$ or 1 according as k is even or odd. We define a pairing $\langle\!\langle\, ,\, \rangle\!\rangle$ between $H^\infty(s_1, s_2, \epsilon)$ and $H^\infty(s_2, s_1, \epsilon) = H^\infty(-s_1, -s_2, \epsilon)$ by

$$\langle\!\langle f_1, f_2\rangle\!\rangle = \int_K f_1(\kappa)\, f_2(\kappa)\, d\kappa.$$

Observe that because we have omitted a complex conjugation over the second factor, this pairing is bilinear in both variables f_1 and f_2. This pairing is G-equivariant by Lemma 2.6.1, because the function $f_1\, f_2$ is in $C(P\backslash G, \delta)$.

Let $f_{k,s} \in H^\infty(s_1, s_2, \epsilon)$ be the vector defined by Eq. (6.14). We define a map

$$\sigma : H^\infty(-s_1, -s_2, \epsilon) \to C^\infty(G)$$

by

$$(\sigma f)(g) = \langle\!\langle \rho(g)\, f_{k,s}, f \rangle\!\rangle\,. \tag{6.29}$$

We observe that σf satisfies Eq. (6.21), and moreover by Eq. (5.22), it also satisfies Eq. (6.22). Hence by Lemma 2.6.3, there exists a holomorphic function Σf on $\mathcal{H}$ such that

$$(\Sigma f|_k\, g)(i) = (\sigma f)(g). \tag{6.30}$$

We will now prove that *there exists a constant $c(l)$ such that if $f_{l,1-s}$ is the vector in $H^\infty(s_2, s_1, \epsilon)$ defined by Eq. (6.14), we have*

$$\Sigma f_{l,1-s} = c(l)\,(z-i)^{-(l+k)/2}\,(z+i)^{(l-k)/2}; \tag{6.31}$$

and moreover, that $c(l) = 0$ unless $l \le -k$. We use the G-equivariance of the pairing $\langle\!\langle\ ,\ \rangle\!\rangle$ to write

$$(\sigma f_{l,1-s})(g) = \langle\!\langle f_{k,s}, \rho(g^{-1})\, f_{l,1-s} \rangle\!\rangle\,.$$

From this it follows that

$$(\sigma f_{l,1-s})(\kappa_\theta g) = e^{-il\theta}\,(\sigma f_{l,1-s})(g).$$

This implies that $\phi = \Sigma f_{l,1-s}$ satisfies

$$\phi|_k\, \kappa_\theta = e^{-il\theta}\,\phi. \tag{6.32}$$

Now we show that a holomorphic function on $\mathcal{H}$ that satisfies Eq. (6.32) is automatically a constant multiple of $(z-i)^{-l-k}\,(z+i)^l$. We make use of the Cayley transform $\mathcal{C}$ defined by Eq. (2.25). If Φ satisfies Eq. (6.32), and if $\psi = \phi|_k\, \mathcal{C}^{-1}$, then it follows from Eq. (2.27) that ψ is a function on the unit disk satisfying

$$\psi\bigg|_k \begin{pmatrix} e^{i\theta} & \\ & e^{-i\theta} \end{pmatrix} = e^{-il\theta}\,\psi.$$

Consideration of the Taylor expansion of Ψ shows immediately that all coefficients but the $(-l-k)/2$-nd must vanish (recall that $l \equiv k \bmod 2$), and that for some constant c,

$$\psi(w) = c\, w^{(-l-k)/2}.$$

Thus $l \le -k$ and ϕ is a constant multiple of $(z-i)^{-(l+k)/2}\,(z+i)^{(l-k)}$. This proves the formula (6.31).

Recall Theorem 2.5.3(iii), which describes the structure of $H(-s_1, -s_2, \epsilon)$. This has an irreducible invariant subspace spanned by the vectors $f_{l,1-s}$ with $2-k \le l \le k-2$, and the quotient is a direct sum of two infinite-dimensional subspaces; taking the preimage of one of these, we see that $H(-s_1, -s_2, \epsilon)$ has a (reducible) invariant subspace spanned by the vectors $f_{l,1-s}$ with $l \ge 2-k$, and these we have shown to be in the kernel of Σ. On the other hand, the image under Σ of the vectors $f_{l,1-s}$ with $l \le -k$ are all square-integrable; indeed, because $|(z-i)/(z+i)| < 1$ on $\mathcal{H}$ and $l+k \le 0$, each is dominated by $|z+i|^{-k}$, which is easily checked to be square-integrable with respect to the inner product (Eq. 6.28) under the assumption that $k \ge 2$. Hence the image under Σ of the K-finite vectors in $H(-s_1, -s_2, \epsilon)$ is contained in $\mathfrak{H}$ defined as above. To see that these span the Hilbert space $\mathfrak{H}$, one may proceed as follows. Transfer any holomorphic function on $\mathcal{H}$ to the unit disk via the Cayley transform; this has a power series expansion. It is clear from the above discussion that this may be regarded as a Fourier series expansion in terms of the vectors $\Sigma f_{l,1-s}$, so the image of the K-finite vectors spans the space $\mathfrak{H}$. It is easily seen that the group action is as given by Eq. (6.27). This completes the proof of (i).

To prove (ii), observe that the correspondence described in Lemma 2.6.3 between functions on G and functions on $\mathcal{H}$ is an isometry; because we have already checked that the image of the K-finite vectors under Σ are square integrable, their images under σ are also. This actually gives rise to a realization of $\mathcal{D}^-(k)$ in the *left*-regular representation of G on $L^2(G)$; however, by composing with $g \to g^{-1}$, this realization may be transferred to the right regular representation. ■

The "limits of discrete series" $\mathcal{D}^{\pm}(1)$ may also be realized in a space of holomorphic functions on $\mathcal{H}$, with the norm

$$|f|^2 = \sup_{y>0} \int_{-\infty}^{\infty} |f(x+iy)|^2 \, dx.$$

However, we do not need this: We have already proved that they are unitarizable because they occur in the induced representation $H(0, 0, 1)$, which is unitary by Theorem 2.6.1.

We make a few remarks about the nature of discrete series representations; for further information, consult the books of Knapp (1986), Wallach (1988), and Warner (1972).

Let $(\pi, \mathfrak{H})$ be a unitary representation of a separable locally compact group G admitting a "large" maximal compact subgroup K. This is a technical hypothesis that we do not need to explain, but suffice it to say that this class of groups contains reductive Lie and p-adic groups. Harish-Chandra proved that the following three conditions are equivalent: (i) $(\pi, \mathfrak{H})$ is equivalent to a subrepresentation of $L^2(G, \chi)$ for some character χ of the center Z of G; (ii) a single K-finite matrix coefficient of $(\pi, \mathfrak{H})$ is square integrable modulo

Z; and (iii) every matrix coefficient of $(\pi, \mathfrak{H})$ is square integrable modulo Z. We have already defined a representation satisfying condition (i) to be *square integrable*. The terms "discrete series representation" and "square integrable representation" are more or less synonymous.

The following theorem is true for reductive Lie groups. We prove it for $GL(2)$ only.

Theorem 2.6.6 *Let $(\pi, \mathfrak{H})$ and $(\pi', \mathfrak{H}')$ be irreducible admissible unitary representations of $G = GL(2, \mathbb{R})^+$. Suppose that these representations are infinitesimally equivalent. Then they are equivalent.*

Proof The assumption that the representations are infinitesimally equivalent means that the spaces $V = \mathfrak{H}_{\text{fin}}$ and $V = \mathfrak{H}'_{\text{fin}}$ of K-finite vectors are isomorphic as $(\mathfrak{g}, K)$-modules. Let $\phi : V \to V'$ be an isomorphism. We show that (after multiplying by a suitable constant) ϕ is an isometry. Let $V = \oplus_k V(k)$ and $V' = \oplus_k V'(k)$ be the decompositions (Eq. 5.2). Choose k so that $V(k)$ is nonzero; by Proposition 2.5.2(iv), it is one dimensional. Let $0 \neq x \in V(k)$. We normalize x so it has length one. We normalize ϕ so that $|\phi(x)| = 1$ also. Now we have decompositions (Eq. 5.3) for both V and V', and because the spaces in Eq. (5.3) are mutually orthogonal (by Exercise 2.1.5 or 2.4.1) in order to establish that ϕ is an isometry, it is clearly sufficient to show that $|R^n x| = |R^n \phi(x)|$ and $|L^n x| = |L^n \phi(x)|$. Indeed, we have, by Eqs. (6.6) and (5.4)

$$|Rx|^2 = \langle Rx, Rx\rangle = -\langle LRx, x\rangle = \left(\lambda + \tfrac{k}{2}(1 + \tfrac{k}{2})\right)\langle x, x\rangle = \lambda + \tfrac{k}{2}(1 + \tfrac{k}{2}).$$

The identical calculation gives the identical value for $|R\,\phi(x)|$, and repeating this argument n times gives $|R^n x| = |R^n \phi(x)|$; similarly, $|L^n x| = |L^n \phi(x)|$. This proves that ϕ is an isometry.

Because $\mathfrak{H}$ and $\mathfrak{H}'$ are the Hilbert space completions of V and V', ϕ induces between these a Hilbert space isomorphism, which we shall also denote as $\phi : \mathfrak{H} \to \mathfrak{H}'$. It remains to be shown that ϕ is an intertwining operator.

Let f be a K-finite vector in $\mathfrak{H}$. Then f is smooth by Proposition 2.4.5, so by Exercise 2.4.2 (b) we have, for all $X \in \mathfrak{gl}(2, \mathbb{R})$

$$\pi(e^X)\, f = \sum_{n=0}^{\infty} \frac{1}{n!} X^n f.$$

Because ϕ is an isometry, we have therefore

$$\phi\big(\pi(e^X)\, f\big) = \sum_{n=0}^{\infty} \frac{1}{n!} \phi(X^n f),$$

and because the restriction of ϕ to V is an isomorphism of $(\mathfrak{g}, K)$-modules, this equals

$$\sum_{n=0}^{\infty} \frac{1}{n!} X^n \phi(f) = \pi'(e^X)\,\phi(f).$$

This proves that

$$\phi\big(\pi(e^X) f\big) = \pi'(e^X)\,\phi(f)$$

when f is K-finite; because the K-finite vectors are dense, it is true for all $f \in \mathfrak{H}$. Because G is generated by elements of the form e^X, we see that ϕ is G-equivariant, as required. ∎

Theorem 2.6.7: Unitary representations of $GL(2, \mathbb{R})^+$***:*** *The following is a complete list of the isomorphism classes of irreducible admissible unitary representations of* $GL(2, \mathbb{R})^+$. *Each of the following infinitesimal equivalence classes of irreducible representations has a unique representative that is a unitary representation.*

(i) The one-dimensional representations $g \mapsto |\det(g)|^{\mu}$, *where* μ *is purely imaginary;*
(ii) The unitary principal series $\mathcal{P}_\mu(\lambda, \epsilon)$, *where* μ *is purely imaginary,* $\epsilon = 0$ *or* 1, *and* λ *is a real number,* $\lambda \geq \frac{1}{4}$*;*
(iii) The complementary series representations $\mathcal{P}_\mu(\lambda, 0)$, *where* μ *is purely imaginary and* $0 < \lambda < \frac{1}{4}$*; and*
(iv) The holomorphic discrete series (when $k \geq 2$*) and limits of discrete series (when* $k = 1$*)* $\mathcal{D}^{\pm}_{\mu}(k)$, *where* μ *is purely imaginary.*

The assumption of admissibility here is actually unnecessary, but this result is sufficient for the applications to automorphic forms in view of the second corollary to Theorem 2.4.3.

Proof The fact that each infinitesimal class contains at most one unitary representative is stated in Theorem 6.14. Given this fact, we may classify the irreducible admissible unitary representations by first classifying the infinitesimal equivalence classes of irreducible admissible representations, that is, by classifying the isomorphism classes of $(\mathfrak{g}, K)$-modules associated with admissible representations, which is accomplished in Theorem 2.5.4, and the discussion following that theorem; then asking which of these infinitesimal equivalence classes has a unitary representative. The latter question is answered in this section, particularly in Theorems 2.6.2, 2.6.3, 2.6.4, Propositions 2.6.4, and 2.6.5, Theorem 2.6.5, and the observation on the unitaricity of the limits of discrete series immediately following the proof of Theorem 2.6.5. ∎

Exercises

Exercise 2.6.1 Prove that though $H(s_1, s_2, \epsilon)$ and $H(s_2, s_1, \epsilon)$ are infinitesimally equivalent when irreducible, they are not in general isomorphic.

[HINT: Assume that $\text{re}(s_1) > \text{re}(s_2)$. It is sufficient to prove that the intertwining operator $M(s) : H^\infty(s_1, s_2, \epsilon) \to H^\infty(s_2, s_1, \epsilon)$ does not extend to an isomorphism of Hilbert spaces, where we recall that $H(s_1, s_2, \epsilon)$ is the completion of $H^\infty(s_1, s_2, \epsilon)$ with respect to the inner product (Eq. 5.14). Because the constants in Eq. (6.15) are dependent on k and because the vectors $f_{k,s}$ and $f_{k,1-s}$ are unit vectors it is clear that $M(s)$ is not an isometry, but it is possible to show more. The constants on the right side of Eq. (6.15) decay as $k \to \pm\infty$, slowly it's true, but fast enough that the image of $H^\infty(s_1, s_2, \epsilon)$ in $H(s_2, s_1, \epsilon)$ is not dense.]

Exercise 2.6.2 Give an alternative proof of the unitaricity of the discrete series (Theorem 2.6.5) based on the method of Theorem 2.6.4.

Exercise 2.6.3 Prove that if two irreducible admissible unitary representations of $GL(2, \mathbb{R})$ are infinitesimally equivalent, they are isomorphic. [HINT: Identify their underlying $(\mathfrak{g}, K)$-modules. Use Schur's Lemma (Exercise 2.4.6(b)) to show that their Hermitian inner products are proportional on the K-finite vectors.]

Exercise 2.6.4 Prove that if λ and μ are complex numbers, there are at most two nonisomorphic irreducible admissible unitary representations on which Z and Δ act by λ and μ. [HINT: Use Exercise 2.6.3 and the classification of the unitarizable $(\mathfrak{g}, K)$-modules in this section.]

2.7 Representations and the Spectral Problem

Now equipped with the benefit of the results of the intervening sections, we return to the discussion of the spectral problem in Section 2.1. Notation is as in Section 2.1, except that here Δ is the self-adjoint extension of the Laplace–Beltrami operator Δ that was defined at the end of Section 2.3. (The original Δ, introduced in Eq. (1.28) was defined only for smooth functions.)

This result gives a very clear picture of the relationship between versions I and II of the spectral problem in Section 2.1 and the representation theory of $GL(2, \mathbb{R})$. It answers the first of the two questions with which we began Section 2.1. (The second question is answered by the systematic use that we made of the elements L and R of $\mathfrak{g}_\mathbb{C}$, $\mathfrak{g} = \mathfrak{gl}(2, \mathbb{R})$ in the proofs throughout!)

Theorem 2.7.1 *Let Γ be a discontinuous subgroup of $G = GL(2, \mathbb{R})^+$ such that $-I \in \Gamma \subset G_1 = SL(2, \mathbb{R})$ and such that $\Gamma\backslash G_1$ is compact. Let χ be a character of Γ. Let $\chi(-1) = (-1)^\epsilon$, where $\epsilon = 0$ or 1.*

(i) The space $\mathfrak{H} = L^2(\Gamma\backslash G, \chi)$ *decomposes into a Hilbert space direct sum of irreducible representations. The spaces* $L^2(\Gamma\backslash\mathcal{H}, \chi, k)$ *each decompose into a Hilbert space direct sum of eigenspaces for* Δ_k.

(ii) Let H *be any subspace of* $\mathfrak{H}$ *that is invariant under the action of* G. *Then* H *is also invariant under the action of* Δ. *Conversely, let* $\lambda \in \mathbb{R}$, *and let* $\mathfrak{H}_\lambda$ *be the* λ*-eigenspace of* Δ *on* $\mathfrak{H}$. *Then* $\mathfrak{H}_\lambda$ *is* G*-invariant.*

(iii) If H *is an irreducible subspace of* $\mathfrak{H}$, *then* Δ *acts as a scalar on* H. *The eigenvector* $\lambda = \lambda(H)$ *of* Δ *on* H *depends only on the isomorphism class of* H. *If* λ *is such an eigenvalue, then* λ *is a real number. Either* $\lambda \geq 0$ *if* $\epsilon = 0$ *and* $\lambda \geq \frac{1}{4}$ *if* $\epsilon = 1$, *or else* $\lambda = \frac{k}{2}(1 - \frac{k}{2})$ *where* $1 \leq k \in \mathbb{Z}$ *and* $k \equiv \epsilon \bmod 2$.

(iv) There is only one finite-dimensional representation of G *that can occur in* $\mathfrak{H}$, *namely, the trivial representation: If* $\chi = 1$, *then the constant function spans a one-dimensional irreducible subspace of* $\mathfrak{H}$, *on which space the eigenvalue* λ *of* Δ *equals* 0. *All other irreducible constituents of* $\mathfrak{H}$ *are infinite dimensional.*

(v) Assume that λ *is not of the form* $\frac{k}{2}(1 - \frac{k}{2})$, *where* $1 \leq k \in \mathbb{Z}$ *and* $\equiv \epsilon \bmod 2$. *There exists a unique irreducible representation* $\mathcal{P}(\lambda, \epsilon)$ *of* G *depending only on* λ *(and not on* Γ*) such that if* H *is an infinite-dimensional irreducible subrepresentation of* $\mathfrak{H}$ *with* $\lambda(H) = \lambda$, *then* $H \cong \mathcal{P}(\lambda, \epsilon)$. *Let* k *be an integer* $\equiv \epsilon \bmod 2$. *The multiplicity of the representation* $\mathcal{P}(\lambda, \epsilon)$ *is equal to the multiplicity of the eigenvalue* λ *in* $L^2(\Gamma\backslash\mathcal{H}, \chi, k)$.

(vi) Assume that λ *is of the form* $\frac{k}{2}(1 - \frac{k}{2})$, *where* $1 \leq k \in \mathbb{Z}$, *and* $k \equiv \epsilon \bmod 2$. *There exist two irreducible representations,* $\mathcal{D}^+(k)$ *and* $\mathcal{D}^-(k)$ *of* G, *depending only on* k *such that if* H *is an infinite-dimensional irreducible representation of* $\mathfrak{H}$ *with* $\lambda(H) = \lambda$, *then either* $H \cong \mathcal{D}^+(k)$ *or* $H \cong \mathcal{D}^-(k)$. *The representations* $\mathcal{D}^\pm(k)$ *have the same multiplicity in* $L^2(\Gamma\backslash G, \chi)$; *this multiplicity equals the dimension of the space of holomorphic modular forms of weight* k *that satisfy* $f(\gamma z) = \chi(\gamma)\,(cz + d)^k\, f(z)$ *for* $\gamma \in \Gamma$.

Proof This is an assembling of previous results. Part (i) is Theorem 2.3.3.

As for (ii), by (i) it is clearly sufficient to show that the Laplacian is a scalar on an irreducible invariant subspace H of $\mathfrak{H}$; indeed, from the way the self-adjoint extension of Δ was defined at the end of Section 2.3, it is sufficient to show that it acts as a scalar on a dense subspace of H. We note that H is admissible by the second corollary to Theorem 2.4.3. Let H_{fin} be the subspace of K-finite vectors ($K = SO(2)$) in H. This is an irreducible admissible $(\mathfrak{g}, K)$-module by the remarks at the end of Section 2.4 and is dense in H by Proposition 2.4.5. Moreover, Δ acts as a scalar on H_{fin} by Proposition 2.5.1. Hence it acts as a scalar on H. This proves (ii).

Parts (iv), (v), and (vi) follow from the classification of the irreducible unitary representations of $GL(2,\mathbb{R})^+$ in Theorem 2.6.7, because an irreducible subspace H of $\mathfrak{H}$ is admissible by the second corollary to Theorem 2.4.3, and is unitary because it inherits an invariant inner product from the L^2 norm on

$L^2(\Gamma\backslash G, \chi)$. We note that according to the definition of $L^2(\Gamma\backslash G, \chi)$, scalar matrices act trivially, which implies that the parameter μ appearing in Theorem 2.6.7 is zero, and in accordance with the convention in Sections 2.5 and 2.6, we suppress it from the notation.

The only point worth commenting on is the connection of the discrete series representations with holomorphic modular forms. The only irreducible admissible unitary representations with eigenvalue $\frac{k}{2}(1-\frac{k}{2})$ (and with $\mu = 0$) are $\mathcal{D}^{\pm}(k)$, where $1 \le k \in \mathbb{Z}$. These are distinguished by the fact that in the notation (Eq. 5.2) if $H \cong \mathcal{D}^+(k)$, then $\dim\ H(k) = 1$ whereas $\dim\ H(-k) = 0$; and if $H \cong \mathcal{D}^-(k)$, then $\dim\ H(-k) = 1$, whereas $\dim\ H(k) = 0$. Thus the multiplicity of $D^+(k)$ in $\mathfrak{H}$ equals the dimension of the $\frac{k}{2}(1-\frac{k}{2})$-eigenspace in $L^2(\Gamma\backslash G, \chi, k)$, or eqivalently by Proposition 2.1.6, it equals the multiplicity of $\frac{k}{2}(1-\frac{k}{2})$-eigenspace in $L^2(\Gamma\backslash\mathcal{H}, \chi, k)$. We show that this eigenspace is isomorphic to a space of modular forms. An element of this space is K-finite and hence is a smooth vector by Proposition 2.4.5, and so we may consider the action of the element $L \in \mathfrak{g}_{\mathbb{C}}$ defined by Eq. (2.23). By Proposition 2.5.2(ii), this is in $H(k-2)$, but according to the definition of $\mathcal{D}^+(k)$ at the end of Section 2.5, $H(k-2) = 0$, so L annihilates the $\frac{k}{2}(1-\frac{k}{2})$-eigenspace of $L^2(\Gamma\backslash\mathcal{H}, \chi, k)$. Now by Eq. (1.29), this implies that any $f \in L^2(\Gamma\backslash\mathcal{H}, \chi, k)$ is annihilated by the operator L_k defined by Eq. (1.2), which implies that $y^{-k/2} f(z)$ satisfies the Cauchy–Riemann equations and hence is holomorphic and is in fact a holomorphic modular form (Exercise 1.7). We see that the multiplicity of $\mathcal{D}^+(k)$ equals the dimension of the $\frac{k}{2}(1-\frac{k}{2})$-eigenspace of $L^2(\Gamma\backslash\mathcal{H}, \chi, k)$, which is equal to the dimension of the space of holomorphic modular forms as described in (vi) above. Because complex conjugation interchanges $L^2(\Gamma\backslash\mathcal{H}, \chi, k)$ and $L^2(\Gamma\backslash\mathcal{H}, \chi, -k)$, it is easy to see that this is equal to the multiplicity of $\mathcal{D}^-(k)$ also. ■

2.8 Whittaker Models

The results of this section will be used in Chapter 3. This section may be read immediately after Section 2.4. We are concerned with smooth functions W on $G = GL(2, \mathbb{R})^+$, which satisfy

$$W\left(\begin{pmatrix} 1 & x \\ & 1 \end{pmatrix} g\right) = \psi(x)\, W(g). \tag{8.1}$$

Let Δ and Z be as in Eqs. (1.29) and (2.18).

We say that a function W satisfying Eq. (8.1) is *of moderate growth* if, for fixed x, u, and θ in the coordinates Eq. (1.19), W is bounded by a polynomial in y as $y \to \infty$. We say it is *rapidly decreasing* if (in the same coordinates) $y^N W \to 0$ as $y \to \infty$ for every $N > 0$. We say that W is *analytic* if it is given by a convergent power series expansion in a neighborhood of every $g \in GL(2, \mathbb{R})^+$.

Proposition 2.8.1 *Let μ and $\lambda \in \mathbb{C}$, and let $k \in \mathbb{Z}$. Let $\mathcal{W}(\lambda, \mu, k)$ be the space of functions W on G satisfying Eq. (8.1) and*

$$W(g\kappa_\theta) = e^{ik\theta}\, W(g)$$

with κ_θ as in Eq. (1.19) such that $\Delta W = \lambda\, W$ and $Z\, W = \mu\, W$ and such that W is of moderate growth. Then $\mathcal{W}(\lambda, \mu, k)$ is one dimensional. A function in this space is actually rapidly decreasing and analytic. The differential operators R and L defined by Eqs. (1.27) and (1.28) map $\mathcal{W}(\lambda, \mu, k)$ into $\mathcal{W}(\lambda, \mu, k + 2)$ and $\mathcal{W}(\lambda, \mu, k - 2)$, respectively.

Proof The character $\psi(x) = e^{iax}$, where a is a nonzero real number. We will assume that $a = 1/2$, and leave the reader to make the minor modifications for other values of a. This normalization has the advantage that the "Whittaker functions" W of the theorem are identical with classical Whittaker functions, which are solutions of the confluent hypergeometric equation. These were introduced by E. T. Whittaker in 1904 and are studied in Whittaker and Watson (1927, Chapter 16).

The condition $Z\, W = \mu\, W$ means that

$$W\left(\begin{pmatrix} u & \\ & u \end{pmatrix} g\right) = u^\mu\, W(g).$$

Thus if we use the coordinates Eq. (1.19)

$$W(g) = u^\mu\, e^{ix/2}\, e^{ik\theta}\, w(y), \qquad w(y) = W\begin{pmatrix} y^{1/2} & \\ & y^{-1/2} \end{pmatrix}.$$

Thus it is sufficient to study the values of $w(y)$. Applying Eq. (1.29), we find that

$$w'' + \left\{ -\frac{1}{4} + \frac{k}{2y} + \frac{\lambda}{y^2} \right\} w = 0. \tag{8.2}$$

This is the confluent hypergeometric equation. Let $s - 1/2$ be a square root of $-\lambda + 1/4$; referring to Whittaker and Watson 16.31, the solutions to this equation are then the linear combinations of $W_{\frac{k}{2}, s-\frac{1}{2}}(y)$ and $W_{-\frac{k}{2}, s-\frac{1}{2}}(-y)$, which are asymptotically $e^{-y/2}\, y^{k/2}$ and $e^{y/2}\, (-y)^{-k/2}$. The assumption of moderate growth excludes the second solution but permits the first. Hence $\mathcal{W}(\lambda, \mu, k)$ is spanned by the function (in the coordinates Eq. (1.19))

$$W_{k,\lambda,\mu}(g) = u^\mu\, e^{ix/2}\, e^{ik\theta}\, W_{\frac{k}{2}, s-\frac{1}{2}}(y). \tag{8.3}$$

Note that by Whittaker and Watson 16.3 and 16.12, this function is rapidly decreasing and analytic.

The action of R and L remains to be considered. Of course any operator in $U\big(\mathfrak{gl}(2, \mathbb{R})\big)$ preserves the property (Eq. 8.1) because such operators commute with left translation; it follows from Eq. (1.30) that if $W \in \mathcal{W}(\lambda, \mu, k)$, then

RW and LW have the appropriate transformation properties with respect to κ_θ; and with Z and Δ being in the center of $U(\mathfrak{gl}(2, \mathbb{R}))$, it is clear that RW and LW have the same eigenvalues for Z and Δ. However, it merits proof that RW and LW are of moderate growth. What needs to be checked is that $(d/dy)\, W_{\frac{k}{2}, s-\frac{1}{2}}(y)$ decays as $y \to \infty$; and in fact, it may be verified that it is legitimate to differentiate the asymptotic expansions of Whittaker and Watson, Section 16.3, term by term. (The actual constant of proportionality between $LW_{k,\lambda,\mu}$ and $W_{k-2,\lambda,\mu}$ is given by the recurrence formula in Whittaker and Watson, Example 3, Chapter 16.) ■

Theorem 2.8.1 *Let (π, V) be an irreducible admissible $(\mathfrak{g}, K)$-module for $G = GL(2, \mathbb{R})^+$ or $GL(2, \mathbb{R})$. Then there exists at most one space $\mathcal{W}(\pi, \psi)$ of smooth K-finite functions W satisfying Eq. (8.1) that are of moderate growth such that the space $\mathcal{W}(\pi, \psi)$ is invariant under the actions of $U(\mathfrak{g})$ and K on $C^\infty(G)$ and such that $\mathcal{W}(\pi, \psi)$ is isomorphic to (π, V) as a $(\mathfrak{g}, K)$-module. Every function in $\mathcal{W}(\pi, \psi)$ is actually rapidly decreasing and analytic.*

The space $\mathcal{W}(\pi, \psi)$ is called the *Whittaker model* of π. This result is also true if $\mathbb{R}$ is replaced by $\mathbb{C}$ (see Jacquet and Langlands, (1970), Theorem 6.3, p. 232).

Proof By Proposition 2.5.1, Δ and Z act by scalars on V. Let λ and μ be their eigenvalues. We decompose V as in Eq. (5.2). When $V(k) \neq 0$, it is one dimensional. If $G = GL(2, \mathbb{R})^+$, this is Proposition 2.5.2(iv), whereas if $G = GL(2, \mathbb{R})$, it follows from the classification of Theorem 2.5.5. If $V(k) \neq 0$, its image under the isomorphism with $\mathcal{W}(\pi, \psi)$ satisfies the conditions in Proposition 2.8.1, and hence is $\mathcal{W}(\lambda, \mu, k)$. Thus $\mathcal{W}(\pi, \psi)$ is the algebraic direct sum of the $\mathcal{W}(\lambda, \mu, k)$ for all k such that $V(k) \neq 0$. This explicit characterization shows that $\mathcal{W}(\pi, \psi)$ is uniquely determined. The rapid decrease and analyticity of the functions in $\mathcal{W}(\lambda, \mu, k)$ is also a consequence of Proposition 2.8.1. ■

* * *

Although Theorem 2.8.1 is sufficient for most of our purposes, we now describe a different version of this result, which on $GL(n)$ is due to Shalika (1974). Let $F = \mathbb{R}$ or $\mathbb{C}$, and let (π, V) be an irreducible admissible unitary representation of $G = GL(n, F)$. Thus V is a Hilbert space. Let V^∞ be the space of smooth vectors in V. Then V^∞ has a natural topology distinct from that inherited from V, with respect to which it is a Fréchet space. If $D \in U(\mathfrak{g})$, then $f \mapsto |\pi(D)\, f|$ is a seminorm on V^∞, where $\pi(D)\, f$ is defined in Eq. (4.1) when $D \in \mathfrak{g}$, extended to all of $U(\mathfrak{g})$ by Proposition 2.2.3, and where $|\ |$ is the Hilbert space norm. The Fréchet topology is then the smallest topology on V^∞, with respect to which these seminorms are continuous. If K is a maximal compact subgroup of G, then the K-finite vectors are dense in V^∞. For further information about the Fréchet topology on the smooth vectors, see

Sections 1.6 and 3.3 of Wallach (1988). Let ψ be a nontrivial additive character of F. We define a character ψ_N of the group $N(F)$ of upper triangular unipotent matrices in G by

$$\psi_N(u) = \psi\left(\sum_{i=1}^{n-1} u_{i,i+1}\right), \qquad u = (u_{ij}) \in N(F).$$

By a *Whittaker functional* on V^∞, we mean a continuous linear functional $\lambda : V^\infty \to \mathbb{C}$ such that $\lambda(\pi(u)\,x) = \psi_N(u)\,\lambda(x)$ for all $u \in N(F)$, $x \in V^\infty$. Shalika (1974) proved the following theorem.

Theorem 2.8.2: Shalika's local multiplicity one theorem *In this situation, the dimension of the space of Whittaker functionals on V^∞ is at most one dimensional.*

Here we omit the proof. We will not need this result. However, it is important in much of the literature. If one is studying an automorphic cuspidal representation with unitary central character, the representation is automatically unitary, and there are advantages to working not with the K-finite vectors (as we do in this book) but with the smooth vectors. On the other hand, for Eisenstein series, it is best to work with the K-finite vectors. Theorem 2.8.2 is also valid for non-Archimedean fields, too – it is Theorem 4.4.1, which we will prove in Chapter 4 when $n = 2$.

2.9 A Theorem of Harish-Chandra

The results of this section will be needed in Chapter 3. This section can be read immediately after Section 2.4. Let G be a semisimple Lie group, K its maximal compact subgroup, and $\mathfrak{g}$ its Lie algebra. Let $\mathcal{Z}$ be the center of $U(\mathfrak{g}_\mathbb{C})$. Then G acts on $C^\infty(G)$ by the right regular representation ρ defined in Section 2.2, and $U(\mathfrak{g}_\mathbb{C})$ acts extending the action (Eq. 2.10) of $\mathfrak{g}$ (and its complexification $\mathfrak{g}_\mathbb{C}$) by means of Proposition 2.2.3. Suppose that $f \in C^\infty(G)$ is both K-finite and $\mathcal{Z}$-finite. (This means that the right translates of f by elements of K span a finite-dimensional vector space, as do the functions Df with $D \in \mathcal{Z}$.) Harish-Chandra (1966) proved that there exists a function $\alpha \in C_c^\infty(G)$ such that the convolution $f * \alpha = f$. We will prove this useful fact if $G = SL(2, \mathbb{R})$.

An alternative exposition of this result of Harish-Chandra, likewise specialized to the case of $SL(2, \mathbb{R})$, may be found in Borel (to appear).

We will require a basic result from the theory of partial differential equations, namely, the analyticity of solutions to elliptic differential equations. We recall that a function on a domain M of $\mathbb{R}^n$ (taking values in $\mathbb{R}$, $\mathbb{C}$, or in a real or complex vector space) is *analytic* if it has a power series representation in a neighborhood of each point. More generally, we may consider a manifold M in which the transition functions between charts are analytic. Such a manifold is called *analytic*, and a function on M is called *analytic* if whenever $\mu : \Omega \to M$ is a chart (with Ω an open set in $\mathbb{R}^n$), the function $f \circ \mu$ is analytic.

Let $x = (x_1, \cdots, x_n)$ be local coordinates on M. We consider a linear differential operator

$$L = \sum_{\substack{\alpha = (\alpha_1, \cdots, \alpha_n) \in \mathbb{Z}_+^n \\ \sum \alpha_i \leq d}} \lambda_\alpha(x) \frac{\partial^{\alpha_1 + \ldots + \alpha_n}}{\partial^{\alpha_1} x_1 \cdots \partial^{\alpha_n} x_n}. \tag{9.1}$$

We form a homogeneous polynomial of degree d with the terms of highest degree:

$$\sum_{\sum \alpha_i = d} \lambda_\alpha(x)\, \xi_1^{\alpha_1} \cdots \xi_n^{\alpha_n}.$$

If this polynomial has no zeros $(\xi_1, \cdots, \xi_n) \neq 0$ for some value of $x \in M$, then we say that L is *elliptic* at x. If it is elliptic for all $x \in M$, then we say it is *elliptic*.

Theorem 2.9.1 *Suppose that L is an elliptic differential operator on the domain $M \subseteq \mathbb{R}^n$ and suppose that the coefficients in L (i.e., the functions λ_α in Eq. (9.1)) are analytic. Then every solution to $L\phi = 0$ is analytic.*

See Friedman (1969), Theorem 1.2, Chapter 3, p. 205), and Bers and Schechter (1964, especially the Appendix on p. 207). Appendix 4 on elliptic equations in Lang (1985) does not prove this analyticity but is useful reading on elliptic equations nonetheless.

Theorem 2.9.2 *Let $G = SL(2, \mathbb{R})$, $K = SO(2)$, let $\mathfrak{g}$ be the Lie algebra of G, and let $\mathcal{Z} = \mathbb{C}[\Delta]$ be the center of $U(\mathfrak{g}_\mathbb{C})$, where Δ is as in Eq. (2.39). Let $f \in C^\infty(G)$ be both K-finite and $\mathcal{Z}$-finite. Then f is analytic, $U(\mathfrak{g}_\mathbb{C})\, f$ is an admissible $(\mathfrak{g}, K)$-module, and there exists $\alpha \in C_c^\infty(G)$ that satisfies*

$$\alpha(kgk^{-1}) = \alpha(g), \qquad k \in K, \tag{9.2}$$

such that $f * \alpha = f$. Suppose furthermore that there exist constants C and N such that $|f(g)| < C\, ||g||^N$, where the height $||g||$ is the Euclidean norm that G inherits from its inclusion in $\mathrm{Mat}_2(\mathbb{R}) \cong \mathbb{R}^4$. *Then if $D \in U(\mathfrak{g}_\mathbb{C})$, Df satisfies a similar equality with the same constant N.*

Harish-Chandra acknowledges the contributions of Jacquet and Borel in arriving at the following beautiful proof. We have adapted the argument to the special case of $SL(2, \mathbb{R})$.

Proof We will work in the complexification $\mathfrak{g}_\mathbb{C}$ of $\mathfrak{g}$. Let H, R, L, W, and Δ be as in Eqs. (2.23), (2.35), and (2.39). Our hypothesis is that $\mathbb{C}[\Delta]\, f$

is a finite-dimensional vector space; also W lies in the Lie algebra of K, so $\mathbb{C}[W] f = \mathbb{C}[H] f$ is also finite dimensional. Because Δ and H commute, it follows that $\mathcal{R} f$ is finite dimensional, where $\mathcal{R}$ is the ring $\mathbb{C}[\Delta, H]$.

We define a $\mathbb{Z}$-grading on $U(\mathfrak{g}_\mathbb{C})$ by giving H degree zero, R degree 1, and L degree -1. Then Δ and H are homogeneous of degree zero, and it is easy to see that $\mathcal{R}$ is the subring of elements of degree zero in this grading. We note that this assignment of degrees is compatible with the commutation relations (see Eq. (2.24))

$$RL - LR = H, \qquad HR - RH = 2R, \qquad HL - LH = -2L, \tag{9.3}$$

which are the only relations between the generators H, R, and L of $U(\mathfrak{g}_\mathbb{C})$. Making use of Eq. (9.3), every element of $U(\mathfrak{g}_\mathbb{C})$ may be written as a linear combination of elements of the form $H^i\, R^j\, L^j$ where i, j, and k are integers, and every element of $\mathcal{R}$ may be written as a linear combination of elements of the form $H^i\, R^j\, L^j$. We may write

$$U(\mathfrak{g}_\mathbb{C}) = \bigoplus_{0 \le i \in \mathbb{Z}} L^i\, \mathcal{R} \oplus \bigoplus_{0 < i \in \mathbb{Z}} R^i\, \mathcal{R}. \tag{9.4}$$

We topologize $C^\infty(G)$ as follows. If X is a compact subset of G and N is a positive integer, we define a seminorm $p_{X,N}$ on $C^\infty(G)$ by

$$p_{X,N}(f) = \max |H^i\, R^j\, L^k\, f(x)|, \qquad x \in X,\ i + j + k \le N.$$

If $\epsilon > 0$, we then define $B(X, N, \epsilon) = \{f | p_{X,N}(f) < \epsilon\}$. We take these sets to be a base of neighborhoods of the identity. With this topology, $C^\infty(G)$ becomes a Fréchet space. (This is identical to the case of $C^\infty(\mathbb{R}^n)$ in Rudin (1991, Section 1.46).) Let V be the smallest closed G-invariant subspace of $C^\infty(G)$ containing f, and let $V_0 = U(\mathfrak{g}_\mathbb{C})\, f$.

If $n \in \mathbb{Z}$, let $V(n)$ be the n-eigenspace of H on V. We will prove that

$$V_0 = \bigoplus_{n=-\infty}^{\infty} \big(V_0 \cap V(n)\big). \tag{9.5}$$

Note that $\phi \in V(n)$ if and only if (with κ_θ as in Eq. (1.19)) we have

$$\phi(g\kappa_\theta) = e^{in\theta}\, \phi(g).$$

Thus we can define a continuous projection $E_n : V \to V$ with image $V(n)$ by

$$E_n\phi = \frac{1}{2\pi} \int_0^{2\pi} e^{-in\theta}\, \phi(g\kappa_\theta)\, d\theta.$$

Because f is K-finite, there exists an N such that

$$f = \sum_{n=-N}^{N} E_n f. \tag{9.6}$$

It follows from Eq. (9.3) that H, R, and L map $V(n)$ into $V(n)$, $V(n+2)$, and $V(n-2)$, respectively. It follows that if $D \in U(\mathfrak{g}_\mathbb{C})$, there exists an M (depending on D) such that

$$Df = \sum_{n=-M}^{M} E_n\, Df \in \sum_{-M}^{M} V(n).$$

Thus Df is K-finite. Now let us note that actually $E_n\, Df \in V_0$. To see this, let p be a polynomial such that $p(k) = 0$ if $-M \leq k \leq M$, $k \neq n$, while $p(n) = 1$. Then $p(H)$ is the identity on $V(n)$ and zero on $V(k)$ with $k \neq n$, $-M \leq k \leq M$, so $p(H)\, Df = E_n\, Df$ and thus $E_n\, Df \in V_0$. Thus we have proved Eq. (9.5).

Next, we show that $V_0 \cap V(n)$ is finite dimensional. Let $\{f_1, \cdots, f_\nu\}$ be a basis for the finite-dimensional space $\mathcal{R} f$. Because each f_k is K-finite, there exists a positive integer M such that each $f_k = \sum_{n=-M}^{M} E_n\, f_k$. Then by Eq. (9.4)

$$V_0 = \sum_{k=1}^{\nu} \sum_{m=-M}^{M} \left[\sum_{0 \leq i \in \mathbb{Z}} \mathbb{C}\, L^i\, E_m\, f_k + \sum_{0 < i \in \mathbb{Z}} \mathbb{C}\, R^i\, E_m\, f_k \right].$$

Now we note that $L_i\, E_m F_k \in V(m-2i)$ and $R^i\, E_n f_k \in V(m+2i)$. Thus

$$V_0 = \bigoplus_{n=-\infty}^{\infty} V_0(n)$$

where

$$V_0(n) = \sum_{\substack{1 \leq k \leq \nu \\ -M \leq m \leq M \\ 0 \leq i \in \mathbb{Z} \\ m-2i=n}} \mathbb{C}\, L^i\, E_m\, f_k + \sum_{\substack{1 \leq k \leq \nu \\ -M \leq m \leq M \\ 0 < i \in \mathbb{Z} \\ m+2i=n}} \mathbb{C}\, R^i\, E_m\, f_k \subseteq V(n).$$

This is evidently finite dimensional, so $V_0 \cap V(n)$ is finite dimensional.

To proceed further, we must show that f is actually analytic. Because f is $\mathcal{Z}$-finite, there exist complex constants $c_0, \cdots, c_{m-1}$ for some m such that

$$(\Delta^m + c_{m-1}\, \Delta^{m-1} + \cdots + c_0) f = 0. \tag{9.7}$$

By Eq. (9.6), it is sufficient to show that each $E_k f$ is analytic. Because Δ commutes with right translation by Proposition 2.2.4, it commutes with E_k, and so $E_k f$ also satisfies Eq. (9.7). Now using the coordinates Eq. (1.19), we see that Δ (given by Eq. (1.29)) has the same effect on $E_k f$ as the weight-k Laplacian Eq. (1.3), and substituting this operator into Eq. (9.7), the highest-order term is just

$$(-1)^m\, y^{2m} \left(\frac{\partial^2}{\partial x^2} + \frac{\partial^2}{\partial y^2} \right)^m.$$

Thus Eq. (9.7) boils down to an elliptic partial differential equation in two dimensions (the coordinate θ in Eq. (1.19) disappears), and its solution $E_k f$ is analytic. Therefore f is analytic.

We will show next that V is the closure of V_0. Suppose it is not. Then by the Hahn–Banach theorem (see Rudin, 1991, 3.3) there exists a nonzero continuous linear functional Λ on V such that $\Lambda(V_0) = 0$. If $F \in C^\infty(G)$, let $\phi_F(g) = \Lambda\big(\rho(g)\, F\big)$, and let $\phi = \phi_f$. It is easy to deduce from the definition (Eq. 2.10) that if $X \in \mathfrak{g}$, then $dX\, \phi_F = \phi_{dX\,F}$, so if $D \in U(\mathfrak{g}_\mathbb{C})$, we have

$$D\phi_F = \phi_{DF}. \tag{9.8}$$

Because f is $\mathcal{Z}$-finite, Eq. (9.8) implies that ϕ is also, and therefore ϕ is analytic by the same argument used to prove that f is analytic. Now observe that if $D \in U(\mathfrak{g}_\mathbb{C})$, Eq. (9.8) implies that $(D\phi)(1) = \Lambda(Df) = 0$ because $Df \in V_0$. Because ϕ is analytic, we may consider its Taylor expansion at 1, and the vanishing of all $(D\phi)(1)$ implies that $\phi = 0$. Thus $\Lambda\big(\rho(g)\, f\big) = \phi(g) = 0$ for all g, a contradiction because the functions $\rho(g)\, f$ span a dense subspace of V, and Λ is not identically zero. This contradiction shows that actually V_0 is dense in V.

Now we can show that actually $V(n) \subset V_0$. Indeed, because V_0 is dense in V, $E_n\, V_0 = V_0 \cap V(n)$ is dense in $E_n\, V = V(n)$. But we have just shown that $V_0 \cap V(n)$ is finite dimensional, so $V_0 \cap V(n) = V(n)$. Thus $V(n)$ is finite dimensional and contained in V_0, and Eq. (9.5) may be rewritten

$$V_0 = \bigoplus V(n). \tag{9.9}$$

This is evidently an admissible $(\mathfrak{g}, K)$-module.

Now let J be the convolution algebra of $\alpha \in C_c^\infty(G)$ that satisfy Eq. (9.2). If $\alpha \in J$, we may write $f * \alpha = \int_G \alpha(x)\, \rho(x^{-1})\, f\, dx$, so $f * \alpha \in V$. It is easy to see using Eq. (9.2) that convolution with α commutes with the action of K, so if $f \in \sum_{-N}^{N} V(n)$, then $f * \alpha$ lies in the same finite-dimensional space $\sum_{-N}^{N} V(n)$. Let X be a sufficiently large relatively compact open subset of G such that the canonical map $C^\infty(G) \to C^\infty(X)$ is injective on $\sum_{-N}^{N} V(n)$. We will show that f can be uniformly approximated on X by elements of $f * J$. (For this it is sufficient to work with the L^∞ norm on $C^\infty(X)$ instead of the Fréchet topology on $C^\infty(G)$.) The finite-dimensionality of $\sum_{-N}^{N} V(n)$ will then imply that actually $f \in f * J$.

We recall the notion of a *Dirac sequence*. A sequence of functions $\{\alpha_n\}$ in $C_c^\infty(G)$ is called a *Dirac sequence* if $\alpha_n(g) > 0$ for all g, $\int_G \alpha_n(g)\, dg = 1$, and if for every open neighborhood Ω of the identity in G, the support of α_n is contained in Ω for all sufficiently large n. It is clear that Dirac sequences exist, and if $\{\alpha_n\}$ is a Dirac sequence, we may replace α_n by the function $g \mapsto \int_G \alpha_n(kgk^{-1})\, dk$ to obtain a Dirac sequence in J.

Now if $\{\alpha_n\} \subset J$ is a Dirac sequence, it is easy to see that $f * \alpha_n$ converges uniformly to f on any compact set, in particular on X. As we have noted, this implies that $f = f * \alpha$ for some $\alpha \in J$.

Finally, let us assume that $|f(g)| < C||g||^N$. We wish to show that if $D \in U(\mathfrak{g}_{\mathbb{C}})$, then Df satisfies a similar estimate with the same N. It is sufficient to check this when $D = X \in \mathfrak{g}_{\mathbb{C}}$. We write $f = f * \alpha$ with $\alpha \in J$. It is easy to see directly from the definition (Eq. 2.10) of the action of X on $C^\infty(G)$ that $dX\,(f * \alpha) = f * (dX\,\alpha)$. From this it follows easily that $dX\,f$ satisfies an estimate similar to that satisfied by f, with the same constant N. ∎

3
Automorphic Representations

There exists a satisfactory theory of *automorphic representations of* $GL(n)$. One may associate an automorphic representation of $GL(1)$ with a Dirichlet character, and one may associate an automorphic representation of $GL(2)$ with a modular form or Maass form. Thus, automorphic representations are attached to classical objects of fundamental importance in number theory.

We will discuss Tate's (1950) thesis, which is essentially the theory of automorphic representations of $GL(1)$, before defining the general notion of an automorphic representation on $GL(n)$. Insofar as we can in a brief account, we will (at least in Section 3.3) give statements that are valid for $GL(n)$, but proofs that are valid only for $n = 2$. In places, we will give proofs that are only complete when the ground field is $\mathbb{Q}$, though the only serious obstacle to filling these gaps is our omission in Chapter 2 of the representation theory of $GL(2, \mathbb{C})$.

In Section 3.2, we will extend our discussion of the "classical" problem of decomposing $L^2\big(\Gamma\backslash PGL(2,\mathbb{R})^+\big)$ to the case when the quotient is noncompact – for example, when $\Gamma = SL(2,\mathbb{Z})$ or one of its congruence subgroups. Our discussion is far from complete. The space falls into two pieces, and we will concentrate on the cuspidal part, $L_0^2\big(\Gamma\backslash PGL(2,\mathbb{R})^+\big)$. After proving the fundamental result of Gelfand, Graev, and Piatetski-Shapiro (1968), that the integral operators on $L_0^2\big(\Gamma\backslash PGL(2,\mathbb{R})^+\big)$ are compact, we will prove that this space falls into irreducible invariant subspaces, each of which occurs with finite multiplicity, and that it contains a dense subspace $\mathcal{A}_0^2\big(\Gamma\backslash PGL(2,\mathbb{R})^+\big)$ consisting of automorphic forms. We will apply Theorem 2.9.2 of Harish-Chandra to prove that if V is an irreducible invariant subspace in the decomposition of $L_0^2\big(\Gamma\backslash PGL(2,\mathbb{R})^+\big)$, then $\mathcal{A}_0^2\big(\Gamma\backslash PGL(2,\mathbb{R})^+\big) \cap V$ is an irreducible admissible $(\mathfrak{g}, K)$-module and hence falls into the classification of Chapter 2. We will prove that the multiplicity of a given irreducible unitary representation in $L_0^2\big(\Gamma\backslash PGL(2,\mathbb{R})^+\big)$ is finite. The orthogonal complement of $L_0^2\big(\Gamma\backslash PGL(2,\mathbb{R})^+\big)$ is also interesting. It does *not* decompose into a direct sum of invariant subspaces, but rather into a direct *integral* parametrized by the Eisenstein series. We will not prove this, but rather concentrate on the cuspidal part of L^2.

In Section 3.3, we define automorphic representations of the adele group $GL(n, A)$. Certain technical points such as admissibility reduce to the classical case of Section 3.2. We will formulate a couple of important results that will be proved in later sections, namely, the tensor product theorem of Flath, and the multiplicity one theorem of Jacquet-Langlands.

The tensor product theorem asserts that an irreducible admissible representation of $G(A)$, where A is the adele ring of a global field F and G is a reductive algebraic group defined over F, decomposes into a "restricted tensor product" of representations of the groups $G(F_v)$. Results of this type may be found in Gelfand, Graev and Piatetski-Shapiro (1968) and Jacquet and Langlands (1970), but Flath (1979) obtained a definitive purely algebraic result in his thesis. What was published in Borel and Casselman (1979) was, unfortunately, only an abbreviated account, and I felt it was a good idea to reconstruct the proof in some detail. An important contribution by Flath was the introduction of a certain ring of distributions on a reductive Lie group, commonly called the *Hecke algebra*. Its significance is that a $(\mathfrak{g}, K)$-module is the same as a module over this ring. A detailed and completely self-contained discussion of this ring recently appeared in Knapp and Vogan (1995), which made the writing of this section much easier for me.

The multiplicity one theorem asserts that no irreducible admissible representation of $GL(n, A)$ occurs with multiplicity greater than one in the cusp forms. As Jacquet and Langlands (1970) showed, this fundamental result is a consequence of the uniqueness of Whittaker models, which is also the key to constructing the L-function. We discuss these fundamental matters in Section 3.5.

Section 3.6 discusses the adelization of classical automorphic forms.

In Section 3.7, we introduce the Eisenstein series. Our treatment of this topic is limited to a couple of specific goals. We prove what we need about the Eisenstein series – notably, their analytic continuation and functional equations – by examining their Fourier expansions. The constant term in the Fourier expansion leads to the intertwining integrals that are important in the purely local theory, and one of our aims in Section 3.7 is to show how this comes about. Our other objective is to establish the facts we need for our discussion of the Rankin–Selberg method in Section 3.8.

Section 3.9 describes the Langlands functoriality conjecture and the L-group. This topic is also discussed in Sections 1.8 and 4.9.

Section 3.10 is a random inclusion. Garrett (1987) showed that it is possible to attach an L-function with a functional equation to three modular forms. To give the reader a taste of the possibilities of the Rankin–Selberg method on higher-rank groups, we briefly study the triple L-function by a method different from Garrett's, that is essentially due to Gelbart and Piatetski-Shapiro (1985).

Starting with Theorem 3.3.3, whose proof is given in Section 3.5, we will begin needing results from Chapter 4. Chapters 3 and 4 can be read in either order, but there is something to be said for supplying some motivation for the local results before delving into their sometimes lengthy proofs. One approach would be to read Chapter 3 before reading Chapter 4, taking the results needed there on faith, then reviewing Chapter 3 again after completing Chapter 4. Alternatively, Chapter 4 (which does not use any results from Chapter 3) could be read first, or they could be approached simultaneously.

Chapters 2 and 3 can also be read in either order. However the reader who begins with Chapter 3 should be aware that Chapter 2 contains much important background material. After reading Sections 2.1, 2.2, 2.3 (through Theorem 2.3.4) and 2.4, the reader should be ready for Chapter 3. We also make use of the result of Section 2.9 in Theorem 3.2.1. Note that Theorem 3.2.1 can be proved without this result for $\mathcal{A}_0(\Gamma\backslash G, \chi, \omega)$, as in Exercise 2.2.5. Section 2.9 can be read immediately after Section 2.4.

3.1 Tate's Thesis

In Section 1.1, we considered the Dirichlet L-functions and their functional equations, a theory that was generalized by Hecke (1918) and (1920) in work that we reviewed in Section 1.7. Hecke's theory was recast by Tate in his 1950 Ph.D. dissertation (See also Cassels and Fröhlich, 1967). This theory is now understood as the theory of automorphic forms on $GL(1)$. We will reproduce Tate's theory here as an introduction to analysis on adele groups. In Proposition 3.1.2, we will make a connection with the classical theory of Dirichlet characters.

The classical theory of Fourier analysis works very nicely in the context of a locally compact Abelian group. Let us recall the basic facts. For reference, see Rudin (1962) and also Katznelson (1967, Chapter VII). In the next few paragraphs, we will write the group law in a locally compact Abelian group additively.

If G is a locally compact Abelian group, by a *quasicharacter* of G we mean a continuous homomorphism $\chi : G \to \mathbb{C}^\times$. If $|\chi(g)| = 1$ for all $g \in G$, then χ is called *unitary*; a unitary quasicharacter is called a *character*. (Some authors use an alternative terminology in which a character is not assumed to be unitary; that is, what we call a quasicharacter is called a character in part of the literature.) Let G^* be the group of characters of G. It is given the topology of uniform convergence on compact sets; with this topology, it becomes another locally compact Abelian group, known as the *(Pontriagin) dual* of G. The dual of a compact group is discrete, and the dual of a discrete group is compact.

For example, we consider the dual groups of the additive groups $\mathbb{R}$, $\mathbb{Z}$, and $\mathbb{R}/\mathbb{Z}$. Every character of $\mathbb{R}$ has the form $\chi_a(t) = e^{2\pi i a t}$ for some $a \in \mathbb{R}$, and $a \mapsto \chi_a$ is an isomorphism of $\mathbb{R}$ onto its dual group (Exercise 3.1.1(f)).

The dual of $\mathbb{Z}$ is $\mathbb{R}/\mathbb{Z}$. To see this, note that $a \mapsto \chi_a|\mathbb{Z}$ gives a homomorphism of $\mathbb{R}$ onto the dual group of $\mathbb{Z}$, but this homomorphism has a kernel, namely $\mathbb{Z}$, and hence induces an isomorphism of $\mathbb{R}/\mathbb{Z}$ onto $\mathbb{Z}^*$. (This illustrates the fact that the dual of a discrete group is compact.)

The dual of $\mathbb{R}/\mathbb{Z}$ is isomorphic to $\mathbb{Z}$. To see this, note that if $a \in \mathbb{Z}$, then χ_a is trivial on $\mathbb{Z}$ and hence induces a character of $\mathbb{R}/\mathbb{Z}$, and $a \mapsto \chi_a$ is then an isomorphism of $\mathbb{Z}$ onto the dual group of $\mathbb{R}/\mathbb{Z}$. (This illustrates the fact that the dual of a compact group is discrete.)

There is a canonical homomorphism $\iota : G \to G^{**}$, namely, if $g \in G$, let $\iota(g)$ be the character $\chi \mapsto \chi(g)$ of G^*. It is a fundamental fact (Pontriagin duality) that ι is an isomorphism of topological groups, so we may identify G with G^{**}. To emphasize the symmetry between G and G^*, we may write $\langle g, g^* \rangle$ for $g^*(g)$ if $g^* \in G^*$ and $g \in G$.

There is on G a translation-invariant Borel measure, unique up to constant multiple, which we shall denote $\int_G dg$. This is the *Haar measure*. We have as usual the Banach spaces $L^p(G)$, $1 \le p \le \infty$, consisting of measurable functions that are bounded with respect to the corresponding norms

$$|f|_p = \left[\int_G |f(g)|^p \, dg \right]^{1/p}, \qquad 1 \le p < \infty,$$

$$|f|_\infty = \inf_{\text{null sets } N \subset G} \; \sup_{g \in G - N} |f(g)|.$$

The space $L^2(G)$ is a Hilbert space.

If $f_1, f_2 \in L^1(G)$, we define the *convolution*

$$(f_1 * f_2)(g) = \int_G f_1(x) \, f_2(g - x) \, dx.$$

This makes $L^1(G)$ into a Banach algebra (without unit).

Let $f : G \to \mathbb{C}$ be an element of $L^1(G)$. We define the *Fourier transform* $\hat{f} : G^* \to \mathbb{C}$ by

$$\hat{f}(x) = \int_G f(y) \, \langle y, x \rangle \, dy. \tag{1.1}$$

It is an element of $C_0(G^*)$, which is defined to be the space of continuous functions on G^* such that for any $\epsilon > 0$ there exists a compact set K such that $|f(x)| < \epsilon$ for $x \in G^* - K$. $C_0(G^*)$ is a closed subspace of $L^\infty(G^*)$.

The *Fourier inversion formula* is the formula

$$\hat{\hat{f}}(x) = f(-x). \tag{1.2}$$

This must be interpreted correctly because it is not necessarily true that $\hat{f}$ is L^1, so $\hat{\hat{f}}$ may not be defined. In any case, it is necessary to choose the Haar measures on G and G^* "compatibly," because otherwise Eq. (1.2) will only be valid up to a constant. Thus the choice of a Haar measure on G forces a Haar measure on G^* such that Eq. (1.2) is valid, although we still have to explain in what sense. We will say that the Haar measures on G and G^* are *dual* if they are matched in this fashion. Assuming that the measures are dual, there are a couple of different approaches to interpreting Eq. (1.2).

The first approach to the Fourier inversion formula involves reinterpreting the Fourier transform, originally defined on $L^1(G)$, as an operator on $L^2(G)$. One shows that the Fourier transform maps $L^1(G) \cap L^2(G)$ onto a dense subspace of $L^2(G^*)$ and moreover is an isometry in the L^2 norm:

$$|f|_2 = |\hat{f}|_2. \tag{1.3}$$

The Fourier transform therefore induces an isometry of $L^2(G)$ onto $L^2(G^*)$. Thus if f is L^2 but not L^1, then the Fourier transform $\hat{f}$ is still defined, but *not* by the formula (1.1). For L^2 functions f, Eq. (1.2) is valid almost everywhere.

The second approach to the Fourier transform is to find a suitable subspace of $C_0(G)$ that is dense in $L^2(G)$, closed under convolution, and invariant under the Fourier transform. Such a space, known as the *Schwartz space*, or *Bruhat–Schwartz space*, was introduced by Schwartz when G is a real vector space and by Bruhat more generally. We will only need it in particular cases. We will describe the Schwartz space $S(G)$ explicitly for the following additive groups: $G = F^n$, where F is any local field (that is, $F = \mathbb{R}, \mathbb{C}$, a finite extension of the p-adic field $\mathbb{Q}_p$, or the field $\mathbb{F}_q((t))$ of formal power series in one variable over a finite field); or $G = A^n$, where A is the adele ring of a global field. We refer to Weil's (1964) paper for Bruhat's general definition of the Schwartz space.

Let F be a local field. If $F = \mathbb{R}$, we define a complex-valued function f on F^n to be *smooth* if it has derivatives of all order. If $F = \mathbb{C}^n$, we may identify $F^n = \mathbb{R}^{2n}$, and we define f to be *smooth* if it is smooth as a function on $\mathbb{R}^{2n}$. If F is non-Archimedean, we define a function f on F^n to be *smooth* if it is locally constant.

The Schwartz space for $\mathbb{R}^n$ is the space of all functions $f : \mathbb{R}^n \to \mathbb{C}$ that are smooth and for which

$$|f|_{\alpha,\beta} = \sup_{x \in \mathbb{R}^n} |x_1^{\alpha_1} \cdots x_n^{\alpha_n}| \frac{\partial^{\beta_1+\ldots+\beta_n} f}{\partial x_1^{\beta_1} \cdots \partial x_n^{\beta_n}}(x)$$

is bounded for all $0 \leq \alpha_i, \beta_i \in \mathbb{Z}$. We topologize $S(\mathbb{R}^n)$ by giving it the smallest topology in which all the seminorms $| \ |_{\alpha,\beta}$ are continuous.

If $F = \mathbb{C}$, we regard $\mathbb{C}^n = \mathbb{R}^{2n}$, for which we have already defined the Schwartz space.

Finally, if F is a non-Archimedean local field, the Schwartz space will be simply the space of compactly supported smooth (i.e., locally constant) functions. The Schwartz space is given the weakest topology in which *every* linear functional is continuous. In the non-Archimedean case, one may ignore the topology on the Schwartz space completely.

Let F be a global field and A its adele ring. If v is a place of F, we will denote by F_v the completion of F at v; if v is non-Archimedean, we will denote by $\mathfrak{o}_v$ the ring of integers in F_v, by $\mathfrak{p}_v$ the maximal ideal of $\mathfrak{o}_v$, by q_v the cardinality of the residue field $\mathfrak{o}_v/\mathfrak{p}_v$, and by ϖ_v a fixed generator of $\mathfrak{p}_v$. Then A is the restricted direct product of the F_v with respect to the $\mathfrak{o}_v$. We assume the reader is familiar with the definition of the adele ring A, its topology, and the fact that $F \subset A$ is a discrete subgroup, with quotient A/F compact.

The Schwartz space $S(A^n)$ of A^n is defined to be the space of all finite linear combinations of functions of the form

$$\Phi(x) = \prod_v \Phi_v(x_v), \qquad x = (x_v) \in A^n,$$

where each $\Phi_v \in S(F_v^n)$ and Φ_v is the characteristic function of $\mathfrak{o}_v^n$ for all but finitely many v.

Let ψ be a nontrivial character of A that is trivial on F. Such a character always exists – compare Exercise 1.2 for a proof in the number field case. Let ψ_v be the composition of ψ with the inclusion $F_v \to A$. Then if $x = (x_v)_v \in A$, we have $\psi_v(x_v) = 1$ for almost all v, and

$$\psi(x) = \prod_v \psi_v(x_v).$$

Every continuous character of A has the form $\psi_a(x) = \psi(ax)$ for some $a \in A$, and $a \mapsto \psi_a$ is an isomorphism of A with its dual group.

Suppose that G is a locally compact group that is identified with its dual group G^* in some fixed manner. For example, $G \cong G^*$ if $G = F$, where F is a local field, or $G = A$, where A is an adele ring. For these examples, the identification is as above, by means of a fixed additive character ψ. In this case, there is a unique normalization of the Haar measure for which the Fourier inversion formula (1.2) is valid. This measure will be called *self-dual*.

Let F be a global field and v a place of F. We will denote by $\int_{F_v} dx$ the self-dual Haar measure on F_v. If $a \in F_v$, we will denote by $|a|_v$ the canonical absolute value of a, defined by the property that the transformation $x \mapsto ax$ multiplies the measure dx by $|a|_v$. (For example, if $F_v \cong \mathbb{R}$, then $|\ |_v$ is the usual absolute value, whereas if $F_v \cong \mathbb{C}$, then $|\ |_v$ is the *square* of the usual absolute value.) We will denote by $\int_{F_v^\times} d^\times x_v$ the Haar measure on the multiplicative group, normalized so that

$$d^\times x_v = m_v \frac{dx_v}{|x_v|_v}, \qquad m_v = \begin{cases} 1 & \text{if } v \text{ is Archimedean;} \\ \left(1 - \frac{1}{q_v}\right)^{-1} & \text{if } v \text{ is non-Archimedean.} \end{cases} \tag{1.4}$$

The significance of this normalization is that for all but finitely many v, it will give $\mathfrak{o}_v$ volume one with respect to the additive measure and will give $\mathfrak{o}_v^\times$ volume one with respect to the multiplicative measure. To be precise, if v is a non-Archimedean place such that the conductor of ψ_v is $\mathfrak{o}_v$ (see Exercise 3.1.1 (b)), then the additive measure on F_v for which $\mathfrak{o}_v$ has volume one is self-dual, because we may exhibit a specific function that is equal to its own Fourier transform, namely the characteristic function of $\mathfrak{o}_v$. With the above choice of measures, $\mathfrak{o}_v^\times$ will also have measure one with respect to $d^\times x_v$. Of course the conductor of ψ_v is $\mathfrak{o}_v$ for almost all v. (Here as always *almost all* when applied to a set of places means *all but finitely many*.)

Similarly, we will denote by $|x| = \prod_v |x_v|_v$ the canonical absolute value on A and by $dx = \prod_v dx_v$ and $d^\times x = \prod_v d^\times x_v$ the corresponding Haar measures on A and $A^\times$. Because the local measures dx_v are self-dual, it follows that the additive measure dx on A is also self-dual.

There is an analogy that warrants a remark: $\mathbb{R}$ is a self-dual group with a discrete subgroup $\mathbb{Z}$ such that $\mathbb{R}/\mathbb{Z}$ is compact, and the dual group of $\mathbb{Z}$ is $\mathbb{R}/\mathbb{Z}$. Thus the relationship between $\mathbb{R}$ and $\mathbb{Z}$ is identical to that between A and F. Many analytic problems (those with number theoretic implications) may be carried out either in the context of the self-dual group $\mathbb{R}$ and its discrete, cocompact subgroup $\mathbb{Z}$, or in the context of the self-dual group A, and its discrete, cocompact subgroup F. The former context is often called classical, the second adelic. Thus the theory of L-functions of Hecke characters and their functional equations, including Theorem 1.1.1 may also be carried out in the adele setting.

What we call a *Hecke character* will be a continuous character of $A^\times/F^\times$, where F is a global field and A its adele ring. It is necessary for us to consider why this notion is equivalent to the notion of a Hecke character, previously introduced in Section 1.7. We do this in Proposition 3.1.2 below for $F = \mathbb{Q}$, leaving the general case for the exercises (cf. Exercise 3.1.4). First we need to prove a property of characters of $A^\times$.

Proposition 3.1.1 *Let F be a global field and A its adele ring. Let χ be a character of $A^\times$. Then there exists a finite set S of places, including all Archimedean ones, such that if $v \notin S$, then χ_v is trivial on $\mathfrak{o}_v^\times$.*

If v is non-Archimedean and χ_v is trivial on $\mathfrak{o}_v^\times$, then we call χ_v *nonramified* at the place v. The content of this proposition is that χ_v is nonramified for all but finitely many v.

Proof Let $A_{\mathrm{f}}^\times \subset A^\times$ be the group of "finite ideles," that is, the restricted direct product of the $F_v^\times$ with respect to the $\mathfrak{o}_v^\times$, as v runs through the non-Archimedean places of F. It is a totally disconnected group on which the restriction of χ is a continuous character, so by Exercise 3.1.1 (a), the kernel of χ contains an open neighborhood of the identity in $A_{\mathrm{f}}^\times$. Any open neighborhood of the identity contains a subset of the form $\prod_{v \notin S} \mathfrak{o}_v^\times$ for some sufficiently large finite set of places S. ■

Now suppose that χ_v is nonramified for some non-Archimedean place v of F. Let ϖ_v be a generator of the maximal ideal $\mathfrak{p}_v$ of $\mathfrak{o}_v$. Then because χ_v is trivial on $\mathfrak{o}_v^\times$, $\chi_v(\varpi_v)$ is independent of the choice of generator ϖ_v. We will denote by $\chi(\mathfrak{p}_v)$ this complex number $\chi_v(\varpi_v)$. It is only defined for v nonramified.

We now show that if $F = \mathbb{Q}$, then the characters of $A^\times/F^\times$ correspond to primitive Dirichlet characters. If v is a non-Archimedean place of $\mathbb{Q}$, let p_v be the corresponding prime number. (It is actually equal to q_v, but we use the notation p_v to emphasize that it is prime.)

We will call $\chi : A^\times/F^\times \to \mathbb{C}$ a character *of finite order* if there exists a positive integer m such that $\chi(x)^m = 1$ for all $x \in A^\times$.

Proposition 3.1.2

(i) *Suppose that $F = \mathbb{Q}$ and that χ is a character of $A^\times/F^\times$. There exists a unique character χ_1 of finite order of $A^\times/F^\times$ and a unique purely imaginary number λ such that $\chi(x) = \chi_1(x)\,|x|^\lambda$.*

(ii) *Suppose that $F = \mathbb{Q}$ and that χ is a character of finite order of $A^\times/F^\times$. There exists an integer N whose prime divisors are precisely the primes p_v such that v is a non-Archimedean place of $\mathbb{Q}$ and χ_v is ramified, and a primitive Dirichlet character χ_0 modulo N such that if v is a non-Archimedean place such that $p_v \nmid N$, then $\chi_0(p_v) = \chi(\mathfrak{p}_v)$. This correspondence $\chi \mapsto \chi_0$ is a bijection between the characters of finite order of $A^\times/F^\times$ and the primitive Dirichlet characters.*

This result allows us to relate the classical theory of Theorem 1.1.1 to the adelic theory to be developed below.

Proof We write $A^\times = \mathbb{R}^\times \times A_\mathfrak{f}^\times$, where $A_\mathfrak{f}^\times$ is the group of finite ideles, that is, the restricted direct product of the $\mathbb{Q}_v^\times$ with respect to the $\mathfrak{o}_v^\times$, as v runs through the non-Archimedean places of $\mathbb{Q}$. Let N be a positive integer. Define $S_0(N)$ to be the set of non-Archimedean places v such that $p_v|N$, and let $S_1(N)$ be the set of non-Archimedean places v such that $p_v \nmid N$. If $v \in S_0(N)$, let

$$U_v(N) = \{x \in \mathfrak{o}_v | x \equiv 1 \bmod N\}.$$

Let

$$U_\mathfrak{f}(N) = \prod_{v \in S_0(N)} U_v(N) \times \prod_{v \in S_1(N)} \mathfrak{o}_v^\times, \qquad U(N) = \mathbb{R}_+^\times \times U_\mathfrak{f}(N).$$

$A_\mathfrak{f}^\times$ is a totally disconnected locally compact group, so by Exercise 3.1.1 (a), any continuous character χ of $A^\times$ is trivial on an open subgroup of $A_\mathfrak{f}^\times$. As N varies, the subgroups $U_\mathfrak{f}(N)$ form a basis of neighborhoods of the identity in $A_\mathfrak{f}^\times$, so there exists some N such that χ is trivial on $U_\mathfrak{f}(N)$. The restriction of χ to $\mathbb{R}_+^\times$ is of the form $|x|^\lambda$ for some purely imaginary λ, and λ is uniquely determined by χ. Hence if we let $\chi_1(x) = \chi(x)\,|x|^{-\lambda}$, the character χ_1 is trivial on $U(N)$. We will show that χ_1 has finite order; this will conclude the proof of (i).

Let

$$V(N) = \mathbb{R}_+^\times \times \prod_{v \in S_0(N)} U_v(N) \times \prod_{v \in S_1(N)} \mathbb{Q}_v^\times \qquad \text{(restricted direct product.)}$$

This is an open subgroup of $A^\times$, and by the "approximation theorem" (Lang, 1986, Theorem 1, p. 35) we have $A^\times = \mathbb{Q}^\times V(N)$. Consequently

$$A^\times/\mathbb{Q}^\times \cong V(N)/\big(\mathbb{Q}^\times \cap V(N)\big). \tag{1.5}$$

To prove (i), it is sufficient, therefore, to show that the restriction of χ_1 to $V(N)$ has finite order. Indeed, χ_1 is trivial on $U(N)\big(\mathbb{Q}^\times \cap V(N)\big)$, and we will show this group to be of finite index in $V(N)$. More precisely, we will show that

$$V(N)/U(N)\big(\mathbb{Q}^\times \cap V(N)\big) \cong (\mathbb{Z}/N\mathbb{Z})^\times. \tag{1.6}$$

Note that Eq. (1.6) will prove (i), because if the restriction χ_1 to $V(N)$ annihilates a subgroup of finite index, it obviously has finite order. Let I_N be the group of all fractional ideals of $\mathbb{Q}$ prime to N, and let P_N be the subgroup of principal fractional ideals $\alpha\mathbb{Z}$, where $\alpha \in \mathbb{Q}^\times \cap V(N)$. If $a \in \mathbb{Z}$ is prime to N, then the ideal $a\mathbb{Z}$ is in I_N, and it is easy to check that $a \mapsto a\mathbb{Z}$ induces an isomorphism

$$(\mathbb{Z}/N\mathbb{Z})^\times \to I_N/P_N. \tag{1.7}$$

On the other hand, we may show that

$$V(N)/U(N)\big(\mathbb{Q}^\times \cap V(N)\big) \cong I_N/P_N \tag{1.8}$$

as follows. If $a = (a_v) \in A^\times$, there is associated with a a fractional ideal of $\mathbb{Q}$, namely

$$\iota(a) := \prod_v \mathfrak{p}_v^{\mathrm{ord}_v(a_v)},$$

where the product is over all non-Archimedean places v. The image of $V(N)$ under ι is I_N, and the kernel of the composite

$$V(N) \xrightarrow{\iota} I_N \xrightarrow{\text{proj}} I_N/P_N$$

is, as is easily checked, just $U(N)\big(\mathbb{Q}^\times \cap V(N)\big)$. Combining Eqs. (1.7) and (1.8) gives (1.6), which, we have noted, proves (i).

Next, we prove (ii). Assume now that $\chi = \chi_1$ has finite order. Let N be as above; composing the restriction χ to $V(N)$ with the isomorphism (Eq. 1.6) gives a character χ_0. Suppose that $v \in S_1(N)$. We check that $\chi_0(p_v) = \chi(\mathfrak{p}_v)$. Indeed, under the isomorphism (Eq. 1.7), the image of p_v in $(\mathbb{Z}/N\mathbb{Z})^\times$ corresponds to the fractional ideal $p_v\mathbb{Z} \in I_N$, and under Eq. (1.8), this corresponds to the coset of $\varpi_v \in V(N)$ in $V(N)/U(N)\big(\mathbb{Q}^\times \cap V(N)\big)$, so $\chi_0(p_v) = \chi(\mathfrak{p}_v)$.

Conversely, given a Dirichlet character χ_0 modulo N, composing with Eq. (1.6) gives a character of $V(N)$ that is trivial on $\mathbb{Q}^\times \cap V(N)$, and by Eq. (1.5), such a character may be reinterpreted as a character of $A^\times/\mathbb{Q}^\times$, which will of course have finite order. The only arbitrariness in these constructions was that in passing from χ to χ_0, we had some freedom to choose N, the only requirement being that χ be trivial on the group $U(N)$. There is a minimal N for which $\chi|_{U(N)}$ is trivial, and it is easy to see that for this N, the corresponding χ_0 is primitive. This completes the proof of (ii). ■

Let F be a global field, A its adele ring. If χ is a character of $A^\times/F^\times$, and if S is a finite set of places including the non-Archimedean ones such that χ_v is nonramified for $v \notin S$, then we define the *partial L-function*

$$L_S(s, \chi) = \prod_{v \notin S} L_v(s, \chi_v),$$

where the *local L-function*

$$L_v(s, \chi_v) = \left(1 - \chi(\mathfrak{p}_v)\, q_v^{-s}\right)^{-1}$$

for nonramified v. If $\chi(x) = \chi_1(x)\,|x|^\lambda$, then

$$L_S(s, \chi) = L_S(s + \lambda, \chi_1),$$

so by Proposition 3.1.2(i), if the ground field is $\mathbb{Q}$, there is no loss of generality in considering characters χ of finite order, and with this proviso, Proposition 3.1.2(ii) shows that L-functions of Dirichlet characters and L-functions of characters of $A^\times/\mathbb{Q}^\times$ are the same class of function. Similarly, if the ground field is an algebraic number field, the L-function of a Hecke character (in the classical sense) is the same as the L-function of a character of $A^\times/F^\times$. (See Exercise 3.1.4.)

Because the classical notion of a Hecke character is equivalent to the notion of a character of $A^\times/F^\times$, we are justified in calling a character of $A^\times/F^\times$ a Hecke character. For the remainder, we will follow Tate in proving the functional equations of L-functions of Hecke characters in an adelic setting.

Throughout the sequel, $\psi = \prod_v \psi_v$ will be a fixed character of A/F.

Proposition 3.1.3: Poisson summation

(i) The volume of A/F is 1 with respect to the self-dual Haar measure dx on A.

(ii) Let Φ be an element of the Schwartz space $S(A)$, and let

$$\hat{\Phi}(x) = \int_A \Phi(y)\,\psi(xy)\,dy$$

be its Fourier transform. Then if $t \in A^\times$,

$$\sum_{\alpha \in F} \Phi(\alpha t) = \frac{1}{|t|} \sum_{\alpha \in F} \hat{\Phi}\left(\frac{\alpha}{t}\right).$$

It is easy to see that if Φ is a Schwartz function, then $\sum_{\alpha\in F}\Phi(\alpha\, t)$ is convergent for all ideles t (Exercise 3.1.7).

Proof Define

$$F(x)=\sum_{\alpha\in F}\Phi\big((x+\alpha)\,t\big).$$

This is a continuous function on the compact Abelian group A/F by Exercise 3.1.7. Consequently, it has a Fourier expansion in terms of the characters of this group, and by Exercise 3.1.3 (c), these have the form $\psi(-\beta x)$ with $\beta\in F$. Thus

$$F(x)=\sum_{\beta\in F}c_\beta\,\psi(-\beta x).$$

We compute c_β by orthogonality. Let V be the volume of A/F.

$$c_\beta=\frac{1}{V}\int\limits_{A/F}F(x)\,\psi(\beta x)\,dx=\frac{1}{V}\int\limits_{A/F}\sum_{\alpha\in F}\Phi\big((x+\alpha)\,t\big)\,\psi(\beta x)\,dx$$

$$=\frac{1}{V}\int\limits_{A/F}\sum_{\alpha\in F}\Phi\big((x+\alpha)\,t\big)\,\psi\big(\beta(x+\alpha)\big)\,dx=\frac{1}{V}\int\limits_{A}\Phi(x\,t)\,\psi(\beta x)\,dx.$$

We make a change of variables, $x\to x/t$, $dx\to|t|^{-1}\,dx$, and this becomes

$$\frac{1}{V\,|t|}\int\limits_{A}\Phi(x)\psi(\beta x/t)\,dx=\frac{1}{V\,|t|}\hat{\Phi}\left(\frac{\beta}{t}\right).$$

Now

$$\sum_{\alpha\in F}\Phi(\alpha t)=F(0)=\sum_{\beta\in F}c_\beta=\frac{1}{V\,|t|}\sum_{\beta\in F}\hat{\Phi}\left(\frac{\beta}{t}\right). \tag{1.9}$$

We must now show that $V=1$; this will prove both (i) and (ii). To see this, we invoke Eq. (1.9) twice, taking $t=1$, say. Because we have chosen the Haar measure dx to be self-dual, we have the Fourier inversion formula (1.2), and we obtain

$$\sum_{\alpha\in F}\Phi(\alpha)=\frac{1}{V^2}\sum_{\alpha\in F}\Phi(-\alpha).$$

This implies $V^2=1$, and because $V>0$, $V=1$, as required. ∎

Let χ be a Hecke character. Recall that, by the definition of a character, $|\chi(x)|=1$ for all $x\in A^\times$. We will also consider quasicharacters of $A^\times/F^\times$; such a quasicharacter has the form $\chi(x)\,|x|^s$ for some Hecke character χ and some complex number s (Exercise 3.1.5). Let $\Phi \in S(A)$. We would like to study integrals of the form

$$\zeta(s, \chi, \Phi) = \int_{A^\times} \Phi(x)\, \chi(x)\, |x|^s\, d^\times x. \tag{1.10}$$

It is sufficient to study this type of integral in the special case where $\Phi(x) = \prod_v \Phi_v(x_v)$ with $\Phi_v \in S(F_v)$, because by definition, a Schwartz function on A is a finite linear combination of functions of this type. If this is the case, we have, at least formally

$$\zeta(s, \chi, \Phi) = \prod_v \zeta_v(s, \chi_v, \Phi_v), \tag{1.11}$$

where

$$\zeta_v(s, \chi_v, \Phi_v) = \int_{F_v^\times} \Phi_v(x)\, \chi_v(x)\, |x|_v^s\, d^\times x. \tag{1.12}$$

Before we can assert this as more than a purely formal identity, we must face the question of convergence.

Proposition 3.1.4 *(i) The local integrals (Eq. 1.12) are convergent if* $\mathrm{re}(s) > 0$. *(ii) There exists a finite set S of places containing the Archimedean ones such that if $v \notin S$, then*

$$\zeta_v(s, \chi_v, \Phi_v) = \left(1 - \chi(\mathfrak{p}_v)\, q_v^{-s}\right)^{-1}. \tag{1.13}$$

Indeed, it is sufficient that if $v \notin S$, then χ_v is nonramified, and Φ_v is the characteristic function of $\mathfrak{o}_v$.
(iii) The integral (Eq. 1.10) is absolutely convergent if $\mathrm{re}(s) > 1$, *in which case the decomposition (Eq. 1.11) is valid.*

Proof First let us show that the local integrals (Eq. 1.12) are absolutely convergent when $\mathrm{re}(s) > 0$. Because χ_v is unitary, the integral (Eq. 1.12) is dominated by

$$\int_{F_v^\times} |\Phi_v(x)|\, |x|_v^s\, d^\times x.$$

We split the integral into two parts. Over

$$\int_{\substack{F_v^\times \\ |x|_v > 1}} |\Phi_v(x)|\, |x|_v^s\, d^\times x,$$

the rapid decay of $\Phi_v(x)$ makes the integral convergent for all s. We must therefore only consider

$$\int_{\substack{F_v^\times \\ |x|_v \le 1}} |\Phi_v(x)|\, |x|_v^s \, d^\times x.$$

In this compact region, $\Phi_v(x)$ is continuous, hence bounded, and so it is sufficient to show that

$$\int_{\substack{F_v^\times \\ |x|_v \le 1}} |x|_v^s \, d^\times x < \infty.$$

If F_v is non-Archimedean, this integral equals

$$\sum_{k=0}^{\infty} \int_{\substack{F_v^\times \\ \mathrm{ord}_v(x) = k}} |x|_v^s \, d^\times x = \sum_{k=0}^{\infty} q^{-ks} < \infty,$$

provided $\mathrm{re}(s) > 0$. If v is a real place, the integral is

$$\int_{-1}^{1} |t|^s \frac{dt}{t} < \infty$$

if $\mathrm{re}(s) > 0$, and if v is complex, we use polar coordinates to write the integral as

$$\frac{1}{2\pi} \int_0^{2\pi} \int_0^1 r^{2s} \frac{dr}{r} \, d\theta < \infty,$$

again assuming that $\mathrm{re}(s) > 0$. This shows that the integrals (Eq. 1.12) are all absolutely convergent.

Let S be a finite set of places such that if $v \notin S$, then v is non-Archimedean, χ_v is nonramified, and Φ_v is the characteristic function of $\mathfrak{o}_v$. We will evaluate $\zeta_v(s, \chi_v, \Phi_v)$ for $v \notin S$.

If $v \notin S$, we have

$$\zeta_v(s, \chi_v, \Phi_v) = \sum_{k=0}^{\infty} \int_{\substack{F_v^\times \\ \mathrm{ord}_v(x) = k}} \chi_v(x)\, |x|_v^s \, d^\times x.$$

Let ϖ_v be a fixed generator of $\mathfrak{p}_v$. Then if $\mathrm{ord}_v(x) = k$, $\chi_v(x)\, |x|_v^s = \left(\chi_v(\varpi_v)\, q_v^{-s}\right)^k$; recall that we have defined, for v nonramified, $\chi(\mathfrak{p}_v) = \chi_v(\varpi_v)$.

Thus

$$\zeta_v(s, \chi_v, \Phi_v) = \sum_{k=0}^{\infty} \left(\chi(\mathfrak{p}_v)\, q^{-s}\right)^k,$$

and summing the geometric series, we obtain Eq. (1.13). It follows by comparison with the Dedekind zeta function of F that the infinite product

$$\prod_v \zeta_v(s, \chi_v, \Phi_v)$$

is absolutely convergent for $\mathrm{re}(s) > 1$, as required, so we have the product decomposition (Eq. 1.11) and the absolute convergence of the integral (Eq. 1.10) in this case. ■

Assume now that $\Phi(g)$ factorizes as $\prod_v \Phi_v(g_v)$.

Proposition 3.1.5: Tate *(i) The local integral (Eq. 1.12) has meromorphic continuation to all $s \in \mathbb{C}$, with no poles in the region* $\mathrm{re}(s) > 0$.
(ii) There exists a meromorphic function $\gamma_v(s, \chi_v, \psi_v)$, independent of the "test function" Φ_v (but dependent on the additive character ψ_v used to define the Fourier transform) such that

$$\zeta_v(1 - s, \chi_v^{-1}, \hat{\Phi}_v) = \gamma_v(s, \chi_v, \psi_v)\, \zeta_v(s, \chi_v, \Phi_v). \tag{1.14}$$

(iii) Let $s_0 \in \mathbb{C}$ be given. We can choose the function Φ_v so that $\zeta_v(s, \chi_v, \Phi_v)$ has neither zero nor pole at $s = s_0$. If v is non-Archimedean, we may even choose the local data so that $\zeta_v(s, \chi_v, \Phi_v)$ is identically equal to 1.

The existence of a meromorphic function γ_v, independent of the test function Φ_v, such that Eq. (1.14) holds, is called the *local functional equation*. See Exercise 4.7.1 for a different proof of this proposition.

Proof We will prove (ii) first. Let us prove the existence of a meromorphic function γ_v such that Eq. (1.14) is true. We will first define γ_v when $0 < \mathrm{re}(s) < 1$, then extend it to all s. By Proposition 3.1.3(i), both functions in the ratio

$$\frac{\zeta_v(1 - s, \chi_v^{-1}, \hat{\Phi}_v)}{\zeta_v(s, \chi_v, \Phi_v)} \tag{1.15}$$

are defined by integrals that are absolutely convergent and are therefore holomorphic in this region. We must show that this ratio is independent of the choice of test function Φ_v. Let Φ_v' be another element of $S(F_v)$. We must prove that

$$\zeta_v(1-s, \chi_v^{-1}, \hat{\Phi}_v)\, \zeta_v(s, \chi_v, \Phi_v') = \zeta_v(1-s, \chi_v^{-1}, \hat{\Phi}_v')\, \zeta_v(s, \chi_v, \Phi_v). \tag{1.16}$$

Indeed, substituting the definitions of the local integrals and the Fourier transform, the left side of Eq. (1.16) equals

$$\int_{F_v^\times} \left[\int_{F_v} \Phi_v(y)\, \psi(xy)\, dy \right] \chi_v(x)^{-1}\, |x|_v^{1-s}\, d^\times x \int_{F_v^\times} \Phi'_v(z)\, \chi_v(z)\, |z|_v^s\, d^\times z$$

$$= \int_{F_v^\times} \int_{F_v} \int_{F_v^\times} \Phi_v(y)\, \Phi'_v(z)\, \psi(xy)\, \chi_v(x)^{-1}\, \chi_v(z)\, |x|_v^{1-s}\, |z|_v^s\, d^\times x\, dy\, d^\times z.$$

For fixed y, make the substitution $x \mapsto y^{-1}x$. This is valid if $y \neq 0$, that is, off a set of measure zero. Using Eq. (1.4), this equals

$$m_v^{-1} \int_{F_v^\times} \int_{F_v^\times} \int_{F_v^\times} \Phi_v(y)\, \Phi'_v(z)\, \chi_v(yz)\, |yz|_v^s\, \psi(x)\, \chi_v(x)^{-1}\, |x|_v^{1-s}\, d^\times x\, d^\times y\, d^\times z.$$

In this symmetrical form, the identity (Eq. 1.16) is now clear.

Next we show that γ_v has meromorphic continuation to all s. The trick is to choose the test function Φ so that $\hat{\Phi}$ vanishes in a neighborhood of zero. (Choose $\hat{\Phi}$ first, then recall that the Schwartz space is invariant under the Fourier transform.) With such a choice, it is clear that the local integral defining $\zeta_v(1-s, \chi_v^{-1}, \hat{\Phi}_v)$ is convergent for all s and defines a holomorphic function. On the other hand, by Proposition 3.1.3(i), $\zeta_v(s, \chi_v, \Phi)$ is convergent and defines a holomorphic function for $\mathrm{re}(s) > 0$. Consequently, this particular choice of test functions represents γ_v as a ratio (Eq. 1.15) of holomorphic functions in the region $\mathrm{re}(s) > 0$. Similarly, we may take Φ_v, instead of $\hat{\Phi}_v$, to vanish in a neighborhood of the identity; this gives the meromorphic continuation of γ_v to $\mathrm{re}(s) < 1$.

Next let us prove (i). If $\mathrm{re}(s) > 0$, we have already observed that the integral defining $\zeta_v(s, \chi_v, \Phi_v)$ is convergent if $\mathrm{re}(s) > 0$, and so we have analytic continuation to this region, with no poles. To obtain the meromorphic continuation to $\mathrm{re}(s) \leq 0$, observe that the analytic continuation of $\zeta_v(1-s, \chi_v^{-1}, \hat{\Phi}_v)$ to $\mathrm{re}(s) > 0$ follows from Eq. (1.14) and the meromorphic continuation of γ_v, which we have already proved. Replacing s by $1-s$, χ_v by χ_v^{-1}, and Φ_v by $\hat{\Phi}_v$, we get the meromorphic continuation of ζ_v to the entire plane.

Finally, let us prove (iii). The point is to choose Φ_v in Eq. (1.12) to be concentrated very near $x = 1$. If Φ_v vanishes in a neighborhood of the identity, then (as we have already observed) the integral (Eq. 1.12) is convergent for all s and hence has no poles anywhere. We may also choose Φ_v to be compactly supported near $x = 1$, so that $\chi_v(x)\, |x|^{s_0}$ has positive real part on the support of Φ_v. In this case, it is clear that the integral is also nonzero at $s = s_0$. If v is non-Archimedean, we may even choose Φ_v to have support contained in $\mathfrak{o}_v^\times$, so that the integrand in Eq. (1.12) is independent of s, in which case, the integral (Eq. 1.12) is simply a constant, which we may normalize to be 1. ■

Next we prove the *global* functional equations.

We will say that a meromorphic function f is *essentially bounded in vertical strips* if whenever $\sigma_1 < \sigma_2$ are real numbers and U is an open set containing all poles of f in the region $\{\sigma_1 < \mathrm{re}(s) < \sigma_2\}$, then f is bounded in the region $\{\sigma_1 < \mathrm{re}(s) < \sigma_2\} - U$.

Proposition 3.1.6 *The function $\zeta(s, \chi, \Phi)$ has meromorphic continuation to all s, and is entire unless the restriction of χ to the subgroup $\mathrm{A}_1^\times$ of ideles of norm one is trivial. If this is the case, then there exists a purely imaginary complex number λ such that $\chi(x) = |x|^\lambda$. In this case, $\zeta(s, \chi, \Phi)$ can have poles at $s = 1 - \lambda$ and $s = -\lambda$ if F is a number field, and at $s = 1 - \lambda + 2\pi ni/\log(q)$ and $s = -\lambda + 2\pi ni/\log(q)$ ($n \in \mathbb{Z}$) if F is a function field and q is the cardinality of the finite ground field. These are the only possible poles. We have the functional equation*

$$\zeta(s, \chi, \Phi) = \zeta(1 - s, \chi^{-1}, \hat{\Phi}). \tag{1.17}$$

The function $\zeta(s, \chi, \Phi)$ is essentially bounded in vertical strips.

Proof For reasons of space, we prove only the case where F is an algebraic number field. We split the integral into two parts. The integral

$$\zeta_1(s, \chi, \Phi) := \int\limits_{\substack{A^\times \\ |x| > 1}} \Phi(x)\, \chi(x)\, |x|^s\, d^\times x$$

is convergent for all s; indeed, we have, in Proposition 3.1.3(iii), already proved that the integral (Eq. 1.10), of which this is a subintegral, is convergent when $\mathrm{re}(s) > 1$, and in the region $|x| > 1$, decreasing $\mathrm{re}(s)$ only improves the convergence. We must consider more carefully the other part of the integral

$$\zeta_0(s, \chi, \Phi) := \int\limits_{\substack{A^\times \\ |x| < 1}} \Phi(x)\, \chi(x)\, |x|^s\, d^\times x.$$

We may write any element of $A^\times$ as αx, where $\alpha \in F^\times$ and x is an element of some fixed set of representatives of the coset space $A^\times/F^\times$. Thus

$$\begin{aligned} \zeta_0(s, \chi, \Phi) &= \sum_{\alpha \in F^\times} \int\limits_{\substack{A^\times/F^\times \\ |x| < 1}} \Phi(\alpha x)\, \chi(\alpha x)\, |\alpha x|^s\, d^\times x \\ &= \int\limits_{\substack{A^\times/F^\times \\ |x| < 1}} \left[\sum_{\alpha \in F^\times} \Phi(\alpha x) \right] \chi(x)\, |x|^s\, d^\times x, \end{aligned}$$

because $\chi(\alpha x) = \chi(x)$ and $|\alpha| = 1$. Adding and subtracting a single term corresponding to $\alpha = 0$, we write this as

$$\int_{\substack{A^\times/F^\times \\ |x|<1}} \left[\sum_{\alpha\in F} \Phi(\alpha x)\right] \chi(x)\,|x|^s\,d^\times x - \Phi(0) \int_{\substack{A^\times/F^\times \\ |x|<1}} \chi(x)\,|x|^s\,d^\times x.$$

Let $A_1^\times$ denote the subgroup of $A^\times$ consisting of ideles of norm 1. Then

$$A^\times/A_1^\times \cong \begin{cases} \mathbb{R}_+^\times & \text{if } F \text{ is a number field;} \\ \mathbb{Z} & \text{if } F \text{ is a function field.} \end{cases}$$

We must now treat the number field case separately from the function field case, and we will consider only the case where F is a number field. Assuming this, we now consider the integral

$$\int_{\substack{A^\times/F^\times \\ |x|<1}} \chi(x)\,|x|^s\,d^\times x = \int_0^1 \int_{\substack{A^\times/F^\times \\ |x|=t}} \chi(x)\,|x|^s\,d^\times x\,\frac{dt}{t}. \tag{1.18}$$

This equals 0 unless the restriction of χ to $A_1^\times$ is trivial. If this is the case, evidently $\chi(x) = |x|^\lambda$ for some complex number λ, and because χ is unitary, λ must be purely imaginary. Assuming this to be the case, Eq. (1.18) equals

$$\mathrm{Vol}(A_1^\times/F^\times) \int_0^1 t^{s+\lambda}\,\frac{dt}{t} = \frac{\mathrm{Vol}(A_1^\times/F^\times)}{s+\lambda}.$$

The volume of the compact space $A_1^\times/F^\times$ is computed in Lang (1970) (stated without proof on Lang's p. 294, but the calculation is essentially carried out on pp. 132–135) or in Tate's (1950) thesis (Tate's theorem 4.3.2). This volume equals

$$\rho_F = \frac{2^{r_1}\,(2\pi)^{r_2}\,h\,R}{\sqrt{|D|}\,w},$$

where r_1 and r_2 are the number of real and complex embeddings of F, respectively; h is the class number, R the regulator, D the discriminant, and w the number of roots of unity.

By the Poisson summation formula (Proposition 3.1.3(ii)), we have

$$\sum_{\alpha\in F} \Phi(\alpha\,x) = \frac{1}{|x|} \sum_{\alpha\in F} \hat{\Phi}\left(\frac{\alpha}{x}\right).$$

Thus

$$\zeta_0(s,\chi,\Phi) = \begin{cases} \displaystyle\int_{\substack{A^\times/F^\times \\ |x|<1}} \left[\sum_{\alpha\in F} \hat{\Phi}\left(\frac{\alpha}{x}\right)\right] \chi(x)\,|x|^{s-1}\,d^\times x & \text{if } \chi|_{A_1^\times} \neq 1; \\ \displaystyle\int_{\substack{A^\times/F^\times \\ |x|<1}} \left[\sum_{\alpha\in F} \hat{\Phi}\left(\frac{\alpha}{x}\right)\right] \chi(x)\,|x|^{s-1}\,d^\times x - \frac{\rho_F\,\Phi(0)}{s+\lambda} & \text{if } \chi(x) = |x|^\lambda. \end{cases}$$

Making the change of variables $x \to x^{-1}$, we have

$$\begin{aligned} &\int_{\substack{A^\times/F^\times \\ |x|<1}} \left[\sum_{\alpha\in F} \hat{\Phi}\left(\frac{\alpha}{x}\right)\right] \chi(x)\,|x|^{s-1}\,d^\times x \qquad (1.19) \\ &\quad = \int_{\substack{A^\times/F^\times \\ |x|>1}} \left[\sum_{\alpha\in F} \hat{\Phi}(\alpha x)\right] \chi(x)^{-1}\,|x|^{1-s}\,d^\times x \\ &\quad = \int_{\substack{A^\times/F^\times \\ |x|>1}} \left[\sum_{\alpha\in F^\times} \hat{\Phi}(\alpha x)\right] \chi(x)^{-1}\,|x|^{1-s}\,d^\times x \\ &\qquad + \hat{\Phi}(0) \int_{\substack{A^\times/F^\times \\ |x|>1}} \chi(x)^{-1}\,|x|^{1-s}\,d^\times x. \end{aligned}$$

Again, the second term vanishes unless the restriction of χ to $A_1^\times$ is trivial; on the other hand, if $\chi(x) = |x|^\lambda$, the second term equals

$$-\frac{\hat{\Phi}(0)\,\rho_F}{1-s-\lambda}.$$

As for the first term in Eq. (1.19), we have $\chi(x)^{-1}\,|x|^{1-s} = \chi(\alpha x)^{-1}\,|\alpha x|^{1-s}$, and so we may collapse the summation with the integration to see that the first term equals

$$\int_{\substack{A^\times \\ |x|>1}} \hat{\Phi}(x)\,\chi(x)^{-1}\,|x|^{1-s}\,d^\times x = \zeta_1(1-s,\chi^{-1},\hat{\Phi}).$$

We see that

$$\zeta(s,\chi,\Phi)=\begin{cases}\zeta_1(s,\chi,\Phi)+\zeta_1(1-s,\chi^{-1},\hat{\Phi}) & \text{if } \chi|_{A_1^\times}\neq 1;\\ \zeta_1(s,\chi,\Phi)+\zeta_1(1-s,\chi^{-1},\hat{\Phi}) & \text{if } \chi(x)=|x|^\lambda.\\ \quad -\rho_F\left\{\frac{\Phi(0)}{s+\lambda}+\frac{\hat{\Phi}(0)}{1-s-\lambda}\right\} & \end{cases} \tag{1.20}$$

Because ζ_1 has analytic continuation to all s, this gives the meromorphic continuation of $\zeta(s,\chi,\Phi)$; the obvious symmetry property of Eq. (1.20) gives rise to the functional equation (1.17).

We note that

$$|\zeta_1(s,\chi,\Phi)|<\int_{\substack{A^\times\\ |x|>1}}|\Phi(x)|\,|x|^{\mathrm{re}(s)}\,d^\times x.$$

Thus $\zeta_1(s,\chi,\Phi)$ is bounded in vertical strips, and it follows from Eq. (1.20) that $\zeta(s,\chi,\Phi)$ is essentially bounded in vertical strips. ■

We may now prove the analytic continuation and functional equations of $L_S(s,\chi)$, which is the main theorem of this section.

Theorem 3.1.1: Hecke and Tate *Let F be a global field, A its adele ring, and let χ be a Hecke character of $A^\times/F^\times$. Let S be a finite set of places of F containing all the Archimedean ones such that if $v\notin S$, then χ_v is nonramified and the conductor of ψ_v is $\mathfrak{o}_v$. Then $L_S(s,\chi)$ has meromorphic continuation to all s, entire unless there exists a complex number λ such that $\chi(x)=|x|^\lambda$, in which case the poles of $L_S(s,\chi)$ are at $s=\lambda$ and $s=1-\lambda$ if F is a number field and at $s=\lambda+2\pi in/\log(q)$ and $s=1-\lambda+2\pi in/\log(q)$ ($n\in\mathbb{Z}$) if F is a function field, and q is the cardinality of its finite field of constants. We have the functional equation*

$$L_S(s,\chi)=\left\{\prod_{v\in S}\gamma_v(s,\chi_v,\psi_v)\right\}L_S(1-s,\chi^{-1}). \tag{1.21}$$

Proof The proof consists of combining the information in Propositions 3.1.4, 3.1.5, and 3.1.6. We choose our Schwartz function Φ so that $\Phi(g)=\prod_v\Phi_v(g_v)$. By Proposition 3.1.4(ii), we have

$$L_S(s,\chi)\prod_{v\in S}\zeta_v(s,\chi_v,\Phi_v)=\zeta(s,\chi,\Phi). \tag{1.22}$$

The right side has meromorphic continuation by Proposition 3.1.6, and the finite product on the left has meromorphic continuation by Proposition 3.1.5(i). Consequently, we obtain the meromorphic continuation of $L_S(s,\chi)$. As for the possible poles, let $s_0\in\mathbb{C}$. If s_0 is not one of the points enumerated in Proposition 3.1.6 as a possible pole of $\zeta(s,\chi,\Phi)$, then we can show that $L_S(s,\chi)$

has no pole at $s = s_0$, because if it did, one of the finite number of factors $\zeta_v(s, \chi_v, \Phi_v)$ would have to have a zero at s_0, and by Proposition 3.1.5(iii), we may choose the Schwartz functions Φ_v so that this does not occur. As for the functional equation, using Eqs. (1.13), (1.5), and (1.14), we see that Eq. (1.22) equals

$$L_S(1-s, \chi^{-1}) \left\{ \prod_{v \in S} \gamma_v(s, \chi_v, \psi_v)\, \zeta_v(s, \chi_v, \Phi_v) \right\}.$$

comparing this expression with Eq. (1.22) yields Eq. (1.21). ■

Although Theorem 3.1.1 is sufficient for many purposes, it is better to define the L-function $L(s, \chi)$ as a product over *all* places, not just $v \notin S$. We will define Euler factor at every place. If v is a non-Archimedean place and χ_v is nonramified, then we have already defined $L_v(s, \chi_v) = (1 - \chi_v(\varpi_v)^{-s})^{-1}$. If v is non-Archimedean and χ_v is ramified, then we define $L_v(s, \chi_v) = 1$. If v is real, $F_v = \mathbb{R}$, then because we are assuming that χ is unitary, we have $\chi_v(x) = \left(x/|x|_v\right)^{\epsilon}$, where $\epsilon = 0$ or 1. We then define $L_v(s, \chi_v) = \pi^{-(s+\epsilon)/2}\,\Gamma\left((s+\epsilon)/2\right)$. Finally, suppose that v is complex, so that $F_v = \mathbb{C}$. Then χ_v has the form

$$\chi_v(x) = |x|_v^{\nu} \left(\frac{x}{\sqrt{|x|_v}}\right)^k, \qquad k \in \mathbb{Z}, \tag{1.23}$$

and because χ_v is unitary, ν is pure imaginary. (We recall that our convention is that $|x|_v$ is the *square* of the "usual" absolute value on the complex numbers.) In this case, we will denote

$$L_v(s, \chi_v) = 2\,(2\pi)^{s+\nu+|k|/2}\,\Gamma\left(s + \nu + \frac{|k|}{2}\right).$$

Proposition 3.1.7 *Suppose that f is a continuous function on the half-closed interval* $(0, 1]$, *which has a convergent series expansion*

$$f(x) = \sum_{\nu \in \Sigma} a(\nu)\, x^{\nu}, \tag{1.24}$$

valid near zero, where $\Sigma \subset \mathbb{R}$ is a discrete set, bounded below. Then

$$\int_0^1 f(x)\, x^s \, \frac{dx}{x} \tag{1.25}$$

has meromorphic continuation to all s, with poles only at the values $s = -\nu$ with $\nu \in \Sigma$. The residue of the pole at $s = -\nu$ is the coefficient $a(\nu)$. Moreover, if $\epsilon > 0$, $N \in \mathbb{R}$, the integral (Eq. 1.25) is bounded in the region $\{\mathrm{re}(s) > -N,\ |s+\nu| > \epsilon\}$.

Proof Let N be given, and let us write

$$f(x) = \sum_{\substack{\nu \in \Sigma \\ \nu < N}} a(\nu)\, x^{\nu} + R(x), \qquad R(x) = O(x^N).$$

Then Eq. (1.25) equals

$$\sum_{\nu<N} \frac{a(\nu)}{s+\nu} + \int_0^1 R(x)\, x^s\, dx. \tag{1.26}$$

The remainder term is analytic for $\mathrm{re}(s) > -N$. This proves the meromorphic continuation to this region with poles only at $s = -\nu$ with $\nu \in \Sigma$. The absolute value of Eq. (1.26) is bounded by

$$\sum_{\nu<N} \frac{|a(\nu)|}{|s+\nu|} + \int_0^1 |R(x)|\, x^{\mathrm{re}(s)}\, dx,$$

and this is bounded in $\{\mathrm{re}(s) > -N + \epsilon,\ |s+\nu| > \epsilon\}$. ■

We say that a nonzero function $e(s)$ of a complex variable s is *of exponential type* if it is of the form $a\, b^s$ for suitable constants $a \in \mathbb{C}^\times$ and $b \in \mathbb{R}$. Note that b being real implies that $e(s)$ is bounded in vertical strips.

Proposition 3.1.8 *Let v be any place of F, and let $s_0 \in \mathbb{C}$. Then $L_v(s, \chi_v)$ has a pole at $s = s_0$ if and only if $\zeta_v(s, \chi_v, \Phi_v)$ has a pole there for some $\Phi_v \in S(F_v)$. In particular, $\zeta_v(s, \chi_v, \Phi_v)/L_v(s, \chi_v)$ is holomorphic for all values of s. If v is non-Archimedean, then $\zeta_v(s, \chi_v, \Phi_v)$ is rational function in q_v^{-s}. There exists a choice of Φ_v such that $\zeta_v(s, \chi_v, \Phi_v)/L_v(s, \chi_v)$ is a function of exponential type.*

In the argument below, we will see that the behavior of $\Phi_v(x)\, \chi_v(x)$ near $x = 0$ determines the poles of $\zeta_v(s, \chi_v, \Phi_v)$. In the process, we will obtain another proof of the meromorphic continuation of these local integrals, which we proved by a different method in Proposition 3.1.5(i).

Proof If v is non-Archimedean, we write (for $\mathrm{re}(s) > 0$)

$$\zeta_v(s, \chi_v, \Phi_v) = \sum_{k=-\infty}^{\infty} q^{ks} \int_{|x|_v = q_v^k} \Phi_v(x)\, \chi_v(x)\, d^\times x. \tag{1.27}$$

The contribution is zero for k large (because $\Phi_v(x) = 0$ off some compact set). On the other hand, if $-k$ is large, then $\Phi_v(x) = \Phi_v(0)$ when $|x|_v = q^k$, so the contribution equals

$$\Phi_v(0) \int_{|x|_v = q_v^k} \chi_v(x)\, d^\times x.$$

This is zero if χ_v is ramified, and so Eq. (1.27) is only a finite sum. Thus $\zeta_v(s, \chi_v, \Phi_v)$ is entire and is rational in q_v^{-s}. If χ_v is nonramified, then at least for k small we have a geometric series that we can evaluate explicitly to verify the proposition.

The Archimedean case will be deduced from Proposition 3.1.7. We note that the poles of $\zeta(s, \chi_v, \Phi_v)$ are the same as the poles of

$$\int_{|x| \le 1} \Phi_v(x)\, \chi_v(x)\, |x|_v^s\, d^\times x, \tag{1.28}$$

because the integral over

$$\int_{|x| > 1} \Phi_v(x)\, \chi_v(x)\, |x|_v^s\, d^\times x$$

is convergent for all s. The poles of the integral (Eq. 1.28) may be determined by Proposition 3.1.7. Assume first that v is a real place. We may split Φ_v into an odd part and an even part and handle these two cases separately. Assuming that Φ_v is either odd or even, the integral (Eq. 1.28) vanishes unless the parity of Φ_v matches that of χ_v. Assuming this, if $\chi_v = 1$, then Φ_v has only even terms in its Taylor expansion, and so the possible poles of Eq. (1.28) are at $0, -2, -4, \cdots$. If χ_v is the sign character, then Φ_v has only odd terms in its Taylor expansion, and the possible poles are at $-1, -3, \cdots$. In either case, these coincide with the poles of $L_v(s, \chi_v)$.

If v is complex, then we use polar coordinates. Let $x = re^{i\theta}$. With χ_v as in Eq. (1.23), we have (up to a constant, coming from the normalization of the additive Haar measure)

$$\zeta_v(s, \chi_v, \Phi_v) = \frac{1}{2\pi} \int_0^\infty \int_0^{2\pi} r^{2\nu+2s}\, e^{ik\theta}\, \Phi_v(re^{i\theta})\, d\theta\, dr/r$$

$$= \int_0^\infty r^{2\nu+2s}\, \phi(r)\, dr/r,$$

where

$$\phi(r) = \frac{1}{2\pi} \int_0^{2\pi} \Phi_v(re^{ik\theta})\, e^{ik\theta}\, d\theta.$$

We consider the Taylor expansion of Φ_v near $x = 0$:

$$\Phi_v(x) = \sum_{n,m\geq 0} a(n,m)\, x^n\, \overline{x}^m,$$

where $\overline{x}$ is the complex conjugate of x. Then

$$\phi(r) = \sum_{m-n=k} a(n,m)\, r^{n+m}.$$

By Proposition 3.1.7, the possible poles of $\zeta_v(s, \chi_v, \Phi_v)$ are at $2s = -2\nu - |k| - 2l$ with $l \in \mathbb{Z}$. Again, these are the poles of the local L-function.

The particular Φ_v such that $\zeta_v(s, \chi_v, \Phi_v)/L_v(s, \chi_v)$ is a function of exponential type may be described as follows. If v is non-Archimedean and χ_v is nonramified, take Φ_v to be the characteristic function of $\mathfrak{o}_v$. If v is non-Archimedean and χ_v is ramified, take

$$\Phi_v(x) = \begin{cases} \chi_v(x)^{-1} & \text{if } x \in \mathfrak{o}^\times, \\ 0 & \text{otherwise.} \end{cases}$$

If v is real, then because χ_v is unitary, it is either trivial, in which case take $\Phi_v(x) = e^{-\pi x^2}$, or it is the sign character, in which case take $\Phi_v(x) = x\, e^{-\pi x^2}$. If v is complex, then with χ_v as in Eq. (1.23), we take

$$\Phi_v(x) = \begin{cases} \overline{x}^{-k}\, e^{-2\pi |x|_v} & \text{if } k > 0, \\ x^{-k}\, e^{-2\pi |x|_v} & \text{if } k < 0, \end{cases}$$

where we recall that $|x|_v$ is the *square* of the usual complex absolute value. ■

Proposition 3.1.9 *Let v be any place of F, and define*

$$\epsilon_v(s, \chi_v, \psi_v) = \frac{\gamma_v(s, \chi_v, \psi_v)\, L_v(s, \chi_v)}{L_v(1-s, \chi_v^{-1})}. \tag{1.29}$$

Then $\epsilon_v(s, \chi_v, \psi_v)$ is a function of exponential type. If v is non-Archimedean, χ_v is nonramified and the conductor of ψ_v is $\mathfrak{o}_v$, then $\epsilon_v(s, \chi_v, \psi_v) = 1$.

The epsilon factors (Eq. 1.29) are sometimes called *root numbers*.

Proof We show first that $\epsilon_v(s, \chi_v, \psi_v)$ has neither zeros nor poles. If $\Phi_v \in S(F_v)$, then it follows from the definitions that

$$\epsilon_v(s, \chi_v, \psi_v) = \frac{\zeta_v(1-s, \chi_v, \hat{\Phi}_v)}{L_v(1-s, \chi_v)} \cdot \frac{L_v(s, \chi_v)}{\zeta_v(s, \chi_v, \Phi_v)}. \tag{1.30}$$

Let s_0 be a fixed complex number. It follows from Proposition 3.1.8 that

$$\zeta_v(1-s, \chi_v, \hat{\Phi}_v)/L_v(1-s, \chi_v)$$

is holomorphic at s_0, and it follows from Proposition 3.1.5(iii) and Proposition 3.1.8 that Φ_v may be chosen so that $\zeta_v(s, \chi_v, \Phi_v)/L_v(s, \chi_v)$ is nonvanishing at s_0. Hence $\epsilon_v(s, \chi_v, \psi_v)$ is holomorphic at s_0, and the same argument applies to its reciprocal, so it is also nonvanishing. If v is non-Archimedean, we've now proved that $\epsilon_v(s, \chi_v, \psi_v)$ is of exponential type because it is a rational function of q_v^{-s} with neither zeros nor poles. If v is non-Archimedean, χ_v is nonramified and the conductor of ψ_v is $\mathfrak{o}_v$, so choose $\Phi_v = \hat{\Phi}_v$ to be the characteristic function of $\mathfrak{o}_v$. Then it follows from Eqs. (1.30) and (1.13) that $\epsilon_v(s, \chi_v, \psi_v) = 1$.

If v is Archimedean, then one proves that $\epsilon_v(s, \chi_v, \psi_v)$ is of exponential type by choosing the test functions Φ_v to be particular functions. In fact, we may use the same test functions as in the proof of the last part of Proposition 3.1.8. We leave the calculation of the Fourier transforms of these test functions to the reader (Exercise 3.1.8), or consult Tate (1950) Section 2.5 for this computation. ■

If v is non-Archimedean and χ_v is ramified, it is possible to express $\epsilon_v(s, \chi_v, \psi_v)$ in terms of Gauss sums. In this case we have $\epsilon_v(s, \chi_v, \psi_v) = \gamma_v(s, \chi_v, \psi_v)$, and Exercise 1.9 gives a formula for this. Again, this calculation is essentially contained in Tate (1950, Section 2.5).

We now define

$$L(s, \chi) = \prod_v L_v(s, \chi_v), \tag{1.31}$$

where the product is over *all* places s of F. This equals the partial L-function $L_S(s, \chi)$, which we've already studied times a finite product, so this product is convergent for $\mathrm{re}(s) > 1$ and has meromorphic continuation to all s. Also, let

$$\epsilon(s, \chi) = \prod_v \epsilon_v(s, \chi_v, \psi_v). \tag{1.32}$$

All but finitely many factors in this product equal 1 by Proposition 4.1.8, and because each factor is of exponential type, so is the product. We will see in the next theorem that although the individual factors in this product depend on the choice of the additive character ψ, $\epsilon(s, \chi)$ is independent of this choice, justifying the omission of ψ from the notation.

Theorem 3.1.2 *The factor $\epsilon(s, \chi)$ is independent of the choice of additive character ψ. The L-function $L(s, \chi)$ has analytic continuation to all s except $s = 0$ or 1, where it can have simple poles as described in Theorem 3.1.1. We have the functional equation*

$$L(s, \chi) = \epsilon(s, \chi)\, L(1 - s, \chi^{-1}). \tag{1.33}$$

The L-function $L(s, \chi)$ is essentially bounded in vertical strips.

Proof The functional equation (1.33) follows by multiplying Eq. (1.21) by $\prod_{v\in S} L_v(s,\chi_v)$ and using Eq. (1.29). It is evident from Eq. (1.33) that $\epsilon(s,\chi)$ is independent of the choice of ψ. It follows from Proposition 3.1.8 and Proposition 3.1.4(ii) that $L(s,\chi)$ has no poles that are not poles of $\zeta(s,\chi,\Phi)$ for some choice of $\Phi\in S(A)$, so the analyticity of $L(s,\chi)$ follows from the analyticity of $\zeta(s,\chi,\Phi)$ as in Proposition 3.1.6.

To see that $L(s,\chi)$ is bounded in vertical strips, we consider the ratio

$$\frac{L(s,\chi)}{\zeta(s,\Phi,\chi)}=\prod_v \frac{L_v(s,\chi)}{\zeta_v(s,\Phi_v,\chi_v)}.$$

All but finitely many factors equal one, and by Proposition 1.8, we may choose the test function Φ_v so that the remaining factors are of exponential type. Hence with this choice of data, the ratio is essentially bounded in vertical strips. On the other hand, $\zeta_v(s,\Phi_v,\chi_v)$ is essentially bounded in vertical strips by Proposition 3.1.6. ∎

Exercises

Exercise 3.1.1 (a) *The "no small subgroups" argument.* Let G be a topological group having a basis of neighborhoods of the identity consisting of open and compact subgroups. (We will call such a group a *totally disconnected locally compact group – see Proposition 4.2.1*. Examples are the F, $F^\times$, or $GL(n,F)$, where F is a non-Archimedean local field.) Prove that the kernel of any continuous homomorphism of $G\to GL(m,\mathbb{C})$ contains an open subgroup. (Let N be a neighborhood of the identity $GL(m,\mathbb{C})$ so small that it does not contain any nontrivial subgroups, and consider the preimage of N in G.)
(b) Let F be a non-Archimedean local field. Let $\mathfrak{o}$ be the ring of integers in F and $\mathfrak{p}$ its maximal ideal. Let ψ be a nontrivial additive character of F. Prove that there exists a unique integer m such that ψ is trivial on $\mathfrak{p}^{-m}$, but not on $\mathfrak{p}^{-m-1}$. We call $\mathfrak{p}^{-m}$ the *conductor* of ψ.
(c) Let k be a finite field, ψ a nontrivial additive character of k, and let ψ_1 be another character. Show that there exists a unique element $a\in k$ such that $\psi_1(x)=\psi(ax)$ for all $x\in k$.
(d) Let ψ_1 be another additive character of the non-Archimedean local field F. Prove that there exists a unique element $a\in F$ such that $\psi_1(x)=\psi(ax)$ for all $x\in F$.

[Hint: Let $\mathfrak{p}^{-m_1}$ be the conductor of ψ_1. Use (c) to prove by induction, beginning with $N=m_1$, that there exists $a_N\in F$ such that $\psi_1(x)=\psi(a_Nx)$ for $x\in\mathfrak{p}^{-N}$. Indeed, if $N=m_1$, take $a_N=0$; and for subsequent N, use (b) to show that you may take $a_N=a_{N-1}+c_N$ for some $c_N\in\mathfrak{p}^{N-1-m}$. Show that $\{a_N\}$ is a Cauchy sequence, and let a be the limit.]
(e) Define $\psi_a(x)=\psi(ax)$, for $a\in F$. Prove that $a\mapsto\psi_a$ is a topological isomorphism of F with its dual group.

(f) Prove that $F = \mathbb{R}$ or $F = \mathbb{C}$ then F may also be identified with its dual group: let $\psi(x) = e^{2\pi i x}$ for $x \in \mathbb{R}$ or $\psi(x) = e^{4\pi i \, \mathrm{re}(x)}$ for $x \in \mathbb{C}$. Show that $a \mapsto \psi_a$ is a topological isomorphism of F with its dual group.

Exercise 3.1.2 Let F be an algebraic number field and A be its adele ring. Suppose that for each place v of F, we select an additive character ψ_v of F_v such that the conductor of ψ_v is $\mathfrak{o}_v$ for almost all v. Then we may define a character $\psi = \prod_v \psi_v$ of A by $\psi(a) = \prod_v \psi_v(a_v)$ for each adele $a = (a_v) \in A$. The aim of this exercise is to show directly how to choose the ψ_v such that ψ is trivial on F. (Of course, this may also be done if F is a function field.)
(a) Assume that F is $\mathbb{Q}$. Define characters $\psi_v : F_v \to \mathbb{C}^\times$ as follows. If $v = \infty$, let $\psi_\infty : \mathbb{Q}_\infty = \mathbb{R} \to \mathbb{C}$ be defined by $\psi_\infty(x) = e^{2\pi i x}$. If $v = p$ is a finite prime, we have $\mathbb{Q}_p = \mathbb{Z}[1/p] + \mathbb{Z}_p$, and so we may write any $x \in \mathbb{Q}_p$ as $y + z$, where $y \in \mathbb{Z}[1/p]$ and $z \in \mathbb{Z}_p$. Then define $\psi_v(x) = e^{-2\pi i y}$. Prove that $\psi = \prod_v \psi_v$ is trivial on $\mathbb{Q}$.
(b) If F is a number field, reduce to the case $F = \mathbb{Q}$ by composition with the trace map.

Exercise 3.1.3 Let F be an algebraic number field and A be its adele ring. Let ψ be an additive character of A that is trivial on F.
(a) Prove that every character of A has the form $\psi_a(x) := \psi(ax)$ for some $a \in A$.
(b) Prove that $a \mapsto \psi_a$ is an isomorphism of A with its dual group.
(c) Prove that the character ψ_a is trivial on the discrete, cocompact subgroup F of A if and only if $a \in F$.
(d) Prove that the dual of the discrete group F is the compact group A/F.

Exercise 3.1.4 In 1.7, we defined the notion of a Hecke character over a totally real field F. Generalize Proposition 3.1.2 to show that there is a bijection between primitive Hecke characters of F and characters of $A^\times/F^\times$, so that if χ_0 is a Hecke character in the sense of Section 1.7, and if χ is the corresponding character of $A^\times/F^\times$, then for nonramified places v, $\chi_v(\varpi_v) = \chi_0(\mathfrak{p}_v)$.

Exercise 3.1.5 (a) Let F be an algebraic number field and A its adele ring. Prove that every quasicharacter of $A^\times/F^\times$ has the form $\chi(x)\,|x|^s$, where χ is unitary (i.e., a Hecke character) and $s \in \mathbb{R}$.
(b) Let F be a non-Archimedean local field. Prove that every quasicharacter of $F^\times$ has the form $\chi(x)\,|x|^s$, where χ is a unitary character of $F^\times$.

Exercise 3.1.6 Let F be an algebraic number field and A its adele ring. Let C be the ideal class group of F, and let χ_0 be a character of C. Show that there exists a character $\chi = \prod_v \chi_v$ of $A^\times/F^\times$ such that for every non-Archimedean place v of F, χ_v is nonramified, and $\chi_v(\varpi_v) = \chi_0(\mathfrak{p}_v)$.

Exercise 3.1.7 Let Φ be a Schwartz function on A. Prove that the sum $\sum_{\alpha \in F} \Phi(x + \alpha)$ converges absolutely and uniformly when x is restricted to any given compact set.

[HINT: Factor $A = A_{\mathfrak{f}} F_\infty$, where $A_{\mathfrak{f}}$ is the restricted direct product of the F_v with v non-Archimedean, and F_∞ is the product of the F_v with v Archimedean. Without loss of generality, assume that there exists a Schwartz function Φ_∞ on the Euclidean space F_∞ and a Schwartz function $\Phi_{\mathfrak{f}}$ on $A_{\mathfrak{f}}$ (that is, a compactly supported, locally constant function) such that when $a \in A$ is factored as $a_\infty a_{\mathfrak{f}}$, $a_\infty \in F_\infty$, $a_{\mathfrak{f}} \in A_{\mathfrak{f}}$, we have $\Phi(a) = \Phi_\infty(a_\infty)\,\Phi_{\mathfrak{f}}(a_{\mathfrak{f}})$. Show that there exists a lattice Λ in the Euclidean space F_∞ such that if $\alpha \in F$, then $\Phi_{\mathfrak{f}}(\alpha_{\mathfrak{f}}) = 0$ unless $\alpha_\infty \in \Lambda$. Thus it is sufficient to show that

$$\sum_{\alpha_\infty \in \Lambda} \Phi_\infty(x_\infty + \alpha_\infty)$$

converges absolutely and uniformly (on compact sets) in x_∞.]

Exercise 3.1.8 Complete the proof of Proposition 3.1.9 by showing that $\epsilon_v(s, \chi_v, \psi_v)$ is of exponential type when v is Archimedean. [HINT: Do this by using particular test functions. If v is real, use $\Phi_v(x) = e^{-\pi x^2}$ or $x\,e^{-\pi x^2}$. If v is complex, refer to Tate (1950), Section 2.5.]

Exercise 3.1.9 Let v be a non-Archimedean place of F, and let $\mathrm{re}(s) < 1$. Prove that if N is sufficiently large, then

$$\gamma_v(s, \chi_v, \psi_v) = \int_{\mathfrak{p}_v^{-N}} |x|^{-s}\, \chi_v^{-1}(x)\, \psi_v(x)\, dx.$$

It is important here that $\int_F dx$ is the self-dual Haar measure. [Take the test function Φ in Eq. (1.14) to be the characteristic function of $1 + \mathfrak{p}^N$.]

3.2 Classical Automorphic Forms and Representations

For the foundations of the spectral theory of automorphic forms on Lie groups, the recent book of Moeglin and Waldspurger (1995) is now the standard reference. The lecture notes of Borel (to appear) are nicely complementary to our discussion. The most important historical landmarks in this area are Selberg (1956) and Langlands' (1976) complete but difficult account. Other useful references are Harish-Chandra (1968) and Borel and Jacquet (1979).

Let $G = GL(2, \mathbb{R})^+$ be the group of 2×2 real matrices with positive determinant acting on the Poincaré upper half plane $\mathcal{H}$ in the usual way by linear fractional transformations. Let $Z(\mathbb{R})$ be the center of G, consisting of scalar matrices, and let $K = SO(2)$ be the standard maximal compact subgroup. We will now define certain spaces of automorphic forms on G. Later we will show how these spaces are related to classical modular and Maass forms.

We now introduce notations that will be in force throughout this section. Let Γ be a discrete subgroup of G. We will assume that Γ is actually contained in $SL(2, \mathbb{R})$. We also assume that $\Gamma\backslash\mathcal{H}$ has finite volume, and that $-I \in \Gamma$, where I is the 2×2 identity matrix. We will assume that $\Gamma\backslash\mathcal{H}$ is noncompact, because the case where it is compact is treated in Section 2.3. (Thus, for example, Γ could be $\Gamma_0(N)$.) Let χ be a character of Γ, and let ω be a character of the center $Z(\mathbb{R})$ of G, consisting of the scalar matrices. We assume that ω and χ agree on -1. (Recall that characters are assumed unitary.)

The group G acts on $C^\infty(G)$ by right translation. This is the representation ρ defined by $\big(\rho(g)\,F\big)(x) = F(xg)$. There is a derived action of the Lie algebra $\mathfrak{g}$, which consists of 2×2 real matrices with Lie bracket $[X, Y] = XY - YX$. This action was denoted $X \mapsto dX$ in Chapter 2 (cf. Eq. (2.10) of Chapter 2), but we shall suppress the d from the notation and write simply

$$XF(g) = \frac{d}{dt}\, F\big(g\, \exp(tX)\big)\big|_{t=0}, \qquad F \in C^\infty(G). \tag{2.1}$$

We recall from Section 2.2 that this representation of $\mathfrak{g}$ extends to an action of the universal enveloping algebra $U(\mathfrak{g})$ or its complexification $U(\mathfrak{g}_{\mathbb{C}})$. In particular, the center $\mathcal{Z}$ of $U(\mathfrak{g}_{\mathbb{C}})$ is a polynomial ring in two variables: $\mathcal{Z} = \mathbb{C}[Z, \Delta]$, where

$$R = \frac{1}{2}\begin{pmatrix} 1 & i \\ i & -1 \end{pmatrix}, \quad L = \frac{1}{2}\begin{pmatrix} 1 & -i \\ -i & -1 \end{pmatrix}, \quad H = -i\begin{pmatrix} 0 & 1 \\ -1 & 0 \end{pmatrix},$$

$$Z = \begin{pmatrix} 1 & 0 \\ 0 & 1 \end{pmatrix}, \tag{2.2}$$

$$\Delta = -\tfrac{1}{4}\,(H^2 + 2RL + 2LR). \tag{2.3}$$

See Eqs. (2.23), (2.18) and (2.39) in Chapter 2. Note that we are using the letter Z in two ways, but this should cause no confusion because the center of G will be denoted $Z(\mathbb{R})$, never simply Z.

The action of G on $\mathcal{H}$ by linear fractional transformations extends to an action on the boundary of $\mathcal{H}$ in the Riemann sphere, which is $\mathbb{R} \cup \{\infty\}$. By a *cusp* of Γ we mean a point of $\mathbb{R} \cup \{\infty\}$ whose stabilizer in Γ contains a nontrivial unipotent matrix. The number of orbits of cusps under the action of Γ is finite, and by abuse of terminology, we also refer to these orbits as *cusps* and say that the number of cusps of Γ is finite. Our assumption that $\Gamma\backslash\mathcal{H}$ is noncompact is equivalent to assuming that Γ has at least one cusp (Exercise 1.2.9).

Let $F \in C^\infty(G)$. We say that F is K-finite if the functions $\rho(k)\,F$, $k \in K$ span a finite-dimensional vector space, and that F is if it is contained in a finite-dimensional $\mathcal{Z}$-invariant subspace.

Let $C(\Gamma\backslash G, \chi, \omega)$ be the space of continuous functions $F : G \to \mathbb{C}$ such that

$$F(\gamma g) = \chi(\gamma)\, F(g), \qquad \gamma \in \Gamma,\ g \in G, \tag{2.4}$$

and

$$F(zg) = \omega(z)\, F(g), \qquad z \in Z(\mathbb{R}), \qquad g \in G. \tag{2.5}$$

Let $C^\infty(\Gamma\backslash G, \chi, \omega)$ denote the space of smooth functions in $C(\Gamma\backslash G, \chi, \omega)$. Let $C_c(\Gamma\backslash G, \chi, \omega)$ and $C_c^\infty(\Gamma\backslash G, \chi, \omega)$ denote the respective subspaces of elements of these two spaces that are compactly supported modulo $Z(\mathbb{R})$. Let $\mathcal{A}(\Gamma\backslash G, \chi, \omega)$ be the subspace of functions in $C^\infty(\Gamma\backslash G, \chi, \omega)$ that are K-finite and $\mathcal{Z}$-finite and such that there exist constants C and N such that

$$|F(g)| < C\,||g||^N, \qquad g \in G, \tag{2.6}$$

where the *height* $||g||$ is the length of the vector $(g, \det g^{-1})$ in the Euclidean space $\mathrm{Mat}_2(\mathbb{R}) \oplus \mathbb{R} = \mathbb{R}^5$. This latter condition we will call the *condition of moderate growth*. As evidence that this is a useful definition, see Theorem 3.2.1 below, where it is shown that the space $\mathcal{A}(\Gamma\backslash G, \chi, \omega)$ is invariant under differentiation by elements of the Lie algebra $\mathfrak{g}$ of G.

Next we define what it means for $F \in \mathcal{A}(\Gamma, \chi, \omega)$ to be *cuspidal*. If a is a cusp of Γ, we will define a notion of F being *cuspidal at a*, and then we will say that F is *cuspidal* if it is cuspidal at every cusp. In this case, we also say that F is a *cusp form*.

Thus we must define what it means for F to be cuspidal at a. Assume first that $a = \infty$. Then Γ contains an element of the form $\tau_r = \left(\begin{smallmatrix}1 & r \\ & 1\end{smallmatrix}\right)$. We say that F is *cuspidal at ∞* if either $\chi(\tau_r) \neq 1$ or

$$\int_0^r F\left(\begin{pmatrix}1 & x \\ & 1\end{pmatrix} g\right) dx = 0. \tag{2.7}$$

(It is easy to see that with this definition, the condition of cuspidality is unchanged if r is replaced by a multiple, so it is not necessary for r to be a generator of the stabilizer of ∞ in Γ.) If a is now an arbitrary cusp, we choose $\xi \in SL(2, \mathbb{R})$ such that $\xi(\infty) = a$. Then $F'(g) = F(\xi g)$ defines an element of $L^2(\Gamma'\backslash G, \chi', \omega)$, where $\Gamma' = \xi^{-1}\Gamma\xi$ and χ' is the character $\chi'(\gamma) = \chi(\xi\gamma\xi^{-1})$ of Γ', and we say that F is *cuspidal at a* if F' is cuspidal at ∞, the latter notion being already defined.

We will denote the space of cusp forms in $\mathcal{A}(\Gamma, \chi, \omega)$ as $\mathcal{A}_0(\Gamma\backslash G, \chi, \omega)$.

We recall the notion of a $(\mathfrak{g}, K)$-module. This is simply a vector space with compatible actions of $\mathfrak{g}$ and K in which every vector is K-finite. (For a precise definition, see the end of Section 2.4.) The $(\mathfrak{g}, K)$-module V is called *admissible* if the isotypic subspace $V(\sigma) = \{v \in V \,|\, \pi(k)\,v = \sigma(k)\,v\}$ is finite dimensional for each character σ of K. (We recall that K is Abelian, so every finite dimensional irreducible representation is a character.)

Theorem 3.2.1 *The spaces $\mathcal{A}(\Gamma\backslash G, \chi, \omega)$ and $\mathcal{A}_0(\Gamma\backslash G, \chi, \omega)$ are stable under the action of $U(\mathfrak{g}_\mathbb{C})$. If $f \in \mathcal{A}(\Gamma\backslash G, \chi, \omega)$, then $U(\mathfrak{g}_\mathbb{C})\,f$ is an admissible $(\mathfrak{g}, K)$-module. If f satisfies Eq. (2.6) and $D \in U(\mathfrak{g}_\mathbb{C})$, then Df satisfies a similar estimate with the same constant N.*

For $\mathcal{A}_0(\Gamma\backslash G, \chi, \omega)$, the use of Harish-Chandra's theorem (Theorem 2.9.2) can be avoided by using Theorem 2.2 below – see Exercise 3.2.6.

Proof If we show that $\mathcal{A}(\Gamma\backslash G, \chi, \omega)$ is stable under the action of $U(\mathfrak{g}_\mathbb{C})$, it will follow that $\mathcal{A}_0(\Gamma\backslash G, \chi, \omega)$ is also stable, because it is easy to see directly that the cuspidal condition is preserved by the action of $\mathfrak{g}$. We note that because of Eq. (2.5), $|f|$ is constant on the cosets of $Z(\mathbb{R})$. It is easy to see that in each coset $Z(\mathbb{R})\,f$, the element g with minimal height $||g||$ is actually in $SL(2,\mathbb{R})$, so the growth condition (Eq. 2.6) is equivalent to the requirement that the restriction of f to $SL(2,\mathbb{R})$ satisfies the same growth condition, where $SL(2, R)$ can be given the same height $||\cdot||$ as in Theorem 2.9.2. Finally, we note that in $\mathcal{Z} = \mathbb{C}[Z, \Delta]$, Z automatically acts by a scalar because of Eq. (2.5). Thus $\mathcal{Z}$-finiteness is equivalent to $\mathbb{C}[\Delta]$-finiteness. Because of these remarks, Theorem 2.9.2 (which was stated for $SL(2,\mathbb{R})$) may be applied and all of our assertions are contained in that result of Harish-Chandra. ∎

We will call elements of $\mathcal{A}(\Gamma\backslash G, \chi, \omega)$ *automorphic forms*. Next we relate this notion of an automorphic form to the classical notions of modular forms and Maass forms.

First we consider modular forms. We call a function ϕ on $\mathcal{H}$ a *(holomorphic) modular form* for Γ with character χ if ϕ is holomorphic, and satisfies

$$\phi\big(\gamma(z)\big) = \chi(\gamma)\,(cz+d)^k\,\phi(z), \qquad \gamma = \begin{pmatrix} a & b \\ c & d \end{pmatrix} \in \Gamma,\ z \in \mathcal{H}, \tag{2.8}$$

and if ϕ is *holomorphic at the cusps* of Γ. The latter condition requires definition. If ∞ is a cusp of Γ, then Γ contains an element of the form $\tau_r = \left(\begin{smallmatrix} 1 & r \\ & 1 \end{smallmatrix}\right)$, $r > 0$. If $\chi(\tau_r) = e^{2\pi i\lambda}$, this means that $\phi(z+r) = e^{2\pi i\lambda}\,\phi(z)$, and so ϕ has a Fourier expansion

$$\phi(z) = \sum_{n\in\mathbb{Z}} a(n+\lambda)\, e^{2\pi i(n+\lambda)z/r}.$$

We say that ϕ is *holomorphic (resp. cuspidal) at* ∞ if the coefficients $a(n+\lambda)$ are zero when $n+\lambda < 0$ (resp. when $n+\lambda \le 0$). Note that this condition is automatic if $\chi(\tau_r) \neq 1$. More generally, if a is a cusp of Γ, let $\xi = \left(\begin{smallmatrix} A & B \\ C & D \end{smallmatrix}\right) \in SL(2,\mathbb{R})$ such that $\xi(\infty) = a$. Then defining

$$\phi'(z) = (Cz+D)^{-k}\,\phi\big(\xi(z)\big),$$

it is not hard to see that ϕ' satisfies Eq. (2.1) with Γ and χ replaced by Γ' and χ', respectively, where $\Gamma' = \xi^{-1}\,\Gamma\,\xi$ and χ' is the character $\chi'(\gamma) = \chi(\xi\gamma\xi^{-1})$ of Γ', and ∞ is a cusp of Γ'. Then we say that ϕ is holomorphic (resp. cuspidal) at a if ϕ' is holomorphic (resp. cuspidal) at ∞. A modular form that is cuspidal at every cusp is called a *cusp form*.

On the other hand, we have also *Maass forms*. To define these, we recall that there is a right action of $GL(2,\mathbb{R})^+$ defined on functions on $\mathcal{H}$ defined as follows. If $g \in GL(2,\mathbb{R})^+$ and $f : \mathcal{H} \to \mathbb{C}$ is a function, let

$$f|_z kg = \left(\frac{c\bar{z}+d}{|cz+d|}\right)^k f\left(\frac{az+b}{cz+d}\right), \qquad g = \begin{pmatrix} a & b \\ c & d \end{pmatrix}. \tag{2.9}$$

(Compare Eq. (1.5) of Chapter 2.)

A Maass form is not holomorphic, but rather is an eigenform of the non-Euclidean Laplacian. Specifically, let $k \in \mathbb{Z}$, and define the *weight k Laplacian* Δ_k by

$$\Delta_k = -y^2\left(\frac{\partial^2}{\partial x^2} + \frac{\partial^2}{\partial y^2}\right) + iky\frac{\partial}{\partial x}.$$

(See Eq. (1.3) of Chapter 2.) Then Δ_k commutes with the action (Eq. 2.9) of $GL(2,\mathbb{R})^+$ (Lemma 2.1.1).

Let us denote the space of smooth functions f on $\mathcal{H}$ that satisfy

$$f|_k\gamma = \chi(\gamma)\, f, \qquad \gamma \in \Gamma \tag{2.10}$$

as $C^\infty(\Gamma\backslash\mathcal{H}, \chi, k)$, and let us denote the space of square integrable functions satisfying Eq. (2.10) as $L^2(\Gamma\backslash\mathcal{H}, \chi, k)$. A *Maass form of weight k* is an element of $C^\infty(\Gamma\backslash\mathcal{H}, \chi, k)$, which is an eigenform for Δ_k with eigenvalue λ, and which is of *moderate growth* at the cusps of Γ. We will now explain the last condition in detail.

First suppose that ∞ is a cusp of Γ; then Γ contains a parabolic element $\tau_r = \left(\begin{smallmatrix}1 & r \\ & 1\end{smallmatrix}\right)$, $r > 0$. If $f(x+iy)$ is bounded by a polynomial in y as $y \to \infty$, we say that f is *of moderate growth at* ∞, and if $f(x+iy) \ll y^{-N}$ for all $N > 0$, we say that f is of *rapid decay* at ∞. If either $\chi(\tau_r) \neq 1$ or

$$\int_0^r f(z+u)\, du = 0$$

for all $z \in \mathcal{H}$, then we say that f is *cuspidal* at ∞. More generally, if $a \in \mathbb{R} \cup \{\infty\}$ is a cusp of Γ, let $\xi \in SL(2,\mathbb{R})$ be such that $\xi(\infty) = a$. Then $f' = f|_k\xi$ is an element of $C^\infty(\Gamma'\backslash\mathcal{H}, \chi', k)$, where $\Gamma' = \xi^{-1}\Gamma\xi$ and $\chi'(\gamma) = \chi(\xi\gamma\xi^{-1})$ for $\gamma \in \Gamma'$. Now ∞ is a cusp of Γ', and we say that f is *of moderate growth at a* or is *cuspidal at a* if f' satisfies the same condition at ∞, these notions already being defined in this case. If f is cuspidal at every cusp, it is called a *cusp form*.

We recall the *Maass operators*, introduced in Eqs. (1.1) and (1.2) of Chapter 2:

$$R_k = iy\frac{\partial}{\partial x} + y\frac{\partial}{\partial y} + \frac{k}{2} = (z - \bar{z})\frac{\partial}{\partial z} + \frac{k}{2}, \tag{2.11}$$

which raises the weight by two, and

$$L_k = -iy\frac{\partial}{\partial x} + y\frac{\partial}{\partial y} - \frac{k}{2} = -(z-\overline{z})\frac{\partial}{\partial \overline{z}} - \frac{k}{2}, \tag{2.12}$$

which lowers the weight by two.

Actually Maass forms with this definition include the holomorphic modular forms, because if f is a modular form of weight $k > 0$, then

$$y^{k/2}\, f(z) \tag{2.13}$$

is a Maass form of weight k (by our definition) with the eigenvalue $\lambda = \frac{k}{2}\left(1-\frac{k}{2}\right)$ of Δ_k. (See Exercise 2.1.7.) In this case, the lowering operator L_k annihilates $y^{k/2}\, f(z)$, and this is equivalent to the Cauchy–Riemann differential equations satisfied by the holomorphic function f.

We may now relate classical modular forms and Maass forms to the $(\mathfrak{g}, K)$-modules of Theorem 3.2.1. In view of Eq. (2.13), we may restrict ourselves to Maass forms of weight k. Let f be a Maass form of weight k. We associate with f a function F on G by

$$F(g) = (f|_k g)(i). \tag{2.14}$$

Then it is easy to see (cf. Proposition 2.1.8) that $F \in C^\infty(\Gamma\backslash G, \chi, \omega)$, where ω is the character of $Z(\mathbb{R}) \cong \mathbb{R}^\times$ that is trivial on the connected component of the identity and agrees with χ on $-I$. Because f is an eigenfunction of Δ_k, it follows from Eq. (1.30) in Chapter 2 that F is eigenfunction of Δ. In particular, it is $\mathcal{Z}$-finite. The function F is K-finite; in fact it satisfies

$$F(g\kappa_\theta) = e^{2\pi i k}\, F(g), \qquad \kappa_\theta = \begin{pmatrix} \cos\theta & \sin\theta \\ -\sin\theta & \cos\theta \end{pmatrix} \tag{2.15}$$

(Exercise 2.1.6). We have given two distinct definitions of moderate growth for f and for F, but it is easy to see that if f and F are related by Eq. (2.14), then these notions are equivalent (Exercise 3.2.2). Therefore $F \in \mathcal{A}(\Gamma\backslash G, \chi, \omega)$. By Theorem 3.2.1, F generates an admissible $(\mathfrak{g}, K)$-module. The effect of the Maass operators R_k and L_k (defined in Eqs. (2.11) and (2.12)) is then seen to arise from the action of R and L (defined in Eq. (2.2)) in the action (Eq. 2.1) of $\mathfrak{g}_\mathbb{C}$ on automorphic forms.

In addition to the spaces $\mathcal{A}(\Gamma\backslash G, \chi, \omega)$ and $\mathcal{A}_0(\Gamma\backslash G, \chi, \omega)$, we will also consider $L^2(\Gamma\backslash G, \chi, \omega)$, the space of square integrable functions satisfying Eqs. (2.4) and (2.5), and its subspace $L_0^2(\Gamma\backslash G, \chi, \omega)$ consisting of cuspidal elements. Here *cuspidality* is defined the same way as for $\mathcal{A}(\Gamma\backslash G, \chi, \omega)$ except that Eq. (2.7) is interpreted in the sense of equality for almost all g.

Let $\phi \in C_c^\infty(G)$. If $f \in L^2(\Gamma\backslash G, \chi, \omega)$, we define

$$\big(\rho(\phi)\, f\big)(g) = \int_G f(gh)\,\phi(h)\, dh. \tag{2.16}$$

(Compare Eq. (3.4) in Chapter 2.) Then $\rho(\phi)$ is an operator on $L^2(\Gamma\backslash G, \chi, \omega)$, leaving $L_0^2(\Gamma\backslash G, \chi, \omega)$ invariant, and ρ is a unitary representation on $L^2(\Gamma\backslash G, \chi, \omega)$ and on $L_0^2(\Gamma\backslash G, \chi, \omega)$ (Exercise 3.2.1). We can rewrite Eq. (2.16) as follows: Let

$$\phi_\omega(g) = \int_{\mathbb{R}^\times} \phi\left(\begin{pmatrix} z & \\ & z \end{pmatrix} g\right) \omega(z)\, dz. \tag{2.17}$$

Then

$$\big(\rho(\phi)\, f\big)(g) = \int_{Z(\mathbb{R})\backslash G} f(gh)\, \phi_\omega(h)\, dh. \tag{2.18}$$

If X is a locally compact Hausdorff space, $C(X)$ denotes the ring of continuous functions on X with the L^∞ norm. It is a closed subspace of $L^\infty(X)$. If Σ is a subset of $C(X)$, then Σ is called *equicontinuous* if for any $\epsilon > 0$ and for any $x \in X$ there exists a neighborhood N of x such that $|f(y) - f(x)| < \epsilon$ for all $y \in N$, $f \in \Sigma$.

Proposition 3.2.1: The Ascoli–Arzéla lemma *Let Y be a compact Hausdorff space, and let $\Sigma \subset C(Y)$ be an equicontinuous set that is bounded in the L^∞ norm. Then the closure of Σ in $C(Y)$ is compact.*

Proof See Rudin (1991, Appendix A5) for the proof. ■

In Proposition 3.2.3, we will encounter *Siegel sets,* which are nicely shaped substitutes for a fundamental domain. Let c and d be positive constants. We will denote by $\mathcal{F}_{c,d}$ the the *Siegel set* of $z = x + iy \in \mathcal{H}$ such that $0 \le x \le d$ and $y \ge c$. We will also denote by $\mathcal{F}_d^\infty$ the set of z such that $0 \le x \le d$ (with no condition on y).

Proposition 3.2.2 *(i) Let $a_1, \cdots, a_h \in \mathbb{R} \cup \{\infty\}$ be representatives of the Γ-orbits of cusps of Γ, and let $\xi_i \in SL(2, \mathbb{R})$ be chosen such that $\xi_i(a_i) = \infty$. If $c > 0$ and $d > 0$ are chosen suitably, the set*

$$\bigcup \xi_i^{-1} \mathcal{F}_{c,d}$$

contains a fundamental domain for Γ.
(ii) Suppose that ∞ is a cusp of Γ. Then if d is large enough, $\mathcal{F}_d^\infty$ contains a fundamental domain for Γ.

The region in (i) has finite volume and can be covered by a finite number of copies of a fundamental domain; the region in (ii) has infinite volume.

Proof First let c be arbitrary. For each $1 \le i \le h$, $\xi_i \Gamma \xi_i^{-1}$ contains a unipotent subgroup generated by an element $\left(\begin{smallmatrix} 1 & \delta_i \\ & 1 \end{smallmatrix}\right)$. If $d \ge \delta_i$, then $\xi_i^{-1}\mathcal{F}_{c,d}$ contains a neighborhood of the cusp a_i in a fundamental domain F of Γ, and so

$F - \bigcup \xi_i^{-1}\mathcal{F}_{c,d}$ is relatively compact in $\mathcal{H}$. We may therefore increase d and decrease c until one of the components $\xi_i^{-1}\mathcal{F}_{c,d}$ swallows up this relatively compact set. Then we have obtained (i). As for (ii), we may assume that $a_1 = \infty$ and $\xi_1 = 1$. Then if d' is large enough, $\mathcal{F}_{d'}^{\infty}$ will contain each of the pieces $\xi_i^{-1}\mathcal{F}_{c,d}$ in (i). Thus we obtain (ii). ■

We have a similar decomposition of the group G: Let us map $G \to \mathcal{H}$ by $\left(\begin{smallmatrix} a & b \\ c & d \end{smallmatrix}\right) \mapsto \frac{ai+b}{ci+d}$; let $\mathcal{G}_{c,d}$ and $\mathcal{G}_d^{\infty}$ be the respective preimages of $\mathcal{F}_{c,d}$ and $\mathcal{F}_d^{\infty}$ in G. If the ξ_i, c, and d are chosen as in Proposition 4.2.2(i), then

$$\bigcup \xi_i^{-1}\mathcal{G}_{c,d}$$

contains a fundamental domain for $\Gamma\backslash G$, and similarly if d is large enough, $\mathcal{G}_d^{\infty}$ contains a fundamental domain. We will also let $\overline{G} = G/Z(\mathbb{R})$ and let $\overline{\mathcal{G}}_{c,d}$ and $\overline{\mathcal{G}}_d^{\infty}$ denote the images of $\mathcal{G}_{c,d}$ and $\mathcal{G}_d^{\infty}$ in $\overline{G}$.

Proposition 3.2.3: Gelfand, Graev and Piatetski-Shapiro *Let $\phi \in C_c^{\infty}(G)$. (i) There exists a constant C depending on ϕ such that for all $f \in L_0^2(\Gamma\backslash G, \chi, \omega)$, we have*

$$\sup_{g \in G} |\rho(\phi)\, f(g)| \leq C\, ||f||_2. \tag{2.19}$$

where

$$||f||_2 = \sqrt{\int_{G/Z(\mathbb{R})} |f(g)|^2\, dg}.$$

(The integrand is well defined because ω is unitary.)
(ii) The restriction of the operator $\rho(\phi)$ to $L_0^2(\Gamma\backslash G, \chi, \omega)$ is a compact operator.

This generalizes Proposition 2.3.1(i). See Gelfand, Graev and Piatetski-Shapiro (1968), Section 6 of Chapter I and Section 4.6 of Chapter III; Godement (1965, 1966a, 1966b and 1970), Langlands (1976, Section 3), and Lang (1975). The following proof is taken from Godement (1966b). Lang (1975) gives an instructive alternative to the Ascoli–Arzela Lemma, proving that the operators $\rho(\phi)$ are Hilbert–Schmidt directly from (i). This implies that they are of trace class (a notion we have not defined), which is also proved by Gelfand, Graev and Piatetski-Shapiro (1968).

Proof Recall that we are assuming that Γ has cusps. Conjugating Γ by an element of $SL(2, \mathbb{R})$ if necessary, we may assume that ∞ is a cusp of Γ and that Γ contains the group

$$\Gamma_{\infty} = \left\{ \begin{pmatrix} 1 & n \\ & 1 \end{pmatrix} \middle| n \in \mathbb{Z} \right\}.$$

We will prove that if $c, d > 0$, then for some constant $C_0 > 0$

$$\sup_{g \in \mathcal{G}_{c,d}} |\rho(\phi) f(g)| \leq C_0 \, ||f||_2. \tag{2.20}$$

We note that this is sufficient to prove (i), because in the general case if $\xi_1, \cdots, \xi_h \in SL(2, \mathbb{R})$ are chosen as in Proposition 3.2.2, then we have a similar estimate on each of the pieces $\xi_i^{-1}\mathcal{G}_{c,d}$, by which we get Eq. (2.19).

We observe that

$$\big(\rho(\phi) f\big)(g) = \int_{Z(\mathbb{R})\backslash G} f(gh) \, \phi_\omega(h) \, dh = \int_{Z(\mathbb{R})\backslash G} f(h) \, \phi_\omega(g^{-1}h) \, dh$$

$$= \int_{\Gamma_\infty \, Z(\mathbb{R})\backslash G} \sum_{\gamma \in \Gamma_\infty} f(\gamma h) \, \phi_\omega(g^{-1}\gamma h) \, dh.$$

Then because $f(\gamma h) = \chi(\gamma) \, f(h)$, we have

$$\big(\rho(\phi) f\big)(g) = \int_{\Gamma_\infty \, Z(\mathbb{R})\backslash G} K(g, h) \, f(h) \, dh, \tag{2.21}$$

where

$$K(g, h) = \sum_{\gamma \in \Gamma_\infty} \chi(\gamma) \, \phi_\omega(g^{-1}\gamma h). \tag{2.22}$$

Now there are two cases, depending on whether or not the restriction of χ to Γ_∞ is trivial. First assume that it is trivial. Then we may argue as follows. Because f is cuspidal, we have

$$\int_{\Gamma_\infty \, Z(\mathbb{R})\backslash G} K_0(g, h) \, f(h) \, dh = 0,$$

where we define

$$K_0(g, h) = \int_{-\infty}^{\infty} \phi_\omega\left(g^{-1} \begin{pmatrix} 1 & x \\ & 1 \end{pmatrix} h\right) \, dx.$$

Thus if f is a cusp form we may write

$$\big(\rho(\phi_\omega) f\big)(g) = \int_{\Gamma_\infty \, Z(\mathbb{R})\backslash G} K'(g, h) \, f(h) \, dh, \tag{2.23}$$

where $K'(g, h) = K(g, h) - K_0(g, h)$. We can estimate $K'(g, h)$ as follows.

Let us write

$$g = \begin{pmatrix} \eta & \\ & \eta \end{pmatrix} \begin{pmatrix} y & x \\ & 1 \end{pmatrix} \kappa_\theta, \qquad h = \begin{pmatrix} \zeta & \\ & \zeta \end{pmatrix} \begin{pmatrix} v & u \\ & 1 \end{pmatrix} \kappa_\sigma, \tag{2.24}$$

where $y, v, \eta, \zeta > 0$, and

$$\kappa_\theta = \begin{pmatrix} \cos\theta & \sin\theta \\ -\sin\theta & \cos\theta \end{pmatrix}.$$

We have

$$K(g,h) = \sum_{n\in\mathbb{Z}} \Phi_{g,h}(n), \qquad \Phi_{g,h}(t) = \phi_\omega\left(g^{-1}\begin{pmatrix} 1 & t \\ & 1 \end{pmatrix} h\right).$$

We may apply the Poisson summation formula ((1.8) in Chapter 1) to obtain

$$K(g,h) = \sum_{n\in\mathbb{Z}} \hat{\Phi}_{g,h}(n),$$

where the Fourier transform

$$\hat{\Phi}_{g,h}(n) = \int_{-\infty}^{\infty} \phi_\omega\left(g^{-1}\begin{pmatrix} 1 & t \\ & 1 \end{pmatrix} h\right) e^{2\pi i n t}\, dt. \tag{2.25}$$

We note that $K_0(g,h) = \hat{\Phi}_{g,h}(0)$, so if we subtract this we obtain

$$K'(g,h) = \sum_{n\neq 0} \hat{\Phi}_{g,h}(n).$$

The key is therefore to estimate Eq. (2.25) when $n \neq 0$ and g is restricted to $\mathcal{G}_{c,d}$. First we note that the change of variables in $t \to t + x - u$ shows that the absolute value of Eq. (2.25) is independent of x and u. After a further change of variables $t \to yt$, we obtain

$$|\hat{\Phi}_{g,h}(n)| = |y|\,|\hat{F}_{\theta,\sigma,y^{-1}v}(yn)|, \tag{2.26}$$

where

$$F_{\theta,\sigma,v}(t) = \phi_\omega\left(\kappa_\theta^{-1}\begin{pmatrix} 1 & t \\ & 1 \end{pmatrix}\begin{pmatrix} v & \\ & 1 \end{pmatrix}\kappa_\sigma\right).$$

We now recall a familiar fact about Fourier transforms: If $F \in C_c^\infty(\mathbb{R})$, the Fourier transform $\hat{F}(y)$ (defined as in Eq. (1.8) in Chapter 1) decays faster than any polynomial as $|y| \to \infty$. To see this, let N be a large integer, and integrate the Fourier transform by parts N times:

$$\widehat{F}(y) = \frac{(-1)^N}{(2\pi i y)^N} \int_{-\infty}^{\infty} F^{(N)}(t)\, e^{2\pi i y t}\, dt,$$

so

$$|\widehat{F}(y)| \leq \frac{1}{(2\pi |y|)^N} \int_{-\infty}^{\infty} |F^{(N)}(t)|\, dt.$$

It follows that there exist constants $C_{\theta,\sigma,v} > 0$ such that

$$|\widehat{F}_{\theta,\sigma,v}(y)| \le C_{\theta,\sigma,v}|y|^{-N},$$

and these constants vary continuously in θ, σ, and v. Now we observe that because ϕ_ω is compactly supported modulo $Z(\mathbb{R})$, it follows from Exercise 3.2.3 that there exists a real constant $B > 1$ such that $F_{\theta,\sigma,v}(t) = 0$ unless $B^{-1} \le v \le B$. Thus in Eq. (2.26), the contribution is zero unless the parameters θ, σ, and $y^{-1}v$ are restricted to a compact set, and hence there exists an absolute constant $C_1 > 0$ such that

$$|\hat{\Phi}_{g,h}(yn)| < C_1\,|y|^{1-N}\,|n|^{-N},$$

and furthermore $\Phi_{g,h}(yn) = 0$ unless $-B \le y^{-1}v \le B$. Consequently, if $N \ge 2$ so the series $\sum_n n^{-N}$ is convergent, we have proved that there exists a constant C_2 depending on ϕ and on N such that

$$|K'(g,h)| \le C_2\,|y|^{1-N} \qquad \text{if } g \in \mathcal{G}_{c,d}. \tag{2.27}$$

Now let us estimate Eq. (2.23) when $g \in \mathcal{G}_{c,d}$. By Proposition 3.2.2(ii), we may replace the domain of integration by $\mathcal{G}_d$ (increasing d if necessary), but we note that the kernel vanishes unless $B^{-1} \le y^{-1}v$, which, if $g \in \mathcal{G}_{c,d}$ so that $c \le y$, implies that $B^{-1}c \le v$. Thus Eq. (2.23) is dominated by

$$C_2\,y^{1-N} \int_{\overline{\mathcal{G}}_{B^{-1}c,d}} |f(h)|\,dh.$$

Now $\overline{\mathcal{G}}_{B^{-1}c,d}$ has finite volume and can be covered by a finite number of copies of a fundamental domain, so this is in turn dominated by the L^1 norm of f, and because the fundamental domain has finite volume, the L^1 norm is dominated by the L^2 norm, by the Cauchy–Schwarz inequality. We have therefore proved Eq. (2.20). As a byproduct, we have found that $\rho(\phi)\,f(g)$ decays more rapidly than any polynomial in y as $y \to \infty$ (Exercise 3.2.5).

We have been assuming that the restriction of χ to Γ_∞ is trivial. If this is not the case, we may still estimate $K(g,h)$ by the same method – it is not necessary to subtract $K_0(g,h)$. We leave the details of this case to the reader.

Regarding the compactness of $\rho(\phi)$, we apply the Ascoli–Arzéla lemma (Proposition 4.2.1). Let Y be the space obtained by compactifying $\Gamma\,Z(\mathbb{R})\backslash G$ by adjoining h cusps. Let Σ be the image of the unit ball in $L^2_0(\Gamma\backslash G,\chi,\omega)$ under $\rho(\phi)$; we extend each $\rho(\phi)\,f$ to Y by making it vanish at the cusps. The set Σ is bounded in the L^∞ norm by Eq. (2.19). We wish to show that Σ is equicontinuous. For this, it is sufficient to bound the derivative

$$\big(X\,\rho(\phi)\,f\big)(g) = \frac{d}{dt}\,\rho(\phi)\,f\big(g\,\exp(tX)\big)\big|_{t=0}$$

uniformly for all f with $||f||_2 < 1$, where X is an element of the Lie algebra $\mathfrak{g}$ of G. It now follows easily from Eq. (2.21) that

$$\big(X\,\rho(\phi)\,f\big)(g) = \rho(\phi_X)\,f(g), \tag{2.28}$$

where

$$\phi_X(g) = \frac{d}{dt}\phi\big(\exp(-tX)\,g\big)\Big|_{t=0}, \tag{2.29}$$

so the same estimates that we found for $\rho(\phi)\,f$ apply also to its derivatives. The family Σ is therefore equicontinuous, and Σ is therefore compact in the L^∞ topology and hence also in the L^2 topology. ■

Theorem 3.2.2 *The space $L^2_0(\Gamma\backslash G, \chi, \omega)$ decomposes into a Hilbert space direct sum of subspaces that are invariant and irreducible under the right regular representation ρ. Let H be such a subspace. The K-finite vectors H_{fin} in H are dense, and every K-finite vector is automatically an element of $C^\infty(\Gamma\backslash G, \chi, \omega)$. The K-finite vectors form an irreducible admissible $(\mathfrak{g}, K)$-module contained in $\mathcal{A}_0(\Gamma\backslash G, \chi, \omega)$.*

Proof The proof that $L^2_0(\Gamma\backslash G, \chi, \omega)$ decomposes into a Hilbert space direct sum of irreducible invariant subspaces is proved identically to Theorem 2.3.3, except that instead of Proposition 2.3.1 one uses Proposition 3.2.4. This proof is left to the reader. Let H be one of these subspaces. Then by Exercise 3.1.5, H decomposes as a Hilbert space direct sum of subspaces H_k, where $\rho(\kappa_\theta)\,f = e^{ik\theta}\,f$ for $f \in H_k$. It follows from Theorem 2.4.3 that each of the spaces H_k is at most one dimensional, so the $(\mathfrak{g}, K)$-module of K-finite vectors is admissible. Finally, we need to show that $H_k \subset \mathcal{A}_0(\Gamma\backslash G, \chi, \omega)$. Let $0 \neq f \in H_k$. By Lemma 2.3.2, there exists $\phi \in C_c^\infty(G)$ such that $\phi(\kappa_\theta g) = \phi(g\kappa_\theta) = e^{-ik\theta}\,\phi(g)$, and $\rho(\phi)\,f \neq 0$. By Proposition 2.3.1, $\rho(\phi)\,f \in H_k$, and because H_k is one dimensional, we may assume without loss of generality that $\rho(\phi)\,f = f$. Now $\rho(\phi)\,f$ is smooth and rapidly decreasing by Exercise 3.2.5 and hence of moderate growth. Thus $f = \rho(\phi)\,f \in \mathcal{A}_0(\Gamma\backslash G, \chi, \omega)$. ■

Harish-Chandra (1968) proved an important finiteness result that is the key to the admissibility of automorphic representations (cf. Theorem 3.3.4 below). Let J be an ideal of finite codimension in $\mathcal{Z}$, and let σ be a character of K. (Because we are specializing to the case where $G = GL(2, \mathbb{R})^+$, irreducible representations of K are one dimensional.) For example, we could consider the ideal generated by $Z - \mu$ and $\Delta - \lambda$, where μ and λ are complex numbers. We may consider the space of automorphic forms in $\mathcal{A}(\Gamma, \chi, \omega)$ that are annihilated by J and, denoting this space by $\mathcal{A}(\Gamma, \chi, \omega, J)$, we may then consider the σ-isotypic component $\mathcal{A}(\Gamma, \chi, \omega, J, \sigma)$. Harish-Chandra proved that this space is finite dimensional. In addition to the proof in Harish-Chandra (1968) and the discussion in Borel and Jacquet (1979), a proof of this result may be found in Borel (to appear). For our purposes, a simpler result is sufficient. Let λ be a complex number, and let $\mathcal{A}(\Gamma, \chi, \omega, \lambda)$ denote the λ-eigenspace of Δ in $\mathcal{A}(\Gamma, \chi, \omega, \lambda)$, and let $\mathcal{A}(\Gamma, \chi, \omega, \lambda, \sigma)$ be the σ-isotypic part of $\mathcal{A}(\Gamma, \chi, \omega, \lambda, \rho)$. We will denote the cuspidal subspaces of each of these by subscripting with 0. Thus

$$\mathcal{A}_0(\Gamma, \chi, \omega, \lambda, \rho) = \mathcal{A}(\Gamma, \chi, \omega, \lambda, \rho) \cap \mathcal{A}_0(\Gamma, \chi, \omega).$$

Theorem 3.2.3 *(i) Let (π, V) be an irreducible admissible unitary representation of $GL(2, \mathbb{R})$. Then the multiplicity of π in the decomposition of $L_0^2(\Gamma\backslash G, \chi, \omega)$ is finite.*
(ii) Let $\lambda \in \mathbb{C}$ and let σ be a character of K. Then $\mathcal{A}_0(\Gamma, \chi, \omega, \lambda, \rho)$ is finite dimensional.

Proof We prove (i). Let σ be a character of K such that $V(\sigma)$ is nonzero, and let $0 \neq \xi \in V(\sigma)$. By Lemma 2.3.2, there exists a $\phi \in C_c^\infty(G)$ such that $\pi(\phi)\xi = \xi$ and such that $\pi(\phi)$ is self-adjoint. Now let us consider the image of ξ under a continuous linear map $T : V \to L_0^2(\Gamma\backslash G, \chi, \omega)$. We have $\rho(\phi)T\xi = T\xi$, so $T\xi$ lies in the 1-eigenspace of the compact operator $\rho(\phi)$, which is finite dimensional by the spectral Theorem 2.3.1 for compact operators. Of course the embedding T is determined by the image of any nonzero vector because π is irreducible. It follows that the space of continuous linear maps $V \to L_0^2(\Gamma\backslash G, \chi, \omega)$ is finite dimensional.

The second part follows from the first. Given the eigenvalue λ and the central character ω, by Exercise 2.6.4 there are only a finite number of isomorphism classes of unitary irreducible admissible representations with central character ω on which Δ acts by λ. (The eigenvalue of Z is determined by ω.) Let Σ be the set of these isomorphism classes. Because we know already that $L_0^2(\Gamma\backslash G, \chi, \omega)$ decomposes into a direct sum of irreducible admissible representations and that the K-finite vectors in each of these are elements of $\mathcal{A}_0(\Gamma\backslash G, \chi, \omega)$, it is clear that $\mathcal{A}_0(\Gamma, \chi, \omega, \lambda, \rho)$ is just the direct sum of the ρ-isotypic components of the irreducible subspaces of $L_0^2(\Gamma\backslash G, \chi, \omega)$ that are isomorphic to an element of Σ. Thus the finite dimensionality follows from the finiteness of Σ together with part (i). ■

Exercises

Exercise 3.2.1 (a) With notations as in this section, show that $C(\Gamma\backslash G, \chi, \omega)$ is dense in $L^2(\Gamma\backslash G, \chi, \omega)$. [HINT: See Proposition 2.1.6.]
(b) Prove that the action ρ on $L^2(\Gamma\backslash G, \chi, \omega)$ is continuous. [HINT: See Proposition 2.1.7.]

Exercise 3.2.2 Verify that the two notions of moderate growth are equivalent when f and F are related by Eq. (2.13). [HINT: Use Proposition 3.2.2.]

Exercise 3.2.3 Let Ω be a compact subgroup of $GL(2,\mathbb{R})$. Prove that there exists a compact subset Ω' of the Borel subgroup $B(\mathbb{R})$ of upper triangular matrices in $GL(2,\mathbb{R})$ such that if $k_1bk_2 \in \Omega$, where $k_1, k_2 \in SO(2)$ and $b \in B(\mathbb{R})$, then $b \in \Omega'$.

Exercise 3.2.4 Complete the proof of Theorem 3.2.2 along the lines of Theorem 2.3.3.

Exercise 3.2.5 Suppose that Γ is a discrete subgroup of $SL(2,\mathbb{R})$ such that $\Gamma\backslash\mathcal{H}$ has finite volume; let χ be a character of Γ, and let ω be a character of $Z(\mathbb{R})^+$. We now define the notion of *rapid decrease at the cusps* of Γ. Suppose first that ∞ is a cusp of Γ. We say that a function $f \in C^\infty(\Gamma\backslash G,\chi,\omega)$ is *rapidly decreasing at* ∞ if for every $N > 0$ there exists a constant $C(N)$ such that for all $g \in \mathcal{G}_{c,d}$ we have $f(g) \leq C(N)\,y^{-N}$, where y is as in Eq. (2.24). More generally, if a is a cusp of Γ, let $\xi \in SL(2,\mathbb{R})$ such that $\xi(\infty) = a$; we say that f is *rapidly decreasing at* a if $f|_\xi$ is rapidly decreasing at ∞. Finally, we say that f is *rapidly decreasing* if it is rapidly decreasing at every cusp.
(a) Prove that if $f \in L^2_0(\Gamma\backslash G,\chi,\omega)$ and $\phi \in C_c^\infty(\Gamma\backslash G,\chi,\omega)$, then $\rho(\phi)f$ is smooth and rapidly decreasing. [HINT: For smoothness, use Eq. (2.28). For rapid decrease, use Eq. (2.27).]
(b) Let Δ be the Laplace–Beltrami operator, defined by Eq. (1.29) in Chapter 2. Show that Δ is a symmetric operator on the space of K-finite elements in $C_c^\infty(\Gamma\backslash G,\chi,\omega)$. That is, if f_1 and $f_2 \in C_c^\infty(\Gamma\backslash G,\chi,\omega)$ are K-finite, then prove that $\langle\Delta f_1, f_2\rangle = \langle f_1, \Delta f_2\rangle$. [HINT: Reduce to the case where f_1 and f_2 both satisfy Eq. (2.15) for the same k. In this case, f_1 and f_2 correspond to Maass forms of weight k and so imitate the proof of Proposition 2.1.4. The hypothesis that f_1 and f_2 are K-finite is unnecessary and is imposed only for reasons of minor convenience.]
(c) State and prove an analog of the corollary to Theorem 2.3.4 for $L^2_0(\Gamma\backslash\mathcal{H},\chi,k)$ when Γ is allowed to have cusps.

Exercise 3.2.6 Prove Theorem 3.2.1 for $\mathcal{A}_0(\Gamma\backslash G,\chi,\omega)$ without making use of Harish-Chandra's Theorem 2.9.2. [HINT: Use Theorem 3.2.2.]

3.3 Automorphic Representations of $GL(n)$

Let F be a number field and A its adele ring. (The case where F is a function field requires only minor modifications.) We will define the notion of an *automorphic representation of* $GL(n,A)$. Let us state at the outset that for us, an "automorphic representation" of $GL(n,A)$ is not really a representation at all – rather it is a representation of $GL(n,A_f)$, where A_f is the ring of finite adeles, which is simultaneously a $(\mathfrak{g}_\infty, K_\infty)$-module, where $\mathfrak{g}_\infty = \prod_{v\in S_\infty}\mathfrak{gl}(n,F_v)$, $K_\infty = \prod_{v\in S_\infty} K_v$, with K_v as in Eq. (3.2) below, where S_∞ is the set of Archimedean places of F. There is no action of $GL(n,A)$ *per se*. In Section 3.5, we will give examples showing how automorphic representations can be associated with classical objects – modular forms, or Maass forms.

It will be our policy in this section to give statements that are valid for $GL(n)$, but to give proofs only for $GL(2)$. Moreover, when convenient we may specialize to the case where $F = \mathbb{Q}$, though the case of a general number field is not much more difficult. For more information, the reader should consult Borel and Jacquet (1979), and Moeglin and Waldspurger (1995).

By an *affine algebraic variety X defined over a field k* (or just an *affine variety*), we mean the locus of some set of polynomial equations with coefficients in k. If R is a commutative ring containing k, we will denote by $X(R)$ or X_R the points of X (i.e., solutions to the given polynomial equations) with coordinates in R. The simplest example of an affine variety is of course the affine n-space $\mathbb{A}^n$, where $\mathbb{A}^n(R) = R^n$; the set of polynomial equations determining this variety may be taken to be the empty set.

Suppose that $f(x_1, \cdots, x_n)$ is a polynomial with coefficients in k. Of course, f defines an affine variety $H \subset \mathbb{A}^n$; indeed, $H(R) \subset R^n$ is the set of solutions to $f(x_1, \cdots, x_n) = 0$. Somewhat more subtle is the fact that (roughly speaking) $\mathbb{A}^n - H$ may be given the structure of an affine algebraic variety. That is, there exists an affine algebraic variety X such that *if R is a field* then $X(R)$ can be identified with $R^n - H(R)$ for any R. The variety X is a subvariety of $\mathbb{A}^{n+1}$; it is the locus of the equation

$$x_{n+1} f(x_1, \cdots, x_n) = 1. \tag{3.1}$$

Obviously, this is an affine algebraic variety, and projection onto the first n coordinates gives a bijection of $X(R)$ with $R^n - H(R)$.

If the commutative ring R is not a field, then $X(R)$ is *not* the same as $\mathbb{A}^n(R) - H(R)$: Instead, it is (alternatively) the set of the set of $(r_1, \cdots, r_n, r_{n+1}) \in R^{n+1}$ that are solutions to Eq. (3.1), or (equivalently) the set of $(r_1, \cdots, r_n) \in R^n$ such that $f(r_1, \cdots, r_n)$ is a *unit* in R. If R is a topological ring such as the adeles, we topologize $X(R)$ as a subset of R^{n+1}. In general, this gives a different topology than the subset topology in R^n. Thus, for example, the group $A^\times$ does not have the subspace topology inherited from the embedding in the adele ring A. Instead, it has the subspace topology inherited from embedding in A^2 via $a \mapsto (a, a^{-1})$.

By an *affine algebraic group* defined over a field k, we mean an affine algebraic variety G defined over k and endowed with a special point $1 \in G(k)$ and a multiplication law $G \times G \to G$ which is given by a polynomial map and which satisfies the appropriate axioms so that $G(R)$ becomes a group for any ring R containing k.

For example, $GL(n)$ is an affine algebraic group: It is the complement of the determinant locus in $\mathrm{Mat}_n = \mathbb{A}^{n^2}$. However, for topological purposes, it should be embedded into $\mathbb{A}^{n^2+1}$, as we have explained above, via $g \mapsto \left(g, \det(g)^{-1}\right)$.

We recall the definition of a *restricted direct product*. Let Σ be some indexing set, and for each $v \in \Sigma$, let there be given a group G_v, and for almost all $v \in \Sigma$, let there be given a subgroup K_v of G_v. Then the restricted direct product of the G_v with respect to the K_v is

$$G = \left\{(a_v)_{v\in\Sigma} \in \prod_v G_v \middle| a_v \in K_v \text{ for almost all } v \in \Sigma\right\}.$$

As always *almost all* when applied to a set of places means *all but finitely many*. If the G_v are locally compact topological groups and the groups K_v are compact, then we may give G the structure of a locally compact topological group as follows. We take as a base of neighborhoods of the identity the products $U = \prod_v U_v$, where each U_v is an open relatively compact neighborhood of of the identity in G_v, and $U_v = K_v$ for almost all v. By the Tychanoff theorem (Kelley (1955), p. 143), such a set U is relatively compact.

The first example of a restricted direct product is of course the adele ring. Let F be a global field, and for each v in the set Σ of places of F, let F_v be the completion of F at v, and if v is non-Archimedean, let $\mathfrak{o}_v$ be the ring of integers in F_v. The adele ring A of F is the restricted direct product of the F_v with respect to the $\mathfrak{o}_v$. The ideles $A^\times$ are the restricted direct product of the $F_v^\times$ with respect to the $\mathfrak{o}_v^\times$.

The group $GL(n, A)$ may be thought of as the restricted direct product of the groups $GL(n, F_v)$ with respect to the groups $GL(n, \mathfrak{o}_v)$, which are maximal compact subgroups for the non-Archimedean v. To see this, if $g = (g_{ij}) \in GL(n, A)$, each matrix entry is an adele, so writing $g_{ij}(v)$, etc., for the v component of the matrix entry g_{ij}, we associate with g the matrices $g_v = \big(g_{ij}(v)\big) \in GL(n, F_v)$. Thus to specify an element of $GL(n, A)$, it is enough to specify $g_v \in GL(n, F_v)$ such that $g_v \in GL(n, \mathfrak{o}_v)$ for all but finitely many places $\mathfrak{o}_v$. Although $GL(n, A) \subset \mathrm{Mat}_n(A)$, it does not have the subspace topology.

The group $GL(n, A)$ is *unimodular*, that is, the left and right invariant Haar measures coincide (Exercise 3.3.3). The group $GL(n, F)$ is a discrete subgroup; this follows from the fact that F is discrete inside of A. On the other hand, it may be shown that $Z(A)\,GL(n, F)\backslash GL(n, A)$ has finite volume. (Recall that Z is the center of $GL(n)$ consisting of scalar matrices.) We will demonstrate this in Proposition 3.3.2 when $n = 2$ and $F = \mathbb{Q}$. First we need the "strong approximation theorem."

Let A_f be the ring of "finite adeles," that is, those adeles (a_v) such that $a_v = 1$ at every Archimedean place v. Let $F_\infty = \prod_{v\in S_\infty} F_v$, where S_∞ is the finite set of Archimedean places of F. We embed F_∞ in A by mapping $(a_v)_{v\in S_\infty}$ to the adele that matches a_v at every infinite place and that is 1 at every finite place. Then $A = F_\infty A_\mathrm{f}$, and if G is an affine algebraic group defined over F, we have $G(A) = G(F_\infty)\,G(A_\mathrm{f})$.

Theorem 3.3.1: Strong approximation theorem *Assume that F is an algebraic number field.*
(i) $SL(n, F_\infty)\,SL(n, F)$ is dense in $SL(n, A)$.
(ii) Let K_0 be a open compact subgroup of $GL(n, A_\mathrm{f})$. Assume that the image

of K_0 *in* $A_{\mathrm{f}}^\times$ *under the determinant map is* $\prod_{v \notin S_\infty} \mathfrak{o}_v^\times$. *Then the cardinality of*

$$GL(n, F)\, GL(n, F_\infty)\backslash GL(n, A)/K_0$$

is equal to the class number of F.

Proof We refer to Humphreys (1980) for the proof of (i). (See also Garrett (1990, Appendix A.3).) Humphreys deduces (i) from (ii) in the case $K_0 = \prod_{v \notin S_\infty} GL(n, \mathfrak{o}_v)$. To deduce the general case of (ii) from (i), use the exact sequence

$$1 \to SL(n, A) \to GL(n, A) \to A^\times \to 1$$

to reduce the problem to the corresponding problem on $A^\times$, that is, (ii) with $n = 1$. Owing to our assumption on the surjectivity of the determinant map, this case coincides with the case proved by Humphreys.

For results on other groups, see Kneser (1966), and the references therein. ∎

We will often use the following fixed compact subgroup of $GL(n, A)$:

$$K = \prod_v K_v, \quad K_v = \begin{cases} O(n) & \text{if } v \text{ is a real place;} \\ U(n) & \text{if } v \text{ is a complex place;} \\ GL(n, \mathfrak{o}_v) & \text{if } v \text{ is a non-Archimedean place.} \end{cases} \tag{3.2}$$

This subgroup is compact by the Tychanoff theorem (Kelley (1955), p. 143). It may be shown that K is maximal among the compact subgroups of $GL(n, A)$ and that every maximal compact subgroup is conjugate to K.

We now show that the Haar volume of $Z(A)\, GL(n, F)\backslash GL(n, A)$ is finite when $n = 2$ and $F = \mathbb{Q}$. Therefore assume temporarily that $F = \mathbb{Q}$. Let K_0 be a compact subgroup of $GL(n, A)$; it is sufficient to show that the volume of $Z(A)\, GL(n, F)\backslash GL(n, A)/K_0$ is finite. We could take $K_0 = K$, but rather than fixing one particular subgroup K_0, we will work with a family of groups $K_0(N)$, obtaining in the process the identity (Eq. 3.3) below, which we will require when we later transfer classical automorphic forms to the adele group.

We fix a positive integer N. Let $K_0(N)$ be the following compact subgroup of $GL(2, A_{\mathrm{f}})$: $K_0(N) = \prod_{v \notin S_\infty} K_0(N)_v$, where $K_0(N)_v = GL(2, \mathfrak{o}_v)$ if $p_v \nmid N$ (p_v is the rational prime corresponding to the non-Archimedean place v), while if $p_v | N$, then $K_0(N)$ is the subgroup of $GL(2, \mathfrak{o}_v)$ of the form $\left(\begin{smallmatrix} a & b \\ c & d \end{smallmatrix}\right)$ where $c \equiv 0$ modulo N in the ring $\mathbb{Z}_v$ of p-adic integers.

Proposition 3.3.1 *Suppose that* A *is the adele ring of* $\mathbb{Q}$. *The inclusion* $SL(2, \mathbb{R}) \to GL(2, A)$ *induces a homeomorphism*

$$\Gamma_0(N)\backslash SL(2, \mathbb{R}) \cong Z(A)\, GL(2, \mathbb{Q})\backslash GL(2, A)/K_0(N). \tag{3.3}$$

Proof Because the class number of $\mathbb{Q}$ is one, part (ii) of the strong approximation theorem implies that $GL(2, A) = GL(2, \mathbb{Q})\, GL(2, \mathbb{R})\, K_0(N)$; because $GL(2, \mathbb{Q})$ contains elements of negative determinant, we have in fact

$GL(2, A) = GL(2, \mathbb{Q})\, GL(2, \mathbb{R})^+\, K_0(N)$. This means that the composition of the canonical maps

$$GL(2, \mathbb{R})^+ \to GL(2, A) \to GL(2, \mathbb{Q})\backslash GL(2, A)/K_0(N)$$

is surjective. We ask when two elements g_∞ and g'_∞ of $GL(2, \mathbb{R})^+$ have the same image in the double coset space. Suppose that $g'_\infty = \gamma g_\infty k_0$ with $\gamma \in GL(2, \mathbb{Q})$ and $k_0 \in K_0(N)$. We factor $\gamma = \gamma_\infty \gamma_{\mathrm{f}}$ with $\gamma_\infty \in GL(2, \mathbb{R})$ and $\gamma_{\mathrm{f}} \in GL(2, A_{\mathrm{f}})$. Then we have $g'_\infty = \gamma_\infty g_\infty$, and $\gamma_{\mathrm{f}} = k_0^{-1}$. The first relation implies that γ_∞ has positive determinant, and the second implies that $\gamma_\infty \in \Gamma_0(N)$. We see that a necessary and sufficient condition for two elements of $GL(2, \mathbb{R})^+$ to have the same image in $GL(2, \mathbb{Q})\backslash GL(2, A)/K_0(N)$ is that they differ on the right by an element of $\Gamma_0(N)$; we have therefore a bijection

$$\Gamma_0(N)\backslash GL(2, \mathbb{R})^+ \cong GL(2, \mathbb{Q})\backslash GL(2, A)/K_0(N).$$

Now let us take the centers into account. By the strong approximation theorem, $A^\times = \mathbb{R}_+^\times \mathbb{Q}^\times \left[\prod_{v \notin S_\infty} \mathfrak{o}_v^\times\right]$, so $Z(A) = Z(\mathbb{R})^+ Z(\mathbb{Q}) \left(Z(A) \cap K_0(N)\right)$, where $Z(\mathbb{R})^+$ is the group of scalar matrices in $Z(\mathbb{R})$ with positive eigenvalue. Thus

$$Z(\mathbb{R})^+\, \Gamma_0(N)\backslash GL(2, \mathbb{R})^+ \cong Z(A)\, GL(2, \mathbb{Q})\backslash GL(2, A)/K_0(N).$$

Now observing that $SL(2, \mathbb{R}) \cong Z(\mathbb{R})^+\backslash GL(2, \mathbb{R})^+$, we obtain Eq. (3.3). ∎

Proposition 3.3.2 *Let A be the adele ring of the global field F. Then the quotient space $GL(n, F)\, Z(A)\backslash GL(n, A)$ has finite measure.*

We will prove this only in a special case. The general case follows from Humphreys (1980, Sections 10.3, 11.1, and 14.4).

Proof (when $F = \mathbb{Q}$ and $n = 2$) If Haar measures on the various groups in Eq. (3.3) (which are all unimodular) are chosen compatibly, then Eq. (3.3) is an isometry, showing that $Z(A)\, GL(2, \mathbb{Q})\backslash GL(2, A)/K_0(N)$ has finite volume because it may be identified with a fundamental domain for the group $\Gamma_0(N)$ acting on $SL(2, \mathbb{R})$, which is known to have finite volume. ∎

Because of Eq. (3.3), one may think of $GL(2, \mathbb{Q})$ and $GL(2, A)$ as being substitutes for $\Gamma_0(N)$ and $GL(2, \mathbb{R})^+$ in the classical theory of automorphic forms on the upper half plane.

Let ω be a unitary Hecke character, that is, a unitary character of $A^\times/F^\times$. Let $L^2\left(GL(n, F)\backslash GL(n, A), \omega\right)$ be the space of all functions ϕ on $GL(n, A)$ that are measurable with respect to Haar measure and that satisfy

$$\phi\left(\begin{pmatrix} z & & \\ & \ddots & \\ & & z \end{pmatrix} g\right) = \omega(z)\, \phi(g), \qquad z \in A^\times, \tag{3.4}$$

$$\phi(\gamma g) = \phi(g), \qquad \gamma \in GL(n, F), \tag{3.5}$$

and that are *square integrable modulo the center*:

$$\int_{Z(A)\,GL(n,F)\backslash GL(n,A)} |\phi(g)|^2\, dg < \infty. \tag{3.6}$$

As usual, we identify functions in L^2 if their difference vanishes off a set of measure zero. We will now define the condition of *cuspidality* that $\phi \in L^2\big(GL(n, F)\backslash GL(n, A), \omega\big)$ may satisfy. For all decompositions $n = r + s$, where $1 \le r, s \le n-1$, let $\mathrm{Mat}_{r\times s}$ denote the additive algebraic group of $r \times s$ matrices. Also, let I_r and I_s denote the $r \times r$ and $s \times s$ identity matrices. The cuspidal condition is that

$$\int_{\mathrm{Mat}_{r\times s}(F)\backslash \mathrm{Mat}_{r\times s}(A)} \phi\left(\begin{pmatrix} I_r & X \\ & I_s \end{pmatrix} g\right) dX = 0 \tag{3.7}$$

almost everywhere. If this is true for all r and s such that $r+s = n$, then we say that ϕ is *cuspidal*. Let $L_0^2\big(GL(n, F)\backslash GL(n, A), \omega\big)$ be the closed subspace of cuspidal elements of the Hilbert space $L^2\big(GL(n, F)\backslash GL(n, A), \omega\big)$.

We let $GL(n, A)$ act on $L^2\big(GL(n, F)\backslash GL(n, A), \omega\big)$ by right translation: that is, we consider the representation

$$\big(\rho(g)\phi\big)(x) = \phi(xg), \qquad g, x \in GL(n, A) \tag{3.8}$$

(Exercise 3.3.1). The cuspidal subspace $L_0^2\big(GL(n, F)\backslash GL(n, A), \omega\big)$ is clearly invariant.

Our aim is to prove an analog of Theorem 3.2.2. To this end, we need an analog of Proposition 3.2.3. We will now define a space of functions on $GL(n, A)$ that we will denote $C_c^\infty\big(GL(n, A)\big)$. This is by definition the space of all finite linear combinations of functions $\pi(g) = \prod_v \phi_v(g_v)$, where for every place v of F, ϕ_v is an element of $C_c^\infty\big(GL(n, F_v)\big)$, and for almost all v, ϕ_v is the characteristic function of $GL(2, \mathfrak{o}_v)$. As in Sections 2.3 and 3.2, if $f \in L^2\big(GL(n, F)\backslash GL(n, A), \omega\big)$, we define

$$\begin{aligned}\big(\rho(\phi)\, f\big)(g) &= \int_{GL(n,A)} \phi(h)\, f(gh)\, dh \\ &= \int_{Z(A)\backslash GL(n,A)} \phi_\omega(h)\, f(gh)\, dh,\end{aligned}$$

where

$$\phi_\omega(g) = \int_{A^\times} \phi\left(\begin{pmatrix} z & & \\ & \ddots & \\ & & z \end{pmatrix} g\right) \omega(z)\, dz.$$

Proposition 3.3.3: Gelfand, Graev, and Piatetski-Shapiro *Let* $\phi \in C_c^\infty\big(GL(n, A)\big)$.
(i) There exists a $C > 0$ *(depending on* ϕ*) such that*

$$\sup_{g \in GL(2,A)} |\rho(\phi) f(g)| \le C \, ||f||_2. \tag{3.9}$$

for all $f \in L_0^2\big(GL(2, F)\backslash GL(2, A), \omega\big)$.
(ii) The operator $\rho(\phi)$ *is compact on* $L_0^2\big(GL(n, F)\backslash GL(n, A), \omega\big)$.

Proof (*when* $n = 2$ *and* $F = \mathbb{Q}$) The proof is a straightforward modification of Proposition 3.2.3. We first define Siegel sets in $GL(2, A)$. We make the assumption that $F = \mathbb{Q}$ so that the class number is one. The modifications needed for $n = 2$ and F an arbitrary global field are very minor and are left to the reader (Exercise 3.3.4).

Let $K = \prod_v K_v$, where $K_\infty = SO(2)$ and $K_v = GL(2, \mathfrak{o}_v)$ for v non-Archimedean. Let c and d be positive numbers. Let $\mathcal{G}_{c,d} \subset GL(2, A)$ be the set of adeles of the form (g_v), where g_∞ is of the form

$$\begin{pmatrix} z & \\ & z \end{pmatrix} \begin{pmatrix} y & x \\ & 1 \end{pmatrix} k_\infty, \qquad z \in \mathbb{R}^\times,\ c \le y,\ 0 \le x \le d,\ k_\infty \in K_\infty,$$

and $g_v \in K_v$ for all non-Archimedean places v. We will also denote by $\overline{\mathcal{G}}_{c,d}$ the image of $\mathcal{G}_{c,d}$ in $Z(A)\backslash GL(2, A)$.

We will show that if $c \le \sqrt{3}/2$ and $d \ge 1$, then $GL(2, A) = GL(2, \mathbb{Q})\, \mathcal{G}_{c,d}$. Indeed, if $g \in GL(2, \mathbb{Q})$, by the strong approximation theorem we can find an element $\gamma_1 \in GL(2, \mathbb{Q})$ such that $\gamma_1 g$ has its v-component in K_v for all $v \notin S_\infty$; then assuming $c \le \sqrt{3}/2$ and $d \ge 1$, it follows from the fact that (in the notation of Section 3.2) $\mathcal{F}_{c,d}$ contains a fundamental domain for $SL(2, \mathbb{Z})$, that we may find $\gamma_0 \in SL(2, \mathbb{Z})$ such that $\gamma_0 \gamma_1 g \in \mathcal{G}_{c,d}$, which shows that $g \in GL(2, \mathbb{Q})\, \mathcal{G}_{c,d}$.

Proceeding now as in the proof of Proposition 3.2.3, we will prove an inequality

$$\sup_{g \in \mathcal{G}_{c,d}} \big|\big(\rho(\phi) f\big)(g)\big| \le C_0 ||f||_2.$$

Let $N(F)$ be the group of upper triangular unipotent matrices with entries in F. As in the proof of Proposition 3.2.3, we have

$$\big(\rho(\phi) f\big)(g) = \int_{N(F)\, Z(A)\backslash GL(2,A)} K'(g, h)\, f(h)\, dh,$$

where

$$K'(g,h) = K(g,h) - K_0(g,h),$$
$$K(g,h) = \sum_{\gamma \in N(F)} \phi_\omega(g^{-1}\gamma h),$$
$$K_0(g,h) = \int_{A/F} \phi_\omega\left(g^{-1}\begin{pmatrix} 1 & x \\ & 1 \end{pmatrix} h\right) dx.$$

Let $g \in \mathcal{G}_{c,d}$. We write

$$g = \begin{pmatrix} \eta & \\ & \eta \end{pmatrix}\begin{pmatrix} y & x \\ & 1 \end{pmatrix}\kappa_g,$$

where $x, y \in \mathbb{R}$ and $c \leq y$ and $\kappa_g \in K$. Let h be an arbitrary element of $GL(2, A)$. We write

$$h = \begin{pmatrix} \zeta & \\ & \zeta \end{pmatrix}\begin{pmatrix} v & u \\ & 1 \end{pmatrix}\kappa_h,$$

where $v \in A^\times$, $u \in A$, and $\kappa_h \in K$. By the Poisson summation formula,

$$K'(g,h) = \sum_{\alpha \in F^\times} \hat{\Phi}_{g,h}(\alpha),$$

where $\Phi_{g,h} : A \to \mathbb{C}$ is the compactly supported continuous function

$$\Phi_{g,h}(x) = \phi_\omega\left(g^{-1}\begin{pmatrix} 1 & x \\ & 1 \end{pmatrix} h\right).$$

(We omit the term $\alpha = 0$ because this is K_0.) The Fourier transform $\hat{\Phi}_{g,h}(\alpha)$ equals

$$\int_A \phi_\omega\left(\kappa_g^{-1}\begin{pmatrix} \eta^{-1}\zeta & \\ & \eta^{-1}\zeta \end{pmatrix}\begin{pmatrix} y^{-1} & \\ & 1 \end{pmatrix}\begin{pmatrix} 1 & t-x+u \\ & 1 \end{pmatrix}\begin{pmatrix} v & \\ & 1 \end{pmatrix}\kappa_h\right)$$
$$\times \psi(\alpha t)\, dt.$$

This equals $\psi\big(\alpha(x-u)\big)\,\omega(\zeta^{-1}\eta)$ times $|y|\,\widehat{F}_{\kappa_g,\kappa_h,y^{-1}v}(\alpha y)$, where

$$F_{\kappa_g,\kappa_h,y}(t) = \phi_\omega\left(\kappa_g^{-1}\begin{pmatrix} 1 & t \\ & 1 \end{pmatrix}\begin{pmatrix} y & \\ & 1 \end{pmatrix}\kappa_h\right).$$

We note that $K\,supp(\phi)\,K \subset GL(2, A)$ is the continuous image of the compact set $K \times supp(\phi) \times K$ and hence is compact, and its intersection with the closed subgroup $B(\mathbb{R})$ is also compact. ($B(\mathbb{R})$ is the Borel subgroup of upper triangular matrices in $GL(2, A)$.) It follows that there exists a compact subset Ω of $A^\times$ such that if $F_{\kappa_g,\kappa_h,y}(t) \neq 0$ for any t, then $y \in \Omega$. To summarize

$$|K'(g,h)| \leq |y| \sum_{\alpha \in F^\times} \widehat{F}_{\kappa_g,\kappa_h,y^{-1}v}(y),$$

where $F_{\kappa_g,\kappa_h,y^{-1}v}(y)$ is a Schwartz function of y, depending continuously on κ_g, κ_h and $y^{-1}v$, and vanishing identically unless $(\kappa_g, \kappa_h, y^{-1}v)$ lies in the compact set $K \times K \times \Omega$. Therefore if $N > 0$ is given, there exists a constant C_N such that

$$|K'(g,h)| \leq C_N\,|y|^{-N},$$

and furthermore $K'(g,h) = 0$ unless $y^{-1}v \in \Omega$. Thus

$$\left|\rho(\phi)(g)\right| \leq C_N\,|y|^{-N} \int_{A/F} \int_{y^{-1}v \in \Omega} \int_K \left| f\left(\begin{pmatrix} v & u \\ & 1 \end{pmatrix} \kappa_h \right) \right| |v|^{-1}\, d\kappa_h\, d^\times v\, du.$$

Hence $|\rho(g)\, f|$ is dominated by the L^1 norm of f, which is in turn dominated by the L^2 norm, which gives us Eq. (3.9).

To show $\rho(\phi)$ is compact, by the Ascoli–Arzéla lemma (Proposition 3.2.1), it suffices to show that the image of the unit ball in $L^2_0\big(GL(2,F)\backslash GL(2,A), \omega\big)$ is equicontinuous. There exists an open subgroup of $GL(2, A_{\mathrm{f}})$ under which ϕ is right invariant, and any element of the image of $\rho(\phi)$ will be right invariant under this same subgroup. It is sufficient, therefore, to show that the functions $(\rho(\phi)\, f)(g)$ are equicontinuous as functions of the infinite component g_∞ of the adele $g \in GL(2, A)$. This is a consequence of a uniform bound for the partial derivative $X\,\rho(\phi)\, f$. The proof of this is identical to the classical case of Proposition 3.2.3 (cf. Eq. (2.28)). ■

Theorem 3.3.2 *The space $L^2_0\big(GL(n,F)\backslash GL(n,A), \omega\big)$ decomposes into a Hilbert space direct sum of irreducible invariant subspaces.*

Proof (*complete when $n = 2$ and $F = \mathbb{Q}$*) The proof is the same as Theorem 2.3.7 or Theorem 3.2.2. ■

We may also define *automorphic forms* on $GL(n, A)$. An *automorphic form with central quasicharacter* ω is a function on $GL(n, A)$ satisfying Eqs. (3.4) and (3.5), which is *smooth*, *K-finite*, *$\mathcal{Z}$-finite*, and *of moderate growth.* We now define these terms.

If F is a function field, then a function f on $GL(n, A)$ is called *smooth* if it is locally constant. If F is a number field, then we call the function f *smooth* if for every $g \in GL(n, A)$, there exists a neighborhood N of g and a smooth function f_g on $GL(n, F_\infty)$ such that for $h \in N$, $f(h) = f_g(h_\infty)$, where we have factored $h = h_\infty h_{\mathrm{f}}$ with $h_\infty \in GL(n, F_\infty)$, $h_{\mathrm{f}} \in A_{\mathrm{f}}$. A function f on $GL(n, A)$ is called *K*-finite if its right translates, by elements of K, span a finite-dimensional vector space. If v is any Archimedean place of F, according to Proposition 2.2.2, an action of $\mathfrak{gl}(n, F_v)$ may be defined on the K-finite vectors by

$$(Xf)(g) = \frac{d}{dt} f\big(g \exp(tX)\big)|_{t=0}.$$

Here $\mathfrak{gl}(n, F_v)$ is the four-dimensional Lie algebra whose underlying space is $\mathrm{Mat}_n(F_v)$, with Lie bracket operation $[X, Y] = XY - YX$.) It may be shown that if f is K-finite, then Xf is defined and is also K-finite (Exercise 3.3.2). This action of $\mathfrak{gl}(n, F_v)$ is extended to the universal enveloping algebra $U\big(\mathfrak{gl}(n, F_v)\big)$ (Proposition 2.2.3). Let $\mathcal{Z}$ be the center of the universal enveloping algebra of $GL(n, F_v)$. The requirement that f be *$\mathcal{Z}$-finite* means that f lies in a finite-dimensional vector space that is invariant by $\mathcal{Z}$. To define *moderate growth*, we need to define a *height function* $||g||$ on $GL(n, A)$. (There are different ways of accomplishing this, but all lead to the same notion of moderate growth.) We embed $GL(n) \to \mathbb{A}^{n^2+1}$ as described above via $g \mapsto \big(g, \det(g)^{-1}\big)$. First we define a local height $||g_v||_v$ on $GL(n, F_v)$ for each place by restricting the height function $(x_1, \cdots, x_{n^2+1}) \mapsto \max_i |x_i|_v$ on $\mathbb{A}^{n^2+1}(F_v)$. We note that $||g_v||_v \geq 1$, and (if v is non-Archimedean) that $||g_v||_v = 1$ if $g_v \in GL(n, \mathfrak{o}_v)$. We then define the global height $||g||$ to be the product of the local heights. We then say that f is *slowly increasing* if there exist constants C and N such that $f(g) < C\,||g||^N$ for all $g \in GL(n, A)$.

We denote by $\mathcal{A}\big(GL(n, F)\backslash GL(n, A), \omega\big)$ the space of automorphic forms with central quasicharacter ω and by $\mathcal{A}_0(GL(n, F)\backslash GL(n, A), \omega)$ the space of *cusp forms*, which are further assumed to satisfy Eq. (3.7) for all decompositions $n = r + s$ with $1 \leq r, s \leq n - 1$. The group $GL(n, A)$ does not quite act on this space: The property of K-finiteness is not preserved by right translation by elements of $GL(n, F_\infty)$. However, $\mathcal{A}\big(GL(n, F)\backslash GL(n, A), \omega\big)$ is stable under right translation by $GL(n, A_\mathfrak{f})$, and it is a $(\mathfrak{g}_\infty, K_\infty)$-module, where $\mathfrak{g}_\infty = \prod_{v \in S_\infty} \mathfrak{gl}(n, F_v)$, $K_\infty = \prod_{v \in S_\infty} K_v$.

Thus $\mathcal{A}\big(GL(n, F)\backslash GL(n, A), \omega\big)$ affords a "representation" of $GL(n, A)$ in the sense that one has a representation of $GL(n, A_\mathfrak{f})$ and a commuting $(\mathfrak{g}_\infty, K_\infty)$-module structure at the infinite places. Taking the term "representation" in this sense, an *automorphic representation* is any irreducible representation of $GL(n, A)$ that can be realized as a subquotient of $\mathcal{A}\big(GL(n, F)\backslash GL(n, A), \omega\big)$. We call ω the *central quasicharacter* of π. (A *subquotient* is a quotient of a submodule.)

Let us define the *restricted tensor product* of an infinite number of vector spaces V_v indexed by a set Σ. For almost all $v \in \Sigma$ (that is, for all but finitely many) let there be given a nonzero $x_v^\circ \in V_v$. Let Ω be the set of all finite subsets S of Σ having the property that if $v \notin S$, then x_v° is defined. We order Ω by inclusion: It is a directed set. If $S, S' \in \Omega$, and $S \subseteq S'$, then we define a homomorphism $\lambda_{S,S'} : \bigotimes_{v \in S} V_v \to \bigotimes_{v \in S'} V_v$, namely, $\lambda_{S,S'}(x)$ is obtained by tensoring $x \in \bigotimes_{v \in S} V_S$ with $\otimes_{v \in S'-S}\, x_v^\circ$. These maps form a direct family, and so we may form the direct limit

$$\varinjlim \bigotimes_{v \in S} V_v.$$

(See Exercise 4.3.2 and Lang (1993, p. 51) for directed sets and direct limits.) We will denote this vector space as $\bigotimes_v V_v$. It is understood that the precise definition of this space depends on assigning the vectors $x_v^\circ \in V_v$. The maps $\lambda_{S,S'}$ are all injective, and so by Exercise 4.3.2, the natural maps

$$\lambda_S : \bigotimes_{v \in S} V_v \to \bigotimes_{v \in \Sigma} V_v$$

are also injective. We will use the following notation: If $x = \otimes_{v \in S} x_v \in \bigotimes_{v \in S} V_v$, we will denote the image of x in $\otimes_{v \in \Sigma} V_v$ as $\otimes_{v \in \Sigma} x_v$, where if $v \in \Sigma - S$, we specify $x_v = x_v^\circ$. An element of this form will be called a *pure tensor*.

Intuitively, we may think of $\bigotimes_{v \in \Sigma} V_v$ as the vector space spanned by all symbols $\otimes x_v$ such that $x_v \in V_v$ for all v, and $x_v = x_v^\circ$ for almost all v (i.e., all but finitely many), subject to the customary condition that $\otimes_v x_v$ is linear in each component. Note that without the saving condition that $x_v = x_v^\circ$ for almost all v, we would have difficulty making this definition.

An important observation, which we will use many times, is that if we specify another collection of vectors $y_v^\circ \in V_v$ for almost all v such that $x_v^\circ \cong y_v^\circ$ for almost all v, then the restricted tensor product of the V_v with respect to the x_v° is essentially the same as the restricted tensor product of the V_v with respect to the y_v° (Exercise 3.3.5). In other words, changing a finite number of the vectors x_v° does not change the restricted tensor product space.

Suppose that for each v, there is given a group G_v and a subgroup K_v. (In practice, Σ will be the set of all places of the global field F, the groups K_v will be as in Eq. (3.2), and we will have either $G_v = GL(n, F_v)$ or $G_v = K_v$.) For each $v \in \Sigma$, let (ρ_v, V_v) be a representation of G_v. Let there be given vectors $\xi_v^0 \in V_v$ for almost all v such that $\rho_v(k_v)\, \xi_v^0 = \xi_v^0$ for all $k_v \in K_v$. Let G be the restricted direct product of the G_v with respect to the K_v. Then we can define a representation $(\otimes_v \rho_v, \otimes_v V_v)$ by

$$\big(\otimes_v \rho_v \big)(g_v)\, \xi_v = \otimes_v \rho_v(g_v)\, \xi_v.$$

Note that the right-hand side is defined!

Lemma 3.3.1 *Let ρ be an irreducible finite-dimensional representation of K. Then there exist finite-dimensional representations ρ_v of K_v such that for almost all v, (ρ_v, V_v) is the one-dimensional trivial representation of K_v, and a vector ξ_v^0 such that (ρ, V) is isomorphic to the representation $\otimes_v \rho_v$ on the restricted tensor product $\otimes_v V_v$ with respect to the V_v.*

Proof Because K_{f} is totally disconnected, by Exercise 3.1.1(c), the kernel of any such representation ρ contains an open subgroup and hence contains K_v

for all but finitely many v. Thus, let S be a finite set of places containing S_∞ such that $\rho(K_v) = 1$ for all $v \notin S$. Then ρ factors through the projection

$$K \to K/\left[\prod_{v\notin S} K_v\right] \cong \prod_{v\in S} K_v.$$

As this is a finite direct product of compact groups, any irreducible representation has the form $\otimes_{v\in S}\rho_v$, where ρ_v is an irreducible representation of K_v. Because the K_v with $v \notin S$ act trivially, we see that the representation ρ is isomorphic to the restricted tensor product as described in the lemma, provided we take (ρ_v, V_v) to be the trivial (one-dimensional) representation for $v \notin S$. ■

Now let (π, V) be a representation of K, and let (ρ, V_ρ) be an irreducible finite-dimensional representation of K. Let $V(\rho)$ be the sum of all K-submodules of V that are isomorphic to K. We call $V(\rho)$ the ρ-*isotypic* part of (V, π).

Let us define the notion of an *admissible representation* of $GL(n, A)$. We apologize in advance, because (as we define it) this will not be a representation of $GL(n, A)$ at all! Just as the action of $GL(n, A)$ on functions does not preserve $\mathcal{A}\big(GL(n, F)\backslash GL(n, A), \omega\big)$ – it does not preserve the property of K-finiteness – $GL(n, A)$ does not act on the space of an admissible representation according to the definition we now give. Let V be a complex vector space that is at once a $(\mathfrak{g}_\infty, K_\infty)$-module and a $GL(n, A_\mathfrak{f})$-module. We will denote both actions by the same letter π and denote the representation by the pair (π, V). We assume that these two actions commute. Because $K = K_\infty \cdot K_\mathfrak{f}$ (with $K_\mathfrak{f} = \prod_{v\notin S_\infty} K_v$), we may combine these actions to obtain a representation π of K. We assume that every vector in V is K-finite, and if ρ is an irreducible finite-dimensional representation of K, then we assume that the ρ-isotypic part $V(\rho)$ of V is finite dimensional.

This definition is the one given by Flath (1979). We will see in Section 3.4 that an admissible representation of $GL(n, A)$ (in this sense) is equivalent to a representation of a certain ring, the *global Hecke algebra* of $GL(n, A)$.

We may also define admissible representations of the groups $GL(n, F_v)$. If v is non-Archimedean, we say that a representation (π, V) of $GL(n, F_v)$ is *smooth* if every vector has an open stabilizer; if furthermore for each irreducible representation (ρ, V_ρ) of $K_v = GL(n, \mathfrak{o}_v)$ the ρ-isotypic part of V is finite dimensional, we say that π is *admissible*. Similarly, if v is Archimedean, we say that a $(\mathfrak{g}_\infty, K_v)$-module V is *admissible* if the ρ-isotypic parts are finite dimensional. If v is non-Archimedean and (π_v, V_v) is an irreducible admissible representation, we say that π_v is *spherical* if V_v contains a nonzero K_v-fixed vector.

Theorem 3.3.3: Tensor product theorem *Let (V, π) be an irreducible admissible representation of $GL(n, A)$. Then there exists for each Archimedean place v of F an irreducible admissible $(\mathfrak{g}_\infty, K_v)$-module (π_v, V_v), and for each non-Archimedean place v there exists an irreducible admissible representation*

(π_v, V_v) of $GL(n, F_v)$ such that for almost all v, V_v contains a nonzero K_v-fixed vector ξ_v^0 such that π is the restricted tensor product of the representations π_v.

So as not to disrupt the continuity of the discussion, we will give the proof of this result, based on Flath (1979), in Section 3.4.

Theorem 3.3.4 *Let (π, V) be an irreducible constituent of the decomposition of $L_0^2(GL(n, F)\backslash GL(n, A), \omega)$ (Theorem 3.3.2). Then π induces an irreducible admissible representation of $GL(n, A)$ on the space of K-finite vectors in V.*

Proof (*when $n = 2$ and $F = \mathbb{Q}$*) We will reduce this to Theorem 3.2.1. We must show that if (ρ, V_ρ) is an irreducible finite dimensional representation of K, then the isotypic part $V(\rho)$ is finite dimensional. By Lemma 3.3.1, we have $\rho = \otimes_v \rho_v$, where each ρ_v is a finite-dimensional irreducible representation of $GL(2, \mathbb{R})^+$ and ρ_v is one dimensional for almost all v. There exists an open subgroup $K_{0,\mathrm{f}}$ of $\prod_{v \notin S_\infty} K_v$ such that every vector in V_ρ is invariant under $K_{0,\mathrm{f}}$.

We will exhibit an isomorphism of $V^{K_{0,\mathrm{f}}}(\rho_\infty)$ (which evidently contains $V(\rho)$) into a finite Cartesian product of spaces of automorphic forms on $GL(2, \mathbb{R})^+$, each of which is finite dimensional by Theorem 3.2.1. By strong approximation, the space $GL(2, A) = GL(2, \mathbb{Q})\, GL(2, \mathbb{R})^+ K_\mathrm{f}$, where $K_\mathrm{f} = \prod_{v \notin S_\infty} K_v$, and $K_{0,\mathrm{f}}$ is of finite index in K_f, so the set of double cosets

$$GL(2, \mathbb{Q})\, GL(2, \mathbb{R})^+ \backslash GL(2, A) / K_{0,\mathrm{f}}$$

is finite. Let $g_1, \cdots, g_h$ be a set of representatives. We may choose the g_i to lie in $GL(2, A_\mathrm{f})$; that is, we may choose them so that $g_{i,\infty} = 1$.

Now let $\phi \in V^{K_{0,\mathrm{f}}}$. We associate with ϕ the h functions Φ_i on $GL(2, \mathbb{R})^+$, defined by $\Phi_i(g_\infty) = \phi(g_\infty g_i)$. We note that if $g \in GL(2, A)$, we can write $g = \gamma\, g_\infty g_i\, k_0$ for $\gamma \in GL(2, \mathbb{Q})$, $g_\infty \in GL(2, \mathbb{R})^+$ and $k_0 \in K_{0,\mathrm{f}}$ with some g_i, so $\phi(g) = \Phi_i(g_\infty)$. Thus ϕ is uniquely determined by the h functions Φ_i. It is sufficient for us to show that each of these lies in a finite-dimensional vector space, and we will deduce this from Theorem 3.2.3.

Let Γ be the projection onto $GL(2, \mathbb{R})^+$ of $GL(2, \mathbb{Q}) \cap \big(GL(2, \mathbb{R})^+ K_{0,\mathrm{f}}\big)$. Because the projection onto $GL(2, \mathbb{R})^+$ of $GL(2, \mathbb{Q}) \cap \big(GL(2, \mathbb{R})^+ K_\mathrm{f}\big)$ is $SL(2, \mathbb{Z})$, the group Γ is of finite index in $GL(2, \mathbb{Z})$. (It is not hard to see that it is a congruence subgroup.) The function Φ_i evidently satisfies $\Phi_i(\gamma\, g_\infty) = \Phi_i(g_\infty)$ for $\gamma \in \Gamma$, $g_\infty \in GL(2, \mathbb{R})^+$, and it is $\mathcal{Z}$-finite. If ϕ is K_∞-finite, then so is Φ_i. It is easy to see that Φ_i inherits the condition of moderate growth from ϕ. Hence if $\omega = \prod_v \omega_v$, we have shown that each $\Phi_i \in \mathcal{A}(\Gamma\backslash GL(2, \mathbb{R})^+, 1, \omega_\infty)$. The ρ_∞-isotypic subspace of this space is finite dimensional by Theorem 3.2.3, so $V^{K_{0,\mathrm{f}}}(\rho_\infty)$ is also finite dimensional.

This proves that the K-finite vectors in V form an admissible representation of $GL(2, A)$. ■

Let F be a non-Archimedean local field and let $\mathfrak{o}$ be its ring of integers. An irreducible admissible representation (π, V) of $GL(n, F)$ is called *spherical* (or *nonramified*) if it contains a $GL(n, \mathfrak{o})$-fixed vector; such a vector is called a *spherical vector*. The spherical vector, if it exists, is unique up to constant multiple by Theorem 4.6.2. The spherical representations are classified in Theorem 4.6.4. We will review this classification in Section 3.5. The importance of the spherical representations comes from the fact that if $\pi = \otimes_v \pi_v$ is automorphic, then π_v is spherical for almost all v. This is one of the conclusions of Theorem 3.3.3. We refer to Section 3.5 and to Section 4.6 for more information on the spherical representations of $GL(2, F)$.

Let F be a non-Archimedean local field. If (π, V) is an admissible representation of $GL(n, F)$, then the *contragredient representation* of $GL(n, F)$ is defined to be the representation $(\hat{\pi}, \widehat{V})$, where $\widehat{V}$ is the space of all *smooth linear functionals* on V: A linear functional $\Lambda : V \to \mathbb{C}$ is called smooth if there exists some open subgroup K_0 such that $\Lambda\big(\pi(k)\, v\big) = \Lambda(v)$ for all $k \in K_0$ and all $v \in K$. If $v \in V$, $\Lambda \in \widehat{V}$, we will write $\langle v, \Lambda \rangle$ instead of $\Lambda(v)$ to emphasize the symmetry between the two spaces. The group action on $\widehat{V}$ is defined by

$$\langle v, \hat{\pi}(g)\, \Lambda \rangle = \big\langle \pi(g^{-1})\, v, \Lambda \big\rangle .$$

The contragredient representation is described in more detail in Section 4.2. There is one deep fact worth mentioning: If π is irreducible, let us define a representation (π_1, V) on the same space as π by $\pi_1(g) = \pi\left({}^{\top}g^{-1}\right)$, then the representation π_1 is equivalent to the contragredient. Moreover, if $n = 2$, and if ω is the central quasicharacter of π, then we may define another representation π_2 of $GL(2, F)$ on V by $\pi_2(g) = \omega\big(\det(g)\big)^{-1}\, \pi(g)$, and this representation is also equivalent to $\hat{\pi}$. These facts are established in Theorem 4.2.2.

More generally, if (π, V) is a smooth representation of $GL(n, F)$, and χ is a character of $F^\times$, there is a representation that we will call $\chi \otimes \pi$. The space of this representation can be taken to be the same vector space V, with the action $(\chi \otimes \pi)(g) = \chi\big(\det(g)\big)\, \pi(g)$. We call the representation $\chi \otimes \pi$ the *twist* of π by χ. The equivalence of π_2 with $\hat{\pi}$ may be expressed by saying that if $n = 2$, and if ω is the central quasicharacter of π, then

$$\hat{\pi} \cong \omega^{-1} \otimes \pi. \tag{3.10}$$

This relation is false on $GL(n)$ if $n > 2$.

We may also define a contragredient representation of an admissible $(\mathfrak{g}, K)$-module (π, V) if $G = GL(n, F)$ where $F = \mathbb{R}$ or $\mathbb{C}$. Now $\widehat{V}$ will be the space of all linear functionals Λ that are zero on the ρ-isotypic part $V(\rho)$ for all but finitely many irreducible representations ρ of K. The action $\hat{\pi}$ of K is given by

$$\langle v, \hat{\pi}(k)\, \Lambda \rangle = \big\langle \pi(k^{-1})\, v, \Lambda \big\rangle ,$$

and the action of $\mathfrak{g}$ is given by

$$\langle v, \hat{\pi}(X)\, \Lambda \rangle = -\, \langle \pi(X)\, v, \Lambda \rangle .$$

The analog of Theorem 4.2.2 remains valid: We can define another $(\mathfrak{g}, K)$-module on the space of V by composing π with the automorphism $g \mapsto {}^{\top}g^{-1}$, and if π is irreducible, the resulting representation is equivalent to $\hat{\pi}$ (Exercise 2.5.8).

Now, let (π, V) be an irreducible admissible representation of $GL(2, A)$. We write $\pi = \otimes_v \pi_v$ by Theorem 3.3.3. Let $\hat{\pi} = \otimes_v \hat{\pi}_v$. This is the *contragredient* of the representation (π, V).

Proposition 3.3.4 *Suppose that (π, V) is an automorphic cuspidal representation of $GL(n, A)$, with central quasicharacter ω. Then so is its contragredient $(\hat{\pi}, \widehat{V})$. If $V \subset \mathcal{A}_0\big(GL(n, F)\backslash GL(n, A), \omega\big)$, then a subspace of $\mathcal{A}_0\big(GL(n, F)\backslash GL(n, A), \omega^{-1}\big)$ affording a representation isomorphic to $\hat{\pi}$ consists of all functions of the form $g \mapsto \phi({}^{\top}g^{-1})$, where $\phi \in V$.*

Because the following argument depends on the proof of Theorem 4.2.2, which is only complete when $n = 2$, the same is true of the present argument. (We are leaving the proof of the Archimedean analog of Theorem 4.2.2 to the reader in the real case, where it can be deduced from the results of Section 2.5 – see Exercise 2.5.8.)

Proof *(Complete if $n = 2$ and F is totally real)* We may regard V as a subspace of $L^2\big(GL(n, F)\backslash GL(n, A), \omega\big)$. Let $\widehat{V}$ be the space of functions of the form

$$\hat{\phi}(g) = \phi\big({}^{\top}g^{-1}\big), \qquad \phi \in V.$$

The representation of $GL(n, A)$ on $\widehat{V}$ is obtained by composing π with the automorphism $g \mapsto {}^{\top}g^{-1}$ of $GL(n, A)$, so this result follows (at least when $n = 2$ and F is totally real) from Theorem 4.2.2 and its Archimedean analog for $(\mathfrak{g}, K)$-modules, which is Exercise 2.5.8 when F is totally real. ■

Let (π, V) be a cuspidal automorphic representation of $GL(n, F)$, so that $V \subset \mathcal{A}_0\big(GL(n, F)\backslash GL(n, A), \omega\big)$, and let χ be a Hecke character of $A^\times/F^\times$. Let V_χ be the space of functions of the form $\phi_\chi(g) = \chi\big(\det(g)\big)\,\phi(g)$, where $\phi \in V$. Then it is easy to see that $V_\chi \subset \mathcal{A}_0\big(GL(n, F)\backslash GL(n, A), \chi^2\omega\big)$, and we obtain on this space an automorphic cuspidal representation of $GL(2, A)$. It is easy to see (Exercise 3.3.6) that if v is a non-Archimedean place, then the v-component of this representation is $\chi_v \otimes \pi_v$. (This would be true at an Archimedean place v too, of course, had we given definitions of the twist $\chi_v \otimes \pi_v$ when (π_v, V_v) is a $(\mathfrak{g}_v, K_v)$-module.) So we call this the representation obtained by *twisting* π by χ, and denote it as $\chi \otimes \pi$.

Theorem 3.3.5 *Let (π, V) be an automorphic cuspidal representation of $GL(2, A)$ with central quasicharacter ω. Then $\hat{\pi} \cong \omega^{-1} \otimes \pi$.*

We will not need this result, though the corresponding non-Archimedean local statement (Eq. 3.10) will play a role in Section 3.5.

Proof We leave the proof to the reader (Exercise 3.3.7). ■

We now state an important result.

Theorem 3.3.6: Multiplicity one *Let (π, V) and (π', V') be two irreducible admissible subrepresentations of $\mathcal{A}_0\big(GL(n, F)\backslash GL(n, A), \omega\big)$. Assume that $\pi_v \cong \pi'_v$ for all Archimedean v and all but finitely many non-Archimedean v. Then $V = V'$.*

This result was first published in this form by Casselman (1973), though the proof we will give for this in Section 3.5 is due to Piatetski-Shapiro (1979). It is only a slight elaboration of arguments from Jacquet and Langlands (1970), which prove the same result under the stronger assumption that $\pi_v \cong \pi'_v$ for *all* v. Actually the hypothesis that $\pi_v \cong \pi'_v$ for Archimedean v is unnecessary. (See Langlands (1980), Lemma 3.1).) This result is sometimes called the *strong multiplicity one* because it is not assumed that $\pi_v \cong \pi'_v$ for every place. The multiplicity one theorem was extended to $GL(n)$ independently by Shalika (1974), Piatetski-Shapiro (1975) and by Gelfand and Kazhdan (1975). The crucial point in the argument is the uniqueness of Whittaker models, and for finite fields, the corresponding result goes back to Gelfand and Graev (1962). Further related papers are Piatetski-Shapiro (1979), Bernstein and Zelevinsky (1976), Jacquet and Shalika (1981) and Cogdell and Piatetski-Shapiro (1994).

We will prove Theorem 3.3.6 in Section 3.5 when $n = 2$.

Exercises

Exercise 3.3.1 Prove the continuity of the representation ρ defined by Eq. (3.8). [HINT: See Exercise 3.2.1.]

Exercise 3.3.2 Suppose that f is a K-finite smooth function on G, where $G = GL(n, \mathbb{R})$ and $K = O(n)$ or else $G = GL(n, \mathbb{C})$ and $G = U(n)$, and let X be an element of the Lie algebra of G. Let $X f$ be defined by Eq. (2.1). Prove that $X f$ is also K-finite. [HINT: See the proof of Proposition 2.4.2.]

Exercise 3.3.3 (a) Let F be a local field, and let dx denote the additive Haar measure on $\mathrm{Mat}_n(F)$. Show that $|\det(x)|^{-n}\,dx$ is both a left and right Haar measure on the group $GL(n, F)$, and conclude that $GL(n, F)$ is unimodular.
(b) Let F be a global field and A its adele ring. Show that $GL(n, A)$ is unimodular.

Exercise 3.3.4 What modifications are needed to prove Proposition 3.3.2 when $n = 2$ and F is an arbitrary global field?

Exercise 3.3.5 Let Σ be an indexing set, and for almost all $v \in S$, let there be given a vector space V_v and vectors x_v° and $y_v^\circ \in V_v$ such that $x_v^\circ = y_v^\circ$ for almost all v. Then the restricted tensor product of the V_v with respect to the x_v° is naturally isomorphic to the restricted tensor product of the V_v with respect to the y_v°. Formulate this remark more precisely, and prove it.

Exercise 3.3.6 Let (π, V) be an automorphic cuspidal representation of $GL(n, A)$, and let χ be a Hecke character of $A^\times/F^\times$. Prove that the v-component of the automorphic representation V_χ is $\chi_v \otimes \pi_v$.

Exercise 3.3.7 Prove Theorem 3.3.5 [Use Eq. (3.10). You will probably need to establish a corresponding result at the Archimedean places also.]

3.4 The Tensor Product Theorem

In this section, we will prove Theorem 3.3.3. Our discussion is based on Flath (1979) and Jacquet and Langlands (1970). We will make free use of results from Chapter 4. Moreover, we will defer to the definite discussion of Knapp and Vogan (1995) in our discussion of in our discussuion of the Hecke algebra associated to reductive Lie group. It is a crucial point (observed by Flath) that a certain algebra may be found whose (smooth) modules are the same as $(\mathfrak{g}, K)$-modules. These algebras are also important for Knapp and Vogan in laying the foundations of the theory of cohomological induction, and their discussion is complete and detailed. There is little point in reproducing it in detail here, so our discussion of the Hecke algebras will be correspondingly brief.

In this section, we will encounter rings both with unit and without unit. If R is a ring with or without unit, we recall that an R-module is *simple* if $M \neq 0$ and if M has no proper nontrivial R-submodules.

Let Ω be a field. In our applications, Ω will be the complex numbers, but the results of this section are true over an arbitrary algebraically closed field Ω. When we need the assumption of algebraic closure, we will make it explicitly, so for the time being Ω can be an arbitrary field. Let R be an Ω-algebra with unit. If M is a module over R, then M is automatically a vector space over Ω, and we have a homomorphism $\phi : R \to \mathrm{End}_\Omega(M)$, namely, $\phi(r)\,m = r \cdot m$ if $r \in R$ and $m \in M$. The kernel of ϕ is a two-sided ideal J of R, and we may naturally identify R/J with its image.

Theorem 3.4.1: Burnside *Assume that Ω is an algebraically closed field. With notation as above, if M is finite dimensional over Ω and simple as an R-module, then the homomorphism $\phi : R \to \mathrm{End}_\Omega(M)$ is surjective, so $R/J \cong \mathrm{End}_\Omega(M)$. Moreover, $\mathrm{End}_R(M)$ is one dimensional over Ω, and consists of just the scalar endomorphisms $m \to \lambda m$ of M, $\lambda \in \Omega$.*

Proof See Lang (1993, Corollary XVII.3.3, p. 648). ■

Throughout the following discussion, tensor product $\otimes$ will always mean tensor product over Ω. Now suppose that A and B are Ω-algebras. Then we may form the tensor product algebra $R = A \otimes B$.

Proposition 3.4.1: Bourbaki (1958, 7.7) *Suppose that A and B are algebras (with unit) over the field Ω. Let $R = A \otimes B$. Let P be a simple R-module that is finite dimensional over Ω. Then there exists a simple A-module M and a simple B-module N such that P is isomorphic to a quotient of $M \otimes N$. Moreover, the isomorphism classes of M and N are uniquely determined.*

Proof We identify A and B with their images in R under the canonical injections $a \mapsto a \otimes 1$ ($a \in A$) and $b \mapsto 1 \otimes b$ ($b \in B$). With this notation, ab ($= ba$) and $a \otimes b$ are synonymous. Because P is finite dimensional over Ω, it contains a simple A-module M – indeed, any nonzero A-module of minimal dimension is obviously simple. Let $N_1 = \mathrm{Hom}_A(M, P)$. We make N_1 a B-module as follows: if $n \in N_1$ and $b \in B$, then $bn \in N_1$ is defined by $(bn)(m) = b \cdot n(m)$. We now define a map $\lambda : M \otimes N_1 \to P$ by $\lambda(m \otimes n) = n(m)$. We claim that λ is an R-module homomorphism. Indeed, if $a \otimes b \in R$, we have

$$\lambda\big((a\otimes b)(m\otimes n)\big) = \lambda\big((am)\otimes(bn)\big) = b{\cdot}n(am) = ba{\cdot}n(m) = (a\otimes b)\lambda(m\otimes n).$$

Now N_1 is nonzero because it contains the identity map $M \to P$ and finite dimensional over Ω, so it contains a simple B-module N. It is a consequence of our definitions that ϕ is nonzero on $M \otimes N$. Because P is simple, the image of ϕ is all of P, proving that P is a quotient of $M \otimes N$.

If $d = \dim_\Omega(N)$, then P is isomorphic, as an A-module, to a direct sum of d copies of M. It is a simple consequence of Schur's lemma that the isomorphism class of M is uniquely determined by that of P. (See Lang (1993, Proposition XVII.1.2, on p. 643).) Similarly, the B-module isomorphism class of N is determined by P. ■

Proposition 3.4.2: Bourbaki (1958, 7.7) *Suppose that the field Ω is algebraically closed. Let A and B be algebras over the field Ω, and let $R = A \otimes B$. If M and N are simple A- and B-modules, respectively, that are finite dimensional over Ω, then $M \otimes N$ is a simple R-module, and every simple R-module that is finite dimensional over Ω has this form for uniquely determined M and N.*

Proof We have homomorphisms $\phi_M : A \to \mathrm{End}_\Omega(M)$ and $\phi_N : B \to \mathrm{End}_\Omega(N)$, given by $\phi_M(a)\, m = a{\cdot}m$ for $a \in A$ and $m \in M$, and $\phi_M(b)\, n = b{\cdot}n$ for $b \in B$ and $n \in N$. By Theorem 3.4.1, these homomorphisms are surjective. Thus to show that $M \otimes N$ is simple as an R-module, it is sufficient to show that it is simple as an $\mathrm{End}_\Omega(M) \otimes \mathrm{End}_\Omega(N)$-module. However, it is easy to see that the natural map $\mathrm{End}_\Omega(M) \otimes \mathrm{End}_\Omega(N) \to \mathrm{End}_\Omega(M \otimes N)$ is surjective, so this is the same as showing that $M \otimes N$ is simple as an $\mathrm{End}_\Omega(M \otimes N)$-module, and this is clear. The rest of the proposition follows easily from Proposition 3.4.1. ∎

We now define the notion of an *idempotented algebra*. This is an Ω-algebra H (usually without unit), together with a set $\mathcal{E}$ of idempotents. It is assumed that if $e_1, e_2 \in \mathcal{E}$, then there exists $e_0 \in \mathcal{E}$ such that $e_0 e_1 = e_1 e_0 = e_1$ and $e_0 e_2 = e_2 e_0 = e_2$. Moreover, it is assumed that for every element $\phi \in \mathcal{H}$, there exists an $e \in \mathcal{E}$ such that $e\phi = \phi e = \phi$. If $e, f \in \mathcal{E}$, we write $e \geq f$ if $e f = f e = f$. Then $\geq$ is a partial ordering on $\mathcal{E}$, which is a *directed set* (See Exercise 4.3.2 and Lang (1993, p. 51) for directed sets and direct limits.) To be strict, the idempotented algebra is the ordered pair $(H, \mathcal{E})$, though by abuse of language we may refer to H itself as the idempotented algebra if the set of idempotents $\mathcal{E}$ is understood. The notion of an idempotented algebra is identical to that of an algebra with *approximate identity* in the terminology of Knapp and Vogan (1995).

If R is a ring (with or without unit) and $e \in R$ is an idempotent, we will denote by $R[e]$ the ring eRe, with unit element e. If M is an R-module, we will denote by $M[e]$ the $R[e]$-module $e \cdot M$.

Let $(H, \mathcal{E})$ be an idempotented Ω-algebra. Suppose that $e \in \mathcal{E}$. Then $H[e]$ is an Ω-algebra (with unit e), and $M[e]$ is an $H(e)$-module. We call M *smooth* if $M = \bigcup_{e \in \mathcal{E}} M[e]$ and *admissible* if it is smooth and each $M[e]$ is finite dimensional over Ω.

The first example of an idempotented algebra is any algebra with a unit – we may take $\mathcal{E} = \{1\}$. If $(H_1, \mathcal{E}_1)$ and $(H_2, \mathcal{E}_2)$ are idempotented Ω-algebras, then the tensor product ring $(H_1 \otimes H_2, \mathcal{E})$ is an idempotented Ω-algebra, with

$$\mathcal{E} = \{e_1 \otimes e_2 | e_1 \in \mathcal{E}_1,\ e_2 \in \mathcal{E}_2\}. \tag{4.1}$$

Suppose that G is a locally compact totally disconnected group, which we assume to be unimodular. Let $\mathcal{H}_G$ be the *Hecke algebra*, namely, the convolution algebra $C_c^\infty(G)$ of locally constant, compactly supported functions. (See Chapter 4, Section 2.) If K_0 is an open compact subgroup of G, then the characteristic function of K_0 divided by the volume of K_0 is an idempotent that we denote ϵ_{K_0}. If K_1 is another open compact subgroup, and if $K_0 \subset K_1$, then $\epsilon_{K_0} * \epsilon_{K_1} = \epsilon_{K_1}$. By Proposition 4.2.1, G has a base of neighborhoods of the identity consisting of open and compact subgroups, and it follows easily that $\mathcal{H}_G$ is an idempotented algebra over $\mathbb{C}$.

Next, let K be a compact Lie group. Let $\mathcal{H}_K$ denote the ring (under convolution) of smooth functions $\phi : K \to \mathbb{C}$ that are K-finite under both left and right translation by elements of K. Such functions are called *matrix coefficients* of K. By the Peter–Weyl theorem (Theorem 2.4.1), $\mathcal{H}_K$ is dense in $C(K)$ (with the L^∞ norm) and in $L^2(K)$. This is an idempotented algebra over $\mathbb{C}$. To see this, we take note of some elementary consequences of the Schur orthogonality relations. Let (π_i, V_i) be representatives of the distinct isomorphism classes of irreducible finite-dimensional unitary representations of K, where $i = 1, 2, 3, \cdots$. Then by Exercise 3.4.1,

$$\mathcal{H}_K \cong \bigoplus_{i=1}^{\infty} \mathrm{End}_{\mathbb{C}}(V_i). \tag{4.2}$$

This is an algebraic direct sum. Corresponding to the identity element in $\mathrm{End}_{\mathbb{C}}(V_i)$ is the idempotent $e_i \in \mathcal{H}_K$ defined by

$$e_i(k) = d_i^{-1} \operatorname{tr}\big(\pi_i(k)\big), \tag{4.3}$$

where $d_i = \dim(V_i)$. If Λ is a finite subset of $\{1, 2, 3, \cdots\}$, let

$$e_\Lambda = \sum_{i \in \Lambda} e_i. \tag{4.4}$$

Let $\mathcal{E}$ be the set of all e_Λ. Then it is clear that from Eq. (4.2) that $(\mathcal{H}_K, \mathcal{E})$ is an idempotented algebra over $\mathbb{C}$.

We note an inconsistency of notation between this section and Section 4.2. In that section, the notation $\mathcal{H}_K$ is used to mean something different – there, $\mathcal{H}_K$ is used to denote the algebra of K-biinvariant elements of $\mathcal{H}_G$.

Proposition 3.4.3 *Let (π, V) be a representation of K that is an algebraic direct sum of finite-dimensional representations. Then we obtain a smooth representation $\pi : \mathcal{H}_K \to \mathrm{End}(V)$ by*

$$\pi(\phi)\, v = \int_K \phi(k)\, \pi(k)\, v \, dk. \tag{4.5}$$

Conversely, if a smooth representation π of $\mathcal{H}_K$ is given, there exists a representation π of K such that Eq. (4.5) is valid.

Proof This is a simple consequence of the decomposition Eq. (4.2) (Exercise 3.4.2). ■

More generally, let G be a reductive Lie group, such as $GL(n, \mathbb{R})$, and let K be a maximal compact subgroup. Then we can form an algebra $\mathcal{H}_G$ that is a substitute for the Hecke algebra. This is the ring that is called the *Hecke algebra* by Flath (1979), and Borel and Jacquet (1979), though Jacquet and Langlands (1970) have a different ring that they call the Hecke algebra. We will call $\mathcal{H}_G$ the *Hecke algebra*. A rather extensive and relevant discussion of Flath's Hecke algebras may be found in Knapp and Vogan (1995), and we will defer to their account for certain facts. Indeed, their discussion should be consulted at this point.

We will use the language of distributions, so note that all the prerequisites on distributions are contained in Appendix B of Knapp and Vogan (1995). A *distribution T with compact support* on G is a linear functional on $C^\infty(G)$,

Vogan or in Section 2.9. The space of distributions with compact support is denote $\mathcal{E}'(G)$ for historical reasons. We will denoted the pairing of $T \in \mathcal{E}'(G)$ with $C^\infty(G)$ by $f \mapsto \langle T, f \rangle$. There are natural left and right actions ρ and λ of G on the space of distributions. (The formulas defining these are given in the different context of a p-adic group by Eq. (3.5) in Chapter 4.) If $x \in G$ and $T \in \mathcal{E}'(G)$, we say that x is in the *support* of T if for every neighborhood N of x there exists a function f vanishing outside of N such that $\langle T, f \rangle \neq 0$. The support of T is compact by Proposition B.15 of Knapp and Vogan, justifying the terminology "distribution with compact support." Distributions with compact support form a ring with respect to convolution (see Knapp and Vogan's equations (1.7a) and (1.7b)). Let $\mathfrak{g}$ be the Lie algebra of G, and let $U(\mathfrak{g}_\mathbb{C})$ be its enveloping algebra. Then $U(\mathfrak{g}_\mathbb{C})$ may be identified with the algebra of distributions in $\mathcal{E}'(G)$ that are supported at the identity. In this identification, we associate with $D \in U(\mathfrak{g}_\mathbb{C})$ the distribution $f \mapsto \langle D, f \rangle := Df(1)$. Thus if $D = X \in \mathfrak{g}$, we have

$$\langle X, f \rangle = \frac{d}{dt} f\left(e^{tX}\right)\big|_{t=0}.$$

With elements of $U(\mathfrak{g}_\mathbb{C})$ reinterpreted as distributions in this way, the multiplication in $U(\mathfrak{g}_\mathbb{C})$ is the same as convolution of distributions by Eq. (1.9) of Knapp and Vogan.

The Hecke algebra $\mathcal{H}_G$ is defined by Flath to be the algebra of compactly supported distributions on G that are supported in K and that are K-finite under both left and right translation. (By Proposition 1.83 of Knapp and Vogan, if such a distribution is K-finite under left translation, it is also K-finite under right translation.) Flath gave a purely algebraic construction of this ring, which we partially review. $\mathcal{H}_G$ contains, in addition to $U(\mathfrak{g}_\mathbb{C})$ the Hecke ring $\mathcal{H}_K$ already defined. An element of $\mathcal{H}_K$ is a smooth function ϕ on K, which is interpreted as a distribution by $f \mapsto \int_K \phi(k)\, f(k)\, dk$. Then every element of $\mathcal{H}_G$ has the form $f * D$ with $f \in \mathcal{H}_K$ and $D \in U(\mathfrak{g}_\mathbb{C})$ by Corollary 1.71 in Knapp and Vogan.

We recall that if R is an algebra over a field Ω and A is a right R-module and B is a left R-module, then the tensor product $A \otimes_R B$ is defined as a vector space over R – the tensor product satisfies $ar \otimes b = a \otimes rb$ if $a \in A, b \in B$ and $r \in R$ (see Knapp and Vogan (1995), Appendix A.1). Let $\mathfrak{k}$ be the Lie algebra of K. Let $\iota : U(\mathfrak{g}_\mathbb{C}) \to U(\mathfrak{g}_\mathbb{C})$ be the antiinvolution such that ${}^\iota X = -X$ when $X \in U(\mathfrak{g}_\mathbb{C})$. This involution is called "transpose" by Knapp and Vogan, and although we will avoid this terminology, it is at least true that if $G = GL(n, \mathbb{R})$, $K = SO(n)$ and $X \in \mathfrak{k}$, then ${}^\iota X$ is the matrix transpose of X. If $X \in \mathfrak{k}$, $f \in \mathcal{H}_K$, and $D \in U(\mathfrak{g}_\mathbb{C})$, then it is easy to see that

$$f * X * D = \rho\left({}^\iota X\right) f * D,$$

so if we regard $\mathcal{H}_K$ as a right $U(\mathfrak{k}_\mathbb{C})$-module by means of the action $(f, D_0) \mapsto \rho\left({}^\iota D_0\right) f$, $D_0 \in U(\mathfrak{k}_\mathbb{C})$, $f \in \mathcal{H}_K$, then this means that $(f, D) \mapsto f * D$ induces

$\rho({}^tD_0)\,f$, $D_0 \in U(\mathfrak{k}_\mathbb{C})$, $f \in \mathcal{H}_K$, then this means that $(f, D) \mapsto f * D$ induces a homomorphism $\mathcal{H}_K \otimes_{U(\mathfrak{k}_\mathbb{C})} U(\mathfrak{g}_\mathbb{C}) \to \mathcal{H}_G$, and according to Corollary 1.71 of Knapp and Vogan, this is a vector space isomorphism.

According to the notes in Knapp and Vogan, Flath had originally introduced $\mathcal{H}_K \otimes_{U(\mathfrak{k}_\mathbb{C})} U(\mathfrak{g}_\mathbb{C}) \to \mathcal{H}_G$ as a purely algebraic object and recognized that it had a ring structure, and the interpretation of this algebra as a convolution algebra of distributions was suggested by Deligne. In order to define the multiplicative structure algebraically, a bit of work is required, and this is the content of Proposition 1.80 of Knapp and Vogan.

Proposition 3.4.4 *Let G be a reductive Lie group, K its maximal compact subgroup, and $\mathfrak{g}$ its Lie algebra. Let V be a $(\mathfrak{g}, K)$-module. Then V is naturally a smooth module for $\mathcal{H}_G$, and moreover every smooth module for $\mathcal{H}_G$ arises in this fashion.*

The content of this proposition is that $(\mathfrak{g}, K)$-modules and smooth modules over $\mathcal{H}_G$ are equivalent categories.

Proof This result is stated at the end of Section 1.4 in Knapp and Vogan (1995). The proof is given in their Section 1.6. Our term "smooth" and their term "approximately unital" are synonymous in this context. ■

Sometimes it is convenient to work not with the full set $\mathcal{E}$ of idempotents, but with a sufficiently large subset. If E is any directed set (such as $\mathcal{E}$), we call a subset E° *cofinal* if for every $x \in E$ there exists an element x' of E° such that $x' \geq x$. The set E° is automatically a directed set in its own right.

Proposition 3.4.5 *Let M be a nonzero module over the idempotented Ω-algebra $(H, \mathcal{E})$, and let $\mathcal{E}^\circ$ be a cofinal subset of $\mathcal{E}$. Then M is a simple $(H, \mathcal{E})$-module if and only if $M[e]$ is either zero or is a simple $H[e]$-module for every $e \in \mathcal{E}^\circ$.*

Proof The proof of this is very similar to the equivalence of (ii) and (iii) in Proposition 4.2.3. We leave the proof to the reader (Exercise 3.4.3). ■

Proposition 3.4.6 *Let M and N be simple admissible modules over the idempotented algebra $(H, \mathcal{E})$. Let $\mathcal{E}^\circ$ be a cofinal subset of $\mathcal{E}$. Then $M \cong N$ if and only if $M[e] \cong N[e]$ as $H[e]$-modules for every $e \in \mathcal{E}^\circ$.*

Proof The proof of this is very similar to that of Proposition 4.2.7 and is left to the reader (Exercise 3.4.4). ■

Proposition 3.4.7 *Let R be a ring (not necessarily with unit), and let e, f be idempotents of R such that $ef = fe = e$. Then $f = e + e'$, where e*

$M[f] = M[e] \oplus M[e']$. Suppose furthermore that R is an algebra over the algebraically closed field Ω and that $M[e]$ is finite dimensional over Ω and simple as an $R[e]$-module. Then $\mathrm{Hom}_{R[e]}\big(M[e], M[f]\big)$ is one dimensional.

The fact that $ee' = e'e = 0$ is sometimes expressed by saying that the idempotents e and e' are *orthogonal*.

Proof Let $e' = f - e$. We have $e'^2 = f^2 - ef - fe + e^2 = f - e - e + e = e'$, so e' is an idempotent. It is easy to check that $ee' = e'e = 0$. If $m \in M[f]$, then $m = fm = em + e'm$, so $M[f] = M[e] + M[e']$. We must show that $M[e] \cap M[e'] = 0$. Let $m \in M[e] \cap M[e']$. Then $em = m$ and $e'm = m$, so $m = em = ee'm = 0$. Thus the sum $M[e] + M[e']$ is direct.

Finally, assume that $M[e]$ is simple as an $R[e]$-module. We note that in the decomposition $M[f] = M[e] \oplus M[e']$, $R[e]$ acts by zero on $M[e']$, so there are no nonzero $R[e]$-homomorphisms $M[e] \to M[e']$, and so

$$\mathrm{Hom}_{R[e]}\big(M[e], M[f]\big) \cong \mathrm{Hom}_{R[e]}\big(M[e], M[e]\big),$$

which (if Ω is algebraically closed) is one dimensional by Burnside's Theorem 3.4.1. ∎

Theorem 3.4.2 *Let $(H_1, \mathcal{E}_1)$ and $(H_2, \mathcal{E}_2)$ be idempotented algebras over an algebraically closed field Ω, and let $(H, \mathcal{E})$ be their tensor product. If M_1 and M_2 are simple admissible modules over H_1 and H_2, respectively, then $M_1 \otimes M_2$ is a simple admissible module over H, and every simple admissible module over H has this form. The isomorphism types of M_1 and M_2 are uniquely determined by that of M.*

Proof Let $\mathcal{E}$ be as in Eq. (4.1), and let M_1 and M_2 be simple admissible modules over H_1 and H_2, respectively. If $e_1 \otimes e_2 \in \mathcal{E}$, $e_1 \in \mathcal{E}_1$, $e_2 \in \mathcal{E}_2$, then

$$(M_1 \otimes M_2)[e_1 \otimes e_2] = M_1[e_1] \otimes M_2[e_2].$$

By Proposition 3.4.5 and the definition of admissibility, $M_1[e_1]$ and $M_2[e_2]$, if nonzero, are simple modules over $H_1[e_1]$ and $H_2[e_2]$, respectively, and are finite dimensional over Ω. Thus by Proposition 3.4.2, $(M_1 \otimes M_2)[e_1 \otimes e_2]$ is a simple module over $(H_1 \otimes H_2)[e_1 \otimes e_2] = H_1[e_1] \otimes H_2[e_2]$ and is finite dimensional over Ω. By Proposition 3.4.5, it follows that $M_1 \otimes M_2$ is simple and admissible.

Now suppose that M is a simple admissible module over $H_1 \otimes H_2$. Then M is nonzero, so there exists $e_1^\circ \otimes e_2^\circ \in \mathcal{E}$ such that $M[e_1^\circ \otimes e_2^\circ] \neq 0$. Fix $e_1^\circ \in \mathcal{E}_1$ and $e_2^\circ \in \mathcal{E}_2$ once and for all, and let $\mathcal{E}_i^\circ = \{e_i \in \mathcal{E}_i \,|\, e_i \geq e_i^\circ\}$, $i = 1, 2$. Because $\mathcal{E}_i$ is a directed set, it is easy to see that $\mathcal{E}_i^\circ$ is cofinal in $\mathcal{E}_i$.

We note that if $e_1 \in \mathcal{E}_1^\circ$ and $e_2 \in \mathcal{E}_2^\circ$, then $M[e_1 \otimes e_2]$ is simple. Indeed, by Proposition 3.4.5, it is sufficient to show it is nonzero. This is because $(e_1^\circ \otimes e_2^\circ)\, M[e_1 \otimes e_2] = M[e_1^\circ \otimes e_2^\circ]$, which is nonzero.

Let $e_1 \in \mathcal{E}_1^\circ$ and $e_2 \in \mathcal{E}_2^\circ$. Because as we have just seen $M[e_1 \otimes e_2]$ is simple, by Proposition 3.4.2, there exist simple $H_i[e_1 \otimes e_2]$-modules $M_i(e_1, e_2)$ $(i = 1, 2)$ such that

$$M(e_1 \otimes e_2) \cong M_1(e_1, e_2) \otimes M_2(e_1, e_2). \tag{4.6}$$

We will show next that for each $e_i \in \mathcal{E}_i^\circ$ there exist simple $H_i[e_i]$-modules $M_i(e_i)$ $(i = 1, 2)$ such that

$$M[e_1 \otimes e_2] \cong M_1(e_1) \otimes M_2(e_2). \tag{4.7}$$

Indeed, this will follow from Eq. (4.6) if we can show that the $H_1[e_1]$-isomorphism class of $M_1(e_1, e_2)$ really only depends on e_1 and that the $H_2[e_2]$-isomorphism class of $M_2(e_1, e_2)$ only depends on e_2. Let us show therefore that if $e_1, f_1 \in \mathcal{E}_1^\circ$ and $e_2 \in \mathcal{E}_2^\circ$, then $M_2(e_1, e_2) \cong M_2(f_1, e_2)$. Because $\mathcal{E}_1^\circ$ forms a directed set, we may find $g_1 \in \mathcal{E}_1^\circ$ such that $g_1 \geq e_1, f_1$, and it is sufficient to prove that $M_2(e_1, e_2) \cong M_2(g_1, e_2)$ and $M_2(f_1, e_2) \cong M_2(g_1, e_2)$. Thus there is no loss of generality in assuming that $e_1 \leq f_1$. We note that if $d_1 = \dim_\Omega M_1(e_1, e_2)$, then as an $H_2[e_2]$-module, $M[e_1 \otimes e_2]$ is a direct sum of d_1 copies of the simple $H_2[e_2]$-module $M_2(e_1, e_2)$. Similarly, $M[f_1 \otimes e_2]$ is a direct sum of some number of copies of the simple $H_2[e_2]$-module $M_2(f_1, e_2)$. By Proposition 3.4.6, $M[f_1 \otimes e_2]$ is a direct sum of $M[e_1 \otimes e_2]$ and $M[e_1' \otimes e_2]$, where $e_1' = f_1 - e_1$. It follows from Lang (1993, Proposition XVII.1.2, p. 643) that $M_2(e_1, e_2)$ and $M_2(f_1, e_2)$ are isomorphic. We have just seen that the isomorphism class of $M_2(e_1, e_2)$ depends only on the choice of $e_2 \in \mathcal{E}_2^\circ$. We therefore fix a particular module in this isomorphism class, which we denote $M_2(e_2)$. Similarly, there exists an $H[e_1]$-module $M_1(e_1)$ that is isomorphic to all $M_1(e_1, e_2)$ with $e_2 \in \mathcal{E}_2^\circ$. Now Eq. (4.7) follows from Eq. (4.6).

Next we note that if $e_1, f_1 \in \mathcal{E}_1$, $e_1 \leq f_1$, then

$$\dim_\Omega \operatorname{Hom}_{H_1[e_1]}\big(M_1(e_1), M_1(f_1)\big) \geq 1. \tag{4.8}$$

Indeed, choose $e_2 \in \mathcal{E}_2^\circ$. As an $H_1[e_1]$-module, $M[e_1 \otimes e_2]$ is isomorphic to a direct sum of multiple copies of $M_1(e_1)$, and $M[f_1 \otimes e_1]$ is isomorphic to a direct sum of multiple copies of $M_1(f_1)$. Thus Eq. (4.8) will follow if we can show the existence of a nonzero $H_1[e_1]$-module homomorphism $M[e_1 \otimes e_2] \to M[f_1 \otimes e_2]$.However, there exists a nonzero $H_1[e_1] \otimes H_2[e_2]$-module homomorphism by Proposition 3.4.7, which is *a fortiori* an $H_1[e_1]$-module homomorphism.

Now let $e_1 \leq f_1$ and $e_2 \leq f_2$, $e_i, f_i \in \mathcal{E}_i$. By Proposition 3.4.7

$$\dim_\Omega \operatorname{Hom}_{H_1[e_1] \otimes H_2[e_2]}\big(M_1(e_1) \otimes M_2(e_2), M_1(f_1) \otimes M_2(f_2)\big) = 1. \tag{4.9}$$

We will deduce from this that $\operatorname{Hom}_{H_i[e_i]}\big(M_i(e_i), M_i(f_i)\big)$ is one dimensional for both $i = 1$ and 2. Indeed, if M_1, M_2, N_1, N_2 are finite-dimensional vector spaces over Ω, then

$$\operatorname{Hom}_\Omega(M_1, N_1) \otimes \operatorname{Hom}_\Omega(M_2, N_2) \cong \operatorname{Hom}_\Omega(M_1 \otimes M_2, N_1 \otimes N_2).$$

In this isomorphism, if $\phi_i \in \mathrm{Hom}_\Omega(M_i, N_i)$ $(i = 1, 2)$, then $\phi_1 \otimes \phi_2 \in \mathrm{Hom}_\Omega(M_1, N_1) \otimes \mathrm{Hom}_\Omega(M_2, N_2)$ corresponds to the map in $\mathrm{Hom}_\Omega(M_1 \otimes M_2, N_1 \otimes N_2)$ (also denoted $\phi_1 \otimes \phi_2$) given by

$$(\phi_1 \otimes \phi_2)(m_1 \otimes m_2) = \phi_1(m_1) \otimes \phi_2(m_2).$$

Now (taking $M_1 = M_1(e_1)$, $M_2 = M_2(e_2)$, $N_1 = M_1(f_1)$, $N_2 = M_2(f_2)$) this isomorphism induces an injective homomorphism

$$\mathrm{Hom}_{H_1[e_1]}\big(M_1(e_1), M_1(f_1)\big) \otimes \mathrm{Hom}_{H_2[e_2]}\big(M_2(e_2), M_2(f_2)\big) \to$$
$$\mathrm{Hom}_{H_1[e_1]\otimes H_2[e_2]}\big(M_1(e_1) \otimes M_2(e_2), M_1(f_1) \otimes M_2(f_2)\big).$$

Thus by Eq. (4.9)

$$\dim \mathrm{Hom}_{H_1[e_1]}\big(M_1(e_1), M_1(f_1)\big) \cdot \dim \mathrm{Hom}_{H_2[e_2]}\big(M_2(e_2), M_2(f_2)\big) \leq 1,$$

so by Eq. (4.8),

$$\dim \mathrm{Hom}_{H_1[e_1]}\big(M_1(e_1), M_1(f_1)\big) = \dim \mathrm{Hom}_{H_2[e_2]}\big(M_2(e_2), M_2(f_2)\big) = 1.$$

Now let $e_1, f_1 \in \mathcal{E}_1^\circ$. We wish to choose a particular nonzero element $\lambda(e_1, f_1)$ of the one-dimensional space $\mathrm{Hom}_{H_1[e_1]}\big(M_1(e_1), M_1(f_1)\big)$ in such a way that when $e_1 \leq f_1 \leq g_1$, we have

$$\lambda(e_1, g_1) = \lambda(f_1, g_1) \circ \lambda(e_1, f_1). \tag{4.10}$$

To accomplish this, we first choose (for all $e_1 \in \mathcal{E}_1^\circ$) $\lambda(e_1^\circ, e_1)$ to be an arbitrary nonzero element of $\mathrm{Hom}_{H_1[e_1^\circ]}\big(M_1(e_1^\circ), M_1(e_1)\big)$. Once these choices are made, we can define $\lambda(e_1, f_1)$ as the unique element of $\mathrm{Hom}_{H_1[e_1]}\big(M_1(e_1), M_1(f_1)\big)$ such that

$$\lambda(e_1^\circ, f_1) = \lambda(e_1, f_1) \circ \lambda(e_1^\circ, e_1). \tag{4.11}$$

To elaborate, if $\lambda(e_1, f_1) \in \mathrm{Hom}_{H_1[e_1]}\big(M_1(e_1), M_1(f_1)\big)$, then both sides of Eq. (4.11) live in the one-dimensional space $\mathrm{Hom}_{H_1[e_1^\circ]}\big(M_1(e_1^\circ), M_1(f_1)\big)$, so they are proportional; and adjusting $\lambda(e_1, f_1)$ by a constant, we may make them equal. Now it is easy to see that Eq. (4.10) is satisfied whenever $e_1 \leq f_1 \leq g_1$ are elements of $\mathcal{E}_1^\circ$.

The identities (Eq. 4.10) imply that the Abelian groups $M_1(e_1)$, $e_1 \in \mathcal{E}_i^\circ$, together with the homomorphisms $\lambda(e_1, f_1)$ form a *directed family*. We may therefore form the direct limit (Lang, (1993), Section 3.10, p. 159):

$$M_1 = \varinjlim M(e_1).$$

This comes equipped with homomorphisms $\lambda(e_1) : M(e_1) \to M_1$. By Exercise 4.3.2, because the maps $\lambda(e_1, f_1)$ are all injective, the maps $\lambda(e_1)$ are injective too, and thus there is no loss of generality in replacing $M(e_1)$ by the equivalent module $M[e_1] \subseteq M$.

Similarly, we may construct a module M_2 of H_2 such that $M_2[e_2] \cong M_2(e_2)$ for all $e_2 \in \mathcal{E}_2^\circ$. By Eq. (4.7) we have, for $e_i \in \mathcal{E}_i^\circ$

$$M[e_1 \otimes e_2] \cong M_1[e_1] \otimes M_2[e_2] \cong (M_1 \otimes M_2)[e_1 \otimes e_2],$$

and by Proposition 3.4.6, it follows that $M \cong M_1 \otimes M_2$.

Lastly, we must show that the isomorphism types of M_1 and M_2 are uniquely determined. Indeed, let $e_1 \in \mathcal{E}_1^\circ$. We have $M[e_1 \otimes e_2^\circ] \cong M_1[e_1] \otimes M_2[e_2^\circ]$. This module is simple and finite dimensional over Ω. By Proposition 3.4.2, the isomorphism class of $M_1[e_1]$ is uniquely determined, and so by Proposition 3.4.6, so is the isomorphism class of M_1 and similarly so is that of M_2. ■

Let G be a totally disconnected, locally compact unimodular group. If (π, V) is a smooth representation of G, then we have a representation of the Hecke algebra $\mathcal{H}_G$, by

$$\pi(\phi)\, v = \int_G \phi(g)\, \pi(g)\, v\, dv, \qquad \phi \in \mathcal{H}_G,\ v \in V.$$

Thus V becomes a smooth module for $\mathcal{H}_G$. Moreover, every smooth module if $\mathcal{H}_G$ is of this type.

Proposition 3.4.8 *Let $\mathcal{H}_G$ be the Hecke algebra of a totally disconnected locally compact unimodular group G. Let V be a smooth module over $\mathcal{H}_G$. Then there exists a smooth representation $\pi : G \to \mathrm{End}_{\mathbb{C}}(V)$ such that $\phi \cdot x = \pi(\phi)\, x$ for $\phi \in \mathcal{H}_G$, $x \in V$.*

Proof Let $x \in V$ and $g \in G$. Because V is a smooth module, there exists a subgroup K_0 of G (depending on x) such that $x \in V[\epsilon_{K_0}]$. If Ω is a subset of G, let ϵ_Ω be the characteristic function of Ω divided by $\mathrm{vol}(\Omega)$. Then we define $\pi(g)\, x = \epsilon_{gK_0} \cdot x$. It is easy to see that this definition is independent of the choice of K_0. Now let $g, h \in G$. We choose subgroups K_0 and K_1 stabilizing x and $\pi(h)\, x$, respectively, so

$$\pi(g)\, \pi(h)\, x = \pi(\epsilon_{gK_1})\, \pi(\epsilon_{hK_0})\, x = \pi(\epsilon_{gK_1} * \epsilon_{hK_0})\, x.$$

We choose K_1 so small that $h^{-1} K_1 h \subseteq K_0$. Then it is easy to verify that $\epsilon_{gK_1} * \epsilon_{hK_0} = \epsilon_{ghK_0}$, and so we deduce that $\pi(gh) = \pi(g)\, \pi(h)$. Thus $\pi(g)$ is a representation. It is simple to verify that π is smooth. ■

Proposition 3.4.9 *Let G_1 and G_2 be locally compact totally disconnected groups. Let (π_i, M_i) be irreducible admissible representations of G_i $(i = 1, 2)$. Then $(\pi_1 \otimes \pi_2, M_1 \otimes M_2)$ is an irreducible admissible representation of $G_1 \times G_2$, and every irreducible admissible representation of $G_1 \times G_2$ is of this type.*

Proof The Hecke algebra $\mathcal{H}_{G_1 \times G_2} \cong \mathcal{H}_{G_1} \otimes \mathcal{H}_{G_2}$. So our result follows from Theorem 3.4.2 and Proposition 3.4.8. ■

Now let $(H, \mathcal{E})$ be an idempotented algebra over the field Ω equipped with an antiinvolution $\iota : H \to H$. This is a linear map of order two such that ${}^{\iota}(xy) = {}^{\iota}y\,{}^{\iota}x$. Let $e^\circ \in \mathcal{E}$. We say that the idempotent e° is *spherical* if there exists an antiinvolution ι of H such that ${}^{\iota}x = x$ for all $x \in H[e^\circ]$. The existence of ι implies that $H[e^\circ]$ is commutative, because if $x, y \in H[e^\circ]$, $xy = {}^{\iota}(xy) = yx$.

Let e° be a spherical idempotent of the idempotented algebra $(H, \mathcal{E})$, and let ι be the corresponding involution. If M is a simple admissible H-module, then $M[e^\circ]$ is a simple module over the commutative ring $H[e^\circ]$ and is finite dimensional as a vector space over $\mathbb{C}$; this implies that it is at most one dimensional by Exercise 3.4.4. Let M be a smooth H-module. If λ is a linear functional on M and $m \in M$, we will write $\langle m, \lambda \rangle$ instead of $\lambda(m)$ for reasons of symmetry. We say that λ is *smooth* if there exists $e \in \mathcal{E}$ such that $\langle em, \lambda \rangle = \langle m, \lambda \rangle$ for all $m \in M$. Let $\hat{M}$ be the vector space of smooth linear functionals on M. Give $\widehat{M}$ the following structure of an H-module by

$$\langle m, \phi\lambda \rangle = \langle {}^{\iota}\phi m, \lambda \rangle .$$

Then $\widehat{M}$ is a smooth H-module, and if M is admissible, then so is $\widehat{M}$. Moreover, if $M[e^\circ]$ is nonzero (hence one dimensional), then $\widehat{M}[e^\circ]$ is also. Indeed, let m° be a nonzero element of $M[e^\circ]$. Then we may define an element $\hat{m}^\circ \in \widehat{M}[e^\circ]$ by

$$e^\circ\, m = \langle m, \hat{m}^\circ \rangle\, m^\circ \tag{4.12}$$

for all $m \in M$.

Theorem 3.4.3 *Let $(H, \mathcal{E})$ be an idempotented algebra over the field Ω, and let e° be a spherical idempotent. Let M and N be simple admissible H-modules such that $M[e^\circ]$ and $N[e^\circ]$ are nonzero. If $M[e^\circ] \cong N[e^\circ]$ as $H[e^\circ]$-modules, then $M \cong N$ as H-modules.*

This is a generalization of Theorem 4.6.3.

Proof Let m° and n° be nonzero elements of $M[e^\circ]$ and $N[e^\circ]$, and let $\widehat{M}^\circ$ and $\hat{n}^\circ$ be the corresponding elements of $\widehat{M}[e^\circ]$ and $\hat{N}[e^\circ]$, respectively. As in the proof of Theorem 4.6.3, it is enough to show that

$$\langle \phi\, m^\circ, \hat{m}^\circ \rangle = \langle \phi\, n^\circ, \hat{n}^\circ \rangle \tag{4.13}$$

for all $\phi \in H$. Let $\phi' = e^\circ \phi e^\circ \in H[e^\circ]$. The left side of Eq. (4.13) equals

$$\langle \phi\, e^\circ\, m^\circ, e^\circ\, \hat{m}^\circ \rangle = \left\langle \phi'\, m^\circ, \hat{m}^\circ \right\rangle ,$$

and because $M[e^\circ] \cong N[e^\circ]$, this equals $\left\langle \phi'\, n^\circ, \hat{n}^\circ \right\rangle$, from whence we get Eq. (4.13). ■

We now consider infinite restricted tensor products of idempotented Ω-algebras. Let $(H_v, \mathcal{E}_v)$ $(v \in \Sigma)$ be an indexed family of idempotented algebras, and in all but finitely many v, we specify in $\mathcal{E}_v$ a distinguished element e_v°. Then we consider the restricted tensor product H of the H_v with respect to the e_v. This is itself an idempotented algebra. Let $\mathcal{E}$ be the set of idempotents of H consisting of the tensors $\otimes_v e_v$, where $e_v \in \mathcal{E}_v$ and $e_v = e_v^\circ$ for all but finitely many v. Then $(H, \mathcal{E})$ is an idempotented Ω-algebra.

We assume that $H_v[e_v^\circ]$ is commutative for almost all v. Suppose v is a place such that $H_v[e_v^\circ]$ is commutative. Let M_v be a simple admissible H_v-module. Then $M_v[e_v^\circ]$ is either zero or simple as an $H_v[e_v^\circ]$-module by Proposition 3.4.5. It is a finite-dimensional vector space over $\mathbb{C}$, and if it has greater dimension than one, we can find a simultaneous eigenspace for $H_v[e_v^\circ]$, contradicting the simplicity of $M_v[e_v^\circ]$. Thus when $H_v[e_v^\circ]$ is commutative, $M_v[e_v^\circ]$ is either one dimensional or zero.

Suppose now that for each $v \in \Sigma$ we have specified a simple admissible H_v-module M_v and that $M_v[e_v^\circ]$ is one dimensional for almost all v, and for almost all v, we specify a nonzero element m_v° of the space $M_v[e_v^\circ]$. Then we may form the restricted tensor product $M = \bigotimes_v M_v$ with respect to the m_v°, and this is naturally a module over H.

We remind the reader that if R_1 and R_2 are rings with unit elements 1_{R_1} and 1_{R_2}, a ring homomorphism $\gamma : R_1 \to R_2$ is always assumed to satisfy $\gamma(1_{R_1}) = 1_{R_2}$.

Proposition 3.4.10 *Let R_v $(v \in \Sigma)$ be a family of rings, each with unit $e_v \in R_v$, and let R be the restricted tensor product of the R_v with respect to the e_v. Let $\gamma : R \to \Omega$ be a ring homomorphism. Then there exist ring homomorphisms $\gamma_v : R_v \to \Omega$ such that $\gamma(\otimes_v r_v) = \prod_v \gamma_v(r_v)$.*

We note that given a family of ring homomorphisms γ_v, if $\otimes_v r_v \in \bigotimes_v R_v$, then $\gamma_v(r_v)$ is one for almost all v, so $\prod_v \gamma_v(r_v)$ is defined.

Proof Let 1_v denote the unit element in R_v. We have a ring homomorphism $i_v : R_v \to R$, defined by $i_v(x_v) = x_v \otimes \left(\otimes_{w \neq v} 1_w \right)$. Let $\gamma_v = \gamma \circ i_v$. Then if $r \in R$, we can write $r = \prod_v i_v(r)$, where all but finitely many terms on the right are one. Then it is clear that $\gamma(r) = \prod_v \gamma_v(r_v)$. ∎

Theorem 3.4.4 *Let $(H_v, \mathcal{E}_v)$ $(v \in \Sigma)$ be an indexed family of idempotented Ω-algebras, and for almost all v, let $e_v^\circ \in \mathcal{E}_v$ be a spherical idempotent. Let $(H, \mathcal{E})$ be the restricted tensor product of the H_v with respect to the e_v°, itself an idempotented Ω-algebra. For each $v \in \Sigma$ let there be specified a simple admissible module M_v, and for almost all v let m_v° be a nonzero element of $M_v[e_v^\circ]$. Let M be the restricted tensor product of the M_v with respect to the m_v°. Then $\otimes_v M_v$ is a simple admissible H-module. Moreover, every simple admissible module is of this type, with uniquely determined modules M_v.*

Proof Let simple admissible modules M_v and elements $m_v^\circ \in M_v[e^\circ]$ as described above be given. We will show that $M = \otimes_v M_v$ is simple and admissible.

Let $e = \otimes e_v \in \mathcal{E}$ be given. Then there exists a finite subset S of Σ such that if $v \in \Sigma - S$, then $e_v = e_v^\circ$, and furthermore $H[e_v]$ is commutative, so $M[e_v]$ is one dimensional. Then

$$M[e] \cong \bigotimes_{v \in S} M_v[e_v]. \tag{4.14}$$

Indeed, this is because $\bigotimes_{v \notin S} M_v[e_v]$ is one dimensional, being spanned by the vector $\otimes_{v \notin S} e_v^\circ$. So tensoring with this vector is an isomorphism

$$\bigotimes_{v \in S} M_v[e_v] \to \bigotimes_{v \in \Sigma} M_v[e_v] = M[e].$$

Now the left side (if nonzero) is simple by Theorem 3.4.2 and Proposition 3.4.4 (applied to M_v). By Proposition 3.4.4 (applied to M), it follows that M is simple.

Now let M be a simple H-module. We must show that $M \cong \bigotimes_v M_v$, where the M_v are simple admissible modules for the H_v, and the tensor product is restricted with respect to elements $m_v^\circ \in M_v$. We will prove this by combining two special cases. Firstly, if the indexing set Σ is finite, the restricted tensor product is of course the same as the ordinary tensor product, and this result follows by iterated applications of Theorem 3.4.2.

We next consider another special case. *We assume that e_v° is a spherical idempotent for all v, and we also assume, with $e = \otimes_v e_v^\circ$, that $M[e] \neq 0$.* This implies that $M[e]$ is one dimensional, and if m denotes a generator, we obtain a ring homomorphism $\gamma : H[e] \to \Omega$ by $hm = \gamma(h)m$, $h \in H[e]$. By Proposition 3.4.10, we may factor γ as $\gamma(\otimes_v h_v) = \prod_v \gamma_v(h_v)$ when $h_v \in H_v[e_v^\circ]$, where γ_v is a homomorphism $H_v[e_v^\circ] \to \Omega$. Now we claim that for each v there exists a simple admissible module M_v of H_v and a nonzero element $m_v \in M_v[e_v^\circ]$ such that $h_v m_v = \gamma_v(h_v) m_v$. Indeed, we may see this by decomposing $H = H_v \otimes H_v'$, where $H_v' = \otimes_{w \in \Sigma, w \neq v} H_w$ (tensor product restricted by the e_v°). By Theorem 3.4.2, there exist simple admissible modules M_v and M_v' for H_v and H_v', respectively, such that $M[e] = M_v[e_v^\circ] \otimes M_v'[e_v']$, where $e_v' = \otimes_{w \neq v} e_v^\circ$. Now consider $N = \otimes_v M_v$, the tensor product restricted with respect to the m_v. It is clear that $N[e] = \otimes_v M_v[e_v] \cong M[e]$ as $H[e]$-modules, and therefore $M \cong \otimes_v M_v$ by Theorem 3.4.3.

We deduce the general case from these two special cases. Choose $e \in \mathcal{E}$ such that $M[e] \neq 0$. Let S be a sufficiently large finite subset of Σ such that if $v \in \Sigma - S$, then e_v° is a spherical idempotent. We represent H as a finite tensor product $\bigotimes_{v \in S} H_v \otimes H'$, where $H' = \bigotimes_{v \in \Sigma - S} H_v$. Using the first special case proved above (Σ finite) we can write $M = \bigotimes_{v \in S} M_v \otimes M'$, where M_v is a simple admissible module for H_v, and M' is a simple admissible module for H'. By using the second "spherical" special case considered above, we obtain the further decomposition $M' = \bigotimes_{v \in \Sigma - S} M_v$. ∎

We may now give the following proof.

Proof of Theorem 3.3.3 Let F be a global field, and let A be its adele ring. Let Σ be the set of all places of F. If $v \in \Sigma$, we have defined a Hecke algebra $\mathcal{H}_{GL(n,F_v)}$ above. If v is non-Archimedean, let e_v° be the characteristic function of $GL(n, \mathfrak{o}_v)$. We normalize the Haar measure on $\mathcal{H}_{GL(n,F_v)}$ so that the volume of $GL(n, \mathfrak{o}_v)$ is one; then e_v° is an idempotent. We claim that it is spherical. Indeed, the transpose map on $GL(n)$ induces an antiinvolution ι on $\mathcal{H}_{GL(n,F_v)}$. It follows easily from the elementary divisor theorem (Theorem 1.4.1) that a complete set of double coset representatives for $GL(n, \mathfrak{o}_v)\backslash GL(n, F)/GL(n, \mathfrak{o}_v)$ consists of diagonal matrices. If $n = 2$, this is checked in Proposition 4.6.2, but there is no difference in the general case. As in Theorem 4.6.1, this implies that the *spherical Hecke algebra* of $GL(n, \mathfrak{o}_v)$-biinvariant functions is commutative. This spherical Hecke algebra is $\mathcal{H}_{GL(n,F_v)}[e_v^\circ]$, and we may reinterpret Theorem 4.6.1 (or its generalization to $GL(n)$) as saying that e_v° is a spherical idempotent. We define the *global Hecke algebra* $\mathcal{H}_{GL(n,A)}$ to be the restricted tensor product of the local Hecke algebras $\mathcal{H}_{GL(n,F_v)}$, where $v \in \Sigma$, the set of all places of F; and the tensor product is restricted with respect to the subalgebras $\mathcal{H}_{GL(n,F_v)}[e_v^\circ]$. (Actually, making use of the Cartan decomposition, one may specify a spherical idempotent at the Archimedean places too, but we do not need this.) In view of Propositions 3.4.4 and 3.4.8, we may reinterpret an irreducible admissible representation of $GL(n, A)$ as a simple admissible module for $\mathcal{H}_{GL(n,A)}$. With this reinterpretation, Theorem 3.3.3 follows immediately from Theorem 3.4.4. ■

Exercises

Exercise 3.4.1 Let K be a compact group, and let (π, V) be an irreducible representation. Let $\langle\ ,\ \rangle$ be an invariant inner product on V.
(a) We define a linear map $\phi : V \otimes V \to \mathrm{End}(V)$ by $\phi(x \otimes y)(z) = \langle z, y\rangle\, x$. Prove that ϕ is an isomorphism, and that

$$\phi(x \otimes y) \circ \phi(z \otimes w) = \langle z, y\rangle\, \phi(x \otimes w).$$

(b) Define a map $\Phi : V \otimes V \to C(K)$ by

$$\Phi(x \otimes y)(g) = \dim(V)\, \langle x, \pi(g)y\rangle\,.$$

Prove that

$$\Phi(x \otimes y) * \Phi(z \otimes w) = \langle z, y\rangle\, \Phi(x \otimes w).$$

Conclude that $\phi(x\otimes y) \to \Phi(x\otimes y)$ is a ring homomorphism $\mathrm{End}(V) \to C(V)$. [Hint: Use the Schur orthogonality relations (Knapp (1986, Section I.5). See also Proposition 2.4.3).]
(c) Prove that $\mathcal{H}_K$ is the algebraic direct sum of the images of the maps defined in (b), as (π, V) ranges through the isomorphism classes of irreducible representations of K. [Hint: Use the Peter–Weyl theorem.]

Exercise 3.4.2 Complete the proof of Proposition 3.4.3.

Exercise 3.4.3 Complete the proof of Proposition 3.4.5. [HINT: See Proposition 4.2.3.]

Exercise 3.4.4 Complete the proof of Proposition 3.4.6. [HINT: See Proposition 4.2.7.]

Exercise 3.4.5 Let R be a commutative algebra over the field Ω, and let M be a simple R-module that is finite dimensional over Ω. Prove that M is one dimensional.

Exercise 3.4.6 Is the assumption in Theorem 3.4.4 that the idempotents e_v° be spherical really necessary? Suppose instead that we assume instead that $\mathcal{H}_v[e_v^\circ]$ is commutative for almost all v. Is the theorem still true?

3.5 Whittaker Models and Automorphic Forms

A fundamental result is the *uniqueness of Whittaker models* – the so-called local multiplicity one theorem. This leads to the proof of the multiplicity one Theorem 3.3.6 and also to the functional equations of the standard L-function of an automorphic cuspidal representation of $GL(2)$.

Many notations will be as in Sections 3.3 and 3.4. We will require some results from Chapter 4 in this section. The reader may take these on faith during a first reading.

Theorem 3.5.1: Local multiplicity one *Let F be a non-Archimedean local field, ψ a nontrivial additive character of F, and let (π, V) be an irreducible admissible representation of $GL(2, F)$. Then up to constant multiple, there exists at most one linear functional Λ on V such that*

$$\Lambda\left(\pi\begin{pmatrix} 1 & x \\ & 1 \end{pmatrix}\xi\right) = \psi(x)\,\Lambda(\xi), \qquad x \in F,\ \xi \in V. \tag{5.1}$$

Proof This is Theorem 4.4.1, and we refer to Chapter 4 for the proof. ■

A nonzero functional satisfying Eq. (5.1) is called a *Whittaker functional* (with respect to ψ).

Our objective will be to show (in Theorem 3.5.4) that the local uniqueness result Theorem 3.5.1 implies a corresponding global uniqueness. The proof is slightly complicated by the presence of Archimedean places. So to emphasize the essential simplicity of the idea, we will prove this global uniqueness first for a function field F. Thus F is a global field such that every place of F is non-Archimedean. Let A be the adele ring of F, and let ψ be an additive character of A/F. Let (π, V) be an irreducible admissible representation of $GL(2, A)$. By a *pre-Whittaker functional*, we mean a functional $\Lambda : V \to \mathbb{C}$

satisfying

$$\Lambda\left(\pi\begin{pmatrix}1 & x\\ & 1\end{pmatrix}v\right) = \psi(x)\,\Lambda(v), \qquad x \in A,\ v \in V. \tag{5.2}$$

By a *Whittaker functional* we mean a nonzero pre-Whittaker functional.

Theorem 3.5.2 *Let F be a function field and A its adele ring, let (π, V) be an irreducible admissible representation of $GL(2, A)$, so that by Theorem 3.3.3, $\pi = \otimes_v \pi_v$, where (π_v, V_v) is an irreducible admissible representation of $GL(2, F_v)$ for each place v of F. This is a restricted tensor product with respect to a family of vectors $\xi_v^\circ \in V_v$ such that ξ_v° is $GL(2, \mathfrak{o}_v)$-invariant for almost all v. If Λ is a Whittaker functional on V, then for each place v of F there exists a Whittaker functional Λ_v on V_v such that $\Lambda_v(\xi_v^\circ) = 1$ for almost all v, and*

$$\Lambda(\otimes_v \xi_v) = \prod_v \Lambda_v(\xi_v). \tag{5.3}$$

(All but finitely many factors on the right are equal to one.) The dimension of the space of pre-Whittaker functionals on V is at most one.

Proof If Λ is nonzero, then it is nonzero on some pure tensor, and so there exists a pure tensor ξ° such that $\Lambda(\xi^\circ) = 1$. We write $\xi^\circ = \otimes_v \xi_v^\circ$. This means changing the vectors ξ_v° at a finite number of places, but such a change does not alter the restricted tensor product (Exercise 3.3.5). If w is any fixed place of F, we have an injection $i_w : V_w \to V$ defined by

$$i_w(\xi_w) = \xi_w \otimes \left(\bigotimes_{v \neq w} \xi_v^\circ\right). \tag{5.4}$$

Let $\Lambda_w : V_w \to \mathbb{C}$ be the composition $\Lambda \circ i_w$. We have $\Lambda_w(\xi_w^\circ) = 1$. It is clear that Λ_w is a Whittaker functional on V_w. We will prove Eq. (5.3) by induction on the cardinality of a finite set S such that $\xi_v = \xi_v^\circ$ for all $v \notin S$; for such ξ_v the formula (5.3) can be rewritten

$$\Lambda(\otimes_v \xi_v) = \prod_{v \in S} \Lambda_v(\xi_v).$$

If S is empty, then $\otimes_v \xi_v = \xi^\circ$, and both sides of Eq. (5.3) equal one. To add a single place w to S, let us assume that $\xi_v = \xi_v^\circ$ for $v \notin S \cup \{w\}$. We note that

$$x_w \to \Lambda\left(x_w \otimes \left(\bigotimes_{v \neq w} \xi_v\right)\right)$$

is a Whittaker functional on V_w, so by Theorem 5.1, it agrees with Λ_w up to a constant multiple. Thus

$$\Lambda\left(x_w \otimes \left(\bigotimes_{v\neq w} \xi_v\right)\right) = c\,\Lambda_w(x_w)$$

for some constant c, and to evaluate c, we take $x_w = \xi_w^\circ$. By induction, we see that the constant c equals

$$\prod_{v\in S} \Lambda_v(\xi_v).$$

Thus we obtain Eq. (5.3) when $\xi_v = \xi_v^\circ$ for $v \notin S \cup \{w\}$.

The uniqueness of Λ now follows from the uniqueness of the functionals Λ_v. ■

In order to prove an analog of Theorem 3.5.2 that is valid for number fields, we must be somewhat more careful about the Archimedean places. We will accomplish this in Theorem 3.5.4. The proof is just an elaboration of the proof of Theorem 3.5.2, so it can be skipped on a first reading of this chapter. However, the reader must first absorb the notion of a *Whittaker model*, which is roughly equivalent to that of a Whittaker functional. Let us begin by noting that Theorem 3.5.1 admits an equivalent formulation.

Theorem 3.5.3: Local multiplicity one, equivalent form *Let F be a non-Archimedean local field, ψ a nontrivial additive character of F, and let (π, V) be an irreducible admissible representation of $GL(2, F)$. Then there exists at most one space $\mathcal{W}$ of functions on $GL(2, F)$ such that if $W \in \mathcal{W}$, then*

$$W\left(\begin{pmatrix} 1 & x \\ & 1 \end{pmatrix} g\right) = \psi(x)\, W(g), \qquad x \in F,\ g \in GL(2, F), \tag{5.5}$$

and such that W is closed under right translation by elements of $GL(2, F)$, and the resulting representation of $GL(2, F)$ is isomorphic to π.

A space of functions $\mathcal{W}$ satisfying these conditions is called a *Whittaker model* for the representation (π, V) (with respect to ψ).

Proof The equivalence of these two formulations is explained in terms of Frobenius reciprocity at the end of Section 4.1 for the analogous result over finite fields. The equivalence of Theorem 3.5.1 and Theorem 3.5.3 is easy enough to explain and important, so we repeat this argument here. If Λ is a Whittaker functional, then we obtain a space $\mathcal{W}$ of functions on $GL(2, F)$ satisfying Eq. (5.5) as follows. For every vector $\xi \in V$, let

$$W_\xi(g) = \Lambda\big(\pi(g)\,\xi\big), \qquad g \in GL(2, F). \tag{5.6}$$

If $g \in GL(2, F)$, it is easy to see that $W_{\pi(g)\xi} = \rho(g)\, W_\xi$, where ρ is the action of $GL(2, F)$ by right translation: $\rho(g)\, W_\xi(h) = W_\xi(hg)$. This means that the space $\mathcal{W} = \{W_\xi | \xi \in V\}$ is closed under right translation and that the representation it affords is isomorphic to π. Conversely, let $\mathcal{W}$ be given, and an isomorphism $\xi \mapsto W_\xi$ of π onto $\mathcal{W}$. Then we define a Whittaker functional Λ by $\Lambda(\xi) = W_\xi(1)$. We see that specifying a Whittaker functional is equivalent to specifying a Whittaker model. ■

We require a result analogous to Theorem 3.5.3 at the Archimedean places. A statement for $(\mathfrak{g}, K)$-modules of $GL(2, \mathbb{R})$ may be found in Theorem 2.8.1. The identical result is true for $GL(2, \mathbb{C})$ – see Jacquet and Langlands (1970, Theorem 5.3, p. 232). The principal difference between the Archimedean result and Theorem 3.5.3 is that it is necessary to assume that the functions in $\mathcal{W}$ are of moderate growth. We prefer a formulation that unites the Archimedean and non-Archimedean cases. Let F be a local field, let $G = GL(2, F)$, let K denote its standard maximal compact subgroup, and let $\mathcal{H}_G$ be the Hecke algebra, defined in Section 3.4. Thus $\mathcal{H}_G$ is $C_c^\infty(G)$ if F is non-Archimedean, whereas if F is Archimedean, $\mathcal{H}_G$ is the convolution algebra of distributions on G that are K-finite and have support contained in K – as explained in Section 3.4, this ring may alternatively be constructed algebraically by amalgamating the ring $\mathcal{H}_K$ of matrix coefficients of K with $U(\mathfrak{g}_\mathbb{C})$, where $\mathfrak{g}_\mathbb{C}$ is the complexified Lie algebra of G. There is a natural action of $\mathcal{H}_G$ on $C_c^\infty(G)$, which we denote by ρ. Thus if G is non-Archimedean, $\phi \in \mathcal{H}_G$, and $f \in C_c^\infty(G)$, then

$$\big(\rho(\phi)\, f\big)(g) = \int_G \phi(h)\, f(gh)\, dh.$$

If G is Archimedean, and if we regard the elements of $\mathcal{H}_G$ as distributions, then the action of $\mathcal{H}_G$ is again by convolution; alternatively, the action of $\mathcal{H}_K$ is by convolution, and the action of $U(\mathfrak{g}_\mathbb{C})$ is by differentiation, as in Section 2.2; these actions are compatible and combine to give the action ρ of $\mathcal{H}_G$ on $C_c^\infty(G)$.

Let V be a simple admissible $\mathcal{H}_G$-module. We will denote the action of $\mathcal{H}_G$ on V by $\pi : \mathcal{H}_G \to \mathrm{End}(V)$. By a *Whittaker model* of (π, V) with respect to a fixed nontrivial additive character ψ of F, we mean a space of functions $\mathcal{W} : G \to \mathbb{C}$ that satisfy

$$W\left(\begin{pmatrix} 1 & x \\ & 1 \end{pmatrix} g\right) = \psi(x)\, W(g), \qquad x \in F,\ W \in \mathcal{W}. \tag{5.7}$$

It is assumed that the functions in $\mathcal{W}$ are smooth (which if F is non-Archimedean just means locally constant) and satisfy a growth condition: for fixed $g \in G$, $W \in \mathcal{W}$, the function

$$W\left(\begin{pmatrix} y & \\ & 1 \end{pmatrix} g\right)$$

is bounded by a polynomial in $|y|$ as $|y| \to \infty$. In addition, it is assumed that there exists a vector space isomorphism $\xi \mapsto W_\xi$ of V onto $\mathcal{W}$ such that if $\xi \in V$ and $\phi \in \mathcal{H}_G$, then

$$W_{\pi(\phi)\xi} = \rho(\phi)\, W_\xi. \tag{5.8}$$

Proposition 3.5.1 *Let F be a local field, ψ a nontrivial additive character of F, and let (π, V) be a simple admissible $\mathcal{H}_G$-module, where $\mathcal{H}_G$ is the Hecke algebra of $G = GL(2, F)$. Then (π, V) has at most one Whittaker model with respect to ψ.*

Proof In view of Propositions 3.4.3 and 3.4.4, this is simply a reformulation of Theorem 3.5.3 if F is non-Archimedean, or of Theorem 2.8.1 if $F = \mathbb{R}$. If $F = \mathbb{C}$, we omit the proof, which may be found in Jacquet and Langlands (1970, Theorem 5.3, p. 232). ■

Proposition 3.5.2 *Let F be a local field, ψ a nontrivial additive character of F, and let (π, V) be a simple admissible $\mathcal{H}_G$-module, where $\mathcal{H}_G$ is the Hecke algebra of $G = GL(2, F)$. Let $\mathcal{W}$ be a Whittaker model of (π, V) with respect to ψ, and let $\xi \mapsto W_\xi$ be an isomorphism of V onto $\mathcal{W}$ satisfying Eq. (5.8). Then there exists $\xi \in V$ such that $W_\xi(1) \neq 0$. If V is non-Archimedean and π is spherical, and if the conductor of ψ is the ring $\mathfrak{o}$ of integers of F, then we may take ξ to be $GL(2, \mathfrak{o})$-invariant.*

Proof If F is non-Archimedean, let W_ξ be a nonzero element of $\mathcal{W}$. Then $W_\xi(g_0) \neq 0$ for some $g_0 \in G$. By the second assertion in Proposition 3.4.3, there exists a representation π of $GL(2, F)$ such that Eq. (4.5) is valid; and we denote by $\rho : GL(2, F) \to \mathrm{End}\big(C^\infty(G)\big)$ the action by right translation. It is easy to see that if $g \in GL(2, F)$, then Eq. (5.8) is equivalent to

$$W_{\pi(g)\xi} = \rho(g)\, W_\xi. \tag{5.9}$$

Thus $W_{\pi(g_0)\xi}(1) = W_\xi(g_0) \neq 0$.

If F is Archimedean, we must argue more carefully. By Proposition 3.4.4, V is a $(\mathfrak{g}, K)$-module, and we make use of the action $\pi : K \to \mathrm{End}(V)$. Let W_ξ be a nonzero element of $\mathcal{W}$. We note that K intersects every connected component of G. (There are two such connected components if $F = \mathbb{R}$, one if $F = \mathbb{C}$.) Applying $\pi(k)$ for some suitable $k \in K$, we may therefore assume that W_ξ does not vanish identically on the connected component of the identity of G. Now W_ξ is analytic. If $F = \mathbb{R}$, this follows from either Theorem 2.8.1 or Theorem 2.9.2. (The complex case is similar.) Because W_ξ is nonzero on the connected component of the identity, analyticity implies that some coefficient in the Taylor expansion of W_ξ at the identity is nonzero, so $DW_\xi(1) \neq 0$ for some $D \in U(\mathfrak{g})$. This equals $W_{D\xi}(1)$, so we may take $W = W_{D\xi}$.

Finally, assume that F is non-Archimedean, that π is spherical, and that the conductor of ψ_v is $\mathfrak{o}_v$. By Proposition 4.5.7, π is a spherical principal series representation, because it is easy to see that the one-dimensional representations factoring through the determinant map do not have Whittaker models. The nonvanishing of $W^\circ(1)$ now follows from the explicit formula (5.30) in Chapter 4. ■

We return now to the global case. Let F be a global field, and let Σ be the set of all places of F. If $v \in \Sigma$, let $\mathcal{H}_v = \mathcal{H}_{GL(2,F_v)}$ be the local Hecke algebra, and let $\mathcal{H} = \mathcal{H}_{GL(2,A)} = \bigotimes_v \mathcal{H}_v$ be the restricted tensor product of the $\mathcal{H}_v$ with respect to the spherical idempotents e_v°, as in the proof of Theorem 3.3.3 given at the end of Section 3.4. Let ψ be a nontrivial additive character of A, trivial on F. Let (π, V) be an irreducible admissible $GL(2, A)$-module, which as in the proof of Theorem 3.3.3 we regard as a simple $\mathcal{H}$-module. We write π as usual as $\otimes_v \pi_v$ with (π_v, V_v) an irreducible admissible $\mathcal{H}_v$-module, and the tensor product is restricted with respect to a system ξ_v° of vectors such that $\xi_v^\circ \in V_v$ is spherical for almost all v.

The definition of a Whittaker model in the global context is almost the same as the local definition. By a *Whittaker model* of π with respect to the nontrivial character ψ of A/F, we mean a space $\mathcal{W}$ of smooth K-finite functions on $GL(2, A)$ satisfying

$$W\left(\begin{pmatrix} 1 & x \\ & 1 \end{pmatrix} g\right) = \psi(x)\, W(g). \tag{5.10}$$

(The notions of smoothness and K-finiteness are introduced in Section 3.3, after Theorem 3.3.2.) It is assumed that the functions $W \in \mathcal{W}$ are *of moderate growth*, that is, $W\left(\begin{pmatrix} y & \\ & 1 \end{pmatrix} g\right)$ is bounded by a polynomial in $|y|$ for large y. If for all $N > 0$ there exists a constant C_N such that $W\left(\begin{pmatrix} y & \\ & 1 \end{pmatrix} g\right) < C_N\, |y|^{-N}$ for $|y|$ sufficiently large, then we say that the function W is *rapidly decreasing*. We assume that the space $\mathcal{W}$ is closed under the action ρ of $\mathcal{H}$ on the K-finite functions and isomorphic as a $\mathcal{H}$-module to V. Thus it is assumed that there exists an isomorphism $\xi \to W_\xi$ of V onto $\mathcal{W}$ such that

$$W_{\pi(\phi)\xi} = \rho(\phi)\, W_\xi, \qquad \phi \in \mathcal{H},\ \xi \in V. \tag{5.11}$$

Theorem 3.5.4: Global uniqueness of Whittaker models *Let (π, V) be an irreducible admissible representation of $GL(2, A)$. With notation as above, (π, V) has a Whittaker model $\mathcal{W}$ with respect to the additive character ψ of A/F if and only if each (π_v, V_v) has a Whittaker model $\mathcal{W}_v$ with respect to the character ψ_v of F_v. If this is the case, then $\mathcal{W}$ is unique and consists of all finite linear combinations of functions of the form $W(g) = \prod_v W_v(g_v)$, where $W_v \in \mathcal{W}_v$, and $W_v = W_v^\circ$ for almost all v, where W_v° is the spherical element of $\mathcal{W}_v$, normalized so that $W_v^\circ(k_v) = 1$ for $k_v \in GL(2, \mathfrak{o}_v)$.*

Proof First suppose that each π_v has a Whittaker model $\mathcal{W}_v$. Then we can define a Whittaker model for π as follows. By Proposition 3.5.2, we may find, for every non-Archimedean v such that π_v is spherical and the conductor of ψ_v is $\mathfrak{o}_v$, an element W_v° of $\mathcal{W}_v$ such that $W_v^\circ(k_v) = 1$ for all $k_v \in GL(2, \mathfrak{o}_v)$ – this is the spherical element of $\mathcal{W}_v$, which is unique up to constant multiple by Theorem 4.6.2. We choose an isomorphism $\xi_v \mapsto W_{v,\xi_v}$ of V_v onto $\mathcal{W}_v$ in such a way that $W_{v,\xi_v^\circ} = W_v^\circ$ for almost all v. Let $\mathcal{W}$ be the space of all finite linear combinations of functions of the form W_ξ, where if $\xi = \otimes_v \xi_v$ is a pure tensor in $V = \otimes_v V_v$ (so that $\xi_v = \xi_v^\circ$ for almost all v), we define

$$W_\xi(g) = \prod_v W_{v,\xi_v}(g_v), \qquad g = (g_v) \in GL(2, A). \tag{5.12}$$

We note that with our definitions, the factor $W_{v,\xi_v}(g_v)$ is one for almost all v. It is clear that $\mathcal{W}$ affords an irreducible admissible representation of $GL(2, A)$ (in the sense of Section 3.3) that is isomorphic to π. It is also clear that $W \in \mathcal{W}$ satisfies Eq. (5.10). We must show that W is rapidly decreasing. It is sufficient to show this when $W = W_\xi$ is as in Eq. (5.12), with $\xi = \otimes_v \xi_v$ a pure tensor. Fix $g = (g_v) \in GL(2, A)$, and let $y = (y_v) \in A^\times$. We know the following about the behavior of

$$W_v\left(\begin{pmatrix} y_v & \\ & 1 \end{pmatrix} g_v\right). \tag{5.13}$$

If v is non-Archimedean, then Eq. (5.13) vanishes for large $|y_v|$; indeed, Eq. (5.13) lies in the space of the Kirillov model of π_v, introduced in Section 4.4 – see Eq. (4.24) in that section in particular – so this follows from Proposition 4.7.2. On the other hand if v is Archimedean, Eq. (5.13) is rapidly decreasing and analytic in the sense of Section 2.8 by Theorem 2.8.2. It follows that W is rapidly decreasing and analytic.

Now let us show that if π has a Whittaker model, it is the one just described. We will denote by $\xi \mapsto W_\xi$ an isomorphism of V onto the Whittaker model $\mathcal{W}$.

Let us show that there exists $\xi \in V$ such that $W_\xi(1) \neq 0$. Let $\xi_1 \in V$ be any nonzero vector. Then $W_{\xi_1}(g_1) \neq 0$ for some $g_1 \in GL(2, A)$. We make use of the action of $GL(2, A_f)$ and the identity

$$W_{\pi(g)\xi}(h) = W_\xi(gh), \tag{5.14}$$

which is valid for $g \in GL(2, A_f)$. Let $g_1 = g_\infty g_f$, where $g_\infty \in GL(2, F_\infty)$ and $g_f \in GL(2, A_f)$. Then

$$W_{\pi(g_f)\xi_1}(g_\infty) = W_{\xi_1}(g_1) \neq 0,$$

so without loss of generality, we may assume that $g_1 \in GL(2, F_\infty)$. Proceeding now as in the proof of the Archimedean case of Proposition 3.5.2, we make use of the action of $K_\infty = \prod_{v \in S_\infty} K_v$. (The notation is as in Section 3.3.) Because K_∞ intersects every connected component of $GL(2, F_\infty)$, we may assume that W_{ξ_1} does not vanish on the connected component of the identity. Then W_{ξ_1} is analytic because it satisfies an elliptic differential equation – the proof of Theorem 5.9.2 extends naturally to the Lie group $GL(2, F_\infty)$, which is a product of copies of the groups $GL(2, \mathbb{R})$ and $GL(2, \mathbb{C})$. Hence W_{ξ_1} has a nonzero coefficient in its Taylor expansion at the identity, so $D\, W_{\xi_1}(1) \neq 0$ for some $D \in U(\mathfrak{g}_\infty)$ (with notation as in Section 3.3); that is, $W_{\pi(D)\xi_1}(1) \neq 0$. This shows that $W_\xi(1) \neq 0$ for some $\xi \in V$.

We may assume without loss of generality that ξ is a pure tensor and that $W_\xi(1) = 1$. Let $\xi = \otimes_v \xi_v^\circ$, where for almost all v, ξ_v° is $GL(2, \mathfrak{o}_v)$-fixed. This involves changing the definitions of the vectors ξ_v° for a finite number of places, but this does not change the meaning of the restricted tensor product (Exercise 3.3.5).

Now for each v, if $\xi_v \in V_v$ and $g_v \in GL(2, F_v) \subset GL(2, A)$, we define

$$W_{v,\xi_v}(g_v) = W_{i_v(\xi_v)}(g_v),$$

where i_v is as in Eq. (5.4). It is easy to see that the space of functions W_{v,ξ_v} form a Whittaker model $\mathcal{W}_v$ for π_v. We have $W_{v,\xi_v^\circ}(1) = 1$. Let S be a finite set of places. Let $A_S = \prod_{v \in S} F_v \subset A$. Thus $GL(2, A_S) \subset GL(2, A)$. We will prove first that if $g \in GL(2, A_S)$, then

$$W_\xi(g) = \prod_{v \in S} W_{v,\xi_v}(g_v). \tag{5.15}$$

We prove Eq. (5.15) by induction on the cardinality of S. If S is empty, both sides of Eq. (5.15) are one, so let w be a fixed particular place in S. For every $\eta_w \in V_w$ and $h_w \in GL(2, F_w)$, let $\Omega_{w,\eta_w}(h_w) = W_\eta(h)$, where

$$\eta = \eta_w \otimes \left(\bigotimes_{v \neq w} \xi_v \right)$$

and $h \in GL(2, A)$ is the adele whose v-component equals g_v if $v \neq w$ or h_w if $v = w$. Then for $\xi_w \in V_w$, the function $\Omega_{w,\xi}$ is of moderate growth and satisfies Eq. (5.5), and we have

$$\Omega_{w,\pi_w(\phi_w)\eta_w} = \rho_w(\phi_w)\, \Omega_{w,\eta_w}, \qquad \phi_w \in \mathcal{H}_w.$$

Thus either the space of functions Ω_{w,η_w} ($\eta \in V_w$) is zero, or it comprises a Whittaker model for the representation π_w. In either case, it follows from Theorem 3.5.3 that there exists a constant c such that

$$\Omega_{w,\eta_w} = c\, W_{w,\eta_w}. \tag{5.16}$$

We evaluate the constant c by taking $\eta_w = \xi_w^\circ$ and evaluating at the identity. Our induction hypothesis shows that

$$c = \prod_{\substack{v \in S \\ v \neq w}} W_{v,\xi_v}(g_v).$$

Now taking $\eta_w = \xi_w$ and evaluating Eq. (5.16) at g_w, we obtain Eq. (5.15).

Finally, we show that if $g = (g_v)$ is any adele and $\xi = \otimes_v \xi_v$ is any pure tensor in V, then

$$W_\xi(g) = \prod_v W_{v,\xi_v}(g_v). \tag{5.17}$$

Indeed, there exists a finite set S of places such that if $v \notin S$, then v is Archimedean, ξ_v is K_v-fixed, and $g_v \in K_v$. Let $h \in GL(2, A_f)$ be the adele whose v-component is g_v when $v \in S$ and 1 when $v \notin S$. Then the left side of Eq. (5.17) equals $W_\xi(h)$, and the right side equals $\prod_{v \in S} W_{v,\xi_v}(g_v)$, so these are equal by Eq. (5.15). This concludes the proof of Eq. (5.17).

We have proved that if π has a Whittaker model, then it is the one described in the statement of the theorem. It is therefore unique. ∎

We now come to an important result: *Automorphic cuspidal representations have Whittaker models.* This remains true on $GL(n)$. It is false for certain other groups such as $Sp(2n)$: Holomorphic Siegel modular forms do not have Whittaker models.

Theorem 3.5.5: Existence of Whittaker models for automorphic representations *Let F be a global field, A its adele ring, and let (π, V) be an automorphic cuspidal representation of $GL(2, F)$, so $V \subset \mathcal{A}_0\big(GL(2, F)\backslash GL(2, A), \omega\big)$, where ω is a character of $A^\times/F^\times$. If $\phi \in V$ and $g \in GL(2, A)$, let*

$$W_\phi(g) = \int_{A/F} \phi\left(\begin{pmatrix} 1 & x \\ & 1 \end{pmatrix} g\right) \psi(-x)\, dx. \tag{5.18}$$

Then the space $\mathcal{W}$ of functions W_ϕ is a Whittaker model for π. We have the "Fourier expansion"

$$\phi(g) = \sum_{\alpha \in F^\times} W_\phi\left(\begin{pmatrix} \alpha & \\ & 1 \end{pmatrix} g\right). \tag{5.19}$$

Proof The function

$$F(x) = \phi\left(\begin{pmatrix} 1 & x \\ & 1 \end{pmatrix} g\right), \qquad x \in A,$$

is continuous and (because ϕ is automorphic) satisfies $F(x + a) = F(x)$ if $a \in F$. Thus it may be regarded as a function on the compact group A/F. It therefore has a Fourier expansion in terms of the characters of A/F. By Exercise 3.1.3, each such character has the form $x \mapsto \psi(ax)$, where $\alpha \in F$. Thus we may write

$$\phi\left(\begin{pmatrix} 1 & x \\ & 1 \end{pmatrix} g\right) = \sum_{\alpha \in F} C(\alpha)\, \psi(\alpha x), \tag{5.20}$$

where the Fourier coefficient

$$C(\alpha) = \int_{A/F} \phi\left(\begin{pmatrix} 1 & x \\ & 1 \end{pmatrix} g\right) \psi(-\alpha x)\, dx.$$

If $\alpha = 0$, then $C(\alpha) = 0$ because ϕ is cuspidal, so we may restrict the summation to $\alpha \in F^\times$. If $\alpha \neq 0$, then because ϕ is automorphic

$$C(\alpha) = \int_{A/F} \phi\left(\begin{pmatrix} \alpha & \\ & 1 \end{pmatrix} \begin{pmatrix} 1 & x \\ & 1 \end{pmatrix} g\right) \psi(-\alpha x)\, dx$$

$$= \int_{A/F} \phi\left(\begin{pmatrix} 1 & \alpha x \\ & 1 \end{pmatrix} \begin{pmatrix} \alpha & \\ & 1 \end{pmatrix} g\right) \psi(-\alpha x)\, dx.$$

Now we make the change of variables $x \mapsto \alpha^{-1} x$, which is unimodular because $\alpha \in F$. Thus

$$C(\alpha) = W_\phi\left(\begin{pmatrix} \alpha & \\ & 1 \end{pmatrix} g\right).$$

Substituting this back into Eq. (5.20) and putting $x = 0$, we obtain Eq. (5.19).

We note that the functions W_ϕ satisfy Eq. (5.10) by construction; they are of moderate growth because ϕ is of moderate growth in the sense of Section 3.3. Let $\mathfrak{g}_\infty$ and K_∞ be as in Section 3.3. It is easy to see that $GL(2, A_\mathfrak{f})$ and $(\mathfrak{g}_\infty, K_\infty)$ act on the space $\mathcal{W}$ of functions W_ϕ and that this action is compatible with the actions on V. Essentially, this is because $GL(2, A_\mathfrak{f})$ and $(\mathfrak{g}_\infty, K_\infty)$ act on both ϕ and W_ϕ by right translation, and W_ϕ is obtained by averaging left translates of ϕ, and left and right translation commute. The formula (5.19) guarantees that the map $\phi \mapsto W_\phi$ is injective. Hence the space $\mathcal{W}$ of functions W_ϕ comprises a Whittaker model for π. ■

Following Piatetski-Shapiro (1979), we may now give the following proof.

Proof of Theorem 3.3.6 when $n = 2$. To fix the ideas, let us first prove Theorem 3.3.7 under the assumption that $\pi_v \cong \pi'_v$ for *every* place V. Thus let V and V' be isomorphic irreducible subrepresentations of $\mathcal{A}_0\big(GL(2, F)\backslash GL(2, A), \omega\big)$. We will prove that $V = V'$. Indeed, the representation π afforded by V has a Whittaker model $\mathcal{W}$ by Theorem 3.5.5; this model is unique by Theorem 3.5.4. Let $\mathcal{W}$ denote this Whittaker space. By Theorem 3.5.5, V consists of the space of all functions ϕ of the form

$$\phi(g) = \sum_{\alpha \in F^\times} W\left(\begin{pmatrix} \alpha & \\ & 1 \end{pmatrix} g\right), \qquad W \in \mathcal{W},$$

and by the same reasoning, V' consists of the same space! And so they are equal.

In the general case, let (π, V) and (π', V') be automorphic representations such that $V, V' \subseteq \mathcal{A}_0(GL(2,F)\backslash GL(2,A),\omega)$, $\pi \cong \bigotimes_v \pi_v$, $\pi' \cong \bigotimes_v \pi'_v$, and $\pi_v \cong \pi'_v$ for all $v \notin S$, where S is a finite set of places and all the places in S are non-Archimedean. Let $\mathcal{W}_v$ and $\mathcal{W}'_v$ be the Whittaker models of π_v and π'_v. For each v, we choose a nonzero function $W_v \in \mathcal{W}_v$. The choice of the W_v is not quite arbitrary, but subject to two conditions. Firstly, we require that for all but finitely many v, W_v must be the unique element of $\mathcal{W}_v$ such that $W_v(k_v) = 1$ when $k_v \in K_v$. This is possible because π_v is spherical for almost all v, as a consequence of Theorem 3.3.3. Secondly, we require that for the finite number of places $v \in S$, the function $W_v\left(\begin{smallmatrix} y & \\ & 1\end{smallmatrix}\right)$ on $F_v^\times$ is compactly supported. This is possible because by Theorem 4.7.1, if σ is an arbitrary compactly supported locally constant function on $F_v^\times$, there exists an element W_v of $\mathcal{W}_v$ such that

$$W_v\begin{pmatrix} y & \\ & 1\end{pmatrix} = \sigma(y), \qquad y \in F_v^\times.$$

We choose functions $W'_v \in \mathcal{W}'_v$ as follows. If $v \notin S$, then $\mathcal{W}'_v = \mathcal{W}_v$, so we choose $W'_v = W_v$. If $v \in S$, then (again using Theorem 4.7.1) we may arrange that

$$W_v\begin{pmatrix} y & \\ & 1\end{pmatrix} = W'_v\begin{pmatrix} y & \\ & 1\end{pmatrix}$$

for all $y \in F_v^\times$. With W_v and W'_v chosen in this way, we define $\phi \in V$ by

$$\phi(g) = \sum_{\alpha\in F^\times} W\left(\begin{pmatrix} \alpha & \\ & 1\end{pmatrix} g\right),$$

where if $g = (g_v) \in GL(2,A)$, $W(g) = \prod_v W_v(g_v)$. Similarly, we define $\phi' \in V'$ from the W'_v. We will show that $\phi = \phi'$.

It is clear from the definitions that $\phi(g) = \phi'(g)$ if g is of the form

$$\begin{pmatrix} y & \\ & 1\end{pmatrix}, \qquad y \in A^\times.$$

Also, $W_v = W'_v$ when v is Archimedean and W and W' are right invariant under some open subgroup K_0 of $GL(2, A_f)$. Most importantly, ϕ and ϕ' are automorphic. Thus we have $\phi(g) = \phi(g')$ if g is of the form

$$\gamma\begin{pmatrix} y & \\ & 1\end{pmatrix} g_\infty k_0, \qquad \gamma \in GL(2,F),\ y \in A^\times,\ g_\infty \in GL(2,F_\infty),\ k_0 \in K_0.$$

However, it follows from strong approximation (Theorem 3.3.1) that every element of $GL(2,A)$ has this form. Thus $\phi = \phi'$. We note that ϕ is nonzero because W is nonzero, and W may be expressed in terms of ϕ by Eq. (5.18). Thus thc intersection of V and V' is nonzero and hence they are equal. ■

Theorems 3.5.4 and 3.5.5, we have seen, lead to the important multiplicity one Theorem 3.3.6, but that is by no means the end of their importance. They are also the key to the construction of the standard L-function of an automorphic cuspidal representation. Specifically, there are (at least) two distinct direct generalizations of Tate's theory, described in Section 3.1, to $GL(n)$. (Tate's theory is the case of $GL(1)$.) These two generalizations are related and give equivalent definitions of the factors γ_v in the local functional equations, as does a third distinct approach to the functional equations (not directly generalizing Tate's work) that is due to Langlands (1971) and Shahidi (1981) and is based on Eisenstein series.

The first generalization of Tate's work is related to the Weil representation and is described in Godement and Jacquet (1972), and in Jacquet (1979). We will not discuss it here. The second approach is based on Whittaker functions and is due to Jacquet and Langlands (1970) when $n = 2$; see Jacquet, Piatetski-Shapiro and Shalika (1979) for this theory (and for its connection with the Jacquet–Godement approach) when $n = 3$ and Cogdell and Piatetski-Shapiro (1994) for general n.

We will follow the Whittaker function approach here. The results of the remainder of this section are strongly analogous to those of Section 3.1. In preparation, we discuss briefly the *spherical principal series representations* of a non-Archimedean local field F with a ring $\mathfrak{o}$ of integers.

Let χ_1 and χ_2 be quasicharacters of $F^\times$. We assume that $\chi_1\chi_2^{-1}$ is not identically equal to either of the quasicharacters $y \mapsto |y|$ or $y \mapsto |y|^{-1}$. We consider the space of all smooth (i.e., locally constant) functions $f : GL(2, F) \to \mathbb{C}$ that satisfy

$$f\left(\begin{pmatrix} y_1 & x \\ & y_2 \end{pmatrix} g\right) = \left|\frac{y_1}{y_2}\right|^{1/2} \chi_1(y_1)\, \chi_2(y_2)\, f(g).$$

Then $GL(2, F)$ acts on this space by right translation, and the resulting representation, denoted $\pi(\chi_1, \chi_2)$, is irreducible, and

$$\pi(\chi_1, \chi_2) \cong \pi(\chi_2, \chi_1). \tag{5.21}$$

These representations are called *principal series representations*. If χ_1 and χ_2 are nonramified, then $\pi(\chi_1, \chi_2)$ is irreducible, in which case it is called a *spherical principal series* or *nonramified principal series* representation. Every spherical representation (i.e., every representation having a $GL(2, \mathfrak{o})$-fixed vector) is either one dimensional or else is isomorphic to $\pi(\chi_1, \chi_2)$ for some nonramified quasicharacters χ_1 and χ_2. (The one dimensional representations cannot occur as constituents of automorphic cuspidal representations because they do not have Whittaker models.) For proofs and further details, see Sections 4.5 and 4.6.

Assume that χ_1 and χ_2 are nonramified. Let $\alpha_1 = \chi_1(\varpi)$ and $\alpha_2 = \chi_1(\varpi)$, where ϖ is a generator of the maximal ideal of $\mathfrak{o}$. We call α_1 and α_2 the *Satake parameters* of $\pi(\chi_1, \chi_2)$. (In view of Eq. (5.21), the order of α_1 and α_2 is irrelevant.) If $\pi(\chi_1, \chi_2)$ is unitary, then either χ_1 and χ_2 are unitary, so that α_1 and α_2 have absolute value one; or else there exists a constant α of absolute value one and a real constant $-\frac{1}{2} < s < \frac{1}{2}$ such that $\alpha_1 = \alpha\, q^s$ and $\alpha_2 = \alpha\, q^{-s}$. This determination of principal series that are unitary is established at the end of Section 4.6. The principal series representations of the first type we call the *tempered principal series representations*; the representations of the second type we call the *complementary series representations*. It is conjectured that the complementary series representations cannot occur as constituents of automorphic representations, but this has not been proved; this is called the *Ramanujan conjecture*.

Deligne (1971) proved the Ramanujan conjecture for holomorphic modular forms as a consequence of his proof of the Weil conjectures. (It should be mentioned that Deligne's work on the Ramanujan conjecture was preceded by the investigations of Ihara (1967), who contributed important ideas.) The proof of the Ramanujan conjecture for modular forms of weight two is simpler than for higher weight and follows from the Eichler–Shimura (Shimura, 1971) theory, which is touched on at the end of the exercises for Section 1.4, together with Weil's proof of the Riemann hypothesis for curves and the clarification of the reduction of modular curves by Igusa. Let us consider briefly the very simplest case where the space $S_2\big(\Gamma_0(N)\big)$ of cusp forms of weight two is one dimensional. In this case, the modular curve $X = X_0(N)$ is an elliptic curve defined over $\mathbb{Q}$. Let $f(z) = \sum_n a(n)\, q^n$ be the normalized Hecke eigenform in $S_2\big(\Gamma_0(N)\big)$. If p is a prime not dividing N, then the curve X has good reduction at p; this means that we may choose its equation $f(x, y) = 0$ to have coefficients in $\mathbb{Z}$, and if we consider the corresponding equation over $\mathbb{Z}/p\mathbb{Z}$, it remains an elliptic curve $X/\mathbb{F}_p$. It is proved in the Eichler–Shimura theory (Shimura 1971, Chapter 7) that the cardinality of the finite group of $\mathbb{F}_p$-rational points of this curve is $1 - a(p) + p$. By the Riemann hypothesis for curves over finite fields (Weil (1940), or Silverman (1986), Theorem V.2.4, p. 136), we have $|a(p)| \le 2\sqrt{p}$. The Satake parameters are the roots of the polynomial

$$X^2 - p^{-1/2}\, a(p)\, X + 1 = 0,$$

so this implies the Ramanujan conjecture for the automorphic representation corresponding to f.

For automorphic representations attached to Maass forms, these ideas of algebraic geometry do not seem to give any progress. However, the Ramanujan conjecture for Maass forms is a consequence of the Langlands conjectures, which are described in Section 1.8, and all partial progress toward the Ramanujan conjectures to date is closely related to partial progress toward the Langlands conjectures. The reason that the Langlands conjecture implies the Ramanujan conjecture is that if π is an automorphic cuspidal representation of $GL(2, A)$, and if π_v is not tempered for some v, then we consider the lift $\vee^r \pi$ of π to $GL(r+1, A)$ described in Prediction 1.8.1. If r is sufficiently large, then π_v will not be unitary. However, an automorphic cuspidal representation of $GL(n, A)$ is automatically unitary, because it inherits a Hermitian inner product from $L^2\big(GL(n, F)\backslash GL(n, A), \omega\big)$ (where ω is its central quasicharacter). This is a contradiction. Thus, for example, the result of Gelbart and Jacquet (1978), establishing the existence of the automorphic representation $\vee^2\pi$ on $GL(3)$, led to the estimate $|\alpha_i| \leq q^{1/4}$, a result also obtained by a different and seemingly unrelated method using Poincaré series by Selberg (1965). If $|\alpha_i| > q^{1/4}$, the lift to $GL(3)$ will not be unitary, which is a contradiction.

The best results for Maass forms toward the Ramanujan conjecture are that the Langlands parameters α_1 and α_2 are bounded by $q^{5/28}$ if the ground field is $\mathbb{Q}$ (Bump, Duke, Hoffstein and Iwaniec 1992), and by $q^{1/5}$ for an arbitrary ground field (Shahidi 1988, 1990a, b). As we will explain in Section 3.7, the Ramanujan conjecture has an Archimedean analog (due to Selberg) – Formulated adelically, Selberg's conjecture is that if v is an Archimedean place, then π_v cannot be a complementary series representation. The Archimedean analog of the result of Bump, Duke, Hoffstein, and Iwaniec (1992) (toward the Selberg conjecture) may be found in Luo, Rudnick, and Sarnak (in press).

Let q be the cardinality of the residue field $\mathfrak{o}/(\varpi)$. We call

$$L(s,\pi) = (1 - \alpha_1 q^{-s})^{-1} (1 - \alpha_2 q^{-s})^{-1} \tag{5.22}$$

the *local L-function* of π. More generally, let ξ be a nonramified character of $F^\times$. Then we define

$$L(s,\pi,\xi) = (1 - \alpha_1 \xi(\varpi) q^{-s})^{-1} (1 - \alpha_2 \xi(\varpi) q^{-s})^{-1}. \tag{5.23}$$

Now let F be a global field, A its adele ring, and let (π, V) be an automorphic cuspidal representation of $GL(2, A)$. We assume that the central quasicharacter ω of π is unitary. This entails no essential loss of generality: If the central quasicharacter of π is not unitary, we simply replace V by the space of functions $|\det(g)|^s \phi(g)$, where $\phi \in V$, which has the effect of replacing the central quasicharacter ω of π by the character $y \mapsto |y|^{2s} \omega(y)$. If s is chosen appropriately, we may make this character unitary. (See Exercise 3.1.5(a).)

We write $\pi = \otimes_v \pi_v$ as usual. Let S be a finite set of places such that if $v \notin S$, then π_v is spherical. If $v \notin S$, let $L_v(s, \pi_v)$ be the local L-function as in Eq. (5.22), and let

$$L_S(s,\pi) = \prod_{v \notin S} L_v(s, \pi_v) \tag{5.24}$$

be the partial L-function. We will show how Theorems 3.5.4 and 3.5.5 lead to a functional equation for $L_S(s, \pi)$.

Let $\phi \in V$. We note that

$$\phi\begin{pmatrix} y & \\ & 1 \end{pmatrix} \tag{5.25}$$

is rapidly decreasing as $|y| \to \infty$; that is, for any $N > 0$ there exists a constant C_N such that

$$\phi\begin{pmatrix} y & \\ & 1 \end{pmatrix} < C_N |y|^{-N} \tag{5.26}$$

for $|y|$ sufficiently large. This is because every term in the Fourier expansion (Eq. 5.19) is rapidly decreasing. Also, we claim that Eq. (5.25) is rapidly decreasing as $|y| \to 0$, that is, that given $N > 0$ there exists a constant C'_N such that

$$\phi\begin{pmatrix} y & \\ & 1 \end{pmatrix} < C'_N\, |y|^N \tag{5.27}$$

when $|y|$ is sufficiently small. This may be seen as follows: Because ϕ is automorphic,

$$\phi\begin{pmatrix} y & \\ & 1 \end{pmatrix} = \phi\left(\begin{pmatrix} & 1 \\ 1 & \end{pmatrix}\begin{pmatrix} y & \\ & 1 \end{pmatrix}\right) = \left(\pi\begin{pmatrix} & 1 \\ 1 & \end{pmatrix}\phi\right)\begin{pmatrix} 1 & \\ & y \end{pmatrix}$$

$$= \omega(y)\left(\pi\begin{pmatrix} & 1 \\ 1 & \end{pmatrix}\phi\right)\begin{pmatrix} y^{-1} & \\ & 1 \end{pmatrix}.$$

So the rapid decrease of Eq. (5.25) as $|y| \to 0$ follows from the rapid decrease of

$$\left(\pi\begin{pmatrix} & 1 \\ 1 & \end{pmatrix}\phi\right)\begin{pmatrix} y & \\ & 1 \end{pmatrix}$$

as $|y| \to \infty$, which has already been established in Eq. (5.26).

Because Eq. (5.25) is rapidly decreasing as $|y| \to \infty$ or 0, the integral

$$Z(s, \phi) = \int_{A^\times/F^\times} \phi\begin{pmatrix} y & \\ & 1 \end{pmatrix} |y|^{s-1/2}\, d^\times y \tag{5.28}$$

is absolutely convergent for all values of s. We substitute Eq. (5.19) into Eq. (5.28), and then we may collapse the integral with the summation to obtain

$$Z(s, \phi) = \int_{A^\times} W_\phi\begin{pmatrix} y & \\ & 1 \end{pmatrix} |y|^{s-1/2}\, d^\times y, \tag{5.29}$$

provided the latter integral is absolutely convergent. For the purpose of determining the convergence of Eq. (5.29), there is no loss of generality in assuming that under the isomorphism $\pi \cong \otimes_v \pi_v$, the vector ϕ corresponds to a pure tensor $\otimes_v \phi_v$. Let $W_v \in \mathcal{W}_v$ be the element of the local Whittaker model corresponding to the vector $\phi_v \in V_v$. Then by Theorem 3.5.4, we may write

$$W(g) = \prod_v W_v(g_v). \tag{5.30}$$

Although the integral (Eq. 5.28) is convergent for all s, the integral (Eq. 5.29) is only convergent if $\mathrm{re}(s)$ is sufficiently large. Write the idele $y = (y_v)$, $y_v \in F_v^\times$; the integrand can be written as

$$\prod_v W_v\begin{pmatrix} y_v & \\ & 1 \end{pmatrix} |y_v|_v^{s-1/2},$$

and so the integral decomposes as an Euler product:

$$Z(s, \phi) = \prod_v Z_v(s, W_v), \tag{5.31}$$

where

$$Z_v(s, W_v) = \int_{F_v^\times} W_v\begin{pmatrix} y_v & \\ & 1 \end{pmatrix} |y_v|_v^{s-1/2} \, d^\times y_v. \tag{5.32}$$

We may now discuss the convergence of the integrals (Eq. 5.29), which will be absolutely convergent provided the integrals (Eq. 5.32) are absolutely convergent for all v and provided the infinite product (Eq. 5.31) is convergent. We will show that the integrals (Eq. 5.32) are convergent if $\mathrm{re}(s) > 1/2$ and that the infinite product (Eq. 5.31) is convergent provided $\mathrm{re}(s) > 3/2$.

Suppose first that v is non-Archimedean. In this case, the function

$$W_v\begin{pmatrix} y & \\ & 1 \end{pmatrix} \tag{5.33}$$

is an element of the Kirillov model of π_v, which is described explicitly in Theorems 4.7.1, 4.7.2 and 4.7.3. It is zero if $|y|$ is large. If $|y|$ is small, the asymptotics of Eq. (5.33) may be described as follows. The representation π_v may be classified as a principal series, a special representation, or as supercuspidal. If it is supercuspidal, then (referring to Theorem 4.7.1) the Jacquet module $J(V)$ is zero, and so Eq. (5.33) is zero when $|y|$ is small. The integral (Eq. 5.32) is therefore convergent for all values of s.

Suppose that π_v is principal series. Then $\pi_v \cong \pi(\chi_1, \chi_2)$, where χ_1 and χ_2 are quasicharacters of $F^\times$, and by Theorem 3.7.5, when y is small, Eq. (5.32) is asymptotic to a linear combination of the quasicharacters $|y|^{1/2}\,\chi_1(y)$ and $|y|^{1/2}\,\chi_2(y)$. Let σ be the real number such that $|\chi_1(y)| = |y|^\sigma$. Because the central quasicharacter $\chi_1\chi_2$ of π_v is unitary, we have $|\chi_2(y)| = |y|^{-\sigma}$. Now the integral (Eq. 5.32) is convergent if $\mathrm{re}(s) > \max(\sigma, -\sigma)$. The representation π_v is unitary, because V inherits an invariant Hermitian inner product from $L^2\big(GL(2, F)\backslash GL(2, A), \omega\big)$. By Theorem 4.6.1, this implies that $|\sigma| < 1/2$, and consequently the integral (Eq. 5.32) is convergent if $\mathrm{re}(s) > 1/2$.

The special representation is handled similarly.

If v is Archimedean, we will only discuss the case where v is real, though the complex case is essentially the same. In this case, the Whittaker function is expressed as a finite linear combination of confluent hypergeometric functions (Eq. (8.3) in Chapter 2). These decay exponentially as $y \to \infty$, whereas the asymptotics at zero are given by Whittaker and Watson (1927, 16.41 and 16.1). To summarize, Eq. (5.33) is a finite linear combination of functions $W_{k/2,\sigma}$, where in the worst case (when π_v is complementary series) σ can be a real number between $-\frac{1}{2}$ and $\frac{1}{2}$ and $W_{k/2,\sigma}$ is asymptotic to a linear combination of $|y|^{\sigma+1/2}$ and $|y|^{-\sigma+1/2}$ near $y = 0$. Again, we have convergence for $\mathrm{re}(s) > 1/2$.

We will show momentarily that for almost all v, $Z_v(s, W_v) = L_v(s, \pi_v)$, where the local L-function is given by Eq. (5.22). Because of Theorem 4.6.7, we have $|\alpha_1|$ and $|\alpha_2| < q^{1/2}$, where q is the residue cardinality at the place v, so the convergence of the product (Eq. 5.31) for $\mathrm{re}(s) > 3/2$ follows by comparison with the infinite product for the Dedekind zeta function of the field F.

We will now evaluate $Z_v(s, W_v)$ when v is "nonramified." To explain a common abuse of terminology, the term "nonramified" is used loosely in the context of any given computation to mean a place where the computation proceeds as typically as possible – one imposes conditions that exclude a finite number of places. In the case at hand, we will say that v is nonramified if v is non-Archimedean, π_v is a spherical principal series, the conductor of the additive character ψ_v is $\mathfrak{o}_v$, the vector ϕ_v is the spherical vector in the representation, and the Whittaker function W_v is normalized so that $W_v(1) = 1$. These conditions will be true for almost all v.

Proposition 3.5.3 *If v is nonramified in the above sense, then for s sufficiently large,*

$$Z_v(s, W_v) = L_v(s, \pi_v) \tag{5.34}$$

Proof With our hypotheses on v, there is an explicit formula for W_v in terms of the Satake parameters α_1 and α_2. Let $m = \mathrm{ord}_v(y)$, and let q_v be the cardinality of the residue field $\mathfrak{o}_v/(\varpi_v)$. Then

$$W_v\begin{pmatrix} y & \\ & 1 \end{pmatrix} = \begin{cases} q^{-m/2}\,\dfrac{\alpha_1^{m+1}-\alpha_2^{m+1}}{\alpha_1-\alpha_2} & \text{if } m \geq 0; \\ 0 & \text{otherwise.} \end{cases} \tag{5.35}$$

This formula is proved in Proposition 4.6.5. Thus we may break the integral (Eq. 5.32) up into a sum over $m = 0$ to ∞ to obtain

$$\sum_{m=0}^{\infty} q^{-m/2}\,\frac{\alpha_1^{m+1} - \alpha_2^{m+1}}{\alpha_1 - \alpha_2}\, q^{m/2-ms}$$

$$= (\alpha_1 - \alpha_2)^{-1}\left[\alpha_1 \sum_{m=0}^{\infty} (\alpha_1 q^{-s})^m - \alpha_2 \sum_{m=0}^{\infty} (\alpha_2 q^{-s})^m\right].$$

Now summing the geometric series, this equals

$$(\alpha_1 - \alpha_2)^{-1}\big(\alpha_1\,(1 - \alpha_1\, q^{-s})^{-1} - \alpha_2\,(1 - \alpha_2\, q^{-s})^{-1}\big)$$

$$= (1 - \alpha_1\, q^{-s})^{-1}\,(1 - \alpha\, q^{-s})^{-1}\,(\alpha_1 - \alpha_2)^{-1}\big(\alpha_1\,(1 - \alpha_2\, q^{-s}) - \alpha_2\,(1 - \alpha_1\, q^{-s})\big).$$

This simplifies to Eq. (5.22). ∎

We can generalize these zeta integrals by "twisting." Let ξ be a unitary Hecke character, that is, a character of $A^\times/F^\times$. In place of Eq. (5.28), consider

$$Z(s,\phi,\xi) = \int_{A^\times/F^\times} \phi\begin{pmatrix} y & \\ & 1 \end{pmatrix} |y|^{s-1/2}\,\xi(y)\,d^\times y. \tag{5.36}$$

Just as with Eq. (5.28), we may substitute the Fourier expansion (Eq. 5.19) for ϕ to obtain

$$Z(s,\phi,\xi) = \int_{A^\times} W_\phi\begin{pmatrix} y & \\ & 1 \end{pmatrix} |y|^{s-1/2}\,\xi(y)\,d^\times y, \tag{5.37}$$

and as with Eq. (5.31), if ϕ corresponds to a pure tensor in $\otimes_v \pi_v$, we have

$$Z(s,\phi,\xi) = \prod_v Z_v(s, W_v, \xi_v), \tag{5.38}$$

where

$$Z_v(s, W_v, \xi_v) = \int_{F_v^\times} W_v\begin{pmatrix} y_v & \\ & 1 \end{pmatrix} |y_v|_v^{s-1/2}\,\xi_v(y_v)\,d^\times y_v. \tag{5.39}$$

Again, these manipulations are justified if $\mathrm{re}(s) > 3/2$.

Suppose now that v is a nonramified place, which in the present context means that it satisfies all the conditions that were imposed for Proposition 3.5.3, and that furthermore ξ_v is nonramified. Then we have

$$Z_v(s, W_v, \xi_v) = L_v(s, \pi_v, \xi_v), \tag{5.40}$$

where now the local L-functions are defined by Eq. (5.23). The proof of this is a straightforward generalization of Proposition 3.5.3, and is left to the reader.

The global zeta integral satisfies a functional equation. Let

$$w_1 = \begin{pmatrix} & 1 \\ -1 & \end{pmatrix}. \tag{5.41}$$

Because ϕ is automorphic, Eq. (5.36) equals

$$\int_{A^\times/F^\times} \phi\left(w_1 \begin{pmatrix} y & \\ & 1 \end{pmatrix}\right) |y|^{s-1/2}\,\xi(y)\,d^\times y$$

$$= \int_{A^\times/F^\times} \phi\left(\begin{pmatrix} 1 & \\ & y \end{pmatrix} w_1\right) |y|^{s-1/2}\,\xi(y)\,d^\times y.$$

We can substitute y^{-1} for y and make use of the invariance of ϕ under the central quasicharacter to obtain

$$\int_{A^\times/F^\times} (\pi(w_1)\phi)\begin{pmatrix} y & \\ & 1 \end{pmatrix} |y|^{-s+1/2} (\xi\omega)^{-1}(y)\, d^\times y.$$

Therefore

$$Z(s, \phi, \xi) = Z\big(1 - s, \pi(w_1)\phi, \xi^{-1}\omega^{-1}\big). \tag{5.42}$$

Here it is important that the global integral (Eq. 5.36) defining $Z(s, \phi, \xi)$ be defined for *all* s. The local integral (Eq. 5.39) defining the local zeta integral $Z_v(s, W_v, \xi_v)$ is only defined for $\mathrm{re}(s)$ sufficiently large. Nevertheless, it does have analytic continuation and a local functional equation analogous to that of Proposition 3.1.5.

Proposition 3.5.4: Local functional equation *The local zeta integral $Z_v(s, W_v, \xi_v)$, defined by Eq. (5.39) for $\mathrm{re}(s)$ sufficiently large, has meromorphic continuation to all s. There exists a meromorphic function $\gamma_v(s, \pi_v, \xi_v, \psi_v)$ such that*

$$Z_v\big(1 - s, \pi_v(w_1)\, W_v, \xi_v^{-1}\omega_v^{-1}\big) = \gamma_v(s, \pi_v, \xi_v, \psi_v)\, Z_v(s, W_v, \xi_v). \tag{5.43}$$

Proof If v is non-Archimedean, this is proved in Theorem 4.7.5. If v is Archimedean, we will confine ourselves to some remarks concerning the proof. One may on the one hand prove this along the lines of Theorem 4.7.5. This requires leaving the domain of $(\mathfrak{g}, K)$-modules and interpreting π_v as the module of K-finite vectors in a true representation of $GL(2, F_v)$, on a Hilbert space for example. Jacquet and Langlands (1970) take an alternative approach to this. Every irreducible admissible representation of $GL(2, F_v)$ that admits a Whittaker model is either in the principal series or else in the discrete series. In either case, π_v has a model in the Weil representation that proves Eq. (5.43) in the course of actually evaluating $\gamma_v(s, \pi_v, \xi_v, \psi_v)$. These arguments, which may be found in Jacquet and Langlands (1970) are essentially the same as the proofs of our Proposition 4.7.7 (which we give in detail in Section 4.8) and Theorem 4.9.1 (whose proof we omit, but which is in Jacquet and Langlands). ■

Let S be a finite set of places of the global field F. Assume that S contains all the Archimedean places and that if $v \notin S$, then π_v is spherical. We define the *partial L-function*

$$L_S(s, \pi) = \prod_{v \notin S} L_v(s, \pi_v). \tag{5.44}$$

If ξ is a Hecke character and ξ_v is nonramified for $v \notin S$, then we can also define:

$$L_S(s, \pi, \xi) = \prod_{v \notin S} L_v(s, \pi_v, \xi_v). \tag{5.45}$$

We now have the following analog of Theorem 3.1.1.

Theorem 3.5.6: Global functional equation *Let π be an automorphic cuspidal representation of $GL(2, A)$, where A is the adele ring of a global field F. Let ξ be a Hecke character of F. Let S be a finite set of places of F containing all the Archimedean ones such that if $v \notin S$, then π_v is spherical and ξ_v is nonramified and the additive character ψ_v has conductor $\mathfrak{o}_v$. Then*

$$L_S(s, \pi, \xi) = \left\{ \prod_{v \in S} \gamma_v(s, \pi_v, \xi_v, \psi_v) \right\} L_S(1 - s, \hat{\pi}, \xi^{-1}). \tag{5.46}$$

Here $\hat{\pi}$ is the contragredient representation, discussed near the end of Section 3.3.

Proof We choose a pure tensor $\phi = \otimes_v \phi_v \in V$ such that ϕ_v is spherical for $v \notin S$ and normalized so that if W_v is then a local Whittaker function corresponding to ϕ_v, then $W_v(1)$ for $v \notin S$. We will prove Eq. (5.46) by evaluating

$$\left\{ \prod_{v \in S} Z_v(s, W_v, \xi_v)^{-1} \right\} Z(s, \phi, \xi) \tag{5.47}$$

in two different ways. We note that this is a meromorphic function of s because the integral (Eq. 5.36) is convergent for all s (hence entire) and because the expression in brackets is meromorphic by Proposition 3.5.4. First, we take $\mathrm{re}(s)$ to be large and note that Eq. (5.47) equals the left side of Eq. (5.46) by Eqs. (5.38) and (5.40).

Next, take $-\mathrm{re}(s)$ to be large and positive. We make use of Eqs. (5.42) and (5.38) to write Eq. (5.47) as

$$\left\{ \prod_{v \in S} Z_v(s, W_v, \xi_v)^{-1} Z_v\big(1 - s, \pi(w_1) W_v, \xi_v^{-1}\omega_v^{-1}\big) \right\}$$
$$\times \prod_{v \notin S} Z_v\big(1 - s, \pi(w_1) W_v, \xi_v^{-1}\omega_v^{-1}\big).$$

Thus we will obtain the right side of Eq. (5.46), provided we show that if $v \notin S$, then

$$Z_v\big(s, \pi(w_1) W_v, \xi_v^{-1}\omega_v^{-1}\big) = L_v(s, \hat{\pi}_v, \xi_v^{-1}). \tag{5.48}$$

Indeed, because v is nonramified, W_v is the spherical vector, and the left side equals $Z_v(s, W_v, \xi_v^{-1}\omega_v^{-1})$, which equals $L_v(s, \pi_v, \omega_v^{-1}\xi_v)$, which equals $L_v(s, \hat{\pi}_v, \xi_v^{-1})$ by Eq. (3.10) or by direct calculation (using the fact that by Proposition 4.5.5, if α_1 and α_2 are the Satake parameters of π_v, then α_1^{-1} and α_2^{-1} are the Satake parameters of $\hat{\pi}_v$). ■

It is desirable to define L-factors for *every* place v. These should have properties similar to the L-factors defined for $GL(1)$ in Section 3.1: we should have $Z_v(s, W_v, \xi_v)/L(s, \pi_v, \xi_v)$ holomorphic for all W_v in the local Whittaker model, and if we define

$$\epsilon_v(s, \pi_v, \chi_v, \psi_v) = \frac{\gamma_v(s, \pi_v, \chi_v, \psi_v)\, L_v(s, \pi_v, \chi_v)}{L_v(1 - s, \hat{\pi}_v, \chi_v^{-1})}, \tag{5.49}$$

then $\epsilon_v(s, \pi_v, \chi_v, \psi_v)$ is a function of exponential type. We note that if v is Archimedean, then L_v is often equal to 1 – this is the case if π_v is supercuspidal or if $\pi_v = \pi(\chi_1, \chi_2)$ and $\chi\chi_1$, $\chi\chi_2$ are both ramified. In these cases, we have

$$\epsilon_v(s, \pi_v, \chi_v, \psi_v) = \gamma_v(s, \pi_v, \chi_v, \psi_v).$$

We describe the factors $L_v(s, \pi_v, \chi_v)$ when F_v is non-Archimedean. If π_v is supercuspidal, then $L_v(s, \pi_v, \chi_v) = 1$. If $\pi_v = \pi(\chi_{1,v}, \chi_{2,v})$ is a principal series representation, then by definition

$$L_v(s, \pi_v, \chi_v) = L(s, \chi_v\chi_{1,v})\, L(s, \chi_v\chi_{2,v}), \tag{5.50}$$

where the L-factors on the right are as defined in Section 3.1. In this case it follows also from Proposition 4.7.7 that

$$\epsilon_v(s, \pi_v, \chi_v, \psi_v) = \epsilon(s, \chi_v\chi_{1,v}, \psi_v)\, \epsilon(s, \chi_v\chi_{2,v}, \psi_v). \tag{5.51}$$

Finally, suppose that $\pi_v = \sigma_v(\chi_{1,v}, \chi_{2,v})$ is a special representation, with $\chi_{1,v}\chi_{2,v}^{-1}(x) = |x|$. Then according to Jacquet and Langlands (1970, Proposition 3.6),

$$L_v(s, \pi_v, \chi_v) = L(s, \chi_v\chi_{1,v}), \tag{5.52}$$

and

$$\epsilon(s, \pi_v, \chi_v, \psi_v) = \epsilon(s, \chi_v\chi_{1,v}, \psi_v)\, \epsilon(s, \chi_v\chi_{2,v}, \psi_v)\, \frac{L(1 - s, \chi_v^{-1}\chi_{1,v}^{-1})}{L(s, \chi_1\chi_2)}. \tag{5.53}$$

3.6 Adelization of Classical Automorphic Forms

We have already touched on the relationship between *classical* automorphic forms (as in Section 3.2), and *adelic* automorphic forms (as in Section 3.3), in the proof of Theorem 3.3.4, where we deduced the admissibility of automorphic representations of $GL(n, A)$ from the corresponding classical result, which is Theorem 3.2.1, at least when $n = 2$ and the ground field is $\mathbb{Q}$, the general case being similar. In this section, we will look a bit more closely at the relationship between classical and adelic automorphic forms.

One procedure for obtaining automorphic representations is to invoke Theorem 3.3.2 to assert the decomposition of $L_0^2\big(GL(n, F)\backslash GL(n, A), \omega\big)$ into irreducible subspaces; then if (V, π) is the space of K-finite vectors in such a space, this is an automorphic representation. An alternative is to take automorphic forms on $GL(n, F_\infty)$ and adelize them, by a process similar to Proposition 3.1.2. We carry this program out when $n = 2$ and $F = \mathbb{Q}$.

Let us begin with a modular form or Maass form of weight k for $\Gamma_0(N)$. As in Section 3.2 – see Eq. (2.14) in particular – we may associate with this modular form or Maass form a function $F : GL(2,\mathbb{R})^+ \to \mathbb{C}$, which is of moderate growth, is an eigenfunction of the Laplace–Beltrami operator Δ (given by Eq. (1.39) in chapter 2), satisfies Eq. (2.15), and, for

$$\gamma = \begin{pmatrix} a & b \\ c & d \end{pmatrix} \in \Gamma_0(N),$$

the identity

$$F(\gamma g) = \chi(d)\, F(g), \tag{6.1}$$

where χ is a Dirichlet character modulo N (*not* assumed to be primitive). We will eventually also assume that F is an eigenfunction of the Hecke operators T_p with $p \nmid N$.

We will transfer the function F to a function ϕ on $GL(2, A)$ by means of Eq. (3.3). If χ is trivial, it is clear how to do this. In the general case, we must compensate for the character χ with a character of $K_0(N)$. We will describe in detail how to do this.

Let ω be the adelization of χ, as described in Proposition 3.1.2. Thus $\omega = \prod_v \omega_v$ is a character of $A^\times/\mathbb{Q}^\times$ of finite order, and if p is a rational prime not dividing N, if v is the place of $\mathbb{Q}$ corresponding to p, and $\varpi_v \in \mathfrak{o}_v$ is a generator of the prime ideal of $\mathfrak{o}_v$, then $\chi(p) = \omega_v(\varpi_v)$. It follows from the proof of Proposition 3.1.2 that ω_∞ is trivial on $\mathbb{R}_+^\times$, that if v is a non-Archimedean place not dividing N, then ω_v is nonramified (that is, is trivial on $\mathfrak{o}_v^\times$), and that if v is a non-Archimedean place dividing N, then ω_v is trivial on the subgroup of $\mathfrak{o}_v^\times$ consisting of elements congruent to the identity modulo N. Moreover, suppose that d is a positive integer prime to N. Then

$$\chi(d) = \prod_{v \notin S_\mathfrak{f}(N)} \omega_v(d_v),$$

where $S_\mathfrak{f}(N)$ is the union of all non-Archimedean places of $\mathbb{Q}$ dividing N. To see this, it is sufficient to assume that d is a prime p not dividing N, in which case the factor on the right, where v is the place of $\mathbb{Q}$ corresponding to p, equals $\chi(p)$, and all other factors are one. Now because $\omega = \prod_v \omega_v$ is trivial on $\mathbb{Q}^\times$, this equals

$$\chi(d) = \prod_{v \in S_\mathfrak{f}(N)} \omega_v^{-1}(d_v). \tag{6.2}$$

Moreover, Eq. (6.2), which we have checked when d is positive, remains true when $d_\infty < 0$, as is easily seen by checking the case $d = -1$.

We define a character λ of $K_0(N)$ by

$$\lambda\begin{pmatrix} a & b \\ c & d \end{pmatrix} = \prod_{v \in S_\mathfrak{f}(N)} \omega_v(d_v).$$

If $g \in GL(2, A)$, we use strong approximation (Theorem 3.3.1) to write $g = \gamma\, g_\infty\, k_0$, where $\gamma \in GL(2, \mathbb{Q})$, $g_\infty \in GL(2, \mathbb{R})^+$, and $k_0 \in K_0(N)$. We then define

$$\phi(g) = F(g_\infty)\, \lambda(k_0). \tag{6.3}$$

Let us check that this is well defined. We must show that if $g_\infty, g'_\infty \in GL(2, \mathbb{R})^+$, $\gamma \in GL(2, \mathbb{Q})$, and $k_0 \in K_0$ such that $g'_\infty = \gamma g_\infty k_0$, then

$$F(g'_\infty) = F(g_\infty)\, \lambda(k_0).$$

Write $\gamma = \gamma_\infty \gamma_f$, where $\gamma_\infty \in GL(2, \mathbb{R})$ and $\gamma_f \in GL(2, A_f)$. Evidently $g'_\infty = \gamma_\infty g_\infty$ and $\gamma_f = k_0^{-1}$. The first relation implies that γ_∞ has positive determinant and the second that it is in $\Gamma_0(N)$. Thus we have

$$F(g'_\infty) = \chi(d_\infty)\, F(g_\infty), \qquad \gamma_\infty = \begin{pmatrix} a_\infty & b_\infty \\ c_\infty & d_\infty \end{pmatrix},$$

so what we require is that $\lambda(k_0) = \chi(d_\infty)$, and because $k_0 = \gamma_f^{-1}$, this follows from Eq. (6.2).

The function ϕ defined by Eq. (6.3) is an automorphic form with a central quasicharacter ω. It must be checked that if $z \in A^\times$, then

$$\phi\left(\begin{pmatrix} z & \\ & z \end{pmatrix} g\right) = \omega(z)\, \phi(g). \tag{6.4}$$

Because

$$A^\times = \mathbb{Q}^\times\, \mathbb{R}_+^\times \prod_{v \text{ non-Archimedean}} \mathfrak{o}_v^\times,$$

it is sufficient to check this in the separate cases where $z \in \mathbb{Q}^\times$, $z \in \mathbb{R}_+^\times$, or when z is a unit in $\mathfrak{o}_v$, where v is non-Archimedean, and the latter case may be subdivided into $v \in S_f(N)$ and $v \notin S_f(N)$. Each case is easy.

The Hecke operators of Section 1.4 (which are extended to congruence subgroups in the exercises there) may be extended to operators on functions satisfying Eq. (6.1). Let Σ denote the set of all primes dividing N, and let $\mathbb{Z}_\Sigma$ be the localization of $\mathbb{Z}$ at the primes in Σ. Thus $\mathbb{Z}_\Sigma$ consists of all fractions r/s in $\mathbb{Q}$ with r and $s \in \mathbb{Z}$ and $N \nmid s$. We extend the character χ to $\mathbb{Z}_\Sigma$ by $\chi(r/s) = \chi(r)/\chi(s)$. Let $G_0(N)$ be the group of all $\left(\begin{smallmatrix} a & b \\ c & d \end{smallmatrix}\right) \in GL(2, \mathbb{Z}_\Sigma)$ with $c \in N\mathbb{Z}_\Sigma$. We define a right action of $G_0(N)$ on functions on $GL(2, \mathbb{R})^+$ by

$$(F|_\chi \alpha)(g) = \chi(d)^{-1}\, F(\alpha g), \qquad \alpha = \begin{pmatrix} a & b \\ c & d \end{pmatrix} \in G_0(N). \tag{6.5}$$

Let $\xi \in G_0(N)$, and let $\xi_1, \cdots \xi_h$ be a complete set of coset representatives for $\Gamma_0(N)\backslash\Gamma_0(N)\, \xi\, \Gamma_0(N)$. Then we define

$$T_\xi F = \sum_{i=1}^{h} F|_\chi \xi_i, \tag{6.6}$$

and if $\xi = \begin{pmatrix} p & \\ & 1 \end{pmatrix}$, where p is a prime not dividing N, we will denote $T_\xi = T_p$. The Hecke operators form a commutative ring of normal operators commuting with the Laplacian and hence may be simultaneously diagonalized.

Theorem 3.6.1 *Suppose that F is an eigenfunction of all the Hecke operators T_p when $p \nmid N$. Then ϕ lies in an irreducible subspace of $L_0^2\big(GL(2, F)\backslash GL(2, A), \omega\big)$.*

Thus every Hecke eigenform gives rise to an automorphic representation.

Proof By Theorem 3.3.2, $L_0^2\big(GL(2, F)\backslash GL(2, A), \omega\big)$ decomposes into a direct sum of irreducible invariant subspaces. Let (π, V) be an irreducible invariant subspace such that ϕ has a nonzero projection on V. We will show that π is uniquely determined by the eigenvalues of the Hecke operators T_p with $p \nmid N$, together with ω. This will show that actually $\phi \in \pi$.

Let S be the set of places of $\mathbb{Q}$ consisting of the Archimedean place, and let every place v correspond to a rational prime dividing N. Note that ϕ is K_v-fixed for every $v \notin S$. This is clear because the character λ is trivial on K_v.

Let p be a rational prime not dividing N, and let $v_p \notin S$ be the place of $\mathbb{Q}$ corresponding to p. Let $\mathcal{H}_p$ be the *Hecke algebra* of compactly supported, locally constant functions on $GL(2, \mathbb{Q}_p)$, which is a ring (without unit) under convolution, and let $\mathcal{H}_p^\circ$ be the subalgebra of functions that are both left and right invariant under $K_p = K_{v_p}$. The *spherical Hecke algebra* $\mathcal{H}_p^\circ$ is a commutative ring (with unit), and its structure is determined in Section 4.6, particularly Proposition 4.6.5. (See also Section 3.9 of this chapter for more discussion of the structure of spherical Hecke algebra.) It is generated by elements that we will denote $\mathbb{T}_p$ and $\mathbb{R}_p$, together with $\mathbb{R}_p^{-1}$, where p is the rational prime (not dividing N) corresponding to the place v. (We are using a different notation from Section 4.6 in order to avoid confusion between $\mathbb{T}_p$ and the closely related "classical Hecke operator" T_p.) Here $\mathbb{T}_p$ is the characteristic function of the double coset

$$K_p \begin{pmatrix} \varpi_p & \\ & 1 \end{pmatrix} K_p,$$

where ϖ_p is the idele whose v_pth component is p and all of whose other components are trivial, and $\mathbb{R}_p$ and $\mathbb{R}_p^{-1}$ are the characteristic functions of

$$K_p \begin{pmatrix} \varpi_p & \\ & \varpi_p \end{pmatrix} \quad \text{and} \quad K_p \begin{pmatrix} \varpi_p & \\ & \varpi_p \end{pmatrix}^{-1},$$

respectively. As in Eq. (6.4) of Chapter 4, we decompose the double coset

$$K_p \begin{pmatrix} \varpi_p & \\ & 1 \end{pmatrix} K_p = \bigcup_{i=1}^{p+1} i_p(\xi_i)\, K_p, \tag{6.7}$$

where $i_p : GL(2, \mathbb{Q}) \to GL(2, A)$ is the map induced by the ring homomorphism, $\mathbb{Q} \to \mathbb{Q}_p \to A$, the first arrow being the inclusion map, and where

$$\xi_i = \begin{pmatrix} p & i \\ & 1 \end{pmatrix} \qquad (i = 1, \cdots, p), \qquad \xi_{i+1} = \begin{pmatrix} 1 & \\ & p \end{pmatrix}.$$

We have an action of the Hecke algebras on automorphic forms, which we denote by ρ, because it is derived from right translation. If ϕ is an automorphic form and $\sigma \in \mathcal{H}_p$, then

$$(\rho(\sigma)\phi)(g) = \int_{GL(2,A)} \sigma(h)\,\phi(gh)\,dh.$$

We will denote $\rho(\mathbb{T}_p)\phi$ as simply $\mathbb{T}_p(\phi)$. With ϕ as in Eq. (6.3), let us evaluate $(\mathbb{T}_p\phi)(g)$, with $g = \gamma g_\infty k_0$, $\gamma \in GL(2, \mathbb{Q})$, $g_\infty \in GL(2, \mathbb{R})^+$, and $k_0 \in K_0(N)$. Because ϕ is right K_p-invariant, Eq. (6.7) implies that

$$(\mathbb{T}_p\phi)(g) = \sum_{i=1}^{p+1} \phi(g\xi_i).$$

By Eq. (6.7), if $1 \le i \le p+1$, there exist $1 \le j \le p+1$ and $k_0' \in K_0(N)$ such that $k_0\, i_p(\xi_i) = i_p(\xi_j)\, k_0'$. Let us factor $\xi_j \in GL(2, \mathbb{Q})$ as $\xi_{j,\infty}\, \xi_{j,\mathrm{f}}$, where $\xi_{j,\infty} \in GL(2, \mathbb{R})^+$, $\xi_{j,\mathrm{f}} \in GL(2, A_\mathrm{f})$. Then

$$g\, i_p(\xi_i) = (\gamma\, \xi_j)\big(\xi_{j,\infty}^{-1}\, g_\infty\big)\big(\xi_{j,\mathrm{f}}^{-1}\, i_p(\xi_j)\, k_0'\big).$$

The three factors here lie in $GL(2, \mathbb{Q})$, $GL(2, \mathbb{R})^+$, and $K_0(N)$, respectively. Therefore by Eq. (6.3)

$$\phi\big(g\, i_p(\xi_i)\big) = F\big(\xi_{j,\infty}^{-1} g_\infty\big)\, \lambda\big(\xi_{j,\mathrm{f}}^{-1} i_p(\xi_j) k_0'\big).$$

Now if v is any non-Archimedean place dividing N, the vth component of $i_p(\xi_j)k_0'$ is the same as the vth component of k_0' or k_0. Thus, remembering the definition of λ, we have

$$\lambda\big(\xi_{j,\mathrm{f}}^{-1} i_p(\xi_j) k_0'\big) = \lambda\big(\xi_{j,\mathrm{f}}^{-1}\big)\, \lambda(k_0).$$

Thus by Eq. (6.5)

$$\phi\big(g\, i_p(\xi_i)\big) = \big(F|_\chi \xi_{j,\infty}^{-1}\big)(g)\, \lambda(k_0).$$

Thus if F is an eigenfunction of the classical Hecke operator T_α, where $\alpha = \begin{pmatrix} p & \\ & 1 \end{pmatrix}$, then ϕ is an eigenfunction of $\mathbb{T}_p$ with the same eigenvalue. Also, ϕ is an eigenfunction of $\mathbb{R}_p$ with eigenvalue $\chi(p)$, by Eq. (6.4). To summarize, ϕ is an eigenvector $\mathcal{H}_p$, and the eigenvalues of the elements of $\mathcal{H}_p$ on ϕ are determined by χ and the eigenvalues of the classical Hecke operators on F.

The projection of $L_0^2\big(GL(2, F)\backslash GL(2, A), \omega\big)$ onto the invariant subspace V is $GL(2, A)$-equivariant, and therefore the image of ϕ in V is an eigenvector of $\mathcal{H}_p$ with the same eigenvalues. These determine the irreducible constituent

π_{v_p} of π by Theorem 4.6.3. Thus π is itself determined by the multiplicity one Theorem 3.3.7. ∎

3.7 Eisenstein Series and Intertwining Integrals

Let F be a global field, A its adele ring, and let other notations be as in Section 3.3. We have in Theorem 3.3.2 a rather satisfactory result on the decomposition of the space $L_0^2\big(GL(n,F)\backslash GL(n,A),\omega\big)$ of cusp forms in $L^2\big(GL(n,F)\backslash GL(n,A),\omega\big)$. We recall that $L^2\big(GL(n,F)\backslash GL(n,A),\omega\big)$ is the space of measurable functions f on $GL(n,A)$ that satisfy

$$f(\gamma g) = f(g), \qquad \gamma \in GL(n,F),\ g \in GL(n,A),$$

$$f\left(\begin{pmatrix} z & & \\ & \ddots & \\ & & z \end{pmatrix} g\right) = \omega(z), \qquad z \in Z(A),\ g \in GL(n,A),$$

and

$$\int\limits_{Z(A)\,GL(n,F)\backslash GL(n,A)} |f(g)|^2\,dg < \infty.$$

The subspace $L_0^2\big(GL(n,F)\backslash GL(n,A),\omega\big)$ is characterized by the vanishing of the constant terms: if $n=2$, this amounts to the condition

$$\int\limits_{A/F} f\left(\begin{pmatrix} 1 & x \\ & 1 \end{pmatrix} g\right) dx = 0, \tag{7.1}$$

or see Eq. (3.7) for the general case. We saw in Theorem 3.3.2 that the Hilbert space $L_0^2\big(GL(n,F)\backslash GL(n,A),\omega\big)$ decomposes into a direct sum of irreducible invariant subspaces. In a sense, this is the most interesting part of $L^2\big(GL(n,F)\backslash GL(n,A),\omega\big)$. It is possible to describe the orthogonal complement of $L_0^2\big(GL(n,F)\backslash GL(n,A),\omega\big)$ rather explicitly, though a full discussion of this question is outside the scope of this book. We will partially discuss this question when $n=2$, leaving the reader to consult other sources, particularly Moeglin and Waldspurger (1995) for the complete story.

Firstly, this orthogonal complement contains the direct sum of the one-dimensional invariant subspaces spanned by the functions $\chi\big(\det(g)\big)$, where χ ranges through the Hecke characters of $A^\times/F^\times$ such that $\chi^2=\omega$. (There are either no such characters or an infinite number, depending on ω.) The direct sum of $L_0^2\big(GL(2,F)\backslash GL(2,A),\omega\big)$ with these one-dimensional spaces is the *discrete spectrum* $L^2_{\text{disc}}\big(GL(2,F)\backslash GL(2,A),\omega\big)$ and is the largest subspace of $L^2\big(GL(2,F)\backslash GL(2,A),\omega\big)$ that can be decomposed into a direct sum of irreducible invariant subspaces. (This is a slight misuse of the word "spectrum" of course.) The orthocomplement of this space is called the *Eisenstein spectrum*, denoted $L^2_{\text{Eis}}\big(GL(2, F)\backslash GL(2, A), \omega\big)$. It does not decompose into a direct sum of irreducible representations, but rather, into a direct integral. We will not prove this decomposition, but will content ourselves with studying to some extent the automorphic forms that underlie it, the *Eisenstein series.*

Before proceeding, we review briefly the spectral decomposition of $L^2(\Gamma\backslash\mathcal{H})$, where $\Gamma = SL(2,\mathbb{Z})$, in order to get some intuition for the adelic case. This space contains, firstly, the constant function and, secondly, the span of the one-dimensional subspaces spanned by the weight-zero Maass cusp forms (Exercise 3.2.5(c)). The Hilbert space direct sum of these one dimensional spaces (spanned by the constant function and the cusp forms) is the "discrete spectrum," $L^2_{\text{disc}}(\Gamma\backslash\mathcal{H})$. Let $L^2_{\text{Eis}}(\Gamma\backslash\mathcal{H})$ be the orthogonal complement of $L^2_{\text{disc}}(\Gamma\backslash\mathcal{H})$.

In order to understand $L^2_{\text{Eis}}(\Gamma\backslash\mathcal{H})$, we recall the Eisenstein series $E(z, s)$ for $SL(2,\mathbb{Z})$ defined in Section 1.6. Recall that

$$E(z, s) = \pi^{-s}\,\Gamma(s)\,\zeta(2s) \sum_{\Gamma_\infty\backslash SL(2,\mathbb{Z})} \text{im}\,\big(\gamma(z)\big)^s. \tag{7.2}$$

These are not square integrable for any s. To understand how big $E(z, s)$ is, we consult the Fourier expansion (Eq. 6.9) in Chapter 1. If $y = \text{im}(z)$ is large, the constant term, which is given by Eq. (6.12) in Chapter 1, by

$$\int_0^1 E(x+iy, s)\,dx = \pi^{-s}\,\Gamma(s)\,\zeta(2s)\,y^s + \pi^{s-1}\,\Gamma(1-s)\,\zeta(2-2s)\,y^{1-s} \tag{7.3}$$

predominates, because the other terms in the Fourier expansion are given by Eq. (6.13) in Chapter 1, and the Bessel function $K_{s-1/2}$ is of rapid decay as $y \to \infty$. The magnitude of the constant term is therefore smallest if $\text{re}(s) = 1/2$. Even in this case, the Eisenstein series is not quite square integrable. However, it is close, as we may see by integrating

$$\int_{\substack{0<x<1\\ 1<y<Y}} \left|y^{1/2}\right|^2 \frac{dx\,dy}{y^2} = \log(Y).$$

Here if $\text{re}(s) = 1/2$, then $|y^{1/2}|$ is the magnitude of the constant term in the Eisenstein series and hence of the Eisenstein series itself, and if Y is large, the region of integration is an approximation to the fundamental domain. The function $\log(Y)$ grows only slowly as $Y \to \infty$. This is why we say that the Eisenstein series is "almost" square integrable when $\text{re}(s) = 1/2$.

The spectral expansion for $SL(2,\mathbb{Z})$ has the following form: If $f \in L^2_{\text{Eis}}(\Gamma\backslash\mathcal{H})$, then f may be expanded as

$$\int_0^\infty E\big(z, \tfrac{1}{2} + it\big)\ \tilde{f}(t)\,\frac{dt}{t} \tag{7.4}$$

for suitable $\tilde{f}$. This spectral decomposition was proved independently by Roelcke (1956) and Selberg (1956). For Godement's proof, see Godement (1966a), Terras (1985, Theorem I, p. 254), or Lang (1985), Chapter 13. See Iwaniec (1995) for further material on the spectral theory of automorphic forms and Eisenstein series.

We could more generally consider (in the notation of Section 2.1) the space $L^2(\Gamma\backslash\mathcal{H}, \chi, k)$, where Γ is an arbitrary discontinuous subgroup of $SL(2, \mathbb{R})$ with finite volume, which without loss of generality can be assumed to contain $-I$; k is an integer, and χ is a character of Γ such that $\chi(-I) = (-1)^k$. In this case, the dimension of the space of Eisenstein series (for a given value of s) is equal to the number of cusps.

If the group Γ and its subgroup $\ker(\chi)$ are congruence subgroups of $SL(2, \mathbb{Z})$, then except for this minor detail (that there are Eisenstein series coming from every cusp of the group) the situation is very analogous to that of $SL(2, \mathbb{Z})$. The poles of the Eisenstein series are at the poles of the constant terms, and for these particular groups, these constant terms are essentially in the form (Eq. 7.3), except that the zeta function can be replaced by a Dirichlet L-function $L(s, \xi)$. Because the only possible pole of $L(s, \xi)$ is the pole at $s = 1$ when ξ is trivial, the Eisenstein series $E(z, s)$ has no "unexpected" poles in the case of a congruence subgroup.

But if Γ is a noncongruence subgroup, or if the Eisenstein series is made with a character χ whose kernel does not contain a congruence subgroup, then the Eisenstein series can have many poles. On the other hand, there may be few cusp forms. This may be seen from the work of Phillips and Sarnak (1985). They consider what happens when a congruence group is deformed into a nonarithmetic one. They found that the cusp forms can "disappear" when the group is deformed, and it was shown in subsequent papers of Deshouillers and Iwaniec (1986), of Deshouillers, Iwaniec, Phillips and Sarnak (1985), and of Luo (1993) (who obtained the best result to date) that this destruction of cusp forms is the rule. The cusp forms that are destroyed when the group is deformed are replaced by poles of Eisenstein series. It now seems clear that for a general noncongruence subgroup, there should be few if any cusp forms, but that the Eisenstein series can have many poles.

A similar situation pertains with the Ramanujan conjecture. Again, we find special features in the "arithmetic" case where Γ and the kernel of χ are congruence subgroups, showing that in the context of general discontinuous groups with cofinite volume, these groups are atypical. Selberg conjectured that in the arithmetic case, the eigenvalue λ of the Laplacian Δ corresponding to a Maass cusp form in $L^2_0(\Gamma\backslash\mathcal{H}, \chi, k)$ is greater than or equal to $\frac{1}{4}$. This may be formulated as the assertion that the principal series representation of $SL(2, \mathbb{R})$ corresponding to this eigenvalue is not in the complementary series. (This conjecture is really only a question about weight-zero Maass forms, as the complementary series representations all have $SO(2)$-fixed vectors.) Expressed this way, we see that Selberg's conjecture is the exact analog of the Ramanujan conjecture, stated in Section 3.5. As we explain there, the Ramanujan conjecture and the Selberg conjecture would follow from the Langlands conjectures and may therefore be assumed to be true, though deep and difficult, and the best result toward the Selberg conjecture is that of Luo, Rudnick and Sarnak (in press).

The depth of the Selberg conjecture may be seen from the fact that it definitely fails in the nonarithmetic case. Indeed, if M is a compact Riemann surface of genus greater than or equal to 2, then M may be realized as $\Gamma\backslash\mathcal{H}$ for suitable Γ – indeed, the universal covering surface of M is conformally equivalent to $\mathcal{H}$, and if we make this identification, the fundamental group $\pi_1(M)$ is realized as a discontinuous group Γ acting on $\mathcal{H}$ (see the exercises to Section 1.3). An eigenvalue of the (weight-zero) Laplacian on $L^2(M)$ that is less than $1/4$ is called an *exceptional eigenvalue*. Randol (1974) proved that there exist compact Riemann surfaces M for which exceptional eigenvalues do exist. See Iwaniec (1984) for further information on this and related topics.

We see that there are important differences between the arithmetic case and the general case for the spectral theory of $\Gamma\backslash\mathcal{H}$. The adelic case of $L^2\big(GL(2,F)\backslash GL(2,A),\omega\big)$ resembles the arithmetic case, so it is from this case that we should draw our intuition.

Our aim in this section is not to give a complete discussion of these topics. In particular, we will not prove the spectral expansion (Eq. 7.4) or its adelic analog. We will not prove that the orthogonal complement of $L^2_{\text{disc}}\big(GL(2,F)\backslash GL(2,A),\omega\big)$ decomposes into a direct integral of irreducible representations coming from the Eisenstein series. We will, however, prove the analytic continuation of the Eisenstein series and show how the constant term in the Eisenstein series leads to the intertwining integrals that play an important role in the local theory (cf. Sections 2.6 and 4.5). We will also briefly discuss the Rankin–Selberg method.

Let F be a global field and A its adele ring. Notations will be as in Section 3.3. Let χ_1 and χ_2 be quasicharacters of $A^\times/F^\times$. We wish to be able to deform these characters continuously, which we accomplish as follows. We start with two unitary characters ξ_1 and ξ_2 of $A^\times/F^\times$ and two complex numbers s_1 and s_2. We will sometimes denote $s = \frac{1}{2}(s_1 - s_2 + 1)$, a useful auxiliary variable. Let $\chi_i(y) = \xi_i(y)\,|y|^{s_i}$ $(i = 1, 2)$. As the complex numbers s_1 and s_2 are varied, the characters χ_1 and χ_2 change continuously. We will be discussing zeros and poles of various objects, and in such a context we would rather be dealing with functions of one complex variable. So when convenient, we will assume that s_1 and s_2 are varied in such a way that $s_1 + s_2$ is held constant.

We now construct an admissible representation $\pi(\chi_1, \chi_2)$ of $GL(2, A)$ in the sense of Section 3.3. The space V_{χ_1,χ_2} of this representation consists of all smooth, K-finite functions on $GL(2, A)$ that satisfy

$$f\left(\begin{pmatrix} y_1 & x \\ & y_2 \end{pmatrix} g\right) = \left|\frac{y_1}{y_2}\right|^{1/2} \chi_1(y_1)\,\chi_2(y_2)\,f(g). \tag{7.5}$$

(The notion of *smoothness* for functions on $GL(n, A)$ was defined in Section 3.3.) The action of $GL(2, F_v)$ (if v is non-Archimedean) or the $(\mathfrak{g}_v, K_v)$-module structure (if v is Archimedean) are by right translation and differentiation. It follows from the Iwasawa decomposition that f is determined by its restriction to K, and consequently, if ρ is any irreducible finite-dimensional representation of K, the dimension of $V(\rho)$ is at most the dimension of ρ, because by the representation theory of compact groups (Peter-Weyl formula and Schur orthogonality), the multiplicity of ρ in $L^2(K)$ is $\dim(\rho)$. So this representation is admissible.

It is easy to see that this representation is the restricted tensor product of the principal series representations $\pi(\chi_{1,v}, \chi_{2,v})$ of the groups $GL(2, F_v)$. These representations, which we introduced in Section 3.5 are studied in detail in Chapter 4 Section 4.5 (if v is non-Archimedean) or in Chapter 2 Section 2.5 (if v is Archimedean).

Suppose that ξ_1 and ξ_2 are fixed and that for every s_1 and s_2 we are given an element $f_{s_1,s_2} \in V_{\chi_1,\chi_2}$. Assume further that for $k \in K$, $f_{s_1,s_2}(k)$ is independent of s_1 and s_2. Then we say that the assignment $(s_1, s_2) \mapsto f_{s_1,s_2}$ comprises a *flat section* of the representations $\pi(\chi_1, \chi_2)$.

Proposition 3.7.1 *Every element of V_{χ_1,χ_2} can be embedded in a unique flat section.*

Proof By the Iwasawa decomposition, every element of $GL(2, A)$ can be written in the form bk, where b is an element of the Borel subgroup $B(A)$ of upper triangular matrices in $GL(2, A)$, and $k \in K$. Therefore any element f of V_{χ_1,χ_2} is determined by its restriction to K, which must satisfy

$$f\left(\begin{pmatrix} y_1 & x \\ & y_2 \end{pmatrix} k\right) = \chi_1(y_1)\, \chi_2(y_2)\, |y_1/y_2|^{1/2}\, f(k) \tag{7.6}$$

when

$$\begin{pmatrix} y_1 & x \\ & y_2 \end{pmatrix} \in K \cap B(A),$$

and conversely, any smooth K-finite function satisfying Eq. (7.7) can be extended to an element of $\pi(\chi_1, \chi_2)$. The key point is that this condition is independent of s_1 and s_2. Indeed, the condition that

$$\begin{pmatrix} y_1 & x \\ & y_2 \end{pmatrix} \in K \cap B(A)$$

can only be satisfied if $|y_1| = |y_2| = 1$, and so $\chi_1(y_1) = \xi_1(y_1)$ and $\chi_2(y_2) = \xi_2(y_2)$. Thus Eq. (7.6) can be rewritten

$$f\left(\begin{pmatrix} y_1 & x \\ & y_2 \end{pmatrix} k\right) = \xi_1(y_1)\, \xi_2(y_2)\, f(k), \quad \begin{pmatrix} y_1 & x \\ & y_2 \end{pmatrix} \in K \cap B(A). \tag{7.7}$$

Thus there is a bijection between the elements of V_{χ_1,χ_2} and the space of smooth functions on K satisfying Eq. (7.7). Because this condition is independent of s_1 and s_2, the proposition is now evident. ∎

The *Eisenstein series* is a $GL(2, A)$-equivariant map

$$\pi(\chi_1, \chi_2) \to \mathcal{A}\big(GL(2, F)\backslash GL(2, A), \omega\big).$$

It is constructed just as the Eisenstein series is in Section 1.6. Let $f \in \pi(\chi_1, \chi_2)$. We define the Eisenstein series

$$E(g, f) = \sum_{\gamma \in B(F)\backslash GL(2,F)} f(\gamma g), \qquad g \in GL(2, A), \tag{7.8}$$

provided the sum is convergent. We note that this is well defined because Eq. (7.5) implies that if $\beta \in B(F)$, then $f(\beta g) = f(g)$. It is clear that if $\gamma \in GL(2, F)$, then $E(\gamma g, f) = E(g, f)$, and it is not hard to see that $g \mapsto E(g, F)$ satisfies the other conditions of an automorphic form ($\mathcal{Z}$-finiteness, K-finiteness, and moderate growth). We will discuss this in Proposition 3.7.3. Thus as a function of g, the Eisenstein series $E(g, f)$ is an element of $\mathcal{A}\big(GL(2, F)\backslash GL(2, A), \omega\big)$, with $\omega = \chi_1\chi_2$. Moreover, the map $f \mapsto E(\cdot, f)$ from $\pi(\chi_1, \chi_2) \to \mathcal{A}\big(GL(2, F)\backslash GL(2, A), \omega\big)$ is clearly $GL(2, A)$-equivariant.

Proposition 3.7.2 *The sum (Eq. 7.8) is convergent if* $\mathrm{re}(s_1 - s_2) > 1$.

Proof if $F = \mathbb{Q}$ Let $\Gamma = SL(2, \mathbb{Z})$, and let $\Gamma_\infty = \Gamma \cap B(\mathbb{Q})$. Using the fact that $\mathbb{Q}$ has class number one, it is easy to see that if $\gamma \in GL(2, \mathbb{Q})$, then there exists $\beta \in B(\mathbb{Q})$ such that $\beta\gamma \in \Gamma$. Consequently, we may write Eq. (7.8) as

$$\sum_{\gamma \in \Gamma_\infty\backslash\Gamma} f(\gamma g).$$

Every element of $\pi(\chi_1, \chi_2)$ is a finite linear combination of elements of the form

$$f(g) = \prod_v f_v(g_v), \qquad g = (g_v) \in GL(2, A), \tag{7.9}$$

where $f_v \in \pi(\chi_{1,v}, \chi_{2,v})$, and for almost all v, f_v is the spherical vector f_v°, normalized so that $f_v^\circ(k_v) = 1$ for $k_v \in K_v$. We may assume without loss of generality that f is of this special type. This assumption is equivalent to assuming that the image of f in the decomposition of Theorem 3.3.3 is a pure tensor.

Now we fix a particular $g = (g_v)$ in Eq. (7.8). For every place v, because K_v is compact, there exists a constant $B_v > 0$ such that $|f(k_v g_v)| \leq B_v$ for all

$k_v \in K_v$ and for almost all v, $B_v = 1$. Let $\gamma \in SL(2,\mathbb{Z})$. We regard $\gamma = (\gamma_v)$ as an adele. For all non-Archimedean places v, $\gamma_v \in K_v$, and so

$$f(\gamma g) \le B\, f_\infty(\gamma_\infty g_\infty), \qquad B = \prod_{v \text{ non-Archimedean}} B_v.$$

Therefore we are reduced to showing that

$$\sum |f_\infty(\gamma_\infty g_\infty)| < \infty.$$

We make use of the fact that $SL(2,\mathbb{R})/SO(2)$ can be identified with the upper half plane and that with this identification, the natural action of $SL(2,\mathbb{R})$ by left translation is the same as the action on $\mathcal{H}$ by linear fractional transformation. Thus let $\gamma = \left(\begin{smallmatrix} a & b \\ c & d \end{smallmatrix}\right)$, and let

$$g_\infty = \begin{pmatrix} u & \\ & u \end{pmatrix} \begin{pmatrix} y^{1/2} & x\,y^{-1/2} \\ & y^{-1/2} \end{pmatrix} \kappa_\theta, \qquad \kappa_\theta = \begin{pmatrix} \cos(\theta) & \sin(\theta) \\ -\sin(\theta) & \cos(\theta) \end{pmatrix}.$$

Then

$$\gamma_\infty g_\infty = \begin{pmatrix} u & \\ & u \end{pmatrix} \begin{pmatrix} y_1^{1/2} & x_1 y_1^{-1/2} \\ & y_1^{-1/2} \end{pmatrix} \kappa_{\theta_1},$$

where, with $z = x + iy$

$$x_1 + iy_1 = \frac{az+b}{cz+d}.$$

We do not need to know the value of κ_{θ_1} precisely. We have

$$|f_\infty(\gamma_\infty g_\infty)| = u^{s_1+s_2}\, y_1^{(s_1-s_2+1)/2}\, |f(\kappa_{\theta_1})| \le u^{s_1+s_2}\, B_\infty \frac{y^s}{|cz+d|^{2s}},$$

where $s = \frac{1}{2}(s_1 - s_2 + 1)$, by Eq. (2.2) in Chapter 1. Thus Eq. (7.8) is convergent if $\mathrm{re}(s) > 1$ by Exercise 1.6.1. ∎

We will now consider the Fourier expansions of the Eisenstein series, along the lines of the proof of Theorem 1.6.1. We will see that the constant term in the Fourier expansion leads naturally to the intertwining integrals that play a role in the local theory (Sections 2.6 and 4.5).

Let $\alpha \in F$. We consider the integral

$$c_\alpha(g,f) = \int_{A/F} E\left(\begin{pmatrix} 1 & x \\ & 1 \end{pmatrix} g\right) \psi(-\alpha x)\, dx. \tag{7.10}$$

We have a Fourier expansion

$$E(g,f) = \sum_{\alpha \in F} c_\alpha(g,f). \tag{7.11}$$

Indeed, we prove Eq. (7.11) by the Fourier inversion formula as follows. For fixed g, the function

$$x \mapsto E\left(\begin{pmatrix} 1 & x \\ & 1 \end{pmatrix} g\right)$$

is a continuous function on the compact group A/F and hence has a Fourier expansion in terms of the characters of A/F. By Exercise 3.1.3, these are the functions of the form $x \mapsto \psi(\alpha x)$ with $\alpha \in F$, and so the Fourier coefficients in this expansion are just the $c_\alpha(g, f)$. Therefore

$$E\left(\begin{pmatrix} 1 & x \\ & 1 \end{pmatrix} g\right) = \sum_{\alpha \in F} c_\alpha(g, f)\, \psi(\alpha x),$$

and putting $x = 0$, we obtain Eq. (7.11).

We are still assuming that $\mathrm{re}(s_1 - s_2) > 1$, so that the sum (Eq. 7.8) is convergent. In Theorem 3.7.1, we will use the Fourier expansion (Eq. 7.11) to obtain the analytic continuation of $E(g, f)$ to other values of s_1 and s_2, so the Fourier expansion will remain valid. In the meantime, assuming that $\mathrm{re}(s_1 - s_2) > 1$, let us analyze the coefficients $c_\alpha(g, f)$ further. A complete set of representatives for $B(F)\backslash GL(2, F)$ are

$$I = \begin{pmatrix} 1 & \\ & 1 \end{pmatrix}, \qquad \text{and} \qquad w_0 \begin{pmatrix} 1 & \lambda \\ & 1 \end{pmatrix}, \qquad (\lambda \in F), \tag{7.12}$$

where

$$w_0 = \begin{pmatrix} & -1 \\ 1 & \end{pmatrix}. \tag{7.13}$$

Thus we may split the sum (Eq. 7.8) into two parts, corresponding to the first and second types of representatives γ in Eq. (7.12). If $\alpha \neq 0$, then the contribution of $\gamma = I$ is zero, and $c_\alpha(g, f)$ equals

$$\sum_{\lambda \in F} \int_{A/F} f\left(w_0 \begin{pmatrix} 1 & x+\lambda \\ & 1 \end{pmatrix} g\right) \psi(-\alpha x)\, dx.$$

Noting that $\psi(\alpha\, x) = \psi\big(\alpha\, (x + \lambda)\big)$, we may collapse the summation with the integration, and we find that

$$c_\alpha(g, f) = \int_A f\left(w_0 \begin{pmatrix} 1 & x \\ & 1 \end{pmatrix} g\right) \psi(-\alpha x)\, dx, \qquad \alpha \neq 0. \tag{7.14}$$

If $\alpha = 0$, the analysis is identical, except that the contribution of $\gamma = I$ cannot be ignored, and we obtain

$$c_0(g, f) = f(g) + \int_A f\left(w_0 \begin{pmatrix} 1 & x \\ & 1 \end{pmatrix} g\right) dx. \tag{7.15}$$

Now assume, that f is a "pure tensor" (Eq. 7.9). Then the integrals in Eqs. (7.14) and (7.15) factorize and may be studied locally. Let us begin with Eq. (7.14). If $\alpha \neq 0$, this integral is analogous to one of the terms in Eq. (5.19). We make the change of variables $x \to \alpha^{-1}x$ and Eq. (7.15) becomes

$$c_\alpha(g, f) = \int_A f\left(\begin{pmatrix} 1 & \\ & \alpha^{-1} \end{pmatrix} w_0 \begin{pmatrix} 1 & x \\ & 1 \end{pmatrix} \begin{pmatrix} \alpha & \\ & 1 \end{pmatrix} g\right) \psi(-x)\, dx.$$

Using Eq. (7.6) and noting that $\chi_2(\alpha) = |\alpha| = 1$ because $\alpha \in F^\times$, we obtain

$$c_\alpha(g, f) = W\left(\begin{pmatrix} \alpha & \\ & 1 \end{pmatrix} g\right), \tag{7.16}$$

where

$$W(g) = \int_A f\left(w_0 \begin{pmatrix} 1 & x \\ & 1 \end{pmatrix} g\right) \psi(-x)\, dx. \tag{7.17}$$

Now using Eq. (7.10), we have

$$W(g) = \prod_v W_v(g_v), \tag{7.18}$$

where $W_v : GL(2, F_v) \to \mathbb{C}$ is the local Whittaker function

$$W_v(g_v) = \int_{F_v} f\left(w_0 \begin{pmatrix} 1 & x \\ & 1 \end{pmatrix} g_v\right) \psi_v(-x)\, dx. \tag{7.19}$$

This type of integral is used in Chapter 4 (cf. Eq. (6.8) in Section 4.6) to study the Whittaker functions of the spherical principal series.

As we have already pointed out, the representation $(\pi, V) = \pi(\chi_1, \chi_2)$ decomposes into a tensor product of the local principal series representations (π_v, V_v), where in the notation of Section 4.5, $(\pi_v, V_v) = \pi(\chi_{1,v}, \chi_{2,v})$. The space V_v consists of all K_v-finite functions $f_v : GL(2, F_v) \to \mathbb{C}$ that satisfy

$$f_v\left(\begin{pmatrix} y_1 & x \\ & y_2 \end{pmatrix} g_v\right) = \chi_{1,v}(y_1)\, \chi_{2,v}(y_2)\, |y_1/y_2|^{1/2} f_v(g_v). \tag{7.20}$$

The action of $GL(2, F_v)$ is by right translation. (If v is Archimedean, then V_v is a $(\mathfrak{g}_v, K_v)$-module.) These principal series representations are studied in detail in Sections 4.5 and 2.5.

The second integral in Eq. (7.15) is the "intertwining integral" that is studied in detail in Sections 4.5 and 2.6. We recall the basic facts here. The integral factorizes just like Eq. (7.18) if we assume that f is a "pure tensor" as in Eq. (7.9). Then

$$\int_A f\left(w_0 \begin{pmatrix} 1 & x \\ & 1 \end{pmatrix} g\right) dx = \prod_v \left(M_v(s) f_v\right)(g_v), \tag{7.21}$$

where

$$\big(M_v(s) f_v\big)(g_v) = \int_{F_v} f_v\left(w_0 \begin{pmatrix} 1 & x \\ & 1 \end{pmatrix} g \right) dx. \tag{7.22}$$

Here the integral (Eq. 7.21) is convergent if $\mathrm{re}(s) > 1/2$. It has "meromorphic continuation" to all s except that it has a pole at $s - 1/2 = \nu$ if $\xi_{1,v}\xi_{2,v}^{-1}(x) = |x|^\nu$. Meromorphic continuation is to be understood as follows: Let f_v vary in a flat section of $\pi(\chi_{1,v}, \chi_{2,v})$ – that is, let the restriction of f_v to K_v be independent of s_1 and s_2. Then $\big(M_v(s) f_v\big)(g_v)$ is a meromorphic function of s_1 and s_2. The analytic continuation of the intertwining integral is proved in Proposition 4.5.7 if v is non-Archimedean and (for K_v-finite vectors) is a consequence of the formula (6.15) in Chapter 2 if v is Archimedean. The intertwining integral is a $GL(2, F_v)$-equivariant map from $\pi(\chi_{1,v}, \chi_{2,v})$ to $\pi(\chi_{2,v}, \chi_{1,v})$, as is shown by an easy calculation (Propositions 4.5.6, 4.5.7 and 2.6.2). If $\pi(\chi_{1,v}, \chi_{2,v})$ is irreducible (which is true unless $(\chi_{1,v}\chi_{2,v}^{-1})(x) = |x|$ or $|x|^{-1}$ if v is non-Archimedean), and if $\chi_{1,v} \neq \chi_{2,v}$ (to avoid the pole of the intertwining integral), then $M_v(s)$ is an isomorphism $\pi(\chi_{1,v}, \chi_{2,v}) \to \pi(\chi_{2,v}, \chi_{1,v})$.

We will "normalize" the Eisenstein series as follows. Let S be a finite set of places such that if $v \notin S$, then v is non-Archimedean, $\xi_{1,v}$ and $\xi_{2,v}$ are nonramified, and the conductor of ψ_v is $\mathfrak{o}_v$. Let

$$E^*(g, f) = L_S\big(2s, \xi_1\xi_2^{-1}\big)\, E(g, f), \tag{7.23}$$

where $s = \frac{1}{2}(s_1 - s_2 + 1)$, and the partial L-function $L_S\big(s, \xi_1\xi_2^{-1}\big)$ is as in Section 3.1. (We will also make use of the complete L-function $L\big(s, \xi_1\xi_2^{-1}\big)$, defined in Section 3.1.) The factor $L_S\big(2s, \xi_1\xi_2^{-1}\big)$ is called the *normalizing factor* of the Eisenstein series. Inclusion of the normalizing factor has the effect of giving the Eisenstein series a more symmetrical functional equation, as we will see later.

Theorem 3.7.1 *Let f vary in a flat section. Then the Eisenstein series $E^*(g, f)$ has meromorphic continuation to all s. It is entire unless there exists a complex number ν such that $\big(\xi_1\xi_2^{-1}\big)(x) = |x|^\nu$ for all x, in which case the only possible pole is at $s + \nu/2 = 1$.*

We recall that when we speak of the Eisenstein series having a pole, we mean that $s_1 + s_2$ is held constant and that we consider it as a function of $s = \frac{1}{2}(s_1 - s_2 + 1)$.

We note that because we are assuming that ξ_1 and ξ_2 are unitary, the parameter ν, if it exists, is purely imaginary. There is no real loss of generality in assuming that $\nu = 0$, so the pole of the Eisenstein series is at $s = 1$.

The analysis of the constant term below is adapted from Gelbart and Shahidi (1988). The discussion of Eisenstein series and the Rankin–Selberg method in Gelltart and Shahidi is instructive and recommended.

We are assuming that F is a number field so that the set S is nonempty. If F is a function field, we could take S to be the empty set, in which case there would also be a pole at $s+\nu/2=0$. To explain this, we will see below that the poles of $E^*(g,f)$ are contained in the set of poles of $L_S(2s,\xi_1\xi_2^{-1})$ and of $L(2s-1,\xi_1\xi_2^{-1})$. Here $L_S(2s,\xi_1\xi_2^{-1})$ has a pole at $2s+\nu=1$, and $L(2s-1,\xi_1\xi_2^{-1})$ has poles at $2s-1+\nu=0$ or 1. (The two apparent poles at $s+\nu/2=1$ cancel.) If S is empty, then $L_S(2s,\xi_1\xi_2^{-1})$ has an additional pole at $s+\nu/2=0$.

Proof (*Complete when F is totally real*) If $\mathrm{re}(s)>1$, then, combining Eqs. (7.11) and (7.16), we may write the Fourier expansion

$$E^*(g,s)=L_S(2s,\xi_1\xi_2^{-1})\,c_0(g,f)+\sum_{\alpha\in F^\times}W^*\left(\begin{pmatrix}\alpha&\\&1\end{pmatrix}g\right),\tag{7.24}$$

where $W^*=L_S(2s,\xi_1\xi_2^{-1})\,W$.

We consider first the "constant term" $L_S(2s,\xi_1\xi_2^{-1})\,c_0(g,f)$. This splits into two parts by Eq. (7.15). The first part is $L_S(2s,\xi_1\xi_2^{-1})\,f(g)$, which can have no poles unless $\xi_1\xi_2^{-1}(y)=|y|^\nu$ for some pure imaginary ν, in which case its possible poles are the poles at $s+\nu/2=0$ and $s+\nu/2=1/2$ of $L_S(2s,\xi_1\xi_2^{-1})$. We will show that the other part

$$L_S(2s,\xi_1\xi_2^{-1})\int_A f\left(w_0\begin{pmatrix}1&x\\&1\end{pmatrix}g\right)dx,\tag{7.25}$$

has analytic continuation to all s, except that it can have simple poles at $s+\nu/2=1/2$ and $s+\nu/2=1$ if $\xi_1\xi_2^{-1}(y)=|y|^\nu$. Eventually, we will show that the poles of the two terms at $s+\nu/2=1/2$ cancel. This integral factorizes into a product over all s, provided we assume that f is chosen as in Eq. (7.9). Thus Eq. (7.25) equals

$$\left[\prod_{v\in S}\int_{F_v}f_v\left(w_0\begin{pmatrix}1&x\\&1\end{pmatrix}g_v\right)dx\right]$$

$$\times\left[\prod_{v\notin S}L_v(2s,\xi_{1,v}\xi_{2,v}^{-1})\int_{F_v}f_v\left(w_0\begin{pmatrix}1&x\\&1\end{pmatrix}g_v\right)dx\right].\tag{7.26}$$

If v is nonramified, then it is proved in Proposition 4.6.7 by a simple calculation that

$$\int_{F_v}f_v\left(w_0\begin{pmatrix}1&x\\&1\end{pmatrix}g_v\right)dx=\frac{L_v(2s-1,\xi_{1,v}\xi_{2,v}^{-1})}{L_v(2s,\xi_{1,v}\xi_{2,v}^{-1})}\,\tilde f_v^\circ(g_v),\tag{7.27}$$

where $\tilde{f}_v^\circ$ is the spherical vector in $\pi(\chi_{2,v}, \chi_{1,v})$, normalized so that $\tilde{f}_v^\circ = 1$ on K_v. Thus the second product in Eq. (7.26) equals

$$L_S(2s-1, \xi_1\xi_2^{-1}) \prod_{v \notin S} \tilde{f}_v^\circ(g_v).$$

All but finitely many factors in this product are one, so the product is convergent. Thus Eq. (7.25) equals

$$L(2s-1, \xi_1\xi_2^{-1}) \left[\prod_{v \in S} \frac{1}{L_v(2s-1, \xi_{1,v}\xi_{2,v}^{-1})} \int_{F_v} f_v\left(w_0 \begin{pmatrix} 1 & x \\ & 1 \end{pmatrix} g_v \right) dx \right]$$

$$\times \prod_{v \notin S} \tilde{f}_v^\circ(g_v). \tag{7.28}$$

We will prove for each of the finite number of places $v \in S$ that

$$\frac{1}{L_v(2s-1, \xi_{1,v}\xi_{2,v}^{-1})} \int_{F_v} f_v\left(w_0 \begin{pmatrix} 1 & x \\ & 1 \end{pmatrix} g_v \right) dx \tag{7.29}$$

has analytic continuation to all s. We split this into two parts. Note that $L_v(2s-1, \xi_{1,v}\xi_{2,v}^{-1})$ can have poles but no zeros, and the integral

$$\int_{|x| \le 1} f_v\left(w_0 \begin{pmatrix} 1 & x \\ & 1 \end{pmatrix} g_v \right) dx$$

is entire because the domain of integration is compact. Thus we are reduced to proving the analytic continuation of

$$\frac{1}{L_v(2s-1, \xi_{1,v}\xi_{2,v}^{-1})} \int_{|x|>1} f_v\left(w_0 \begin{pmatrix} 1 & x \\ & 1 \end{pmatrix} g_v \right) dx.$$

We make use of the decomposition

$$w_0 \begin{pmatrix} 1 & x \\ & 1 \end{pmatrix} = \begin{pmatrix} x^{-1} & -1 \\ & x \end{pmatrix} \begin{pmatrix} 1 & \\ x^{-1} & 1 \end{pmatrix}, \tag{7.30}$$

and because $f_v \in \pi(\chi_{1,v}, \chi_{2,v})$, we see that what we must prove is the analytic continuation of

$$\frac{1}{L_v(2s-1, \xi_{1,v}\xi_{2,v}^{-1})} \int_{|x|>1} |x|^{-2s} (\xi_{1,v}^{-1}\xi_{2,v})(x)\, f_v\left(\begin{pmatrix} 1 & \\ x^{-1} & 1 \end{pmatrix} g_v \right) dx. \tag{7.31}$$

If v is non-Archimedean, then because f_v is locally constant, there exists an $N > 0$ such that

$$f_v\left(\begin{pmatrix} 1 & \\ x^{-1} & 1 \end{pmatrix} g_v \right) = f_v(g_v) \tag{7.32}$$

if $|x| > q^N$, an observation which we will need later. Whether v is Archimedean or non-Archimedean, we note that dx is a constant multiple of $|x|\, d^\times x$ and make the variable change $x \mapsto x^{-1}$ to see that Eq. (7.32) is a constant times

$$\frac{1}{L_v\left(2s-1, \xi_{1,v}\xi_{2,v}^{-1}\right)} \int_{|x|<1} |x|^{2s-1} \left(\xi_{1,v}\xi_{2,v}^{-1}\right)(x)\, d^\times x,$$

whose holomorphy was proved in Section 3.1 in the analysis of Eq. (1.28). This completes the proof that Eq. (7.29) is entire.

The possible poles of $c_0(g, f)$ are at the poles of $L_S(2s, \xi_{1,v}\xi_{2,v}^{-1})\, f(g)$ and at the poles of Eq. (7.25). Thus there are possible poles at $2s + \nu = 1$ (from the first term) and at $2s - 1 + \nu = 0, 1$. The poles at $2s + \nu = 1$ will eventually be shown to cancel, so $c_0(g, f)$ actually has a pole at $s + \nu/2 = 1$ only. We will postpone this point to the end of the proof, leaving open for the moment the possibility that there is a pole at $s + \nu/2 = 1/2$ also. As we noted in the remark after the statement of the theorem, if the set S is empty (which can only happen in the function field case) there would also be a pole at $s + \nu/2 = 0$.

It is proved in the general theory of Eisenstein series that the poles of an Eisenstein series are contained in the poles of its various constant terms. (For the case at hand, there is only one constant term, $c_0(g, f)$.) So if we made use of the general theory, we could now conclude that $E^*(g, f)$ has analytic continuation to all $s+\nu/2$ except 1 and perhaps $1/2$. We will proceed differently, continuing to analyze Eq. (7.24). We look next at the second term, the Whittaker function. We will first show that $W^*(g)$ has analytic continuation to all s. Then we will look at the implications of summing over $F^\times$ as in Eq. (7.24).

First, let v be a place outside S. Then the integral W_v is studied in Section 4.6, particularly Theorem 4.6.5. (The factor $(1 - q^{-1}\alpha_1\alpha_2^{-1})^{-1}$ in that theorem is $L_v(2s, \xi_{1,v}\xi_{2,v}^{-1})$, with $\alpha_i = q^{-s_i}\, \xi_{i,v}(\varpi_v)$.) It is found that $L_v(2s, \xi_{1,v}\xi_{2,v}^{-1})\, W_v$ has analytic continuation to all s, where $\pi(\chi_{1,v}, \chi_{2,v})$ is irreducible; by Theorem 4.5.1, this is all s except $s + \nu/2 = 0$ or 1. Moreover, f_v is equal to the normalized spherical vector f_v° for all but finitely many v. When this is the case, $L_v\left(2s, \xi_{1,v}\xi_{2,v}^{-1}\right) W_v$ is the spherical vector W_v° in the Whittaker model of $\pi(\chi_{1,v}, \chi_{2,v})$, normalized so that $W_v^\circ(g_v) = 1$ when $g_v \in K_v$. Indeed, it is obvious that W_v is a K_v-fixed vector in the Whittaker model of $\pi(\chi_{1,v}, \chi_{2,v})$, and the normalization is contained in Theorem 4.6.5. More precisely, that theorem contains the following explicit formula for W_v, which was already used in this chapter in the proof of Proposition 3.5.3:

$$\frac{1}{L_v(2s, \xi_{1,v}\xi_{2,v}^{-1})} W_v\left(\begin{pmatrix} 1 & x \\ & 1 \end{pmatrix}\begin{pmatrix} y_1 & \\ & y_2 \end{pmatrix} k\right)$$

$$= \begin{cases} \psi_v(x)\, (\alpha_1\alpha_2)^{m_2}\, q^{-m/2}\, \frac{\alpha_1^{m+1} - \alpha_2^{m+1}}{\alpha_1 - \alpha_2} & \text{if } m \geq 0, \\ 0 & \text{otherwise,} \end{cases} \tag{7.33}$$

when $k \in K_v$, where $m_1 = \text{ord}(y_1)$, $m_2 = \text{ord}(y_2)$, and $m = m_1 - m_2$, and where $\alpha_i = q^{-s_1}\, \xi_i(\varpi)$ $(i = 1, 2)$. This formula must be interpreted by taking the limit when $s+\nu/2 = 1/2$, that is when $\alpha_1 = \alpha_2$. The representation remains irreducible at this point, and the formula (7.33) is interpreted by the limit

$$\lim_{\alpha_1 \to \alpha_2} \frac{\alpha_1^{m+1} - \alpha_2^{m+1}}{\alpha_1 - \alpha_2} = (m+1)\,\alpha_2^m.$$

It is also easy to see what happens in the exceptional cases where $s+\nu/2 = 0$ or 1. If $s+\nu/2 = 1$, then the integral (Eq. 7.19) is absolutely convergent – indeed, it is absolutely convergent when $\mathrm{re}(s) = \mathrm{re}(s+\nu/2) > 1/2$ by comparison with the integral (Eq. 5.16) in Chapter 4, whose convergence is discussed in Proposition 4.5.6. Thus when $s+\nu/2 = 1$, the formula (7.33) remains valid. When $s+\nu/2 = 0$, the integral (Eq. 7.19) is not convergent, but the formula (7.33) gives the analytic continuation to this point.

It is worth noting that the right side of Eq. (7.33) is symmetric in α_1 and α_2. This symmetry is established in Proposition 4.6.8 prior to proving the explicit formula (7.33), and it relates the values of Eq. (7.19) outside the region of convergence to the values inside. But all we need from Eq. (7.33) for the present is that it be an analytic function of s_1 and s_2 defined for all values and that it equal one when $g_v \in K_v$.

Thus for $v \notin S$, the function $L_v\left(2s, \xi_{1,v}\xi_{2,v}^{-1}\right) W_v(g_v)$ has analytic continuation to all s, and equals 1 for almost all v. If we write

$$W^*(g) = \prod_{v \in S} W_v(g_v) \times \left[\prod_{v \notin S} L_v\left(2s, \xi_{1,v}\xi_{2,v}^{-1}\right) W_v(g_v)\right], \tag{7.34}$$

each factor of the product in brackets has analytic continuation to all s and equals 1 for almost all v. We will show for the finite number of places in S that $W_v(g_v)$ has analytic continuation to all s. If v is non-Archimedean, we proceed as in the analysis of Eq. (7.29). There exists a constant q^N such that if $|x| > q^N$, then Eq. (7.32) is true. Thus using Eq. (7.32), we see (after splitting the integral into two parts) that Eq. (7.19) equals

$$\int_{|x| \le q^N} f\left(w_0 \begin{pmatrix} 1 & x \\ & 1 \end{pmatrix} g_v\right) \psi_v(-x)\,dx$$
$$+ f_v(g_v) \int_{|x| > q^N} |x|^{-2s} \left(\xi_{1,v}^{-1}\xi_{2,v}\right)(x)\,\psi_v(-x)\,dx.$$

The first integral is over a compact set and hence defines an entire function of s. As for the second integral, if k is sufficiently large, the integral over $|x| = q^k$ vanishes, owing to the presence of the additive character. This is because if $\mathfrak{a}$, is a fractional ideal on which ψ is nontrivial, if k is sufficiently large, we may break $\{x \mid |x| = q^k\}$ up into cosets of $\mathfrak{a}$, on each of which $|x|^{-2s}\left(\xi_{1,v}^{-1}\xi_{2,v}\right)(x)$ is constant, whereas the integral of $\psi(-x)$ is zero. Hence the sum is essentially finite, and we get the analytic continuation to all s.

Next, suppose that v is a real place. Then because f_v is K-finite, we may write

$$f_v\begin{pmatrix} \cos(\theta) & \sin(\theta) \\ -\sin(\theta) & \cos(\theta) \end{pmatrix} = \sum_{k=-N}^{N} a_k\, e^{ik\theta},$$

where the summation is over even integers if $\xi_{1,v}\xi_{2,v}(-1) = 1$ and over odd integers if $\xi_{1,v}\xi_{2,v}(-1) = -1$. For the purpose of proving analytic continuation, there is no loss of generality in assuming that all but one of the a_k vanish. Thus we may assume

$$f_v\begin{pmatrix} \cos(\theta) & \sin(\theta) \\ -\sin(\theta) & \cos(\theta) \end{pmatrix} = e^{ik\theta}. \tag{7.35}$$

Let us write

$$g_v = \begin{pmatrix} \zeta & \\ & \zeta \end{pmatrix} \begin{pmatrix} \eta^{1/2} & \xi\eta^{-1/2} \\ & \eta^{-1/2} \end{pmatrix} \begin{pmatrix} \cos(\phi) & \sin(\phi) \\ -\sin(\phi) & \cos(\phi) \end{pmatrix},$$

where η and ζ are positive. Using Eq. (7.35)

$$f_v\left(w_0\begin{pmatrix} 1 & x \\ & 1 \end{pmatrix} g_v\right)$$

$$= (\chi_1^{-1}\chi_2)(\eta^{1/2})\,|\eta|^{-1}\,(\chi_1\chi_2)(\zeta)\,e^{ik\phi}\, f_v\left(w_0\begin{pmatrix} 1 & \eta^{-1}(x+\xi) \\ & 1 \end{pmatrix}\right).$$

Thus making the change of variables $x \to \eta x - \xi$, we are reduced to proving the analytic continuation of

$$\int_{\mathbb{R}} f_v\left(w_0\begin{pmatrix} 1 & x \\ & 1 \end{pmatrix}\right) \psi_v(\eta x)\, dx. \tag{7.36}$$

We have

$$w_0\begin{pmatrix} 1 & x \\ & 1 \end{pmatrix} = \begin{pmatrix} \Delta^{-1} & -x\Delta^{-1} \\ & \Delta \end{pmatrix} \begin{pmatrix} x\Delta^{-1} & -\Delta^{-1} \\ \Delta^{-1} & x\Delta^{-1} \end{pmatrix},$$

where $\Delta = \sqrt{1+x^2}$. Because by Eq. (7.36)

$$f_v\begin{pmatrix} x\Delta^{-1} & -\Delta^{-1} \\ \Delta^{-1} & x\Delta^{-1} \end{pmatrix} = \left(\frac{x-i}{|x-i|}\right)^k.$$

We see that Eq. (7.36) equals

$$\int_0^\infty (x^2+1)^{-s-k/2}\,(x-i)^k\, e^{i\lambda\eta x}\, dx,$$

where λ is the real number such that $\psi_v(x) = e^{i\lambda x}$. To prove the analytic continuation of this integral, we employ the *Hankel contour* that we encountered in Exercise 1.1.8. Assume first that $\lambda > 0$. In this case, when $\mathrm{re}(s)$ is large, we can use Cauchy's theorem to deform the path of integration so that it begins at $i\infty$, circles the singularity at $x = i$ once in the clockwise direction, then returns to $i\infty$. The integral is now convergent for all values of x owing to the rapid decay of $e^{i\lambda x}$ in the direction of $i\infty$, proving that Eq. (7.36) is an entire function of s_1 and s_2. Moreover, if we are careful to keep the path of integration entirely in the upper half plane, this integral representation shows that Eq. (7.36) is of rapid decay as $\eta \to \infty$, a fact that we will need shortly. If $\lambda < 0$, we proceed similarly, but begin at $-i\infty$. This proves the analytic continuation of W^* to all s_1 and s_2.

We leave the analytic continuation of $W_v(g_v)$ when v is complex to the reader (Exercise 3.7.2).

We have shown that W^* has analytic continuation to all s_1 and s_2. Now we must prove the same for

$$\sum_{\alpha \in F} W^*\left(\begin{pmatrix} \alpha & \\ & 1 \end{pmatrix} g_v\right), \tag{7.37}$$

for this is what occurs in Eq. (7.24). Let v be a non-Archimedean place of F. Then there exists a constant C_v such that if $y_v \in F_v^\times$ and $|y_v| > C_v$, then

$$W_v\left(\begin{pmatrix} y_v & \\ & 1 \end{pmatrix} g_v\right) = 0.$$

Indeed, referring to Eq. (4.24), in Chapter 4, this function of y_v lies in the Kirillov model of $\pi(\chi_{1,v}, \chi_{2,v})$, so this vanishing is asserted in Proposition 4.7.2. By Eq. (7.33), we may take $C_v = 1$ for almost all v. In Eq. (7.37), we may restrict ourselves to nonzero $\alpha \in F$ that satisfy $|\alpha_v| \leq C_v$ for all non-Archimedean places v. If we embed F in the Euclidean space $F_\infty = \prod_{v \in S_\infty} F_v$, this means that α is restricted to a lattice. The convergence of the sum (Eq. 7.37) now follows from the fact that the local Whittaker functions W_v are rapidly decreasing, a fact we noted as following from their representation by a Hankel integral.

The series Eq. (7.24) is thus absolutely convergent except where the constant term has poles, and it is easily seen that the convergence is uniform in s_1 and s_2. This proves the meromorphic continuation. There is one point that we have left for the end, which is (in the case where $\xi_1\xi_2^{-1}(y) = |y|$) that there is no pole at $s + \nu/2 = 1/2$. At this point, each of the two terms in $c_0(g, f)$ has a pole. We claim that these poles cancel.

The two terms in $c_0(g, f)$ are

$$L_S(2s, \xi_1\xi_2^{-1})\, f(g)$$

and

$$L_S(2s, \xi_1\xi_2^{-1}) \int_A f\left(w_0 \begin{pmatrix} 1 & x \\ & 1 \end{pmatrix} g\right) dx.$$

Let us first consider a value of s such that $\mathrm{re}(s) = \mathrm{re}(s + \nu/2)$ is slightly bigger than 1/2, so that these live, respectively, in $\pi(\chi_1, \chi_2)$ and $\pi(\chi_2, \chi_1)$. (The justification for our statement that the second term is in $\pi(\chi_2, \chi_1)$ is in Proposition 4.5.6 and Proposition 2.6.2.) When we move s to the point $(1-\nu)/2$, the two characters χ_1 and χ_2 become equal, so these two terms have poles, and their residues are thus in the same space $\pi(\chi_1, \chi_1)$.

To see that the residues cancel, we note that if the Eisenstein series has a pole at this point, its residue $R(g)$ must be automorphic; that is, it satisfy $R(\gamma g) = R(g)$ for $\gamma \in GL(2, F)$. We have seen that Eq. (7.37) has no pole for any value of s. This means that only the constant term $c_0(g, f)$ contributes to $R(g)$, and so $R \in \pi(\chi_1, \chi_1)$. Because $w_0 \in GL(2, F)$, with y an arbitrary idele

$$R\left(w_0 \begin{pmatrix} y & \\ & y^{-1} \end{pmatrix} g\right) = R\left(\begin{pmatrix} y & \\ & y^{-1} \end{pmatrix} g\right) = |y|\, R(g).$$

On the other hand, this equals

$$R\left(\begin{pmatrix} y^{-1} & \\ & y \end{pmatrix} w_0\, g\right) = |y|^{-1}\, R(w_0\, g) = |y|^{-1}\, R(g).$$

This shows that the function R is identically zero. ■

Proposition 3.7.3 *With notation as in Theorem 3.7.1, let s_1 and s_2 be complex numbers such that $E(g, f)$ does not have a pole. Then $E(g, f)$ lies in $\mathcal{A}\big(GL(2, F)\backslash GL(2, A), \omega\big)$, where the central quasicharacter $\omega = \chi_1\chi_2$.*

Proof It is clear that $E(\gamma g, f) = E(g, f)$ when $\gamma \in GL(2, F)$; indeed, this is true by construction if $\mathrm{re}(s) > 1$, and if s is arbitrary, both sides of this identity are equal by the uniqueness of analytic continuation. The function $E(g, f)$ inherits the properties of K-finiteness and $\mathcal{Z}$-finiteness from f. The only condition remaining to be checked is moderate growth. This follows from the Fourier expansion (Eq. 7.24), where the first term $L_S\big(2s, \xi_1\xi_2^{-1}\big)\, c_0(g, f)$ is of moderate growth, and the second term (involving the Whittaker function) is of rapid decay. ■

A key property of the Eisenstein series is that they are orthogonal to the cusp forms.

Proposition 3.7.4 *Let (π, V) be an irreducible cuspidal automorphic representation of $GL(2, A)$ with central character ω. Let other notations be as in Theorem 3.7.1, except that we assume that $\chi_1\chi_2 = \omega$. Then if $\phi \in V$ and $f \in \pi(\chi_1, \chi_2)$, we have*

$$\int_{Z(A)\, GL(2,F)\backslash GL(2,A)} \overline{\phi(g)}\, E(g, f)\, dg = 0. \tag{7.38}$$

Proof The assumption that $\chi_1\chi_2 = \omega$ guarantees that if z lies in the center $Z(A)$ of $GL(2, A)$, then

$$\overline{\phi(zg)}\, E(zg, f) = \overline{\phi(g)}\, E(g, f),$$

so the integral is well defined. We note that the integral is absolutely convergent, because $\phi(g)$ is rapidly decreasing whereas $E(g, f)$ is of moderate growth, so the integrand is rapidly decreasing and hence is integrable.

First assume that $\mathrm{re}(s) > 1$. We may then *unfold* the Eisenstein series, analogous to the proof of Proposition 1.6.1. This means substituting the series definition of the Eisenstein series (which is now absolutely convergent by Proposition 3.7.2) and pursuing the consequences. The left side equals

$$\int_{Z(A)\,GL(2,F)\backslash GL(2,A)} \sum_{\gamma\in B(F)\backslash GL(2,F)} \overline{\phi(g)}\, f(\gamma g)\, dg$$

$$= \int_{Z(A)\,GL(2,F)\backslash GL(2,A)} \sum_{\gamma\in B(F)\backslash GL(2,F)} \overline{\phi(\gamma g)}\, f(\gamma g)\, dg$$

$$= \int_{Z(A)\,B(F)\backslash GL(2,A)} \overline{\phi(g)}\, f(g)\, dg.$$

We have collapsed the integration with the summation in a typical manipulation. We factor the Borel subgroup as a semidirect product, $B = NT$, where T is the torus of diagonal matrices and N is the group of upper triangular unipotent matrices. We write the last integral as

$$\int_{Z(A)\,T(F)\,N(F)\backslash GL(2,A)} \overline{\phi(g)}\, f(g)\, dg$$

$$= \int_{Z(A)\,N(A)\,T(F)\backslash GL(2,A)} \int_{N(F)\backslash N(A)} \overline{\phi(ng)}\, f(ng)\, dn\, dg$$

$$= \int_{Z(A)\,N(A)\,T(F)\backslash GL(2,A)} \int_{N(F)\backslash N(A)} \overline{\phi(ng)}\, dn\, f(g)\, dg,$$

where we have used the fact that $f(ng) = f(g)$ when $n \in N(A)$. The inner integral now vanishes because ϕ is cuspidal.

Next, we consider the case of arbitrary s. We begin with $\mathrm{re}(s)$ large, and move $s = \frac{1}{2}(s_1 - s_2 + 1)$ to its desired value by varying s_1 and s_2 and holding $s_1 + s_2$ fixed so that the condition $\chi_1\chi_2 = \omega$ (needed for Eq. (7.38) to be well defined) is valid at all times. The left side is analytic in s, because the integrand is analytic, and the integral is absolutely and uniformly convergent. Because it is zero for large s, it is zero for all s. ■

Proposition 3.7.5 *Let the notation be as in Theorem 3.7.1, except assume that $\xi_1 = \xi_2 = 1$, and let $s_2 = -s_1$, so that the central character of $E(g, f)$ is one and the Eisenstein series $E(g, f)$ has its pole at $s = 1$. Then the residue of the Eisenstein series is a constant, independent of g. The section f may be chosen so that this constant is nonzero.*

Proof We return to the formulas in the proof of Theorem 3.7.1. Evidently, the residue of $E(g, f)$ is $L_S(2, \xi_1\xi_2^{-1})^{-1}$ times the residue of $E^*(g, s)$, so it is sufficient to show that the latter is constant. We saw in the proof of Theorem 3.7.1 that the second term in Eq. (7.24) has no pole, and writing the first term as a sum of two terms, $L_S(2s, \xi_1\xi_2^{-1}) f(g)$ has no pole at $s = 1$, so the residue of the Eisenstein series is the residue of Eq. (7.25). We rewrite this as Eq. (7.28). This is an element of $\pi(\chi_2, \chi_1)$, where $\chi_1(x) = |x|^{s_1}$ and $\chi_2(x) = |x|^{-s_1}$; at the value where the pole occurs, we will have $s_1 = 1/2$, so $s = 1$. The function $L(2s - 1, \xi_1\xi_2^{-1})$ is just the Dedekind zeta function $\zeta_F(s) = \prod_v \zeta_{F,v}(s)$ of the field F, and up to a constant, the residue is

$$\left[\prod_{v\in S} \frac{1}{\zeta_{F,v}(1)} \int_{F_v} f_v\left(w_0 \begin{pmatrix} 1 & x \\ & 1 \end{pmatrix} g_v\right)\right] \prod_{v\notin S} \tilde{f}_v^\circ(g_v). \tag{7.39}$$

We claim that for each v, the v-component of this function is a constant multiple of the K_v-fixed vector in $\pi(\chi_{2,v}, \chi_{1,v})$, which is unique up to constant multiple. If $v \notin S$, this is true by the definition of $\tilde{f}_v^\circ$. If $v \in S$, this follows from the fact that the image of the intertwining integral (when $\chi_{1,v}(x) = |x|^{1/2}$ and $\chi_{2,v}(x) = |x|^{-1/2}$) is the unique one-dimensional invariant subspace of $\chi_{1,v}(x) = |x|^{1/2}$ and $\chi_{2,v}(x) = |x|^{-1/2}$. If v is non-Archimedean, this fact is contained in Exercise 4.5.4. If v is Archimedean, this fact follows from Proposition 2.6.3, because $k = 0$ is the only even integer such that the right side of Eq. (6.15) in Chapter 2 does not have a zero at $s = 1$, so the image of $M(1)$ is (in the notation of that proposition) just the one-dimensional linear span of $f_{0,0}$.

With $\chi_{1,v}(x) = |x|^{1/2}$ and $\chi_{2,v}(x) = |x|^{-1/2}$, the K_v-fixed vector in $\pi(\chi_{2,v}, \chi_{1,v})$ is just the constant function, identically equal to one. We have therefore proved that the residue of Eq. (7.25), and hence the residue of the Eisenstein series itself, is constant. It is clear from Exercise 4.5.4 that we may choose each of the functions f_v so that the local factors in Eq. (7.40) are nonzero, so f may be chosen so that the residue of the Eisenstein series is a nonzero constant. ■

The following proposition and its proof show that although the Eisenstein series are orthogonal to the cusp forms, even if our interest is principally with the cusp forms, we cannot ignore the Eisenstein series. In order to prove a result about cusp forms, we make use of the Eisenstein series.

Proposition 3.7.6 *Let (π, V) be an automorphic cuspidal representation of $GL(2, A)$ with trivial central character, and let ϕ be a cusp form in V. Then*

$$\int_{Z(A)\,GL(2,F)\backslash GL(2,A)} \phi(g)\,dg = 0.$$

Proof By Proposition 3.7.5, we may find an Eisenstein series $E(g, f)$, where f is a flat section of $\pi(\chi_1, \chi_2)$, with $\chi_1(x) = \chi_2(x)^{-1} = |x|^{s-1/2}$ such that the residue of $E(g, f)$ at $s = 1$ is the constant function identically equal to one on $GL(2, A)$. It is not hard to see from the Fourier expansion (Eq. 7.24) and from the rapid decay of $\phi(g)$ that $(s-1)\,E(g, f)\,\phi(g)$ then converges uniformly in g to $\phi(g)$. We are therefore justified in writing

$$\int_{Z(A)\,GL(2,F)\backslash GL(2,A)} \phi(g)\,dg = \lim_{s\to 1}\,(s-1) \int_{Z(A)\,GL(2,F)\backslash GL(2,A)} \phi(g)\,E(g, s)\,dg.$$

However, the integral on the right is identically equal to zero by Proposition 3.7.4. ∎

In the following proposition, it is crucial that for each place v of F, we take the additive Haar measure $\int_{F_v} dx$ to be self-dual with respect to the additive character ψ_v. With this normalization, the Haar measure $\prod_v dx_v$ on the adele group A induces on the compact quotient A/F an additive measure such that A/F has volume 1, as was proved in Proposition 3.1.3(i).

Proposition 3.7.7 *Let $M_v(s) : \pi(\chi_{1,v}, \chi_{2,v}) \to \pi(\chi_{2,v}, \chi_{1,v})$ be the intertwining integral defined by Eq. (7.23) or its analytic continuation, and let $\widetilde{M}_v(1-s) : \pi(\chi_{2,v}, \chi_{1,v}) \to \pi(\chi_{1,v}, \chi_{2,v})$ be the corresponding intertwining integral in the opposite direction. Then $\widetilde{M}_v(1-s) \circ M_v(s)$ acts on $\pi(\chi_{1,v}, \chi_{2,v})$ by a scalar; this scalar is*

$$\gamma_v\big(2-2s, \xi_{1,v}^{-1}\xi_{2,v}, \psi_v\big)\,\gamma_v\big(2s, \xi_{1,v}\,\xi_{2,v}^{-1}, \psi_v\big). \tag{7.40}$$

Proof We will prove this if v is non-Archimedean or real and omit the complex case. If v is non-Archimedean, we will prove this in Proposition 4.5.10. If v is real, we argue as follows. We consider first the effect of changing the additive character ψ_v. If we replace this by $x \mapsto \psi_v(ax)$, then the additive Haar measure is multiplied by $|a|^{1/2}$, so the left side of Eq. (7.40) is multiplied by $|a|$. On the other hand, this change of character multiplies $\gamma_v(s, \chi_v, \psi_v)$ by $|a|^{s-1/2}$, so the right side of Eq. (7.40) is also multiplied by $|a|$. Hence there is no loss of generality in assuming that $\psi_v(x) = e^{2\pi i x}$, and then the self-dual measure $\int_{\mathbb{R}} dx$ is just the Lebesgue measure.

Let $\epsilon = 0$ or 1 such that the unitary character $\xi_{1,v}\xi_{2,v}^{-1}$ of $\mathbb{R}$ is $\big(x/|x|\big)^{\epsilon}$. We make use of the test function $\Phi_v(x) = x^\epsilon\, e^{-\pi x^2}$, so that $\hat{\Phi}_v = i^\epsilon\, \Phi$

$$\zeta_v(s, \Phi_v) = \pi^{-(s+\epsilon)/2}\, \Gamma\left(\frac{s+\epsilon}{2}\right), \qquad \gamma_v(s, \chi_v, \psi_v) = i^\epsilon\, \pi^{s-1/2}\, \frac{\Gamma\left(\frac{1+\epsilon-s}{2}\right)}{\Gamma\left(\frac{s+\epsilon}{2}\right)}.$$

Thus Eq. (7.40) equals

$$\begin{cases} \pi\, \dfrac{\Gamma\left(\frac{1}{2}-s\right)\Gamma\left(s-\frac{1}{2}\right)}{\Gamma(s)\,\Gamma(1-s)} = \dfrac{\pi}{s-\frac{1}{2}} \tan(\pi s), & \text{if } \epsilon = 0, \\ -\pi\, \dfrac{\Gamma(1-s)\,\Gamma(s)}{\Gamma\left(s+\frac{1}{2}\right)\Gamma\left(\frac{3}{2}-s\right)} = -\pi\, \cot(\pi s), & \text{if } \epsilon = 1, \end{cases}$$

where we have used the identities

$$\Gamma(s)\,\Gamma(1-s) = \frac{\pi}{\sin(\pi s)}, \qquad \Gamma\left(s-\tfrac{1}{2}\right)\Gamma\left(s+\tfrac{1}{2}\right) = \frac{\pi}{\cos(\pi s)}. \tag{7.41}$$

On the other hand, by Proposition 2.6.3, $\widetilde{M}_v(1-s) \circ M_v(s)$ is the operator that multiplies the vector $f_{k,s}$ (in the notation of that proposition) by

$$(-1)^k\, \frac{\Gamma(s)\,\Gamma(1-s)\,\Gamma\left(s-\frac{1}{2}\right)\Gamma\left(\frac{1}{2}-s\right)}{\Gamma\left(s+\frac{k}{2}\right)\Gamma\left(1-s+\frac{k}{2}\right)\Gamma\left(s-\frac{k}{2}\right)\Gamma\left(1-s-\frac{k}{2}\right)}.$$

Using Eq. (7.41) and the periodicity of trigonometric functions, this value is independent of k so long as k is an integer congruent to ϵ modulo 2, which of course it must be. We may now verify the proposition by taking $k = \epsilon$. ■

Next we consider the functional equations of the Eisenstein series. We will now normalize the Eisenstein series more carefully than in Theorem 3.7.1 by using the complete L-series $L\left(s, \xi_1\xi_2^{-1}\right)$ rather than just the partial L-function $L_S\left(s, \xi_1\xi_2^{-1}\right)$. Thus instead of Eq. (7.23), we will consider

$$E^{\#}(g, f) = L\left(2s, \xi_1\xi_2^{-1}\right) E(g, f). \tag{7.42}$$

Also, having fixed a flat section f of $\pi(\chi_1, \chi_2)$, we need a particular section $\tilde{f}$ of $\pi(\chi_2, \chi_1)$. We assume that f factorizes as in Eq. (7.9), and we define

$$\tilde{f}(g) = \prod_v \tilde{f}_v(g_v),$$

where (with S and $\tilde{f}_v^\circ$ as in the proof of Theorem 3.7.1) we have

$$\tilde{f}_v = \begin{cases} \tilde{f}_v^\circ & \text{if } v \notin S; \\ \dfrac{L_v\left(2s, \xi_{1,v}\xi_{2,v}^{-1}\right)}{\gamma_v\left(2-2s, \xi_{1,v}^{-1}\xi_{2,v}, \psi_v\right) L_v\left(2-2s, \xi_{1,v}^{-1}\xi_{2,v}\right)}\, M_v(s)\, f_v & \text{if } v \in S. \end{cases} \tag{7.43}$$

If $v \notin S$, the section $\tilde{f}_v$ of $\pi(\chi_2, \chi_1)$ is obviously flat. If $v \in S$, the section is not flat, but it is a finite linear combination, with meromorphic coefficients, of flat sections by Exercise 4.5.10. (This exercise assumes that v is non-Archimedean, but the Archimedean case is identical because we are restricting ourselves to

K_v-finite vectors.) So $\tilde{f}$ is a finite linear combination, with meromorphic coefficients, of flat sections. We define

$$E^{\#}(g, \tilde{f}) = L\left(2-2s, \xi_1^{-1}\xi_2\right) E(g, \tilde{f}). \tag{7.44}$$

Theorem 3.7.2 *The Eisenstein series $E^{\#}(g, f)$ satisfies the functional equation*

$$E^{\#}(g, f) = E^{\#}(g, \tilde{f}). \tag{7.45}$$

Proof The proof will consist of showing that the two Eisenstein series in Eq. (7.45) have the same constant term. Given this, $E^{\#}(g, f) - E^{\#}(g, \tilde{f})$ is a cusp form and hence it is orthogonal to itself, because by Proposition 3.7.4, cusp forms are orthogonal to Eisenstein series. Hence it is zero.

By Eq. (7.15)

$$\int_{A/F} E^{\#}\left(\begin{pmatrix} 1 & x \\ & 1 \end{pmatrix} g\right) dx = L\left(2s, \xi_1\xi_2^{-1}\right) f(g) + L\left(2s, \xi_1\xi_2^{-1}\right) \times \int_{A} f\left(w_0 \begin{pmatrix} 1 & x \\ & 1 \end{pmatrix} g\right) dx. \tag{7.46}$$

We will prove that the second term here equals

$$L(2-2s, \xi_1^{-1}\xi_2)\, \tilde{f}(g),$$

where $\tilde{f}$ is defined in Eq. (7.43). Indeed, using Eq. (7.27), this second term equals

$$L_S\left(2s-1, \xi_1\xi_2^{-1}\right) \left[\prod_{v\in S} L_v\left(2s, \xi_{1,v}\xi_{2,v}^{-1}\right) M_v(s)\, f_v(g_v)\right] \prod_{v\notin S} \tilde{f}_v^{\circ}(g_v),$$

so using Eq. (1.21), this equals

$$L_S\left(2-2s, \xi_1^{-1}\xi_2\right) \left[\prod_{v\in S} \frac{L_v\left(2s, \xi_{1,v}\xi_{2,v}^{-1}\right)}{\gamma_v\left(2-2s, \xi_{1,v}^{-1}\xi_{2,v}, \psi_v\right)} M_v(s)\, f_v(g_v)\right] \prod_{v\notin S} \tilde{f}_v^{\circ}(g_v),$$

and so Eq. (7.46) equals

$$L\left(2s, \xi_1\xi_2^{-1}\right) f(g) + L\left(2-2s, \xi_1^{-1}\xi_2\right) \tilde{f}(g).$$

Applying the same calculation to the second Eisenstein series gives

$$\int_{A/F} E^{\#}\left(\begin{pmatrix} 1 & x \\ & 1 \end{pmatrix} g, \tilde{f}\right) dx = L\left(2-2s, \xi_1^{-1}\xi_2\right) \tilde{f}(g) + L\left(2s, \xi_1\xi_2^{-1}\right) \tilde{\tilde{f}}(g),$$

where (analogous to Eq. (7.43)) we have

$$\tilde{\tilde{f}}(g_v) = \prod_v \tilde{\tilde{f}}_v(g_v),$$

with

$$\tilde{\tilde{f}}_v = \begin{cases} f_v^\circ & \text{if } v \notin S; \\ \dfrac{L_v\left(2-2s, \xi_{1,v}^{-1}\xi_{2,v}\right)}{\gamma_v\left(2s, \xi_{1,v}\xi_{2,v}^{-1}, \psi_v\right) L_v\left(2s, \xi_{1,v}\xi_{2,v}^{-1}\right)} \widetilde{M}_v(1-s)\, \tilde{f}_v & \text{if } v \in S. \end{cases} \tag{7.47}$$

But now it follows from Proposition 3.7.7 that $\tilde{\tilde{f}} = f$. We have shown that the two Eisenstein series in Eq. (7.45) have equal constant terms, and as we have noted, this implies that they are equal. ■

Exercises

Exercise 3.7.1 Discuss the possible poles of $E^\#(g, f)$.

Exercise 3.7.2 Complete the proof of Theorem 3.7.1 by proving the analytic continuation of $W_v(g_v)$ when v is a complex place.

3.8 The Rankin–Selberg Method

We return to the *Rankin–Selberg method*, which was introduced in a classical context in Section 1.6, and reconsidered from a philosophical point of view in Section 1.8. A basic reference for the Rankin–Selberg method for $GL(2) \times GL(2)$ is Jacquet (1972).

We will attach an L-function to a pair of automorphic forms on $GL(2)$ and make a start on studying its analytic properties. Our purpose in including this material is not to give a complete theory (for this, see Jacquet (1972)), but simply to introduce the reader to this important L-function, and also to certain important typical manipulations. We will omit the proof of one important fact (Proposition 3.8.3), in order to give the reader an incentive to consult Jacquet (1972).

There are further instructive remarks in Gelbart and Shahidi (1988). For a survey of the Rankin–Selberg method on higher-rank groups, though somewhat out of date, see Bump (1989), but note that paper needs a correction in its discussion of the uniqueness of Whittaker functions on $GL(n, \mathbb{R})$. It is asserted there that a spherical Whittaker function can be characterized by its eigenvalues under the center of the universal enveloping algebra, together with a growth condition. This is correct, but it is asserted (in Section 2.1 of Bump (1989)) that this follows from the multiplicity one theorem of Shalika (1974). However, it is not clear that this is the case. A better reference is Wallach (1983), which contains results of the type needed in Bump (1989). See Bump and Huntley (1994) for further comments on the asymptotics of Whittaker functions.

Let (π_1, V_1) and (π_2, V_2) be automorphic cuspidal representations of $GL(2, A)$, with unitary central characters ω_1 and ω_2. We fix a set of places S such that if $v \notin S$, then v is non-Archimedean, π_1 and π_2 are both spherical, and the conductor of the additive character ψ_v is $\mathfrak{o}_v$. If $v \notin S$, let $\alpha_1(v)$ and $\alpha_2(v)$ be the Satake parameters of $\pi_{1,v}$, and let $\beta_1(v)$ and $\beta_2(v)$ be the Satake parameters of $\pi_{2,v}$. The *Rankin–Selberg L-function* $L_S(s, \pi_1 \times \pi_2)$ is the Euler product

$$\prod_{v \notin S} L_v(s, \pi_{1,v} \times \pi_{2,v}), \tag{8.1}$$

where

$$L_v(s, \pi_{1,v} \times \pi_{2,v}) = \prod_{i=1}^{2} \prod_{j=1}^{2} \left(1 - \alpha_i(v)\, \beta_j(v)\, q_v^{-s}\right)^{-1}. \tag{8.2}$$

The Rankin–Selberg method proves a functional equation for this L-function. We will show that $\pi_1 \cong \hat{\pi}_2$ if and only if this L-function has a pole at $s = 1$, a result that has many applications.

Let ξ_1 and ξ_2 be unitary Hecke characters of $A^\times / F^\times$ such that $\xi_1 \xi_2 \omega_1 \omega_2 = 1$. The character ξ_1 will turn out to be slightly more important than ξ_2, so we will sometimes write simply ξ for ξ_1. We will assume that $\xi_{1,v}$ and $\xi_{2,v}$ are nonramified when $v \notin S$. This can always be arranged: For example, we may take $\xi_1 = 1$ and $\xi_2 = (\omega_1 \omega_2)^{-1}$. If $v \notin S$, then $\omega_{1,v}$ and $\omega_{2,v}$ are nonramified because $\pi_{1,v}$ and $\pi_{2,v}$ are nonramified. Let s be a complex number, and let $s_1 = s - 1/2$ and $s_2 = -s + 1/2$, so that $s = \frac{1}{2}(s_1 - s_2 + 1)$ as in Section 3.7. Let $\chi_i(x) = \xi_i(x)\, |x|^{s_i}$ $(i = 1, 2)$, and let f be a flat section of $\pi(\chi_1, \chi_2)$. We assume that $f(g) = \prod_v f_v(g_v)$ as in Eq. (7.9), and we choose the local sections so that $f_v = 1$ on $K_v = GL(2, \mathfrak{o}_v)$ if $v \notin S$. We can do this because $\xi_{1,v}$ and $\xi_{2,v}$ are nonramified. We consider the Eisenstein series $E^*(g, f)$, defined as in Section 3.7. Let $\phi_1 \in V_1$ and $\phi_2 \in V_2$ be cusp forms, chosen such that in the tensor product decompositions, $\pi_i = \otimes_v \pi_{i,v}$, ϕ_1 and ϕ_2 correspond to pure tensors, and such that ϕ_1 and ϕ_2 are spherical at every place $v \notin S$. This is possible because $\pi_{1,v}$ and $\pi_{2,v}$ are spherical for all such v.

The cusp forms ϕ_1 and ϕ_2 are rapidly decreasing, whereas E^* is of moderate growth, so $\phi_1(g)\, \phi_2(g)\, E^*(g, f)$ is rapidly decreasing. In particular, it is bounded. Moreover, because the product of the central (quasi-) characters of ϕ_1, ϕ_2, and E^* is one, $\phi_1(g)\, \phi_2(g)\, E^*(g, f)$ is invariant under the center of $GL(2, A)$. By Proposition 3.3.2, the integral

$$\int_{Z(A)\, GL(2,F) \backslash GL(2,A)} \phi_1(g)\, \phi_2(g)\, E^*(g, f)\, dg \tag{8.3}$$

is absolutely convergent for all values of s, apart from the possible pole of $E^*(g, f)$. Moreover, Eq. (8.2) is a meromorphic function of s, holomorphic except at the pole of $E^*(g, f)$ (if it has one). Generalizing Eq. (8.1), we will study

$$L_S(s, \pi_1 \times \pi_2, \xi) = \prod_{v \notin S} L_v(s, \pi_{1,v} \times \pi_{2,v}, \xi_v), \tag{8.4}$$

where

$$L_v(s, \pi_{1,v} \times \pi_{2,v}, \xi_v) = \prod_{i=1}^{2} \prod_{j=1}^{2} \left(1 - \alpha_i(v)\, \beta_j(v)\, \xi_v(\varpi)\, q_v^{-s}\right)^{-1}. \tag{8.5}$$

(Recall that $\xi = \xi_1$.) We will show that the integral (Eq. 8.3) "represents" this L-function, much as the integral (Eq. 5.28) "represents" the standard L-function $L_S(s, \pi)$.

As in Section 3.3, we have the Whittaker functions

$$W_1(g) = \int_{A/F} \phi_1\left(\begin{pmatrix} 1 & x \\ & 1 \end{pmatrix} g\right) \psi(-x)\, dx,$$

$$W_2(g) = \int_{A/F} \phi_2\left(\begin{pmatrix} 1 & x \\ & 1 \end{pmatrix} g\right) \psi(x)\, dx, \tag{8.6}$$

Note that we have used the inverse of the additive character ψ in the Whittaker function of ϕ_2. By Theorem 3.5.4, the Whittaker functions decompose as products:

$$W_i(g) = \prod_v W_{i,v}(g_v), \qquad i = 1, 2, \tag{8.7}$$

where $W_{i,v}$ is an element of the local Whittaker model $\mathcal{W}_{i,v}(\pi_{i,v})$, and if $v \notin S$, then $W_{i,v} = W_{i,v}^{\circ}$ is the spherical Whittaker function, normalized so that $W_{i,v}^{\circ}(g_v) = 1$ when $g_v \in K_v = GL(2, \mathfrak{o}_v)$. Let B denote the Borel subgroup of $GL(2)$ consisting of upper triangular matrices. We factor $B = TN$, where T is the maximal torus of diagonal matrices and N is the group of upper triangular unipotent matrices. Let T_1 denote the subgroup of T, consisting of matrices whose lower right entry is 1. Let $\delta_B : B(A) \to \mathbb{C}$ or $B(F_v) \to \mathbb{C}$ denote the modular quasicharacter. We will study the local integrals

$$Z_v(s, W_{1,v}, W_{2,v}, f_v) = \int_{K_v} \int_{T_1(F_v)} W_{1,v}(tg)\, W_{2,v}(tg)\, f_v(tg)\, \delta_B(t)^{-1}\, dt\, dg, \tag{8.8}$$

where K_v is the standard maximal compact subgroup, defined in Eq. (3.2). Later we will relate these integrals to Eq. (8.3).

Proposition 3.8.1 *(i) The integral (Eq. 8.8) is absolutely convergent if* $\mathrm{re}(s) > 1$. *It has meromorphic continuation to all* s.
(ii) If $v \notin S$, *then*

$$L_v(2s, \xi_v^2\, \omega_1\, \omega_2)\, Z_v(s, W_{1,v}, W_{2,v}, f_v) = L_v(s, \pi_{1,v} \times \pi_{2,v}, \xi_v). \tag{8.9}$$

The reader should bear in mind that with our conventions, $\xi = \xi_1$ and $\xi^2 \omega_1 \omega_2 = \xi_1 \xi_2^{-1}$.

Proof Because $W_{1,v}$, $W_{2,v}$, and f_v are all K_v-finite, Eq. (8.8) is a finite linear combination of functions of the form

$$\int_{T_1(F_v)} W_{1,v}(tg)\, W_{2,v}(tg)\, f_v(tg)\, \delta_B(t)^{-1}\, dt.$$

The convergence of the integral is analogous to that of Eq. (5.32), which is discussed immediately after that formula. For the meromorphic continuation, it follows from Exercise 4.7.2(a) that the integrand is the Mellin transform of a finite linear combination of functions of the type considered in Exercise 4.7.2(b), and so we obtain the meromorphic continuation of the local integrals.

Assuming that $v \notin S$, we prove Eq. (8.9). Note that $GL(2, F_v) = B(F_v)\, K_v$, where as usual $K_v = GL(2, \mathfrak{o}_v)$ is the standard maximal compact subgroup, so we may replace the integration over $B(F_v)\backslash GL(2, F_v)$ with an integration over K_v. However, if $v \notin S$, then on our hypotheses $W_{1,v}$, $W_{2,v}$, and f_v are all right invariant under K_v, so we may omit this integration altogether and write

$$\begin{aligned} Z_v(s, W_{1,v}, W_{2,v}, f_v) &= \int_{T_1(F_v)} W_{1,v}(t)\, W_{2,v}(t)\, f_v(t)\, \delta_B(t)^{-1}\, dt \\ &= \int_{F_v^\times} W_{1,v}\begin{pmatrix} y & \\ & 1 \end{pmatrix} W_{2,v}\begin{pmatrix} y & \\ & 1 \end{pmatrix} \xi_{1,v}(y)\, |y|^{s-1}\, d^\times y. \end{aligned}$$

Now we recall the explicit formula (5.35), whose proof is in Proposition 4.5.10. We may write our last integral as

$$\sum_{k=0}^{\infty} \int_{\mathfrak{o}_v^\times \varpi^k} W_{1,v}\begin{pmatrix} y & \\ & 1 \end{pmatrix} W_{2,v}\begin{pmatrix} y & \\ & 1 \end{pmatrix} \xi_{1,v}(y)\, |y|^{s-1}\, d^\times y = \sum_{k=0}^{\infty} A(k)\, B(k)\, q^{-ks},$$

where in terms of the Satake parameters

$$A(k) = \frac{\alpha_1(v)^{k+1} - \alpha_2(v)^{k+1}}{\alpha_1(v) - \alpha_2(v)}, \qquad B(k) = \frac{\beta_1(v)^{k+1} - \beta_2(v)^{k+1}}{\beta_1(v) - \beta_2(v)}.$$

We use Lemma 1.6.1, with $x = \xi_{1,v}(\varpi_v)\, q^{-s}$, noting that

$$\alpha_{1,v}(\varpi_v)\, \alpha_{2,v}(\varpi_v)\, \beta_{1,v}(\varpi_v)\, \beta_{2,v}(\varpi_v) = (\omega_{1,v}\omega_{2,v})(\varpi_v) = (\xi_{1,v}\, \xi_{2,v})^{-1}(\varpi_v),$$

and because $\xi_v = \xi_{1,v}$ and $\xi_v^2 \omega_{1,v}\omega_{2,v} = \xi_{1,v}\xi_{2,v}^{-1}$, we obtain Eq. (8.9). ■

Proposition 3.8.2 *We have*

$$\int_{Z(A)\, GL(2,F)\backslash GL(2,A)} \phi_1(g)\, \phi_2(g)\, E^*(g, f)\, dg$$

$$= \left\{ \prod_{v \in S} Z_v(s, W_{1,v}, W_{2,v}, f_v) \right\} L_S(s, \pi_1 \times \pi_2, \xi_1). \tag{8.10}$$

Proof For the purpose of explaining the Rankin–Selberg unfolding, we will temporarily work with the "unnormalized" Eisenstein series $E(g, f)$ instead of $E^*(g, f)$. Assume that $\mathrm{re}(s) > 1$, so that the sum (Eq. 7.8) is absolutely convergent. Then we may substitute the definition of the Eisenstein series and "unfold" the sum to see that

$$\int_{Z(A)\,GL(2,F)\backslash GL(2,A)} \phi_1(g)\,\phi_2(g)\,E(g,f)\,dg$$

$$= \int_{Z(A)\,B(F)\backslash GL(2,A)} \phi_1(g)\,\phi_2(g)\,f(g)\,dg$$

$$= \int_{B(A)\backslash GL(2,A)} \int_{Z(A)\,B(F)\backslash B(A)} \phi_1(bg)\,\phi_2(bg)\,f(bg)\,db\,dg \qquad (8.11)$$

Then we may identify the quotient $Z\backslash B$ with T_1N. If we use the coordinates tn on the group $T_1(A)\,N(A)$, with $t_1 \in T_1$ and $n \in N$, then it is easy to see that the left Haar measure is $dt\,dn$, where dt and dn are Haar measures on $T_1(A)$ and $N(A)$ – indeed, one easily checks the left invariance of this measure by $T_1(A)$ and $N(A)$ separately. Noting that $f(tng) = f\left(tnt^{-1} \cdot tg\right) = f(tg)$ because $tnt^{-1} \in N(A)$, we may thus write Eq. (8.10) as

$$\int_{B(A)\backslash GL(2,A)} \int_{T_1(F)\backslash T_1(A)} \int_{N(F)\backslash N(A)} \phi_1(tng)\,\phi_2(tng)\,f(tg)\,dn\,dt\,dg. \qquad (8.12)$$

We have, by Theorem 3.5.5, the Fourier expansions

$$\phi_i(g) = \sum_{\alpha \in T_1(F)} W_i(\alpha g), \qquad i = 1, 2. \qquad (8.13)$$

We substitute the Fourier expansion of ϕ_1, then invoke the automorphicity of ϕ_2 to write Eq. (8.10) as

$$\int_{B(A)\backslash GL(2,A)} \int_{N(F)\backslash N(A)} \int_{T_1(F)\backslash T_1(A)} \sum_{\alpha \in T_1(F)} W_1(\alpha tng)\,\phi_2(tng)\,f(tg)\,dt\,dn\,dg$$

$$= \int_{B(A)\backslash GL(2,A)} \int_{T_1(F)\backslash T_1(A)} \int_{N(F)\backslash N(A)} \sum_{\alpha \in T_1(F)} W_1(\alpha tng)\,\phi_2(\alpha tng)\,f(\alpha tg)\,dn\,dt\,dg$$

$$= \int_{B(A)\backslash GL(2,A)} \int_{T_1(A)} \int_{N(F)\backslash N(A)} W_1(tng)\,\phi_2(tng)\,f(tg)\,dn\,dt\,dg.$$

Now we make the change of variables $n \to t^{-1}nt$, which multiplies the measure

dn by $\delta_B(t)^{-1}$, where the modular quasicharacter on $B(A)$ is

$$\delta_B\begin{pmatrix} t_1 & \\ & 1 \end{pmatrix} = |t_1|.$$

The integral equals

$$\int_{B(A)\backslash GL(2,A)} \int_{T_1(A)} \int_{N(F)\backslash N(A)} W_1(ntg)\,\phi_2(ntg)\,f(tg)\,\delta_B(t)^{-1}\,dn\,dt\,dg$$

$$= \int_{B(A)\backslash GL(2,A)} \int_{T_1(A)} W_1(tg) \int_{N(F)\backslash N(A)} \psi(n)\,\phi_2(ntg)\,dn\,f(tg)\,\delta_B(t)^{-1}\,dt\,dg$$

$$= \int_{B(A)\backslash GL(2,A)} \int_{T_1(A)} W_1(tg)\,W_2(tg)\,f(tg)\,\delta_B(t)^{-1}\,dt\,dg. \tag{8.14}$$

We note that by the Iwasawa decomposition, $GL(2, A) = B(A)\,K$, where K is as in Eq. (3.2), and so we may replace the integration over $B(A)\backslash GL(2, A)$ with an integration over $K = \prod_v K_v$. Substituting into the resulting integral the decompositions (Eq. 8.7) and (Eq. 7.9), we see that Eq. (8.13) equals

$$\prod_v Z_v(s, W_{1,v}, W_{2,v}, f_v). \tag{8.15}$$

Now multiplying by $L_S(2s, \xi_1\xi_2^{-1})$ and using Eq. (8.9), we obtain Eq. (8.10). ■

Proposition 3.8.3 *Let s_0 be a complex number, and let v be a place of F. The data $W_{1,v}$, $W_{2,v}$, and f_v may be chosen so that $Z_v(s, W_{1,v}, W_{2,v}, f_v)$ does not have a zero at $s = s_0$.*

We will not prove this fact, which will remain a gap in our exposition. Actually, if v is non-Archimedean, the proof is not at all difficult and one may choose the flat section f_v so that $Z_v(s, W_{1,v}, W_{2,v}, f_v)$ is identically one. For the proof see Lemma 14.7.5 on p. 19 of Jacquet (1972). On the other hand, if v is Archimedean, the proof is more difficult. A fairly simple argument, somewhat parallel to this lemma in the non-Archimedean case may be given if instead of working with $(\mathfrak{g}, K)$-modules $\pi_{1,v}$ and $\pi_{2,v}$, one makes use of the fact that π_1 and π_2 are unitary representations and takes the representatives $W_{1,v}$ and $W_{2,v}$ to be smooth vectors in their Whittaker models. (However, it is essential that f_v be a K_v-finite vector, because the theory of Eisenstein series is problematical if one tries to work with smooth vectors.) Jacquet (1972) in his Sections 17 and 18 gives precise results for the local integrals $Z_v(s, W_{1,v}, W_{2,v}, f_v)$. The nonvanishing then follows from the fact that though the gamma function may have poles, it has no zeros. ■

Proposition 3.8.4 *The Rankin–Selberg L-function*

$$L_S(s,\pi_1\times\pi_2,\xi)=\prod_{v\notin S}L_v(s,\pi_{1,v}\times\pi_{2,v},\xi_v)$$

has meromorphic continuation to all s. It can have a pole only where $E^*(g,f)$ has a pole for some choice of data.

Proof The meromorphic continuation of $L_S(s,\pi_1\times\pi_2,\xi_1)$ follows from Eq. (8.10), because the left side is absolutely and uniformly (in s) convergent apart from the poles of $E^*(g,f)$, and the finite product in braces is meromorphic by Proposition 3.8.1(i). To show that it has no poles apart from the possible poles of $E^*(g,f)$, it is sufficient to show that for a fixed place v and $s_0\in\mathbb{C}$, the data $W_{1,v}$, $W_{2,v}$, and f_v may be chosen so that $Z_v(s,W_{1,v},W_{2,v},f_v)$ does not have a zero at s_0, and this is Proposition 3.8.3. ■

Proposition 3.8.5 *Suppose that (π_1,V_1) and (π_2,V_2) are automorphic cuspidal representations of $GL(2,A)$. If $L_S(s,\pi_1\times\pi_2)$ has a pole at $s=1$, then $\pi_1=\hat{\pi}_2$.*

Proof We take the character $\xi=\xi_1$ to be trivial. Then $E(g,f)$ has a pole at $s=1$ by Proposition 3.8.4, so by Theorem 3.7.1, $\xi_1\xi_2^{-1}=1$. Thus $\xi_2=1$ also. Because $\xi_1\xi_2\omega_1\omega_2=1$, we see that the central characters of π_1 and π_2 are inverses of each other.

We choose the local data so that the integrals $Z_v(s,W_{1,v},W_{2,v},f_v)$ are nonzero at $s=1$ (by Proposition 3.8.3). Let $\phi_1\in V_1$ and $\phi_2\in V_2$ be the automorphic forms corresponding to this choice of data. As in the proof of Proposition 3.8.4, $E(g,f)$ has a pole at $s=1$, and by Proposition 3.7.5, the residue of $E(g,f)$ is a constant, independent of g. Consequently, the fact that Eq. (8.10) has a pole implies that

$$\int_{Z(A)\,GL(2,F)\backslash GL(2,A)}\phi_1(g)\,\phi_2(g)\,dg\neq 0.$$

Now we have obtained a nondegenerate bilinear pairing $V_1\times V_2\to\mathbb{C}$, namely, integration over the fundamental domain $Z(A)\,GL(2,F)\backslash GL(2,A)$. This proves that π_1 and π_2 are contragredients of each other. ■

When $\pi_1=\hat{\pi}_2$, the L-function $L_S(s,\pi_1\times\pi_2)$ does have a pole at $s=1$, and in fact we may factor $L_S(s,\pi_1\times\pi_2)=\zeta_{F,S}(s)\,L_S(s,\pi_1,\mathrm{Ad})$, where $\zeta_{F,s}$ is the partial Dedekind zeta function of F, and $L_S(s,\pi_1,\mathrm{Ad})$ is the *adjoint square L-function*

$$L_S(s,\pi_1,\mathrm{Ad}^\circ)=\prod_v L_v(s,\pi_{1,v},\mathrm{Ad}^\circ),$$

$$L_v(s,\pi_{1,v},\mathrm{Ad}^\circ)=\prod_v\left(1-\alpha_1(v)\,\alpha_2(v)^{-1}\,q_v^{-s}\right)^{-1}$$
$$\times\left(1-q_v^{-s}\right)^{-1}\left(1-\alpha_1(v)^{-1}\,\alpha_2(v)\,q_v^{-s}\right)^{-1}.$$

(The notation Ad° will be explained in Section 3.9.) It follows from results of Shimura (1971) and Gelbart and Jacquet (1978) that this L-function is entire. This result is obtained by a different application of the Rankin–Selberg method, namely, the adjoint square L-function is expressed as the Rankin–Selberg convolution of an automorphic form in the space of π_1 with a *theta function*, which belongs to an automorphic representation of the metaplectic double cover of $GL(2)$. We will look at the adjoint square L-function and the related from the point of view of the Langlands conjectures in the next section.

Let us close this section with a brief discussion of what Jacquet (1972) accomplishes. Some of this is analogous to the local theory that we have covered for the standard L-functions on $GL(1)$ (in Section 3.1) and on $GL(2)$ (in Section 3.5). There are *local functional equations*. For these, the reader should consult Gelbart and Shahidi (1988), and Exercise 7.3 of Chapter 4, where we outline a proof of the local functional equations at a non-Archimedean place.

Jacquet defines local L-functions $L_v(s, \pi_{1,v} \times \pi_{2,v})$ at *every* place, not just the nonramified ones, and epsilon factors $\epsilon_v(s, \pi_{1,v} \times \pi_{2,v}, \psi_v)$, that are, like the factors in Proposition 4.1.9, of exponential type. Then the functional equation takes a form analogous to Eq. (1.33).

In the final section of his work, Jacquet (1972) carries out the program that we outlined in Chapter 1 for proving the existence of base change lifts. (We will change his notation in explaining what he does.) If (π, V) is an automorphic cuspidal representation of $GL(2, A)$, where A is the adele ring of a global field F and E/F is a quadratic extension, then Jacquet defines an irreducible admissible representation Π of $GL(2, A_E)$, where A_E is the adele ring of E. (Each local factor Π_w is described in terms of π_v, where w is a place of E dividing the place v of F.) To prove that Π is automorphic, it is necessary to prove a functional equation for $L(s, \Pi, \chi)$, where χ is a Hecke character of $A_E^\times/E^\times$ – the converse theorem of Jacquet and Langlands (1970) then allows one to conclude that Π is automorphic. The functional equations for $L(s, \Pi, \chi)$ are proved by means of the Rankin–Selberg method. An auxiliary automorphic representation $\pi(\chi)$ of $GL(2, A)$ (analogous to the Maass forms that we constructed in Theorem 1.9.1) is constructed in Theorem 12.2, Jacquet and Langlands' and also by means of the converse theorem. Then the functional equation of $L(s, \Pi, \chi)$ is obtained by means of the Rankin–Selberg convolution $L\big(s, \pi \times \pi(\chi)\big)$. This program is described in more detail in Section 1.8.

3.9 The Global Langlands Conjectures

The Langlands conjectures were first set forth in Langlands (1970c). The Yale Monograph (Langlands, 1971) contains some of the supporting evidence that Langlands gave for the conjectures, and in the intervening years, much more evidence has appeared. Still, the conjectures seem far from proof.

Our aim is to state the Langlands functoriality conjecture, at least for $GL(n)$, a bit more formally than in Section 1.8, but still less formally than in the standard reference, which is Borel (1979). We will *not* give a precise definition of the L-group, except in the case of $GL(n)$. Instead, we will describe the L-group and its use in impressionistic terms, hoping that our less formal discussion will serve as a useful guide to Borel (1979). See also Arthur and Gelbart (1991), Gelbart (1977, 1984), and Kudla (1994).

In this section, we will state many facts without proof. We will assume in a few places more knowledge of algebraic groups than in the rest of the book. The reader who is only interested in $GL(n)$ should be able to read this section by ignoring the more technical parts.

We recall some definitions on affine algebraic groups, referring to Borel (1979), Borel and Tits (1965), Bruhat and Tits (1972), Humphreys (1975), Satake (1971), Serre (1965), Springer (1981) and Tits (1966), and other papers in Borel and Mostow (1966) and Borel and Casselman (1979) for further information. Let G_m denote the "multiplicative group," which is the algebraic group such that $G_m(F) \cong F^\times$ for every field F. A *torus* is an algebraic group isomorphic to a product of copies of G_m. A connected affine algebraic group is called *reductive* if it has no nontrivial normal unipotent subgroups, and *semisimple* if it is reductive and furthermore has no nontrivial normal tori. If G is reductive, then we define a *Borel subgroup* to be a maximal connected solvable subgroup, and it is a theorem that all Borel subgroups are conjugate. We recall that the algebraic groups G and G' are *isogenous* if there exists an algebraic group $\tilde{G}$ and surjective homomorphisms $\tilde{G} \to G$ and $\tilde{G} \to G'$ with finite kernel. We call a semisimple algebraic group *almost simple* if it is connected and is not isogenous to a product of semisimple groups of lower dimension.

Suppose that the torus T is defined over a field F. Then T is called *split* if there exists an isomorphism of T with a product of copies of G_m that is defined over the ground field F. See Section 4.1 for examples of split and nonsplit tori. The reductive group G is called *split* if it has a maximal torus that is split its ground field or *quasisplit* if it has a Borel subgroup defined over the ground field. Every split group is quasisplit.

If G is a split almost simple algebraic group defined over F, then it is isogenous to one of the following groups: $SL(n)$, or

$$SO(n) = \{g \mid g\,J\,{}^\top g = J\}, \qquad J = \begin{pmatrix} & & 1 \\ & \cdot^{\cdot^{\cdot}} & \\ 1 & & \end{pmatrix},$$

or

$$Sp(2n) = \left\{ g \mid g \begin{pmatrix} & -J \\ J & \end{pmatrix} {}^\top g = \begin{pmatrix} & -J \\ J & \end{pmatrix} \right\},$$

or one of the exceptional groups, G_2, F_4, E_6, E_7, or E_8.

Let F be a non-Archimedean local field, $\mathfrak{o}$ its ring of integers, and ϖ a generator of the maximal ideal of $\mathfrak{o}$. Let G be a reductive algebraic group defined over F. In contrast with the case of $G = GL(n, F)$, it is false in general that the maximal compact subgroups of $G(F)$ are conjugate. For example, in $SL(2, F)$, there are two conjugacy classes of maximal compact subgroups, namely $SL(2, \mathfrak{o})$, and the group

$$\begin{pmatrix} \varpi & \\ & 1 \end{pmatrix} SL(2, \mathfrak{o}) \begin{pmatrix} \varpi & \\ & 1 \end{pmatrix}^{-1}.$$

However, there are certain distinguished maximal compact subgroups, called *special maximal compact subgroups.* We will not define these, but note that if $G = SL(n)$, $SO(n)$, or $Sp(2n)$, then $SL(n, \mathfrak{o})$, $SO(n, \mathfrak{o})$, or $Sp(2n, \mathfrak{o})$, respectively, is a special maximal compact subgroup. An irreducible admissible representation of $G(F)$ is called *spherical* if it has a fixed vector with respect to a special maximal compact subgroup.

Let F be either a local or global field. The *connected L-group* ${}^L G^\circ$ is a complex Lie group that is canonically associated with F. If F is a non-Archimedean local field and the group G is split, then the semisimple conjugacy classes in ${}^L G^\circ$ naturally parametrize the spherical representations of $G(F)$. For example, suppose that $G = GL(n)$. Then ${}^L G^\circ = GL(n, \mathbb{C})$. We will explain in this particular example how the semisimple conjugacy classes of $GL(n, \mathbb{C})$ (that is, those conjugacy classes containing diagonal matrices) parametrize the spherical representations of $GL(n, F)$.

Let $\chi_1, \cdots, \chi_n$ be quasicharacters of $F^\times$. Let χ be the character of the Borel subgroup $B(F)$ consisting of upper triangular matrices in $GL(n, F)$, defined by

$$\chi \begin{pmatrix} y_1 & * & \cdots & * \\ & y_2 & & * \\ & & \ddots & \vdots \\ & & & y_n \end{pmatrix} = \chi_1(y_1) \cdots \chi_n(y_n). \tag{9.1}$$

Let $\mathcal{B}(\chi_1, \cdots, \chi_n)$ be the representation of $GL(n, F)$ induced from this representation of $B(F)$.

We recall that a *composition series* of a representation (π, V) consists of a sequence of submodules

$$1 = V_0 \subsetneq V_1 \subsetneq \cdots \subsetneq V_m = V$$

such that the quotients V_{i+1}/V_i are irreducible. The composition series (if it exists) is usually not uniquely determined. However, if

$$1 = V_0' \subsetneq V_1' \subsetneq \cdots \subsetneq V_{m'}' = V$$

is another composition series, the *Jordan–Hölder theorem* asserts that $m = m'$, and the quotients V_{i+1}/V_i are isomorphic to the quotients V_{i+1}'/V_i', though their order may be changed. These are called the *composition factors* of V. The proof in the general case is exactly parallel to the case of groups in Lang (1993, Section I.3).

Proposition 3.9.1 *The representation* $\mathcal{B}(\chi_1, \cdots, \chi_n)$ *is irreducible unless* $\chi_i \chi_j^{-1}(x) = |x|$ *or* $|x|^{-1}$ *for some* $i \neq j$*. If* $\mathcal{B}(\chi_1, \cdots, \chi_n)$ *is irreducible, then its isomorphism class is independent of the order of the* χ_i*. Even if it is reducible, it has a composition series. In the reducible case, if the* χ_i *are permuted, the isomorphism class of* $\mathcal{B}(\chi_1, \cdots, \chi_n)$ *may change, but its composition factors do not.*

If $\mathcal{B}(\chi_1, \cdots, \chi_n)$ is irreducible, we denote it as $\pi(\chi_1, \cdots, \chi_n)$. We will prove this proposition in the special case $n = 2$ in Section 4.5. For the general case, this result is due to Jacquet. For proofs, the reader should consult Bernstein and Zelevinsky (1977) or Casselman (unpublished manuscript).

An instructive case is the example where $n = 2$, $\chi_1(x) = |x|^{1/2}$, and $\chi_2(x) = |x|^{-1/2}$. There are two composition factors, the trivial representation and the special representation. In $\mathcal{B}(\chi_1, \chi_2)$, the trivial representation is a quotient and the special representation is a subrepresentation; these are reversed for $\mathcal{B}(\chi_2, \chi_1)$.

In the general case, we will assume this without proof and explore the consequences. An irreducible admissible representation of $GL(n, F)$ will be called *spherical* if it admits a $GL(n, \mathfrak{o})$-fixed vector, where $\mathfrak{o}$ is the ring of integers in F.

Proposition 3.9.2 *If* $\chi_1, \cdots, \chi_n$ *are nonramified, then the representation* $\mathcal{B}(\chi_1, \cdots, \chi_n)$ *contains a* $GL(n, \mathfrak{o})$*-fixed vector. It has a unique composition factor that is spherical. Every spherical representation of* $GL(n, F)$ *arises in this fashion.*

If $n = 2$, we prove this in Section 4.6.

Thus if π is a spherical representation of $GL(n, F)$, there exist nonramified quasicharacters $\chi_1, \cdots, \chi_n$ of $F^\times$ such that π is a composition factor of $\mathcal{B}(\chi_1, \cdots, \chi_n)$. We parametrize the isomorphism class of this representation by the conjugacy class of

$$\begin{pmatrix} \alpha_1 & & \\ & \ddots & \\ & & \alpha_n \end{pmatrix} \in {}^L G^\circ,$$

where $\alpha_1 = \chi_1(\varpi), \cdots, \alpha_n = \chi_n(\varpi)$ are the *Satake parameters* of π. This parametrization is closely related to the *Satake isomorphism*, which is a structure theorem for the *spherical Hecke algebra* $\mathcal{H}^\circ$ of compactly supported, locally constant $GL(n, \mathfrak{o})$-biinvariant functions, which is a ring under convolution. This ring is commutative and is isomorphic to the ring of symmetric elements in the polynomial ring $\mathbb{C}[X_1, \cdots, X_n, X_1^{-1}, \cdots, X_n^{-1}]$. Under this isomorphism, if $\phi \in \mathcal{H}^\circ$ corresponds to the symmetric polynomial p_α, then for each spherical representation (π, V), if $\xi \in V$ is the $GL(n, \mathfrak{o})$-fixed vector (uniquely determined up to constant multiple), then $\pi(\phi)\xi = \lambda_\phi \xi$, where $\phi \mapsto \lambda_\phi$ is a homomorphism of the Hecke algebra into $\mathbb{C}$, and in fact

$$\lambda_\phi = p_\phi(\alpha_1, \cdots, \alpha_n) \tag{9.2}$$

in terms of the Satake parameters. If $n = 2$, the Satake isomorphism is equivalent to Proposition 4.6.6. For the general case, see Satake (1963), Macdonald (1971, 1979, Chapter V).

We will not give a general definition of the connected L-group ${}^LG^\circ$ – for this, see Borel (1979). However, we give a few more examples. The connected L-groups of almost simple split groups are then given by the following table, with $GL(n)$, which is not semisimple, thrown in for good measure.

G	Cartan type of G	${}^LG^\circ$	Cartan type of ${}^LG^\circ$
$SL(n+1, F)$	A_n	$PGL(n+1, \mathbb{C})$	A_n
$PGL(n+1, F)$	A_n	$SL(n+1, \mathbb{C})$	A_n
$GL(n+1, F)$	A_n	$GL(n+1, \mathbb{C})$	A_n
$SO(2n+1, F)$	B_n	$Sp(2n, \mathbb{C})$	C_n
$Sp(2n, F)$	C_n	$SO(2n+1, \mathbb{C})$	B_n
$SO(2n, F)$	D_n	$SO(2n, \mathbb{C})$	D_n
	G_2		G_2
	F_4		F_4
	E_6		E_6
	E_7		E_7
	E_8		E_8

The root system of ${}^LG^\circ$ is in each case obtained from the root system of G by interchanging long and short roots, so that the Cartan classification types B_n and C_n are interchanged. This rule determines the group ${}^LG^\circ$ up to isogeny. Within this isogeny class, the isomorphism type of ${}^LG^\circ$ is then determined by a further rule: The finite center of ${}^LG^\circ$ is the fundamental group (in the sense of algebraic groups) of G and *vice versa*.

Let us fix a large finite Galois field extension Ω/F. We could work over the algebraic closure of F, but this leads to some problems because we wish to consider Frobenius elements. The choice of Ω is somewhat arbitrary – it must be large enough to accommodate all the fields that occur in whatever problem we are trying to solve. In particular, the group G must split over Ω. On the other hand, if F is non-Archimedean, we will ignore the case where Ω/F is ramified. The *L-group* ${}^LG$ is a semidirect product of ${}^LG^\circ$ with the Galois group $\mathrm{Gal}(\Omega/F)$, where Ω is the algebraic closure. Thus we have a split short exact sequence

$$1 \to {}^LG^\circ \to {}^LG \to \mathrm{Gal}(\Omega/F) \to 1. \tag{9.3}$$

If G is split, then ${}^LG$ is simply the direct product of ${}^LG^\circ$ with $\mathrm{Gal}(\Omega/F)$. To describe the L-group more generally, we assume first that G is quasisplit. Let G' be a split group that is isomorphic to G over Ω. Then the connected L-group ${}^LG^\circ$ is the same as that of G', but the group extension (Eq. 9.3) is different. By definition, G being quasisplit means that there is a Borel subgroup B of G defined over F; let T be a maximal torus contained in B that is defined, though not necessarily split, over F. By a *positive root* of G, we mean a rational character $T \to G_m$ defined over Ω that occurs in the representation Ad on the Lie algebra of the maximal unipotent subgroup U of B. A *simple positive root* is one that is not expressible as a product of other positive roots. The *(absolute) Dynkin diagram* Δ is a graph whose vertices are in one-to-one correspondence with the simple positive roots. The Galois group $\mathrm{Gal}(\Omega/F)$ acts on the positive roots and hence on the Dynkin diagram. The Dynkin diagram ${}^L\Delta$ of ${}^LG^\circ$ is obtained from the Dynkin diagram of G by reversing the arrows that point from long roots to short roots, and so we have an action of $\mathrm{Gal}(\Omega/F)$ on the Dynkin diagram of the complex analytic group ${}^LG^\circ$. Let ${}^L\widetilde{G}^\circ$ be the universal covering group of ${}^LG^\circ$. There exists a homomorphism from $\mathrm{Aut}({}^L\Delta)$ to $\mathrm{Aut}({}^L\widetilde{G}^\circ)$, and in fact $\mathrm{Aut}({}^L\widetilde{G}^\circ)$ is the semidirect product of this (finite) group of "outer" automorphisms by the (connected) group of inner automorphisms. In any case, we see that we have a homomorphism $\mathrm{Gal}(\Omega/F) \to \mathrm{Aut}({}^L\widetilde{G}^\circ)$, and the automorphisms in the image of this homomorphism stabilize the finite kernel of the projection map ${}^L\widetilde{G}^\circ \to {}^LG^\circ$, so in fact we have a homomorphism $\mathrm{Gal}(\Omega/F) \to \mathrm{Aut}({}^LG^\circ)$. Thus we may form the semidirect product ${}^LG$ as in Eq. (9.3).

If G is not quasisplit, then at least there exists a quasisplit group G'' that differs from G by inner twisting, and we define the L-group of G to be the L-group of G''.

We have stated that if the group G is split, then the spherical representations of $G(F)$ are parametrized by the semisimple conjugacy classes in ${}^LG^\circ$. This is closely related to a structure theorem for the spherical Hecke algebra $\mathcal{H}^\circ$ of locally constant, compactly supported K-biinvariant functions on ${}^LG^\circ$, where K is a special maximal compact subgroup. In the split case, $\mathcal{H}^\circ$ is isomorphic to the ring generated by the characters of irreducible analytic finite-dimensional representations of ${}^LG^\circ$. This is Langlands' formulation of the Satake isomorphism. Suppose that π is a spherical representation and that α is the semisimple conjugacy class of ${}^LG^\circ$ parametrizing π. Then π determines a character of $\mathcal{H}^\circ$, as is explained in Section IV.6—see (IV.6.1). In view of the Satake isomorphism, we should therefore exhibit a character of the ring generated by the characters of irreducible analytic finite-dimensional representations of ${}^LG^\circ$, and this character is evaluation at α. In the example of $GL(n)$, we see that Eq. (9.2) is a special case of this.

If G is not split, these assertions require modification. Suppose that G is quasisplit and splits over an unramified extension E/F. We will also assume that Ω is unramified over F and that $E \subseteq \Omega$. Let Φ be a Frobenius element in $\mathrm{Gal}(\Omega/F)$ and consider the coset ${}^LG^\circ \cdot \Phi$ in ${}^LG$. Langlands' formulation of the Satake isomorphism in this case amounts to the assertion that the spherical Hecke algebra is isomorphic to the ring of functions on this coset that are the restrictions of characters of irreducible representations of ${}^LG$. The spherical representations of $G(F)$ are parametrized by the semisimple conjugacy classes of ${}^LG$ in this coset.

Next let F be a global field, A its adele ring, and G an reductive algebraic group over F. Again, there is associated with G a connected L-group ${}^LG^\circ$ and an L-group ${}^LG$, and their descriptions are exactly as in the case of a local field. Let (π, V) be an automorphic cuspidal representation of $G(A)$. Then π is a restricted tensor product of irreducible admissible representations (π_v, V_v) of the groups $G(F_v)$. Let S be a finite set of places including the Archimedean ones such that π_v is spherical for all $v \notin S$, and such that the field Ω that occurs in the definition of the L-group is unramified over F. If v is any place of F, let Ω_v be the completion of Ω with respect to one of the places of Ω dividing v. (These places are all conjugate under the action of $\mathrm{Gal}(\Omega/F)$.) We use the field Ω_v in the definition of the local L-group ${}^LG_v$, which was previously denoted ${}^LG$, but which must now be distinguished from the L-group of G over the global field F. There is a canonical homorphism ${}^LG_v \to {}^LG$ that is "dual" to the inclusion $G(F) \to G(F_v)$.

Let $r : {}^LG \to GL(m, \mathbb{C})$ be a homomorphism. If $v \notin S$, then π_v is spherical and Ω_v is unramified over F_v. Let α_v be the semisimple conjugacy class in ${}^LG_v$ that parametrizes π_v. Let $r_v : {}^LG_v \to GL(m, \mathbb{C})$ be the composite of r with the canonical map ${}^LG_v \to {}^LG$, and let

$$L_v(s, \pi_v, r_v) = \det\left(I - q_v^{-s}\, r_v(\alpha_v)\right)^{-1},$$

where q_v is the cardinality of the residue field at v. Let

$$L_S(s, \pi, r) = \prod_{v \notin S} L_v(s, \pi_v, r_v).$$

Langlands (1970c, 1971) proved that this Euler product, known as a *Langlands L-function*, is convergent when $\mathrm{re}(s)$ is sufficiently large and conjectured that it should have a functional equation.

If G is split, then we may obtain a large subclass of the Langlands L-functions by ignoring the Galois part of the L-group altogether. In this case, ${}^LG = {}^LG^\circ \times \mathrm{Gal}(\Omega/F)$, and every analytic representation of ${}^LG$ has the form $r_0 \otimes r_1$, where r_0 is a representation of ${}^LG^\circ$ and r_1 is a representation of $\mathrm{Gal}(\Omega/F)$. We consider the subclass of Langlands L-functions in which r_1 is trivial.

Thus assume that G is split, and let $r : {}^LG^\circ \to GL(m, \mathbb{C})$ be a representation. If v is any place where π_v is unramified, then π_v is parametrized by a semisimple conjugacy class $\alpha_v \in {}^LG^\circ$, and so if S is any finite set of places such that π is nonramified for all $v \notin S$, we may define

$$L_S(s, \pi, r) = \prod_{v \notin S} L_v(s, \pi_v, r),$$

where

$$L_v(s,\pi_v,r_v)=\det\left(I-q_v^{-s}\,r(\alpha_v)\right)^{-1}.$$

This is really just a restatement of the previous definition.

It is possible that there may be two automorphic representations π and π' of $G(A)$ such that the L-functions

$$L_S(s,\pi,r)=L_S(s,\pi',r)$$

are equal for all representations of the L-group (where S is sufficiently large) and, moreover, that when H is another reductive algebraic group over F, and σ is automorphic representation of $H(A)$, so that $\pi'\otimes\sigma$ is an automorphic representation of $(G\times H)(A)$, and r_1 is a complex analytic representation of ${}^LH$, we have

$$L_S(s,\pi\otimes\sigma,r\otimes r_1)=L_S(s,\pi'\otimes\sigma,r\otimes r_1).$$

Thus there is no way to tell π and π' apart using L-functions. In this case, π and π' are called *L-indistinguishable*. Examples already occur when $G=SL(2)$. However, when $G=GL(n)$, it is a consequence of the strong multiplicity one theorem that if π and π' are L-indistinguishable, they are equal. The equivalence classes of L-functions under the relation of L-indistinguishability are called *L-packets*. An L-packet is finite, and it is conjectured that every L-packet contains a representation having a Whittaker model.

Langlands went beyond the conjectured functional equations to formulate his *functoriality conjecture*. Suppose that G_1 and G_2 are reductive groups, and suppose that G_2 is quasisplit. Suppose that a homomorphism $\phi:{}^LG_1\to{}^LG_2$ of complex analytic groups is given, which is the identity map on Galois groups. Actually, it is not necessary to take the same universal field Ω in both L-groups – but if ${}^LG_1={}^LG_1^\circ\cdot\mathrm{Gal}(\Omega_1/F)$ and ${}^LG_2={}^LG_2^\circ\cdot\mathrm{Gal}(\Omega_2/F)$, we must take $\Omega_1\supseteq\Omega_2$, and ϕ must induce the canonical map $\mathrm{Gal}(\Omega_1/F)\to\mathrm{Gal}(\Omega_2/F)$ on Galois groups. Such a map is called an *L-homomorphism*. Langlands conjectured that if π_1 is any automorphic representation of $G_1(A)$, then there exists an automorphic representation π_2 of $G_2(A)$ such that

$$L_S(s,\pi_2,r)=L_S(s,\pi_1,r\circ\phi),$$

where $r:{}^LG_2\to\mathbb{C}$ is any analytic representation. We express this situation by saying that π_2 is a *functorial lifting* of the automorphic representation π_1. The functoriality conjecture implies the conjectured functional equations for Langlands L-functions, because if $r:{}^LG\to GL(m,\mathbb{C})$ is a representation, then by functoriality we may lift any automorphic representation of $G(A)$ to an automorphic representation of $GL(m,A)$ (whose L-group is $GL(m,\mathbb{C})$), then use one of the known constructions such as that of Godement and Jacquet (1972) to prove the functional equation of the standard L-function of the lift, which agrees with the Langlands L-function $L_S(s,\pi,r)$.

Let us illustrate these conjectures by considering some important L-functions on $GL(n)$, the Rankin–Selberg square, the symmetric square, the exterior square, and the adjoint square.

If π_1 and π_2 are automorphic representations of $GL(n, A)$ and $GL(m, A)$, then $\pi_1 \otimes \pi_2$ is naturally an automorphic representation of $GL(n, A) \times GL(m, A)$, so we may consider L-functions attached to a pair of automorphic representations (of $GL(n, A)$ and $GL(m, A)$). The connected L-group of $GL(n) \times GL(m)$ is $GL(n, \mathbb{C}) \times GL(m, \mathbb{C})$, which has an nm-dimensional representation, namely the tensor product $\otimes : GL(n, \mathbb{C}) \times GL(m, \mathbb{C}) \to GL(nm, \mathbb{C})$. The corresponding L-function has the form

$$\prod_v \prod_{i=1}^{n} \prod_{j=1^m} \left(1 - \alpha_i(v)\,\beta_j(v)\,q_v^{-s}\right)^{-1},$$

where $\alpha_i(v)$ are the Satake parameters of $\pi_{1,v}$, and $\beta_j(v)$ are the Satake parameters of $\pi_{2,v}$. It is customary to denote this L-function as $L_S(s, \pi_1 \times \pi_2)$, and if $n = m = 2$, we studied this L-function in Section 3.8.

Let π be an automorphic representation of $GL(2, A)$, and let $\hat{\pi}$ be its contragredient. Then we will show that the Langlands conjectures imply the pole of the L-function $L_S(s, \pi \times \hat{\pi})$ at $s = 1$, which we saw in Section 3.8. Indeed, let $\alpha_1, \cdots, \alpha_n$ be the Satake parameters of π at a nonramified place v. Then $\alpha_1^{-1}, \cdots, \alpha_n^{-1}$ are the Satake parameters of $\hat{\pi}$. If $n = 2$, this follows from Proposition 4.5.5, and the extension of this proposition to $GL(n)$ is straightforward. We may express this by saying that $L(s, \hat{\pi})$ is the Langlands L-function of π with respect to the contragredient representation $r : {}^LGL(n)^\circ = GL(n, \mathbb{C}) \to GL(n, \mathbb{C})$, that is, $r(g) = {}^\top g^{-1}$. Thus $L(s, \pi \times \hat{\pi})$ is the Langlands L-function of π with respect to the representation of $GL(n, \mathbb{C})$ obtained by tensoring the standard representation $GL(n, \mathbb{C}) \to GL(n, \mathbb{C})$ (i.e., the identity map) with its contragredient. It is not hard to see that this n^2-dimensional representation is equivalent to the adjoint representation of $GL(n, \mathbb{C})$ on its Lie algebra $\mathrm{Mat}_n(\mathbb{C})$, namely, $\mathrm{Ad}(g)X = gXg^{-1}$, where $X \in \mathrm{Mat}_n(\mathbb{C})$, $g \in GL(n, \mathbb{C})$. This representation splits into two pieces, namely, the one-dimensional span of the identity matrix in $\mathrm{Mat}_n(\mathbb{C})$ and the matrices of trace zero. Let Ad° be the $n^2 - 1$-dimensional representation of $GL(n, \mathbb{C})$ on the elements of trace zero in $\mathrm{Mat}_n(\mathbb{C})$. The decomposition of the adjoint representation of $GL(n, \mathbb{C})$ into a one-dimensional subspace (affording the trivial representation) and an $n^2 - 1$-dimensional subspace (affording the representation Ad°) means that the tensor product representation $L_S(s, \pi \times \hat{\pi})$ factors as a product of the partial Dedekind zeta function $\zeta_{F,S}(s)$ and $L_S(s, \pi, \mathrm{Ad}^\circ)$. Because $\zeta_{F,S}(s)$ has a pole at $s = 1$, $L_S(s, \pi \times \hat{\pi})$ will also have a pole at $s = 1$ so long as $L_S(s, \pi, \mathrm{Ad}^\circ)$ has no zero. But conjecturally, this is the same as the Langlands L-function of an automorphic representation of $GL(n^2-1)$, namely, the functorial lift of π with respect to the L-homomorphism $\mathrm{Ad}^\circ : GL(n, \mathbb{C}) = {}^LGL(n)^\circ \to GL(n^2 - 1, \mathbb{C}) = {}^LGL(n^2 - 1)^\circ$, and it is a general theorem of Jacquet and Shalika (1976) that such an L-function can have no zero on the line $\mathrm{re}(s) = 1$.

Next we consider $L_S(s, \pi \times \pi)$. We will see that this L-function is also reducible. The analysis of this L-function is similar to that of the L-function $L_S(s, \pi \times \hat{\pi})$, except that now we tensor the standard representation $GL(n, \mathbb{C}) \to GL(n, \mathbb{C})$ not with its contragredient, but with itself. Thus $L_S(s, \pi \times \pi)$ is the Langlands L-function associated with the n^2-dimensional representation $g \mapsto g \otimes g$ of ${}^LGL(n)^\circ = GL(n, \mathbb{C})$ acting on $\mathbb{C}^n \otimes \mathbb{C}^n$. This representation has two invariant irreducible subspaces, namely, the space $\vee^2\mathbb{C}^N$ of symmetric tensors, spanned by elements of the form $x \otimes y + y \otimes x$, and the space $\wedge^2\mathbb{C}^n$ of *skew-symmetric tensors*, spanned by elements of the form $x \otimes y - y \otimes x$. The representations of $GL(n, \mathbb{C})$ on these two spaces are the $\frac{1}{2}n(n+1)$-dimensional *symmetric square* representation and the $\frac{1}{2}n(n-1)$-dimensional *exterior square* representation, respectively. We will denote the corresponding L-functions as $L_S(s, \pi, \vee^2)$ and $L_S(s, \pi, \wedge^2)$. We have, therefore

$$L_S(s, \pi \times \pi) = L_S(s, \pi, \vee^2)\, L_S(s, \pi, \wedge^2).$$

Because (by the same reasoning that $L_S(s, \pi, \mathrm{Ad}^\circ)$ cannot have a zero at $s = 1$) neither $L_S(s, \pi, \vee^2)$ nor $L_S(s, \pi, \wedge^2)$ can have a zero at $s = 1$, we see that if $\pi \cong \hat{\pi}$, so that $L_S(s, \pi \times \pi)$ has a pole at $s = 1$, either the $L_S(s, \pi, \vee^2)$ or $L_S(s, \pi, \wedge^2)$ should have a simple pole – one or the other but not both.

The exterior square and symmetric square L-functions may both be studied by means of the Rankin–Selberg method. There are two distinct Rankin–Selberg constructions of the exterior square L-function. These may be found in Jacquet and Shalika (1990b) and in Bump and Friedberg (1990). The construction of the symmetric square L-function is considered in Shimura (1975), Gelbart and Jacquet (1978), Patterson and Piatetski-Shapiro (1989), and Bump and Ginzburg (1992).

Exercises

Exercise 3.9.1 We have seen that if π is a self-contragredient automorphic cuspidal representation of $GL(n, A)$, then $L_S(s, \pi \times \pi)$ has a pole at $s = 1$, and that one or the other, but not both of $L_S(s, \pi, \vee^2)$ or $L_S(s, \pi, \wedge^2)$ can have a pole. Moreover, it is known that if n is odd, then $L_S(s, \pi, \wedge^2)$ can have no pole at $s = 1$ because it is represented by a Rankin–Selberg integral of "Hecke type" (not involving an Eisenstein series), which has no pole. This is the final theorem of Jacquet and Shalika (1990b). (See also Bump and Friedberg (1990).) Consider the L-group inclusions

$${}^LSO(2n)^\circ = SO(2n, \mathbb{C}) \to {}^LGL(2n, \mathbb{C})^\circ = GL(2n, \mathbb{C}),$$

$${}^LSO(2n+1)^\circ = Sp(2n, \mathbb{C}) \to {}^LGL(2n, \mathbb{C})^\circ = GL(2n, \mathbb{C}),$$

$${}^LSp(2n)^\circ = SO(2n+1, \mathbb{C}) \to {}^LGL(2n+1, \mathbb{C})^\circ = GL(2n+1, \mathbb{C}).$$

(Here, as in the text, $SO(m)$ denotes the split form of the orthogonal group.) There should be functorial liftings from automorphic representations on the groups $SO(2n)$, $SO(2n+1)$, and $Sp(2n)$ to $GL(2n)$, $GL(2n)$, and $GL(2n+1)$, respectively. Assume that the Langlands conjectures are true. Show that if π is a lift from any of these groups, then it is self-contragredient and that if π is a lift from $SO(2n)$ or $Sp(2n)$, then $L_S(s, \pi, \vee^2)$ has a pole, whereas if π is a lift from $SO(2n+1)$, then $L_S(s, \pi, \wedge^2)$ has a pole. It is conjectured that every self-contragredient automorphic form on $GL(m)$ is a lift from a classical group.

This situation is exactly analogous to that found with the *Asai L-function*. Let E/F be a quadratic extension of global fields, and let A_E and A be the adele groups of E and F, respectively. There exist functorial liftings from automorphic representations $GL(2, A)$ and from the unitary group $U(2)$ attached to the extension E/F to the automorphic representations on $GL(2, A_E)$. The fact that there are two distinct liftings manifests itself first in the case of $SL(2, \mathbb{Z})$, from Doi and Naganuma (1969) and Naganuma (1973). It is shown, using the trace formula, that the Galois invariant automorphic representations of $GL(2, A)$ are precisely the unions of the images of these two liftings. See Saito (1975), Shintani (1979), Langlands (1980) and Arthur and Clozel (1989). There is an L-function approach to distinguishing the images of the two liftings, precisely analogous to the use of the symmetric and exterior square L-functions to distinguish the images of the two liftings from orthogonal groups to $GL(2n)$. The original paper of Asai (1977) (which, like Doi and Naganuma (1969) acknowledges Shimura for his suggestion) explains this situation very clearly. The poles of the Asai L-function are related to certain "periods" of the automorphic form, which are important in both the algebraic geometry of Hilbert modular surfaces and in emerging philosophies about automorphic forms. See Harder, Langlands, and Rapoport (1986).

3.10 The Triple Convolution

There exist many examples where we may prove a functional equation for a Langlands L-function by integrating a cusp form against an Eisenstein series. Such an integral is called a *Rankin–Selberg integral*. There is no systematic theory that explains which Langlands L-functions may be represented as Rankin–Selberg integrals. We refer to Bump (1989) for a survey of the literature prior to 1986. We will give here just one example to whet the appetite. In this section, we will depart from our policy of giving complete proofs, because we will make use of an Eisenstein series on the group $GSp(4)$. Considerations of space preclude us from proving the analytic continuation and functional equation of this Eisenstein series. Moreover, we will refer to Garrett (1987) for some algebra in connection with our "nonramified" computation at the end.

Let F be a global field and A its adele ring. Let (π_0, V_0), (π_1, V_1), and (π_2, V_2) be three automorphic representations on $GL(2, A)$. We may regard $\pi_0 \otimes \pi_1 \otimes \pi_2$ as an automorphic representation of the group $G(A)$, where

$G = GL(2) \times GL(2) \times GL(2)$. The connected L-group of G is $GL(2, \mathbb{C}) \times GL(2, \mathbb{C}) \times GL(2, \mathbb{C})$, which has a natural eight-dimensional representation, namely, $(g_1, g_2, g_3) \to g_1 \otimes g_2 \otimes g_3$. Suppose that v is a nonramified place, which in the present context means that v is non-Archimedean, that $\pi_{0,v}$, $\pi_{1,v}$, and $\pi_{2,v}$ are nonramified, and that ψ_v has conductor equal to the ring $\mathfrak{o}_v$ of units in F_v, where, as usual, $\psi = \prod_v \psi_v$ is a fixed nontrivial additive character of A/F. Let $\alpha_1(v)$ and $\alpha_2(v)$ be the Satake parameters of π_0, let $\beta_1(v)$ and $\beta_2(v)$ be the Satake parameters of π_1, and let $\gamma_1(v)$ and $\gamma_2(v)$ be the Satake parameters of π_2. Then the local Langlands L-function of $\pi_0 \otimes \pi_1 \otimes \pi_2$ with respect to this L-group representation is

$$L_v(s, \pi_{0,v} \times \pi_{1,v} \times \pi_{2,v}) = \prod_{i,j,k} (1 - \alpha_i(v)\, \beta_j(v)\, \gamma_k(v)\, q_v^{-s})^{-1},$$

and we would like to study the "triple L-function"

$$L_S(s, \pi_0 \times \pi_1 \times \pi_2) = \prod_{v \notin S} L_v(s, \pi_{0,v} \times \pi_{1,v} \times \pi_{2,v})$$

where, as usual, S is a finite set of places such that every $v \notin S$ is nonramified in the above sense. We will study these L-functions in this section, assuming for simplicity that the central quasicharacters of π_0, π_1, and π_2 are trivial. Our goal is simply to give the reader some flavor of the possibilities of the Rankin–Selberg method on higher-rank groups. We will see that although all three L-functions are only on $GL(2)$, the construction of the L-function involves the intervention of a higher-rank group, namely, $GSp(4)$. We will assume for simplicity that the central characters of all three representations π_i are trivial. We will exhibit a Rankin–Selberg integral, which we will unfold, and show that it represents the triple L-function.

It is a consequence of the Langlands conjectures that one should similarly have an analytic continuation and functional equation for an L-function formed with any number of automorphic representations. For two automorphic forms, we studied this problem in Section 3.8. For three automorphic representations, the analytic properties of the triple L-function were established by Garrett (1987). Garrett's construction involved forming an Eisenstein series on $GSp(6)$, restricting it to the subgroup $PGL(2) \times PGL(2) \times PGL(2)$, and integrating it against the three automorphic forms. This construction was modified by Piatetski-Shapiro and Rallis (1987) to give a twisted convolution associated with an automorphic form on $GL(2)$ over a cubic field. Piatetski-Shapiro and Rallis (1987) and Ikeda (1992) used this same convolution to give precise results about the poles of the triple-product L-function. A particularly interesting problem is the algebraicity of the special values of this L-function when π_0, π_1, and π_2 correspond to holomorphic modular forms. This has been the subject of several deep investigations by Garrett and Harris (1993), Gross and Kudla (1992), and Harris and Kudla (1991).

The integral found by Garrett is not, however, the only known Rankin–Selberg integral that represents the triple convolution. Gelbart and Piatetski-Shapiro (1985) found Rankin–Selberg integrals representing automorphic representations of $SO(2n) \times GL(n)$. If $n = 2$, in view of the isogeny of $SO(4)$ and $PGL(2) \times PGL(2)$, this integral gives the triple convolution. Unlike Garrett's integral, the latter method does not adapt to give the twisted convolution, and it is Garrett's integral that has been used to study the L-function in the literature cited above. The work of Gelbart and Piatetski-Shapiro does, however, extend in another way. Ginzburg, Piatetski-Shapiro, Rallis, and Soudry have generalized the work of Gelbart and Piatetski-Shapiro to give Rankin–Selberg constructions for L-functions on $SO(n) \times GL(m)$ for any n and m. (See Ginzburg (1990), Soudry (1993), and Ginzburg, Piatetski-Shapiro, and Rallis (preprint) for this theory.) Taking $n = 4$, this yields triple convolutions associated with automorphic forms on $GL(2) \times GL(2) \times GL(m)$.

Shahidi's (1981, 1988) method may also be used to prove these functional equations. There exist Eisenstein series on a group of Cartan classification type D_m that involve the triple L-functions on $GL(2) \times GL(2) \times GL(m-2)$. Other triple convolutions, of type $GL(2) \times GL(3) \times GL(m)$, where $m = 3, 4$, or 5 may be obtained from Shahidi's method by means of the exceptional groups E_6, E_7, and E_8. These L-functions have not yet been achieved as Rankin–Selberg convolutions.

The convolution to be described here is adapted from the construction of Gelbart and Piatetski-Shapiro (1985). Their integral is on the group $SO(5)$. However, we will take advantage of the isogeny between the groups $SO(5)$ and $PGSp(4)$ to write the integral on the symplectic group.

Let $GSp(4)$ denote the algebraic subgroup of $GL(4)$ consisting of matrices g that satisfy

$$g\,J\,{}^{\top}g = \mu\,J$$

for some scalar *similitude factor* μ, where

$$J = \begin{pmatrix} &&&-1\\ &&-1&\\ &1&&\\ 1&&& \end{pmatrix}.$$

We call $GSp(4)$ the group of *symplectic similitudes*. Let P be the parabolic subgroup consisting of elements of the form

$$p = \begin{pmatrix} y_1y_2 & * & * & *\\ & a & b & *\\ & c & d & *\\ &&& y_2^{-1} \end{pmatrix}. \tag{10.1}$$

In order for this matrix to be a symplectic similitude, we must have $ad-bc = y_1$.

Let (π_0, V_0) be an automorphic cuspidal representation of $GL(2)$, whose central character we assume to be trivial.

Let us factor the adele ring $A = F_\infty A_f$ as in Section 3.3. Let H be a reductive affine algebraic group, which in our application will be $GSp(4)$. We call the function f on $H(A)$ *smooth* (resp. *analytic*) if for every $g \in H(A)$, there exists a neighborhood N of g and a smooth (resp. analytic) function f_g defined in a neighborhood of the image of g in $H(F_\infty)$ such that for $h \in N$, $f(h) = f_g(h_\infty)$, where we have factored $h = h_\infty h_f$ with $h_\infty \in H(F_\infty)$, $h_f \in H(A_f)$. (If F is a function field, we define f to be *smooth* or *analytic* if it is locally constant. There is no distinction between these two concepts in the function field case.) We define the *global Hecke algebra* $\mathcal{H}_{H(A)}$ just as in the proof of Theorem 3.3.3 at the end of Section 3.4. The global Hecke algebra acts on smooth or analytic functions on $H(A)$. If $\sigma \in \mathcal{H}_{H(A)}$ and f is a smooth function on $H(A)$, we will denote by $\rho(\sigma) f$ the image of f under this action. It is analytic if f is analytic.

For the remainder of this section, H will be the algebraic group $GSp(4)$. Let δ be the modular quasicharacter of $P(A)$, so if p is as in Eq. (10.1), we have $\delta(p) = |y_1^2 y_2^4|$ and let s be a complex number. We will define a certain representation of $H(A)$ – in the sense of Section 3.3 – which is to be thought of as *induced* from $P(A)$. We will first define the space $V(s)$ of this representation, which we will denote $\mathrm{Ind}\left(\pi_0 \otimes \delta^s\right)$. In order to be an element of $V(s)$, we ask that $f : H(A) \to \mathbb{C}$ be analytic and K-finite, where $K = \prod_v K_v$ is a maximal compact subgroup of $H(A)$, with $K_v = H(\mathfrak{o}_v)$ when v is non-Archimedean, or the intersection of $H(F_v)$ with $O(4)$ or $U(4)$ when v is real or complex. Moreover, we require that if $\sigma \in \mathcal{H}_{H(A)}$, then $\rho(\sigma) f$ satisfies

$$\big(\rho(\sigma)\, f\big)(p) = \delta(p)^s\, \phi\begin{pmatrix} a & b \\ c & d \end{pmatrix} \tag{10.2}$$

for all $p \in P(A)$, with a, b, c, and d as in Eq. (10.1), where ϕ is a cusp form in V_0, depending of course on σ. The action of $\mathcal{H}_{H(A)}$ on $V(s)$ is induced by the action ρ of $\mathcal{H}_{H(A)}$ on analytic functions. Let $s \mapsto f_s$ be a flat section of the spaces $V(s)$. Thus we assume that the restriction of f_s to K is independent of s. By Theorem 3.4.4, the space $\mathrm{Ind}\left(\pi_0 \otimes \delta^s\right)$ decomposes as a restricted tensor product. We will assume that f_s is a pure tensor in this decomposition.

Let $\mathcal{W}_{\pi_0}$ be the Whittaker model of π_0. We may again define an induced representation $\mathrm{Ind}\left(\mathcal{W}_{\pi_0} \otimes \delta^s\right)$ in exactly the same way that we defined $\mathrm{Ind}\left(\pi_0 \otimes \delta^s\right)$, except that instead of ϕ being in V_0 in Eq. (10.2), it must be in $\mathcal{W}_{\pi_0}$. As in the proof of Theorem 3.3.3, it follows from Theorem 3.4.4 that the space $\mathrm{Ind}\left(\mathcal{W}_{\pi_0} \otimes \delta^s\right)$, which is of course isomorphic to $\mathrm{Ind}\left(\pi_0 \otimes \delta^s\right)$, decomposes as a restricted tensor product of spaces $\mathrm{Ind}\left(\mathcal{W}_{\pi_0,v} \otimes \delta^s\right)$, and these local spaces admit the following description. If v is non-Archimedean, this is the space of all locally constant functions $W_{s,v}$ on $H(F_v)$ such that for every $g \in H(F_v)$, if

$p \in P(F_v)$ is as in Eq. (10.1), we have

$$W_{s,v}(pg) = \delta(p)^s\, W\begin{pmatrix} a & b \\ c & d \end{pmatrix},$$

where $W \in \mathcal{W}_{\pi_{0,v}}$, the Whittaker model of $\pi_{0,v}$. Of course the function W depends on g as well as f; if v is Archimedean, then $W_{s,v}$ is analytic, and for every σ in the local Hecke algebra $\mathcal{H}_{H(F_v)}$, we have

$$\big(\rho(\sigma)\, W_{s,v}\big)(pg) = \delta(p)^s\, W\begin{pmatrix} a & b \\ c & d \end{pmatrix},$$

where $W \in \mathcal{W}_{\pi_{0,v}}$, and of course W depends on σ. If W_s is a pure tensor in the decomposition, then by Theorem 3.5.4, we may write

$$W_s(g) = \prod_v W_{s,v}(g_v), \qquad W_{s,v} \in \operatorname{Ind}\big(\mathcal{W}_{\pi_{0,v}}(\pi_0) \otimes \delta^s\big), \tag{10.3}$$

where for almost all places the restriction of $W_{s,v}$ to K_v is one.

If f_s is a smooth section of $\operatorname{Ind}\big(\pi_0 \otimes \delta^s\big)$, then we may define an element

$$W_s(g) = \int_{A/F} f_s\begin{pmatrix} 1 & & & \\ & 1 & x & \\ & & 1 & \\ & & & 1 \end{pmatrix} \psi(-x)\, dx$$

of $\operatorname{Ind}\big(\mathcal{W}_{\pi_0} \otimes \delta^s\big)$. Then we claim that

$$f_s(g) = \sum_{\alpha \in F^\times} W_s\begin{pmatrix} \alpha & & & \\ & \alpha & & \\ & & 1 & \\ & & & 1 \end{pmatrix}. \tag{10.4}$$

Indeed, it follows from the proof of Eq. (5.19) that if

$$F(g) = f_s(g) - \sum_{\alpha \in F^\times} W_s\left(\begin{pmatrix} \alpha & & & \\ & \alpha & & \\ & & 1 & \\ & & & 1 \end{pmatrix}\right),$$

then $F(g)$ is an analytic function on $H(A)$ and that $\rho(\sigma)\, F(1) = 0$ for every $\sigma \in \mathcal{H}_{H(A)}$. This implies that $F = 0$.

Let $E(h, s)$ be an Eisenstein series on $H(A)$ constructed from π_0 relative to the parabolic subgroup P. More precisely, if $\operatorname{re}(s)$ is sufficiently large, the summation

$$E(h, s) = L_S(4s - 1, \pi_0, \vee^2) \sum_{\xi \in P(F)\backslash H(F)} f_s(\xi h)$$

is convergent, where $L_S(s, \pi_0, \vee^2)$ is the symmetric square L-function defined in the previous section. Here $L_S(4s-1, \pi_0, \vee^2)$ is the *normalizing factor* for the Eisenstein series, analogous to the factor $L_S(2s, \xi_1\xi_2^{-1})$ in Section 3.7. To see that $L_S(4s-1, \pi_0, \vee^2)$ is the correct normalizing factor, one consults the tables in Langlands (1971). Note that Langlands' s is actually our $s - 1/2$, because he has normalized the Eisenstein series so that the functional equation is $s \to -s$. The particular example at hand is (iv) on his p. 48 with $n = 1$, so that $a_1 = 4$. The Langlands L-function $\xi\big(a_1(s-1/2)+1, \tilde{\pi}, \phi\big)$, in his notation, is $L_S\big(4s-1/2, \pi_0, \vee^2\big)$ in our notation. The Eisenstein series so defined is convergent for sufficiently large s and has meromorphic continuation to all s and a functional equation with respect to $s \to 1-s$.

Let Z be the center of H, and let G be the group of all elements (g_1, g_2) in $GL(2) \times GL(2)$ such that $\det(g_1) = \det(g_2)$, embedded in H via

$$\left(\begin{pmatrix} a_1 & b_1 \\ c_1 & d_1 \end{pmatrix}, \begin{pmatrix} a_2 & b_2 \\ c_2 & d_2 \end{pmatrix} \right) \to \begin{pmatrix} a_2 & & & b_2 \\ & a_1 & b_1 & \\ & c_1 & d_1 & \\ c_2 & & & d_2 \end{pmatrix}.$$

Let $\phi_1 \in V_1$ and $\phi_2 \in V_2$ be two further automorphic forms on $GL(2)$. Recall that we are assuming for simplicity that the central characters of (π_2, V_2) and (π_3, V_3) are trivial. We now consider the integral

$$\int_{Z(A)\,G(F)\backslash G(A)} E\big((g_1, g_2), s\big)\, \phi_1(g_1)\, \phi_2(g_2)\, d(g_1, g_2). \tag{10.5}$$

To unfold this convolution, we take s to be large and substitute the definition of $E(h, s)$ as a sum over a particular complete set of coset representatives for $P(F)\backslash H(F)$, which may be described as follows. Let ξ run through a complete set of coset representatives for $P(F)\backslash H(F)/G(F)$, and for fixed ξ, let G^ξ be the algebraic group $G \cap \xi^{-1} P \xi$. For each ξ, let ξ' run through a complete set of coset representatives for $G^\xi(F)\backslash G(F)$. Then $\xi\xi'$ runs through a complete set of coset representatives for $P(F)\backslash H(F)$. Thus Eq. (10.5) equals

$$L_S(4s-1, \pi_0, \vee^2) \\ \times \sum_{\xi \in P(F)\backslash H(F)/G(F)} \int_{Z(A)G^\xi(F)\backslash G(A)} f_s\big(\xi(g_1, g_2)\big)\, \phi_1(g_1)\, \phi_2(g_2)\, d(g_1, g_2).$$

At this point, we need to know the contents of the set $P(F)\backslash H(F)/G(F)$. To achieve this, we represent a coset in $P\backslash H$ by the bottom row of a representative. This is a row vector. Two vectors represent the same coset if and only if they are proportional. Thus we see that $P\backslash H$ is really a projective four space. The action of G on the right on $P\backslash H$ is transparent in this interpretation, and we see that there are three orbits, corresponding to the vectors $(0, 0, 0, 1)$, $(0, 0, 1, 0)$, and $(0, 0, 1, 1)$. Therefore there are three double cosets in $P(F)\backslash H(F)/G(F)$, and representatives are

$$\begin{pmatrix} 1 & & & \\ & 1 & & \\ & & 1 & \\ & & & 1 \end{pmatrix}, \quad \begin{pmatrix} 0 & 1 & & \\ -1 & 0 & & \\ & & 0 & -1 \\ & & 1 & 0 \end{pmatrix}, \quad \begin{pmatrix} 1 & & & \\ -1 & 1 & & \\ & & 1 & \\ & & 1 & 1 \end{pmatrix}.$$

The contributions of the first two representatives are zero. For example, let us show that the first representative, $\xi = 1$, contributes zero. The contribution is

$$L_S(4s-1, \pi_0, \vee^2) \int_{Z(A)G^1(F)\backslash G(A)} f_s(g_1, g_2)\, \phi_1(g_1)\, \phi_2(g_2)\, d(g_1, g_2).$$

For any $x_2 \in A$, this equals

$$L_S(4s-1, \pi_0, \vee^2)$$
$$\times \int_{Z(A)G^1(F)\backslash G(A)} f_s\left(g_1, \begin{pmatrix} 1 & x_2 \\ & 1 \end{pmatrix} g_2\right) \phi_1(g_1)\, \phi_2\left(\begin{pmatrix} 1 & x_2 \\ & 1 \end{pmatrix} g_2\right) d(g_1, g_2).$$

Because

$$f_s\left(g_1, \begin{pmatrix} 1 & x_2 \\ & 1 \end{pmatrix} g_2\right) = f_s(g_1, g_2),$$

the contribution equals

$$L_S(4s-1, \pi_0, \vee^2)$$
$$\times \int_{A/F} \int_{Z(A)G^1(F)\backslash G(A)} f_s(g_1, g_2)\, \phi_1(g_1)\, \phi_2\left(\begin{pmatrix} 1 & x_2 \\ & 1 \end{pmatrix} g_2\right) d(g_1, g_2),$$

which vanishes because ϕ_2 is cuspidal. We leave it to the reader to check that the vanishing of the contribution of the second ξ follows from the cuspidality of ϕ_1. Now let

$$\xi = \begin{pmatrix} 1 & & & \\ -1 & 1 & & \\ & & 1 & \\ & & 1 & 1 \end{pmatrix}.$$

Then G^ξ is the set of elements of P of the form

$$\begin{pmatrix} y & & & x_2 \\ & y & x_1 & \\ & & 1 & \\ & & & 1 \end{pmatrix}.$$

Invoking Eq. (10.4), we see that the contribution of this ξ (which, we have noted, is the only nonzero contribution) to Eq. (10.5) equals

$$L_S(4s-1,\pi_0,\vee^2)\int_{Z(A)U(F)\backslash G(A)} W_s\big(\xi(g_1,g_2)\big)\,\phi_1(g_1)\,\phi_2(g_2)\,d(g_1,g_2),$$

where U is the subgroup of G consisting of elements of the form

$$\begin{pmatrix}1&&&x_2\\&1&x_1&\\&&1&\\&&&1\end{pmatrix}.$$

This equals

$$L_S(4s-1,\pi_0,\vee^2)\int_{Z(A)U(A)\backslash G(A)}\int_{(A/F)^2} W_s\left(\xi\left(\begin{pmatrix}1&x_1\\&1\end{pmatrix}g_1,\begin{pmatrix}1&x_2\\&1\end{pmatrix}g_2\right)\right)$$

$$\times\,\phi_1\left(\begin{pmatrix}1&x_1\\&1\end{pmatrix}g_1\right)\phi_2\left(\begin{pmatrix}1&x_1\\&1\end{pmatrix}g_2\right)dx_1\,dx_2\,d(g_1,g_2).$$

Now because

$$\xi\begin{pmatrix}1&&&x_2\\&1&x_1&\\&&1&\\&&&1\end{pmatrix}\xi^{-1}=\begin{pmatrix}1&&-x_2&x_2\\&1&x_1+x_2&-x_2\\&&1&\\&&&1\end{pmatrix},$$

we have

$$W_s\left(\xi\left(\begin{pmatrix}1&x_1\\&1\end{pmatrix}g_1,\begin{pmatrix}1&x_2\\&1\end{pmatrix}g_2\right)\right)=e(x_1+x_2)\,W_s\big(\xi(g_1,g_2)\big),$$

and so Eq. (10.5) equals

$$L_S(4s-1,\pi_0,\vee^2)\int_{Z(A)U(A)\backslash G(A)}\int_{(A/F)^2} e(x_1+x_2)\,W_s\big(\xi(g_1,g_2)\big)$$

$$\times\,\phi_1\left(\begin{pmatrix}1&x_1\\&1\end{pmatrix}g_1\right)\phi_2\left(\begin{pmatrix}1&x_1\\&1\end{pmatrix}g_2\right)dx_1\,dx_2\,d(g_1,g_2)$$

$$=L_S(4s-1,\pi_0,\vee^2)$$

$$\times\int_{Z(A)U(A)\backslash G(A)} W_s\big(\xi(g_1,g_2)\big)\,W_{\phi_1}(g_1)\,W_{\phi_2}(g_2)\,d(g_1,g_2).$$

Here W_{ϕ_1} and W_{ϕ_2} are the Whittaker functions formed with the *opposite* character $\overline{\psi}$. We assume that ϕ_1 and ϕ_2 are pure tensors in the decompositions

$\pi_i = \otimes \pi_{i,v}$, so that by Theorem 3.5.4, we may write

$$W_{\phi_i}(g) = \prod_v W_{\pi_i,v}(g_v),$$

where $W_{\pi_i,v} \in \mathcal{W}_{\pi_i}$, and for all $v \notin S$, we may choose the local component of ϕ_i to be the spherical vector. Thus if $v \notin S$, we have $W_{\pi_i,v} = W^\circ_{\pi_i,v}$, the spherical vector in the Whittaker model, normalized so that $W^\circ_{\pi_i,v}(g_v) = 1$ when $g_v \in GL(2, \mathfrak{o}_v)$.

Our last integral evidently has a Euler product. We now evaluate the nonramified factors.

Let v be a nonramified place. The representation $\mathrm{Ind}\left(\mathcal{W}_{\pi_0,v} \otimes \delta^s\right)$ has a unique spherical vector, $W^\circ_{s,v}$, normalized so that the restriction of $W^\circ_{s,v}$ to $H(\mathfrak{o})$ is one. In the decomposition (Eq. 10.3), we must have $W_{s,v} = W^\circ_{s,v}$ for all $v \notin S$. It is easy to give a formula for $W^\circ_{s,v}$. Let $W^\circ_{\pi_0,v}$ be the spherical vector in $\mathcal{W}_{\pi_0,v}$, normalized so that $W^\circ_{\pi_0,v} = 1$ on $GL(2, \mathfrak{o}_v)$. If $g \in H(F_v)$ is factored as pk, where $p \in P(F_v)$ and $k \in K_v$, then

$$W^\circ_{s,v}(pk) = W^\circ_{\pi_0,v}\begin{pmatrix} a & b \\ c & d \end{pmatrix} \delta(p)^s.$$

The local factor is

$$L_v(4s-1, \pi_0, \vee^2) \int_{Z(F_v)U(F_v)\backslash G(F_v)} W^\circ_{s,v}\big(\xi(g_1, g_2)\big)$$
$$\times\, W_{\pi_1,v}(g_1)\, W_{\pi_2,v}(g_2)\, d(g_1, g_2). \tag{10.6}$$

Now let B be the Borel subgroup of $GL(2)$ consisting of upper triangular matrices, and let $B' = (B \times B) \cap G$. Also, let $K = GL(2, \mathfrak{o})$ be the standard maximal compact subgroup of $GL(2, F_v)$. Let $K' = (K \times K) \cap G$. The integrand in Eq. (10.6) is invariant under K', and so by the Iwasawa decomposition, Eq. (10.6) is equal to

$$L_v(4s-1, \pi_0, \vee^2) \int_{Z(F_v)U(F_v)\backslash B'(F_v)} W^\circ_{s,v}\big(\xi(b_1, b_2)\big)\, W_{\pi_1}(b_1)\, W_{\pi_2}(b_2)\, d(b_1, b_2),$$

where $d(b_1, b_2)$ is the left invariant Haar measure on $Z(F_v)\backslash B'(F_v)$. Let us choose coordinates on $Z(F_v)\backslash B'(F_v)$ as follows: We will represent a typical coset by a matrix

$$\begin{pmatrix} 1 & & & x_2 \\ & 1 & x_1 & \\ & & 1 & \\ & & & 1 \end{pmatrix} \begin{pmatrix} y_1 y_2 & & & \\ & y_1 & & \\ & & 1 & \\ & & & y_2^{-1} \end{pmatrix}.$$

In our previous notation, this is to be identified with

$$(b_1, b_2) = \left(\begin{pmatrix} y_1 & x_1 \\ & 1 \end{pmatrix}, \begin{pmatrix} y_1 y_2 & x_2 y_2^{-1} \\ & y_2^{-1} \end{pmatrix} \right).$$

Now the left Haar measure in these coordinates is

$$d(b_1, b_2) = |y_1 y_2|^{-2}\, dx_1\, dx_2\, d^\times y_1\, d^\times y_2.$$

The integrand in Eq. (10.6) is also independent of x_1 and x_2, and so Eq. (10.6) equals

$$\begin{aligned} &L_v(4s-1, \pi_0, \vee^2) \\ &\times \int_{F_v^\times}\int_{F_v^\times} W_{s,v}^\circ\left(\begin{pmatrix} 1 &&& \\ -1 & 1 && \\ && 1 & \\ && 1 & 1 \end{pmatrix} \begin{pmatrix} y_1y_2 &&& \\ & y_1 && \\ && 1 & \\ &&& y_2^{-1} \end{pmatrix} \right) \\ &\times W_{\pi_1,v}\begin{pmatrix} y_1 & \\ & 1 \end{pmatrix} W_{\pi_2,v}\begin{pmatrix} y_1y_2 & \\ & y_2^{-1} \end{pmatrix} |y_1y_2|^{-2}\, d^\times y_1\, d^\times y_2. \end{aligned}$$

Using the assumption that the central character of π_2 is trivial, we have

$$W_{\pi_2,v}\begin{pmatrix} y_1y_2 & \\ & y_2^{-1} \end{pmatrix} = W_{\pi_2,v}\begin{pmatrix} y_1y_2^2 & \\ & 1 \end{pmatrix}.$$

We break the integral into two parts. Firstly, suppose that $y_2 \in \mathfrak{o}$, so that $|y_2| \le 1$. Then because f_s is right invariant by $H(\mathfrak{o})$, we have

$$\begin{aligned} &W_{s,v}^\circ\left(\begin{pmatrix} 1 &&& \\ -1 & 1 && \\ && 1 & \\ && 1 & 1 \end{pmatrix} \begin{pmatrix} y_1y_2 &&& \\ & y_1 && \\ && 1 & \\ &&& y_2^{-1} \end{pmatrix} \right) \\ &= W_{s,v}^\circ\left(\begin{pmatrix} 1 &&& \\ -1 & 1 && \\ && 1 & \\ && 1 & 1 \end{pmatrix} \begin{pmatrix} y_1y_2 &&& \\ & y_1 && \\ && 1 & \\ &&& y_2^{-1} \end{pmatrix} \begin{pmatrix} 1 &&& \\ y_2 & 1 && \\ && 1 & \\ && -y_2 & 1 \end{pmatrix} \right) \\ &= W_{\pi_0,v}\begin{pmatrix} y_1 & \\ & 1 \end{pmatrix} |y_1^2 y_2^4|^s. \end{aligned}$$

Thus we have the contribution

$$\begin{aligned} &L_v(4s-1, \pi_0, \vee^2) \\ &\times \int_{\mathfrak{o}}\int_{F_v^\times} W_{\pi_0,v}\begin{pmatrix} y_1 & \\ & 1 \end{pmatrix} W_{\pi_2,v}\begin{pmatrix} y_1 & \\ & 1 \end{pmatrix} W_{\pi_2,v}\begin{pmatrix} y_1y_2^2 & \\ & 1 \end{pmatrix} \\ &\times |y_1y_2|^{-2}\, |y_1^2y_2^4|^s\, d^\times y_1\, d^\times y_2. \end{aligned}$$

On the other hand, if $y_2 \notin \mathfrak{o}$, so that $|y_2| > 1$, then

$$W^\circ_{s,v}\left(\begin{pmatrix} 1 & & & \\ -1 & 1 & & \\ & & 1 & \\ & & 1 & 1 \end{pmatrix}\begin{pmatrix} y_1y_2 & & & \\ & y_1 & & \\ & & 1 & \\ & & & y_2^{-1} \end{pmatrix}\begin{pmatrix} y_2^{-1} & -1 & & \\ 1 & & & \\ & & -y_2^{-1} & 1 \\ & & -1 & \end{pmatrix}\right)$$

$$= W_{\pi_0,v}\begin{pmatrix} y_1y_2^2 & \\ & 1 \end{pmatrix} |y_1^2|^s.$$

Thus we have the contribution

$$L_v(4s-1, \pi_0, \vee^2)$$
$$\times \int_{F_v^\times - \mathfrak{o}} \int_{F_v^\times} W_{\pi_0,v}\begin{pmatrix} y_1y_2^2 & \\ & 1 \end{pmatrix} W_{\pi_1,v}\begin{pmatrix} y_1 & \\ & 1 \end{pmatrix} W_{\pi_2,v}\begin{pmatrix} y_1y_2^2 & \\ & 1 \end{pmatrix}$$
$$\times |y_1y_2|^{-2}\, |y_1^2|^s\, d^\times y_1\, d^\times y_2.$$

In this integral, we make the substitutions $y_1 \to y_1y_2^2$, $y_2 \to y_2^{-1}$, and we obtain the contribution

$$L_v(4s-1, \pi_0, \vee^2)$$
$$\times \int_{\mathfrak{o} - \mathfrak{o}^\times} \int_{F_v^\times} W_{\pi_0,v}\begin{pmatrix} y_1 & \\ & 1 \end{pmatrix} W_{\pi_1,v}\begin{pmatrix} y_1y_2^2 & \\ & 1 \end{pmatrix} W_{\pi_2,v}\begin{pmatrix} y_1 & \\ & 1 \end{pmatrix}$$
$$\times |y_1y_2|^{-2}\, |y_1^2y_2^4|^s\, d^\times y_1\, d^\times y_2.$$

Note the symmetry between the two contributions!

Now suppose that $\alpha_1(v)$ and $\alpha_2(v)$ are the parameters of $\pi_{0,v}$. Because we are assuming that the central character is trivial, we have $\alpha_1(v)\alpha_2(v) = 1$. Similarly, let $\beta_1(v)$, $\beta_2(v) = \beta_1(v)^{-1}$ and $\gamma_1(v)$, $\gamma_2(v) = \gamma_1(v)^{-1}$ be the parameters of $\pi_{1,v}$ and $\pi_{2,v}$, respectively. We have proved that the local factor is equal to

$$L_v(4s-1, \pi, \vee^2)$$
$$\times \int_{\mathfrak{o}} \int_{F^\times} W_{\pi_0,v}\begin{pmatrix} y_1 & \\ & 1 \end{pmatrix} W_{\pi_1,v}\begin{pmatrix} y_1 & \\ & 1 \end{pmatrix} W_{\pi_2,v}\begin{pmatrix} y_1y_2^2 & \\ & 1 \end{pmatrix}$$
$$\times |y_1y_2|^{-2}\, |y_1^2y_2^4|^s\, d^\times y_1\, d^\times y_2 + L_v(4s-1, \pi, \vee^2)$$
$$\times \int_{\mathfrak{o} - \mathfrak{o}^\times} \int_{F^\times} W_{\pi_0,v}\begin{pmatrix} y_1 & \\ & 1 \end{pmatrix} W_{\pi_1,v}\begin{pmatrix} y_1y_2^2 & \\ & 1 \end{pmatrix} W_{\pi_2,v}\begin{pmatrix} y_1 & \\ & 1 \end{pmatrix}$$
$$\times |y_1y_2|^{-2}\, |y_1^2y_2^4|^s\, d^\times y_1\, d^\times y_2.$$

It is interesting to note that although the integral that we have chosen to represent the triple L-function is different from that of Garrett (1987), at this stage the local calculation is identical to his. The algebra required to establish this (based on Eq. (5.35)) is routine but tedious – substituting Eq. (5.35) for the three nonramified Whittaker functions and expanding out, one arrives at sixteen geometric series, which are convergent for $\mathrm{re}(s)$ large; summing these and simplifying, one obtains for the local factor

$$\prod_{i,j,k}(1-\alpha_i(v)\,\beta_j(v)\,\gamma_j(v)\,q_v^{-2s+1/2})^{-1},$$

which is $L_v\left(2s-\frac{1}{2},\pi_{0,v}\times\pi_{1,v}\times\pi_{2,v}\right)$. Or see Garrett (1987, Theorem 4.2). Note that this change of variables does not change the functional equation: If $s'=2s-1/2$, then as $s\to 1-s$, we have $s'\to 1-s'$.

4

Representations of $GL(2)$ Over a p-adic Field

In this chapter, we will discuss the representation theory of $GL(2)$ over a p-adic field, a cornerstone of the modern theory of automorphic forms, emphasizing techniques that are applicable to $GL(n)$.

We begin by looking at the representation theory of $GL(2)$ over a finite field, where we encounter an essential tool, Mackey theory, which is the calculus of intertwining operators between induced representations. We will see that the irreducible representations of $G = GL(2, F)$ when F is finite are roughly parametrized by the characters of maximal tori in G. The representations parametrized by maximal *split* tori are induced representations, those parametrized by *nonsplit* tori must be constructed by some other method. A convenient method of accomplishing this is afforded by the Weil representation, which we study in detail.

In the rest of the chapter, we will adapt the results of Section 4.1 to representations of $GL(2, F)$ where F is local. In Section 4.2, we introduce the categories of smooth and admissible representations and establish their basic properties. In Section 4.3, we introduce some tools, sheaves, and distributions, which are needed to extend Mackey theory to locally compact groups, as in Bruhat's thesis (1956, 1961). We follow Bernstein and Zelevinsky (1976) in emphasizing these tools. In Section 4.4, we prove the uniqueness of Whittaker models, a fundamental result that was applied in Section 3.5 to the multiplicity one theorem for automorphic forms and to the construction of L-functions. We also introduce the Jacquet module, a functor adjoint to parabolic induction, and establish its basic properties. In Section 4.5, we construct the principal series representations by induction from a Borel subgroup, and in Section 4.6, we study the spherical representations, which are those that admit a vector invariant by the maximal compact subgroup. In Section 4.7, we prove the "local functional equations," which were needed in Chapter 3 in the study of the L-function of an automorphic representation. Following Jacquet and Langlands (1970), Section 4.8 constructs many supercuspidal representations by means of the Weil representation, and Section 4.9 describes the local Langlands correspondence. This is a classification of the irreducible representations of $GL(n, F)$, which is proved if $n = 2$ or 3 or in the case of positive characteristic but is still conjectural in characteristic zero if $n > 3$. In Section 4.8, we show that if the residue characteristic is odd, then the Langlands correspondence (which we will not prove or even precisely state) shows that the picture is the same as over finite fields: The irreducible representations are roughly parametrized by the characters of tori. This is false in the case of even residue characteristic.

4.1 Representations of $GL(2)$ Over a Finite Field

There are parallels between the representation theory of $GL(n)$ or any other reductive algebraic group in different contexts: One may work over a finite field, a local field, or the adele ring of a global field. Because many of the most important concepts persist at each level, we will begin with the representation theory of $GL(2)$ over a finite field. In this section, we will give complete proofs using methods that generalize (with modification) to local fields and adele rings. For further readings on this subject, we recommend three books that have influenced our treatment, Piatetski-Shapiro (1983), Howe and Moy (1985), and Zelevinsky (1981).

All representations in this section will be finite-dimensional representations over the complex numbers. If (π, V) is a representation of G, then by definition V is a complex vector space and $\pi : G \to GL(V)$ is a homomorphism. We also refer to V as a *G-module*. If (π, V) and (π', V') are representations, then an *intertwining operator* or *G-module homomorphism* is a linear map $\phi : V \to V'$ such that $\phi \circ \pi(g) = \pi'(g) \circ \phi$ for all $g \in G$. We will also use the term *equivariant* or *G-equivariant* to indicate that a map $\phi : V \to V'$ satisfies $\phi \circ \pi(g) = \pi'(g) \circ \phi$.

We denote the space of G-module homomorphisms interchangeably as $\mathrm{Hom}_G(V, V')$ or $\mathrm{Hom}_G(\pi, \pi')$.

If (π, V) is an irreducible representation of the finite group G, then by Schur's lemma, the center of G acts by scalars. In particular, if $G = GL(n, F)$ and F is a finite field, the center of G consists of the group of scalar matrices, isomorphic to $F^\times$, and hence there exists a character $\omega : F^\times \to \mathbb{C}^\times$, known as the *central character* of π, such that

$$\pi \begin{pmatrix} z & & \\ & \ddots & \\ & & z \end{pmatrix} v = \omega(z)\, v \tag{1.1}$$

for all $z \in F^\times, v \in V$.

Let (π, V) be a representation of the finite group G. Let V^* be the dual space of V; if $v \in V$ and $l \in V^*$, we will sometimes denote $\langle v, l \rangle = l(v)$ to emphasize the symmetry between V and V^*. We then have a representation $\hat{\pi}$ on V^*, called the *contragredient representation* of π, defined by

$$\langle v, \hat{\pi}(g)\, l \rangle = \left\langle \pi(g^{-1})\, v, l \right\rangle .$$

If $G = GL(n, F)$, the contragredient representation has the following useful realization.

Proposition 4.1.1 *Let $G = GL(n, F)$, F finite, and let (π, V) be a representation of G.*

(i) Define a representation (π_1, V) on the same space by $\pi_1(g) = \pi\left({}^\top g^{-1}\right)$. Then $\hat{\pi} \cong \pi_1$.
(ii) Suppose that $n = 2$ and that (π, V) is irreducible. Let ω be the central character of π. Define another representation (π_2, V) on the same space by

$$\pi_2(g) = \omega\big(\det(g)\big)^{-1}\,\pi(g).$$

Then $\hat{\pi} \cong \pi_2$.

Proof The idea for both parts is to compare the characters of the two representations in question. We note that if χ is the character of π, then $g \mapsto \chi(g^{-1}) = \overline{\chi(g)}$ is the character of $\hat{\pi}$. For (i), we make use of the fact that in $GL(n, F)$, every matrix is conjugate to its transpose (Exercise 1.3); because the character is preserved under conjugation, we see that the character of π_1 is also $g \mapsto \chi(g^{-1})$, and so the representations are isomorphic.

For (ii), let $g = \left(\begin{smallmatrix} a & b \\ c & d \end{smallmatrix}\right) \in GL(2, F)$ have determinant D. We note that

$$g^{-1} = \frac{1}{D}\begin{pmatrix} d & -b \\ -c & a \end{pmatrix} = \begin{pmatrix} D^{-1} & \\ & D^{-1} \end{pmatrix}\begin{pmatrix} & 1 \\ -1 & \end{pmatrix}{}^\top g \begin{pmatrix} & 1 \\ -1 & \end{pmatrix}^{-1},$$

from which we see that $\hat{\pi}$ and π_2 have the same character. ■

Let (π, V) be a representation of a subgroup H of G. We will take the following definition of the *induced representation* (π^G, V^G) of G. Let V^G be the vector space of all functions $f : G \to V$ that satisfy $f(hg) = \pi(h)\,f(g)$ for $h \in H$, $g \in G$. The action of G on such functions is by *right translation*: We define

$$\big(\pi^G(g)\,f\big)(x) = f(xg). \tag{1.2}$$

Now suppose that (τ, U) is a representation of G and (π, V) is a representation of H. Because $H \subset G$, we may regard U as an H-module by restricting the action; let us denote this H-module as U_H. The *Frobenius reciprocity law* amounts to a natural isomorphism

$$\mathrm{Hom}_G(U, V^G) \cong \mathrm{Hom}_H(U_H, V). \tag{1.3}$$

In this isomorphism, we let $\phi \in \mathrm{Hom}_G(U, V^G)$ correspond with $\phi' \in \mathrm{Hom}_H(U_H, V)$, where (for $u \in U$) we have $\phi'(u) = \phi(u)(1)$ and, conversely, $\phi(u)(g) = \phi'(gu)$ (Exercise 4.1.1).

An important tool is the *Mackey theory*, which is the science of intertwining operators between induced representations. Let G be a finite group, let H_1 and H_2 be subgroups, and let (π_1, V_1) and (π_2, V_2) be representations of H_1 and H_2, respectively. An important problem is to "calculate" $\mathrm{Hom}_G(V_1^G, V_2^G)$. It will turn out that this space can be identified with something fairly concrete and amenable to computation.

Let Δ be a map $G \to \mathrm{Hom}_{\mathbb{C}}(V_1, V_2)$, and let $f : G \to V_1$ be a map. We define the convolution $\Delta * f : G \to V_2$ to be the map

$$(\Delta * f)(x) = \frac{1}{|G|} \sum_{g \in G} \Delta(xg^{-1})\big(f(g)\big).$$

(Here $|G|$ is the cardinality of G.)

Proposition 4.1.2: Mackey *The space* $\mathrm{Hom}_G(V_1^G, V_2^G)$ *is isomorphic as a vector space to the space of functions* $\Delta : G \to \mathrm{Hom}_{\mathbb{C}}(V_1, V_2)$ *such that for* $h_1 \in H_1$, $h_2 \in H_2$, *and* $g \in G$, *we have*

$$\Delta(h_2 g h_1) = \pi_2(h_2) \circ \Delta(g) \circ \pi_1(h_1). \tag{1.4}$$

Given such a function Δ, *the corresponding intertwining operator* $L \in \mathrm{Hom}_G(V_1^G, V_2^G)$ *is given by* $L(f_1) = \Delta * f_1$ *for* $f_1 \in V_1^G$.

It is possible to refine this result into an "algorithm" for computing $\mathrm{Hom}_G(V_1^G, V_2^G)$. For this, we refer to Exercise 4.1.2. Bruhat (1956, 1961), extended this technique to allow the computation of intertwining operators between induced representations of Lie groups and p-adic groups – for this, the function Δ becomes a distribution.

Proof Let $\mathcal{D}$ be the vector space of functions Δ satisfying Eq. (1.4).

It follows readily from the definitions that given such a Δ, if $f_1 \in V_1^G$, then $\Delta * f_1 \in V_2^G$, and $L : V_1^G \to V_2^G$ defined by $L(f_1) = \Delta * f_1$ is an intertwining operator. Hence we have a linear map $\mathcal{D} \to \mathrm{Hom}_G(V_1^G, V_2^G)$.

Conversely, we construct an inverse mapping $\mathrm{Hom}_G(V_1^G, V_2^G) \to \mathcal{D}$. Let us define a collection $f_{g,v}$ of elements of V_1^G indexed by $g \in G$, $v \in V_1$. We define

$$f_{g,v}(x) = \begin{cases} \pi(h)\, v & \text{if } x = hg, h \in H_1; \\ 0 & \text{if } x \notin H_1 g. \end{cases}$$

One easily checks that (with $L \in \mathrm{Hom}_G(V_1^G, V_2^G)$ defined by $L(f_1) = \Delta * f_1$ as above) we have, for $v \in V_1$, the identity

$$\Delta(g)(v) = [G : H_1]\, L(f_{g^{-1},v})(1). \tag{1.5}$$

We leave it to the reader to check that for any $L \in \mathrm{Hom}_G(V_1^G, V_2^G)$, Eq. (1.5) defines an element of $\mathcal{D}$ and that the two maps between $\mathcal{D}$ and $\mathrm{Hom}_G(V_1^G, V_2^G)$ defined above are inverses of each other. ■

We will use Proposition 4.1.2 to study the representation theory of $GL(2, F)$, where $F = \mathbb{F}_q$ is the finite field with q elements, where q is a power of some prime p. The methods we will use are applicable in the case where F is a local field but, we hasten to caution, *not without modification.*

Let F be a field and T an algebraic group defined over F. (For this discussion, let F be any field. We will presently return to the case where F is finite.) We call T a *torus* if over an algebraic closure of F it becomes isomorphic to a direct product of copies of the multiplicative group. The torus is called *split* if the isomorphism is defined over F (rather than just over the algebraic closure), so that $T(F)$ is itself isomorphic to a direct product of copies of $F^\times$.

For example, inside of $GL(2)$ there is the subgroup T_s consisting of diagonal elements. Obviously, if Ω is any field containing F, $T(\Omega) \cong (\Omega^\times)^2$, so T_s is a torus. It is *maximal* in the sense that it cannot be embedded into any properly larger torus. It is also clearly split. It may be shown that any maximal split torus is conjugate to this one.

To give examples of nonsplit tori, suppose that E/F is a quadratic extension. We can find an algebraic subgroup T_a of $GL(2)$ such that $T_a(F) \cong E^\times$. To do this, let us assume for simplicity that the characteristic of E is not two. Then $E = F(\sqrt{D})$ for some $D \in E$. We have

$$E^\times = \{x + y\sqrt{D} | x, y \in F, \text{not both zero}\}.$$

Let T_a be the affine algebraic group consisting of matrices

$$\left\{ \begin{pmatrix} x & y \\ Dy & x \end{pmatrix} \middle| x^2 - Dy^2 \neq 0 \right\}.$$

Then $T_a(F) \cong E^\times$; indeed, we associate $x + y\sqrt{D} \in E^\times$ with x and $y \in F$ with the matrix $\left(\begin{smallmatrix} x & y \\ Dy & x \end{smallmatrix}\right) \in T_a(F)$.

Why is T_a a torus? Suppose that Ω is an algebraic closure of F or indeed any field containing E. Then we claim that there is an algebraic isomorphism $\mu : T_a(\Omega) \cong T_s(\Omega)$. Indeed, the isomorphism is given by

$$\mu \begin{pmatrix} x & y \\ Dy & x \end{pmatrix} = \begin{pmatrix} x + y\sqrt{D} & 0 \\ 0 & x - y\sqrt{D} \end{pmatrix}.$$

To show that this is an isomorphism, we note that $x + y\sqrt{D} = \xi$ and $x - y\sqrt{D} = \eta$ are equivalent to

$$x = \tfrac{1}{2}(\xi + \eta), \qquad y = \tfrac{1}{\sqrt{D}}(\xi - \eta).$$

Thus the groups T_a and T_s, though not isomorphic over F, become isomorphic over an algebraic extension. These are regarded as *forms* of the same group.

Returning to the case where $F = \mathbb{F}_q$ is a finite field, the basic paradigm is that *the irreducible representations of $GL(2, F)$ are (roughly) parametrized by the characters of its maximal tori.*

Let us first show how certain irreducible representations of $GL(2,F)$ are parametrized by characters of the split torus $T = T_s$ of diagonal matrices. Let χ_1 and χ_2 be characters of $F^\times$. Then we get a character χ of $T(F)$ defined by

$$\chi\begin{pmatrix} y_1 & \\ & y_2 \end{pmatrix} = \chi_1(y_1)\,\chi_2(y_2).$$

Let B be the "Borel subgroup" consisting of upper triangular matrices. Thus $B = TN$, where N is the group of upper triangular unipotent matrices. We extend χ to a character of $B(F)$ by letting $N(F)$ lie in the kernel. (Note: we will use the same letter N to denote the norm mapping $K \to F$, where K is a quadratic extension of F; we hope that this causes no confusion.) Thus

$$\chi\begin{pmatrix} y_1 & x \\ & y_2 \end{pmatrix} = \chi_1(y_1)\,\chi_2(y_2). \tag{1.6}$$

Let $\mathcal{B}(\chi_1, \chi_2)$ be the representation of $GL(2,F)$ induced from this character of $B(F)$.

We remind the reader of the *Bruhat decomposition* for $GL(2)$. This is the identity

$$GL(2,F) = B(F) \cup B(F)\,w_0\,B(F), \quad \text{(disjoint)}, \quad w_0 = \begin{pmatrix} & -1 \\ 1 & \end{pmatrix}. \tag{1.7}$$

Note that $B(F)$ is really a double coset $B(F)\,1\,B(F)$, so Eq. (1.7) tells us that the space of double cosets $B(F)\backslash GL(2,F)/B(F)$ has cardinality two and that $\{1, w_0\}$ are a complete set of representatives for these double cosets. To prove Eq. (1.7), a matrix $\left(\begin{smallmatrix} a & b \\ c & d \end{smallmatrix}\right)$ lies in $B(F)$ if $c = 0$; otherwise

$$\begin{pmatrix} a & b \\ c & d \end{pmatrix} = \begin{pmatrix} 1 & a/c \\ & 1 \end{pmatrix} w_0 \begin{pmatrix} c & d \\ & Dc^{-1} \end{pmatrix} \in B(F)\,w_0\,B(F), \qquad D = ad - bc.$$

Lemma 4.1.1 *Let χ_1, χ_2, μ_1 and μ_2 be characters of $F^\times$. Then*

$$\dim \operatorname{Hom}_{GL(2,F)}\big(\mathcal{B}(\chi_1,\chi_2), \mathcal{B}(\mu_1,\mu_2)\big) = e_1 + e_2,$$

where

$$e_1 = \begin{cases} 1 & \textit{if } \chi_1 = \mu_1 \textit{ and } \chi_2 = \mu_2; \\ 0 & \textit{otherwise,} \end{cases}$$

and

$$e_2 = \begin{cases} 1 & \textit{if } \chi_1 = \mu_2 \textit{ and } \chi_2 = \mu_1; \\ 0 & \textit{otherwise.} \end{cases}$$

Proof Let χ and μ be the characters of $B(F)$ obtained from χ_1, χ_2 and from μ_1, μ_2 by Eq. (1.6). We regard χ and μ as representations of $B(F)$ acting on the one-dimensional space $\mathbb{C}$; we may identify $\operatorname{Hom}_{\mathbb{C}}(\mathbb{C},\mathbb{C})$ with $\mathbb{C}$ also.

Thus using Proposition 4.1.2, we must compute the dimension of the space of functions $\Delta : GL(2, F) \to \mathbb{C}$ such that

$$\Delta(b_2 g b_1) = \mu(b_2)\, \Delta(g)\, \chi(b_1), \qquad b_i \in B(F). \tag{1.8}$$

It follows from the Bruhat decomposition (Eq. 1.7) that Δ is determined by its values on 1 and w_0. We show that $\Delta(1) = 0$ if $e_1 \neq 1$. Indeed, if $\chi_1 \neq \mu_1$ or $\chi_2 \neq \mu_2$, we can find $t \in T(F)$ such that $\chi(t) \neq \mu(t)$; then $\Delta(1) = \Delta(t \cdot 1 \cdot t^{-1}) = \chi(t)\, \Delta(1)\, \mu(t)^{-1}$, which clearly implies that $\Delta(1) = 0$. Similarly, one shows that $\Delta(w_0) = 0$ if $e_2 \neq 1$.

If $e_1 = 1$, we may construct a function Δ_1 satisfying Eq. (1.8) and having support just in the double coset $B(F) = B(F)\, 1\, B(F)$ by

$$\Delta_1(g) = \begin{cases} \chi(g) = \mu(g) & \text{if } g \in B(F); \\ 0 & \text{otherwise.} \end{cases}$$

If $e_1 = 0$, we take Δ_1 to be zero. Similarly, if $e_2 = 1$, we define a function Δ_2 satisfying Eq. (1.8) and having support within the other double coset, defined by

$$\Delta_2(g) = \begin{cases} \chi(b_1)\mu(b_2) & \text{if } g = b_2 w_0 b_1,\ b_i \in B(F); \\ 0 & \text{otherwise,} \end{cases}$$

and $e_2 \neq 0$ implies that this is well defined. If $e_2 = 0$, we take $\Delta_2 = 0$. Now an arbitrary function satisfying Eq. (1.8) can clearly be expressed as a linear combination of Δ_1 and Δ_2, so we see that the dimension of the space of such functions is indeed $e_1 + e_2$. ■

Theorem 4.1.1 *Let χ_1, χ_2, μ_1, and μ_2 be characters of $F^\times$. Then $\mathcal{B}(\chi_1, \chi_2)$ is an irreducible representation of degree $q+1$ of $GL(2, F)$ unless $\chi_1 = \chi_2$, in which case it is the direct sum of two irreducible representations having degrees 1 and q. We have*

$$\mathcal{B}(\chi_1, \chi_2) \cong \mathcal{B}(\mu_1, \mu_2)$$

if and only if either

$$\chi_1 = \mu_1 \text{ and } \chi_2 = \mu_2 \tag{1.9}$$

or else

$$\chi_1 = \mu_2 \text{ and } \chi_2 = \mu_1. \tag{1.10}$$

We will see in the proof of this that the irreducible representation of dimension one contained in $\mathcal{B}(\chi, \chi)$ is the character $g \mapsto \chi\big(\det(g)\big)$. The other q-dimensional representation is obtained by tensoring this character with the q-dimensional irreducible subrepresentation of $\mathcal{B}(1, 1)$; this q-dimensional representation is called the *Steinberg representation.* In the theory of $GL(2, F)$, where F is p-adic, the representation analogous to the Steinberg representation is called the *special representation*. By abuse of terminology, we may also call the q-dimensional irreducible representation contained in $\mathcal{B}(\chi, \chi)$ a Steinberg representation, although this is not truly correct if $\chi \neq 1$. The irreducible representations $\mathcal{B}(\chi_1, \chi_2)$ with $\chi_1 \neq \chi_2$ are called representations of the *principal series*.

Proof First we apply Lemma 4.1.1 with $\chi_1 = \mu_1$ and $\chi_2 = \mu_2$. We see that the dimension of the space of endomorphisms

$$\mathrm{End}_{GL(n,F)}\big(\mathcal{B}(\chi_1, \chi_2)\big) = \begin{cases} 1 & \text{if } \chi_1 \neq \chi_2; \\ 2 & \text{if } \chi_1 = \chi_2. \end{cases}$$

Now if (π, V) is a representation of a finite group G, and if V is a direct sum of distinct irreducible representations $\pi_1, \cdots, \pi_h$ with multiplicities $d_1, \cdots, d_h$, then the dimension $\mathrm{End}_G(V)$ is $\sum d_i^2$; see Exercise 4.1.1 (b). Hence $\mathcal{B}(\chi_1, \chi_2)$ is irreducible if $\chi_1 \neq \chi_2$, and otherwise, it is the direct sum of two irreducible representations, because $2 = 1^2 + 1^2$ is the only way of writing $\epsilon_1 + \epsilon_2 = 2$ as a sum of squares!

Because the index of $B(F)$ in $GL(2, F)$ is $q + 1$, the degree of $\mathcal{B}(\chi_1, \chi_2)$ is $q + 1$. If $\chi_1 = \chi_2$, we may exhibit an invariant subspace of dimension one, namely, the function $f(g) = \chi\big(\det(g)\big)$ clearly satisfies $f(bg) = \chi(b) f(g)$ for $b \in B(F)$ and hence lies in the space of $\mathcal{B}(\chi_1, \chi_2)$. Hence one of the two irreducible components of $\mathcal{B}(\chi_1, \chi_2)$ is one dimensional and affords the character $g \mapsto \chi\big(\det(g)\big)$ of $GL(2, F)$. The other component is therefore q dimensional.

Once we know that $\mathcal{B}(\chi_1, \chi_2)$ is usually irreducible, the question of when $\mathcal{B}(\chi_1, \chi_2) \cong \mathcal{B}(\mu_1, \mu_2)$ is answered by Lemma 4.1.1: If a nonzero intertwining operator between these spaces exists, then it is an isomorphism. (The case $\chi_1 = \chi_2$ must be handled separately.) Lemma 4.1.1 describes the possible cases where this can occur. ■

We wish to know explicitly the isomorphism $\mathcal{B}(\chi_1, \chi_2) \cong \mathcal{B}(\chi_2, \chi_1)$ when $\chi_1 \neq \chi_2$. Unraveling the definitions gives this explicit intertwining operator: If f lies in the space of $\mathcal{B}(\chi_1, \chi_2)$, then consider

$$(Tf)(g) = \sum_{x \in F} f\left(\begin{pmatrix} & -1 \\ 1 & \end{pmatrix}\begin{pmatrix} 1 & x \\ & 1 \end{pmatrix} g\right). \tag{1.11}$$

Then Tf lies in the space of $\mathcal{B}(\chi_2, \chi_1)$, and T is a nonzero intertwining operator. Analogous intertwining operators occur in the theory of $GL(2, F)$ when F is local; their introduction may be motivated by the theory of the constant terms of Eisenstein series. We study these intertwining operators in Sections 2.6, 3.7 and 4.5.

We have mentioned that characters of the other torus $T_{\mathrm{a}}(F)$ also give rise to representations of $GL(2, F)$. These cannot be constructed by induction, so another procedure is required. The construction of the irreducible representations of $GL(n, F)$ with F as a finite field was solved by Green (1955). The most satisfactory method of parametrizing representations of finite groups of Lie type by characters of their tori is due to Deligne and Lusztig (1976). However, this work uses deep methods from algebraic geometry and has not been generalized to groups over local fields. We will use an alternative method that works well for $GL(2, F)$ when F is finite or local and even constructs automorphic representations of $GL(2)$ over the adele ring of a global field. Its generalizations, though important, do not, however, solve the corresponding problem of parametrizing irreducible representations of $GL(n, F)$ by characters of its tori when $n > 2$.

The method that we will use is called the *Weil representation*, introduced (in far greater generality than we will use) in the great paper of Weil (1964). As we will explain it in this section, the Weil representation will appear as an *ad hoc* unmotivated miracle. This is because we are suppressing the underlying mechanism of the Heisenberg group and finite "Stone-Von Neumann theorem" in the interests of expedience – we refer to the exercises and to Section 4.8 for this aspect.

Let E be a two-dimensional commutative semisimple algebra over F. There are precisely two nonisomorphic possibilities for E: either $E = F \oplus F$ with F embedded diagonally (the "split case"), or E is the unique quadratic field extension of F (the "anisotropic case"). First we construct a representation of $SL(2, F)$, which we will later extend to $GL(2, F)$. We also fix a torus T_0 of F: It is T_s in the split case and T_a in the anisotropic case. Thus $T_0(F) \cong E^\times$. We fix a nontrivial character ψ of the additive group F. Although the representations of $SL(2, F)$ that we construct depend on ψ, we will suppress this dependence from our notation.

We require generators and relations for $SL(2, F)$.

Lemma 4.1.2 *Let F be a field. Let S be the group generated by elements $t(y)$, $n(z)$, and w_1 with $y \in F^\times$ and $z \in F$, subject to the following relations:*

$$t(y_1)\,t(y_2) = t(y_1 y_2); \qquad n(z_1)\,n(z_2) = n(z_1 + z_2); \tag{1.12}$$

$$t(y)\,n(z)\,t(y)^{-1} = n(y^2 z); \qquad w_1 t(y) w_1 = t(-y^{-1}); \tag{1.13}$$

and

$$w_1\,n(z)\,w_1 = t(-z^{-1})\,n(-z)\,w_1\,n(-z^{-1}), \quad (z \neq 0). \tag{1.14}$$

Then S is isomorphic to $SL(2, F)$; in this isomorphism

$$t(y) \mapsto \begin{pmatrix} y & \\ & y^{-1} \end{pmatrix}, \qquad n(z) \mapsto \begin{pmatrix} 1 & z \\ & 1 \end{pmatrix}, \qquad w_1 \mapsto \begin{pmatrix} & 1 \\ -1 & \end{pmatrix}.$$

We recall what it means for S to be the group generated by $t(y)$, $n(z)$, and w_1 subject to Eqs. (1.12–1.14). This means that S contains elements $t(y)$, $z(z)$, and w_1 as described and that if S' is another group containing elements $t'(y)$, $z'(z)$, and w_1' satisfying the same relations (1.12–1.14), for example

$$t'(y_1)\,t'(y_2) = t'(y_1 y_2), \qquad n'(z_1)\,n'(z_2) = n'(z_1 + z_2),$$

etc., then there exists a unique homomorphism $\phi : S \to S'$ such that $t'(y) = \phi\big(t(y)\big), n'(z) = \phi\big(n(z)\big)$, and $w_1' = \phi(w_1)$. We call this property the *universal property* of S, and by a typical purely formal argument, the universal property is sufficient to describe the group up to isomorphism. A group S with this universal property may be constructed as follows: Let $\tilde{S}$ be the free group with generators $\tilde{t}(y)$ ($y \in F^\times$), $\tilde{n}(z)$ ($z \in F$), and $\tilde{w}_1$ (cf. Lang (1993)). Let R be the smallest normal subgroup of $\tilde{S}$ that contains

$$\tilde{t}(y_1)\,\tilde{t}(y_2)\,\tilde{t}(y_1 y_2)^{-1}, \qquad \tilde{n}(z_1)\,\tilde{n}(z_2)\,\tilde{n}(z_1+z_2)^{-1};$$
$$\tilde{t}(y)\,\tilde{n}(z)\,\tilde{t}(y)^{-1}\,\tilde{n}(y^2 z)^{-1}; \qquad \tilde{w}_1 \tilde{t}(y) \tilde{w}_1\,\tilde{t}(-y^{-1})^{-1};$$
$$\tilde{w}_1\,\tilde{n}(z)\,\tilde{w}_1\,\tilde{n}(-z^{-1})^{-1}\,\tilde{w}_1^{-1}\,\tilde{n}(-z)^{-1}\,\tilde{t}(-z^{-1})^{-1}, \quad (z \neq 0).$$

Then it is easy to see that S'/R has the required universal property.

Proof Let

$$t'(y) = \begin{pmatrix} y & \\ & y^{-1} \end{pmatrix}, \quad n'(z) = \begin{pmatrix} 1 & z \\ & 1 \end{pmatrix}, \quad w_1' = \begin{pmatrix} & 1 \\ -1 & \end{pmatrix} \in SL(2, F).$$

These elements satisfy the relations (1.12–1.14), so by the universal property of S, there exists a homomorphism $\phi : S \to SL(2, F)$ such that $\phi\big(t(y)\big) = t'(y)$, etc., and what we must show is that ϕ is an isomorphism. We will accomplish this by constructing an inverse map $\psi : SL(2, F) \to S$. We define

$$\psi\begin{pmatrix} a & b \\ c & d \end{pmatrix} = \begin{cases} n(a/c)\,t(-c^{-1})\,w_1\,n(d/c) & \text{if } c \neq 0; \\ t(a)\,n(b/a) & \text{if } c = 0. \end{cases}$$

We check that ψ is a homomorphism; that is, if

$$\begin{pmatrix} a & b \\ c & d \end{pmatrix}\begin{pmatrix} A & B \\ C & D \end{pmatrix} = \begin{pmatrix} \alpha & \beta \\ \gamma & \delta \end{pmatrix},$$

we must check that

$$\psi\begin{pmatrix} a & b \\ c & d \end{pmatrix}\psi\begin{pmatrix} A & B \\ C & D \end{pmatrix} = \psi\begin{pmatrix} \alpha & \beta \\ \gamma & \delta \end{pmatrix}. \tag{1.15}$$

There are several cases; we will check this if c, C, and γ are all nonzero and leave the remaining cases to the reader. The left side equals

$$\begin{aligned} &n(a/c)\,t(-c^{-1})\,w_1\,n(d/c)\,n(A/C)\,t(-C^{-1})\,w_1\,n(D/C) \\ &= n(a/c)\,t(-c^{-1})\,w_1\,n(d/c)\,n(A/C)\,w_1\,t(-C)\,n(D/C) \\ &= n(a/c)\,t(-c^{-1})\,w_1\,n(\gamma c^{-1} C^{-1})\,w_1\,t(-C)\,n(D/C). \end{aligned}$$

Using Eq. (1.14), this equals

$$n(a/c)\,t(-c^{-1})\,t(-cC/\gamma)\,n(-\gamma/cC)\,w_1\,n(-cC/\gamma)\,t(-C)\,n(D/C).$$

Making use of Eq. (1.13) and the identities

$$a-\frac{C}{\gamma}=\frac{a(cA+dC)-C(ad-bc)}{\gamma}=\frac{c\alpha}{\gamma},$$

$$D-\frac{c}{\gamma}=\frac{D(cA+dC)-c(AD-BC)}{\gamma}=\frac{C\delta}{\gamma},$$

this may be simplified to obtain Eq. (1.15).

Now let us note that $\psi\circ\phi$ is the identity map on S; indeed, because ϕ and ψ are homomorphisms, it is sufficient to check this on generators, which is easily done. It follows that ϕ is injective. It is easy to see that $SL(2, F)$ is generated by the image of ϕ, and so ϕ is an isomorphism. ∎

Let $x\to\overline{x}$ be the automorphism $(\xi,\eta)\to(\eta,\xi)$ of E if $E=F\oplus F$, or let $x\to\overline{x}$ be the nontrivial Galois automorphism of E/F if E is a field. Let tr, $N: E\to F$ be the trace and norm maps, so $\mathrm{tr}(x)=x+\overline{x}$, $N(x)=x\,\overline{x}$. (We use the same letter N to denote the algebraic group of 2×2 upper triangular unipotent matrices in $GL(2)$; it should always be clear from the context which use we have in mind.) If Φ is a function on E, the *Fourier transform* $\Phi(x)$ is defined by

$$\widehat{\Phi}(x)=\epsilon\,q^{-1}\sum_{y\in E}\Phi(y)\,\psi\big(\mathrm{tr}(\overline{x}y)\big),$$

where

$$\epsilon=\begin{cases}1 & \text{in the split case,}\\ -1 & \text{in the anisotropic case.}\end{cases}$$

Now let W be the q^2-dimensional space of all complex-valued functions on E.

Proposition 4.1.3: Weil representation for SL(2,F) *There exists a representation $\omega: SL(2, F)\to End(W)$ such that*

$$\left(\omega\begin{pmatrix}a & \\ & a^{-1}\end{pmatrix}\Phi\right)(x)=\Phi(ax),$$

$$\left(\omega\begin{pmatrix}1 & z\\ & 1\end{pmatrix}\Phi\right)(x)=\psi\big(z\,N(x)\big)\,\Phi(x),$$

and

$$\left(\omega\begin{pmatrix} & 1\\ -1 & \end{pmatrix}\Phi\right)(x)=\widehat{\Phi}(x).$$

Proof To use Lemma 4.1.2, one must verify consistency relations corresponding to Eqs. (1.12), (1.13), and (1.14). Actually only Eq. (1.14) is difficult; we will prove it and leave the remaining cases to the reader. It is worth remarking that if $y = 1$, the second formula in Eq. (1.13) is the Fourier inversion formula

$$\hat{\hat{\Phi}}(x) = \Phi(-x).$$

We must prove that

$$\omega\begin{pmatrix} & 1 \\ -1 & \end{pmatrix}\omega\begin{pmatrix} 1 & a \\ & 1 \end{pmatrix}\omega\begin{pmatrix} & 1 \\ -1 & \end{pmatrix}$$
$$= \omega\begin{pmatrix} -a^{-1} & \\ & -a \end{pmatrix}\omega\begin{pmatrix} 1 & -a \\ & 1 \end{pmatrix}\omega\begin{pmatrix} & 1 \\ -1 & \end{pmatrix}\omega\begin{pmatrix} 1 & -a^{-1} \\ & 1 \end{pmatrix}. \qquad (1.16)$$

For $x \in F$, it is easily checked that the number of $y \in E$ such that $N(y) = x$ is

$$\begin{cases} q + (q-1)\epsilon & \text{if } x = 0; \\ q - \epsilon & \text{if } x \neq 0. \end{cases}$$

This implies that $\sum_{y \in E} \psi\big(N(y)\big) = \epsilon q$. More generally, we have, for $a \in F^\times$, $b \in E$

$$\sum_{y \in E} \psi\big(a\,N(y) + \mathrm{tr}(\overline{b}y)\big) = \epsilon q\, \psi\big(-a^{-1}N(b)\big). \qquad (1.17)$$

To prove this, find $A \in E$ such that $N(A) = a$. Completing the square

$$\sum_y \psi\big(a\,N(y) + \mathrm{tr}(\overline{b}y) + a^{-1}N(b)\big) = \sum_y \psi\big((Ay + b/\overline{A})(\overline{A}\overline{y} + \overline{b}/A)\big),$$

and on making a change of variables, this equals $\sum_{y \in E} \psi\big(N(y)\big) = \epsilon q$. Thus Eq. (1.17) follows from the special case, already established, where $a = 1$ and $b = 0$.

Now, unfolding the definitions on the left side of Eq. (1.16), we have

$$\left(\omega\begin{pmatrix} & 1 \\ -1 & \end{pmatrix}\omega\begin{pmatrix} 1 & a \\ & 1 \end{pmatrix}\omega\begin{pmatrix} & 1 \\ -1 & \end{pmatrix}\Phi\right)(x)$$
$$= \frac{1}{q^2}\sum_{y,z \in E} \psi\big(a\,N(y)\big)\,\psi\big(\mathrm{tr}(\overline{x}y)\big)\,\psi\big(\mathrm{tr}(\overline{z}y)\big)\,\Phi(z),$$

which by Eq. (1.17) equals $\epsilon q^{-1}\sum_z \psi\big(-a^{-1}N(x+z)\big)\,\Phi(z)$. On the other hand

$$\left(\omega\begin{pmatrix} -a^{-1} & \\ & -a \end{pmatrix}\omega\begin{pmatrix} 1 & -a \\ & 1 \end{pmatrix}\omega\begin{pmatrix} & 1 \\ -1 & \end{pmatrix}\omega\begin{pmatrix} 1 & -a^{-1} \\ & 1 \end{pmatrix}\Phi\right)(x)$$
$$= \epsilon q^{-1}\sum_z \psi\big(-a^{-1}N(x)\big)\,\psi\big(-a^{-1}N(z)\big)\,\psi\big(-a^{-1}\,\mathrm{tr}(x\overline{z})\big)\,\Phi(z).$$

The two expressions are equal, so the defining relations of the Weil representation are compatible. ■

Now let χ be a character of $E^\times$. Because $T_0(F) \cong E^\times$, this is the same as a character of $T_0(F)$; in view of our philosophy of parametrizing the irreducible representations of $GL(2, F)$ by characters of its tori, we will associate a representation $\big(\pi(\chi), W(\chi)\big)$ of $GL(2, F)$ with χ. *We assume that χ does not factor through $N : E^\times \to F^\times$.* Let $E_1^\times$ be the subgroup of elements x of $E^\times$ such that $N(x) = 1$. Our assumption is equivalent to assuming that $\chi | E_1^\times$ is nontrivial. Without this assumption, we could still construct the representation $\pi(\chi)$, but it would not be irreducible.

Let

$$W(\chi) = \{\Phi \in W | \Phi(yx) = \chi(y)^{-1}\,\Phi(x) \text{ for } y \in E_1^\times\}.$$

Evidently

$$\dim\ W(\chi) = \begin{cases} q-1 & \text{if } T_0 = T_{\text{a}}; \\ q+1 & \text{if } T_0 = T_{\text{s}}. \end{cases}$$

One easily checks that the space $W(\chi)$ is stable under the Weil representation of $SL(2, F)$. We extend the action of $SL(2, F)$ on $W(\chi)$ to a representation of $GL(2, F)$ by letting

$$\left(\omega\begin{pmatrix} a & \\ & 1\end{pmatrix}\Phi\right)(x) = \chi(b)\,\Phi(bx),$$

where, given $a \in F^\times$, $b \in E^\times$ is chosen so that $N(b) = a$. It must be checked that this is consistent with the definitions in Proposition 4.1.3: What must be shown is that

$$\omega\begin{pmatrix} a & \\ & 1\end{pmatrix}\omega(g)\,\omega\begin{pmatrix} a & \\ & 1\end{pmatrix}^{-1} = \omega\left(\begin{pmatrix} a & \\ & 1\end{pmatrix} g \begin{pmatrix} a & \\ & 1\end{pmatrix}^{-1}\right)$$

when g is one of the generators in $SL(2, F)$; we leave this verification to the reader.

Suppose that $T_0 = T_{\text{s}}$. Then χ is a character of $E = F \oplus F$, so there exist characters χ_1 and χ_2 of $F^\times$ such that $\chi(x_1, x_2) = \chi_1(x_1)\,\chi_2(x_1)$. By abuse of notation, we will also use the same letter χ to denote the character (Eq. 1.6).

Proposition 4.1.4 *In this situation, the Weil representation*

$$\big(\pi(\chi), W(\chi)\big) \cong \mathcal{B}(\chi_1, \chi_2).$$

Proof Our hypothesis that χ is nontrivial on the elements of norm one in $E = F \oplus F$ means that $\chi_1 \neq \chi_2$, so $\mathcal{B}(\chi_1, \chi_2)$ is irreducible by Theorem 4.1.1

and isomorphic to $\mathcal{B}(\chi_2, \chi_1)$. Define $\tau : W(\chi) \to \mathbb{C}$ by $\tau(\Phi) = \Phi(1, 0)$. We easily check that

$$\tau\left(\omega\begin{pmatrix} y_1 & x \\ & y_2 \end{pmatrix}\Phi\right) = \chi_2(y_1)\,\chi_1(y_2)\,\tau(\Phi).$$

It follows that we may define an intertwining operator $L : W(\chi) \to \mathcal{B}(\chi_2, \chi_1)$ by

$$(L\Phi)(g) = \tau\big(\omega(g)\,\Phi\big).$$

To see that this is an isomorphism, observe that the two spaces in question have the same dimension $q + 1$ and that $\mathcal{B}(\chi_2, \chi_1) \cong \mathcal{B}(\chi_1, \chi_2)$ is irreducible. ∎

We call a representation (π, V) of $GL(2, F)$ *cuspidal* if there exists no nonzero linear functional l on V such that

$$l\left(\pi\begin{pmatrix} 1 & x \\ & 1 \end{pmatrix}v\right) = l(v)$$

for all $v \in V$, $x \in F$.

We note that over a finite field, the cuspidality of l is equivalent to assuming that V has no $N(F)$-fixed vector. To see this, we recall that if H is any finite group and (τ, U) is a representation of H, the multiplicity of the trivial representation in τ equals the multiplicity of the trivial representation in its contragredient $\hat{\tau}$; the asserted equivalence then follows by taking $H = N(F)$ and taking τ to be the restriction of π to $N(F)$. Over a local field, however, these two conditions are not equivalent, and the condition as we have stated it is the correct one for defining a *supercuspidal* representation.

Proposition 4.1.5 *Let (π, V) be any cuspidal representation of $GL(2, F)$. Then the dimension of V is a multiple of $q - 1$.*

Proof Every additive character of F has the form $\psi_a(x) = \psi(ax)$ for some $a \in F$; we recall a proof of this basic fact. If $a \neq b$, then $\psi_a \neq \psi_b$ because by assumption $\psi \neq 1$, so $\psi(\lambda) \neq 1$ for some $\lambda \in F$; and if $x = (a - b)^{-1}\lambda$, we have $\psi_a(x)/\psi_b(x) = \psi(\lambda) \neq 1$. We see that $a \mapsto \psi_a$ is an injection of F into the group of characters of the finite Abelian group F, and because these groups have the same cardinality, this is also a surjection.

We decompose the dual space V^*, on which $GL(2, F)$ acts by the contragredient representation, into isotypic spaces according to the characters of $N(F)$; thus $V^* = \oplus_{a \in F} V^*(a)$, where we define

$$V^*(a) = \left\{ l \in V^* | \hat{\pi}\begin{pmatrix} 1 & x \\ & 1 \end{pmatrix} l = \psi(ax)\, l \right\}.$$

The hypothesis of cuspidality implies that $V^*(0)$ is zero. Now, one checks easily that the action of $F^\times$ on l defined by

$$t : l \mapsto \hat{\pi}\begin{pmatrix} t & \\ & 1 \end{pmatrix} l$$

permutes the spaces $V^*(a)$ with $a \neq 0$ transitively; hence these spaces have the same dimension. Because there are $q - 1$ of them, we see that the dimension of V is a multiple of $q - 1$. ■

Now suppose that $T_0 = T_{\text{a}}$.

Proposition 4.1.6 *In this situation, the Weil representation $\big(\pi(\chi), W(\chi)\big)$ is cuspidal. It is irreducible.*

Proof We use the characterization (explained above) that (π, V) is cuspidal if and only if V contains no nonzero $N(F)$ fixed vectors. Suppose on the contrary that Φ_0 is such a vector. If Φ is any element of $V(\chi)$, then $\Phi(0) = 0$ – this is a consequence of our assumption that χ restricted to $E_1^\times$ is nontrivial. On the other hand, if $0 \neq x \in E$, then choose $z \in F$ so that $\psi\big(z\,N(x)\big) \neq 1$. Then because

$$\Phi_0(x) = \left(\omega\begin{pmatrix} 1 & z \\ & 1 \end{pmatrix}\Phi_0\right)(x) = \psi\big(z\,N(x)\big)\,\Phi_0(x),$$

we have $\Phi_0(x) = 0$. Thus only the zero vector is fixed by $N(F)$. Therefore the representation $V(\chi)$ is cuspidal.

To see that $\pi(\chi)$ is irreducible, we note that it is of dimension $q - 1$; a nontrivial invariant subspace also affords a cuspidal representation and hence must equal V by Proposition 4.1.5. ■

We have now constructed all the irreducible representations of $GL(2, F)$ with F finite, as may be checked by showing that the sums of the squares of the degrees of the representations that we have constructed equal the order of the group. Moreover, the techniques that we have used were chosen for their amenability to generalization to local fields and adele rings.

The last significant topic that we must discuss is the notion of a *Whittaker model* of a representation, a matter that attains supreme significance in the theory of automorphic forms. Let (π, V) be an irreducible representation of $GL(2, F)$. Let ψ_N be the character of $N(F)$ defined by

$$\psi_N\begin{pmatrix} 1 & x \\ & 1 \end{pmatrix} = \psi(x).$$

Let $\mathcal{G}$ be the representation of $GL(2, F)$ induced from the character ψ_N of $N(F)$. We say that a representation (π, V) of a finite group G is *multiplicity free* if in its decomposition into irreducibles, no irreducible representation occurs with multiplicity greater than one. We will show that $\mathcal{G}$ is multiplicity free. First, we need a criterion.

Proposition 4.1.7 *Let G be a finite group, and let (π, V) be a representation of G. Then π is multiplicity free if and only if the ring $\mathrm{End}_G(V)$ of G-module endomorphisms of V is commutative.*

Proof Let (π, V) be an irreducible representation of a finite group G. Suppose that $V \cong \bigoplus d_i V_i$ is the decomposition of V into irreducibles; d_i is the multiplicity of the representation V_i. Because by Schur's lemma

$$\mathrm{Hom}_G(V_i, V_j) \cong \begin{cases} \mathbb{C} & \text{if } i = j, \\ 0 & \text{otherwise,} \end{cases}$$

the endomorphism ring

$$\mathrm{End}_G(V) \cong \bigoplus \mathrm{Mat}_{d_i}(\mathbb{C}).$$

Clearly, this is commutative if and only if no $d_i > 1$. ■

Theorem 4.1.2: Uniqueness of Whittaker models *The representation $\mathcal{G}$ is multiplicity free. Every irreducible representation of $GL(2, F)$ that is not one dimensional occurs in $\mathcal{G}$ with multiplicity precisely one.*

The following proof, due to Gelfand and Graev (1962), generalizes to $GL(n)$. It is based on the same idea used in Theorem 1.4.2 to prove the classical Hecke algebras are commutative, and the same idea recurs in Theorem 2.2.3 and later in this chapter in Theorem 4.6.1. One shows that a certain ring is commutative by checking that it is invariant under an involution, that is, an antiautomorphism of order two. We will explain why this theorem is called "uniqueness of Whittaker models" after the proof.

Proof By Proposition 4.1.2, $\mathrm{End}_{GL(2,F)}(\mathcal{G})$ is isomorphic to the ring of functions $\Delta : GL(2, F) \to \mathbb{C}$ that satisfy

$$\Delta(n_2 g n_1) = \psi_N(n_2)\, \Delta(g)\, \psi_N(n_1); \tag{1.18}$$

the multiplication in this ring (corresponding to the composition of intertwining operators) is convolution. Just as in the proof of Theorem 4.1.1, the issue is to determine those double cosets in $N(F)\backslash GL(2, F)/N(F)$ that could support a nonzero function Δ satisfying Eq. (1.18). It may be deduced easily from the Bruhat decomposition that the double cosets are represented by monomial matrices, that is, matrices with one nonzero entry in each row and column. Moreover, it is easy to see that the double coset containing a diagonal matrix cannot support such a function unless the eigenvalues of this matrix are equal.

Thus the double cosets on which Δ satisfying Eq. (1.18) is nonzero have the following representatives:

$$\begin{pmatrix} a & \\ & a \end{pmatrix}; \quad \text{or} \quad \begin{pmatrix} & b \\ c & \end{pmatrix}. \tag{1.19}$$

Define an involution $\iota : GL(2, F) \to GL(2, F)$ by

$$\iota(g) = \begin{pmatrix} & 1 \\ 1 & \end{pmatrix} {}^{\top}g \begin{pmatrix} & 1 \\ 1 & \end{pmatrix}.$$

Thus $\iota^2(g) = g$ and $\iota(g_1 g_2) = \iota(g_2)\,\iota(g_1)$. Note that ι is the identity on the subgroup N. Consequently, ι induces an antiautomorphism of order two of the ring of functions Δ satisfying Eq. (1.18). However, from Eq. (1.19), we see that these functions are supported on double cosets that have representatives that are fixed by ι, and consequently, the antiautomorphism that we have constructed is actually the identity map! This proves that the endomorphism ring is commutative, as required for the application of Proposition 4.1.7.

We have also asserted that every irreducible representation of degree greater than one occurs in $\mathcal{G}$. We will not belabor this point; however, for the principal series representations $\mathcal{B}(\chi_1, \chi_2)$ with $\chi_1 \neq \chi_2$ (and indeed also for the Steinberg representations) this point may be checked using Proposition 4.1.2; for the cuspidal representation, the existence of a Whittaker model may be deduced from the considerations of Propositions 4.1.5 and 4.1.6, together with Frobenius reciprocity. ■

If (π, V) is an irreducible representation that can be embedded into $\mathcal{G}$, we call its image a *Whittaker model* of π. Thus a *Whittaker model* of π is a space $\mathcal{W}(\pi)$ of functions $W : GL(2, F) \to \mathbb{C}$ having the property that

$$W\left(\begin{pmatrix} 1 & x \\ & 1 \end{pmatrix} g\right) = \psi(x)\, W(g).$$

The space of functions is invariant under right translation (because this is how $GL(2, F)$ acts on the induced representation $\mathcal{G}$) and affords a representation of $GL(2, F)$ isomorphic to π. The content of Theorem 4.1.2 is that the Whittaker model is unique if it exists, which it does provided $\dim\ V > 1$.

The Whittaker model is an image of a $GL(2, F)$-module homomorphism from V to $\mathcal{G}$, which is induced from the character ψ_N of $N(F)$; by Frobenius reciprocity (Eq. 1.3), this is the same as an $N(F)$-module homomorphism $V \to \mathbb{C}$, where $\mathbb{C}$ is given the structure of an $N(F)$-module by means of the character ψ_N. Such a map $V \to \mathbb{C}$ is called a *Whittaker functional*. Thus a *Whittaker functional* is a linear functional $L : V \to \mathbb{C}$ with the property that

$$L\left(\pi\begin{pmatrix} 1 & x \\ & 1 \end{pmatrix} v\right) = \psi(x)\, L(v)$$

for $x \in F$, $v \in V$. Theorem 4.1.2 thus implies that the space of Whittaker functionals on V is at most one dimensional.

It may be worth unravelling the definitions in the proof of Eq. (1.3) to make the correspondence between Whittaker models and Whittaker functionals explicit. Let there be given a Whittaker model of π; then we may identify V with $\mathcal{W}(\pi)$, and if we make this identification, the Whittaker functional is evaluation at the identity: $W \mapsto W(1)$ is the Whittaker functional on $\mathcal{W}(\pi)$. Conversely, let there be given a Whittaker functional L on V. Then the Whittaker model is the space of functions of the form

$$W_v(g) = L\big(\pi(g)\,v\big), \qquad v \in V.$$

For further discussion of this point, see the proof of Theorem 3.5.3.

Exercises

Exercise 4.1.1: Frobenius reciprocity (a) Verify that the correspondence described in the text following Eq. (1.3) is an isomorphism.
(b) Let (π_1, V_1) and (π_2, V_2) be representations of the finite group G, and let χ_1, χ_2 be their respective characters. Define their inner product as usual by

$$\langle \chi_1, \chi_2 \rangle_G = \frac{1}{|G|} \sum_{x \in G} \chi_1(x)\, \overline{\chi_2(x)}.$$

(Here $|G|$ is the cardinality of G.) Prove that

$$\langle \chi_1, \chi_2 \rangle_G = \dim\, \mathrm{Hom}_G(\pi_1, \pi_2).$$

[HINT: Note that both sides in this equation are bilinear and hence reduce to the case where π_1 and π_2 are irreducible.]
(c) Now let $H \subset G$, let (π, V) be a representation of H, and let (τ, U) be a representation of G. Let χ and σ be their respective characters, let χ^G be the character of the induced representation (π^G, V^G) of G, and let σ_H be the restriction of σ to H. Prove that

$$\left\langle \chi^G, \sigma \right\rangle_G = \langle \chi, \sigma_H \rangle_H \,.$$

Exercise 4.1.2: Mackey's theorem Let the notation be as in Proposition 4.1.1. Let $x_1, \cdots, x_r$ be a set of double coset representatives for $H_2\backslash G/H_1$. Let $S_i = H_2 \cap x_i H_1 x_i^{-1}$. Define two representations $(\pi_{1,i}, V_1)$ and $(\pi_{2,i}, V_2)$ of S_i on the vector spaces V_1 and V_2 by $\pi_{1,i}(s) = \pi_1(x_i^{-1} s x_i)$ and $\pi_{2,i}(s) = \pi_2(s)$.
(a) Prove that if $\Delta : G \to \mathrm{Hom}_{\mathbb{C}}(V_1, V_2)$ satisfies Eq. (1.3), then $\Delta(x_i) : V_1 \to V_2$ is an intertwining operator from $\pi_{1,i}$ to $\pi_{2,i}$.

(b) Prove that

$$\mathrm{Hom}_G(V_1^G, V_2^G) \cong \bigoplus_{i=1}^{r} \mathrm{Hom}_{S_i}(\pi_{1,i}, \pi_{2,i}).$$

Exercise 4.1.3 Let F be any field. Prove that in $GL(n, F)$, every matrix is conjugate to its transpose. [HINT: Use Corollary XIV.2.3 in Lang (1993) to reduce to the case where F is algebraically closed; then use the Jordan canonical form.]

THE FINITE STONE-VON NEUMANN THEOREM. Let H be a finite *two-step nilpotent group*, also known as a *Heisenberg group*. This means that if Z is the center of H, we assume that $\overline{H} = H/Z$ is Abelian. If A is any subgroup of H containing Z, we will denote $\overline{A} = A/Z \subseteq \overline{H}$; and if $x \in H$, we will denote by $\overline{x}$ its image in H. Because $\overline{H}$ is Abelian, $\overline{A}$ is normal in $\overline{H}$, and consequently, A is normal in H. Let χ_0 be a character of Z. If $\overline{x}, \overline{y} \in \overline{H}$, then $xyx^{-1}y^{-1} \in Z$ does not depend on the representatives $x,\ y \in H$ for $\overline{x}, \overline{y} \in \overline{H}$. Thus we may define

$$\langle \overline{x}, \overline{y} \rangle = \chi_0(xyx^{-1}y^{-1}).$$

Exercise 4.1.4 Show that the pairing

$$(\overline{x}, \overline{y}) \mapsto \langle \overline{x}, \overline{y} \rangle$$

is bilinear and skew symmetric; that is, prove that

$$\langle \overline{x}_1\overline{x}_2, \overline{y} \rangle = \langle \overline{x}_1, \overline{y} \rangle \langle \overline{x}_2, \overline{y} \rangle, \tag{1.20}$$

$$\langle \overline{x}, \overline{y}_1\overline{y}_2 \rangle = \langle \overline{x}, \overline{y}_1 \rangle \langle \overline{x}, \overline{y}_2 \rangle, \tag{1.21}$$

$$\langle \overline{x}, \overline{x} \rangle = 1, \tag{1.22}$$

and

$$\langle \overline{x}, \overline{y} \rangle = \langle \overline{y}, \overline{x} \rangle^{-1}. \tag{1.23}$$

Let G be a finite Abelian group, and let G^* be its character group. If $g \in G$, $g^* \in G^*$, we will denote $\langle g, g^* \rangle = g^*(g)$ to emphasize the symmetry between the two. It is well known that G^* is a finite Abelian group isomorphic to G, though not canonically. However, there is a canonical isomorphism $G \cong G^{**}$, and we identify these two groups. If $A \subseteq G$, we denote

$$A^{\perp} = \left\{ g^* \in G^* | \langle a, g^* \rangle = 1 \text{ for all } a \in A \right\}.$$

Thus $A^{\perp}$ consists of all characters of G that vanish on A, and we may identify $A^{\perp} = (G/A)^*$. We have

$$A^{\perp\perp} = A. \tag{1.24}$$

If A and B are subgroups of G, then clearly

$$(AB)^\perp = A^\perp \cap B^\perp,$$

and using Eq. (1.24), this implies that

$$(A \cap B)^\perp = A^\perp B^\perp. \tag{1.25}$$

Returning to the two-step nilpotent group H, we say that χ_0 is *generic* if the pairing $\langle\, ,\, \rangle$ is nondegenerate, that is, if every character of $\overline{H}$ has the form

$$\overline{x} \mapsto \langle \overline{x}, \overline{y} \rangle$$

for a unique $\overline{y} \in \overline{H}$. In this case, we may use the pairing to identify $\overline{H}$ with its dual group. With the above notation, each subgroup $\overline{A}$ of $\overline{H}$ (corresponding to a subgroup A of H containing Z) gives rise to another subgroup $\overline{A}^\perp$.

We will assume in the remaining exercises that χ_0 is generic. Making this assumption, we say that a subgroup $\overline{A} \subseteq \overline{H}$ is *isotropic* if $\langle \overline{x}, \overline{y} \rangle = 1$ for all $\overline{x}, \overline{y} \in \overline{A}$. We say that $\overline{A}$ is *polarizing* if $\overline{A} = \overline{A}^\perp$; that is, $\langle \overline{x}, \overline{y} \rangle = 1$ for all $\overline{y} \in \overline{A}$ if and only if $\overline{x} \in \overline{A}$. Thus a polarizing subgroup is automatically isotropic. (If A is the preimage of $\overline{A}$ in H, we also say that A is *isotropic* or *polarizing* if $\overline{A}$ has the same corresponding property.)

Exercise 4.1.5 Prove that a maximal isotropic subgroup is polarizing. Thus any isotropic subgroup may be embedded in a polarizing subgroup. In particular, because the trivial subgroup is isotropic, polarizing subgroups exist. [HINT: Suppose that $\overline{A}$ is a maximal isotropic subgroup. To show $\overline{A}$ is isotropic, let $\overline{x} \in \overline{A}^\perp$; one must show that $\overline{x} \in \overline{A}$. Let $\overline{B}$ be the cyclic subgroup generated by $\overline{x}$. By the maximality of A, it is sufficient to show that $\overline{A}\,\overline{B}$ is isotropic.]

Exercise 4.1.6 Let $\overline{A}$ be an isotropic subgroup of $\overline{H}$, and let A be its preimage in H. Then χ_0 may be extended to a character of A. [HINT: Let $Z_0 \subseteq Z$ be the kernel of χ_0; prove that A/Z_0 is Abelian. Then use the fact that if $G_1 \subset G_2$ are finite Abelian groups, then a character of G_1 may always be extended to a character of G_2.]

Exercise 4.1.7 Let H be a two-step nilpotent group, and let χ_0 be a generic character of its center Z. Let $\overline{A}$ and $\overline{B}$ be polarizing subgroups of $\overline{H}$, and let A and B be their preimages in H. Let χ_A and χ_B be characters of A and B extending χ_0, and let π_A and π_B be the representations of H induced from these characters of A and B.

(a) Show that A, B, and BA are normal subgroups of H and that there exists a unique (double) coset $xBA = BxA \in B\backslash H/A$ such that (choosing a representative x of the coset) the characters $s \mapsto \chi_A(x^{-1}sx)$ and $s \mapsto \chi_B(s)$ coincide on $A \cap B$. [HINT: The condition $\chi_B(s) = \chi_A(x^{-1}sx)$ for $s \in A \cap B$ is equivalent to $(\chi_B\chi_A)(s) = \langle s, x \rangle$, which determines x modulo $(A \cap B)^\perp$, which equals AB by Eq. (1.25).]

(b) Prove that

$$\dim\ \mathrm{Hom}_H(\pi_A, \pi_B) = 1.$$

[HINT: Use Mackey theory in the form of Exercise 4.1.2, together with (a).]

Exercise 4.1.8: The Stone–Von Neumann Theorem Let H be a finite two-step nilpotent group, and let χ_0 be a generic character of its center Z. Then there exists a unique isomorphism class of irreducible representations of H with central character χ_0. Such a representation may be constructed as follows: Let A be any polarizing subgroup of H, and let χ_A be any extension of χ_0 to A. Then the representation π of H induced from this character of A is of this class.

Projective representations and covering groups Let G be a group, and let A be an Abelian group. By a *central extension* of G by A we mean a group $\tilde{G}$ together with a short exact sequence

$$1 \mapsto A \mapsto \tilde{G} \mapsto G \mapsto 1$$

such that the image of A is contained in the center of $\tilde{G}$. We say that two extensions $\tilde{G}_1$ and $\tilde{G}_2$ of G by A are *equivalent* if there exists an isomorphism $\tilde{G}_1 \to \tilde{G}_2$ making the following diagram commute:

$$\begin{array}{ccccccccc} 1 & \longrightarrow & A & \longrightarrow & \tilde{G}_1 & \longrightarrow & G & \longrightarrow & 1 \\ & & \| & & \downarrow & & \| & & \\ 1 & \longrightarrow & A & \longrightarrow & \tilde{G}_2 & \longrightarrow & G & \longrightarrow & 1 \end{array}$$

We recall the definition of the cohomology group $H^2(G, A)$ (where G acts trivially on A). Let $Z^2(G, A)$ be the multiplicative group of all maps ("2-cocycles") $\sigma : G \times G \to A$ satisfying the *cocycle relation*

$$\sigma(g_1, g_2)\,\sigma(g_1g_2, g_3) = \sigma(g_1, g_2g_3)\,\sigma(g_2, g_3). \tag{1.26}$$

If $\phi : G \to A$ is any map, then

$$\sigma(g_1, g_2) = \phi(g_1)\,\phi(g_2)\,\phi(g_1g_2)^{-1}$$

satisfies Eq. (1.26); let $B^2(G, A)$ be the subgroup of $Z^2(G, A)$ of elements of this type ("coboundaries"). Then $H^2(G, A) = Z^2(G, A)/B^2(G, A)$. If $\sigma \in Z^2(G, A)$, we denote its class in $H^2(G, A)$ by $[\sigma]$.

Given a two-cocyle $\sigma \in Z^2(G, A)$, we construct a central extension $\tilde{G}_\sigma$ as follows. As a set, $\tilde{G}_\sigma$ is the Cartesian product $G \times A$. The group law is given by

$$(g, a)(g', a') = \big(gg', aa'\sigma(g, g')\big). \tag{1.27}$$

The cocycle condition (Eq. 1.26) implies that this multiplication satisfies the associative law. The homomorphisms $A \to \tilde{G}_\sigma$ and $\tilde{G}_\sigma \to G$ are given by $a \mapsto (1, a)$ and $(g, a) \mapsto g$.

Exercise 4.1.9 Show that the extension $\tilde{G}_\sigma$ depends only on the cohomology class of σ in $H^2(G, A)$ and, moreover, that any central extension $\tilde{G}$ is equivalent to some $\tilde{G}_\sigma$ for a unique class $[\sigma] \in H^2(G, A)$. Hence there is a bijection between the classes of central extensions of G by A and the elements of $H^2(G, A)$.

By a *projective representation* of a group G, we mean a homomorphism of G into $PGL(n, \mathbb{C})$ for some n, where we recall that $PGL(n, \mathbb{C})$ is $GL(n, C)$ modulo its center. Choosing a representative $\rho(g)$ of the image of $g \in G$ in $GL(n, \mathbb{C})$ under this homomorphism, we find that

$$\rho(gg') = \sigma(g, g')^{-1} \rho(g) \rho(g')$$

for some complex number σ. It follows from the associative law in G that σ satisfies the cocycle relation (Eq. 1.26) and hence determines an extension $\widetilde{G} = \widetilde{G}_\sigma$. One checks then that $\tilde{\rho} : \widetilde{G} \to GL(n, \mathbb{C})$ defined by

$$\tilde{\rho}(g, a) = a\, \rho(g), \qquad g \in G,\ a \in \mathbb{C}^\times$$

is a representation. Thus, *a projective representation of G gives rise to a cohomology class in $H^2(G, \mathbb{C}^\times)$ and to a true representation of a central extension of G by $\mathbb{C}^\times$.*

The Weil representation We return now to the setting of Exercise 4.1.8. Let H be a two-step nilpotent group, and let χ_0 be a generic character of the center Z of H. Let G be a group of automorphisms of H that fix Z. We may then construct a projective representation ω of G as follows. By the Stone–Von Neumann theorem (Exercise 4.1.8), there exists a representation (π, W) of H, unique up to isomorphism, having central character χ_0. If $g \in G$, then we obtain another such representation (π_g, W) acting on the same space by $\pi_g(h) = \pi({}^g h)$. By the uniqueness of π, these representations are isomorphic, which means that there is an intertwining map $\omega(g) : W \to W$ from (π, W) to (π_g, W), and $\omega(g)$ is unique up to constant by Schur's lemma, because π and π_g are irreducible.

Exercise 4.1.10 Prove that ω is a projective representation of G and that the defining property of ω can be written

$$\pi({}^g h) = \omega(g)\, \pi(h)\, \omega(g)^{-1}. \tag{1.28}$$

We make this projective representation explicit in a special case and relate it to the Weil representation as described in the text.

Let F be a finite field with q elements, and assume that q is *odd*. We will arrive at a reinterpretation of the Weil representation. Let V be a vector space over F, and let $B : V \times V \to F$ be a nondegenerate symmetric bilinear form. Let ψ be a nontrivial additive character of F. Define a two-step nilpotent group H as follows. As a set, $H = V \oplus V \oplus F$, with multiplication

$$(v_1, v_2, x)(v_1', v_2', x') = \big(v_1 + v_1', v_2 + v_2', x + x' + B(v_1, v_2') - B(v_1', v_2)\big).$$

The center Z of H consists of the subgroup of elements of the form $(0, 0, z)$

with $z \in F$, and we define a character χ_0 of Z by

$$\chi_0(0, 0, z) = \psi(z).$$

Let A be the subgroup of elements of the form $(v_1, 0, x)$.

Exercise 4.1.11 Prove that χ_0 is generic, and that the subgroup A is polarizing. Exercise 4.1.8 shows that there exists a unique irreducible representation of H with central character χ_0 and that this representation can be obtained by extending χ_0 to A in an arbitrary way and then inducing; we induce the extension

$$(v_1, 0, x) \mapsto \psi(x)$$

and call the resulting representation ρ. Thus ρ acts on the space W_ρ of functions $\phi : H \to \mathbb{C}$ satisfying $\phi(ah) = \chi(a)\phi(h)$ for $a \in A$, and $\rho(h)\phi(h') = \phi(h'h)$. We will use a slightly different model of the representation, however.

Exercise 4.1.12 Show that if Φ is an arbitrary $\mathbb{C}$-valued function on V, there exists a unique element ϕ of V_ρ such that $\phi(0, v, 0) = \Phi(v)$. Let W be the space of all complex-valued functions on V, and let (π, W) be the representation determined by the condition that that $\Phi \mapsto \phi$ is an isomorphism. Verify that

$$\pi(u, 0, 0)\Phi(v) = \psi\big(-2B(v, u)\big)\,\Phi(v), \tag{1.29}$$

$$\pi(0, u, 0)\Phi(v) = \Phi(u + v). \tag{1.30}$$

Exercise 4.1.13 Let G_1 be the group $SL(2, F)$. Verify that there is an action of G_1 on H defined by

$${}^g(v_1, v_2, x) = (av_1 + bv_2, cv_1 + dv_2, x), \qquad g = \begin{pmatrix} a & b \\ c & d \end{pmatrix}.$$

Hence by Exercise 4.1.10 there is a projective representation ω_1 of $SL(2, F)$, where $\omega_1 : SL(2, F) \to GL(W)$ is defined up to constant multiple by Eq. (1.28). Let $\chi : F^\times \to \{\pm 1\}$ be the unique quadratic character of the cyclic group $F^\times$. Verify that

$$\left(\omega_1 \begin{pmatrix} 1 & x \\ & 1 \end{pmatrix} \Phi\right)(v) = \psi\big(x\,B(v, v)\big)\Phi(v), \tag{1.31}$$

$$\left(\omega_1 \begin{pmatrix} a & \\ & a^{-1} \end{pmatrix} \Phi\right)(v) = \chi(a)^{\dim V}\,\Phi(av), \tag{1.32}$$

and

$$\omega_1 \begin{pmatrix} & 1 \\ -1 & \end{pmatrix} \Phi = \hat{\Phi}, \tag{1.33}$$

where the Fourier tranform is defined by

$$\hat{\Phi}(v) = \epsilon q^{-\dim(V)/2} \sum_{u \in V} \Phi(u)\,\psi\big(2B(u, v)\big), \tag{1.34}$$

and ϵ is a constant to be chosen later.

[HINT: At the moment, the value of ϵ is irrelevant because $\omega_1(g)$ is only determined up to a constant multiple. For the same reason, the factor $\chi(a)$ on the right side of Eq. (1.32) is unnecessary, but the inclusion of ϵ in Eq. (1.34) and $\chi(a)$ in Eq. (1.32) will become important in Exercise 4.1.14, where we show that the projective representation ω_1 is actually a true representation. To prove Eqs. (1.31–1.33), it is necessary to verify Eq. (1.28), and it is sufficient to check this for h of the form $(u, 0, 0)$ and $(0, u, 0)$, because these generate H.]

At this point, we only know that ω_1 is a projective representation. As such it determines a cocycle in $H^2\big(SL(2, F), \mathbb{C}^\times\big)$. If that cocycle is trivial, then we know that the values of $\omega_1(g)$ can be adjusted by complex constants so as to obtain a bona fide representation. (Actually for most values of q, the *Schur multiplier* $H^2\big(SL(2, F), \mathbb{C}^\times\big)$ is known to be trivial; however, we do not need this fact.)

Exercise 4.1.14 (a) Prove that there exists a fourth root of unity ϵ_0 such that for $a \in F^\times$, the Gauss sum

$$\sum_{x \in F} \psi(ax^2) = \chi(a)\, \epsilon_0 \sqrt{q},$$

and $\epsilon_0^2 = \chi(-1)$.

[HINT: Let χ be the quadratic character of $F^\times$. Then the argument of Exercise 1.1.3 shows that the Gauss sum equals $G = \sum_{y \in F} \chi(y)\, \psi(ay)$. This is a Gauss sum of the type encountered in Section 1.1. Its absolute value is $q^{1/2}$ by the proof of Eq. (1.5) in Chapter 1. To see that ϵ_0 is a fourth root of unity, note that $\overline{G} = \sum_{y \in F} \chi(y)\, \psi(-ay)$, and making the change of variables $y \to -a^2 y$ shows that $\overline{G} = \chi(-a^2)\, G = \chi(-1)\, G$.]

(b) Show that there exists a fourth root of unity ϵ such that

$$\sum_{v \in V} \psi\big(a\, B(v, v)\big) = \epsilon\, \chi(a)\, q^{\dim(V)/2},$$

and $\epsilon^2 = \chi(-1)^{\dim V}$.

[HINT: After choosing an orthogonal basis of V, the sum reduces to a product of sums of the type in (a).]

Exercise 4.1.15 Prove that if ϵ is chosen as in Exercise 4.1.14, the projective representation ω_1 with the formulas in Exercise 4.1.13 is a true representation. [HINT: Imitate the proof of Proposition 4.1.3, and note that a crucial step is supplied by Exercise 4.1.14.]

Exercise 4.1.16 As in the text, let K be a two-dimensional semisimple algebra over F, and let $V = K$ with $B(x, y) = \frac{1}{2}\operatorname{tr}(x\overline{y})$. Verify that the Weil representation as constructed in these exercises agrees with the construction in the text.

Now let $G_2 = O(V)$ be the orthogonal group of all endomorphisms $g : V \to V$ such that $B(gv_1, gv_2) = B(v_1, v_2)$. Then G_2 induces an automorphism of

H fixing Z, so again there is a (projective) representation ω_2 of G_2 on W. This representation is readily made explicit: It is simply

$$\big(\omega_2(g)\Phi\big)(v) = \Phi(g^{-1}v).$$

It is clearly a true representation. The actions ω_1 and ω_2 commute with each other, and so we may regard W as a module for $SL(2, F) \times O(V)$. If π is a representation of $O(V)$, then $\mathrm{Hom}_{O(V)}(\pi, W)$ is a representation of $SL(2)$. There is a strong tendency for this representation to be irreducible if it is nonzero. The same phenomenon is true if we reverse the roles of $SL(2)$ and $O(V)$. Furthermore, everything may be adapted without difficulty to the pair $Sp(2n) \times O(V)$. (The symplectic group $Sp(2)$ is just $SL(2)$.) Over a local field, this phenomenon, precisely formulated, is known as *Howe duality*.

The philosophy of cusp forms Harish-Chandra (1970) emphasized that the way Eisenstein series are built up from cusp forms had an analog – parabolic induction – for representations of linear groups over finite or local fields. The *philosophy of cusp forms* refers to this analogy.

Let G be a reductive algebraic group. Suppose first that G is defined over a global field F with adele ring A. In the theory of Eisenstein series (Langlands (1976), Moeglin and Waldspurger (1993)), it is found that automorphic forms on $G(A)$ that are orthogonal to cusp forms are built up from cusp forms on the Levi factors of parabolic subgroups by the formation of Eisenstein series. On the other hand, if F is a finite or local field, representations that are not cuspidal (or discrete series) are built up from cuspidal (or discrete series) representations by parabolic induction.

If $G = GL(2)$, all these theories are treated in some detail in this book. In this case, the only parabolic subgroup is the Borel subgroup, so matters are somewhat more simple than in the general case. The theory of Eisenstein series on $GL(n)$ is beyond the scope of this book. However, it may be helpful to the reader about to approach this subject to have in mind the analogous theory of parabolic induction in the case of finite fields. This, at least, we cover in these exercises.

A word of caution: It is true for both finite and local fields that representations are built up from cuspidal representations (or discrete series) by parabolic induction. Nevertheless, there are some essential differences between the two theories. The case of a finite field is a good starting point for understanding the process, so long as one bears these differences in mind. For the local case, see Kudla (1994) for statements and Bernstein and Zelevinsky (1977) for proofs.

As we said, the essential paradigm in the representation theory of finite fields is that the representations are built up from basic building blocks, which are the *cuspidal representations*. Let G be a connected reductive algebraic group over a finite field F. Let B be a *Borel subgroup* of G, which is by definition a maximal Zariski-connected solvable (algebraic) subgroup. A *parabolic subgroup* of G is a proper (algebraic) subgroup of G containing B or one of its conjugates. If P is a parabolic subgroup, its *Levi subgroup* (determined up to conjugacy) is a maximal reductive subgroup M of P. The unipotent radical U of P is its (uniquely determined) maximal unipotent subgroup. We have a semidirect product $P = MU$. For further details about these concepts, the reader may consult the works on algebraic groups that are listed near the beginning of Section 3.9.

If $G = GL(n)$, we may take V to be the standard Borel subgroup B of upper triangular nonsingular matrices. The parabolic subgroups then have the following explicit description. Let $\lambda_1, \cdots, \lambda_h$ be positive integers whose sum is n. Then the (ordered) sequence $\lambda = (\lambda_1, \cdots, \lambda_h)$ is called an *ordered partition* of n. We call the ordered partition *proper* if $h > 1$. If λ is an ordered partition, we consider the group of all matrices of the form

$$\begin{pmatrix} A_1 & * & \cdots & * \\ 0 & A_2 & & * \\ \vdots & & \ddots & \vdots \\ 0 & 0 & \cdots & A_h \end{pmatrix},$$

where each A_i is a nonsingular $\lambda_i \times \lambda_i$ block matrix. We call this group P_λ. It is a parabolic subgroup if λ is proper and every parabolic subgroup is conjugate to a subgroup of this type. We call the subgroups P_λ *standard* parabolic subgroups. The *Levi subgroup* M_λ of P_λ is the subgroup with the blocks above the diagonal all zero, so

$$M_\lambda \cong GL(\lambda_1) \times \cdots \times GL(\lambda_h).$$

The unipotent radical U_λ consists of the subgroup where each A_i is the $\lambda_i \times \lambda_i$ identity matrix. We call a subgroup of M_λ *parabolic* if it is a proper subgroup that is the product of subgroups of the groups $GL(\lambda_i)$, each of which is either $GL(\lambda_i)$ or a parabolic subgroup.

Let (π, V) be an irreducible representation of $GL(n, F)$. We say that π is *cuspidal* if for every proper ordered partition λ of n, there exists no nontrivial $U_\lambda(F)$-invariant linear functional on V, that is, no linear functional $\Lambda : V \to \mathbb{C}$ such that $\Lambda\big(\pi(u)\, v\big) = \Lambda(v)$ for all $v \in V$, $u \in U_\lambda(F)$.

If $\lambda = (\lambda_1, \cdots, \lambda_h)$ is an ordered partition of n, and if (π_i, V_i) is a cuspidal representation of $GL(\lambda_i, F)$ for each $1 \le i \le h$, then $\pi = \otimes\pi_i$ is naturally a representation of $M_\lambda(F)$, and we call a representation of $M_\lambda(F)$ of this type *cuspidal.* We extend this representation to a representation of $P_\lambda(F)$ by letting $U_\lambda(F)$ act trivially and then we induce. The resulting representation is said to be obtained from the cuspidal representation π of $M_\lambda(F)$ by *parabolic induction.* We will see in the following exercises that the resulting representation is often, though not always, irreducible. The following exercises show how every irreducible representation of $GL(n, F)$ is either cuspidal or is a quotient of one parabolically induced from a cuspidal representation of $M_\lambda(F)$. See Carter (1985), Howe (1985), and Zelevinsky (1981) for further details.

Exercise 4.1.17 Let (π, V) be an irreducible representation of $GL(n, F)$, where F is a finite field. Prove that π is cuspidal if and only if π does not occur in any representation parabolically induced from a cuspidal representation of the Levi subgroup $M_\lambda(F)$ of a parabolic subgroup. [HINT: Use Frobenius reciprocity.]

Given Exercise 4.1.18, it is important to understand when parabolically induced representations are irreducible, how they decompose when reducible, and when they can be isomorphic. This is essentially an exercise in Mackey theory. The *Bruhat decomposition* is an essential tool in analyzing the double coset spaces that arise.

Exercise 4.1.18: The Bruhat decomposition for GL(n) See, for example, Carter (1972, Chapter 8) for the axiomatics of *Tits' systems* (after J. Tits) and the Bruhat decomposition.

(a) Let W be the group of permutation matrices in $GL(n)$, isomorphic to the symmetric group S_n. Let B be the Borel subgroup of $GL(n)$, consisting of upper triangular matrices. Show that

$$GL(n) = \bigcup_{w\in W} B\,w\,B \qquad \text{(disjoint).}$$

(b) Let $\lambda = (\lambda_1, \cdots, \lambda_h)$ and $\mu = (\mu_1, \cdots, \mu_k)$ be ordered partitions of n. Let $W_\lambda = W \cap M_\lambda \cong S_{\lambda_1} \times \cdots S_{\lambda_h}$. According to (a), the canonical map $W \to GL(n) \to B\backslash GL(n)/B$ is a bijection. Show that the composition

$$W \to GL(n) \to B\backslash GL(n)/B \to P_\mu\backslash GL(n)/P_\lambda$$

induces a bijection

$$W_\mu\backslash W/W_\lambda \cong P_\mu\backslash GL(n)/P_\lambda.$$

Exercise 4.1.19 Let $\lambda = (\lambda_1, \cdots, \lambda_h)$ and $\mu = (\mu_1, \cdots, \mu_k)$ be ordered partitions of n. Let $\pi = \pi_1 \otimes \ldots \otimes \pi_h$ and $\theta = \theta_1 \otimes \ldots \otimes \theta_k$ be cuspidal representations of $M_\lambda(F)$ and $M_\mu(F)$, respectively, and let (Π, V_Π) and (Θ, V_Θ) be the representations of $GL(n, F)$ parabolically induced from π and θ. Show that $\mathrm{Hom}_{GL(n,F)}(\Pi, \Theta) = 0$ unless $k = h$, in which case the dimension of $\mathrm{Hom}_{GL(n,F)}(\Pi, \Theta)$ is equal to the number of permutations σ of $\{1, 2, \cdots, h\}$ such that $\lambda_{\sigma(i)} = \mu_i$ and $\pi_{\sigma(i)} \cong \theta_i$ as a $GL(n, \mu_i)$-module for each $i = 1, \cdots, h$.

[HINT: By Mackey's theorem (Proposition 4.1.2), we must compute the dimension of the space of functions $\Delta : GL(n, F) \to \mathrm{Hom}_{\mathbb{C}}(V_\pi, V_\theta)$ that satisfy

$$\Delta(p_\mu\, g\, p_\lambda) = \theta(p_\mu) \circ \Delta(g) \circ \pi(p_\lambda), \qquad \mathfrak{p}_\mu \in P_\mu(F),\ p_\lambda \in P_\lambda(F).$$

Show that any such function may be expressed uniquely as a sum of such functions, each of which is supported on a single double coset in $P_\mu(F)\backslash GL(n, F)/P_\lambda(F)$. Thus let us study Δ on the assumption that it is supported on a single such double coset, which by Exercise 4.18 may be assumed to be of the form $P_\mu(F)\, w\, P_\lambda(F)$, where $w \in W$. Let $\phi = \Delta(w) : V_\pi \to V_\theta$. Show, as follows, that $M_\mu(F) \subseteq wM_\lambda(F)w^{-1}$. If not, $M_\mu(F) \cap wU_\lambda(F)w^{-1}$ is the unipotent radical of the (not necessarily standard) parabolic subgroup $M_\mu(F) \cap wP_\lambda(F)w^{-1}$ of $M_\mu(F)$. If $n \in M_\mu(F) \cap wU_\lambda(F)w^{-1}$ and $v \in V_\pi$, then

$$\theta(n)\,\phi(v) = \theta(n)\,\phi\big(\pi(w^{-1}nw)\,v\big) = \theta(n)\,\Delta(w)\,\pi\left(w^{-1}n^{-1}w\right)v = \phi(v),$$

and because θ is cuspidal, this implies $\phi(v) = 0$. Thus ϕ is the zero map which is a contradiction. Similarly, show $M_\lambda(F) \subseteq w^{-1}M_\mu(F)w$, and therefore $M_\mu(F) = wM_\lambda(F)w^{-1}$. Hence there exists a permutation σ of $\{1, \cdots, h\}$ such that $\lambda_{\sigma(i)} = \mu_i$. Now the isomorphism $m \to wmw^{-1}$ of M_λ onto $M_\mu(F)$ gives V_θ an $M_\lambda(F)$-module structure. Show that with this structure, $\phi : V_\pi \to V_\theta$ is an $M_\lambda(F)$-module homomorphism. By Schur's lemma, at most one such π can exist, and its existence is contingent upon the isomorphisms $\pi_{\sigma(i)} \cong \theta_i$.]

Exercise 4.1.20 (a) Suppose that $\lambda = (\lambda_1, \cdots, \lambda_h)$ is an ordered partition of n. Let $\pi = \pi_1 \otimes \ldots \otimes \pi_h$ be a cuspidal representation of $M_\lambda(F)$. Show that the representation (Π, V_Π) is irreducible unless there exists some $i \neq j$ such that $\lambda_i = \lambda_j$ and $\pi_i \cong \pi_j$ as $GL(\lambda_i, F)$-modules. [HINT: Take $\Pi = \Theta$ in Exercise 4.1.19.]
(b) Suppose furthermore that $\mu = (\mu_1, \cdots, \mu_k)$ is another ordered partition of n, and that $\theta = \otimes\theta_i$ is a cuspidal representation of $M_\mu(F)$. Assuming that the representations parabolically induced from π and θ are irreducible, show that they are isomorphic if and only if the μ_i and θ_i are the λ_i and π_i rearranged.

These exercises go a long way toward reducing the classification of the representations of $GL(n, F)$ to the classification of the cuspidal representations. What is needed to complete this program is to study the decomposition of Π when $\lambda = (\lambda_0, \cdots, \lambda_0)$ and each $\pi_i \cong \pi_0$ is isomorphic to a fixed $GL(\lambda_0, F)$-module. This question is intimately related to the representation theory of the symmetric groups. See Zelevinsky (1981) for a full discussion.

If we permute the factors M_{λ_i} and their representations π_i, the isomorphism class of the parabolically induced representation does not change. (This is clear from Exercise 4.1.19, at least when the parabolically induced representation is irreducible.) This isomorphism is analogous to the functional equations of Eisenstein series. In the case of $GL(2)$, we hope this analogy is clear from the discussions in Section 3.7 and Section 4.5 of this chapter.

4.2 Smooth and Admissible Representations

Besides Jacquet and Langlands (1970), who first emphasized the appropriate notion of an *admissible representation* and laid foundations suitable for the theory of automorphic forms on $GL(2)$, references for the representation theory of p-adic groups include Godement (1970) and Bernstein and Zelevinsky (1976). There is also the survey article by Cartier (1979). Casselman (unpublished book) is extremely useful if you can find it.

In this section, F will always denote a non-Archimedean local field, $\mathfrak{o}$ its ring of integers, $\mathfrak{p}$ the unique maximal ideal of $\mathfrak{o}$, and ϖ a generator of $\mathfrak{p}$. Let q

be the cardinality of the residue field $\mathfrak{o}/\mathfrak{p}$. We will denote by $v : F \to \mathbb{Z} \cup \{\infty\}$ the valuation, defined by $v(0) = \infty$ and $v(\varpi^r u) = r$ if $u \in \mathfrak{o}^\times$. We will denote by $\int_F dx$ the additive Haar measure on F normalized so the volume of $\mathfrak{o}$ is one, and by $\int_{F^\times} d^\times x$ the multiplicative Haar measure on $F^\times$ normalized so the volume of $\mathfrak{o}^\times$ is one. Thus $d^\times x = (1 - q^{-1})^{-1} |x|^{-1} dx$.

We recall that by a *neighborhood base* at a point x of a topological space X, we mean a collection $\mathcal{U}$ of neighborhoods of x such that every neighborhood of x contains an element of $\mathcal{U}$. By a *totally disconnected locally compact* topological space we mean a Hausdorff topological space X such that each point of X has a neighborhood base consisting of sets that are both open and compact. Thus F in its usual topology is totally disconnected and locally compact, and indeed so is any closed subset of affine space F^n. For example, $GL(n, F)$ and its closed subgroups are totally disconnected locally compact topological groups.

Proposition 4.2.1 *If G is a totally disconnected locally compact topological group, and if U is any neighborhood of the identity, then U contains a compact and open subgroup. Moreover, if G is compact, this subgroup may be taken to be normal.*

Thus the compact and open subgroups of G form a neighborhood base at identity, as do the compact and open normal subgroups if G is compact.

Proof See Montgomery and Zippen (1955, Sections 2.3 and 2.2). ■

We do not really need this proposition (whose proof is not difficult) because our applications will be to particular groups such as $GL(n, F)$ for which a neighborhood base at the identity consisting of compact and open subgroups can be exhibited explicitly; for example, if $G = GL(n, F)$, we may take the groups $K(\varpi^n)$ $(n = 0, 1, 2, 3, \cdots)$ consisting of all elements of $GL(n, \mathfrak{o})$ congruent to the identity modulo ϖ^n; or if G is the compact group $GL(n, \mathfrak{o})$, we may use the same subgroups, which are then normal. Although we therefore do not really need this proposition, it is an important fact to be aware of.

Let G be a totally disconnected locally compact group, and let (π, V) be a representation. We assume no topology on V – it is just a complex vector space of possibly infinite dimension on which G acts. We say that π is *smooth* if for any $v \in V$, the stabilizer $\{g \in G | \pi(g)\, v = v\}$ is open. (Smoothness is a substitute for a topological condition.) If π is smooth, and if furthermore for any open subgroup $U \subset G$, the space V^U of vectors stabilized by U is finite dimensional, then π is called *admissible*. We will study the admissible representations of $GL(n, F)$.

If G is a locally compact group, recall that there exist left-invariant and right-invariant measures on G, called *left* and *right Haar measures*, respectively. These may or may not coincide: If they are the same, then G is called *unimodular*. This condition is satisfied for most of the groups that we will encounter, such as $GL(n, F)$, any unipotent, compact, or Abelian group, etc.

(See Exercises 4.2.3–4.2.5) The only nonunimodular groups that play an important role in the theory are the *parabolic* subgroups, which we now define. The (standard) *Borel subgroup* $B(F)$ is the group of all upper triangular matrices in $GL(n, F)$. (Any conjugate of $B(F)$ is also called a Borel subgroup.) A *parabolic subgroup* is a proper subgroup of $GL(n, F)$ containing a conjugate of $B(F)$ – a *standard* parabolic is a parabolic subgroup containing $B(F)$ itself. Thus $B(F)$ is itself a parabolic subgroup, and it is not unimodular. To see this, let $d_L b$ and $d_R b$ denote left and right Haar measures on $B(F)$. Then if

$$b = \begin{pmatrix} 1 & x_{12} \cdots & x_{1n} \\ & 1 \cdots & x_{2n} \\ & \ddots & \vdots \\ & & 1 \end{pmatrix} \begin{pmatrix} y_1 & & & \\ & y_2 & & \\ & & \ddots & \\ & & & y_n \end{pmatrix},$$

we have

$$d_R b = \prod_{1 \le i < j \le n} dx_{ij} \prod_{i=1}^{n} d^\times y_i, \qquad d_L b = |y_1|^{1-n} |y_2|^{3-n} \cdots |y_n|^{n-1} d_R b. \tag{2.1}$$

(Exercise 4.2.1.) Thus we see that G is not unimodular. We denote by δ_B the *modular quasicharacter* of $B(F)$, which is the ratio of right and left Haar measures; thus

$$d_R b = \delta_B(b)\, d_L b, \qquad \delta_B(b) = |y_1|^{n-1} |y_2|^{n-3} \cdots |y_n|^{1-n}. \tag{2.2}$$

More generally, if G is a locally compact group, we define the modular quasicharacter δ_G by

$$d_R g = \delta_G(g)\, d_L g. \tag{2.3}$$

Let Γ be a compact totally disconnected group; in particular, Γ can be finite. We will denote by $\widehat{\Gamma}$ the set of equivalence classes of finite-dimensional irreducible representations of Γ whose kernel is open and hence of finite index (because Γ is assumed compact – see Exercise 4.2.2.) By abuse of notation, if $\rho \in \widehat{\Gamma}$, we may sometimes denote by (ρ, V_ρ) a representation in the equivalence class of ρ.

Suppose now that Γ is a finite group and that (π, V) is a representation of Γ on a possibly infinite-dimensional vector space. If $\rho \in \widehat{\Gamma}$, let $V(\rho)$ be the sum of all invariant subspaces of V that are isomorphic as Γ-modules to V_ρ. We call $V(\rho)$ the *ρ-isotypic subspace*. Because Γ is finite, $\widehat{\Gamma}$ is also finite. We have

$$V = \bigoplus_{\rho \in \widehat{\Gamma}} V(\rho). \tag{2.4}$$

This decomposition is a consequence of Theorem XVII.4.4 on p. 653 of Lang (1993), together with the semisimplicity of the group algebra, "Maschke's theorem," which in Lang is Theorem XVIII.1.2 on p. 666.

We will generalize the decomposition (Eq. 2.4) to smooth representations of a totally disconnected locally compact group G. We recall from Proposition 4.2.1 that G has a compact open subgroup K, which we fix, and that the compact and open normal subgroups of K form a neighborhood base at the identity in K and hence in G. Let $\rho \in \widehat{K}$. The kernel of ρ is a compact and open normal subgroup K_ρ of K, so ρ might as well be considered a representation of the finite group K/K_ρ. Let (π, V) be a smooth representation of G. Generalizing our previous notation for finite groups, we let $V(\rho)$ denote the sum of all K-invariant irreducible subspaces that are isomorphic, as K-modules, to ρ. We again call $V(\rho)$ the *ρ-isotypic subspace* of V.

Proposition 4.2.2 *Let (π, V) be a smooth representation of G. Then*

$$V = \bigoplus_{\rho \in \widehat{K}} V(\rho) \qquad \text{(algebraic direct sum).} \tag{2.5}$$

The representation π is admissible if and only if each of the $V(\rho)$ is finite dimensional.

Proof We show first that V is the sum of the spaces $V(\rho)$. Indeed, If $v \in V$, then because π is smooth, v is fixed by some compact open subgroup K_0 of K, which by Proposition 4.2.1 may be chosen to be normal. Then, by Eq. (2.4) with $\Gamma = K/K_0$

$$v \in V^{K_0} = \bigoplus_{\rho \in \widehat{\Gamma}} V(\rho) \subseteq \sum_{\rho \in \widehat{K}} V(\rho).$$

Thus V is the sum of the spaces $V(\rho)$.

We now show that the sum (Eq. 2.5) is direct. Suppose it is not. Then there is a relation $\sum_{\rho \in S} c_\rho v_\rho = 0$, where S is a finite subset of $\widehat{K}$, $v_\rho \in V(\rho)$, and the constants c_ρ are not all zero. Let K_0 be the intersection of the kernels of the ρ that appear in this relation. We obtain a contradiction to the directness of the sum (Eq. 2.4), with $\Gamma = K/K_0$.

Suppose now that π is admissible. To see that $V(\rho)$ is finite dimensional, note that $V(\rho)$ is contained in the space $V^{\ker(\rho)}$ of vectors fixed by the kernel of ρ; because $\ker(\rho)$ is open, this space is finite dimensional by the definition of admissibility. On the other hand, if π is not smooth, then V^{K_0} is infinite dimensional for some open normal subgroup K_0 of K. But V^{K_0} decomposes into the finite direct sum of $V(\rho)$ with $\rho \in \widehat{K/K_0}$; because $\widehat{K/K_0}$ is finite, one of these spaces must be infinite dimensional. ■

Let (π, V) be a smooth representation of G. If $\hat{v} : V \to \mathbb{C}$ is a linear functional, we write $\langle v, \hat{v} \rangle$ for $\hat{v}(v)$ when $v \in V$. We say that $\hat{v}$ is *smooth* if there exists an open neighborhood U of the identity in G such that $\langle \pi(g)\, v, \hat{v} \rangle = \langle v, \hat{v} \rangle$ whenever $g \in U$, $v \in V$. Let $\widehat{V}$ be the space of smooth linear functionals on V.

We define a representation $(\hat{\pi}, \widehat{V})$, called the *contragredient representation*, by letting g act on $\widehat{V}$ by

$$\langle v, \hat{\pi}(g)\hat{v}\rangle = \left\langle \pi(g^{-1})\, v,\, \hat{v}\right\rangle. \tag{2.6}$$

It is not hard to check that the linear functional $\hat{\pi}(g)\,\hat{v}$ on V defined by Eq. (2.6) is smooth and hence lies in $\widehat{V}$. It is clear that $(\hat{\pi}, \widehat{V})$ is a smooth representation of G.

The dual space of an infinite direct sum of vector spaces is the direct *product* of the dual spaces. Thus $V^* = \prod V(\rho)^*$. However, it is not difficult to see that a linear functional on V is smooth if and only if it is zero on all but finitely many of the summands in Eq. (2.5). Therefore $\widehat{V}$ may be identified with the direct sum of the $V(\rho)^*$:

$$\widehat{V} = \bigoplus_{\rho} V(\rho)^*. \tag{2.7}$$

In view of Eq. (2.7), the contragredient of an admissible representation is admissible because the dual of the finite-dimensional vector space $V(\rho)$ is again finite dimensional. Moreover, because for a finite-dimensional vector space $V(\sigma)$ we have $V(\sigma)^{**} \cong V(\sigma)$, we see that if (π, V) is admissible, then $\hat{\hat{\pi}} \cong \pi$.

If X is a totally disconnected topological space, we call a function f on X (taking values in $\mathbb{C}$ or in some complex vector space) *smooth* if it is locally constant. Let $\mathcal{H}$ be the space of smooth compactly supported complex-valued functions on G. We assume G to be unimodular, and make $\mathcal{H}$ an algebra (without unit) under convolution:

$$(\phi_1 * \phi_2)(g) = \int_G \phi_1(gh^{-1})\, \phi_2(h)\, dh.$$

This is the *Hecke algebra*. If K_0 is a compact open subgroup of G, let $\mathcal{H}_{K_0}$ denote the subspace of K_0-biinvariant functions in $\mathcal{H}$. It is clear that $\mathcal{H}_{K_0}$ is closed under convolution and hence forms a subring of $\mathcal{H}$, but unlike $\mathcal{H}$, it has an identity element ϵ_{K_0}, where

$$\epsilon_{K_0}(g) = \begin{cases} \mathrm{vol}(K_0)^{-1} & \text{if } g \in K_0; \\ 0 & \text{otherwise.} \end{cases} \tag{2.8}$$

Here the volume is taken with respect to left Haar measure.

We note that many of the results in this section are derived from the fact that $\mathcal{H}$ is an *idempotented algebra,* a notion defined in Section 3.4. A number of the results in this section and Section 4.6 are special cases of more general results for smooth modules over an arbitrary idempotented algebra, whose more general statements can be found in Section 3.4.

Also, let us point out that there is an inconsistency of notation between this chapter and Section 3.4. In that section, the notation $\mathcal{H}_K$ was used to denote the algebra of matrix coefficients of a compact group K.

If $\phi \in \mathcal{H}$, we define an endomorphism $\pi(\phi)$ of V by

$$\pi(\phi)\,v = \int_G \phi(g)\,\pi(g)\,v\,dg. \tag{2.9}$$

It is easily checked that

$$\pi(\phi_1 * \phi_2) = \pi(\phi_1) \circ \pi(\phi_2),$$

so π is a representation of the ring $\mathcal{H}$.

The appearance of integration here should not conceal the fact that the theory that we are considering is purely algebraic in nature; the use of the Haar measure is just a convenience. Topological and analytic considerations could be eliminated completely. Thus, for example, the integral (Eq. 2.9) may be replaced by a finite sum. To see this, note that v is fixed by an open subgroup K_0 and the support ϕ is a finite union of left cosets $g_i\,K_0$ of K_0. If small enough K_0 is chosen, we may assume that ϕ is constant on these cosets; then Eq. (2.9) is replaced by the finite sum

$$\frac{1}{\mathrm{vol}(K_0)} \sum_i \phi(g_i)\,\pi(g_i)v.$$

We recall that if R is a ring (perhaps without unit), an R-module is called *simple* or *irreducible* if it is nonzero and has no proper nonzero submodules. If H is a subgroup of G, let V^H be the space of H-fixed vectors. Let K_0 be a compact open subgroup of G. The space V^{K_0} is clearly stable under $\mathcal{H}_{K_0}$ and is a $\mathcal{H}_{K_0}$-module. Also, if (π, V) is a representation of a group G, it is called *irreducible* if it has no G-invariant subspaces. The irreducible representations that we encounter will always be admissible (not just smooth).

Proposition 4.2.3 *Let (π, V) be a smooth representation of G. Assume that V is nonzero. The following three conditions are equivalent.*

(i) The representation π is irreducible.
(ii) V is simple as a $\mathcal{H}$-module.
(iii) V^{K_0} is either zero or simple as a $\mathcal{H}_{K_0}$-module for all open subgroups K_0 of G.

Compare Proposition 3.4.5.

Proof A G-invariant subspace is clearly $\mathcal{H}$ invariant, so the implication that (ii)$\Rightarrow$(i) is trivial. Conversely, suppose that $W \subset V$ is an $\mathcal{H}$-invariant subspace; we show it is G invariant. If not, there exists $g \in G$, $w \in W$ such that $\pi(g)\,w \notin W$. Now w is fixed by some neighborhood N of the identity, so letting $\phi \in \mathcal{H}$ be the characteristic function of gN divided by its volume, we see that $\pi(\phi)\,w = \pi(g)\,w \notin W$. This contradicts our assumption that W is $\mathcal{H}$ invariant; we have proved that (i) and (ii) are equivalent.

The proof that (iii)$\Rightarrow$(ii) is also simple. Suppose that W is a nonzero proper $\mathcal{H}$-invariant subspace. Then because $V = \bigcup V^{K_0}$, we can find K_0 small enough that W^{K_0} is a nonzero proper subspace of V^{K_0}, contradicting (iii).

The only difficult part of this proposition is (ii)$\Rightarrow$(iii). To prove this, let $W_0 \subset V^{K_0}$ be a proper, nonzero $\mathcal{H}_{K_0}$ submodule. We will show that $\pi(\mathcal{H})W_0 \cap V^{K_0} = W_0$. Note that this is sufficient to contradict (ii), for it shows that $\pi(\mathcal{H})W_0$ is a proper nonzero subspace of V, and it is clear that it is $\mathcal{H}$ invariant. By definition, $\pi(\mathcal{H})W_0$ is the set of vectors that are finite sums $\sum_i \pi(\phi_i)\, w_i$ with $\phi_i \in \mathcal{H}$, $w_i \in W_0$. We assume that such a vector lies in V^{K_0}; we must show it is in W_0. Because $w_i \in W_0 \subset V^{K_0}$, $\pi(\epsilon_{K_0})(w_i) = w_i$, and because $\sum_i \pi(\phi_i) w_i \in V^{K_0}$, we have $\pi(\epsilon_{K_0}) \sum_i \pi(\phi_i) w_i = \sum_i \pi(\phi_i) w_i$. Thus $\sum_i \pi(\phi_i) w_i = \sum_i \pi(\epsilon_{K_0} * \phi_i * \epsilon_{K_0}) w_i$. However, $\epsilon_{K_0} * \phi_i * \epsilon_{K_0} \in \mathcal{H}_{K_0}$, and because W_0 is $\mathcal{H}_{K_0}$ stable, this shows that $\sum_i \pi(\phi_i) w_i \in W_0$, as required. ■

If (π_1, V_1) and (π_2, V_2) are representations of a group G, we recall that an *intertwining operator* is a linear map $T : V_1 \to V_2$ such that $T \circ \pi_1(g) = \pi_2(g) \circ T$ for all $g \in G$.

Proposition 4.2.4: Schur's lemma *Let (π, V) be an irreducible admissible representation of the totally disconnected locally compact group G. Let $T : V \to V$ be an intertwining operator for π. Then there exists a complex number c such that $T(v) = cv$ for all $v \in V$.*

Proof Let K_0 be an open and compact subgroup small enough that V^{K_0} is nonzero. Then V^{K_0} is finite dimensional because π is assumed admissible and it is clearly preserved under T. Thus T has an eigenvalue c on V^{K_0}. If $I : V \to V$ denotes the identity transformation, the kernel of $T - cI$ is a nonzero G-invariant subspace, and because V is irreducible, it must be all of V. ■

An immediate consequence of this is that the center of G acts by scalars on the irreducible admissible representation (π, V). Thus if $G = GL(n, F)$, where F is a non-Archimedean local field, there exists a quasicharacter ω of $F^\times$, the so-called *central quasicharacter*, such that the center $Z(F)$ of $GL(n, F)$ acts by Eq. (1.1).

The central quasicharacter may or may not be unitary. If it is not, then by our conventions it is not a character.

Proposition 4.2.5 *Let (π, V) be an admissible representation of the totally disconnected locally compact group G, and let $(\hat{\pi}, \widehat{V})$ be the contragredient. Let K_0 be an open and compact subgroup of G. Then the canonical pairing between V and $\widehat{V}$ induces a nondegenerate pairing between V^{K_0} and $\widehat{V}^{K_0}$.*

Proof Let $0 \neq v \in V^{K_0}$. We will show that there exists a vector $\hat{v} \in \widehat{V}^{K_0}$ such that $\langle v, \hat{v} \rangle \neq 0$. (By symmetry, this is sufficient because $\hat{\widehat{V}} \cong V$, so the same reasoning shows that the pairing is nondegenerate on the other side also.)

Choose $\hat{v}_1 \in \widehat{V}$ such that $\langle v, \hat{v}_1 \rangle \neq 0$. Because $v \in V^{K_0}$, we have $v = \pi(\epsilon_{K_0})\, v$. Now

$$\langle v, \hat{v}_1 \rangle = \left\langle \pi(\epsilon_{K_0})\, v, \hat{v}_1 \right\rangle = \left\langle v, \hat{\pi}(\epsilon_{K_0})\, \hat{v}_1 \right\rangle,$$

so we may take $\hat{v} = \hat{\pi}(\epsilon_{K_0})\, \hat{v}_1$, which is a K_0-fixed vector in $\widehat{V}$. ■

As with representations of finite groups, the *character* of an admissible representation of a totally disconnected locally compact group G is an important tool. The character is not defined as a function, but as a *distribution*. (It is a theorem of Harish-Chandra, which we will not prove, that if G is a reductive p-adic group then the character, *a priori* a distribution, is actually a locally integrable function defined on a dense subset of G.) Let X be a locally compact totally disconnected space. We recall that a function on X is *smooth* if it is locally constant. Let $C_c^\infty(X)$ be the ring of smooth compactly supported functions $X \to \mathbb{C}$. A *distribution* on X is a linear functional on $C_c^\infty(X)$. We will denote by $\mathfrak{D}(X)$ the space of distributions on X. In contrast with distributions on manifolds, there is no assumption of continuity.

We will define the character as a distribution on G, that is, a linear functional on $\mathcal{H} = C_c^\infty(G)$. First we need some preliminaries on the trace. Let U be a finite-dimensional vector space, and let $f : U \to U$ be an endomorphism. Let U_0 be any subspace containing the image of f. Then f induces an endomorphism $f|U_0 : U_0 \to U_0$. We observe that

$$\text{the trace of } f \text{ equals the trace of } f|U_0. \tag{2.10}$$

To see this, we pick a basis $u_1, \cdots, u_n$ of U such that the first d elements are a basis U_0, where $d = \dim(U_0)$. With respect to this basis, the matrix of f has the form

$$\begin{pmatrix} A & B \\ 0 & 0 \end{pmatrix},$$

where $A, B, 0$, and 0 are blocks of shape $d \times d$, $d \times (n-d)$, $(n-d) \times d$, and $(n-d) \times (n-d)$, respectively, and A is the matrix of $f|U_0$. Thus Eq. (2.10) is evident. Now, suppose that V is a possibly infinite-dimensional vector space and $f : V \to V$ is an endomorphism. We recall that f is said to have *finite rank* if its image is finite dimensional. In this case, we may define the trace of f as follows. Let U be any finite-dimensional subspace of V containing the image of f. Then f induces an endomorphism $f|U$ of U, and we define the trace $\operatorname{tr}(f)$ of f to be the trace of $f|U$. It is a simple consequence of Eq. (2.10) that this definition is independent of the choice of U, so long as U contains the image of f.

Now let (π, V) be an admissible representation of G. By definition, a distribution is a linear functional on the Hecke algebra $\mathcal{H} = C_c^\infty(G)$. If $\phi \in \mathcal{H}$, then because ϕ is locally constant and compactly supported, $\phi \in \mathcal{H}_{K_0}$ for some compact open subgroup K_0. The endomorphism $\pi(\phi)$ defined by Eq. (2.9) has its image in the K_0-fixed vectors, which form a finite-dimensional vector space by the definition of admissibility. Thus $\pi(\phi)$ has finite rank, and we define a distribution $\chi : \mathcal{H} \to \mathbb{C}$ by $\chi(f) = \operatorname{tr}\big(\pi(f)\big)$.

Proposition 4.2.6 *Let R be an algebra over a field k of characteristic zero. Let E_1 and E_2 be simple R-modules that are finite dimensional as vector spaces over k. For each $\phi \in R$, multiplication by ϕ induces endomorphisms $\lambda_i(\phi)$ of E_i $(i = 1, 2)$. Assume that $\operatorname{tr}\big(\lambda_1(\phi)\big) = \operatorname{tr}\big(\lambda_2(\phi)\big)$ for all $\phi \in R$. Then the R-modules E_1 and E_2 are isomorphic.*

Proof See Lang (1993, Corollary XVII.3.8 on p. 650). Actually, Lang proves this results for semisimple modules; an examination of the proof shows that for *simple* modules, the hypothesis of characteristic zero is unnecessary. ■

Proposition 4.2.7 *Let (π_1, V_1) and (π_2, V_2) be irreducible admissible representations of the totally disconnected locally compact group G. If $V_1^{K_1} \cong V_2^{K_1}$ as $\mathcal{H}_{K_1}$-modules for every open and compact subgroup K_1 of G, then $\pi_1 \cong \pi_2$.*

Compare Proposition 3.4.6. A stronger result may be proved by adapting the proof of Theorem 4.6.3. If there exists an open compact subgroup K such that V_1^K and V_2^K are nonzero (hence simple $\mathcal{H}_K$-modules by Proposition 4.2.3), and if these $\mathcal{H}_K$-modules are isomorphic, then $V_1 \cong V_2$.

Proof Let us fix K_0 small enough that $V_1^{K_0}$ and $V_2^{K_0}$ are nonzero. Then by hypothesis, $V_1^{K_0}$ and $V_2^{K_0}$ are isomorphic as $\mathcal{H}_{K_0}$-modules. We fix an isomorphism $\sigma_0 : V_1^{K_0} \to V_2^{K_0}$. (By Schur's lemma, σ_0 is determined up to scalar multiple, because by Proposition 4.2.3, $V_1^{K_0}$ and $V_2^{K_0}$ are irreducible.) We observe that if K_1 is an open subgroup of K_0, then σ_0 can be extended uniquely to a $\mathcal{H}_{K_1}$-module isomorphism $\sigma_{K_1} : V_1^{K_1} \to V_2^{K_1}$. Indeed, the existence of an isomorphism σ_{K_1} is our hypothesis; we observe that $V_i^{K_0} = \pi_i(\epsilon_{K_0})\, V_i^{K_1}$, and because σ_{K_1} is a ring homomorphism, we see that

$$\sigma_{K_1}(V_1^{K_0}) = \sigma_{K_1}\left(\pi_1(\epsilon_{K_0})\, V_1^{K_1}\right) = \pi_2(\epsilon_{K_0})\left(\sigma_{K_1}\, V_1^{K_1}\right) = \pi_2(\epsilon_{K_0})\, V_2^{K_1},$$

which equals $V_2^{K_0}$. Thus σ_{K_1} maps $V_1^{K_0}$ into $V_2^{K_0}$ and induces a $\mathcal{H}_{K_0}$ isomorphism of these spaces. By the uniqueness of σ_0, σ_{K_1} restricted to $V_1^{K_0}$ agrees with σ_0 up to a scalar, which we can normalize to equal one. The uniqueness of σ_{K_1} follows from the irreducibility of $V_1^{K_1}$ and $V_2^{K_1}$ just as with σ_0. It follows from a repetition of this argument that if K_2 is an open subgroup of K_1 then σ_{K_2} agrees with σ_{K_1} on $V_1^{K_1}$. We may therefore define a map $\sigma : V_1 \to V_2$ as follows. If $v \in V_1$, then because π_1 is smooth, $v \in V_1^{K_1}$ for some K_1, and we define $\sigma(v) = \sigma_{K_1}(v)$. It follows from what goes before that this map is independent of the choice of K_1, and it is clear that it is a $\mathcal{H}$-isomorphism.

Thus we are reduced to showing that if $\sigma : V_1 \to V_2$ is a $\mathcal{H}$-module isomorphism, then it is an intertwining operator. To see this, let $g \in G$, $v \in V_1$. We can find a compact open subgroup K_1 of G such that $v \in V_1^{K_1}$ and $\sigma(v) \in V_2^{K_1}$. Let $\phi \in \mathcal{H}_{K_1}$ be the characteristic function of the coset gK_1, divided by the volume of K_1. Then it is clear that $\pi_1(\phi)\,v = \pi_1(g)\,v$ and $\pi_2(\phi)\,\sigma(v) = \pi_2(g)\,\sigma(v)$. Thus

$$\sigma\big(\pi_1(g)\,v\big) = \sigma\big(\pi_1(\phi)\,v\big) = \pi_2(\phi)\,\sigma(v) = \pi_2(g)\,\sigma(v).$$

We see that σ is indeed an intertwining operator, and the representations π_1 and π_2 are therefore equivalent. ■

Theorem 4.2.1 *Let (π_1, V_1) and (π_2, V_2) be irreducible admissible representations of the totally disconnected locally compact group G. If the characters of π_1 and π_2 agree, then the two representations are equivalent.*

Proof If the characters of π_1 and π_2 agree, then the hypotheses of Proposition 4.2.6 are satisfied with $k = \mathbb{C}$, $R = \mathcal{H}_{K_1}$, $E_1 = V_1^{K_1}$, and $E_2 = V_2^{K_1}$. Thus $V_1^{K_1} \cong V_2^{K_1}$ as $\mathcal{H}_{K_1}$-modules, so by Proposition 4.2.7, $\pi_1 \cong \pi_2$. ■

We will use this result to prove the analog of Proposition 4.1.1.

Theorem 4.2.2 *Let $G = GL(n, F)$, where F is a non-Archimedean local field, and let (π, V) be an irreducible admissible representation of G.*

(i) Define a representation (π_1, V) on the same space by $\pi_1(g) = \pi\left({}^{\top}g^{-1}\right)$. Then $\hat{\pi} \cong \pi_1$.
(ii) Suppose that $n = 2$. Let ω be the central quasicharacter of π. Define another representation (π_2, V) on the same space by $\pi_2(g)\,v = \omega\big(\det(g)\big)^{-1}\,\pi(g)$. Then $\hat{\pi} \cong \pi_2$.

Let G be a group acting on a totally disconnected locally compact space X. We assume that if $g \in G$ then the transformation $x \mapsto gx$ of X is a homomorphism. Then G acts on $C_c^\infty(X)$ and on $\mathfrak{D}(X)$ in the obvious way: If $\phi \in C_c^\infty(X)$, we define $(g\phi)(x) = \phi(g^{-1}x)$, and if $D \in \mathfrak{D}(X)$, we define $(gD)(\phi) = D(g^{-1}\phi)$. In particular, if $G = GL(n, F)$, we may take $X = G$ acting by conjugation to obtain an action of G on $\mathfrak{D}(G)$. Similarly, the transformation $g \mapsto {}^{\top}g$ of g by transpose induces an automorphism of $\mathfrak{D}(G)$.

We will deduce Theorem 4.2.2 from the following theorem.

Theorem 4.2.3 *Let $G = GL(n, F)$, where F is a non-Archimedean local field, and let D be a distribution on G that is invariant under conjugation. Then D is also invariant under transpose.*

This result is eminently plausible because in $GL(n, F)$, every matrix is conjugate to its transpose (Exercise 4.1.3), so any conjugation-invariant element of $C_c^\infty(G)$ is also transpose invariant. Nevertheless, the proof is not so simple. For $n = 2$, we will give the proof in the following section. (For $n > 2$, the proof will be left incomplete.) Assuming Theorem 4.2.3, we will show how to deduce Theorem 4.2.2.

Proof of Theorem 4.22 assuming Theorem 4.23 It is easy to see that the character of an irreducible admissible representation of $GL(n, F)$ is conjugation invariant; because we are assuming Theorem 4.2.2, it is therefore transpose invariant. Let χ be the character of π; if $\phi \in C_c^\infty(G)$, let $\phi'(g) = \phi(g^{-1})$, and let $\phi''(g) = \phi({}^\top g^{-1})$. It is easy to see that $\pi_1(\phi) = \pi(\phi'')$, so the character of π_1 is $\phi \mapsto \chi(\phi'')$, which equals $\chi(\phi')$ because χ is transpose invariant.

On the other hand, it is easy to see that $\pi(\phi)$ and $\hat{\pi}(\phi')$ are adjoints of each other and hence have equal trace. This shows that π_1 and $\hat{\pi}$ have the same character and hence are isomorphic by Theorem 4.2.1. This proves Theorem 4.2.2(i).

As for Theorem 4.2.2(ii), it follows from the identity (valid for $g \in GL(2, F)$)

$$ {}^\top g^{-1} = \begin{pmatrix} \det(g) & \\ & \det(g) \end{pmatrix}^{-1} w^{-1} g w, \qquad w = \begin{pmatrix} & -1 \\ 1 & \end{pmatrix} $$

that $\pi(w)$ is an intertwining operator from (π_1, V) to (π_2, V). Thus the representations π_1 and π_2 are equivalent. ■

Proposition 4.2.8 *Let π be an admissible representation of $GL(n, F)$. Then π is irreducible if and only if $\hat{\pi}$ is irreducible.*

Proof It is sufficient to show that π is irreducible if and only if the representation π_1 in Theorem 4.2.2(i) is irreducible. But clearly a subspace is invariant under π if and only if it is invariant under π_1. ■

We end this section with a discussion of the representations of $F^\times$, which is of course isomorphic to $GL(1, F)$ or of its k-fold Cartesian product $(F^\times)^k$. Because $(F^\times)^k$ is a Abelian, one might think that an admissible representation would decompose into a direct sum of one-dimensional invariant subspaces. This is not the case, however. Consider the homomorphism $\rho : F^\times \to GL(2, \mathbb{C})$ given by

$$ \rho(a) = \begin{pmatrix} 1 & \log|a| \\ & 1 \end{pmatrix}. $$

Evidently, this representation is admissible and has a unique one-dimensional invariant subspace, so it is not irreducible, yet there is no complementary one-dimensional invariant subspace.

Nevertheless we have the following result.

Proposition 4.2.9 *Let (π, V) be an admissible representation of $(F^\times)^k$. Then V, if nonzero, contains a one-dimensional invariant subspace.*

Proof Because (π, V) is smooth, $V = \bigoplus_\chi V_\chi$, where χ runs through $\widehat{(\mathfrak{o}^\times)^k}$ and where V_χ is the space of vectors v satisfying $\pi(u)\, v = \chi(u)\, v$ with u in the compact subgroup $(\mathfrak{o}^\times)^k$. Because π is admissible, the spaces V_χ are finite dimensional. Let χ be such that V_χ is nonzero. Let $p_r = (1, \cdots, 1, \varpi, 1, \cdots, 1) \in (F^\times)^k$, where the ϖ is in the rth position, $1 \le r \le k$. Then the operators $\pi(p_r)$ commute with the action of $(\mathfrak{o}^\times)^k$, and thus leave the finite-dimensional space V_χ invariant, and they commute with each other, so they have a common one-dimensional eigenspace. This space is invariant under a set of generators of $(F^\times)^k$ and hence meets our requirements. ■

Exercises

Exercise 4.2.1 Verify that the measures (Eq. 2.1) are left and right Haar measures for $B(F)$.

Exercise 4.2.2 Let G be a compact totally disconnected group. Show that a subgroup of G is open if and only if it has finite index. Hence there can be at most one topology with respect to which G is compact and totally disconnected.

Exercise 4.2.3 Prove that a compact group is unimodular.

Exercise 4.2.4 Let F be a local field. Let $d_a g$ denote the additive Haar measure on $\mathrm{Mat}_n(F)$. Prove that the measure $dg = |\det(g)|^{-n}\, d_a g$ on $GL(n, F)$ is both left and right invariant, and conclude that $GL(n, F)$ is unimodular.

Exercise 4.2.5 Let N be the group of upper triangular unipotent matrices in $GL(n, F)$, where F is a local field. Show that if

$$x = \begin{pmatrix} 1 & x_{12} & \cdots & x_{1n} \\ & 1 & & \vdots \\ & & \ddots & x_{n-1,n} \\ & & & 1 \end{pmatrix},$$

then (denoting by dx_{ij} additive Haar measure on F)

$$\prod_{1 \le i < j \le n} dx_{ij}$$

is the Haar measure on $N(F)$, and conclude that $N(F)$ is unimodular.

Exercise 4.2.6 Would Proposition 4.2.9 be true for an admissible representation of an arbitrary totally disconnected locally compact group?

Exercise 4.2.7 Let $V = \oplus_{n \in \mathbb{Z}} \mathbb{C}$ be the direct sum of an infinite number of copies of $\mathbb{C}$. Thus an element $(a_n) \in V$ consists of a sequence of complex numbers indexed by $n \in \mathbb{Z}$ such that $a_n = 0$ for all but finitely many n. We have a representation $\rho : \mathbb{Z} \to \mathrm{End}(V)$ in which $\rho(m)$ is the shift operator that sends (a_n) to (a_{n+m}). Describe the invariant subspaces of V.

Exercise 4.2.8 Prove that the contragredient of an irreducible admissible representation of $GL(n, F)$ is irreducible.

Exercise 4.2.9 Prove that if (π, V) is a finite-dimensional irreducible admissible representation of $GL(2, F)$, then V is one dimensional. [HINT: By the "no small subgroups argument" (Exercise 3.1.1(a)) the kernel of π contains an open subgroup. Show that an open normal subgroup of $GL(2, F)$ contains $SL(2, F)$.]

Exercise 4.2.10 We call a representation *indecomposable* if it does not decompose into a direct sum of irreducible representations. Before Proposition 4.2.9 we gave an example of an indecomposable irreducible admissible representation of $F^\times$ of dimension two. Show that any two-dimensional indecomposable admissible representation of $F^\times$ is obtained by tensoring this example with a quasicharacter of $F^\times$.

4.3 Distributions and Sheaves

To complete the proof of Theorem 4.2.2, we need some foundational material on sheaves and distributions. Our treatment of these topics is based on the excellent paper of Bernstein and Zelevinsky (1976).

Let X be a totally disconnected locally compact space. We recall from the previous section that by *distribution* on X, we mean a linear functional on $C_c^\infty(X)$, the space of smooth compactly supported functions on X, and that $\mathfrak{D}(X)$ be the space of distributions on X. We note that (unlike the theory of distributions on manifolds) there is no assumption that the functional be continuous. Distribution theory is therefore much less analytic in the case of a totally disconnected locally compact space than in the case of a manifold.

Proposition 4.3.1 *Let X be a totally disconnected locally compact space, and let $C \subseteq X$ be a closed subset. Then we have exact sequences*

$$0 \to C_c^\infty(X - C) \to C_c^\infty(X) \to C_c^\infty(C) \to 0 \tag{3.1}$$

and

$$0 \to \mathfrak{D}(C) \to \mathfrak{D}(X) \to \mathfrak{D}(X - C) \to 0. \tag{3.2}$$

Proof The maps in Eq. (3.1) are clear: We obtain a map $C_c^\infty(X - C) \to C_c^\infty(X)$ by extending a function $f \in C_c^\infty(X - C)$ by zero on C to a function in $C_c^\infty(X)$, and the map $C_c^\infty(X) \to C_c^\infty(C)$ is just a restriction of functions to C. Regarding the exactness of Eq. (3.1), the only point that is not clear is the surjectivity of restriction to C. Let $f \in C_c^\infty(C)$. Because f is compactly supported and locally constant, there exist disjoint open and compact sets $U_i \subseteq C$ and $a_i \in \mathbb{C}$ such that $f(x) = a_i$ if $x \in U_i$ and $f(x) = 0$ off $\bigcup U_i$. Let V_i be open and compact subsets of X such that $U_i = V_i \cap C$. We may arrange that the V_i are disjoint by successively replacing V_i by $V_i - \bigcup_{j<i} V_j$, which are also open and compact and give the U_i on intersection with C. Then we extend the function f to X by letting $f(x) = a_i$ if $x \in V_i$ and $f(x) = 0$ off $\bigcup V_i$.

This proves the exactness of Eq. (3.1), and the exactness of Eq. (3.2) follows by dualizing. ■

Let G be a locally compact totally disconnected compact group. We define actions of G on G, $C_c^\infty(G)$, and $\mathfrak{D}(G)$ by left and right translation:

$$\rho(g)\,x = xg^{-1}, \qquad \lambda(g)\,x = gx, \qquad g, x \in G; \tag{3.3}$$

$$\big(\rho(g)\,f\big)(x) = f(xg), \quad \big(\lambda(g)\,f\big)(x) = f(g^{-1}x), \quad f \in C_c^\infty(G), \tag{3.4}$$

if $g,\ x \in G$, and

$$\big(\rho(g)\,T\big)(f) = T\big(\rho(g^{-1})f\big), \qquad \big(\lambda(g)\,T\big)(f) = T\big(\lambda(g^{-1})f\big) \tag{3.5}$$

for $T \in \mathfrak{D}(G)$, $f \in C_c^\infty(G)$. (The arrangement of -1s in these formulas is determined by the requirement that $\rho(g_1g_2) = \rho(g_1) \circ \rho(g_2)$ and $\lambda(g_1g_2) = \lambda(g_1) \circ \lambda(g_2)$.)

Proposition 4.3.2 *Let G be a locally compact totally disconnected group, and let ξ be a character of G. Suppose that T is a distribution on G that satisfies $\lambda(h)\,T = \xi(h)^{-1}\,T$ for $h \in G$. Then there exists a constant c such that*

$$T(f) = c \int_G \xi(h)\,f(h)\,dh, \tag{3.6}$$

where dh denotes the left Haar measure.

Of course, a similar statement (which we will also sometimes need) is true for the right Haar measure.

Proof We define another distribution by $f \mapsto T(\xi^{-1}f)$. (By the "no small subgroups" argument – Exercise 3.1.1(c) – ξ and hence ξf are locally constant.) Replacing T by this distribution, we see that there is no loss of generality in assuming that $\xi = 1$. We therefore assume that $\lambda(h)\,T = T$ for all h.

Let K be a fixed open compact subgroup of G. (Such subgroups exist by Proposition 4.5.1.) We will show that Eq. (3.6) is valid for all f with $c = T(\epsilon_K)$, where as in Eq. (2.8), ϵ_K is the characteristic function of K divided by the volume of K with respect to the left Haar measure.

If $f \in C_c^\infty(G)$, let $S(f) = \{h \in G | \lambda(h)f = f\}$. Then $S(f)$ is a compact open subgroup of G, and if K_0 is any open subgroup of $S(f)$, let h_i ($i = 1, \cdots, r$) be representatives for those cosets in $K_0\backslash G$ (finite in number) on which f does not vanish, and let $a_i = f(h_i)$. Then

$$f = \mathrm{vol}(K_0) \sum_{i=1}^{r} a_i \lambda(h_i)\,\epsilon_{K_0}. \tag{3.7}$$

Using the hypothesis that T is invariant under left translation, we now have

$$T(f) = \mathrm{vol}(K_0)\left[\sum_{i=1}^{r} a_i\right] T(\epsilon_{K_0}). \tag{3.8}$$

Apply this formula with $f = \epsilon_K$, so that $S(f) = K$ and K_0 can be any open subgroup of K. The index of K_0 in K equals $\mathrm{vol}(K)/\mathrm{vol}(K_0)$, and so Eq. (3.8) implies that

$$T(\epsilon_{K_0}) = T(\epsilon_K) = c. \tag{3.9}$$

Now let f be general. Substituting Eq. (3.9) back into Eq. (3.8), where there is no harm in assuming that $K_0 \subseteq K$, we obtain Eq. (3.6). ■

Proposition 4.3.3 *Let G be a topological group and H a closed subgroup. Then the quotient space G/H (given the quotient topology) is Hausdorff, and the projection map $p : G \to G/H$ is both continuous and open. If G is locally compact and totally disconnected, then so is G/H.*

The quotient topology is by definition the topology in which $U \subset G/H$ is open if and only if $p^{-1}(U)$ is open in G.

Proof We recall that a space X is Hausdorff if and only if the diagonal in $\Delta \subset X \times X$ is closed. We review the reason for this criterion. Suppose that Δ is closed so its complement is open. If (P, Q) is not on the diagonal, in other words if $P \neq Q$, then (P, Q) has a neighborhood Ω not intersecting the diagonal. By the definition of the product topology, Ω contains an open set of the form $\Omega_P \times \Omega_Q$, where Ω_P and Ω_Q are open neighborhoods of P and Q; because $\Omega \cap \Delta = \emptyset$, Ω_P and Ω_Q are disjoint. Thus unraveling what it means for the diagonal to be closed, one arrives at *precisely* the definition of a Hausdorff space.

In the case at hand, let $X = G/H$. Then $X \times X$ is homeomorphic to the quotient $(G \times G)/(H \times H)$. The preimage of the diagonal is the image of all pairs (g_1, g_2) in $G \times G$ such that $g_1 g_2^{-1} \in H$. This is closed because H is closed. Because the preimage of Δ in $G \times G$ is closed, Δ is closed. Thus G/H is Hausdorff.

To see that $p : G \to G/H$ is an open map, note that if $U \subset G$ is open, then $p^{-1}p(U)$ is the union of H-translates of U and hence is an open set; therefore $p(U)$ is open in the quotient topology.

Now assume that G is locally compact and totally disconnected. It follows from the definition of the quotient topology (and the fact that p is open) that the image under p of a basis of the topology of G is a basis of the topology of G/H. The compact and open subsets of G form a basis of the topology; the image under p of a compact and open set is open (because p is a continuous open map). Therefore their images form a basis of the topology of G/H consisting of compact and open sets, so G/H is locally compact and totally disconnected. ■

Let us assume now that G is a locally compact group, that H is a closed subgroup, and (as in Proposition 4.3.3) that $p : G \to G/H$ is the projection map. We have a mapping $\Lambda : C_c(G) \to C_c(G/H)$ defined by

$$\Lambda(\phi)\big(p(g)\big) = \int\limits_H \phi(gh)\, dh,$$

where $\int_H dh$ denotes the left Haar measure on G. (Recall that if X is a locally compact topological space, $C_c(X)$ is the space of continuous compactly supported complex-valued functions on X.) Note that this is well defined independent of the choice of representative g. $\Lambda(\phi)$ is compactly supported because its support is contained in $p\big(\mathrm{supp}(\phi)\big)$, the continuous image of a compact set. If G and therefore also H are totally disconnected, and if ϕ is smooth (locally constant), then so is $\Lambda(\phi)$; indeed, if ϕ is left-invariant by the open subgroup K of G, then $\Lambda(\phi)$ is left-invariant by the same subgroup. Thus Λ takes $C_c^\infty(G)$ into $C_c^\infty(G/H)$. We will denote by $\Lambda^\infty : C_c^\infty(G) \to C_c^\infty(G/H)$ the restriction of Λ.

Proposition 4.3.4 *In this situation, the maps $\Lambda : C_c(G) \to C_c(G/H)$ and (if G and therefore also H are totally disconnected) $\Lambda^\infty : C_c^\infty(G) \to C_c^\infty(G/H)$ are surjective.*

Proof The surjectivity of Λ is true if G is assumed only to be locally compact and H is assumed to be closed – we omit the proof, and refer to Hewitt and Ross (1979, Theorem 15.21, p. 204). We will assume that G is totally disconnected, and prove the surjectivity of Λ^∞. Let $\phi \in C_c(G/H)$, and let $U \subset G/H$ be the support of ϕ. Then U is open and compact. Its preimage $p^{-1}(U)$ in G may not be compact, but we will prove the existence of an open and compact set $V \subseteq G$ such that $p(V) = U$. Indeed, U has an open cover $\{V_i | i \in I\}$ by open and compact sets. Each set $p(V_i)$ is open, and the sets $p(V_i)$ cover U. Because U is compact, the indexing set I has a finite subset J such that the sets $p(V_j)$ cover U. Let $V \subset G$ be the union of the V_j, $j \in J$; then V is open and compact and $p(V) = U$.

Let ψ_0 be the characteristic function of V. Then $\Lambda(\psi_0)$ and ϕ have the same support. Thus if we define

$$\psi(x) = \begin{cases} \phi\big(p(x)\big)/\Lambda(\psi_0)\big(p(x)\big) & \text{if } x \in V; \\ 0 & \text{otherwise,} \end{cases}$$

we have $\Lambda(\psi) = \phi$. Note that ψ is smooth. ■

If $g \in G$, we define $\lambda(g) : C_c(G/H) \to C_c(G/H)$ by

$$\big(\lambda(g)\phi\big)\big(p(x)\big) = \phi\big(g^{-1}p(x)\big).$$

We call a measure $\int_{G/H} dx$ on G/H *left-invariant* if

$$\int_{G/H} \big(\lambda(g)\phi\big)(x)\,dx = \int_{G/H} \phi(x)\,dx, \qquad g \in G, \phi \in C_c(G/H).$$

The subspace $C_c^\infty(G/H)$ is invariant under the action λ, and we similarly call a distribution in $\mathfrak{D}(G/H)$ *left-invariant* if $D\big(\lambda(g)\phi\big) = D(\phi)$ for $g \in G$, $\phi \in C_c^\infty(G/H)$.

Proposition 4.3.5 *Let G be a locally compact group and H a closed subgroup. Assume that G and H are both unimodular.*

(i) Let $\int_H dh$ and $\int_G dg$ denote Haar measures on H and G, respectively. If $\phi \in C_c(G)$ such that $\int_H \phi(gh)\,dh = 0$ for all $g \in G$, then $\int_G \phi(g)\,dg = 0$.
(ii) There exists a left-invariant positive regular Borel measure $\int_{G/H} dx$ on $C_c(G/H)$. This measure is finite on compact sets and may be normalized so that

$$\int_{G/H} (\Lambda\phi)(x)\,dx = \int_G \phi(g)\,dg. \tag{3.10}$$

(iii) Suppose that G and therefore also H are totally disconnected. If $D \in \mathfrak{D}(G/H)$ is a left-invariant distribution, then there exists a constant c such that

$$D(\phi) = c \int_{G/H} \phi(x)\,dx.$$

If G and H are not unimodular, the statement must be modified. We refer to Hewitt and Ross (1979, Theorem 15.24, p. 206) for a definitive theorem. We will make use of the Riesz representation theorem in the following proof, which was already discussed in detail in the proof of Proposition 2.1.5.

Proof We first prove (i). By the surjectivity of Λ in Proposition 4.3.4, there exists a function $f \in C_c(G)$ such that $\Lambda(f)$ is equal to one on the support of ϕ. Then by Fubini's theorem,

$$\begin{aligned}
\int_G \phi(g)\,dg &= \int_G \int_H \phi(g)\,f(gh)\,dh\,dg \\
&= \int_H \int_G \phi(g)\,f(gh)\,dg\,dh \\
&= \int_H \int_G \phi(gh^{-1})\,f(g)\,dg\,dh.
\end{aligned}$$

Now the transformation $h \to h^{-1}$ interchanges left and right Haar measures on H, but because H is unimodular, these are the same. Thus if we interchange the G and H integrations again and replace h by h^{-1}, the inner integral with respect to h vanishes, so $\int_G \phi(g)\, dg$ is zero.

Now we have a positive linear functional on $C_c(G/H)$ defined by $\Lambda(\phi) \mapsto \int_G \phi(g)\, dg$. Part (i) shows that this is well defined. It therefore follows from the Riesz representation theorem that there exists a positive regular Borel measure that is finite on compact sets satisfying Eq. (3.10), which gives us (ii).

Assume that G is totally disconnected, and let D be a left invariant distribution on G/H. Then applying Proposition 4.3.2 to $D \circ \Lambda^\infty$ shows that $D \circ \Lambda^\infty$ is, up to constant multiple, $\int_G dg$; combined with the surjectivity of Λ^∞ and Eq. (3.10), this implies (iii). ■

Proposition 4.3.6 *Let X and Y be totally disconnected locally compact spaces. Then*

$$C_c^\infty(X \times Y) \cong C_c^\infty(X) \otimes C_c^\infty(Y);$$

under this identification $\phi_1 \otimes \phi_2 \in C_c^\infty(X) \otimes C_c^\infty(Y)$ *is associated with the function* $(x, y) \mapsto \phi_1(x)\phi_2(y)$ *on* $X \times Y$.

Proof This follows easily from the readily verified fact that any compact open subset of $X \times Y$ is a finite union of sets of the form $U \times V$, where U and V are compact open subsets of X and Y, respectively; so an element of $C_c^\infty(X \times Y)$ is a finite linear combination of characteristic functions of sets of this type. ■

Following Bernstein and Zelevinsky, we wish to introduce distributions in the slightly more general context of a *sheaf*.

Let X be a topological space, and let $\mathfrak{T}$ be the topology of X, that is, the set of open sets. We recall that a *base* $\mathfrak{T}_0$ for the topology is a subset of $\mathfrak{T}$ with the property that if $x \in U \in \mathfrak{T}$, there exists $U_0 \in \mathfrak{T}_0$ such that $x \in U_0 \subseteq U$. Of course, specifying a base determines the topology. By a *presheaf* (of Abelian groups) *on X with the base* $\mathfrak{T}_0$, we mean the following data. Let there be given, for every $U \in \mathfrak{T}_0$, an Abelian group $\mathcal{F}(U)$, and when $U \supseteq V$ are elements of $\mathfrak{T}_0$, we assume a *restriction homomorphism* $\rho_{U,V} : \mathcal{F}(U) \to \mathcal{F}(V)$ to be defined such that $\rho_{U,U}$ is the identity map and such that when $U \supseteq V \supseteq W$ are elements of $\mathfrak{T}_0$, we have $\rho_{V,W} \circ \rho_{U,V} = \rho_{U,W}$. If $\emptyset \in \mathfrak{T}_0$, we assume that $\mathcal{F}(\emptyset) = 0$. Specifying a presheaf requires several data, but by abuse of terminology, we speak of *the presheaf* $\mathcal{F}$. Of course presheaves of rings, vector spaces, etc., are defined in the same way.

Let $\mathcal{F}_1$ and $\mathcal{F}_2$ be presheaves on X with the base $\mathfrak{T}_0$. By a *morphism* $\phi : \mathcal{F}_1 \to \mathcal{F}_2$ in the category presheaves with base $\mathfrak{T}_0$, we mean the following data. For every $U \in \mathfrak{T}_0$, let there be specified a homomorphism $\phi_U : \mathcal{F}_1(U) \to \mathcal{F}_2(U)$; we require that if $U \supseteq V$ are elements of $\mathfrak{T}_0$, then $\phi_V \circ \rho_{U,V} = \rho_{U,V} \circ \phi_U$.

We call the presheaf a *sheaf* if the following further *sheaf axiom* is satisfied.

Sheaf axiom *Let $V \in \mathfrak{T}_0$, and let $\{V_i | i \in I\}$ be an open cover, $V_i \in \mathfrak{T}_0$. Suppose there is given for every i in the indexing set an element $f_i \in \mathcal{F}(V_i)$, and suppose that whenever $V_{ij} \in \mathfrak{T}_0$ such that $V_{ij} \subseteq V_i \cap V_j$, we have $\rho_{V_i,V_{ij}}(f_i) = \rho_{V_j,V_{ij}}(f_j)$. Then there exists a unique $f \in \mathcal{F}(V)$ such that $\rho_{V,V_i}(f) = f_i$.*

The *stalk* $\mathcal{F}_x$ of a presheaf $\mathcal{F}$ at a point of $x \in X$ is defined as follows: If $x \in X$, then the elements U of $\mathfrak{T}_0$ containing x, partially ordered with respect to reverse inclusion (so $U \leq V$ if $U \supseteq V$), form a directed set (cf. Lang 1993, I.10 and III.10). By Lang's Theorem III.10.1 (on p. 160) there exists a direct limit

$$\mathcal{F}_x = \varinjlim \mathcal{F}(U).$$

(See also Exercise 4.3.2 and the discussion that precedes it.)

If $x \in U \in \mathfrak{T}_0$, let $\rho_{U,x} : \mathcal{F}(U) \to \mathcal{F}_x$ be the canonical projection. Let

$$\widehat{\mathcal{F}} = \bigcup_{x \in X} \mathcal{F}_x \qquad \text{(disjoint union).}$$

The space $\widehat{\mathcal{F}}$ is called the *étale space* associated with the presheaf $\mathcal{F}$. Let $U \subset X$ be open. (We do not insist that $U \in \mathfrak{T}_0$.) By a *section* of $\widehat{\mathcal{F}}$ over U, we mean a mapping $s : U \to \widehat{\mathcal{F}}$ such that for $x \in U$, $s(x) \in \mathcal{F}_x$ and such that there exists an open cover $\{U_i\}$ of U with $U_i \in \mathfrak{T}_0$ and $f_i \in \mathcal{F}(U_i)$ such that $s(x) = \rho_{U_i,x}(f_i)$ for all $x \in U_i$.

Proposition 4.3.7 *Let $\mathcal{F}$ be a presheaf on X with base $\mathfrak{T}_0$. Then there exists a sheaf $\mathcal{F}^+$ together with a morphism $\phi : \mathcal{F} \to \mathcal{F}^+$ such that any morphism of $\mathcal{F}$ into a sheaf factors through ϕ.*

Proof Let $\widehat{\mathcal{F}}$ be the étale space associated with the presheaf $\mathcal{F}$. If $U \in \mathfrak{T}_0$, let $\mathcal{F}^+(U)$ be the Abelian group of all sections of $\widehat{\mathcal{F}}$ over U. If $U \supseteq V$, $U, V \in \mathfrak{T}_0$, we define the restriction map $\rho^+_{U,V} : \mathcal{F}^+(U) \to \mathcal{F}^+(V)$ by letting $\rho^+_{U,V}(s)$ be the composition of s with the inclusion map $V \to U$; to see that this is a section, we note that because s is a section, by definition there exists an open cover $\{U_i\}$ of U with $U_i \in \mathfrak{T}_0$ and $f_i \in \mathcal{F}(U_i)$ such that $s(x) = \rho_{U_i,x}(f_i)$ for all $x \in U_i$. We may find an open cover $\{V_j\}$ of V with $V_j \in \mathfrak{T}_0$ such that each V_j is contained in some U_i, and let $g_j = \rho_{U_i,V_j}(f_i)$. Then the elements $g_j \in V_j$ show that $\rho^+_{U,V}(s)$ satisfies the definition of a section.

One verifies rather easily that $\mathcal{F}^+$ is a sheaf. To construct the morphism $\phi : \mathcal{F} \to \mathcal{F}^+$, if $U \in \mathfrak{T}_0$, and if $f \in \mathcal{F}(U)$, define a section $s_f : U \to \bigcup_{x \in U} \mathcal{F}_x$ by $s_f(x) = \rho_{U,x}(f)$. (To see that this satisfies the defining property of sections, one may take the open cover $\{U_i\}$ to consist of just U itself, and $f_i = f$.) Then $\phi_U(f) = s_f$ defines a map $\phi_U : \mathcal{F}(U) \to \mathcal{F}^+(U)$.

If $\psi : \mathcal{F} \to \mathcal{F}'$ is another morphism of $\mathcal{F}$ into a sheaf, we define a morphism $\Psi : \mathcal{F}^+ \to \mathcal{F}'$ as follows. If $s \in \mathcal{F}^+(U)$, then by the defining property of sections, there exists an open cover $\{U_i\} \subseteq \mathfrak{T}_0$ of U and $f_i \in \mathcal{F}(U_i)$ such that

$\rho_{U_i,x}(f_i) = s(x)$ for $x \in U_i$. Consider $f_i' = \psi_{U_i}(f_i) \in \mathcal{F}'(U_i)$. We will show that

$$\rho_{U_i,U_i\cap U_j}(f_i') = \rho_{U_j,U_i\cap U_j}(f_j'). \tag{3.11}$$

Assuming this, it follows from the sheaf axiom for $\mathcal{F}'$ that there exists a unique $f \in \mathcal{F}'(U)$ such that $\rho'_{U,U_i}(f') = f_i'$. We define $\Psi_U(s) = f'$. One may verify that $\Psi : \mathcal{F}^+ \to \mathcal{F}'$ is the unique morphism of presheaves such that $\psi = \Psi \circ \phi$.

It remains for us to verify Eq. (3.11). Let $x \in U_i \cap U_j$. The morphism ψ induces a homomorphism $\psi_x : \mathcal{F}_x \to \mathcal{F}_x'$ on the stalks. We find that

$$\rho_{U_i\cap U_j,x}\big(\rho_{U_i,U_i\cap U_j}(f_i') - \rho_{U_j,U_i\cap U_j}(f_j')\big) = \psi_x\big(\rho_{U_i,x}(f_i) - \rho_{U_j,x}(f_j)\big) = \psi_x(0),$$

which vanishes, because $\rho_{U_i,x}(f_i) = s(x) = \rho_{U_j,x}(f_j)$. Now Eq. (3.11) follows from Exercise 4.3.3. ■

We note that $\mathcal{F}^+$ is defined by a universal property and hence is characterized up to isomorphism. It is called the *sheafification* of $\mathcal{F}$. It is a formal consequence of the universal property in Proposition 4.3.7 that if $\mathcal{F}$ is itself a sheaf, then $\mathcal{F} \cong \mathcal{F}^+$. Consequently, every sheaf may be identified with the sheaf of sections of its étale space.

Suppose that $s : U \to \widehat{\mathcal{F}}$ is a section. We define the support $\mathrm{supp}(s)$ of s to be $\{x \in U | s(x) \neq 0\}$.

Proposition 4.3.8 *Let $s : U \to \widehat{\mathcal{F}}$ be a section of the sheaf $\mathcal{F}$. Then the support* $\mathrm{supp}(s)$ *is closed.*

Proof Suppose that $x \notin \mathrm{supp}(s)$. By the definition of a section, x has a neighborhood $W \in \mathfrak{T}_0$ with an element $f \in \mathcal{F}(W)$ such that $s(y) = \rho_{W,y}(f)$ for all $y \in W$. Because by assumption $\rho_{W,x}(f) = s(x) = 0$, it follows from Exercise 4.3.2 that there exists $V \in \mathfrak{T}_0$ such that $x \in V \subseteq W$ with $\rho_{W,V}(f) = 0$. Then for all $y \in V$, $s(y) = \rho_{V,x} \circ \rho_{W,V}(f) = \rho_{V,x}(0) = 0$, so V does not intersect $\mathrm{supp}(s)$. This shows that the support of s is closed. ■

We remark that the category of sheaves on X with base $\mathfrak{T}_0$ is equivalent to the category of sheaves on X in the usual sense, that is, to the category of sheaves on X with base $\mathfrak{T}$. Indeed, if $\mathcal{F}$ is a sheaf with base $\mathfrak{T}_0$ and if U is an arbitrary open set, then one defines $\mathcal{F}^+(U)$ to be the group of sections $s : U \to \bigcup_{x\in U} \mathcal{F}_x$ defined for an arbitrary open set exactly as in the case where $U \in \mathfrak{T}_0$ in the proof of Proposition 4.3.7. Thus $\mathcal{F}^+$ becomes a sheaf in the usual sense. (See Exercise 4.3.1.)

Now, for the remainder of our discussion of sheaves, we specialize to the case of interest. *For the remainder of the section, we will assume that X is a totally disconnected locally compact space and that $\mathfrak{T}_0 = \mathfrak{T}_c$ is the set of all open compact subsets of X.* Our assumption on X means that this is a base to the topology. We will work with the categories of presheaves and sheaves on

X with base $\mathfrak{T}_c$. This is a very special case, as is made clear by the following proposition.

Proposition 4.3.9 *Let X be a totally disconnected locally compact space, and let $\mathcal{F}$ be a sheaf on X with base $\mathfrak{T}_c$. Then there exists an Abelian group $\mathcal{F}_c$ and sheaf $\mathcal{F}_1$ isomorphic to $\mathcal{F}$ such that for every $U \in \mathfrak{T}_c$, $\mathcal{F}_1(U) \subseteq \mathcal{F}_c$ and such that if $U \supseteq V$ are elements of $\mathfrak{T}_c$, then $\mathcal{F}_1(V) \subseteq \mathcal{F}_1(U)$ and, if $f \in \mathcal{F}_1(V)$, then the image of f under the restriction map $\rho_{U,V} : \mathcal{F}_1(U) \to \mathcal{F}_1(V)$ is f itself, whereas the image of f under the restriction map $\rho_{U,U-V} : \mathcal{F}_1(U) \to \mathcal{F}_1(U - V)$ is zero. Moreover, we may assume that $\mathcal{F}_c$ is the union of the spaces $\mathcal{F}(U)$ with $U \in \mathfrak{T}_c$.*

To clarify the meaning of this proposition, let us consider the special case of the sheaf $C^\infty(X)$. If U is a compact and open set of X, we embed $C^\infty(U)$ in $C^\infty(X)$ by extending a function in $C^\infty(U)$ by zero off U. Then it is clear that if $U \supseteq V$, then $C^\infty(V) \subset C^\infty(U)$, and that if $f \in C^\infty(V)$ is extended to U and then restricted back to V, it is unchanged, but if extended to U and then restricted to $U - V$, it is zero. The meaning of the proposition is that *any* sheaf on X (with base $\mathfrak{T}_c$) is isomorphic to a sheaf of this type. Of course, this type of topological space is very special, and usually one would not expect a sheaf to be realizable in this way.

Proof Because we may replace $\mathcal{F}$ by an isomorphic sheaf, we take $\mathcal{F}$ to be the sheaf of sections of an étale space $\widehat{\mathcal{F}}$. Let $\mathcal{F}_c$ be the space of all compactly supported sections in $\mathcal{F}(X)$. If $U \in \mathfrak{T}_c$, we embed $\mathcal{F}(U)$ in $\mathcal{F}_c$ by extending the section $s : U \to X$ by zero off U. Because $X - U$ is open, this is a section. All the assertions of the proposition are now easily checked. ■

Let $C^\infty = C_X^\infty$ be the sheaf of smooth (that is, locally constant) complex-valued functions on X. Thus (consistent with notation previously introduced) if U is an open set in X, $C^\infty(U)$ is the space of locally constant functions on U. In view of our emphasis on sheaves and presheaves with base $\mathfrak{T}_c$, we really only need $C^\infty(U)$ when U is open and compact. We will denote by $C_c^\infty(U)$ the subring (without unit, if U is noncompact) of elements of $C^\infty(X)$ having compact support contained in U.

Note that C^∞ is a sheaf of rings. Thus (X, C^∞) is a *ringed space*, which is, by definition, a topological space endowed with a sheaf of rings. If $U \in \mathfrak{T}_c$, we will denote by $\mathbf{1}_U$ the characteristic function of U. Because U is both open and closed, this is an element of $C^\infty(X)$.

If $(X, \mathcal{R})$ is a ringed space, a *sheaf of modules over* $(X, \mathcal{R})$ (or, *over* $\mathcal{R}$) consists of a sheaf $\mathcal{M}$ such that for each $U \in \mathfrak{T}_c$, $\mathcal{M}(U)$ is a module over $\mathcal{R}(U)$, for which the following compatibility with restriction applies. If $U \supseteq V$ are elements of $\mathfrak{T}_c$, then the restriction map $\mathcal{R}(U) \to \mathcal{R}(V)$ automatically makes any $\mathcal{R}(V)$-module also a $\mathcal{R}(U)$-module. Then we insist that the restriction map

$\mathcal{M}(U) \to \mathcal{M}(V)$ be a homomorphism of $\mathcal{R}(U)$-modules. For example, we have the following proposition.

Proposition 4.3.10 *Let X be a totally disconnected locally compact space, and let $\mathcal{M}$ be a sheaf of vector spaces over X with base $\mathfrak{T}_c$. Then $\mathcal{M}$ is naturally a sheaf of modules for C^∞.*

Proof We may identify $\mathcal{M}$ with the sheaf of sections of its étale space $\widehat{\mathcal{M}}$. Let $U \in \mathfrak{T}_c$, let $s : U \to \widehat{\mathcal{M}}$ be a section, and let $f \in C^\infty(U)$. Then $x \to f(x)\,s(x)$ is defined, and because f is locally constant, it is easy to see that this is a section; thus $\mathcal{M}(U)$ becomes a module over $C^\infty(U)$. ■

We call a $C_c^\infty(X)$-module or $C^\infty(X)$-module M *cosmooth* if for every $x \in M$, there exists an open compact subset U of X such that $\mathbf{1}_U \cdot x = x$. For example, if $\mathcal{F}$ is a sheaf of modules over C^∞, then the Abelian group $\mathcal{F}_c$ introduced in Proposition 4.3.9 is naturally a $C^\infty(X)$-module, and it is easy to see that it is cosmooth (Exercise 4,3.4(a)).

Actually, there is no difference between cosmooth modules for the two rings $C_c^\infty(X)$ and $C^\infty(X)$:

Proposition 4.3.11 *Let M be a cosmooth $C_c^\infty(X)$-module. Then the action of M may be extended in a unique way to make M a $C^\infty(X)$-module.*

Proof To extend the action of $C^\infty(X)$ on M to $C^\infty(X)$, let $x \in M$. Let U be an open compact set such that $x = \mathbf{1}_U \cdot x$. Assuming the action can be extended, let $f \in C^\infty(X)$; we must have $f \cdot x = f\mathbf{1}_U \cdot x = f' \cdot x$, where $f' = f\mathbf{1}_U$. Because $f' \in C_c^\infty(X)$, this shows that the extension of the action, if it exists, is unique. To show that the extension exists, we define $f \cdot x = f' \cdot x$ with f and f' as above; it is not hard to check that $f \cdot x$ is independent of the choice of U and that this definition makes M a $C^\infty(X)$-module. ■

Proposition 4.3.12 *Let M be a cosmooth $C_c^\infty(X)$-module. We associate a presheaf $\mathcal{M}$ with base $\mathfrak{T}_c$ with M in the following way. If $U \in \mathfrak{T}_c$, let $\mathcal{M}(U) = \mathbf{1}_U \cdot M$. If $U \supseteq V$, $U, V \in \mathfrak{T}_c$, we define a restriction map $\rho_{U,V} : \mathcal{M}(U) \to \mathcal{M}(V)$ by $\rho_{U,V}(m) = \mathbf{1}_V \cdot m$. Then $\mathcal{M}$ is a sheaf.*

Thus we may associate a sheaf of modules over the ringed space (X, C^∞) with an arbitrary cosmooth module of $C_c^\infty(X)$.

Proof We note that because $\mathbf{1}_U$ is an idempotent, we have

$$\mathcal{M}(U) = \{m \in M | \mathbf{1}_U \cdot m = m\},$$

a characterization that is easier to work with. The verification that $U \mapsto \mathcal{M}(U)$ is a presheaf follows simply from the fact that if $U \supset V$ are open sets, then $\mathbf{1}_V \cdot \mathbf{1}_U = \mathbf{1}_V$.

To verify that the presheaf $\mathcal{M}$ set is a sheaf, we must verify the sheaf axiom. Thus assume that $\{U_i | i \in I\}$ is an open cover of $U \in \mathfrak{T}_c$ with $U_i \in \mathfrak{T}_c$ and with $m_i \in \mathcal{M}(U_i)$ such that $\rho_{U_i, U_i \cap U_j}(m_i) = \rho_{U_j, U_i \cap U_j}(m_j)$ for all i, j. We will show that there exists a unique element $m \in U$ such that $\rho_{U,U_i}(m) = m_i$.

We will show that if $j, k \in I$, we may replace U_j and U_k by $U_0 = U_j \cup U_k$. Specifically, we will show that there exists a unique element $m_0 \in \mathcal{M}(U_j \cup U_k)$ such that $\rho_{U_0,U_j}(m_0) = m_j$ and $\rho_{U_0,U_k}(m_0) = m_k$. Furthermore, if $l \in I$, we will show that $\rho_{U_0, U_0 \cap U_l}(m_0) = \rho_{U_l, U_0 \cap U_l}(m_l)$. This means that we may successively replace pairs of sets in the cover $\{U_i\}$ by their unions. Because U is compact, we will (in a finite number of steps) reach a state where U itself is an element of the cover; at this point, the truth of the sheaf axiom becomes transparent.

Let $m_{jk} = \mathbf{1}_{U_j \cap U_k} \cdot m_j = \mathbf{1}_{U_j \cap U_k} \cdot m_k$. Because $m_j = \mathbf{1}_{U_j} \cdot m_j$, and because $\mathbf{1}_{U_j}\mathbf{1}_{U_k} = \mathbf{1}_{U_j \cap U_k}$, we have $\mathbf{1}_{U_k} \cdot m_j = m_{jk}$, and similarly $\mathbf{1}_{U_j} \cdot m_k = m_{jk}$. Also $\mathbf{1}_{U_j} \cdot m_{jk} = \mathbf{1}_{U_k} \cdot m_{jk} = \mathbf{1}_{U_0} \cdot m_{jk} = \mathbf{1}_{U_j \cap U_k} \cdot m_{jk} = m_{jk}$. Now let us define $m_0 = m_j + m_k - m_{jk}$. Note that $\mathbf{1}_{U_0} = \mathbf{1}_{U_j} + \mathbf{1}_{U_k} - \mathbf{1}_{U_j \cap U_k}$. We have $\mathbf{1}_{U_j} \cdot m_0 = m_j$, $\mathbf{1}_{U_k} \cdot m_0 = m_k$ and $\mathbf{1}_{U_j \cap U_k} \cdot m_0 = m_{jk}$. Hence $\mathbf{1}_{U_0} \cdot m_0 = m_0$ and $m_0 \in \mathcal{M}(U_0)$ satisfy $\rho_{U_0,U_j}(m_0) = m_j$ and $\rho_{U_0,U_k}(m_0) = m_k$. As for uniqueness, if $\rho_{U_0,U_i}(m_0') = m_i$ $(i = j, k)$ also, then note that

$$m_0' = \mathbf{1}_{U_0} \cdot m_0' = \mathbf{1}_{U_j} \cdot m_0' + \mathbf{1}_{U_k} \cdot m_0' - \mathbf{1}_{U_j}\mathbf{1}_{U_k} \cdot m_0' = m_j + m_k - m_{jk} = m_0.$$

The remaining assertion that $\rho_{U_0, U_0 \cap U_l}(m_0) = \rho_{U_l, U_0 \cap U_l}(m_l)$ may now be checked as follows. Our assumption is that $\mathbf{1}_{U_l} \cdot m_j = \mathbf{1}_{U_j} \cdot m_l$ and that $\mathbf{1}_{U_l} \cdot m_k = \mathbf{1}_{U_k} \cdot m_l$. Furthermore, we have $\mathbf{1}_{U_l} \cdot m_{jk} = \mathbf{1}_{U_l}\mathbf{1}_{U_k} \cdot m_j = \mathbf{1}_{U_j}\mathbf{1}_{U_k} \cdot m_l = \mathbf{1}_{U_j \cap U_k} \cdot m_l$. Therefore

$$\mathbf{1}_{U_l} \cdot m_0 = \mathbf{1}_{U_l} \cdot (m_j + m_k - m_{jk}) = (\mathbf{1}_{U_j} + \mathbf{1}_{U_k} - \mathbf{1}_{U_j \cap U_k}) \cdot m_l = \mathbf{1}_{U_0} \cdot m_l,$$

as required.

This completes the proof that $\mathcal{M}$ is a sheaf. ■

Let us summarize the situation. We have been working with categories of sheaves with base $\mathfrak{T}_c$, but by Exercise 4.3.1, this is equivalent to the category of sheaves in the usual sense. By Proposition 4.3.10, a sheaf of vector spaces over the totally disconnected locally compact space X is automatically a sheaf of modules over C^∞. Given such a sheaf $\mathcal{F}$, the Abelian group $\mathcal{F}_c$ constructed in Proposition 4.3.9 is naturally a cosmooth module over $C_c^\infty(X)$ (Exercise 4.3.4(a)), and conversely, given a cosmooth module over $C_c^\infty(X)$, Proposition 4.3.12 constructs a sheaf of modules over C^∞. These two constructions are inverses of each other: It is easy to see that if M is a cosmooth module over $C^\infty(X)$ and $\mathcal{M}$ is the associated sheaf of modules over C^∞, then $\mathcal{M}_c$ is isomorphic to M (Exercise 4.3.4(b)); conversely, one can see that if $\mathcal{F}$ is a sheaf of modules over C^∞, then the sheaf associated with $\mathcal{F}_c$ is isomorphic to $\mathcal{F}$. Therefore the following three categories are equivalent: the category of sheaves of vector spaces over X, the category of sheaves of C^∞-modules, and the category of cosmooth modules over the ring $C_c^\infty(X)$ or, by Proposition 4.3.11, over $C^\infty(X)$. (We refer to Mac Lane (1971) for the notion of equivalence of categories.)

Proposition 4.3.13 *Let M be a cosmooth module over $C_c^\infty(X)$, and let $\mathcal{M}$ be the corresponding sheaf constructed in Proposition 4.3.12. Let $x \in X$. Then the fiber $\mathcal{M}_x$ is isomorphic to $M/M(x)$, where $M(x)$ is the submodule of all $m \in M$ such that $\mathbf{1}_U \cdot m = 0$ for some $U \in \mathfrak{T}_c$ containing x; an equivalent condition is that $m = \mathbf{1}_V \cdot f$ for some $V \in \mathfrak{T}_c$ such that $x \notin V$.*

Proof If $U \in \mathfrak{T}_c$ contains x, let $\pi_U : M(U) \to M/M(x)$ be the composition of the inclusion $M(U) \to M$ with the projection $M \to M/M(x)$. We claim that if $U \supseteq V$ are elements of $\mathfrak{T}_c$ containing x, then $\pi_V \circ \rho_{U,V} = \pi_U$. Indeed, if $m \in M(U)$, then $\pi_V \circ \rho_{U,V}(m) - \pi_U(m)$ is the image in $M/M(x)$ of $\mathbf{1}_V \cdot m - m$. However, this element is in $M(x)$ because $\mathbf{1}_V \cdot (\mathbf{1}_V \cdot m - m) = 0$.

Now by the universal property of the direct limit, there exists a homomorphism $\sigma : \mathcal{M}_x \to M/M(x)$ such that $\sigma \circ \rho_{U,x} = \pi_U$ of $x \in U \in \mathfrak{T}_c$. It is clear that σ is surjective, and injectivity is an easy consequence of Exercise 4.3.2. This proves that $\mathcal{M}_x \cong M/M(x)$.

To see that $\mathbf{1}_U \cdot m = 0$ for some $U \in \mathfrak{T}_c$ containing x if and only if $\mathbf{1}_V \cdot m = m$ for some $V \in \mathfrak{T}_c$ such that $x \notin V$, first let U be given. Let W be a sufficiently large open and compact set containing both x and U such that $\mathbf{1}_W \cdot m = m$; such a set exists because M is cosmooth. Then we may take $V = W - U$, and similarly if V is given. ■

Now we may generalize the notion of a distribution. Let $\mathcal{F}$ be a sheaf of C^∞ modules on X, and let $\mathcal{F}_c$ be the corresponding cosmooth module of compactly supported sections. By an *$\mathcal{F}$-distribution on X*, we mean a complex linear functional on the vector space $\mathcal{F}_c$. Let $\mathfrak{D}(X, \mathcal{F})$ denote the space of $\mathcal{F}$-distributions on X.

A subset Z of a topological space X is called *locally closed* if it is the intersection of an open set with a closed set. Let $\mathcal{F}$ be a sheaf on X. Then we may construct a sheaf $\mathcal{F}_Z$ on a locally closed subspace Z as follows. Let $\widehat{\mathcal{F}}$ be the étale space of $\mathcal{F}$. We identify $\mathcal{F}$ with the sheaf of sections of $\widehat{\mathcal{F}}$. If U is an open set in Z, then let $\mathcal{F}_Z(U)$ be the set of maps $s : U \to \widehat{\mathcal{F}}$ such that for $x \in U$, $s(x) \in \mathcal{F}_x$ and such that there exists a neighborhood V of x and a section $s' \in \mathcal{F}(V)$ that agrees with s on $V \cap U$. It is very easy to see that $\mathcal{F}_Z$ is a sheaf and that the fiber $\mathcal{F}_{Z,x}$ is isomorphic to $\mathcal{F}_x$. Thus the étale space of $\mathcal{F}_Z$ is just the restriction of the étale space of $\mathcal{F}$ to Z.

Now suppose that Z is closed in X. In this case, we have a map $\mathcal{F}_c \to (\mathcal{F}_Z)_c$; namely, if $s \in \mathcal{F}_c$ is a compactly supported section in $\mathcal{F}(X)$, its restriction to Z is a compactly supported section of $\mathcal{F}_Z$. Similarly, we have a map $(\mathcal{F}_{X-Z})_c \to \mathcal{F}_X$; namely, any compactly supported section of $\mathcal{F}_{X-Z}$ may be extended to $\mathcal{F}$ by making it zero on Z.

Proposition 4.3.14 *Let $\mathcal{F}$ be a sheaf of C^∞-modules on the totally disconnected locally compact space X. We have exact sequences*

$$0 \to (\mathcal{F}_{X-Z})_c \to \mathcal{F}_c \to (\mathcal{F}_Z)_c \to 0 \tag{3.12}$$

and

$$0 \to \mathfrak{D}(Z, \mathcal{F}_Z) \to \mathfrak{D}(X, \mathcal{F}) \to \mathfrak{D}(X - Z, \mathcal{F}_{X-Z}) \to 0. \tag{3.13}$$

Proof The proof is identical to the proof of Eqs. (3.1) and (3.2). Again, the only subtle part is the surjectivity of the restriction $\mathcal{F}_c \to (\mathcal{Z})_c$, and the proof is the same as in Proposition 4.3.1. ■

Now suppose that G is a group acting on the totally disconnected locally compact space X, and let $\mathcal{F}$ be a sheaf on X. We suppose that the action of g extends to an action on the pair $(X, \mathcal{F})$, so that for $g \in G$ and $U \in \mathfrak{T}_c$, we have an isomorphism $\mathcal{F}(U) \to \mathcal{F}(gU)$. Then g evidently acts also on $\mathcal{F}_c$ and on $\mathfrak{D}(X, \mathcal{F})$. If Z is a closed subset that is stable under the action of G, we also have actions on $(\mathcal{F}_Z)_c$ and $\mathfrak{D}(Z, \mathcal{F}_Z)$.

Proposition 4.3.15: Bernstein and Zelevinsky *Let X and Y be totally disconnected locally compact spaces, and let $p : X \to Y$ be a continuous mapping. Let $\mathcal{F}$ be a sheaf on X. Suppose that G is a group acting on X and on its sheaf $\mathcal{F}$. Assume that the action satisfies $p(g \cdot x) = p(x)$ for $g \in G$, $x \in X$. Let χ be a character of G.*

(i) Let $y \in Y$, and let $Z = p^{-1}(y)$. Let $\mathcal{F}_c(\chi)$ (resp. $(\mathcal{F}_Z)_c(\chi)$) be the submodule of $\mathcal{F}_c$ (resp. $(\mathcal{F}_Z)_c$) generated by elements of the form $g \cdot f - \chi(g)^{-1} f$ for $f \in \mathcal{F}_c$ (resp. $f \in (\mathcal{F}_Z)_c$) and $g \in G$. Then $M = \mathcal{F}_c/\mathcal{F}_c(\chi)$ is a cosmooth C_Y^∞-module; let $\mathcal{G}$ be the corresponding sheaf on Y constructed by Proposition 4.3.12. If $y \in Y$, then the stalk $\mathcal{G}_y$ is isomorphic to $(\mathcal{F}_Z)_c/(\mathcal{F}_Z)_c(\chi)$.

(ii) Assume that there are no nonzero distributions D in $\mathfrak{D}\big(p^{-1}(y), \mathcal{F}_{p^{-1}(y)}\big)$ that satisfy

$$g\, D = \chi(g)\, D, \qquad (g \in G), \tag{3.14}$$

for any $y \in Y$. Then there are no nonzero distributions in $\mathfrak{D}(X, \mathcal{F})$ satisfying Eq. (3.14).

Proof We first prove (i). By Proposition 4.3.14, the canonical homomorphism $\mathcal{F}_c \to (\mathcal{F}_Z)_c$ is surjective. We show that its kernel consists of the submodule L generated by elements of the form $(\phi \circ p) \cdot f$, where $\phi \in C_c^\infty(Y)$ vanishes at y, and $f \in \mathcal{F}_c$. It is clear that L is contained in this kernel. Conversely, let $f \in \mathcal{F}_c$ be in the kernel. Then $\operatorname{supp}(f)$ is compact and disjoint from Z, so its image under p is a compact subset of Y that does not contain y. Hence there is an open and compact subset of Y containing $p\big(\operatorname{supp}(f)\big)$ but not y. If ϕ is the characteristic function of this set, then $f = (\phi \circ p) \cdot f$, showing that $f \in L$.

Because $(\mathcal{F}_Z)_c \cong \mathcal{F}_c/L$, we see that $(\mathcal{F}_Z)_c/(\mathcal{F}_Z)_c(\chi)$ is isomorphic to $\mathcal{F}_c$ modulo the submodule generated by L and by $\mathcal{F}_c(\chi)$. On the other hand, by Proposition 4.3.13, the stalk $\mathcal{G}_y$ is isomorphic to $M = \mathcal{F}_c/\mathcal{F}_c(\chi)$ modulo $M(y)$, where $M(y)$ is the image of L in M; so $\mathcal{G}_y \cong (\mathcal{F}_Z)_c/(\mathcal{F}_Z)_c(\chi)$. This proves (i).

We note that (ii) is an immediate consequence of (i). Indeed, the action of g on $\mathcal{F}$-distributions has the effect $(gD)(f) = D(g^{-1}\cdot f)$, so $D \in \mathfrak{D}(X, \mathcal{F})$ satisfies Eq. (3.14) if and only if it annihilates $\mathcal{F}_c(\chi)$. Hence it is sufficient to show $M = 0$, or equivalently, that $\mathcal{G}$ is the zero sheaf, and this will follow if each stalk $\mathcal{G}_y$ is zero. Because (denoting $Z = p^{-1}(y)$ as in (i)) $\mathcal{G}_y \cong (\mathcal{F}_Z)_c/(\mathcal{F}_Z)_c(\chi)$, the assumption that there are no $\mathcal{F}_Z$-distributions in $\mathfrak{D}(Z, \mathcal{F}_Z)$ is precisely the condition that this stalk be zero. ■

We now have enough tools at our disposal to prove Theorem 4.2.3 (and hence Theorem 4.2.2) when $n = 2$. A proof of the general case along these lines could also be given. However, it is better to develop a bit more machinery in order to prove this result. In any case, we must omit the proof of Theorems 4.2.2 and 4.2.3 when $n > 2$ for limitations of space and refer to the paper of Bernstein and Zelevinsky (1976) for this.

Proof of Theorem 4.2.3 when $n = 2$ Let F be a non-Archimedean local field. Let D be a conjugation-invariant distribution on $GL(2, F)$. Let ${}^{\top}D$ be its image under transpose. Replacing D by $D - {}^{\top}D$, we may assume that ${}^{\top}D = -{}^{\top}D$, in which case our problem is to show that $D = 0$.

This motivates us to introduce the following group. Let $\{1, \tau\}$ be a cyclic group of order two, with $\tau^2 = 1$. We have a homomorphism from this group into $\mathrm{Aut}\big(GL(2, F)\big)$ that sends τ to the automorphism $g \mapsto {}^{\top}g^{-1}$. Therefore there exists a group G, which is a semidirect product of $GL(2, F)$ and $\{1, \tau\}$, which contains copies of both groups, with $GL(2, F)$ as a normal subgroup of index two, and $\tau g \tau^{-1} = {}^{\top}g^{-1}$ for $g \in GL(2, F)$. (See Lang 1993, Exercise I.12(c), p. 76) for the construction of the semidirect product.) The group G acts on $GL(2, F)$, with its subgroup $GL(2, F)$ acting by conjugation and τ acting by transpose: If $x \in GL(2, F)$, then ${}^{g}x = gxg^{-1}$ for $g \in GL(2, F)$, whereas ${}^{\tau}x = {}^{\top}x$. Let χ be the character of G that is 1 on $GL(2, F)$ and -1 on the nonidentity coset. Thus our task is to show that a distribution D on $GL(2, F)$ that satisfies

$$g \cdot D = \chi(g)\, D \tag{3.15}$$

zero.

We will denote by $GL(2, F)_{\mathrm{reg}}$ the subset of $g \in GL(2, F)$ having distinct eigenvalues and by $GL(2, F)_{\mathrm{sing}}$ the subset of g having equal eigenvalues. We note that $GL(2, F)_{\mathrm{sing}}$ is the locus of the discriminant $\mathrm{tr}(g)^2 - 4\det(g)$ of the characteristic polynomial of g, and is hence a closed set. We have, by Eq. (3.2), an exact sequence

$$1 \to \mathfrak{D}\big(GL(2, F)_{\text{sing}}\big) \to \mathfrak{D}\big(GL(2, F)\big) \to \mathfrak{D}\big(GL(2, F)_{\text{reg}}\big) \to 0. \quad (3.16)$$

We will first show that the image of D in $\mathfrak{D}\big(GL(2, F)_{\text{reg}}\big)$ is zero. Then exactness will show that D lives in $\mathfrak{D}\big(GL(2, F)_{\text{sing}}\big)$, which we will need to further analyze. Thus we first consider a distribution D_{reg} in $\mathfrak{D}\big(GL(2, F)_{\text{reg}}\big)$ satisfying Eq. (3.15).

We will apply Proposition 4.3.15 with $X = GL(2, F)_{\text{reg}}$. We need to construct a suitable space Y to parametrize the conjugacy classes. We take

$$Y = \{(x, y) \in F \oplus F | x^2 \neq 4y\},$$

and let $p : X \to Y$ be defined by $p(g) = \big(\text{tr}(x), \det(y)\big)$. This map is clearly continuous, and the fibers are exactly the regular conjugacy classes in $GL(2, F)_{\text{reg}}$. Proposition 4.3.15 shows that it is sufficient for our purposes to show that there are no nonzero distributions satisfying Eq. (3.15) on a single conjugacy class $Z = p^{-1}(y)$. Let $x \in GL(2, F)_{\text{reg}}$ be a representative of this conjugacy class, and let $H \in GL(2, F)$ be the centralizer of x. We obtain a bijection between $\sigma : GL(2, F)/H \to Z$ by $\sigma(gH) = gxg^{-1}$. Giving $GL(2, F)/H$ the quotient topology, we claim that this is a homeomorphism. To see this, note that if $\Omega \subset Z$ is an open set, its preimage in G under $g \mapsto gxg^{-1}$ is a union of open sets and hence is open, and because by Proposition 4.3.3 the canonical map $GL(2, F) \to GL(2, F)/H$ is open, it follows that $\sigma^{-1}(\Omega)$ is open. Thus σ is continuous. To show that its inverse is continuous, because σ is equivariant under the actions of $GL(2, F)$ on both $GL(2, F)/H$ and Z, it is sufficient to check that σ^{-1} is continuous near the identity. Thus it is necessary to check that if gxg^{-1} is near x, then g is near an element of H. To accomplish this, note that multiplying g by an element of the center, we may assume that $q^{-1} \leq |\det(g)| \leq 1$, where q is the cardinality of the residue field. Let K be the field obtained from F by adjoining the eigenvalues of x. There exists an element γ of $GL(2, K)$ such that $x_1 = \gamma x \gamma^{-1}$ is diagonal, and so $H = \gamma^{-1} T \gamma \cap GL(2, F)$, where T is the group of diagonal matrices in $GL(2, K)$. Now if $g_1 \in GL(2, K)$ satisfies $q^{-1} \leq |\det(g_1)| \leq 1$ and $g_1 x_1 g_1^{-1}$ is near x_1, where x_1 is a diagonal matrix with distinct eigenvalues, then it is easy to check that the off-diagonal entries of g_1 are small. Applying this to $g_1 = \gamma g \gamma^{-1}$, we see that g_1 is near an element of T, and so g is near an element of Z. Therefore σ is a homeomorphism.

Now we may regard a $GL(2, F)$-invariant distribution on Z as an invariant distribution on the homogeneous space $GL(2, F)/H$. Because (as noted in the preceding paragraph) H is conjugate in $GL(2, K)$ to a subgroup of T, it is Abelian and hence unimodular. Therefore Proposition 4.3.5(iii) is applicable, and we see that (up to a constant) such a distribution is just integration with respect to the invariant measure on $GL(2, F)$. This distribution is transpose invariant, so we see that a distribution on a regular conjugacy class that satisfies

Eq. (3.15) is zero. By Proposition 4.3.15, this implies that the image of D in $\mathfrak{D}\big(GL(2, F)_{\mathrm{reg}}\big)$ is zero, and therefore by the exactness of Eq. (3.16), D resides in $\mathfrak{D}\big(GL(2, F)_{\mathrm{sing}}\big)$.

Now we divide $GL(2, F)_{\mathrm{sing}}$ into two pieces. Let C be the center of $GL(2, F)$, consisting of scalar matrices. Then we have an exact sequence

$$0 \to \mathfrak{D}(C) \to \mathfrak{D}\big(GL(2, F)_{\mathrm{sing}}\big) \to \mathfrak{D}\big(GL(2, F)_{\mathrm{sing}} - C\big) \to 0.$$

We may show that the image of D in $\mathfrak{D}\big(GL(2, F)_{\mathrm{sing}} - C\big)$ is zero by the same method. This time a parameter space Y for the conjugacy classes in $GL(2, F)_{\mathrm{sing}} - C$ is just $F^\times$ because each conjugacy class has a unique representative of the form

$$g(a) = \begin{pmatrix} a & 1 \\ & a \end{pmatrix},$$

and we fiber $GL(2, F)_{\mathrm{sing}} - C$ over $F^\times$ by $g(a) \mapsto a$. Then Proposition 4.3.5(iii) again shows that the distribution is zero, provided we show that (for conjugacy classes of this type) a distribution supported on a single conjugacy class that satisfies Eq. (3.15) is zero; again, one interprets the conjugacy class as a homogeneous space and applies Proposition 4.3.15. Finally one is reduced to distributions on the center, and because the center is fixed pointwise by transposition, we see that a nonzero distribution satisfying Eq. (3.15) is not possible. ∎

Exercises

For further exercises on sheaves, we recommend the those in Section II.1 of Hartshorne (1977).

Exercise 4.3.1 Let $\mathfrak{T}$ be the topology of the topological space X, and let $\mathfrak{T}_0 \subseteq \mathfrak{T}$ be a base for the topology. Let $\mathcal{F}$ be a sheaf on X with base $\mathfrak{T}_0$. Show that there is a sheaf $\mathcal{F}'$ on X with base $\mathfrak{T}$, together with isomorphisms $\sigma_U : \mathcal{F}(U) \to \mathcal{F}'(U)$ for $U \in \mathfrak{T}_0$ such that if $U \supset V$ are elements of $\mathfrak{T}_0$, we have $\rho_{U,V} \circ \sigma_U = \sigma_V \circ \rho_{U,V}$, and that $\mathcal{F}'$ is unique up to isomorphism. [Hint: Define $\mathcal{F}'(U)$ to be the space of sections of the étale space $\widehat{\mathcal{F}}$ of $\mathcal{F}$.]

Because of this, the category of sheaves on X with base $\mathfrak{T}_0$ is independent (in the sense of equivalence of categories) of the choice of $\mathfrak{T}_0$. There is no corresponding independence for the category of presheaves.

We recall that a *directed set* I is a partially ordered set with the property that if $i, j \in I$, then there exists $k \in I$ such that $k \geq i$ and $k \geq j$. A *directed family* of Abelian groups indexed by I consists of the following data. For each $i \in I$, there is an Abelian group G_i, and if $i \leq j$, there is a homomorphism $\phi_{ij} : G_i \to G_j$. (We may also say that the morphisms ϕ_{ij} form a *directed family*.) It is assumed that ϕ_{ii} is the identity map, and that $\phi_{jk} \circ \phi_{ij} = \phi_{ik}$ when $i < j < k$. A *direct limit* of this directed family of Abelian groups then consists of a group G, and for each $i \in I$, a homomorphism $\phi_i : G_i \to G$ such that if

$i < j$, then $\phi_j \circ \phi_{ij} = \phi_i$. It is assumed that if H is any Abelian group, and if a family of homomorphisms $\psi_i : G_i \to H$ is given such that $\psi_j \circ \phi_{ij} = \psi_i$ when $i < j$, then there exists a unique homomorphism $\psi : G \to H$ such that $\psi_i = \phi_i \circ \psi$.

As usual for objects characterized by a universal property, the direct limit is determined up to isomorphism by this property if it exists; this is a purely formal categorical argument. It is a theorem that the direct limit exists. (Theorem III.10.1 on p. 160 of Lang (1993).)

Exercise 4.3.2 In this situation, prove that if $g_i \in G_i$ satisfies $\phi_i(g_i) = 0$, then $\phi_{ij}(g_i) = 0$ for some $j > i$. [HINT: One method of proof would be to deduce this from the explicit construction of the direct limit as a quotient of $\bigoplus G_i$ in the proof of the theorem just quoted from Lang (1993).]

Exercise 4.3.3 Let $\mathcal{F}$ be a sheaf on X, and let $U \subset X$ be open. Let $f \in \mathcal{F}(U)$ have the property that $\rho_{U,x}(f) = 0$ in $\mathcal{F}_x$ for all $x \in U$. Prove that $f = 0$. [HINT: By Exercise 4.3.2, for each $x \in U$ there exists an open neighborhood U_x (in $\mathfrak{T}_0$ if $\mathcal{F}$ a sheaf with basis $\mathfrak{T}_0$) of x such that $\rho_{U,U_x}(f) = 0$. Regard the sets U_x as forming an open cover of U and use the sheaf axiom.]

Exercise 4.3.4 Let X be a totally disconnected locally compact space.
(a) Let $\mathcal{F}$ be a sheaf of modules over C^∞. Show that the Abelian group $\mathcal{F}_c$ constructed in Proposition 4.3.6 is naturally a module over $C_c^\infty(X)$ and that it is cosmooth.
(b) Let M be a cosmooth module over $C^\infty(X)$, and let $\mathcal{M}$ be the associated sheaf of modules over C^∞, constructed in Proposition 4.3.9. Show that $\mathcal{M}_c$ is isomorphic to M.
(c) Conversely, let $\mathcal{F}$ be a sheaf of modules over C^∞ (with base $\mathfrak{T}_0$, the set of open and compact sets of X). Show that the sheaf associated with $\mathcal{F}_c$ by Proposition 4.3.9 is isomorphic to $\mathcal{F}$.

Exercise 4.3.5 (a) Prove that if G is a compact topological group and H is an open subgroup, then H has finite index in G.
(b) Prove that if G is a locally compact topological group and H is an open subgroup, then H has nonzero volume with respect to the left Haar measure.

4.4 Whittaker Models and the Jacquet Functor

Let F be a non-Archimedean local field, and let ψ be a nontrivial additive character of F. Let $\mathfrak{o}$, $\mathfrak{p}$, q, and ϖ be as in Section 4.2. We define a character ψ_N of the group $N(F)$ of upper triangular unipotent matrices in $GL(n, F)$ by

$$\psi_N(u) = \psi\left(\sum_{i=1}^{n-1} u_{i,i+1}\right), \qquad u = (u_{ij}) \in N(F).$$

Let (π, V) is a smooth representation of $GL(n, F)$. By a *Whittaker functional* on V, we mean a linear functional $\lambda : V \to \mathbb{C}$ such that $\lambda\big(\pi(u)\,x\big) =$

$\psi_N(u)\,\lambda(x)$ for all $u \in N(F)$, $x \in V$. Note that λ is *not* assumed to be smooth (and in practice won't be.)

Theorem 4.4.1 *Let (π, V) be an irreducible admissible representation of $GL(n, F)$. Then the dimension of the space of Whittaker functionals on V is at most one.*

We will prove this result, referred to as the "uniqueness of Whittaker models" or the "local multiplicity one theorem," in this section when $n = 2$. We will set up the proof to some extent for the general case. We define actions ρ and λ of $GL(n, F)$ on $GL(n, F)$, on $C_c^\infty\big(GL(n, F)\big)$, and on $\mathfrak{D}\big(GL(n, F)\big)$ by Eqs. (3.3–3.5). We will be concerned with distributions $\Delta \in \mathfrak{D}\big(GL(n, F)\big)$ that satisfy

$$\lambda(u)\,\Delta = \psi_N(u)^{-1}\,\Delta, \qquad \rho(u)\,\Delta = \psi_N(u)\,\Delta, \qquad (u \in N(F)), \tag{4.1}$$

which is the analog of Eq. (1.18).

We define an involution $\iota : GL(n, F) \to GL(n, F)$ by

$$\iota(g) = w^{0\top} g w^0, \qquad w^0 = \begin{pmatrix} & & 1 \\ & \cdot^{\cdot^{\cdot}} & \\ 1 & & \end{pmatrix}.$$

Then ι also induces involutions on $C_c^\infty\big(GL(n, F)\big)$ and $\mathfrak{D}\big(GL(n, F)\big)$. We note that $N(F)$ is mapped to itself by ι and that $\psi_N({}^\iota u) = \psi_N(u)$ for $u \in N(F)$. (If $n = 2$, then ${}^\iota u = u$, but this is not true in general.)

Theorem 4.4.2 *Let $\Delta \in \mathfrak{D}\big(GL(n, F)\big)$ satisfy Eq. (4.1). Then Δ is stable under ι.*

The relevance of this theorem to Theorem 4.4.1 is clear from the proof of Theorem 4.1.2. In this section, we will prove Theorem 4.4.2 when $n = 2$, and then we will show (for general n) how Theorem 4.4.2 implies Theorem 4.4.1. Our discussion of some aspects of the proofs is modeled on Soudry (1987).

Proof of Theorem 4.4.2 when $n = 2$ Although we will eventually specialize to the case $n = 2$, let us at least reformulate the statement of the theorem for general n. We may replace Δ by $\Delta - {}^\iota\Delta$. This distribution satisfies, in addition to Eq. (4.1), the condition

$${}^\iota\Delta = -\Delta, \tag{4.2}$$

and so our task is to show that a distribution satisfying Eqs. (4.1) and (4.2) is zero.

We introduce a group G that is a certain semidirect product. G will have as a subgroup of index two the group $N(F) \times N(F)$. G also contains an element $\mathcal{I}$ that satisfies $\mathcal{I}^2 = 1$, and $\mathcal{I}(u_1, u_2)\mathcal{I}^{-1} = ({}^\iota u_2^{-1}, {}^\iota u_1^{-1})$ for $(u_1, u_2) \in N(F) \times N(F)$. We define a character χ of G by $\chi(u_1, u_2) = \psi_N(u_1)^{-1}\,\psi_N(u_2)$,

$\chi(\mathcal{I}) = -1$. Now let us define actions (to be denoted σ) of G $GL(n, F)$, $C_c^\infty(GL(n, F))$, and $\mathfrak{D}(GL(n, F))$ by

$$\sigma(u_1, u_2) = \lambda(u_1)\,\rho(u_2), \qquad \sigma(\mathcal{I}) = \iota.$$

The conditions (4.1) and (4.2) may then be summarized in the single condition

$$\sigma(g)\,\Delta = \chi(g)\,\Delta. \tag{4.3}$$

We will show (when $n = 2$) that a distribution satisfying Eq. (4.3) is zero.

Specializing now to the case where $n = 2$, the proof of Theorem 4.1.2 gives a hint as to how Theorem 4.4.2 is to be proved: Roughly speaking, what we expect is that a distribution satisfying Eq. (4.1) has its support restricted to the double cosets of the form (Eq. 1.19). Then because these double cosets are stable under ι, it will follow that such a distribution must be stable under ι.

We will denote by $B(F)$ the Borel subgroup consisting of all upper triangular matrices in $GL(2, F)$. Let $X = B(F)\,w^0\,B(F)$ be the open cell in the Bruhat decomposition (Eq. 1.7). By Proposition 4.3.1, we have an exact sequence

$$0 \to \mathfrak{D}(B(F)) \to \mathfrak{D}(GL(2, F)) \to \mathfrak{D}(X) \to 0. \tag{4.4}$$

We will first show that the image in $\mathfrak{D}(X)$ of Δ (assumed to satisfy Eq. (4.3)) is zero. Thus we consider a distribution Δ in $\mathfrak{D}(X)$ that satisfies Eq. (4.3). We have a continuous mapping $p : X \to Y$, where $Y = F^\times \oplus F^\times$ is given by

$$p\begin{pmatrix} a & b \\ c & d \end{pmatrix} = \big(c, (ad - bc)/c\big).$$

Note that the fibers of this map are invariant under the action σ of G; they are precisely the double cosets

$$N(F)\begin{pmatrix} & b_0 \\ c_0 & \end{pmatrix} N(F). \tag{4.5}$$

Applying Proposition 4.3.15 to the sheaf C^∞, we see that it is sufficient to show that there are no nonzero distributions Δ on the single double coset (Eq. 4.5) that satisfy Eq. (4.3). But Eq. (4.3) is homeomorphic to $N(F) \times N(F)$ under the map

$$(u_1, u_2) \to u_1 \begin{pmatrix} & b_0 \\ c_0 & \end{pmatrix} u_2^{-1}.$$

Thus we may transfer a distribution satisfying Eq. (4.1) to a distribution on $N(F)\times N(F)$ by means of this homeomorphism, and applying Proposition 4.3.2, we learn (if f is a smooth and compactly supported function on the double coset (Eq. 4.5)) that

$$D(f) = c \int\limits_{N(F)\times N(F)} \psi_N(u_1)\,\psi_N(u_2)\,f\left(u_1 \begin{pmatrix} & b_0 \\ c_0 & \end{pmatrix} u_2\right) du_1\,du_2.$$

But this distribution is easily seen to be invariant under ι, so if Eq. (4.2) is also satisfied, the distribution is zero. This shows that a distribution $\Delta \in \mathfrak{D}(X)$ satisfying Eq. (4.3) is zero.

Now by the exactness of Eq. (4.4), if $\Delta \in \mathfrak{D}(GL(2, F))$ satisfies Eq. (4.3), then Δ resides in $\mathfrak{D}(B(F))$. We must therefore show that a distribution in $\mathfrak{D}(B(F))$ that satisfies Eq. (4.3) is zero. Let $Y_1 = F^\times \times F^\times$, and let $p : B(F) \to Y_1$ be the map

$$p\begin{pmatrix} a & b \\ & d \end{pmatrix} = (a, d).$$

The fibers of this map are stable under the action σ of G. We have a homeomorphism $N(F) \to p^{-1}(a, d)$ via $u \mapsto u\delta$, where $\delta = \begin{pmatrix} a & \\ & d \end{pmatrix}$. Using this homeomorphism, we transfer the distribution to $\mathfrak{D}(N(F))$, where we may apply Proposition 4.3.2; then the first equation in Eq. (4.1) implies that there exists a constant c_1 such that for $\phi \in \mathfrak{D}(p^{-1}(a, d))$ we have

$$\Delta(\phi) = c_1 \int_{N(F)} \phi(u\delta)\, \psi_N(u)\, du. \tag{4.6}$$

Similarly, using the analog of Proposition 4.3.2 for right translation and the second equation in Eq. (4.1), we obtain

$$\Delta(\phi) = c_2 \int_{N(F)} \phi(u\delta)\, \psi_N(\delta^{-1}u\delta)\, du.$$

(Here du is of course the *right* Haar measure, but the group $N(F)$ is unimodular.) If $a \neq d$, then these two equations imply that $c_1 = c_2 = 0$. If not, we find that we may choose u such that $c_1\psi_N(u) \neq c_2\psi_N(\delta^{-1}u\delta)$, and then taking a test function ϕ that is the characteristic function of a small neighborhood of u gives a contradiction. On the other hand, if $a = d$, then it is easy to see that the distribution (Eq. 4.6) is invariant under ι, so Eq. (4.2) implies that $c_1 = 0$. ∎

We may now give the proof of Theorem 4.4.1. The following proof is complete assuming Theorems 4.4.2 and 4.2.2. These have already been proved if $n = 2$.

Proof of Theorem 4.4.1 We define another representation (π', V) by $\pi'(g) = \pi({}^t g^{-1})$. We note that $\pi(w^0)$ is an intertwining operator between π' and the representation π_1 in Theorem 4.2.2(i), so π' is isomorphic to the contragredient representation $(\hat{\pi}, \hat{V})$. It follows that there exists a bilinear pairing $\langle\ \rangle : V \times V \to \mathbb{C}$ such that

$$\langle \pi(g)\xi, \eta\rangle = \langle \xi, \pi({}^t g)\eta\rangle\,. \tag{4.7}$$

We recall that the space $\hat{V}$ of the contragredient representation consists of all smooth linear functionals on V. Thus if Λ is a smooth linear functional on V, there exists an element $[\Lambda]$ of V that satisfies

$$\langle \xi, [\Lambda] \rangle = \Lambda(\xi). \tag{4.8}$$

As in Section 4.2, let $\mathcal{H}$ be the Hecke algebra, which is the space $C_c^\infty(G)$, given the structure of an algebra (without unity) under convolution; if $\phi \in \mathcal{H}$, $\pi(\phi)$ is defined as in Section 4.2. If Λ is any linear functional on V, and if $\phi \in \mathcal{H}$, we define a linear functional $\Lambda * \phi$ on V by

$$(\Lambda * \phi)(\xi) = \Lambda\big(\pi(\phi)\,\xi\big) = \int_G \Lambda\big(\pi(g)\,\xi\big)\,\phi(g)\,dg. \tag{4.9}$$

Note that even if Λ is not smooth, the integrand is smooth and compactly supported. It is easy to see that $\Lambda * \phi$ is smooth and that if $\phi_1, \phi_2 \in \mathcal{H}$, then

$$\Lambda * (\phi_1 * \phi_2) = (\Lambda * \phi_1) * \phi_2.$$

We prove now that if Λ is any linear functional on V, and if $\phi \in \mathcal{H}$, then

$$\pi(g)[\Lambda * \phi] = [\Lambda * \rho({}^\iota g^{-1})\phi]. \tag{4.10}$$

To see this, we take the inner product with an arbitrary $\xi \in V$:

$$\begin{aligned}\langle \xi, \pi(g)[\Lambda * \phi]\rangle &= \langle \pi({}^\iota g)\xi, [\Lambda * \phi]\rangle = (\Lambda * \phi)\big(\pi({}^\iota g)\xi\big)\\ &= \int_G \Lambda\big(\pi(h)\,\pi({}^\iota g)\,\xi\big)\,\phi(h)\,dh\\ &= \int_G \Lambda\big(\pi(h)\,\xi\big)\,\phi(h\,{}^\iota g^{-1})\,dh = \big\langle \xi, [\Lambda * \rho({}^\iota g^{-1})\phi]\big\rangle,\end{aligned}$$

which gives us Eq. (4.10). ■

We show next that if L is a *smooth* linear functional, then

$$[L * \phi] = \pi({}^\iota\phi)[L]. \tag{4.11}$$

Indeed, we take the inner product with an arbitrary vector ξ:

$$\langle \xi, [L * \phi]\rangle = (L * \phi)(\xi) = L\big(\pi(\phi)\,\xi\big),$$

whereas by Eq. (4.7)

$$\langle \xi, \pi({}^\iota\phi)[L]\rangle = \langle \pi(\phi)\xi, [L]\rangle = L\big(\pi(\phi)\,\xi\big).$$

Because these are equal for all ξ, we obtain Eq. (4.11).

Next, we prove that if Λ is a Whittaker functional, and if $u \in N(F)$, then

$$[\Lambda * \lambda(u)\phi] = \psi_N(u)\,[\Lambda * \phi]. \tag{4.12}$$

Again, this is proved by taking the inner product with an arbitrary element of V:

$$\langle \xi, [\Lambda * \lambda(u)\phi] \rangle = \big(\Lambda * \lambda(u)\phi\big)(\xi) = \int_G \Lambda\big(\pi(g)\,\xi\big)\,\phi(u^{-1}g)\,dg$$

$$= \int_G \Lambda\big(\pi(u)\,\pi(g)\,\xi\big)\,\phi(g)\,dg$$

$$= \psi_N(u)\,(\Lambda * \phi)(\xi) = \psi_N(u)\,\langle \xi, [\Lambda * \phi] \rangle\,.$$

Now suppose that Λ_1 and Λ_2 are two Whittaker functionals. We will show that they are proportional. We define a distribution Δ on G by

$$\Delta(\phi) = \Lambda_2([\Lambda_1 * \phi]). \tag{4.13}$$

It follows from Eqs. (4.10) and (4.12) that Eq. (4.1) is satisfied. Hence by Theorem 4.4.2, we have $\Delta = {}^{\iota}\Delta$.

Lemma 4.4.1 *If $\phi \in \mathcal{H}$ such that $\Lambda_1 * \phi = 0$, then also $\Lambda_2 * \phi = 0$.*

Proof We claim that $\Lambda_1 * \rho(g)\phi = 0$ for all $g \in G$. Indeed, it is sufficient to show that $[\Lambda_1 * \rho(g)\phi] = 0$, and this is an immediate consequence of Eq. (4.10). Next, we claim that

$$\Lambda_2\big([\Lambda_1 * \lambda(g)\,{}^{\iota}\phi]\big) = 0 \tag{4.14}$$

for all g. Indeed

$$0 = \Lambda_2\big([\Lambda_1 * \rho(g)\phi]\big) = \Delta\big(\rho(g)\phi\big) = \Delta\big({}^{\iota}(\rho(g)\phi)\big) = \Lambda_2\big([\Lambda_1 * \lambda({}^{\iota}g^{-1})\,{}^{\iota}\phi]\big).$$

Replacing g by ${}^{\iota}g^{-1}$ gives Eq. (4.14). Now we observe that if $\sigma \in \mathcal{H}$, then

$$(\sigma * {}^{\iota}\phi)(g) = \int_G \sigma(h)\,\big(\lambda(h)\,{}^{\iota}\phi\big)(g)\,dh,$$

so multiplying Eq. (4.14) by $\sigma(g)$ and integrating shows that

$$\Lambda_2([\Lambda_1 * \sigma * {}^{\iota}\phi]) = 0 \tag{4.15}$$

for all σ. Noting that $\Lambda_1 * \sigma$ is smooth, we may use Eq. (4.11) to rewrite this as

$$\Lambda_2\big(\pi(\phi)\,[\Lambda_1 * \sigma]\big) = 0. \tag{4.16}$$

Next we show that if Λ is any nonzero linear functional on V, then

$$V = \{[\Lambda * \sigma] | \sigma \in \mathcal{H}\}\,. \tag{4.17}$$

Indeed, the right side of Eq. (4.17) is an invariant subspace of V, by Eq. (4.10). Because V is assumed to be irreducible, we need only show that the right side of Eq. (4.17) contains a nonzero vector. Let $\xi \in V$ such that $\Lambda(\xi) \neq 0$. The stabilizer in G of V is an open set because π is smooth. Let ϕ be a function with support contained in this open set such that $\int \phi(g)\, dg = 1$; then it is easy to see that $(\Lambda * \phi)(\xi) = \Lambda(\xi) \neq 0$, and we get Eq. (4.17). Now Eqs. (4.16), (4.17), and (4.9) imply that $\Lambda_2 * \phi = 0$, so we have Lemma 4.4.1. □

We may now define a map $T : V \to V$ by

$$T([\Lambda_1 * \phi]) = [\Lambda_2 * \phi]. \tag{4.18}$$

This is well defined by Lemma 4.4.1; moreover, it is defined on all of V by Eq. (4.17). It is an intertwining operator because by Eq. (4.10)

$$T\big(\pi(g)[\Lambda_1 * \phi]\big) = T\big([\Lambda_1 * \rho({}^t g^{-1})\phi]\big) = [\Lambda_2 * \rho({}^t g^{-1})\phi] = \pi(g)[\Lambda_2 * \phi].$$

By Schur's lemma, there exists a constant c such that $T(\xi) = c\,\xi$ for all $\xi \in V$. We may now prove that $\Lambda_2 = c\,\Lambda_1$. Indeed, let $\xi \in V$ be any vector, and let ϕ be a smooth function with compact support contained in the (open) stabilizer of ξ such that $\int \phi(g)\, dg = 1$. Then $(\Lambda_1 * \phi)(\xi) = \Lambda_1(\xi)$, and $(\Lambda_2 * \phi)(\xi) = \Lambda_2(\xi)$. Then we have

$$\Lambda_2(\xi) = (\Lambda_2 * \phi)(\xi) = \langle \xi, [\Lambda_2 * \phi] \rangle = c\,\langle \xi, [\Lambda_1 * \phi] \rangle = c\,\Lambda_1(\xi).$$

This completes the proof of Theorem 4.4.1. ■

After some preliminaries, we will prove a result complementary to Theorem 4.4.1: We will show (for $GL(2)$ only – the result is false for $GL(n)$) that an irreducible admissible representation, if not one dimensional, always has a Whittaker model.

We now introduce the important *Jacquet functor*. Although for simplicity we will restrict ourselves to $GL(2)$, the concepts are important for $GL(n)$ and more general groups. Let (π, V) be a smooth representation of the Borel subgroup $B(F)$ of $GL(2, F)$. In most cases, we will be interested in representations (π, V) that are representations of $GL(2, F)$ restricted to $B(F)$; however, only the $B(F)$-module structure is involved in the definition of the Jacquet module, which we now give. First we define V_N to be the vector subspace of V generated by elements of the form

$$\pi(u)\, v - v, \qquad u \in N(F),\ v \in V.$$

Let $T(F)$ be the torus of diagonal elements in $GL(2, F)$. Because if $u \in N(F)$, $t \in T(F)$, we have

$$\pi(t)\,\big(\pi(u)\, v - v\big) = \pi(u')\, v' - v', \qquad u' = tut^{-1},\ v' = \pi(t)\, v.$$

Thus the group V_N is invariant under $T(F)$, and consequently the quotient $J(V) = V/V_N$ is a $T(F)$-module. Let $\pi_N : T(F) \to \operatorname{End}\big(J(V)\big)$ be the action of $T(F)$ on $J(V)$. It is obvious that $\big(\pi_N, J(V)\big)$ is a smooth representation. $J(V)$ is called the *Jacquet module* of V.

Proposition 4.4.1 *Let $x \in V$. Then $x \in V_N$ if and only if for sufficiently large n*

$$\int_{\mathfrak{p}^{-n}} \pi\begin{pmatrix} 1 & x \\ & 1 \end{pmatrix} v \, dx = 0. \tag{4.19}$$

Here $\mathfrak{p}$ denotes the maximal ideal in the ring $\mathfrak{o}$ of integers in F.

Proof Let

$$v = \pi\begin{pmatrix} 1 & \xi \\ & 1 \end{pmatrix} w - w$$

be a typical generator of V_N. If $\xi \in \mathfrak{p}^{-n}$, then the integral (Eq. 4.19) vanishes, because the change of variables $x \to x - \xi$ shows that

$$\int_{\mathfrak{p}^{-n}} \pi\begin{pmatrix} 1 & x \\ & 1 \end{pmatrix} \pi\begin{pmatrix} 1 & \xi \\ & 1 \end{pmatrix} w \, dx = \int_{\mathfrak{p}^{-n}} \pi\begin{pmatrix} 1 & x \\ & 1 \end{pmatrix} w \, dx.$$

This shows that V_N is contained in the space of all v that satisfy Eq. (4.19). Conversely, suppose that v satisfies Eq. (4.19) for some n. Let $m > -n$ be such that w is fixed by $\pi\left(\begin{smallmatrix} 1 & x \\ & 1 \end{smallmatrix}\right)$ when $x \in \mathfrak{p}^m$. We note that the integrand in Eq. (4.19) is constant on the cosets of x modulo $\mathfrak{p}^m$. Hence Eq. (4.19) may be rewritten

$$\sum_{\alpha \in \mathfrak{p}^{-n}/\mathfrak{p}^m} \pi\begin{pmatrix} 1 & \alpha \\ & 1 \end{pmatrix} v = 0.$$

Thus

$$\begin{aligned} v &= v - q^{-(n+m)} \sum_{\alpha \in \mathfrak{p}^{-n}/\mathfrak{p}^m} \pi\begin{pmatrix} 1 & \alpha \\ & 1 \end{pmatrix} v \\ &= \sum_{\alpha \in \mathfrak{p}^{-n}/\mathfrak{p}^m} \left[q^{-(n+m)} v - \pi\begin{pmatrix} 1 & \alpha \\ & 1 \end{pmatrix} q^{-(n+m)} v \right]. \end{aligned}$$

This latter expression shows that $v \in V_N$. ■

We recall that a functor is called *exact* if it takes exact sequences to exact sequences. It is sufficient to check that it takes *short* exact sequences to short exact sequences; this follows from the following observation: In order for a sequence

$$M' \xrightarrow{f} M \xrightarrow{g} M''$$

to be exact, it is necessary and sufficient for the homomorphisms f and g to be factorizable as $f = f_1 \circ f_2$, $g = g_1 \circ g_2$, where

$$0 \longrightarrow N' \xrightarrow{f_1} M \xrightarrow{g_2} N'' \longrightarrow 0,$$

$$M' \xrightarrow{f_2} N' \longrightarrow 0, \qquad 0 \longrightarrow N'' \xrightarrow{g_1} M''$$

are exact. (Take $N' = \mathrm{im}(f)$ and $N'' = \mathrm{im}(g)$.) From this criterion it is clear that a functor preserving short exact sequences preserves all exact sequences.

Any homomorphism $V \to W$ between $B(F)$-modules induces a homomorphism $J(V) \to J(W)$ of $T(F)$-modules. Thus the Jacquet module is a functor from the category of smooth $B(F)$-modules to the category of smooth $T(F)$-modules; it is therefore referred to as the *Jacquet functor*.

Proposition 4.4.2 *The Jacquet functor is exact.*

Proof Let

$$0 \longrightarrow V' \xrightarrow{i} V \xrightarrow{p} V'' \longrightarrow 0 \tag{4.20}$$

be a short exact sequence of $B(F)$-modules. First we show that the induced sequence

$$0 \longrightarrow V'_N \xrightarrow{i_N} V_N \xrightarrow{p_N} V''_N \longrightarrow 0$$

is exact. Without loss of generality, we may assume that V' is a submodule of V, and so i_N is the inclusion of a submodule, so the injectivity of i_N is clear. The surjectivity of p_N also easy: Consider a typical generator $\pi''(u)\,v'' - v''$, $u \in N(F)$, $v'' \in V''$. (Here of course π'' is the representation of $B(F)$ on V''.) Because p is surjective, we may find $v \in V$ with $p(v) = v''$. Then p_N maps $\pi(u)\,v - v$ into $\pi''(u)\,v'' - v''$.

Thus the only nontrivial point is exactness at V_N. Suppose that $v \in V_N$ and $p(v) = 0$. As before, regarding V' as a submodule of V, we see that $v \in V' \cap V_N$, and so exactness at V_N amounts to showing that if V' is a submodule of V, then $V'_N = V' \cap V_N$. However, this is clear from the characterization of Proposition 4.4.1. We see that Eq. (4.20) is exact.

Now we have a commutative diagram

$$\begin{array}{ccccccccc}
0 & \longrightarrow & V'_N & \xrightarrow{i_N} & V_N & \xrightarrow{p_N} & V''_N & \longrightarrow & 0 \\
 & & \downarrow j' & & \downarrow j & & \downarrow j'' & & \\
0 & \longrightarrow & V' & \xrightarrow{i} & V & \xrightarrow{p} & V'' & \longrightarrow & 0
\end{array}$$

where j, j', and j'' are the inclusion maps. By the Snake lemma Lang (1993, Section III.9, p. 157), we have an exact sequence

$$\ker(j'') \to \mathrm{coker}(j') \to \mathrm{coker}(j) \to \mathrm{coker}(j'') \to 0.$$

Here $\ker(j'') = 0$, $\text{coker}(j') = J(V')$, $\text{coker}(j) = J(V)$, and $\text{coker}(j'') = J(V'')$, so we obtain the exactness of the functor J. ■

We need a variant of this construction, the "twisted" Jacquet functor. Fixing as before a nontrivial additive character of F, let us define, for (π, V) a smooth representation of $B(F)$, the vector subspace $V_{N,\psi}$ to be the vector space generated by elements of the form $\pi(u)\,v - \psi_N(u)\,v$ with $u \in N(F)$, $v \in V$; and let $J_\psi(V) = V/V_{N,\psi}$. We note that unlike $J(V)$, $J_\psi(V)$ is not a module for $T(F)$; it is only a module for the center $Z(F)$ of $GL(2, F)$. J_ψ is a functor from the category of $B(F)$-modules to the category of $Z(F)$-modules.

Proposition 4.4.3 *The functor J_ψ is exact.*

Proof The proof is identical to that of Proposition 4.4.2, except that the criterion of Proposition 4.4.1 must be modified: Instead, we find that $v \in J_{N,\psi}(V)$ if and only if for sufficiently large N

$$\int_{\mathfrak{p}^{-N}} \overline{\psi(x)}\pi\begin{pmatrix} 1 & x \\ & 1 \end{pmatrix} v\,dx = 0. \tag{4.21}$$

Apart from this, the proofs are identical. ■

Proposition 4.4.4 *If (π, V) is an irreducible admissible representation of $GL(2, F)$, then* $\dim\ J_\psi(V) \leq 1$.

Proof The module $J_\psi(V)$ has an immediate interpretation in terms of Whittaker functionals: It is clear from the definitions that a linear functional on V is a Whittaker functional if and only if it annihilates $V_{N,\psi}$, and consequently the space of Whittaker functionals on V may be identified with the algebraic dual space of $J_\psi(V)$. Thus Proposition 4.4.4 is simply a paraphrase of Theorem 4.4.1. ■

Now let V be a smooth $B(F)$-module. We will associate with V a certain sheaf. We recall from Exercise 3.1.1 (e) that F is isomorphic to its own dual group. (See, for example, Rudin (1962) for the duality theory of locally compact Abelian groups.) The isomorphism of F with its dual depends upon the choice of a nontrivial additive character ψ of F, a choice which we have made; then $a \in F$ corresponds to the character $\psi_a(x) = \psi(ax)$. The dual group is given the topology of uniform convergence on compact sets, and the isomorphism of F with its dual is then a topological isomorphism. The Fourier inversion formula takes compactly supported locally constant functions to compactly supported locally constant functions, but interchanges the two multiplicative structures (convolution and pointwise multiplication) on $C_c^\infty(F)$.

To be precise, if $\phi \in C_c^\infty(F)$, the Fourier transform of ϕ is defined by

$$\hat{\phi}(x) = \int_F \phi(y)\,\psi(xy)\,dy.$$

There is a close relation between the choice of the additive character and the normalization of the Haar measure: The additive character forces a unique choice of the additive Haar measure dx such that the Fourier inversion formula has the form

$$\hat{\hat{\phi}}(x) = \phi(-x).$$

We will assume the Haar measure to be normalized in this way. The Fourier transform of the convolution $\phi_1 * \phi_2$ is $\hat{\phi}_1\hat{\phi}_2$, whereas the Fourier transform of $\phi_1\phi_2$ is $\hat{\phi}_1 * \hat{\phi}_2$. The first of these assertions is proved by a simple change of variables, and then the second is a formal consequence of the first, together with the Fourier inversion formula.

To recapitulate, let $\left(C_c^\infty(F), \cdot\right)$ and $\left(C_c^\infty(F), *\right)$ denote the rings (without unit) of smooth compactly supported functions on F with the two possible multiplications, pointwise and convolution; then the Fourier transform is an isomorphism between these two rings.

Returning to the $B(F)$-module (π, V), we use the isomorphism $F \cong N(F)$ to construct a representation ρ of $N(F)$ on V. Thus $\rho : N(F) \to \mathrm{End}(V)$ is given by

$$\rho(x)\,v = \pi\begin{pmatrix} 1 & -x \\ & 1 \end{pmatrix} v.$$

In accordance with Eq. (2.9), we have a corresponding representation (also denoted ρ) of the Hecke algebra $\left(C_c^\infty(F), *\right)$ on V. We then use the Fourier transform to transfer this to a representation of $\left(C_c^\infty(F), \cdot\right)$. Thus if $\phi \in C_c^\infty(F)$ and $v \in V$, we define

$$\phi \cdot v = \rho(\hat{\phi})\,v = \int_F \hat{\phi}(x)\,\pi\begin{pmatrix} 1 & -x \\ & 1 \end{pmatrix} v\,dx.$$

Let us observe that with this structure, the $C_c^\infty(F)$-module V is cosmooth in the sense of Section 4.3. To see this, let $\mathfrak{p}^n$ be the conductor of ψ; that is, $\mathfrak{p}^n$ is the largest fractional ideal on which ψ is trivial. Then, denoting by $\mathbf{1}_X$ the characteristic function of a subset X of F, the Fourier transform of $\mathbf{1}_{\mathfrak{p}^{-k}}$ is $(\mathrm{Vol}\,\mathfrak{p}^{-k})\,\mathbf{1}_{\mathfrak{p}^{n+k}}$. Now for every vector $v \in V$, if k is sufficiently large, $\rho(x)\,v = v$ for $x \in \mathfrak{p}^{n+k}$. Then

$$\mathbf{1}_{\mathfrak{p}^{-k}}\,v = \mathrm{Vol}(\mathfrak{p}^{-k})\,\mathrm{Vol}(\mathfrak{p}^{n+k})\,v = v,$$

because (as may be easily shown) $\mathrm{Vol}(\mathfrak{p}^{-k})\,\mathrm{Vol}(\mathfrak{p}^{n+k}) = 1$ with respect to the self-dual Haar measure. This shows that V satisfies the definition of a cosmooth module, and there is hence a sheaf, which we will denote $\mathcal{S}(V)$ on F associated with V by Proposition 4.3.12.

Let us compute the stalks of the sheaf $\mathcal{S}(V)$. We recall that if $a \in F$, then $\psi_a : F \to \mathbb{C}^\times$ is the character $\psi_a(x) = \psi(ax)$.

Proposition 4.4.5 *Let V be a smooth $B(F)$-module, and let $a \in F$. Then the stalk*

$$\mathcal{S}(V)_a \cong \begin{cases} J(V) & \text{if } a = 0; \\ J_{\psi_a}(V) \cong J_\psi & \text{if } a \neq 0. \end{cases}$$

More specifically, the projection map $V \to \mathcal{S}(V)_a$ is surjective, and its kernel is V_N (so $S(V) \cong J(N)$) if $a = 0$, and V_{N,ψ_a} (so $S(V) \cong J_{\psi_a}(V)$) if $a \neq 0$, and

$$\pi(g_a), \qquad g_a = \begin{pmatrix} a & \\ & 1 \end{pmatrix}$$

maps V_{N,ψ_a} onto $V_{N,\psi}$, so $J_{\psi_a}(V) \cong J_\psi(V)$.

Proof The stalks are described in Proposition 4.3.13: If $a \in F$, then $\mathcal{S}(V)_a$ is V modulo the subgroup consisting of elements v that satisfy $\mathbf{1}_U \cdot v = 0$ whenever U is a sufficiently small neighborhood of a. It is sufficient to consider U of the form $a + \mathfrak{p}^k$ for k large, because if $a + \mathfrak{p}^k \subseteq U$ then $\mathbf{1}_{a+\mathfrak{p}^k} \cdot \mathbf{1}_U = \mathbf{1}_{a+\mathfrak{p}^k}$. We find that

$$\hat{\mathbf{1}}_{a+\mathfrak{p}^k}(x) = \psi(ax)\,\mathrm{Vol}(\mathfrak{p}^k)\,\mathbf{1}_{\mathfrak{p}^{n-k}}(x).$$

Thus the condition of Proposition 4.3.13 for v to be in the kernel of the canonical map $V \mapsto \mathcal{S}(V)_a$ amounts to the condition that

$$\int_{\mathfrak{p}^{n-k}} \psi(ax)\,\pi\begin{pmatrix} 1 & -x \\ & 1 \end{pmatrix} dx = 0$$

for sufficiently large k, a condition that we recognize as Eq. (4.21) if $a \neq 0$ (with ψ replaced by the character ψ_a) or Eq. (4.19) if $a = 0$. Now it is easy to see that $\pi(g_a)V_{N,\psi_a} = V_{N,\psi}$, so $\pi(g_a)$ induces an isomorphism $J_{\psi_a}(V) \cong J_\psi(V)$, which gives us Proposition 4.4.5. ∎

We recall that if X is a Hausdorff topological space and P is a point in X, then a *skyscraper sheaf* $\mathcal{F}$ on X supported on P is a sheaf whose only nonzero stalk is $\mathcal{F}_P$. If $\mathcal{F}_P$ is an Abelian group A, then the sheaf $\mathcal{F}$ is uniquely determined by A: We have (for open $U \subseteq X$)

$$\mathcal{F}_P(U) = \begin{cases} A & \text{if } P \in U; \\ 0 & \text{otherwise.} \end{cases}$$

Theorem 4.4.3 *Let (π, V) be a smooth representation of $GL(2, F)$ that has no Whittaker functional. Then π factors through the determinant map $GL(2, F) \to F^\times$. In particular, if π is irreducible and admissible, it is one dimensional.*

This result is *not* true without modification for $GL(n)$. There are infinite-dimensional irreducible admissible representations of $GL(n, F)$ that do not admit Whittaker models. Nevertheless, there is a sense (made precise by the notion of Gelfand–Kirillov dimension) in which such a representation is a "smaller" and more degenerate one having a Whittaker model. Representations having Whittaker models are often called *generic*.

Proof We have noted that (immediately from the definitions) a linear functional on V is a Whittaker functional if and only if it factors through $J_\psi(V)$. Hence by Proposition 4.4.5, our assumption implies that the stalks $\mathcal{S}(V)_a$ are zero when $a \neq 0$, so $\mathcal{S}(V)$ is a skyscraper sheaf supported at zero, and its stalk there is $J(V)$. Because V is itself equal to the space of all compactly supported sections of $\mathcal{S}(V)$, we see that the canonical map $V \to \mathcal{S}(V)_0 = J(V)$ is an isomorphism. Thus $J_N = 0$, which (from the definition of J_N) means that $\pi(u)\, v = v$ for all $u \in N(F)$, and thus $N(F)$ acts trivially.

It follows that all conjugates of $N(F)$ act trivially also. But by Exercise 4.4.2, $N(F)$ and its conjugates generate $SL(2, F)$, and so $SL(2, F)$ acts trivially on V. Thus the representation π factors through the determinant map $GL(2, F) \to F^\times$. Because the irreducible representations of the locally compact Abelian group $F^\times$ are all one dimensional, $\dim(V) = 1$. ■

Proposition 4.4.6 *Let (π, V) be an irreducible admissible representation of $GL(2, F)$. Assume that V is infinite dimensional. Then there is no nonzero vector $v \in V$ that is invariant under all of $N(F)$.*

Proof Because the stabilizer of v is open, it contains an element of $SL(2, F) - B(F)$. Thus by Exercise 4.4.2, v is stabilized by $SL(2, F)$. By Schur's lemma (Proposition 4.2.4), the one-dimensional subspace spanned by v is also fixed by the center $Z(F)$. First assume that the characteristic of F is not two. If $g \in GL(2, F)$, the coset of g in $GL(2, F)/SL(2, F)\, Z(F)$ is determined by the class of $\det(g)$ in $F^\times/(F^\times)^2$, and (assuming the characteristic of F is not two) there are only finitely many such classes – four if the residue characteristic of F is not equal to two. Therefore there are only finitely many such classes, and it follows that the space spanned by $\pi(g)\, v$, $g \in GL(2, F)$ is finite dimensional. This contradicts the irreducibility and infinite dimensionality of V if $v \neq 0$.

If the characteristic of F is two, we must modify this argument slightly. The one-dimensional subspace spanned by v is stabilized by $SL(2, F)$, $Z(F)$, and some open subgroup, and these together generate a subgroup of finite index, so the translates of v span a finite-dimensional space as before. ■

Proposition 4.4.7 *Let* (π, V) *be an irreducible admissible representation of* $GL(2, F)$*, and let* $\Lambda : V \to \mathbb{C}$ *be a nonzero Whittaker functional on* V*. If* $0 \neq v \in V$*, then there exists* $a \in F^\times$ *such that*

$$\Lambda\left(\pi\begin{pmatrix} a & \\ & 1 \end{pmatrix} v\right) \neq 0. \tag{4.22}$$

Proof We will show that if Eq. (4.22) vanishes for all a, then v is $N(F)$-invariant and hence zero by Theorem 4.4.3. We make use of the sheaf $S(V)$ of Proposition 4.4.5. It is easy to see that if $0 \neq a \in F$, the kernel of

$$v \mapsto \Lambda\left(\begin{pmatrix} a & \\ & 1 \end{pmatrix} v\right)$$

contains V_{N,ψ_a}; because Λ is nonzero, this map is nonzero. By Proposition 4.7, V_{N,ψ_a} is of codimension one, so the kernel of this map is all of V_{N,ψ_a}. Thus if Eq. (4.22) fails for all a, the vector v has zero image in $S(V)_a$ for all $0 \neq a \in F$, and if $x \in F$, then clearly

$$v' = v - \pi\begin{pmatrix} 1 & x \\ & 1 \end{pmatrix} v$$

has the same property; moreover, v' has zero image in $S(V)_0$, and because V is identified with the space of all compactly supported sections of the sheaf $S(V)$ by the construction of Proposition 4.3.12, it follows that $v' = 0$. Because this is true for all x, it follows from Proposition 4.4.6 that $v = 0$. ■

We now introduce two important models of an irreducible admissible representation (π, V) of $GL(2, F)$. (A *model* is a particular realization of a representation in a particular concrete space, often a subspace of an induced representation.) We assume that V is infinite dimensional, so that by Theorem 4.4.3, there exists a nonzero Whittaker functional Λ on V.

First, there is the *Whittaker model.* This was previously introduced in the context of a finite field subsequent to Theorem 4.1.2, and the reader should consult those paragraphs for some additional remarks. The space $\mathcal{W}$ of the Whittaker model consists of functions W_v on $GL(2, F)$ $(v \in V)$ of the form

$$W_v(g) = \Lambda\big(\pi(g)\, v\big). \tag{4.23}$$

$W_{\pi(g)\,v}(h) = W_v(hg)$; the space $\mathcal{W}$ is closed under the action of $GL(2, F)$ by right translation, and the resulting representation is isomorphic to (π, V).

The *Kirillov model* of (π, V) is the space $\mathcal{K}$ of functions $\phi_v : F^\times \to \mathbb{C}$ defined (for $v \in V$) by

$$\phi_v(a) = W_v\begin{pmatrix} a & \\ & 1 \end{pmatrix}. \tag{4.24}$$

By Proposition 4.4.7, the map $v \mapsto \phi_v$ is nonzero. It is not so easy to describe the action of an arbitrary element of $GL(2, F)$ on $\mathcal{K}$, but at least the action of

$B(F)$ is easy to describe. Because the map $v \mapsto \phi_v$ is an isomorphism of V with $\mathcal{K}$, let us identify V with $\mathcal{K}$, so that the action of $GL(2, F)$ on $\mathcal{K}$ is still denoted by π. Then we find that if $\phi \in \mathcal{K}$

$$\pi\begin{pmatrix} a & \\ & 1 \end{pmatrix}\phi(x) = \phi(ax), \qquad \pi\begin{pmatrix} 1 & b \\ & 1 \end{pmatrix}\phi(x) = \psi(bx)\phi(x). \tag{4.25}$$

Of course if ω is the central character of π, the center $Z(F)$ of $GL(2, F)$ acts by Eq. (1.1), so the action of the Borel subgroup on the Kirillov model is completely explicit. We will determine the space of functions $\mathcal{K}$ explicitly in Section 4.7.

The remainder of this section will be devoted to completing the elementary theory of the Jacquet functor. We begin by introducing a couple of specific open and compact subgroups of $GL(2, F)$. Let $\mathfrak{a}$ be an ideal of $\mathfrak{o}$, and (with $K = GL(2, \mathfrak{o})$ as before) let

$$K_0(\mathfrak{a}) = \left\{ \begin{pmatrix} a & b \\ c & d \end{pmatrix} \in K \;\middle|\; c \equiv 0 \bmod \mathfrak{a} \right\},$$

$$K_1(\mathfrak{a}) = \left\{ \begin{pmatrix} a & b \\ c & d \end{pmatrix} \in K \;\middle|\; c \equiv 0,\ a \equiv d \equiv 1 \bmod \mathfrak{a} \right\}.$$

The group $K_0(\mathfrak{p})$ is known as the *Iwahori subgroup*.

These groups admit a decomposition known as the *Iwahori factorization*. Let us define (for an ideal $\mathfrak{a}$ of $\mathfrak{o}$)

$$N(\mathfrak{a}) = \left\{ \begin{pmatrix} 1 & x \\ & 1 \end{pmatrix} \;\middle|\; x \in \mathfrak{a} \right\},$$

$$N_-(\mathfrak{a}) = \left\{ \begin{pmatrix} 1 & \\ x & 1 \end{pmatrix} \;\middle|\; x \in \mathfrak{a} \right\},$$

and

$$T(\mathfrak{a}) = \left\{ \begin{pmatrix} t_1 & \\ & t_2 \end{pmatrix} \;\middle|\; t_1, t_2 \in \mathfrak{o}^\times,\ t_1, t_2 \equiv 1 \bmod \mathfrak{a} \right\}.$$

Then if the ideal $\mathfrak{a}$ is a proper ideal, we have

$$K_0(\mathfrak{a}) = N_-(\mathfrak{a})\, T(\mathfrak{o})\, N(\mathfrak{o}), \tag{4.26}$$

and

$$K_1(\mathfrak{a}) = N_-(\mathfrak{a})\, T(\mathfrak{a})\, N(\mathfrak{o}). \tag{4.27}$$

To see this, suppose that

$$k = \begin{pmatrix} a & b \\ c & d \end{pmatrix} \in K_0(\mathfrak{a}).$$

Because we are assuming that $\mathfrak{a}$ is a proper ideal, this implies that a is a unit in $\mathfrak{o}$. Thus we have

$$k = n_- \beta, \qquad n_- = \begin{pmatrix} 1 & \\ c/a & 1 \end{pmatrix} \in N_-(\mathfrak{a}),$$

where β is an upper triangular element of $K_0(\mathfrak{a})$, and hence β factors uniquely as $t\,n$, where $t \in T(\mathfrak{o})$ and $n \in N(\mathfrak{o})$. Furthermore, if $k \in K_1(\mathfrak{a})$, then $t \in T(\mathfrak{a})$. We see that an arbitrary element k of $K_0(\mathfrak{a})$ (resp. of $K_1(\mathfrak{a})$) can be factored uniquely as n_-tn with $n_- \in N_-(\mathfrak{a})$, $t \in T(\mathfrak{o})$ (resp. $t \in T(\mathfrak{a})$) and $n \in N(\mathfrak{o})$. It is not hard to see that the multiplication maps

$$N_-(\mathfrak{a}) \times T(\mathfrak{o}) \times N(\mathfrak{o}) \to K_0(\mathfrak{a})$$

and

$$N_-(\mathfrak{a}) \times T(\mathfrak{a}) \times N(\mathfrak{o}) \to K_1(\mathfrak{a})$$

are homeomorphisms. We have the following integration formulas. Let $K_0 = K_0(\mathfrak{a})$ or $K_1(\mathfrak{a})$, so $K_0 = N_-(\mathfrak{a})\, T_0\, N(\mathfrak{o})$, where, $T_0 = T(\mathfrak{o})$ or $T(\mathfrak{a})$ depending on whether $K_0 = K_0(\mathfrak{a})$ or $K_1(\mathfrak{a})$. Then if ϕ is an integrable function on K_0, we have

$$\int_{K_0} \phi(k)\, dk = \int_{N_-(\mathfrak{a})} \int_{T_0} \int_{N(\mathfrak{o})} \phi(n_- t_0 n)\, dn\, dt_0\, dn_-, \tag{4.28}$$

where for convenience we have normalized the various Haar measures so that the volume of all four compact subgroups is one. To see this, we note that the measure

$$\int_{N_-(\mathfrak{a})} \int_{T_0} dt\, dn_-$$

is a Haar measure on $N_-(\mathfrak{a})\, T_0$ by Proposition 2.1.5 (taking $P = N_-(\mathfrak{a})$ and $K = T_0$). Then Eq. (4.26) follows from another application of the same proposition with $P = N_-(\mathfrak{a})\, T_0$ and $K = N(\mathfrak{o})$. Finally, we note that the factors in the Iwahori factorization may be taken in the opposite order: Equally, we may write

$$K_0(\mathfrak{a}) = N(\mathfrak{o})\, T(\mathfrak{o})\, N_-(\mathfrak{a}) \tag{4.29}$$

and

$$K_1(\mathfrak{a}) = N(\mathfrak{o})\, T(\mathfrak{a})\, N_-(\mathfrak{a}), \tag{4.30}$$

and with notations as in Eq. (4.28),

$$\int_{K_0} \phi(k)\, dk = \int_{N_-(\mathfrak{a})} \int_{T_0} \int_{N(\mathfrak{o})} \phi(n t_0 n_-)\, dn\, dt_0\, dn_-. \tag{4.31}$$

Lemma 4.4.2: Jacquet *Let (π, V) be a smooth representation of $GL(2, F)$. Let $K_0 = K_0(\mathfrak{a})$ or $K_1(\mathfrak{a})$, where $\mathfrak{a}$ is a proper ideal of $\mathfrak{o}$, so that K_0 has an Iwahori factorization $K_0 = N_-(\mathfrak{a})\, T_0\, N(\mathfrak{o})$, where $T_0 = T(\mathfrak{o})$ or $T(\mathfrak{a})$. Then V^{K_0} and $V^{N_-(\mathfrak{a})T_0}$ have the same image in the Jacquet module $J(V)$.*

Proof Let $p : V \to J(V)$ be the projection map. Let $x \in V^{N_-(\mathfrak{a})T_0}$. Then $x_1 = \int_{K_0} \pi(k)\, x\, dk$ lies in V^{K_0}; we will show that $p(x_1) = p(x)$. Note that this is sufficient because it shows that $p\big(V^{N_-(\mathfrak{a})T_0}\big) \subseteq p(V^{K_0})$, the other inclusion being obvious. By Eq. (4.31), we have

$$x_1 = \int\limits_{N_-(\mathfrak{a})} \int\limits_{T_0} \int\limits_{N(\mathfrak{o})} \pi(n)\pi(t_0 n_-)\, x\, dn\, dt_0\, dn_-,$$

or, because x is fixed by t_0 and n_-

$$x_1 = \int\limits_{N(\mathfrak{o})} \pi(n)\, x\, dn.$$

Because $p(x) = p\big(\pi(n)\, x\big)$ by the definition of $J(V)$, it follows that $p(x_1) = p(x)$. ∎

Theorem 4.4.4 *(i) Let (π, V) be a smooth representation of $GL(2, F)$. Let $K_0 = K_0(\mathfrak{a})$ or $K_1(\mathfrak{a})$, where $\mathfrak{a}$ is a proper ideal of $\mathfrak{o}$, so that K_0 has an Iwahori factorization $K_0 = N_-(\mathfrak{a})\, T_0\, N(\mathfrak{o})$, where $T_0 = T(\mathfrak{o})$ or $T(\mathfrak{a})$. Then the projection map $p : V \to J(V)$ induces a surjection of V^{K_0} onto $J(V)^{T_0}$.*
(ii) If (π, V) is an admissible representation of $GL(2, F)$, then the representation of $T(F)$ on $J(V)$ is also admissible.

In greater generality than we have stated here, part (ii) is due to Harish-Chandra and part (i) to Jacquet. The following simultaneous proof of both assertions (which works without essential modification in the most general context) is taken from Casselman (unpublished), where it is attributed to Borel.

Proof We will prove (i) while simultaneously showing that the dimension of $J(V)^{T_0}$ is bounded above by the dimension of V^{K_0}. Thus $J(V)^{T_0}$ is finite dimensional if π is admissible, and because we may take $T_0 = T(\mathfrak{a})$ to be an arbitrarily small neighborhood of the identity in $T(F)$, this will prove the admissibility of $J(V)$.

We note that any T_0-fixed vector in $J(V)$ may be pulled back to a T_0-fixed vector in V; indeed, if $p(x)$ is T_0-fixed and $t \in T_0$, then we note that $p\big(\pi(t)\, x\big) = \pi_N(t)\, p(x) = p(x)$, where $\pi_N : T(F) \to \mathrm{End}\big(J(V)\big)$ is the action of $T(F)$ on the Jacquet module, so integrating, we find that

$$p(x_1) = p(x), \qquad x_1 = \frac{1}{\mathrm{Vol}(T_0)} \int\limits_{T_0} \pi(t)\, x\, dt,$$

and $x_1 \in V^{T_0}$.

Thus if $\overline{U}$ is any finite-dimensional subspace of $J(V)^{T_0}$, we may find a subspace $U \subseteq V^{T_0}$ that is mapped isomorphically onto $\overline{U}$ by p. Now U is stabilized by $N_-(\mathfrak{p}^n)$ for some sufficiently large n, and so U is stabilized by $N_-(\mathfrak{p}^n)\, T_0$. The ideal $\mathfrak{a} = \mathfrak{p}^m$ for some m. We note that

$$\pi(d)\, N_-(\mathfrak{p}^n)T_0\, \pi(d)^{-1} = N_-(\mathfrak{a})T_0, \qquad d = \begin{pmatrix} \varpi^{n-m} & \\ & 1 \end{pmatrix},$$

so $\pi(d)U$ is stabilized by $N_-(\mathfrak{a})T_0$. By Lemma 4.4.2, it follows that $\pi_N(d)p(U) = p\big(\pi(d)U\big) \subseteq p(V^{K_0})$. Consequently, the dimension of $p(U)$ is bounded by dim V^{K_0}. Thus we may as well take $U = J(V)^{T_0}$, and we have proved that there exists $d \in T(F)$ such that $\pi_N(d)\, J(V)^{T_0} \subseteq p(V^{K_0})$. However, because $T(F)$ is Abelian, $\pi_N(d)$ commutes with the action of T_0 on $J(V)$, so this means that $J(V)^{T_0} \subseteq p(V^{K_0})$. The opposite inclusion is also clear. ■

Exercises

Exercise 4.4.1 (Waldspurger 1980, Proposition 9, p. 31.) Let T be a maximal torus of $GL(2)$ defined over F. (T may or may not be split – the result is valid in both cases.) Let (π, V) be an irreducible admissible representation of $GL(2, F)$. Use the involution method to prove that there exists at most one linear functional $L : V \to \mathbb{C}$ such that $L\big(\pi(t)\, v\big) = L(v)$ for $t \in T(F)$.

Exercise 4.4.2 Prove that if F is a field and $\gamma \in SL(2, F) - B(F)$, then γ and $N(F)$ generate $SL(2, F)$. [HINT: Make use of the identity

$$\begin{pmatrix} a & b \\ c & d \end{pmatrix} = \begin{pmatrix} 1 & a/c \\ & 1 \end{pmatrix} \begin{pmatrix} & -c^{-1} \\ c & \end{pmatrix} \begin{pmatrix} 1 & d/c \\ & 1 \end{pmatrix}$$

if $\left(\begin{smallmatrix} a & b \\ c & d \end{smallmatrix}\right) \in SL(2, F) - B(F)$.]

4.5 The Principal Series Representations

Let G be a totally disconnected locally compact group, let H be a closed subgroup, and let (π, V) be a smooth representation of H. We will define not one but *two* induced representations of G. Let V^G be the space of functions $f : G \to V$ such that the following properties hold:
(i) We have

$$f(hg) = \delta_G(h)^{-1/2}\, \delta_H(h)^{1/2}\, \pi(h)\, f(g) \tag{5.1}$$

for $h \in H$, $g \in G$, where δ_G and δ_H are the modular quasicharacters of G and H, and
(ii) there exists an open subgroup K_0 of G such that $f(gk) = f(g)$ for all $g \in G$ when $k \in K_0$.

A function satisfying (ii) is automatically smooth (locally constant), and if $H\backslash G$ is compact, it is not hard to show that this condition is equivalent to smoothness. We have a representation $\pi^G : G \to \mathrm{End}\left(V^G\right)$ on V^G, in which $\left(\pi^G(g)\, f\right)(x) = f(xg)$. It is clear (because left and right translations commute) that $\pi^G(g)\, f$ satisfies (i), and if f satisfies (ii) for a certain group K_0, then $\pi^G(g)\, f$ satisfies (i) with the group gK_0g^{-1}. The condition (ii) may be paraphrased by the assertion that the "vector" f is fixed by the subgroup K_0 in π^G, and so this representation is smooth. This is the *(ordinary) induced representation*, and we will also use the notation $\mathrm{Ind}_H^G(\pi) = \left(\pi^G, V^G\right)$.

On the other hand, there is also the notion of *compact induction*. Let V_c^G be the space of functions in V^G that, in addition to (i) and (ii) are *compactly supported modulo* H. This means that the image in $H\backslash G$ of the support of f (which is both open and closed) is compact. Let $\pi_c^G(g)$ be the restriction of $\pi^G(g)$ to this space. Then $\left(\pi_c^G, V_c^G\right)$ is also a smooth representation. The two notions agree if $G\backslash H$ is compact.

Proposition 4.5.1: Frobenius reciprocity *Let G be a totally disconnected locally compact group and H a closed subgroup. Let (π, V) and (σ, W) be smooth representations of H and G, respectively. Then there is a natural isomorphism* $\mathrm{Hom}_G\left(\sigma, \pi^G\right) \cong \mathrm{Hom}_H\left(\sigma, \pi \otimes \delta_G^{-1/2}\delta_H^{1/2}\right)$.

Here $\pi \otimes \delta_G^{-1/2}\delta_H^{1/2} : H \to \mathrm{End}(V)$ is the representation π "twisted" (i.e., multiplied) by $\delta_G^{-1/2}\,\delta_H^{1/2}$:

$$\left(\pi \otimes \delta_G^{-1/2}\delta_H^{1/2}\right)(h) = \delta_G^{-1/2}(h)\,\delta_H^{1/2}(h)\,\pi(h).$$

Proof The proof is similar to the case of a finite group G (Exercise 4.1.1). The isomorphism is explicitly as follows. Let $\Phi : W \to V^G$ be an intertwining map. Then we obtain a map $\phi : W \to V$ by $\phi(w) = \Phi(w)(1)$. It is not hard to check that ϕ is an intertwining operator when V is given the H-module structure $\pi \otimes \delta_G^{-1/2}\delta_H^{1/2}$. Conversely, if $\phi : W \to V$ is given, we may define $\Phi : W \to V^G$ by $\Phi(w)(g) = \phi\left(\sigma(g)\, w\right)$. One checks that $\Phi(w) \in V^G$; property (i) is a formal verification, whereas property (ii) follows from the smoothness of σ. It is also easily verified that Φ is an intertwining operator. Finally, one checks that the two maps $\phi \mapsto \Phi$ and $\Phi \mapsto \phi$ are inverses of each other. (Compare Exercise 4.1.1.) ■

An important example for us is when $G = GL(2, F)$, F a non-Archimedean local field, and $H = B(F)$ is the Borel subgroup consisting of upper triangular matrices in G. If χ_1 and χ_2 are two quasicharacters of $F^\times$, then we define a quasicharacter χ of $B(F)$ by

$$\chi\begin{pmatrix} y_1 & * \\ & y_2 \end{pmatrix} = \chi_1(y_1)\,\chi_2(y_2). \tag{5.2}$$

Let

$$\mathcal{B}(\chi_1, \chi_2) = \mathrm{Ind}_{B(F)}^{GL(2,F)}(\chi).$$

The representations $\mathcal{B}(\chi_1, \chi_2)$, when irreducible, are called *principal series* representations. We will see later that they are usually (though not always) irreducible. If $\mathcal{B}(\chi_1, \chi_2)$ is irreducible, its isomorphism class is denoted $\pi(\chi_1, \chi_2)$.

We have mentioned that the two notions of induction agree if H is cocompact. Let us now observe that this is the case when $G = GL(2, F)$ and $H = B(F)$.

Proposition 4.5.2: Iwasawa decomposition *Let $G = GL(n, F)$, where F is a non-Archimedean local field, let $B(F)$ be the Borel subgroup consisting of upper triangular matrices, and let $K = GL(n, \mathfrak{o})$, where $\mathfrak{o}$ is the ring of integers in F. Then $G = B(F)\,K$, and $B(F)\backslash G$ is compact.*

The group K is a maximal compact subgroup of G, and every maximal subgroup of G is conjugate to K (Exercise 4.5.1).

Proof We prove this by induction on n. Assuming that it is true for $n = 1$, let $g \in GL(n, F)$. We will find $k \in K$ such that gk is upper triangular. The first step is to find $k_1 \in K$ such that gk_1 has zeros in its bottom row below the main diagonal. Let

$$(x_1, x_2, \cdots, x_n)\, w$$

be the bottom row of g^{-1}, where w is a permutation matrix chosen so that x_n has minimal valuation among the x_i. Then we take

$$k_1 = w^{-1} \begin{pmatrix} 1 & & & & \\ 0 & 1 & & & \\ 0 & 0 & 1 & & \\ \vdots & & & \ddots & \\ -x_1x_n^{-1} & -x_2x_n^{-1} & -x_3x_n^{-1} & \cdots & 1 \end{pmatrix}.$$

Now gk_1 has the form

$$\begin{pmatrix} g_{n-1} & * \\ 0 & x_n \end{pmatrix},$$

where $g_{n-1} \in GL(n-1, F)$. By induction, we may find $k' \in GL(n-1, \mathfrak{o})$ such that $g_{n-1}k'$ is upper triangular. Then if

$$k = k_1 \begin{pmatrix} k' & \\ & 1 \end{pmatrix},$$

we find that gk is upper triangular, proving that $G = B(F)\,K$.

The group K is a closed and bounded subset of $\mathrm{Mat}_n(F) \cong F^{n^2}$ and hence is compact. We consider the composition $K \to GL(n, F) \to B(F)\backslash GL(n, F)$.

This is a continuous surjective map, hence its image $B(F)\backslash GL(n, F)$ is compact. ■

We now prepare to discuss the Whittaker functionals on the representations $\mathcal{B}(\chi_1, \chi_2)$ using the tools of Section 4.3. We will eventually show that $\mathcal{B}(\chi_1, \chi_2)$ is (usually) irreducible or (always) has a unique irreducible constituent in its Jordan–Hölder series that is not one dimensional, and from this, together with the results of Section 4.4, it follows that the representation has a unique Whittaker model. However, we will find it convenient to show directly that the representation has at most one Whittaker model (without proving existence) and to use this fact as a tool to deduce the irreducibility.

The reason that a direct approach to irreducibility will not work merits comment. In Section 4.1, we proved the irreducibility of the principal series representation $(\pi, V) = \mathcal{B}(\chi_1, \chi_2)$ for $GL(2, F)$ when F is finite by showing that $\mathrm{Hom}_{GL(2,F)}(V, V)$ is one dimensional. In a more general setting, if (π, V) is a unitary representation of some group G, it is true that if $\mathrm{Hom}_G(V, V)$ is one-dimensional then V is necessarily irreducible. However, the representation $\mathcal{B}(\chi_1, \chi_2)$ is only unitary for certain choices of χ_1 and χ_2. To give an example, consider (for F a local field) the representations $\mathrm{Ind}_{B(F)}^{GL(2,F)}(\delta^{1/2})$ and $\mathrm{Ind}_{B(F)}^{GL(2,F)}(\delta^{-1/2})$, where $\delta = \delta_{B(F)}$ is the modular quasicharacter of $B(F)$. We will see later that if (π, V) is one of these representations, then $\mathrm{Hom}_{GL(2,F)}(V, V)$ is one dimensional, yet the representations are reducible. This is possible because they are nonunitary. Because of this phenomenon, a more cautious approach to the irreducibility of the principal series is necessary.

Let $(\pi, V) = \mathcal{B}(\chi_1, \chi_2)$. We will make use of a map $P : C_c^\infty(GL(2, F)) \to V$, defined by

$$(P\phi)(g) = \int_{B(F)} \phi(b^{-1}g)\,(\delta^{1/2}\chi)(b)\,db, \qquad \phi \in C_c^\infty(GL(2, F)). \tag{5.3}$$

Here, as usual, $\int_{B(F)} db$ is the left Haar integral on $B(F)$, and (because there is only one nonunimodular group in the picture) δ will always denote $\delta_{B(F)}$. It is easy to check that $P\phi \in V$. With λ and ρ denoting left and right translation as in Eq. (3.4), we have

$$P\big(\lambda(b)^{-1}\phi\big) = \big(\delta^{-1/2}\chi\big)(b)\,P(\phi), \qquad b \in B(F). \tag{5.4}$$

This is seen immediately from the definitions after rewriting Eq. (5.3) in terms of the right Haar measure and making a change of variables. On the other hand, it is equally clear that

$$P\big(\rho(g)\phi\big) = \pi(g)\,P\phi, \qquad g \in GL(2, F). \tag{5.5}$$

Proposition 4.5.3 *The map P is surjective.*

Proof Let $f \in V$. We define

$$\phi(g) = \begin{cases} f(g) & \text{if } g \in K; \\ 0 & \text{otherwise.} \end{cases}$$

We will show that $P\phi = \mathrm{Vol}\big(K \cap B(F)\big)\, f$. Indeed, it follows from the Iwasawa decomposition (Proposition 4.5.2) that an element of V is determined by its restriction to K, so it is sufficient to show that $P\phi$ and $\mathrm{Vol}\big(K \cap B(F)\big)\, f$ agree on K. Now if $g \in K$, then the integrand on the right in Eq. (5.3) equals

$$\begin{cases} \phi(g) & \text{if } b \in K \cap B(F); \\ 0 & \text{otherwise,} \end{cases}$$

and the assertion follows. This proves the surjectivity. ■

Proposition 4.5.4 *The representation $\mathcal{B}(\chi_1, \chi_2)$ admits at most one Whittaker functional.*

In fact, it may be shown that it always admits exactly one. The technique of proof is, by now, a familiar one.

Proof With notations as above, let $\Lambda : V \to \mathbb{C}$ be a Whittaker functional. Then we define a distribution Δ on $GL(2, F)$ by

$$\Delta(\phi) = \Lambda(P\phi), \qquad \phi \in C_c^\infty\big(GL(2, F)\big). \tag{5.6}$$

It follows from Eqs. (5.4) and (5.5) that

$$\lambda(b)\, \Delta = \big(\delta^{-1/2}\chi\big)(b)\, \Delta, \qquad b \in B(F), \tag{5.7}$$

and

$$\rho(n)\, \Delta = \psi_N(n)^{-1}\, \Delta, \qquad n \in N(F). \tag{5.8}$$

In view of the surjectivity of P (Proposition 4.5.3), it is sufficient to show that the space of distributions satisfying Eqs. (5.7) and (5.8) is one dimensional.

It follows from the Bruhat decomposition that there are two double cosets in $B(F)\backslash GL(2, F)/N(F)$ having as representatives the identity and the element

$$w_0 = \begin{pmatrix} & -1 \\ 1 & \end{pmatrix}.$$

The corresponding double cosets are $B(F) = B(F)\, 1\, N(F)$ and $GL(2, F) - B(F) = B(F)\, w_0\, N(F)$. By Proposition 4.3.1, we have therefore a short exact sequence

$$0 \to \mathfrak{D}\big(B(F)\big) \to \mathfrak{D}\big(GL(2, F)\big) \to \mathfrak{D}\big(GL(2, F) - B(F)\big) \to 0. \tag{5.9}$$

First we analyze the image of Δ in $\mathfrak{D}\big(GL(2, F) - B(F)\big)$, which we show is uniquely determined up to constant multiple.

Thus we show that if $\Delta_1 \in \mathfrak{D}\big(GL(2, F) - B(F)\big)$ satisfies Eqs. (5.7) and (5.8), then there exists a constant c such that

$$\Delta_1(\phi) = c \int_{B(F)} \int_{N(F)} \phi\big(bw_0 n^{-1}\big)\, \psi_N(n)\, \big(\delta^{1/2}\chi^{-1}\big)(b)\, db\, dn \qquad (5.10)$$

for $\phi \in C_c^\infty\big(GL(2, F) - B(F)\big)$. To see this, we note that the map $(b, n) \mapsto bw_0 n^{-1}$ is a diffeomorphism of $B(F) \times N(F)$ onto $GL(2, F) - B(F)$. We therefore may define a distribution T on $B(F) \times N(F)$ by $T(\Phi) = \Delta_1(\phi)$, where $\Phi \in C_c^\infty\big(B(F) \times N(F)\big)$ is defined by $\Phi(b, n) = \phi\big(bw_0 n^{-1}\big)$; our assertion follows by applying Proposition 4.3.2 to this distribution T.

Next we show that a distribution $\Delta_2 \in \mathfrak{D}\big(B(F)\big)$ that satisfies Eqs. (5.7) and (5.8) is zero. Indeed, it follows from Proposition 4.3.2 that if Δ_2 satisfies Eqs. (5.8), then there exists a constant c such that

$$\Delta_2(\phi) = c \int_{B(F)} \phi(b)\, \big(\delta^{1/2}\chi^{-1}\big)(b)\, db, \qquad \phi \in C_c^\infty\big(B(F)\big). \qquad (5.11)$$

But then $\rho(n)\, \Delta_2 = \Delta_2$ for $n \in N(F)$, which together with Eq. (5.8) implies that $c = 0$.

Now we may show that the space of $\Delta \in \mathfrak{D}\big(GL(2, F)\big)$ satisfying Eqs. (5.7) and (5.8) is one dimensional. Indeed, given two such distributions, their images in $\mathfrak{D}\big(GL(2, F) - B(F)\big)$ are proportional. Thus there is a nontrivial linear combination of the two that has zero image in $\mathfrak{D}\big(GL(2, F) - B(F)\big)$, and by the exactness of Eq. (5.9) this linear combination lies in the image of $\mathfrak{D}\big(B(F)\big)$; but a distribution in this space satisfying Eqs. (5.7) and (5.8) is zero, so the two distributions are proportional. ■

Proposition 4.5.5 *Let χ_1 and χ_2 be quasicharacters of $F^\times$. The contragredient of $\mathcal{B}(\chi_1, \chi_2)$ is $\mathcal{B}\big(\chi_1^{-1}, \chi_2^{-1}\big)$.*

Proof Let us denote $(\pi, V) = \mathcal{B}(\chi_1, \chi_2)$ and $(\pi', V') = \mathcal{B}\big(\chi_1^{-1}, \chi_2^{-1}\big)$. We note that if $f \in V$ and $f' \in V'$, then ff' satisfies Eq. (6.1) of Chapter 2 with $P = B(F)$. Thus by Lemma 2.6.1, if we define a pairing $\langle\ ,\ \rangle$ on $V \times V'$ by

$$\langle f, f' \rangle = \int_K f(k)\, f'(k)\, dk,$$

then

$$\big\langle \pi(g)\, f, \pi'(g)\, f' \big\rangle = \big\langle f, f' \big\rangle$$

when $g \in GL(2, F)$. Moreover, it is clear that if $f' \in V'$, then the linear functional $l_{f'} : V \to \mathbb{C}$ defined by $l_{f'}(f) = \big\langle f, f' \big\rangle$ is smooth, and if $l_{f'} = 0$ then $f' = 0$. Thus $f' \mapsto l_{f'}$ is an injection

$$\mathcal{B}\big(\chi_1^{-1}, \chi_2^{-1}\big) \to \widehat{\mathcal{B}(\chi_1, \chi_2)}. \qquad (5.12)$$

We must show it is also a surjection. By symmetry, we also have an injection

$$\mathcal{B}(\chi_1, \chi_2) \to \widehat{\mathcal{B}(\chi_1^{-1}, \chi_2^{-1})}. \tag{5.13}$$

Now it is not hard to see from the discussion of the contragredient representation following Proposition 4.2.2 that an injection $V \to W$ of admissible modules induces a surjection $\hat{W} \to \hat{V}$. Because $\hat{\hat{W}} \cong W$, the surjection induced from Eq. (5.13) coincides with the injection (Eq. 5.12), proving that Eq. (5.12) is indeed an isomorphism. ■

Lemma 4.5.1 *Let χ_1 and χ_2 be quasicharacters of $F^\times$.*
(i) If $\mathcal{B}(\chi_1, \chi_2)$ admits a one-dimensional invariant subspace, then $(\chi_1\chi_2^{-1})(y) = |y|^{-1}$ for all $y \in F$.
(ii) If $\mathcal{B}(\chi_1, \chi_2)$ admits a one-dimensional quotient representation, then $(\chi_1\chi_2^{-1})(y) = |y|$ for all $y \in F$.

Proof We first prove (i). Suppose that $\mathcal{B}(\chi_1, \chi_2) = (\pi, V)$ and that $f \in V$ spans a one-dimensional invariant subspace. Then there exists a quasicharacter $\rho : GL(2, F) \to \mathbb{C}$ such that $\pi(g)\, f = \rho_0(g)\, f$. Evidently, ρ_0 is trivial on the commutator subgroup of $GL(2, F)$, which is $SL(2, F)$, so ρ_0 factors through the determinant map, and thus we may write $\pi(g)\, f = \rho\big(\det(g)\big)\, f$ for some quasicharacter ρ of $F^\times$. Combining this with the definition of V, we see that

$$f(bg) = \big(\delta^{1/2}\chi\big)(b)\, \rho\big(\det(g)\big)\, f(1), \qquad b \in B(F), g \in GL(2, F),$$

where χ is as in Eq. (5.2). Because $f \neq 0$, this shows that $f(1) \neq 0$. Now taking

$$b = g^{-1} = \begin{pmatrix} y & \\ & y^{-1} \end{pmatrix},$$

we have $\big(\delta^{1/2}\chi\big)(b) = 1$, which implies (i).

To prove (ii), we note that if V has a one-dimensional irreducible quotient, then the contragredient $\hat{V}$ has a one-dimensional irreducible subrepresentation, so (ii) follows from (i) in view of Proposition 4.5.5. ■

Theorem 4.5.1 *Let χ_1 and χ_2 be quasicharacters of $F^\times$. Then $\mathcal{B}(\chi_1, \chi_2)$ is irreducible except in the following two cases:*
(i) If $(\chi_1\chi_2^{-1})(y) = |y|^{-1}$ for all $y \in F$, then $\mathcal{B}(\chi_1, \chi_2)$ has a one-dimensional invariant subspace and the quotient representation is irreducible.
(ii) If $(\chi_1\chi_2^{-1})(y) = |y|$, then $\mathcal{B}(\chi_1, \chi_2)$ has an irreducible invariant subspace of codimension one.

Proof Let $(\pi, V) = \mathcal{B}(\chi_1, \chi_2)$. Suppose that $0 \subsetneq V' \subsetneq V$ is a nontrivial invariant subspace, and let $V'' = V/V'$; let π' and π'' be the representations of

$GL(2, F)$ on V' and V'', respectively. By the exactness of the twisted Jacquet functor (Proposition 4.4.3), we have an exact sequence

$$0 \to J_\psi(V') \to J_\psi(V) \to J_\psi(V'') \to 0.$$

By Proposition 4.5.4, $J_\psi(V)$ is (at most) one dimensional, so one of $J_\psi(V')$ or $J_\psi(V'')$ is zero.

First suppose that $J_\psi(V') = 0$. Then by Theorem 4.4.3, π' factors through $\det : GL(2, F) \to F^\times$. Because V' is admissible, it has an invariant one-dimensional subspace by Proposition 4.2.9, and so by Lemma 4.5.1, we have $(\chi_1\chi_2^{-1})(y) = |y|^{-1}$. In this case, there is an obvious one-dimensional invariant subspace, namely, let us write $\chi_1(y) = \chi(y)\,|y|^{-1/2}$, so that $\chi_2(y) = \chi(y)\,|y|^{1/2}$. Then the function $f(g) = \chi\big(\det(g)\big)$ spans an invariant one-dimensional subspace. In this situation, we are in case (i) of the theorem.

The other case where $J_\psi(V'') = 0$ we may treat by dualizing. By Proposition 4.5.5, the contragredient module $\hat{V} \cong \mathcal{B}(\chi_1^{-1}, \chi_2^{-1})$, and we have a short exact sequence

$$0 \to \hat{V}'' \to \hat{V} \to \hat{V}' \to 0,$$

where now the representation $\hat{\pi}''$ on $\hat{V}''$ factors through the determinant map $GL(2, F) \to F^\times$ (because π'' does), and so case (i) of the theorem, already discussed applies to $\hat{V}$. We see that $(\chi_1^{-1}\chi_2)(y) = |y|^{-1}$, and so we are in case (ii) of the theorem.

The irreducibility of (in case (i)) the quotient representation by the invariant subspace or (in case (ii)) the invariant subrepresentation of codimension one we leave as an exercise for the reader (Exercise 4.5.2). It may be handled by considerations similar to the irreducibility assertions that we have already proved. ■

Next we consider the question of when two of the representations $\mathcal{B}(\chi_1, \chi_2)$ can be isomorphic.

Theorem 4.5.2 *Suppose that there exists a nonzero intertwining map $\mathcal{B}(\chi_1, \chi_2) \to \mathcal{B}(\mu_1, \mu_2)$. Then either $\chi_1 = \mu_1$ and $\chi_2 = \mu_2$, or $\chi_1 = \mu_2$ and $\chi_2 = \mu_1$.*

Proof We will denote $(\pi, V) = \mathcal{B}(\chi_1, \chi_2)$. By Frobenius reciprocity (Proposition 4.5.1) there exists a nonzero $B(F)$-intertwining map $\Lambda : V \to \mathbb{C}$, where $\mathbb{C}$ is given a $B(F)$-module structure by means of the quasicharacter $\delta^{1/2}\mu$ of $B(F)$, where

$$\mu\begin{pmatrix} y_1 & * \\ & y_2 \end{pmatrix} = \mu_1(y_1)\,\mu_2(y_2).$$

Thus $\Lambda\big(\pi(b)\,v\big) = \big(\delta^{1/2}\mu\big)(b)\Lambda(v)$, $v \in V$, $b \in B(F)$. Let $P : C_c^\infty\big(GL(2, F)\big) \to V$ be as in Eq. (5.3). We define a distribution Δ on $GL(2, F)$ by

$\Delta(\phi) = \Lambda\big(P(\phi)\big)$. Because P is surjective (Proposition 4.5.3), Δ is nonzero. As in the proof of Proposition 4.5.4, we have (with χ as in Eq. (5.2))

$$\lambda(b)\,\Delta = \big(\delta^{-1/2}\chi\big)(b)\,\Delta, \tag{5.14}$$

$$\rho(b)\,\Delta = \big(\delta^{-1/2}\,\mu^{-1}\big)(b)\,\Delta. \tag{5.15}$$

Now making use of the short exact sequence (Eq. 5.9), we see that there exists a nonzero distribution in either $\mathfrak{D}\big(B(F)\big)$ or $\mathfrak{D}\big(GL(2,F) - B(F)\big)$ satisfying both Eqs. (5.14) and (5.15).

First suppose that there exists a nonzero distribution $\Delta \in \mathfrak{D}\big(GL(2,F) - B(F)\big)$ that satisfies Eqs. (5.14) and (5.15). We note that Eq. (5.15) implies that for $n \in N(F)$, we have $\rho(n)\,\Delta = \Delta$. Thus, analogous to Eq. (5.10), we have (after adjusting Δ by a nonzero constant)

$$\Delta(\phi) = \int\limits_{B(F)}\int\limits_{N(F)} \phi\big(bw_0n^{-1}\big)\,\big(\delta^{1/2}\chi^{-1}\big)(b)\,db\,dn.$$

Now let $t \in T(F)$, and apply Eq. (5.15) with $b = t$. We see that

$$\begin{aligned}\big(\delta^{-1/2}\,\mu^{-1}\big)(t)\,\Delta(\phi) &= \big(\rho(t)\Delta\big)(\phi)\\ &= \int\limits_{N(F)}\int\limits_{B(F)} \phi\big(bw_0n^{-1}t^{-1}\big)\big(\delta^{1/2}\chi^{-1}\big)(b)\,db\,dn.\end{aligned}$$

After rewriting this in terms of the right Haar measure on $B(F)$ and making the change of variables $n \mapsto t^{-1}nt$, $dn \mapsto \delta(t)^{-1}\,dn$, $b \to bw_0tw_0^{-1}$, we obtain

$$\big(\delta^{-1/2}\mu^{-1}\big)(t) = \delta(t)^{-1}\,\big(\delta^{-1/2}\chi^{-1}\big)\big(w_0tw_0^{-1}\big).$$

Because $\delta(t) = \delta\big(w_0tw_0^{-1}\big)^{-1}$, this reduces to $\mu(t) = \chi\big(w_0tw_0^{-1}\big)$, and so in this case, $\chi_1 = \mu_2$ and $\chi_2 = \mu_1$.

On the other hand, suppose there exists a nonzero distribution $\Delta \in \mathfrak{D}\big(B(F)\big)$ satisfying Eqs. (5.14) and (5.15). As noted in the proof of Proposition 4.5.4, Eq. (5.14), which is the same as Eq. (5.7), implies that Δ is given by Eq. (5.11). Rewriting the integral in terms of the right Haar measure, a change of variables shows that for $b \in B(F)$

$$\rho(b)\,\Delta = \big(\delta^{-1/2}\chi^{-1}\big)(b)\,\Delta.$$

Comparison with Eq. (5.15) then shows that $\chi_1 = \mu_1$ and $\chi_2 = \mu_2$. ■

Theorem 4.5.2 (together with the evidence of Theorem 4.1.1 over a finite field) suggests that there should be isomorphisms between $\mathcal{B}(\chi_1, \chi_2)$ and $\mathcal{B}(\chi_2, \chi_1)$ when these are irreducible. In order to prove that these isomorphisms actually exist, we will define *intertwining integrals* that are generalizations of Eq. (1.11). The Archimedean analogs of these integrals have already been considered in Chapter 2, Section 2.6, and in Chapter 3, Section 3.7, we saw how these intertwining integrals arise naturally in the constant terms of Eisenstein series. The intertwining integrals may or may not (depending on χ_1 and χ_2) be convergent. When they diverge, they must be interpreted by a process of analytic continuation. To this end, we must be able to parametrize the quasicharacters χ_1 and χ_2 by complex parameters that can be analytically continued. Let us therefore fix two (unitary) characters ξ_1 and ξ_2, and we will take $\chi_i(y) = |y|^{s_i}\xi_i(y)$ $(i = 1, 2)$. Let $(\pi_{s_1,s_2}, V_{s_1,s_2}) = \mathcal{B}(\chi_1, \chi_2)$, and let $(\pi'_{s_2,s_1}, V'_{s_2,s_1}) = \mathcal{B}(\chi_2, \chi_1)$. For the first part of our discussion, s_1 and s_2 may be considered fixed, and so we will denote $(\pi_{s_1,s_2}, V_{s_1,s_2})$ and $(\pi'_{s_2,s_1}, V'_{s_2,s_1})$ more simply by (π, V) and (π', V'), respectively. Let $f \in V$. We tentatively define $Mf : GL(2, F) \to \mathbb{C}$ by

$$Mf(g) = \int_F f\left(\begin{pmatrix} & -1 \\ 1 & \end{pmatrix}\begin{pmatrix} 1 & x \\ & 1 \end{pmatrix} g\right) dx, \tag{5.16}$$

if convergent.

Proposition 4.5.6 *If* $\mathrm{re}(s_1 - s_2) > 0$, *the integral (Eq. 5.16) is absolutely convergent,* $Mf \in V'$, *and* $M : V \to V'$ *is a nonzero intertwining map. Thus if* $\mathcal{B}(\chi_1, \chi_2)$ *and* $\mathcal{B}(\chi_2, \chi_1)$ *are both irreducible, they are isomorphic.*

Proof We have

$$f\left(\begin{pmatrix} & -1 \\ 1 & \end{pmatrix}\begin{pmatrix} 1 & x \\ & 1 \end{pmatrix} g\right) = f\left(\begin{pmatrix} x^{-1} & -1 \\ & x \end{pmatrix}\begin{pmatrix} 1 & \\ x^{-1} & 1 \end{pmatrix} g\right)$$

$$= |x|^{-1}(\chi_1^{-1}\chi_2)(x)\, f\left(\begin{pmatrix} 1 & \\ x^{-1} & 1 \end{pmatrix} g\right). \tag{5.17}$$

Now because f is smooth (locally constant), there exists a constant N such that if $|x| > q^N$, then

$$f\left(\begin{pmatrix} 1 & \\ x^{-1} & 1 \end{pmatrix} g\right) = f(g). \tag{5.18}$$

Hence the absolute convergence of Eq. (5.16) is equivalent to the convergence of

$$\int_{|x|>q^N} |x|^{-1}|(\chi_1^{-1}\chi_2)(x)|\, dx = \int_{|x|>q^N} |x|^{-s_1+s_2-1}\, dx,$$

and it is not hard to see that this integral converges if $\mathrm{re}(s_1 - s_2) > 0$. This proves the assertion of absolute convergence.

Let us show now that $Mf \in V'$. First we verify condition (i) in the definition of the induced representation that was given at the beginning of this section. We must show that

$$Mf\left(\begin{pmatrix} 1 & x \\ & 1 \end{pmatrix} g\right) = Mf(g) \tag{5.19}$$

and

$$Mf\left(\begin{pmatrix} y_1 & \\ & y_2 \end{pmatrix} g\right) = \left|\frac{y_1}{y_2}\right|^{1/2} \chi_2(y_1)\,\chi_1(y_2)\,Mf(g). \tag{5.20}$$

We can absorb the x on the left side of Eq. (5.19) into the variable of integration in the definition of Mf, and after a change of variables, we obtain $Mf(g)$, so Eq. (5.19) is easy. The left side of Eq. (5.20) equals

$$\int_F f\left(\begin{pmatrix} y_2 & \\ & y_1 \end{pmatrix}\begin{pmatrix} & -1 \\ 1 & \end{pmatrix}\begin{pmatrix} 1 & y_2 y_1^{-1} x \\ & 1 \end{pmatrix} g\right) dx.$$

Making the change of variable $x \mapsto y_1 y_2^{-1} x$, so that $dx \mapsto |y_1 y_2^{-1}|\,dx$, and using the fact that $f \in V$, we obtain Eq. (5.20). On the other hand, it is clear that if f satisfies property (ii) in the definition of the induced representation with a certain group K_0, then Mf satisfies this property (ii) with the same K_0. We see that $Mf \in V'$.

Because the operations of right translation and left translation commute, and because Mf is obtained by integrating left translates of f, it is clear that $M\big(\pi(g)\,f\big) = \pi'(g)\,Mf$, so M is an intertwining operator.

It remains to be shown that $Mf \neq 0$ for some $f \in V$. This may be accomplished as follows. We define

$$f(g) = \left|\frac{y_1}{y_2}\right|^{1/2} \chi_1(y_1)\,\chi_2(y_2) \tag{5.21}$$

if

$$g = \begin{pmatrix} y_1 & z \\ & y_2 \end{pmatrix}\begin{pmatrix} & -1 \\ 1 & \end{pmatrix}\begin{pmatrix} 1 & x \\ & 1 \end{pmatrix}, \qquad y_1, y_2 \in F^\times,\ z \in F,\ x \in \mathfrak{o},$$

while $F(g) = 0$ if g is not of this form. Of course, if g has a representation as in Eq. (5.21), that representation is unique. Also y_1, y_2, x, and z depend continuously on g, so this f is smooth. It is clear that f satisfies (i) in the definition of an induced representation, so $f \in V$. Now if $g = 1$, the integral (Eq. 5.16) is compactly supported (regardless of s_1 and s_2) and equals 1, so $Mf(1) = 1$ for all s_1 and s_2. Thus $Mf \neq 0$. ■

Suppose that $\mathcal{B}(\chi_1, \chi_2)$ is irreducible, that is, is not one of the two exceptional cases of Theorem 4.5.1. Then clearly $\mathcal{B}(\chi_2, \chi_1)$ is also not one of the two exceptional cases, so it is isomorphic also. Proposition 4.5.6 shows that if $\mathrm{re}(s_1 - s_2) > 0$, then $\mathcal{B}(\chi_1, \chi_2) \cong \mathcal{B}(\chi_2, \chi_1)$, whereas if $\mathrm{re}(s_1 - s_2) < 0$, we obtain the same result by interchanging the roles of χ_1 and χ_2. This leaves only the case $\mathrm{re}(s_1 - s_2) = 0$. This turns out to be the most important case! For example, if s_1 and s_2 are pure imaginary, then by Theorem 2.6.1, $\mathcal{B}(\chi_1, \chi_2)$ is unitary, and although there are other cases where $\mathcal{B}(\chi_1, \chi_2)$ is unitary (the p-adic complementary series), the Ramanujan conjecture asserts that only when

$\mathrm{re}(s_1) = \mathrm{re}(s_2) = 0$ can $\mathcal{B}(\chi_1, \chi_2)$ occur as a constituent in an automorphic cuspidal representation. (Such a constituent is automatically unitary.)

The next stage in our analysis is to extend the intertwining integral to all χ_1 and χ_2 by a process of analytic continuation. Actually, we will see that the "analytic continuation" of Mf has a "pole" when $\chi_1 = \chi_2$, but in this case, we obviously do not need the intertwining integral M to see that $\mathcal{B}(\chi_1, \chi_2) \cong \mathcal{B}(\chi_2, \chi_1)$!

We begin by noting that by the Iwasawa decomposition, $f \in V$ is determined by its restriction to $K = GL(2, \mathfrak{o})$, which must satisfy

$$f\left(\begin{pmatrix} y_1 & x \\ & y_2 \end{pmatrix} k\right) = \xi_1(y_1)\,\xi_2(y_2)\,f(k) \tag{5.22}$$

when $y_1, y_2 \in \mathfrak{o}^\times, x \in \mathfrak{o}, k \in K$. Note that this condition is independent of s_1 and s_2. We now vary s_1 and s_2 while keeping ξ_1 and ξ_2 fixed.

Let V_0 be the space of all smooth functions on K that satisfy Eq. (5.22). If $f_0 \in V_0$, then for fixed s_1 and s_2, there exists a unique extension f_{s_1,s_2} of f_0 to V_{s_1,s_2}. We will refer to $(s_1, s_2) \mapsto f_{s_1,s_2}$ as a *flat section* of the family of representations $(\pi_{s_1,s_2}, V_{s_1,s_2})$. In other words, a flat section specifies for each s_1, s_2 a representative f_{s_1,s_2} such that the restriction of f_{s_1,s_2} to K is independent of s_1 and s_2.

Proposition 4.5.7 *Fix $f_0 \in V_0$, and let $(s_1, s_2) \mapsto f_{s_1,s_2}$ be the corresponding flat section. For fixed $g \in GL(2, F)$, the integral $Mf_{s_1,s_2}(g)$, originally defined by Eq. (5.16) for $\mathrm{re}(s_1 - s_2) > 0$, has analytic continuation to all s_1, s_2, where $\chi_1 \neq \chi_2$, and defines a nonzero intertwining operator $V_{s_1,s_2} \to V'_{s_2,s_1}$.*

We note that for $\xi_1, \xi_2, f_0 \in V_0$, and g fixed, $f_{s_1,s_2}(g)$ is just a complex number depending on two complex numbers s_1 and s_2, so the notion of analytic continuation makes sense here. We note that flatness of the section is *not* preserved by the intertwining integral M. This is not a problem: If f_{s_1,s_2} is a flat section of the representations V_{s_1,s_2}, then even though Mf_{s_1,s_2} is not a flat section of the representations V'_{s_2,s_1}, it is a finite linear combination with analytic coefficients of flat sections – see Exercise 4.5.10.

Proof The proof of Proposition 4.5.7 closely parallels the proof of Proposition 4.5.6. We choose a positive integer N so that Eq. (5.18) is true when $|x| > q^N$. Then making use of Eq. (5.17), we have (assuming first that $\mathrm{re}(s_1 - s_2) > 0$, so that the integral (Eq. 5.16) is absolutely convergent)

$$Mf_{s_1,s_2}(g) = \int\limits_{|x| \le q^N} f_{s_1,s_2}\left(\begin{pmatrix} & -1 \\ 1 & \end{pmatrix}\begin{pmatrix} 1 & x \\ & 1 \end{pmatrix} g\right) dx$$

$$+ \int\limits_{|x| \ge q^{N+1}} |x|^{-s_1+s_2-1}\left(\xi_1^{-1}\xi_2\right)(x)\,dx\; f_{s_1,s_2}(g). \tag{5.23}$$

In the first term, the domain of integration is compact, so analytic continuation is clear. In the second term, there are two cases. If $\xi_1\xi_2^{-1}$ is ramified, then for each m

$$\int_{|x|=q^m} (\xi_1^{-1}\xi_2)(x)\, dx = 0,$$

so the second integral is identically zero. If on the other hand, $\xi_1\,\xi_2^{-1}$ is nonramified, then there exists $\alpha \in \mathbb{C}$ of absolute value one (because ξ_1 and ξ_2 are assumed to be unitary) such that $(\xi_1\,\xi_2^{-1})(y) = \alpha^{\mathrm{ord}(y)}$, where $\mathrm{ord} : F^\times \to \mathbb{Z}$ is the valuation on F. The second integral may be written

$$\sum_{m=N+1}^{\infty} \int_{|x|=q^m} |x|^{-s_1+s_2}\,(\xi_1^{-1}\xi_2)(x)\,\frac{dx}{|x|}\, f_{s_1,s_2}(g)$$

$$= \mathrm{vol}(\mathfrak{o}^\times)\, f_{s_1,s_2}(g) \sum_{m=N+1}^{\infty} \left(\alpha\, q^{-s_1+s_2}\right)^m.$$

If $\mathrm{re}(s_1 - s_2) > 0$, this equals $\left(\alpha q^{-s_1+s_2}\right)^{N+1}\left(1 - \alpha\, q^{-s_1+s_2}\right)^{-1}$. We see that Eq. (5.23) has analytic continuation to all s_1 and s_2, except where $\left(\alpha\, q^{-s_1+s_2}\right)^{-1} = 1$. The latter condition is equivalent to $\chi_1 = \chi_2$. This proves the analytic continuation of the integral (Eq. 5.16) to all values of s_1 and s_2 for which $\chi_1 \neq \chi_2$.

We need to verify that the analytically continued integral $M : V_{s_1,s_2} \to V'_{s_2,s_1}$ remains an intertwining operator. This means that

$$M\left(\pi_{s_1,s_2}(g)\, f_{s_1,s_2}\right)(h) - \left(\pi'_{s_2,s_1}(g)\, Mf_{s_1,s_2}\right)(h) = 0 \tag{5.24}$$

for all $g, h \in GL(2, F)$. But Eq. (5.24) is true if $\mathrm{re}(s_1 - s_2) > 0$ by Proposition 4.5.6. Because the left side is defined by analytic continuation, it remains true for general s_1 and s_2.

We must show that the analytically continued intertwining integral is nonzero. We make use of the flat section (Eq. 5.21). As noted in the proof of Proposition 4.5.6, the integral defining $Mf(1)$ is compactly supported and hence convergent for all s_1 and s_2, and equals 1. If Mf_{s_1,s_2} is defined by means of the integral when $\mathrm{re}(s_1 - s_2) > 0$ and by analytic continuation for other values of s_1 and s_2, this remains true, so $Mf_{s_1,s_2} \neq 0$. ■

Thus we have the following complement to Theorem 4.5.2.

Theorem 4.5.3 *Assume that $\mathcal{B}(\chi_1, \chi_2)$ is irreducible. Then $\mathcal{B}(\chi_1, \chi_2) \cong \mathcal{B}(\chi_2, \chi_1)$.*

Proof It follows from Theorem 4.5.1 that if $\mathcal{B}(\chi_1, \chi_2)$ is irreducible, then so is $\mathcal{B}(\chi_2, \chi_1)$. If $\chi_1 \neq \chi_2$, the analytically continued integral M is a nonzero intertwining integral between two irreducible representations, so these are isomorphic. On the other hand, if $\chi_1 = \chi_2$, there is nothing to prove. ■

If χ_1 and χ_2 are quasicharacters of $F^\times$ such that $\mathcal{B}(\chi_1, \chi_2)$ is irreducible, then the principal series representation $\mathcal{B}(\chi_1, \chi_2)$ – or rather, its isomorphism class, as strictly speaking, $\mathcal{B}(\chi_1, \chi_2)$ refers to one particular realization of the representation – is denoted $\pi(\chi_1, \chi_2)$. If $\mathcal{B}(\chi_1, \chi_2)$ is reducible, then we have seen that it has two composition factors in its Jordan–Hölder series, a one-dimensional factor and an infinite-dimensional factor; depending on whether $(\chi_1\chi_2^{-1})(y) = |y|$ or $|y|^{-1}$, the infinite-dimensional composition factor may be either a subrepresentation or a quotient. In either case, the infinite-dimensional representation is denoted $\sigma(\chi_1, \chi_2)$ and is called a *special* or *Steinberg representation*, though the latter terminology is historically incorrect and is only justified by the analogy with the case of the finite field. The one-dimensional quotient is denoted $\pi(\chi_1, \chi_2)$.

We may summarize our results by asserting that the representations $\pi(\chi_1, \chi_2)$ and $\sigma(\chi_1, \chi_2)$ (the latter only defined for certain χ_1, χ_2) are irreducible, that $\pi(\chi_1, \chi_2) \cong \pi(\chi_2, \chi_1)$ and $\sigma(\chi_1, \chi_2) \cong \sigma(\chi_2, \chi_1)$, and that there are no other isomorphisms between these representations. There *are* other irreducible admissible representations of $GL(2, F)$ – the supercuspidals – which we will study in Section 4.7.

Our next objective in this section is to determine the Jacquet modules of the $\mathcal{B}(\chi_1, \chi_2)$.

Proposition 4.5.8 *Let (ρ, V) be a two-dimensional smooth representation of $F^\times$. Then either there exist quasicharacters ξ and ξ' of $F^\times$ such that ρ is isomorphic to the representation*

$$t \mapsto \begin{pmatrix} \xi(t) & \\ & \xi'(t) \end{pmatrix} \tag{5.25}$$

or else there exists a quasicharacter ξ of $F^\times$ such that ρ is isomorphic to the representation

$$t \mapsto \xi(t) \begin{pmatrix} 1 & v(t) \\ & 1 \end{pmatrix}, \tag{5.26}$$

where $v : F^\times \to \mathbb{Z} \subset \mathbb{C}$ is the valuation map.

Proof By Proposition 4.2.9, there exists a vector $x \in V$ that spans a one-dimensional invariant subspace; thus there exists a quasicharacter $\xi : F^\times \to \mathbb{C}$ such that

$$\rho(t)\,x = \xi(t)\,x, \qquad t \in F^\times.$$

The quotient space $V/\mathbb{C}x$ is one dimensional, so $F^\times$ acts by a quasicharacter ξ' on this space. Thus if y is another vector in V whose image in $V/\mathbb{C}x$ is nonzero, we have

$$\rho(t)\, y = \xi'(t)\, y + \lambda(t)\, x,$$

where $\lambda(t)$ is some complex number. Let us elucidate the nature of the function λ. Because $\rho(tu)\, y = \rho(t)\, \rho(u)\, y$, we must have

$$\lambda(tu) = \xi'(u)\, \lambda(t) + \lambda(u)\, \xi(t). \tag{5.27}$$

Suppose first that $\xi \neq \xi'$. Then there exists $u \in F^\times$ such that $\xi(u) \neq \xi'(u)$. Because Eq. (5.27) must be symmetrical in t and u, we have

$$\lambda(t)[\xi(u) - \xi'(u)] = \lambda(u)[\xi(t) - \xi'(t)],$$

and therefore $\lambda(t) = C\big(\xi(t) - \xi'(t)\big)$ with $C = \lambda(u)\big(\xi(u) - \xi'(u)\big)^{-1}$. Thus if $z = y - Cx$, the matrix of $\rho(t)$ with respect to the basis $\{x,\ z\}$ is Eq. (5.25).

On the other hand, if $\xi = \xi'$, then Eq. (5.27) implies that $t \mapsto \lambda(t)/\xi(t)$ is a homomorphism of $F^\times$ into the additive group $\mathbb{C}$. The kernel of this homomorphism must contain $\mathfrak{o}^\times$, because the image of $\mathfrak{o}^\times$ is a compact subgroup of $\mathbb{C}$, and the only such group is $\{0\}$. Thus there exists a constant c such that $\lambda(t) = c\,\xi(t)\, v(t)$, and in this case ρ is isomorphic to the representation Eq. (5.26) if $c \neq 0$ or to Eq. (5.25) if $c = 0$. ■

Theorem 4.5.4 *Let χ_1 and χ_2 be quasicharacters of $F^\times$, and let χ and χ' be quasicharacters of $T(F)$ defined by*

$$\chi\begin{pmatrix} t_1 & \\ & t_2 \end{pmatrix} = \chi_1(t_1)\, \chi_2(t_2), \qquad \chi'\begin{pmatrix} t_1 & \\ & t_2 \end{pmatrix} = \chi_2(t_1)\, \chi_1(t_2). \tag{5.28}$$

Then the representation of $T(F)$ on the Jacquet module of $\mathcal{B}(\chi_1, \chi_2)$ is equivalent to the following two-dimensional complex representation:

$$t \mapsto \begin{cases} \begin{pmatrix} \delta^{1/2}\chi(t) & \\ & \delta^{1/2}\chi'(t) \end{pmatrix} & \text{if } \chi_1 \neq \chi_2; \\[2ex] \delta^{1/2}\chi(t) \begin{pmatrix} 1 & v(t_1/t_2) \\ & 1 \end{pmatrix} & \text{if } \chi_1 = \chi_2. \end{cases}$$

Here $v : F^\times \to \mathbb{Z} \subset \mathbb{C}$ is the valuation map.

Proof Let $(\pi, V) = \mathcal{B}(\chi_1, \chi_2)$. By Exercise 4.5.3, $J(V)$ is exactly two dimensional. We consider first the action of $T_1(F)$, the group of elements in $T(F)$ of the form $\begin{pmatrix} a & \\ & 1 \end{pmatrix}$. This group is isomorphic to $F^\times$, so its two-dimensional representations are of the form described by Proposition 4.5.8; and because $Z(F)$ acts by scalars, and because $T(F) = T_1(F)\, Z(F)$, it follows that either there exist quasicharacters ξ and ξ' of $T(F)$ such that the representation π_N of

$T(F)$ on $J(V)$ is isomorphic to the representation

$$t \mapsto \begin{pmatrix} \delta^{1/2}\xi(t) & \\ & \delta^{1/2}\xi'(t) \end{pmatrix}, \tag{5.29}$$

or else there exists a quasicharacter ξ of $T(F)$ such that π_N is isomorphic to the representation

$$\delta^{1/2}\xi(t) \begin{pmatrix} 1 & v(t_1/t_2) \\ & 1 \end{pmatrix}. \tag{5.30}$$

The two cases may be distinguished easily because, if η_1 and η_2 are quasicharacters of $F^\times$ and if $\eta : T(F) \to \mathbb{C}^\times$ is the quasicharacter

$$\eta\begin{pmatrix} t_1 & \\ & t_2 \end{pmatrix} = \eta_1(t_1)\,\eta_2(t_2),$$

which is extended to $B(F)$ as usual to be trivial on $N(F)$, we have

$$\begin{aligned} \mathrm{Hom}_{T(F)}\left(J(V), \delta^{1/2}\eta\right) &\cong \mathrm{Hom}_{B(F)}\left(V, \delta^{1/2}\eta\right) \\ &\cong \mathrm{Hom}_{GL(2,F)}\left(V, \mathcal{B}(\eta_1, \eta_2)\right), \end{aligned} \tag{5.31}$$

where the first step is clear from the definition of $J(V)$, and the second step is the Frobenius reciprocity, Proposition 5.1. Suppose first that $\chi_1 \neq \chi_2$. Then by Proposition 4.5.7, Eq. (5.31) is nonzero if either $\eta = \chi$ or $\eta = \chi'$, and this is only possible if π_N is of the form (Eq. 5.29) with $\xi = \chi$ and $\xi' = \chi'$. On the other hand, if $\chi_1 = \chi_2$, then by Theorem 4.5.2, Eq. (5.31) is nonzero if and only if $\eta_1 = \eta_2 = \chi_1 = \chi_2$, in which case it is one dimensional by Schur's lemma (Proposition 4.2.4) because $\mathcal{B}(\chi_1, \chi_2)$ is irreducible by Theorem 4.5.1. This is possible only if π_N is of the form (Eq. 5.30) with $\xi = \chi$. ■

Our last task in this chapter is to understand the composition of intertwining integrals. Specifically, we have intertwining operators $M : \mathcal{B}(\chi_1, \chi_2) \to \mathcal{B}(\chi_2, \chi_1)$, defined by Eq. (5.16) or its analytic continuation, and $M' : \mathcal{B}(\chi_2, \chi_1) \to \mathcal{B}(\chi_1, \chi_2)$, defined the same way. The composite $M' \circ M$ is a scalar by Schur's lemma, because $\mathcal{B}(\chi_1, \chi_2)$ is irreducible for most χ_1 and χ_2. The problem is to calculate this scalar.

This calculation is important for multiple reasons. We required this knowledge in our discussions of the functional equations of Eisenstein series in Section 3.7. To give another example, Kazhdan and Patterson (1984, Corollary I.8) and Banks (1994) use precise information about the composition of intertwining integrals to prove the irreducibility of the image of an intertwining integral. Moreover, the calculation accomplishes the determination of the Plancherel measure for $GL(2, F)$, as was known to Harish-Chandra.

In the discussion that follows, it is essential that $\int_F dx$ be the self-dual Haar measure with respect to a fixed additive character ψ of F.

Our approach to calculating $M' \circ M$ will be based on the uniqueness of Whittaker models. We have a Whittaker functional $\Lambda : \mathcal{B}(\chi_1, \chi_2) \to \mathbb{C}$, defined by

$$\Lambda f = \int_F f\left(w_0 \begin{pmatrix} 1 & x \\ & 1 \end{pmatrix}\right) \psi(-x)\, dx, \tag{5.32}$$

where by Proposition 4.5.6, the integral is absolutely convergent if $\mathrm{re}(s_1 - s_2) > 0$, by comparison with Eq. (5.16). Moreover, this integral has analytic continuation to all s_1 and s_2. Analytic continuation is to be understood as in Proposition 4.5.7: If (in our previous notation) we let $f_{s_1,s_2} \in V_{s_1,s_2}$ vary in a flat section, then we claim that $\Lambda f_{s_1,s_2}$ is an analytic function of s_1 and s_2 and that its analytic continuation defines a Whittaker function for all s_1 and s_2. Indeed

$$\lim_{N\to\infty} \int_{\mathfrak{p}^{-N}} f\left(w_0 \begin{pmatrix} 1 & x \\ & 1 \end{pmatrix}\right) \psi(-x)\, dx \tag{5.33}$$

provides such an analytic continuation. To see this, one makes use of Eq. (5.17), the point being that the value of $\int_{\mathfrak{p}^{-N}} |x|^{-1} \left(\chi_1^{-1}\chi_2\right)(x)\, \psi(x)\, dx$ stabilizes when N is sufficiently large. We define $\Lambda' : \mathcal{B}(\chi_2, \chi_1) \to \mathbb{C}$ by the same formula.

If χ is any additive character of F, we have defined in Proposition 3.1.5 certain factors $\gamma(s, \chi, \psi)$. (We suppress the subscripts v from the notation because we are now working completely in the context of a local field.)

Proposition 4.5.9 *With notations as above, we have*

$$\Lambda' \circ M = \xi_1 \xi_2^{-1}(-1)\, \gamma\left(1 - s_1 + s_2, \xi_1^{-1}\xi_2, \psi\right) \Lambda. \tag{5.34}$$

The following proof was worked out with Banks (1994), who used a similar technique to normalize some more difficult intertwining integrals. An equivalent calculation may be found in Shahidi (1981, Sections 3.1 and 3.2).

Proof We note that $\Lambda' \circ M$ is a Whittaker functional on $\mathcal{B}(\chi_1, \chi_2)$, so by the uniqueness Theorem 4.4.1, there exists, for each fixed χ_1 and χ_2, a constant λ such that

$$(\Lambda' \circ M)\, f = \lambda\, \Lambda\, f, \tag{5.35}$$

for all $f \in V_{s_1,s_2}$, and if we let f vary in a flat section, $(\Lambda' \circ M)\, f$ and $\Lambda\, f$ are both meromorphic functions, so λ is itself a meromorphic function of s_1 and s_2.

By uniqueness of analytic continuation, there is no loss of generality in assuming that $\mathrm{re}(s_1 - s_2) > 0$. In this case, we may use Eqs. (5.16) and (5.33) to write

$$\Lambda' M\, f = \lim_{N\to\infty} \int_{\mathfrak{p}^{-N}} \int_F f\left(w_0 \begin{pmatrix} 1 & y \\ & 1 \end{pmatrix} w_0 \begin{pmatrix} 1 & x \\ & 1 \end{pmatrix}\right) \psi(-x)\, dy\, dx.$$

The integral is absolutely convergent by Proposition 4.5.6 and the compactness of $\mathfrak{p}^{-N}$, so we are justified in interchanging the order of integration to write this as

$$\lim_{N\to\infty} \int_F \int_{\mathfrak{p}^{-N}} f\left(\begin{pmatrix} y^{-1} & -1 \\ & y \end{pmatrix}\begin{pmatrix} & -1 \\ 1 & x - y^{-1} \end{pmatrix}\right) \psi(-x)\, dx\, dy$$

$$= \lim_{N\to\infty} \int_F \int_{\mathfrak{p}^{-N}} \phi(x - y^{-1})\, (\chi_1^{-1}\chi_2)(y)\, \psi(-x)\, dx\, \frac{dy}{|y|},$$

where we have denoted

$$\phi(x) = f\left(w_0 \begin{pmatrix} 1 & x \\ & 1 \end{pmatrix}\right). \tag{5.36}$$

We note that if ϕ is any locally constant function on F such that $\phi(x)$ is constant when $|x|$ is sufficiently large, then we can find a function $f \in V_{s_1,s_2}$ such that Eq. (5.36) is valid. In particular, ϕ could be any Schwartz function. Comparing with Eq. (5.33), we have proved that if $\phi \in C_c^\infty(F)$, then

$$\lim_{N\to\infty} \int_F \int_{\mathfrak{p}^{-N}} \phi(x - y^{-1})\, (\chi_1^{-1}\chi_2)(y)\, \psi(-x)\, dx\, \frac{dy}{|y|}$$

$$= \lambda \lim_{N\to\infty} \int_{\mathfrak{p}^{-N}} \phi(x)\, \psi(-x)\, dx.$$

We make the change of variables $y \to y^{-1}$ to obtain

$$\lim_{N\to\infty} \int_F (\xi_1\xi_2^{-1})(y)\, |y|^{s_1 - s_2 - 1}\, \psi(-y) \int_{\mathfrak{p}^{-N}} \phi(x - y)\, \psi(-(x - y))\, dx\, dy$$

$$= \lambda \lim_{N\to\infty} \int_{\mathfrak{p}^{-N}} \phi(x)\, \psi(-x)\, dx. \tag{5.37}$$

Now let us take ϕ to be the characteristic function of $\mathfrak{p}^M$, where M is chosen large enough that ψ is trivial on $\mathfrak{p}^M$. Then as long as $N \geq -M$, it is easy to see that

$$\int_{\mathfrak{p}^{-N}} \phi(x - y)\, \psi(-(x - y))\, dx$$

equals $\mathrm{vol}(\mathfrak{p}^M)$ times the characteristic function of $\mathfrak{p}^{-N}$, so the left side of Eq. (5.37) equals

$$\lim_{N\to\infty} \mathrm{vol}(\mathfrak{p}^M) \int_{\mathfrak{p}^{-N}} (\xi_1\xi_2^{-1})(y)\, |y|^{s_1 - s_2 - 1}\, \psi(-y)\, dy,$$

whereas the right side equals $\lambda\,\mathrm{vol}(\mathfrak{p}^M)$. Therefore

$$\lambda = \lim_{N\to\infty} \int_{\mathfrak{p}^{-N}} (\xi_1\xi_2^{-1})(y)\,|y|^{s_1-s_2-1}\,\psi(-y)\,dy,$$

which equals $\xi_1\xi_2^{-1}(-1)\,\gamma\big(1-s_1+s_2,\xi_1^{-1}\xi_2,\psi\big)$ by Exercise 3.1.10. (Make a change of variables $x \to -x$ in that exercise to see the factor $\xi_1\xi_2^{-1}(-1)$.) ■

Proposition 4.5.10 *Let $M : \mathcal{B}(\chi_1,\chi_2) \to \mathcal{B}(\chi_2,\chi_1)$ and $M' : \mathcal{B}(\chi_2,\chi_1) \to \mathcal{B}(\chi_1,\chi_2)$ be the intertwining integrals defined by Eq. (5.16) or its analytic continuation, where $\int_F dx$ is the Haar measure self-dual with respect to the additive character ψ. Then $M' \circ M : \mathcal{B}(\chi_1,\chi_2) \to \mathcal{B}(\chi_1,\chi_2)$ is the scalar*

$$\gamma\big(1-s_1+s_2,\xi_1^{-1}\xi_2,\psi\big)\,\gamma\big(1+s_1-s_2,\xi_1\xi_2^{-1},\psi\big). \tag{5.38}$$

Proof We know that $M' \circ M$ is a scalar, because for χ_1 and χ_2 in general position, $\mathcal{B}(\chi_1,\chi_2)$ is irreducible, so this follows from Schur's lemma. To evaluate the scalar, we compose $M' \circ M$ with Λ and use Proposition 4.5.9 twice:

$$\begin{aligned}\Lambda \circ M' \circ M &= \xi_1^{-1}\xi_2(-1)\,\gamma\big(1+s_1-s_2,\xi_1\xi_2^{-1},\psi\big)\,\Lambda' \circ M\\ &= \xi_1^{-1}\xi_2(-1)\,\gamma\big(1+s_1-s_2,\xi_1\xi_2^{-1},\psi\big)\\ &\quad\times \xi_1\xi_2^{-1}(-1)\,\gamma\big(1-s_1+s_2,\xi_1^{-1}\xi_2,\psi\big)\,\Lambda,\end{aligned}$$

so evidently $M' \circ M$ is just multiplication by Eq. (5.38). ■

Exercises

Exercise 4.5.1 Prove that if $G = GL(n,F)$, where F is a non-Archimedean local field and $\mathfrak{o}$ is its ring of integers, then the group $K = GL(n,\mathfrak{o})$ is a maximal compact subgroup of G and that every compact subgroup of G is conjugate to a subgroup of K.

Exercise 4.5.2 (a) If χ_1 and χ_2 are quasicharacters of $F^\times$ such that $(\chi_1\chi_2^{-1})(y) = |y|^{-1}$, prove the irreducibility of the quotient of $\mathcal{B}(\chi_1,\chi_2)$ by its one-dimensional invariant subspace, as asserted in Theorem 4.5.1(i).
(b) If $(\chi_1\chi_2^{-1})(y) = |y|$, prove the irreducibility of invariant subspace of codimension one in $\mathcal{B}(\chi_1,\chi_2)$, as asserted in Theorem 4.5.1(ii).

Exercise 4.5.3 Prove that the Jacquet module of $(\pi, V) = \mathcal{B}(\chi_1,\chi_2)$ is always exactly two dimensional.
[HINTS: This is a little subtle when $\chi_1 = \chi_2$. The arguments outlined below are valid even in this case. It is easiest to count the linear functionals on $J(V)$, which are the same as linear functionals $L : V \to \mathbb{C}$ that annihilate V_N, or

equivalently, that satisfy $L(\pi(n)\phi) = L(\phi)$ when $n \in N(F)$, $\phi \in V$. Imitate the proof of Theorem 4.5.2 to show that the space of such functionals is at most two dimensional. To exhibit two linearly independent functionals, define $L_1(\phi) = \phi(1)$ and

$$L_2(\phi) = \int_F \left[\phi\left(\begin{pmatrix} & -1 \\ 1 & \end{pmatrix}\begin{pmatrix} 1 & x \\ & 1 \end{pmatrix}\right) - h(x)\phi(1)\right] dx,$$

$$h(x) = \begin{cases} |x|^{-1}(\chi_1^{-1}\chi_2)(x) & \text{if } |x| > 1, \\ 0 & \text{if } |x| \leq 1, \end{cases}$$

and note that the latter integral is compactly supported by Eq. (5.17). To show that these functionals are linearly independent, exhibit f_1 and $f_2 \in V$ such that $L_1(f_1) \neq 0$, $L_1(f_2) = 0$, and $L_2(f_2) \neq 0$. You may take f_2 to be defined by Eq. (5.21). Define $f_1(bk) = (\delta^{1/2}\chi)(b)$ if $b \in B(F)$, $k \in K_1(\mathfrak{a})$ and $f_1(g) = 0$ if $g \notin B(F)K_1(\mathfrak{a})$; here $K_1(\mathfrak{a})$ is defined as in Section 4.4, and $\mathfrak{a}$ is a fractional ideal chosen so that f_1 is well defined.]

Exercise 4.5.4 Suppose that χ_1 and χ_2 are characters of $F^\times$ such that $(\chi_1\chi_2^{-1})(y) = |y|^{-1}$, so that $\mathcal{B}(\chi_1, \chi_2)$ is reducible. Prove that the Jacquet modules of $\pi(\chi_1, \chi_2)$ and $\sigma(\chi_1, \chi_2)$ are one dimensional and that the characters of $T(F)$ that they afford are $\delta^{1/2}\chi$ and $\delta^{1/2}\chi'$, respectively, where χ and χ' are as in Eq. (5.28). Show that the image of the intertwining integral $\mathcal{B}(\chi_2, \chi_1) \to \mathcal{B}(\chi_1, \chi_2)$ is the subrepresentation of $\mathcal{B}(\chi_1, \chi_2)$ isomorphic to the one-dimensional representation $\pi(\chi_1, \chi_2)$, and that the image of the intertwining integral $\mathcal{B}(\chi_1, \chi_2) \to \mathcal{B}(\chi_2, \chi_1)$ is the subrepresentation of $\mathcal{B}(\chi_2, \chi_1)$ isomorphic to $\sigma(\chi_1, \chi_2)$.

Exercise 4.5.5 Simple Mackey Theory Let G be a totally disconnected locally compact group, let H_1 be an open subgroup, let H_2 be a closed subgroup, and let (σ_1, U_1) and (σ_2, U_2) be smooth representations of H_1 and H_2, respectively; let (π_1, V_1) be the representation of G obtained from (σ_1, U_1) by compact induction, and let (π_2, V_2) be the representation of G obtained from (σ_2, U_2) by ordinary induction. Then by Frobenius reciprocity (Proposition 4.5.1) $\mathrm{Hom}_G(V_1, V_2) \cong \mathrm{Hom}_{H_2}(V_1, U_2)$. In this simplest case, the finite case of Mackey Theory (Exercise 4.1.2) has a straightforward generalization, without any need for distributions.

(a) Prove that $\mathrm{Hom}_{H_2}(V_1, U_2)$ is isomorphic as a vector space to the space of all functions $\Delta : G \to \mathrm{Hom}_\mathbb{C}(U_1, U_2)$ that satisfy Eq. (1.4).

[Hint: If $u \in U_1$ and $g \in G$, define $f_{g,u} \in V_1$ as in the proof of Proposition 4.1.2; because V_1 is obtained by compact induction, every element of V_1 is a finite linear combination of these; If $L \in \mathrm{Hom}_{H_2}(V_1, U_2)$, we associate with L the map $\Delta : G \to \mathrm{Hom}_\mathbb{C}(U_1, U_2)$ defined by $\Delta(g)(u) = L(f_{g^{-1},u})$.]

(b) If $\gamma \in G$, let $S_\gamma = H_2 \cap \gamma H_1 \gamma^{-1}$. Define two representations $\sigma_{1,\gamma}$ and $\sigma_{2,\gamma}$ of S_γ by $\sigma_{1,\gamma}(s) = \sigma_1(\gamma^{-1}s\gamma)$ and $\sigma_{2,\gamma}(s) = \sigma_2(s)$. Prove that $\mathrm{Hom}_{H_2}(V, V_2)$ is isomorphic to the direct sum of the spaces $\mathrm{Hom}_{S_\gamma}(\sigma_{1,\gamma}, \sigma_{2,\gamma})$ as γ runs

through a complete set of double coset representatives for $H_2\backslash G/H_1$. [HINT: the proof is identical to Exercise 4.1.2.]
(c) We may obtain an equivalent form of this result by conjugating the group S_γ. Let $S^\gamma = H_1 \cap \gamma^{-1} H_2 \gamma$, and let σ_1^γ and σ_2^γ be the representations $\sigma_1^\gamma(s) = \sigma_1(s)$ and $\sigma_2^\gamma(s) = \sigma_2(\gamma s \gamma^{-1})$. Then

$$\mathrm{Hom}_{S_\gamma}(\sigma_{1,\gamma}, \sigma_{2,\gamma}) \cong \mathrm{Hom}_{S^\gamma}(\sigma_1^\gamma, \sigma_2^\gamma).$$

Putting it all together

$$\mathrm{Hom}_G(V_1, V_2) \cong \bigoplus_{\gamma \in H_2\backslash G/H_1} \mathrm{Hom}_{S_\gamma}(\sigma_{1,\gamma}, \sigma_{2,\gamma}) \cong \bigoplus_{\gamma \in H_2\backslash G/H_1} \mathrm{Hom}_{S^\gamma}(\sigma_1^\gamma, \sigma_2^\gamma).$$

Exercise 4.5.6 Let G be a totally disconnected locally compact group, and let H be an open subgroup. Let (π_0, V_0) be a unitarizable smooth representation of H, and let (π, V) be the representation of G obtained by compact induction. Prove that (π, V) is unitarizable.
[HINT: If $\langle\, ,\, \rangle$ is an H-invariant positive definite Hermitian inner product on V_0, define a G-invariant positive definite Hermitian inner product on V by

$$\langle\langle f_1, f_2 \rangle\rangle = \int_G \langle f_1(g), f_2(g) \rangle \, dg. \Big]$$

Exercise 4.5.7 Let $B_1(F)$ be the group of elements of $GL(2, F)$ of the form $\left(\begin{smallmatrix} a & b \\ & 1 \end{smallmatrix}\right)$. Let $\mathcal{K}$ be the Kirillov representation of $B_1(F)$, defined in Eq. (4.25). Show that $\mathcal{K}$ is isomorphic to the representation obtained from the character ψ_N of the group $N(F)$ (also defined in Section 4.4) by compact induction.
[HINT: If $\phi \in \mathcal{K}$, show that there exists a unique element ϕ' of this induced representation such that

$$\phi(a) = \begin{pmatrix} a & \\ & 1 \end{pmatrix}. \Big]$$

Exercise 4.5.8: Transitivity of induction Let G be a totally disconnected locally compact group, and let $H_1 \subset H_2 \subset G$ be closed subgroups. Let (σ, V) be a smooth representation of H_1. We can induce σ from H_1 to G in one step, or we can induce it first from H_1 to H_2, then from H_2 to G. Show that the resulting representations are isomorphic. Prove the corresponding statement for compact induction under the hypothesis that H_2 is cocompact.
[HINT: If $\phi \in (V^{H_2})^G$, then ϕ is a map $G \to V^{H_2}$. Consider $\phi' : G \to V$ defined by $\phi'(g) = \phi(g)(1)$. Prove that $\phi \mapsto \phi'$ is an isomorphism $(V^{H_1})^G \cong V^G$.]

Exercise 4.5.9: Twisting Let (π, V) be an admissible representation of $GL(2, F)$, and let χ be a quasicharacter of $F^\times$. Define another admissible representation $\chi \otimes \pi$ acting on the same space V by

$$(\chi \otimes \pi)(g)\, v = \chi\big(\det(g)\big)\, \pi(g)\, v, \qquad g \in GL(2, F),\ v \in V.$$

Prove that if χ_1 and χ_2 are quasicharacters of $F^\times$, then

$$\chi \otimes \mathcal{B}(\chi_1, \chi_2) \cong \mathcal{B}(\chi\chi_1, \chi\chi_2).$$

Exercise 4.5.10 Let f_{s_1,s_2} be a flat section of $\mathcal{B}(\chi_1, \chi_2)$, where $\chi_i(x) = \xi_i(x)\,|x|^{s_i}$ $(i = 1, 2)$. Show that there exist a finite number of flat sections $\tilde{f}^j_{s_2,s_1}$ of $\mathcal{B}(\chi_2, \chi_1)$ and meromorphic functions $\phi_j(s_1, s_2)$ that are holomorphic except at values of s_1 and s_2 such that $\chi_1 = \chi_2$ such that

$$Mf_{s_1,s_2} = \sum_j \phi_j(s_1, s_2)\,\tilde{f}^j_{s_2,s_1}.$$

[HINT: Without loss of generality, we may assume that there exists an irreducible admissible representation ρ of $K = GL(2, \mathfrak{o})$ such that $f_{s_1,s_2} \in V_{s_1,s_2}(\rho)$, where notations are as in the text. Let $f^j_{s_2,s_1}$ be an orthonormal basis of $V'_{s_2,s_1}(\rho)$ consisting of flat sections, and let $\phi_j(s_1, s_2)$ be obtained by taking the inner product of Mf_{s_1,s_2} with $f^j_{s_2,s_1}$.]

4.6 Spherical Representations

As usual, in this section F will be a non-Archimedean local field and $\mathfrak{o}$, $\mathfrak{p}$, ϖ will be the ring of integers in F, the maximal ideal in $\mathfrak{o}$, and an arbitrarily chosen generator of $\mathfrak{p}$, respectively. We choose the Haar measure on $GL(2, F)$ so that its maximal compact subgroup $K = GL(2, \mathfrak{o})$ has volume one. An irreducible admissible representation (π, V) of $GL(2, F)$ is called *spherical* (or *nonramified*) if it contains a K-fixed vector, where $K = GL(2, \mathfrak{o})$. A nonzero element of V^K is called a *spherical vector*. Spherical representations are extremely important, because an automorphic representation decomposes into a restricted tensor product of local representations, and these are spherical at all but a finite number of places.

Nearly everything in this section works without essential modification on $GL(n)$.

Proposition 4.6.1 *Let (π, V) be a spherical representation of $GL(2, F)$. Then the contragredient representation $(\hat{\pi}, \hat{V})$ is also spherical.*

Proof It is sufficient to show that the representation (π_1, V) of Theorem 4.2.2(i) has a spherical vector. But because K is stable under transpose, a vector in V that is spherical for π is also spherical for π_1, so this is clear. ■

If (π, V) is any smooth representation of $GL(2, F)$, the space V^K of K-fixed vectors is, as we have seen in Section 4.2, a module for the convolution algebra $\mathcal{H}_K$ of compactly supported K-biinvariant functions, a ring that we call the *spherical Hecke algebra*. We will eventually determine the structure of this ring. Our first aim will be to show that it is commutative, then we draw some conclusions.

We note an inconsistency between the notation in this chapter and that of Chapter 3, particularly Section 3.4, where the notation $\mathcal{H}_K$ was used to denote the algebra of the following Proposition.

Proposition 4.6.2: p-adic Cartan decomposition *A complete set of double coset representatives for $K\backslash GL(2, F)/K$ consists of diagonal matrices*

$$\begin{pmatrix} \varpi^{n_1} & \\ & \varpi^{n_2} \end{pmatrix},$$

where $n_1 \geq n_2$ are integers.

This result is closely related to Proposition 1.4.2, and indeed the proof is the same.

Proof Let $g \in GL(2, F)$. We may find $N \in \mathbb{Z}$ such that $\varpi^{-N}g$ has coefficients in $\mathfrak{o}$. In the elementary divisor theorem, as stated in Section 1.4 prior to Proposition 1.4.2, we take $R = \mathfrak{o}$, $\Lambda_1 = \mathfrak{o}^2$, and Λ_2 to be the sublattice of Λ_1 spanned by the rows of $\varpi^{-N}g$. The elementary divisor theorem then asserts that there exists a basis ξ_1, ξ_2 of $\mathfrak{o}^2$ and $D_1, D_2 \in \mathfrak{o}$ such that $D_2|D_1$ and such that $D_1\xi_1$, $D_2\xi_2$ is a basis of Λ_2. Multiplying D_1 or D_2 by a unit does not affect this property, so without loss of generality, we may assume that $\varpi^N D_1 = \varpi^{n_1}$ and $\varpi^N D_2 = \varpi^{n_2}$ are powers of ϖ. The matrix ξ with rows ξ_1 and ξ_2 is then an element of $GL(2, \mathfrak{o})$, and the rows $\text{diag}(D_1, D_2)\,\xi$ span the same lattice as $\varpi^{-n_2}g$, so $k\,\varpi^{-N}g = \text{diag}(D_1, D_2)\,\xi$ for some $k \in GL(2, \mathfrak{o})$. We see that the matrices ξ and $\text{diag}(\varpi^{n_1}, \varpi^{n_2})$ generate the same double coset, as required.

To show that distinct diagonal matrices $\text{diag}(\varpi^{n_1}, \varpi^{n_2})$ with $n_1 \geq n_2$ generate disjoint double cosets, observe that n_1 and n_2 can be reconstructed from any representative ξ of the double coset: For the fractional ideal, (ϖ^{n_2}) is the greatest common divisor of the entries of ξ, and $(\varpi^{n_1+n_2})$ is the fractional ideal generated by $\det(\xi)$ (Exercise 4.6.1). ■

Theorem 4.6.1 *The spherical Hecke algebra $\mathcal{H}_K$ is commutative.*

The proof (based on Gelfand's involution method) is similar in structure to the proofs of Lemma 1.4.3, Theorem 2.2.3, or Theorem 4.1.2. The identical argument works for $GL(n)$ without modification.

Proof Matrix transposition is a measure-preserving antiinvolution on $GL(2, F)$, which induces an involution on $\mathcal{H}_K$; specifically, if $\phi \in \mathcal{H}_K$, let ${}^\iota\phi(g) = \phi({}^\top g)$. Then it is simple to check that ${}^\iota(\phi_1 * \phi_2) = {}^\iota\phi_2 * {}^\iota\phi_1$. However, $\mathcal{H}_K$ has a basis over $\mathbb{C}$ consisting of the characteristic functions of the double cosets of K, and by Proposition 4.6.2, these are invariant under transpose; therefore ι is the identity map. Because the identity map is thus an antiinvolution, it follows that $\mathcal{H}_K$ is commutative. ■

Theorem 4.6.2 *Let (π, V) be an irreducible admissible representation of $GL(2, F)$. Then V^K is at most one dimensional, and the space of linear functionals L on V satisfying*

$$L\big(\pi(k)\, v\big) = L(v), \qquad k \in K,\ v \in V$$

is also at most one dimensional.

Proof By Proposition 4.2.3, V^K (if nonzero) is a finite-dimensional simple module for $\mathcal{H}_K$. Because $\mathcal{H}_K$ is commutative by Theorem 4.6.1, it follows that V^K can be at most one dimensional. The second assertion follows from the first, for such an L would be a K-fixed vector in the contragredient representation. (By Proposition 4.2.8, the contragredient of an irreducible admissible representation is irreducible.) ■

Suppose that (π, V) is an irreducible admissible spherical representation, and let $v_K \in V$ be a spherical vector. Then if $\phi \in \mathcal{H}_K$, $\pi(\phi)\, v_K$ is also spherical, so by Theorem 4.6.2,

$$\pi(\phi)\, v_K = \xi(\phi)\, v_K \tag{6.1}$$

for some complex number $\xi(\phi)$. It is clear that ξ is a character of $\mathcal{H}_K$. Our aim is to show that the isomorphism class of (π, V) is determined by ξ. We call $\xi : \mathcal{H}_K \to \mathbb{C}$ the *character of $\mathcal{H}_K$ associated with the spherical representation* (π, V).

Let k be a field, and let R be an algebra over k. Let M be an R-module that is a finite-dimensional vector space over R. By a *matrix coefficient* of M, we mean a function $c : R \to k$ of the form $c(r) = L(r \cdot x)$, where $x \in M$, and $L : M \to k$ is a linear functional.

Proposition 4.6.3 *Let R be an algebra over a field k. If two simple R-modules that are finite dimensional over k have a nonzero matrix coefficient in common, they are isomorphic.*

If k has characteristic zero, then the hypothesis of simplicity may be weakened to semisimplicity. (Modify the proof of Corollary XVII.3.8 on p. 659 of Lang (1993).)

Proof Suppose that M_1 and M_2 are R-modules and that $c(r)$ is a nonzero matrix coefficient on M_1. Then there exist vectors $x_i \in M_i$ and linear functionals $L_i : M_i \to k$ such that $c(r) = L_i(r \cdot x_i)$. Let $u \in R$ such that $c(u) \neq 0$. By Theorem XVII.3.7 on p. 650 of Lang (1993), if M_1 and M_2

are nonisomorphic, there exists $e \in R$ that acts as the identity on M_1 and zero on M_2. Then $c(ue) = L_1(ue \cdot x_1) = L_1(u \cdot x_1) = c(u) \neq 0$, whereas $c(ue) = L_2(ue \cdot x_2) = L_2(i \cdot 0) = 0$. This is a contradiction, so $M_1 \cong M_2$. ■

Theorem 4.6.3 *Let (π_1, V_1) and (π_2, V_2) be irreducible admissible spherical representations. Suppose that the characters of $\mathcal{H}_K$ associated with π_1 and π_2 by Eq. (6.1) are equal. Then $\pi_1 \cong \pi_2$.*

Compare Theorem 3.4.3. See the remark to Proposition 4.2.7.

Proof By Proposition 4.2.7, it is sufficient to show that if K_1 is an open subgroup of $GL(2, F)$, then $V_1^{K_1} \cong V_2^{K_1}$ as $\mathcal{H}_{K_1}$-modules. It is clearly sufficient to prove this when $K_1 \subseteq K$.

By Proposition 4.6.1, the contragredient representations $\hat{V}_1$ and $\hat{V}_2$ are also spherical. Let $v_i \in V_i$ and $\hat{v}_i \in \hat{V}_i$ denote spherical vectors. We normalize them so that

$$\langle v_1, \hat{v}_1 \rangle = \langle v_2, \hat{v}_2 \rangle = 1.$$

This is possible by Proposition 4.2.5. We will show that

$$\langle \pi_1(\phi)\, v_1, \hat{v}_1 \rangle = \langle \pi_2(\phi)\, v_2, \hat{v}_2 \rangle \tag{6.2}$$

for any $\phi \in \mathcal{H}$. Define a map $P : \mathcal{H} \to \mathcal{H}$ by $P(\phi) = \epsilon_K * \phi * \epsilon_K$, where ϵ_K is the characteristic function of K. Because we are normalizing the Haar measure on $GL(2, F)$ so that K has volume one, ϵ_K is an idempotent, and $P \circ P = P$. Thus $\mathcal{H}$ decomposes as the direct sum of the image and kernel of P, and it is clear that the image of P is precisely $\mathcal{H}_K$. It is thus sufficient to prove Eq. (6.2) in the two cases where $\phi \in \mathcal{H}_K$ or where $P(\phi) = 0$.

If $\phi \in \mathcal{H}_K$, then both sides of Eq. (6.2) clearly equal $\xi(\phi)$, where ξ is the character of $\mathcal{H}_K$ associated with π_1 or π_2; we recall that the two associated characters are assumed to be equal. Thus if $\phi \in \mathcal{H}_K$, equation (6.2) is equivalent to our hypothesis.

On the other hand, if $P(\phi) = 0$, we have $\pi_1(\epsilon_K)\, v_1 = v_1$ and $\hat{\pi}_1(\epsilon_K)\, \hat{v}_1 = \hat{v}_1$, so the left side of Eq. (6.2) equals

$$\langle \pi_1(\phi)\, \pi_1(\epsilon_K)\, v_1, \hat{\pi}_1(\epsilon_K)\, \hat{v}_1 \rangle = \langle \pi_1(\epsilon_K)\, \pi_1(\phi)\, \pi_1(\epsilon_K)\, v_1, \hat{v}_1 \rangle.$$

Because $P(\phi) = 0$, we see that the left side of Eq. (6.2) vanishes, and the right side vanishes for the same reason. This completes the proof of Eq. (6.2).

Now for fixed $K_1 \subseteq K$, the restriction of the left side of Eq. (6.2) to $\phi \in \mathcal{H}_{K_1}$ defines a nonzero matrix coefficient of $V_1^{K_1}$ and the right side to a nonzero matrix coefficient of $V_2^{K_1}$. (It is nonzero because it equals one when $\phi = \epsilon_K$.) These matrix coefficients are equal, so by Proposition 4.6.3, $V_1^{K_1}$ and $V_2^{K_1}$ are isomorphic as $\mathcal{H}_{K_1}$-modules. It follows thus from Proposition 4.2.7 that the representations π_1 and π_2 are isomorphic. ■

To proceed further, we need to know the precise structure of the Hecke algebra. If k is a nonnegative integer, let $T(\mathfrak{p}^k) \in \mathcal{H}_K$ be the characteristic function of the set of all $g \in \mathrm{Mat}_2(\mathfrak{o})$ such that the ideal generated by $\det(g)$ in $\mathfrak{o}$ is $\mathfrak{p}^k$. Also, let $R(\mathfrak{p}) \in \mathcal{H}_K$ be the characteristic function of

$$K\begin{pmatrix} \varpi & \\ & \varpi \end{pmatrix}K = K\begin{pmatrix} \varpi & \\ & \varpi \end{pmatrix}.$$

Proposition 4.6.4 *If $k \geq 1$, we have*

$$T(\mathfrak{p})\,T(\mathfrak{p}^k) = T(\mathfrak{p}^{k+1}) + q\,R(\mathfrak{p})\,T(\mathfrak{p}^{k-1}). \tag{6.3}$$

This is essentially the same as Eq. (4.12) in Chapter 1.

Proof We note that $T(\mathfrak{p}) * T(\mathfrak{p}^k)$, $T(\mathfrak{p}^{k+1})$, and $R(\mathfrak{p}) * T(p^{k-1})$ are all supported on the double cosets whose determinants generate the ideal $\mathfrak{p}^{k+1}$, so it is sufficient to verify the equality of these Hecke operators on the matrices

$$\begin{pmatrix} \varpi^{k+1-r} & \\ & \varpi^r \end{pmatrix}, \qquad r \in \mathbb{Z}.$$

Moreover, they all vanish off $\mathrm{Mat}_r(\mathfrak{o})$, so we may assume that $0 \leq r \leq k+1$.

It is clear (from Exercise 4.6.1, for example) that $T(\mathfrak{p})$ is the characteristic function of the double coset

$$K\begin{pmatrix} \varpi & \\ & 1 \end{pmatrix}K = \begin{pmatrix} 1 & \\ & \varpi \end{pmatrix}K \cup \bigcup_{b \bmod \mathfrak{p}} \begin{pmatrix} \varpi & b \\ & 1 \end{pmatrix}K \qquad \text{(disjoint)}. \tag{6.4}$$

Thus if $g \in GL(2, F)$, we have

$$\begin{aligned}
&\left(T(\mathfrak{p}) * T(\mathfrak{p}^k)\right)(g) \\
&\quad = \int_{GL(2,F)} T(\mathfrak{p})(h)\,T(\mathfrak{p}^k)(h^{-1}g)\,dh \\
&\quad = \int_{K\begin{pmatrix} \varpi & \\ & 1 \end{pmatrix}K} T(\mathfrak{p}^k)(h^{-1}g)\,dh \\
&\quad = T(\mathfrak{p}^k)\left(\begin{pmatrix} 1 & \\ & \varpi \end{pmatrix}^{-1} g\right) + \sum_{b \bmod \mathfrak{p}} T(\mathfrak{p}^k)\left(\begin{pmatrix} \varpi & b \\ & 1 \end{pmatrix}^{-1} g\right),
\end{aligned}$$

because the integrand is constant when h is in a fixed right coset of K and each coset has volume one.

From this it is easily verified that

$$\left(T(\mathfrak{p}) * T(\mathfrak{p}^k)\right)\begin{pmatrix} \varpi^{k+1-r} & \\ & \varpi^r \end{pmatrix} = \begin{cases} 1 & \text{if } r = 0 \text{ or } k+1; \\ q+1 & \text{if } 1 \le r \le k+1. \end{cases}$$

On the other hand

$$T(\mathfrak{p}^{k+1})\begin{pmatrix} \varpi^{k+1-r} & \\ & \varpi^r \end{pmatrix} = 1 \qquad (0 \le r \le k+1),$$

and

$$q\,R(\mathfrak{p})\,T(\mathfrak{p}^{k-1})\begin{pmatrix} \varpi^{k+1-r} & \\ & \varpi^r \end{pmatrix} = \begin{cases} 0 & \text{if } r = 0 \text{ or } k+1; \\ q & \text{if } 1 \le r \le k+1. \end{cases}$$

Equation (6.3) follows. ■

Proposition 4.6.5 *The spherical Hecke algebra $\mathcal{H}_K$ is generated by $T(\mathfrak{p})$, $R(\mathfrak{p})$, and $R(\mathfrak{p})^{-1}$.*

Proof By Proposition 4.6.2, a basis for $\mathcal{H}_K$ as a complex vector space consists of the characteristic functions of the double cosets

$$K\begin{pmatrix} \varpi^n & \\ & \varpi^m \end{pmatrix}K, \qquad n \ge m.$$

This equals $R(\mathfrak{p})^m$ times the characteristic function of

$$K\begin{pmatrix} \varpi^{n-m} & \\ & 1 \end{pmatrix}K$$

or $T(\mathfrak{p}^{n-m}) - R(\mathfrak{p})\,T(\mathfrak{p}^{n-m-1})$. Thus it is sufficient to show that the algebra generated by $T(\mathfrak{p})$ and $R(\mathfrak{p})$ contains $T(\mathfrak{p}^k)$ for all $k \ge 0$, and this follows from Eq. (6.3) by induction on k. ■

Let χ_1 and χ_2 be nonramified quasicharacters of $F^\times$, and let $(\pi, V) = \mathcal{B}(\chi_1, \chi_2)$. We will be most interested in the case where $(\chi_1\chi_2^{-1})(x) \neq |x|$ or $|x|^{-1}$. In this case, π is therefore irreducible by Theorem 4.5.1 and is denoted $\pi(\chi_1, \chi_2)$. Let χ be the character (Eq. 5.2) of $B(F)$. We note that V contains a K-fixed vector, which we will denote by $\phi_{K,\chi}$ or simply as ϕ_K when χ is fixed, namely

$$\phi_{K,\chi}(bk) = \phi_K(bk) = (\delta^{1/2}\chi)(b), \qquad b \in B(F),\ k \in K, \tag{6.5}$$

where δ is the modular character of $B(F)$. Note that this is well defined, because if $bk = b'k'$ with $b' \in B(F)$ and $k' \in K$, then $b = b'u$, where $u = (b')^{-1}b = k'k^{-1} \in B(F) \cap K$, and because χ is nonramified, $(\delta^{1/2}\chi)(u) = 1$. The representation π (if irreducible) is thus spherical; we refer to ϕ_K as the *normalized spherical vector* in V. We will show as a consequence of Theorem 4.6.3 that a spherical irreducible admissible representation of $GL(2, F)$, if infinite dimensional, is isomorphic to one of these *nonramified* or *spherical principal series* representations $\pi(\chi_1, \chi_2)$. The finite-dimensional irreducible spherical representations are of course also easy to classify: We know (Exercise 4.2.9) that a finite-dimensional irreducible admissible representation of $GL(2, F)$ is one-dimensional and is of the form $g \mapsto \chi\big(\det(g)\big)$, where χ is a quasicharacter of $F^\times$; evidently such a representation is spherical if and only if χ is nonramified.

To save chalk, if (π, V) is a representation and $v \in V$, we will write $T(\mathfrak{p}^k)\, v$ or $R(\mathfrak{p})\, v$ instead of $\pi\big(T(\mathfrak{p}^k)\big)\, v$ or $\pi\big(R(\mathfrak{p})\big)\, v$.

Proposition 4.6.6 *Let ϕ_K be the normalized spherical vector in $\pi(\chi_1, \chi_2)$, where χ_1 and χ_2 are nonramified quasicharacters of $F^\times$. Let $\alpha_1 = \chi_1(\varpi)$ and $\alpha_2 = \chi_2(\varpi)$. Then $T(\mathfrak{p})\, \phi_K = \lambda\, \phi_K$ and $R(\mathfrak{p})\, \phi_K = \mu\, \phi_K$, where*

$$\lambda = q^{1/2}(\alpha_1 + \alpha_2), \qquad \mu = \alpha_1\alpha_2. \tag{6.6}$$

Proof Because $T(\mathfrak{p})\, \phi_K$ is automatically spherical, it is a multiple of ϕ_K and therefore equals $\lambda\, \phi_K$ for some $\lambda \in \mathbb{C}$; similarly, $R(\mathfrak{p})\phi_K = \mu\, \phi_K$ for some $\mu \in \mathbb{C}$, and because $R(\mathfrak{p})$ has an inverse in $\mathcal{H}_K$, we know that $\mu \neq 0$.

Because $\phi_K(1) = 1$, we have

$$\lambda = \big(T(\mathfrak{p})\phi_K\big)(1) = \int_{K\begin{pmatrix} \varpi & \\ & 1 \end{pmatrix}K} \phi_K(g)\, dg$$

$$= \sum_{\gamma \in K\begin{pmatrix} \varpi & \\ & 1 \end{pmatrix}K/K} \int_K \phi(\gamma k)\, dk = \sum_{\gamma \in K\begin{pmatrix} \varpi & \\ & 1 \end{pmatrix}K/K} \phi_K(\gamma).$$

Using the representatives in Eq. (6.4), this equals

$$(\delta^{1/2}\chi)\begin{pmatrix} 1 & \\ & \varpi \end{pmatrix} + q(\delta^{1/2}\chi)\begin{pmatrix} \varpi & \\ & 1 \end{pmatrix} = q^{1/2}(\alpha_1 + \alpha_2),$$

where $\alpha_i = \chi_i(\varpi)$. Similarly, $\mu = \alpha_1\alpha_2$. ■

Theorem 4.6.4 *Let (π, V) be an irreducible admissible representation of $GL(2, F)$ that is spherical. Then either V is one dimensional, and there exists a nonramified quasicharacter χ of $F^\times$ such that $\pi(g)\, v = \chi\big(\det(g)\big)\, v$ for $g \in GL(2, F)$, $v \in V$, or else π is a spherical principal series representation.*

Proof Let ξ be the character of $\mathcal{H}_K$ associated with the spherical representation (π, V) by Eq. (6.1). Let λ and μ be the eigenvalues of $T(\mathfrak{p})$ and $R(\mathfrak{p})$ on the one-dimensional space of spherical vectors in V. Because $R(\mathfrak{p})$ is invertible in $\mathcal{H}_K$, μ is nonzero. Let α_1 and α_2 be the roots of the quadratic polynomial

$X^2 - q^{-1/2}\lambda X + \mu = 0$, and let χ_1 and χ_2 be the nonramified quasicharacters of $F^\times$ such that $\chi_i(\varpi) = \alpha_i$. Then by Proposition 4.6.6, $T(\mathfrak{p})$ and $R(\mathfrak{p})$ have the same eigenvalues λ and μ on the one-dimensional space of spherical vectors in $\pi(\chi_1, \chi_2)$. By Proposition 4.6.5, it follows that the character of the Hecke algebra associated with the spherical representation $\pi(\chi_1, \chi_2)$ is ξ.

Assuming that $\mathcal{B}(\chi_1, \chi_2)$ is irreducible, by Theorem 4.6.3, (π, V) is isomorphic to this representation. Suppose that $\mathcal{B}(\chi_1, \chi_2)$ is not irreducible. Then by Theorem 4.5.1, we have $\alpha_1\alpha_2^{-1} = q^{\pm 1}$, and interchanging α_1 and α_2, we may assume that $\alpha_1\alpha_2^{-1} = q$ so that $\mathcal{B}(\chi_1, \chi_2)$ has a one-dimensional invariant subspace. By Theorem 4.6.3, (π, V) is isomorphic to the representation of $GL(2, F)$ on this one-dimensional representation, and we find that $\pi(g)\, v = \chi\big(\det(g)\big)\, v$, where $\chi(y) = |y|\,\chi_1(y)$. ■

With χ_1 and χ_2 nonramified, let $(\pi, V) = \mathcal{B}(\chi_1, \chi_2)$ and $(\pi', V') = \mathcal{B}(\chi_2, \chi_1)$. In Section 4.5, we defined an intertwining operator $M : V \to V'$ by Eq. (5.16) if (with $\alpha_i = \chi_i(\varpi)$) we have $|\alpha_1| < |\alpha_2|$ or by analytic continuation of Eq. (5.16) in general. Next we compute the effect of the intertwining operator on the K-fixed vector $\phi_{K,\chi}$. This is a calculation that comes up often, for example, in the calculation of the constant terms of Eisenstein series. (There is a close relationship between Eq. (6.7) below and the second term in Eq. (6.12) in Chapter 1.) We will assume for the remainder of this section that the Haar measure on F is normalized so that $\mathfrak{o}$ has volume one.

Proposition 4.6.7 *We have*

$$M\phi_{K,\chi} = \frac{1 - q^{-1}\alpha_1\alpha_2^{-1}}{1 - \alpha_1\alpha_2^{-1}}\, \phi_{K,\chi'}. \tag{6.7}$$

Proof It is clear that $M\phi_{K,\chi}$ is a spherical vector in V', so it equals some constant multiple of $\phi_{K,\chi'}$. The constant equals $(M\phi_{K,\chi})(1)$, and it is this that we must compute. We may assume that $|\alpha_1| < |\alpha_2|$, so that M is defined by the integral (Eq. 5.16); then we must evaluate

$$\int_F \phi_{K,\chi}\left(\begin{pmatrix} & -1 \\ 1 & \end{pmatrix}\begin{pmatrix} 1 & x \\ & 1 \end{pmatrix}\right) dx.$$

If $x \in \mathfrak{o}$, the integrand equals 1, whereas if $m > 0$ and $x \in \mathfrak{p}^{-m} - \mathfrak{p}^{-(m-1)}$, by Eq. (5.17) with $g = 1$, the integrand equals $q^{-m}\alpha_1^m\alpha_2^{-m}$, and the volume of $\mathfrak{p}^{-m} - \mathfrak{p}^{-(m-1)}$ equals $q^m\,(1 - q^{-1})$. Summing a geometric series yields

$$1 + (1 - q^{-1})\,\alpha_1\alpha_2^{-1}(1 - \alpha_1\alpha_2^{-1})^{-1},$$

and after simplifying, we obtain Eq. (6.7). ■

There are two special functions on $GL(2, F)$ that are associated with a spherical representation (π, V). The first is the *spherical Whittaker function*, the second is the *spherical function*. The method of computation that we will employ is due to Casselman and is found in two papers, Casselman (1980) and Casselman and Shalika (1980). Though we follow this method in the simplest case of $GL(2)$, we should point out that the original papers are far more general.

Let us begin with the formula for the spherical Whittaker function. Let $(\pi, V) = \mathcal{B}(\chi_1, \chi_2)$, where χ_1 and χ_2 are nonramified. This representation admits a Whittaker model by Proposition 4.5.4. The spherical Whittaker function is simply the spherical vector W_0 in the Whittaker model. Thus if ϕ_K is the spherical vector and Λ the Whittaker functional, $W_0(g) = \Lambda\big(\pi(g)\phi_K\big)$. Recall that Λ and ϕ_K are unique up to constant multiple by Theorems 4.4.1 and 4.6.2 (or, in the case of ϕ_K, by more elementary considerations having to do with the Iwasawa decomposition, Proposition 4.5.2), so W_0 is unique up to a constant multiple, and a logical normalization would be to make $W_0(1) = 1$. However, there is another logical normalization that we use.

With notation as above, we call the quasicharacter χ of $T(F)$ or $B(F)$ defined by Eq. (5.2) *regular* if $\chi_1 \neq \chi_2$ and *dominant* if $|\alpha_1/\alpha_2| < 1$. We define a Whittaker functional Λ on $\mathcal{B}(\chi_1, \chi_2)$ by

$$\Lambda(f) = \int_F f\left(w_0 \begin{pmatrix} 1 & x \\ & 1 \end{pmatrix}\right) \psi(-x)\, dx, \qquad w_0 = \begin{pmatrix} & -1 \\ 1 & \end{pmatrix}, \tag{6.8}$$

where ψ is a nonzero additive character of F. (We saw in Section 3.7 – cf. Eq. (7.17) in particular – that this type of integral emerges naturally in the Fourier expansions of Eisenstein series.) The integral is absolutely convergent if χ is dominant, by comparison with Eq. (5.16) (Proposition 4.5.6). We may analytically continue it to all χ by interpreting it to mean

$$\Lambda(f) = \lim_{k\to\infty} \int_{\mathfrak{p}^{-k}} f\left(w_0 \begin{pmatrix} 1 & x \\ & 1 \end{pmatrix}\right) \psi(-x)\, dx. \tag{6.9}$$

Let us show that this makes sense. By Eq. (5.17), we see that if k is sufficiently large, then

$$\int_{\mathfrak{p}^{-(k+1)}-\mathfrak{p}^{-k}} f\left(w_0 \begin{pmatrix} 1 & x \\ & 1 \end{pmatrix}\right) \psi(-x)\, dx$$

$$= \int_{\mathfrak{p}^{-(k+1)}-\mathfrak{p}^{-k}} \psi(-x)\, dx\; |q|^{-(k+1)} (\chi_1^{-1}\chi_2)\, (\varpi)^{-(k+1)}\, f(1),$$

and it is clear that this is zero for k sufficiently large. Thus the limit Eq. (6.9) exists, and it is easy to see that it defines a Whittaker functional. We then define $W_0(g) = \Lambda\big(\pi(g)\,\phi_K\big)$, and the problem is to find an explicit formula for this Whittaker function. It is important to note that (for fixed g) $W_0(g)$ is a holomorphic function of α_1 and α_2.

We note that although this simple argument works well to prove the analytic continuation of the Whittaker functionals on $GL(2)$, for higher rank groups, a different approach is preferrable. Bernstein (1985) showed how integrals defining unique functionals can be analytically continued as a result of a general theorem. This beautiful idea is discussed by Gelbart and Piatetski-Shapiro (1985), who give examples but unfortunately omit the short proof of this theorem.

Note that

$$W_0\left(\begin{pmatrix} 1 & x \\ & 1 \end{pmatrix}\begin{pmatrix} z & \\ & z \end{pmatrix} gk\right) = \psi(x)\,\omega(z)\,W_0(g)$$

for $x \in F$, $z \in F^\times$ and $k \in K$, where $\omega = \chi_1\chi_2$ is the central quasicharacter of $\mathcal{B}(\chi_1, \chi_2)$. Thus it is sufficient to calculate $W_0(g)$ as g runs through a set of coset representatives for $N(F)Z(F)\backslash GL(2, F)/K$; by the Iwasawa decomposition, Proposition 4.5.2, it is thus sufficient to compute

$$W_0\begin{pmatrix} \varpi^n & \\ & 1 \end{pmatrix} \tag{6.10}$$

for $n \in \mathbb{Z}$. We will do this under the assumption that *the conductor of* ψ *is* $\mathfrak{o}$, that is, that ψ is trivial on $\mathfrak{o}$ but not on $\mathfrak{p}^{-1}$. This simplification entails no loss of generality. Indeed, if we fix ψ, an arbitrary nontrivial additive character would have the form $x \mapsto \psi(ax)$ for some $a \in F^\times$, and it is easy to see that

$$\int_F \phi_K\left(w_0\begin{pmatrix} 1 & x \\ & 1 \end{pmatrix} g\right)\psi(-ax)\,dx = |a|^{-1/2}\,\chi_2(a)\,W_0\left(\begin{pmatrix} a & \\ & 1 \end{pmatrix} g\right).$$

Thus there is no real loss of generality in our assumption that the conductor of ψ is one.

We begin by calculating $W_0(1)$. We find that

$$\int_{\mathfrak{o}} f\left(w_0\begin{pmatrix} 1 & x \\ & 1 \end{pmatrix}\right)\psi(-x)\,dx = 1,$$

because the argument of f is in K for all values of x, whereas (using Eq. (5.17))

$$\int_{\mathfrak{p}^{-1}-\mathfrak{o}} f\left(w_0\begin{pmatrix} 1 & x \\ & 1 \end{pmatrix}\right)\psi(-x)\,dx = q^{-1}\alpha_1\alpha_2^{-1}\int_{\mathfrak{p}^{-1}-\mathfrak{o}}\psi(-x)\,dx,$$

which equals $-q^{-1}\alpha_1\alpha_2^{-1}$, and similarly

$$\int_{\mathfrak{p}^{-(k+1)}-\mathfrak{p}^{-k}} f\left(w_0\begin{pmatrix} 1 & x \\ & 1 \end{pmatrix}\right)\psi(-x)\,dx = 0$$

if $k \geq 1$. Thus

$$W_0(1) = 1 - q^{-1}\alpha_1\alpha_2^{-1}. \tag{6.11}$$

We could try to calculate $W_0(g)$ by this method for arbitrary g, but the calculations become messy. The method that we will use is not the shortest. (See Exercise 4.6.2 for a shorter method.) The method that we will follow has certain other advantages; it illustrates an important technique (due to Casselman) that works well for higher-rank groups and gives information about special functions corresponding to other unique models. Thus we will obtain the Macdonald formula for the "spherical function" as a byproduct of our considerations. For this reason, we will take a slightly circuitous path to the final formula for $W_0(g)$.

There are a couple of things we can prove immediately, however. We can prove that

$$W_0(a_m) = 0, \qquad a_m = \begin{pmatrix} \varpi^m & \\ & 1 \end{pmatrix} \qquad \text{if } m < 0. \tag{6.12}$$

Indeed, if $x \in \mathfrak{o}$, then $\begin{pmatrix} 1 & x \\ & 1 \end{pmatrix} \in K$, so the left side of Eq. (6.12) equals

$$W_0\left(a_m \begin{pmatrix} 1 & x \\ & 1 \end{pmatrix}\right) = W_0\left(\begin{pmatrix} 1 & \varpi^m x \\ & 1 \end{pmatrix} a_m\right) = \psi(\varpi^m x)\, W_0(a_m).$$

If $m < 0$ then (because we are assuming that the conductor of ψ is $\mathfrak{o}$) we can choose $x \in \mathfrak{o}$ so that $\psi(\varpi^m x) \neq 1$, which gives us Eq. (6.12).

The other thing that we can prove immediately is the following proposition.

Proposition 4.6.8 *For fixed g, regard $W_0(g)$ as an analytic function of α_1 and α_2. Then*

$$(1 - q^{-1}\alpha_1\alpha_2^{-1})^{-1}\, W_0(g) \tag{6.13}$$

is invariant under the interchange of α_1 and α_2.

Proof This is a reflection of Eq. (6.11) and the uniqueness of Whittaker models. We may assume that $\alpha_1\alpha_2^{-1} \neq q^{-1}$ or q. In this case, the representation $\mathcal{B}(\chi_1, \chi_2)$ is irreducible by Theorem 4.5.1. By Theorem 4.4.1 (see also the discussion of Whittaker models following the proof of Theorem 4.1.2) this representation has a unique Whittaker model, that is, there is a unique space $\mathcal{W}$ of functions on $GL(2, F)$, invariant under right translation, satisfying

$$W\left(\begin{pmatrix} 1 & x \\ & 1 \end{pmatrix} g\right) = \psi(x)\, W(g),$$

and isomorphic as a module over $GL(2, F)$ to $\mathcal{B}(\chi_1, \chi_2)$. Moreover, by Theorem 4.6.2 (or more elementary considerations regarding the Iwasawa decomposition), the K-fixed vector in $\mathcal{W}$ is unique up to constant multiple, and it is unique without qualification if we normalize it to equal one at the identity. This normalized Whittaker function is of course Eq. (6.13). Now by Theorem 4.5.3, $\mathcal{B}(\chi_1, \chi_2) \cong \mathcal{B}(\chi_2, \chi_1)$, so this space, and the normalized spherical vector within it, are invariant under the interchange of α_1 and α_2. ■

Let $K_0(\mathfrak{p})$ be the Iwahori subgroup of K, introduced in Section 4.4, and let $J(V)$ be the Jacquet module.

Proposition 4.6.9 *Suppose that χ defined by Eq. (5.2) is regular, and let $(\pi, V) = \mathcal{B}(\chi_1, \chi_2)$. Then the composition*

$$V^{K_0(\mathfrak{p})} \xrightarrow{\text{inclusion}} V \xrightarrow{\text{projection}} J(V) \tag{6.14}$$

is an isomorphism.

Proof First let us note that $V^{K_0(\mathfrak{p})}$ is at most two dimensional. To see this, we need the following decomposition, which is derived from the Bruhat decomposition:

$$GL(2, F) = B(F)\, 1\, K_0(\mathfrak{p}) \cup B(F)\, w_0\, K_0(\mathfrak{p}) \qquad \text{(disjoint)}. \tag{6.15}$$

To prove this, we first note that by taking the Bruhat decomposition (Eq. 1.7) for $GL(2)$ over the finite field $\mathfrak{o}/\mathfrak{p}$ and pulling it back under the canonical homomorphism $K \to GL(2, \mathfrak{o}/\mathfrak{p})$, we obtain

$$K = K_0(\mathfrak{p}) \cup K_0(\mathfrak{p})\, w_0\, K_0(\mathfrak{p}), \tag{6.16}$$

and we can use the Iwahori factorization (Eq. 4.29) to write $K_0(\mathfrak{p}) = \big(K_0(\mathfrak{p}) \cap B(F)\big)\, N_-(\mathfrak{p})$; because $w_0^{-1} N_-(\mathfrak{p}) w_0 \subset K_0(\mathfrak{p})$, this means that actually

$$K = K_0(\mathfrak{p}) \cup \big(K_0(\mathfrak{p}) \cap B(F)\big)\, w_0\, K_0(\mathfrak{p}),$$

and combined with the Iwasawa decomposition (Proposition 4.5.2), we see that

$$GL(2, F) = B(F)\, 1\, K_0(\mathfrak{p}) \cup B(F)\, \big(K_0(\mathfrak{p}) \cap B(F)\big)\, w_0 K_0(\mathfrak{p}),$$

which gives us Eq. (6.15). (The two double cosets are disjoint because otherwise we would have $w_0 \in B(F)\, K_0(\mathfrak{p})$, and it is easy to see that this is false.)

We will naturally call a vector *Iwahori-fixed* if it is in $V^{K_0(\mathfrak{p})}$. From Eq. (6.15), an Iwahori-fixed vector is determined by its values on 1 and w_0, so $V^{K_0(\mathfrak{p})}$ is at most two dimensional. On the other hand, $J(V)$ is two dimensional by Exercise 4.5.3, so it is enough to show that Eq. (6.14) is surjective. By Theorem 4.4.4, the image of Eq. (6.14) is $J(V)^{T_0}$, where $T_0 = T(\mathfrak{o})$ in the notation of Section 4.4. But as a $T(F)$-module, because χ is regular

$$J(V) \cong \chi\delta^{1/2} \oplus \chi'\delta^{1/2},$$

where

$$\chi'\begin{pmatrix} t_1 & \\ & t_2 \end{pmatrix} = \chi_2(t_1)\, \chi_1(t_2). \tag{6.17}$$

(This follows readily from Frobenius reciprocity and the results of Section 4.5. See Exercise 4.5.4.) Because χ_1 and χ_2 are nonramified, it follows that all of $J(V)$ is $T(\mathfrak{o})$-invariant, so Eq. (6.14) is indeed surjective. ■

Now let us introduce two linear functionals L_0 and L_1 on V as follows. As in Exercise 4.5.3, L_1 is evaluation at the identity; L_0 is obtained by applying the intertwining integral $M : \mathcal{B}(\chi_1, \chi_2) \to \mathcal{B}(\chi_2, \chi_1)$ defined in Section 4.5, then evaluating at the identity. Thus

$$L_1\phi = \phi(1), \qquad L_0\phi = (M\phi)(1).$$

We note that M (and hence L_0) is defined so long as χ is regular.

It is clear that L_1 and L_0 both annihilate the kernel V_N (in the notation of Section 4.4) of the projection map $V \to J(V)$. Thus they may be regarded as linear functionals on either $J(V)$ or on V^{K_0}. We have, for $t \in T(F)$

$$L_1\big(\pi(t)\phi\big) = \phi(t) = (\delta^{1/2}\chi)(t)\, L_1(\phi), \tag{6.18}$$

and similarly (with $(\pi', V') = \mathcal{B}(\chi_2, \chi_1)$, and χ' as in Eq. (6.17))

$$L_0\big(\pi(t)\phi\big) = M\big(\pi(t)\phi\big) = \pi'(t)\, M(\phi) = (\delta^{1/2}\chi')(t)\, L_0(\phi). \tag{6.19}$$

It is clear that L_1 and L_0 are nonzero (using the fact that $M : V \to V'$ is an isomorphism for L_0), and because we are assuming that χ is regular, the characters χ and χ' are linearly independent, so Eqs. (6.18) and (6.19) imply that the linear functionals L_0 and L_1 are linearly independent. By Proposition 4.6.9, this is true whether they are regarded as linear functionals on $J(V)$ or on $V^{K_0(\mathfrak{p})}$.

We may therefore find a basis $\{\phi_0, \phi_1\}$ of the Iwahori fixed vectors such that $L_0(\phi_0) = L_1(\phi_1) = 1$, $L_1(\phi_0) = L_0(\phi_1) = 0$. This is the *Casselman basis.*

Now let us exhibit some Iwahori fixed vectors. Let $m \geq 0$. With a_m as in Eq. (6.12), define

$$F_m(g) = \int_{\mathfrak{o}} \phi_K\left(g \begin{pmatrix} 1 & x \\ & 1 \end{pmatrix} a_m\right) dx. \tag{6.20}$$

Proposition 4.6.10 *If $0 \leq m \in \mathbb{Z}$, the vector $F_m \in V$ is Iwahori fixed.*

Proof It is clear that the vector

$$\int_{K_0(\mathfrak{p})} \pi(ka_m)\phi_K \, dk$$

is Iwahori fixed; we will show that this equals F_m. Indeed, by the Iwahori factorization (Eq. 4.31)

$$\int_{K_0(\mathfrak{p})} \big(\pi(ka_m)\phi_K\big)(g)\, dk = \int_{N_-(\mathfrak{p})} \int_{T_0} \int_{N(\mathfrak{o})} \phi_K(gnt_0n_-a_m)\, dn\, dt_0\, dn_-.$$

Because $m \geq 0$, it is easy to see that $a_m^{-1}t_0n_-a_m \in K$, so the integrand is independent of t_0 and n_-, and this equals Eq. (6.20). ■

Because ϕ_0 and ϕ_1 are a basis of $V^{K_0(\mathfrak{p})}$, we may write

$$F_m = c_0(m)\,\phi_0 + \phi_1(m)\,\phi_1. \tag{6.21}$$

We have

$$c_1(m) = L_1(F_m) = F_m(1) = \phi_K(a_m) = (\delta^{1/2}\chi)(a_m) = q^{-m/2}\,\alpha_1^m. \tag{6.22}$$

Similarly, because L_0 factors through the intertwining map $M : V \to V'$, we have

$$c_0(m) = q^{-m/2}\,\alpha_2^m\,M(\phi_K)(1). \tag{6.23}$$

Indeed, the left side here equals $L_1(F_m)$ or

$$\int_{N(F)} F_m(w_0 n)\,dn = \int_{N(F)}\int_{N(\mathfrak{o})} \phi_K(w_0 n n' a_k)\,dn'\,dn.$$

After a change of variables, the integral over n' may be omitted, and this equals simply $(M\phi_K)(a_m) = (\delta^{1/2}\chi')(a_m)\,(M\phi_K)(1)$, where χ' is as in Eq. (6.17), from whence we get Eq. (6.23).

Now we can obtain some nontrivial information about $W(g)$. Specifically, we can show that if $m \geq 0$, then $W(a_m)$ is a linear combination (with coefficients independent of m) of $q^{-m/2}\alpha_1^m$ and $q^{-m/2}\alpha_2^m$. This will follow from the formula

$$W(a_m) = \int_F F_m\left(w_0\begin{pmatrix}1 & x\\ & 1\end{pmatrix}\right)\psi(-x)\,dx, \tag{6.24}$$

valid when χ is dominant. (The general case follows by analytic continuation.) To prove Eq. (6.24), substitute the definition of F_m on the right, interchange the order of integration, and make a change of variables to drop the integration over $\mathfrak{o}$ as in the proof of Eq. (6.23); one then obtains the integral representing $W(a_m)$.

Now substituting Eq. (6.21) into Eq. (6.24) and making use of Eq. (6.22) and Eq. (6.23), we see that

$$W(a_m) = C_1\,q^{-m/2}\,\alpha_1^m + C_0\,q^{-m/2}\,\alpha_2^m, \tag{6.25}$$

where

$$C_1 = \int_F \phi_1\left(w_0\begin{pmatrix}1 & x\\ & 1\end{pmatrix}\right)\psi(-x)\,dx \tag{6.26}$$

and

$$C_0 = (M\phi_K)(1)\int_F \phi_0\left(w_0\begin{pmatrix}1 & x\\ & 1\end{pmatrix}\right)\psi(-x)\,dx. \tag{6.27}$$

The strategy now is to evaluate Eq. (6.27) directly. Then the value of Eq. (6.26) will follow from the functional equation, Proposition 4.6.8. It is difficult to evaluate Eq. (6.26) directly; indeed, there is a simple formula for ϕ_0 but no corresponding simple formula for ϕ_1. But we are saved from this difficulty by the functional equation. This is characteristic of Casselman's method.

Referring to the Iwahori–Bruhat decomposition Eq. (6.15), write $g \in GL(2, F)$ as $g = bk$, where $b \in B(F)$ and $k \in K$. We will show that

$$\phi_0(g) = \begin{cases} (\delta^{1/2}\chi)(b) & \text{if } g \in B(F)\, w_0\, K_0(\mathfrak{p}), \\ 0 & \text{otherwise.} \end{cases} \tag{6.28}$$

It is clear that Eq. (6.28) is well defined and is an element of V. We must therefore show that L_1 applied to Eq. (6.28) gives zero, whereas L_0 gives 1. Because $1 \notin B(F)\, w_0\, K_0(\mathfrak{p})$, it is clear that L_0 applied to Eq. (6.28) gives zero. On the other hand, it is easy to see that

$$w_0 \begin{pmatrix} 1 & x \\ & 1 \end{pmatrix} \in B(F)\, w_0 K_0(\mathfrak{p})$$

if and only if $x \in \mathfrak{o}$, and from this it is easy to see that the integral defining L_0 when applied to Eq. (6.28) gives one. This proves Eq. (6.28), and given Eq. (6.28), the integral

$$\int_F \phi_0\left(w_0 \begin{pmatrix} 1 & x \\ & 1 \end{pmatrix} \right) \psi(-x)\, dx$$

also equals one, because the integrand equals 1 if $x \in \mathfrak{o}$ and zero if $x \notin \mathfrak{o}$. Thus actually

$$C_0 = \big(M\phi_K\big)(1). \tag{6.29}$$

Theorem 4.6.5 *We have the following explicit formula for the spherical Whittaker function. If a_m is as in Eq. (6.12)*

$$(1 - q^{-1}\alpha_1\alpha_2^{-1})^{-1}\, W_0(a_m) = \begin{cases} q^{-m/2}\, \frac{\alpha_1^{m+1} - \alpha_2^{m+1}}{\alpha_1 - \alpha_2} & \text{if } m \geq 0, \\ 0 & \text{otherwise.} \end{cases} \tag{6.30}$$

We note that the "Schur polynomial" $(\alpha_1^{m+1} - \alpha_2^{m+1})/(\alpha_1 - \alpha_2)$ that occurs here is the value of the character of an irreducible representation (the mth symmetric power representation) of $GL(2, \mathbb{C})$ on the conjugacy class

$$\begin{pmatrix} \alpha_1 & \\ & \alpha_2 \end{pmatrix}.$$

This is no coincidence, for it is known that the values of spherical Whittaker functions (on higher rank groups) are given by the characters of irreducible representations applied to a conjugacy class in the "L-group" that parametrizes the representation. This formula, which was first conjectured in a letter by Langlands to Godement, was proved in varying degrees of generality by Shintani (1976), Kato (preprint), Kazhdan (unpublished), and Casselman and Shalika, (1980).

Proof By Eq. (6.12), we may assume that $m \geq 0$. By Eqs. (6.25), (6.29), and (6.7), the left side of Eq. (6.30) equals

$$C_1(1 - q^{-1}\alpha_1\alpha_2^{-1})^{-1} q^{-m/2} \alpha_1^m + (1 - \alpha_1\alpha_2^{-1}) q^{-m/2} \alpha_2^m. \tag{6.31}$$

The value of C_1 is now determined by the requirement that Eq. (6.31) be its invariant under the interchange of α_1 and α_2 (Proposition 4.6.8). After minor simplification, we obtain Eq. (6.30). ■

If (π, V) is an irreducible admissible representation of $GL(2, F)$, a *matrix coefficient* of π is a function of the form $g \mapsto \langle \pi(g)\, v, \hat{v} \rangle$, where $v \in V$ and $\hat{v} \in \hat{V}$. A particularly important matrix coefficient is the *spherical function.* Assume that (π, V) is spherical and by Proposition 4.6.1, so is $(\hat{\pi}, \hat{V})$. Then we may take v and $\hat{v}$ to be spherical vectors, normalized so $\langle v, \hat{v} \rangle = 1$. Then $\sigma(g) = \langle \pi(g)\, v, \hat{v} \rangle$ is called the *spherical function.*

In order to make this definition, we must show that $\langle v, \hat{v} \rangle \neq 0$. To see this, we note that by Theorem 4.6.4, (π, V) is either one dimensional or a spherical principal series representation. If V is one dimensional, there exists a quasicharacter χ of $F^\times$ such that $\pi(g)\, v = \chi\big(\det(g)\big)\, v$, and the spherical function is easily seen to equal $\chi\big(\det(g)\big)$. Hence we may assume that $(\pi, V) = \mathcal{B}(\chi_1, \chi_2)$ for nonramified characters χ_1 and χ_2. Now we exhibit the spherical vector in $\hat{V}$ in explicit form; this is the linear functional

$$\phi \mapsto \int_K \phi(k)\, dk.$$

It is clear that this is a K-invariant smooth linear functional on V, hence it is a constant multiple of the spherical vector $\hat{v} \in \hat{V}$. Moreover, it is clear that the value of this functional on the normalized spherical vector (Eq. 6.5) is one. As a byproduct of these considerations, we also obtain

$$\sigma(g) = \int_K \big(\pi(g)\phi_K\big)(k)\, dk = \int_K \phi_K(kg)\, dk. \tag{6.32}$$

The spherical function σ is biinvariant under K and satisfies

$$\sigma\left(\begin{pmatrix} z & \\ & z \end{pmatrix} g\right) = \omega(z)\,\sigma(g).$$

Consequently, we will have complete information on its values if we know them on a complete set of coset representatives for $KZ(F)\backslash GL(2,F)/K$. By the p-adic Cartan decomposition (Proposition 4.6.1), it is therefore sufficient to compute $\sigma(a_m)$ with $m \geq 0$.

Theorem 4.6.6: The Macdonald formula *If $m \geq 0$, we have*

$$\sigma(a_m) = \frac{1}{1+q^{-1}} q^{-m/2} \left[\alpha_1^m \frac{1-q^{-1}\alpha_2\alpha_1^{-1}}{1-\alpha_2\alpha_1^{-1}} + \alpha_2^m \frac{1-q^{-1}\alpha_1\alpha_2^{-1}}{1-\alpha_1\alpha_2^{-1}} \right]. \tag{6.33}$$

See Macdonald (1971, 1979), and Casselman (1980).

Proof We begin by noting that

$$\sigma(a_m) = \int_K F_m(k)\, dk. \tag{6.34}$$

To see this, substitute the definition (Eq. 6.20) of F_m, interchange the order of integration, and make a change of variables to see that the x integration is superfluous; compare (6.32). Now by Eq. (6.21), Eqs. (6.22), (6.23), and (6.7), this equals

$$\left[\int_K \phi_0(k)\, dk \right] q^{-m/2} \alpha_2^m \frac{1-q^{-1}\alpha_1\alpha_2^{-1}}{1-\alpha_1\alpha_2^{-1}} + \left[\int_K \phi_1(k)\, dk \right] q^{-m/2} \alpha_1^m. \tag{6.35}$$

Now we claim that

$$\int_K \phi_0(k)\, dk = \frac{1}{1+q^{-1}}. \tag{6.36}$$

Indeed, by definition of ϕ_0, the left side equals the volume of

$$K \cap \big(B(F)\, w_0\, K_0(\mathfrak{p})\big) = K_0(\mathfrak{p})\, w_0\, K_0(\mathfrak{p}).$$

We claim that the coset space $K/K_0(\mathfrak{p})$ has cardinality $q+1$; indeed, the canonical map $K \to GL(2,\mathbb{F}_q)$ induces a bijection of $K/K_0(\mathfrak{p}) \cong GL(2,\mathbb{F}_q)/B(\mathbb{F}_q)$, and this is the projective line over $\mathbb{F}_q$, which has $q+1$ elements. Thus the volume of $K_0(\mathfrak{p})$ is $1/(q+1)$. Now of these $q+1$ cosets, by Eq. (6.16), all but the identity coset lies in $K_0(\mathfrak{p})\, w_0\, K_0(\mathfrak{p})$, so the volume of this double coset is $q/(1+q) = 1/(1+q^{-1})$, which gives us Eq. (6.36). Now we see that $\sigma(a_m)$ is a sum of two terms, one a multiple (independent of m) of α_1^m, the other of α_2^m; the second term agrees with the second term in Eq. (6.33); and the formula will follow if we show that $\sigma(g)$ is invariant under the interchange of α_1 and α_2. This follows from the same idea as Proposition 4.6.8: By the uniqueness of the spherical vectors in V and $\hat{V}$, (π, V) has a unique matrix coefficient that is both left and right invariant by K. Because, for α_1 and α_2 in general position, the isomorphism class of V is invariant when they are interchanged, we see that $\sigma(g)$ is likewise invariant. ∎

* * *

As a final topic, we will determine the unitarizable principal series representations. We recall that if V is a complex vector space, a real bilinear pairing $\langle\,,\,\rangle$ on V is called *sesquilinear* if it is complex linear in the first variable and antilinear in the second variable. Further it is called *Hermitian* if it satisfies

$$\langle x, y\rangle = \overline{\langle y, x\rangle}. \tag{6.37}$$

A representation (π, V) of a group G is *unitarizable* if there exists a positive definite G-invariant Hermitian pairing on V. In this case, we may complete V to a Hilbert space and obtain a unitary representation.

Proposition 4.6.11 *Suppose that χ_1 and χ_2 are unitary characters of $F^\times$. Then the representation $\mathcal{B}(\chi_1, \chi_2)$ is unitarizable.*

Proof Indeed, by Lemma 2.6.1 (with $P = B(F)$ and $K = GL(2, \mathfrak{o})$), the pairing

$$\langle f_1, f_2\rangle = \int_K f_1(k)\,\overline{f_2(k)}\,dk$$

is a $GL(2, F)$-invariant Hermitian pairing. It is clearly positive definite. ∎

In addition to the representations of Proposition 4.6.11, there exist other principal series representations that are unitary, the so-called *complementary series representations*. We encountered the complementary series representations of $GL(2,\mathbb{R})$ in Section 2.6.

Proposition 4.6.12 *Suppose that $\mathcal{B}(\chi_1, \chi_2)$ admits an invariant nondegenerate Hermitian pairing. Then either χ_1 and χ_2 are unitary, or $\chi_1 = \overline{\chi}_2^{-1}$.*

Proof There exists a $GL(2, F)$-invariant antilinear map $\mathcal{B}(\chi_1, \chi_2) \to \mathcal{B}(\overline{\chi}_1, \overline{\chi}_2)$, namely, complex conjugation. Hence if there exists a nondegenerate invariant Hermitian pairing $\langle\,,\,\rangle$ on $\mathcal{B}(\chi_1, \chi_2)$, $(f_1, f_2) \mapsto \langle f_1, \overline{f_2}\rangle$ is a nondegenerate $GL(2, F)$-invariant bilinear pairing

$$\mathcal{B}(\chi_1, \chi_2) \times \mathcal{B}(\overline{\chi}_1, \overline{\chi}_2) \to \mathbb{C}.$$

Thus $\mathcal{B}(\overline{\chi}_1, \overline{\chi}_2)$ is isomorphic to the contragredient representation of $\mathcal{B}(\chi_1, \chi_2)$, so by Proposition 4.5.5,

$$\mathcal{B}(\overline{\chi}_1, \overline{\chi}_2) \cong \mathcal{B}(\overline{\chi}_1^{-1}, \overline{\chi}_2^{-1}).$$

By Theorem 4.5.2, we have either $\chi_1 = \overline{\chi}_1^{-1}$ and $\chi_2 = \overline{\chi}_2^{-1}$, or else $\chi_1 = \overline{\chi}_2^{-1}$. ∎

By Proposition 4.6.12, the problem of classifying the principal series that are unitary but that are not induced from unitary data (as in Proposition 4.6.11) is reduced to deciding when $\mathcal{B}(\chi, \overline{\chi}^{-1})$ is unitary. By Exercise 3.1.5(b), if χ is any quasicharacter of $F^\times$, there exists a unitary character χ_0 of $F^\times$ and a real number s such that $\chi(y) = \chi_0(y)\,|y|^s$. Then $\overline{\chi(y)}^{-1} = \chi_0(y)\,|y|^{-s}$, so by Exercise 4.5.9

$$\mathcal{B}(\chi, \overline{\chi}^{-1}) \cong \chi_0 \otimes \mathcal{B}(\chi_s, \chi_s^{-1}),$$

where $\chi_s(y) = |y|^s$. It is easy to see that a representation π is unitarizable if and only if its twist $\chi_0 \otimes \pi$ by a unitary character of $F^\times$ is unitary (see Exercise 4.5.9 for the definition of the twisted representation), so we are reduced to determining when the spherical representation $\mathcal{B}(\chi_s, \chi_s^{-1})$ is unitarizable.

Proposition 4.6.13 *Suppose that s is a real number not equal to 0 or 1, so that that $\mathcal{B}(\chi_s, \chi_s^{-1})$ is irreducible. Then $\mathcal{B}(\chi_s, \chi_s^{-1})$ is unitarizable if and only if $-\frac{1}{2} < s < \frac{1}{2}$.*

Proof Because we are assuming that $(\pi, V) = \mathcal{B}(\chi_s, \chi_s^{-1})$ is irreducible, Theorem 4.5.1 implies that $s \neq \pm\frac{1}{2}$, and in view of the isomorphism $\mathcal{B}(\chi_s, \chi_s^{-1}) \cong \mathcal{B}(\chi_s^{-1}, \chi_s)$ (and the fact that when $s = 0$ we already know from Proposition 4.6.11 that $\mathcal{B}(\chi_s, \chi_s^{-1})$ is unitary) there is no loss of generality in assuming that $s < 0$.

There can be at most one nondegenerate $GL(2, F)$-invariant sesquilinear pairing on V because V is irreducible; indeed, specifying such a nondegenerate sesquilinear pairing $\langle\,,\,\rangle$ is equivalent to specifying an isomorphism $\lambda : V \to V^*$, where V^* is the vector space of antilinear forms on V, endowed with the group action $\big(\pi^*(g)\Phi\big)(v) = \Phi\big(\pi(g^{-1})\,v\big)$. Specifically, $\lambda(v)\,w = \langle v, w\rangle$. Such an isomorphism is determined up to isomorphism by Schur's lemma (Proposition 4.2.4).

The nondegenerate sesquilinear pairing comes from making the proof of Proposition 4.6.12 explicit as follows. We have an isomorphism

$$M_s : \mathcal{B}(\chi_s, \chi_s^{-1}) \to \mathcal{B}(\chi_s^{-1}, \chi_s)$$

defined as in Section 4.5 by by analytically continuing the integrals (Eq. 5.16). (The parameter s will be allowed to vary, so we retain the subscript in our notation M_s.) We combine the intertwining operator M_s with complex conjugation and with the pairing of Proposition 4.5.5 to obtain the following nondegenerate sesquilinear pairing:

$$\langle f_1, f_2\rangle = \int_K (M_s\, f_1)(k)\,\overline{f_2(k)}\,dk. \tag{6.38}$$

We will show momentarily that this pairing is Hermitian. Then we must determine when some constant multiple of this pairing is positive definite.

Let us consider Eq. (6.38) as f_1 and f_2 vary in flat sections of $\mathcal{B}(\chi_s, \chi_s^{-1})$, as defined in Section 4.5 – this means that as s varies, the restrictions of f_1 and f_2 to K are independent of s. Then it is clear from Proposition 4.5.7 that $\langle f_1, f_2 \rangle$ is an analytic function of s (except perhaps at $s = 0$).

Referring to Eq. (6.15), we may define an Iwahori fixed vector in V by

$$f_0(g) = \begin{cases} \delta^{s+1/2}(b) & \text{if } g = bk,\, b \in B(F),\, k_0 \in K_0(\mathfrak{p}), \\ 0 & \text{if } g \in B(F)\, w_0\, K_0(\mathfrak{p}). \end{cases}$$

We note that as s varies, f_0 forms a flat section. We will calculate $\langle f_0, f_0 \rangle$. First we assume that $s > 0$, so that the integral (Eq. 5.16) is convergent. In the integral (Eq. 6.38), the integrand is zero unless $k \in K_0(\mathfrak{p})$, but if $k \in K_0(\mathfrak{p})$ then by using Eq. (5.16), the integrand equals

$$\int_F f_0\left(\begin{pmatrix} & -1 \\ 1 & \end{pmatrix}\begin{pmatrix} 1 & x \\ & 1 \end{pmatrix}\right) dx.$$

Thus

$$\langle f_0, f_0 \rangle = \mathrm{Vol}\big(K_0(\mathfrak{p})\big) \int_F f_0\left(\begin{pmatrix} & -1 \\ 1 & \end{pmatrix}\begin{pmatrix} 1 & x \\ & 1 \end{pmatrix}\right) dx.$$

It is easy to see that

$$\begin{pmatrix} & -1 \\ 1 & \end{pmatrix}\begin{pmatrix} 1 & x \\ & 1 \end{pmatrix} \in B(F)\, K_0(\mathfrak{p})$$

if and only if $x \notin \mathfrak{o}$. If $x \notin \mathfrak{o}$, then

$$\begin{pmatrix} & -1 \\ 1 & \end{pmatrix}\begin{pmatrix} 1 & x \\ & 1 \end{pmatrix} = \begin{pmatrix} x^{-1} & -1 \\ & x \end{pmatrix}\begin{pmatrix} 1 & \\ x^{-1} & 1 \end{pmatrix},$$

so

$$f_0\left(\begin{pmatrix} & -1 \\ 1 & \end{pmatrix}\begin{pmatrix} 1 & x \\ & 1 \end{pmatrix}\right) = |x|^{-1-2s}.$$

Thus

$$\langle f_0, f_0 \rangle = \frac{1}{q+1} \int_{|x|>1} |x|^{-1-2s}\, dx = \frac{1-q^{-1}}{1+q} \frac{q^{-2s}}{1-q^{-2s}}. \tag{6.39}$$

This expression remains valid after analytic continuation to $s < 0$. If $s < 0$, this expression is negative. We deduce from this that the pairing $\langle\, ,\, \rangle$ is Hermitian. Indeed, the left and right sides of Eq. (6.37) are invariant sesquilinear pairings on V, and we have noted that such a pairing is determined up to constant multiple. Taking $x = y = f_0$, because $\langle f_0, f_0 \rangle$ is real, that constant is one; therefore the pairing is Hermitian. Also, because Eq. (6.39) is negative, if a constant multiple of $\langle\, ,\, \rangle$ is to be positive definite, then $\langle\, ,\, \rangle$ is itself negative definite. We must determine when this is the case.

We next consider $\langle \phi_K, \phi_K \rangle$, where ϕ_K is the standard spherical vector defined by Eq. (6.5). It follows immediately from Eq. (6.7), with $\alpha_1 = q^{-s}$, $\alpha_2 = q^s$, that

$$\langle \phi_K, \phi_K \rangle = \frac{1 - q^{-1-2s}}{1 - q^{-2s}}.$$

This is negative if $-\frac{1}{2} < s < 0$, but positive if $s < -\frac{1}{2}$. This shows that the representation $\mathcal{B}(\chi_s, \chi_{-s})$ cannot be unitarizable if $s < -\frac{1}{2}$. It remains to be shown that it *is* unitarizable if $-\frac{1}{2} < s < 0$.

The idea of the proof is that we will begin with the case where $s = 0$, where by Proposition 4.6.11 we know that the representation $\mathcal{B}(\chi_s, \chi_s^{-1})$ is unitary, and obtain the general case by deforming the representation. Unfortunately, as we noted in connection with Proposition 4.5.7, the intertwining integral M_s has a pole at $s = 0$. We consider instead the modified intertwining operator

$$M_s^* = (1 - q^{-2s})\, M_s.$$

When $s = 0$, the integral M_s^* remains defined, as may be seen by adapting the proof of Proposition 4.5.7, and M_0^* is an endomorphism of an irreducible representation, so by Schur's lemma (Proposition 4.2.4), it is a scalar. The particular scalar may be determined by considering the effect on the spherical vector, so by Eq. (6.7), we see that M_s^* is the scalar $1 - q^{-1}$. Similarly, we consider the Hermitian inner product

$$\langle\, ,\, \rangle^* = (1 - q^{-2s})\, \langle\, ,\, \rangle.$$

This is defined even if $s = 0$, and it is positive definite when $s = 0$, by Proposition 4.6.11.

Let ρ be an irreducible admissible representation of K. We will show that the restriction of $\langle\, ,\, \rangle^*$ to the isotypic part $V(\rho)$ is positive definite. Because π is smooth, this will imply that $\langle\, ,\, \rangle^*$ is positive definite. First let $s = 0$. In this case, M_0 maps $V(\rho)$ to $V(\rho)$, and these spaces $V(\rho)$ for different ρ are orthogonal, so the restriction of $\langle\, ,\, \rangle^*$ to each $V(\rho)$ must be positive definite.

We recall that a necessary and sufficient condition for an $n \times n$ Hermitian matrix to be positive definite is that its eigenvalues all be positive. We refer $V(\rho)$ to a basis of smooth sections, whose restrictions to K are orthonormal in $L^2(K)$. With respect to this basis, $\langle\, ,\, \rangle^*$ is given by a Hermitian matrix, all of whose eigenvalues are positive when $s = 0$. As we vary s, the matrix of $\langle\, ,\, \rangle^*$ deforms through a set of Hermitian matrices, as we have shown. Its eigenvalues are continuous functions of s and (being the eigenvalues of Hermitian matrices) are real, so if it is not positive definite for all $-\frac{1}{2} < s < \frac{1}{2}$, there will be some value of s in this range for which M_s^* has zero as an eigenvalue. Thus

M_s^* has a nontrivial kernel, which is a contradiction, because it follows from Proposition 4.5.7 that M_s^* is an isomorphism for s in this range. ■

Theorem 4.6.7 *The representation $\mathcal{B}(\chi_1, \chi_2)$ is unitary if and only if either χ_1 and χ_2 are unitary, or else there exists a unitary character χ_0 and a real number $-\frac{1}{2} < s < \frac{1}{2}$ such that $\chi_1(y) = \chi_0(y)\,|y|^s$ and $\chi_2(y) = \chi_0(y)\,|y|^{-s}$.*

Proof This is just a summary of what we have proved in Propositions 4.6.11, 4.6.12 and 4.6.13. ■

Exercises

Exercise 4.6.1 Prove that if $g \in GL(2, \mathfrak{o})$, and if $n_1 \geq n_2$ are integers, then g lies in the double coset

$$K \begin{pmatrix} \varpi^{n_1} & \\ & \varpi^{n_2} \end{pmatrix} K$$

if and only if the fractional ideal generated by the coefficients of g is $\mathfrak{p}^{n_2}$ (this is the "greatest common divisor" of these matrix entries) and the fractional ideal generated by the determinant of g is $\mathfrak{p}^{n_1+n_2}$.

Exercise 4.6.2 This exercise gives an alternative approach to Eq. (6.30). Use Eq. (6.4) to show that if W_0 is a spherical Whittaker function, and if

$$w_m = W_0\begin{pmatrix} \varpi^m & \\ & 1 \end{pmatrix},$$

$$w'_m = \big(T(\mathfrak{p})W_0\big)\begin{pmatrix} \varpi^m & \\ & 1 \end{pmatrix},$$

then for $m \geq 0$, we have $w'_m = q\,w_{m+1} + \alpha_1\alpha_2\,w_{m-1}$. Thus by Eq. (6.6), we have for $m \geq 0$

$$q^{1/2}(\alpha_1 + \alpha_2)\,w_m = q\,w_{m+1} + \alpha_1\alpha_2\,w_{m-1}.$$

Combine this with the fact that $w_{-1} = 0$ and deduce that

$$w_m = q^{-m/2}\,\frac{\alpha_1^{m+1} - \alpha_2^{m+1}}{\alpha_1 - \alpha_2}\,w_0.$$

4.7 Local Functional Equations

The local functional equation of this section is the analog of Proposition 3.1.5. It is used in the global theory of L-functions described in Chapter 3, Section 5. The fact that a uniqueness principle (Theorem 4.7.4) underlies the global functional equations is typical of L-function constructions.

We continue the notations of section 5.4 and 4.5. If (π, V) is an irreducible admissible representation of $GL(2, F)$, let $(\pi_N, J(V))$ be the Jacquet module, introduced in Section 4.4. We begin by showing that the Jacquet module controls the asymptotics of the functions in the Kirillov model of V.

Proposition 4.7.1 *Let (π, V) be an irreducible admissible representation of $GL(2, F)$. The dimension of the Jacquet module of V is at most two dimensional. If it is nonzero, then π is isomorphic to a subrepresentation of $\mathcal{B}(\chi_1, \chi_2)$.*

Proof This is trivial if $J(V)$ is zero dimensional, so assume that it is not. By Theorem 4.4.4, $J(V)$ is admissible as a $T(F)$-module, so its contragredient is also admissible as a $T(F)$-module. By Proposition 4.2.9, the contragredient of $J(V)$ has a $T(F)$-invariant one-dimensional subspace, which means that there exists a quasicharacter χ of $T(F)$ and a linear functional L on $J(V)$ such that $L(\pi_N(t)\, v) = (\delta^{1/2}\chi)(t)\, L(v)$ for $v \in J(V)$, $t \in T(F)$. We regard L as a functional on V that is trivial on the kernel V_N of the projection $V \to J(V)$; then (extending χ to a character of $B(F)$ that is trivial on $N(F)$) we have $L(\pi(b)\, v) = (\delta^{1/2}\chi)(b)\, L(v)$ for $v \in V$, $b \in B(F)$. By Frobenius reciprocity (Proposition 4.5.1), L corresponds to a nonzero homomorphism $V \to \mathcal{B}(\chi_1, \chi_2)$, where χ_1 and χ_2 are the characters of $F^\times$ related to χ as in Eq. (5.2). Because V is irreducible, this homomorphism is injective.

Now by the exactness of the Jacquet functor (Proposition 4.4.2) $J(V)$ is isomorphic to a submodule of the Jacquet module of $\mathcal{B}(\chi_1, \chi_2)$, which is two dimensional by Exercise 4.5.3. ■

Proposition 4.7.2 *Let (π, V) be an irreducible representation of $GL(2, F)$. Assume that V is infinite dimensional, so that by Theorem 4.4.3, it has a Kirillov model; identify (π, V) with its Kirillov model, so that V is a space of functions on $F^\times$ on which the Borel subgroup acts by Eq. (4.25) and (denoting by ω the central quasicharacter of π) by Eq. (1.1). Then if $\phi \in V$, ϕ is locally constant, and there exists a positive constant C such that $\phi(y) = 0$ if $|y| > C$. If ϕ lies in the kernel V_N of the projection of V onto its Jacquet module $J(V)$, then there exists an $\epsilon > 0$ such that $\phi(y) = 0$ when $|y| < \epsilon$.*

Proof Because π is smooth, ϕ is stabilized by an open subgroup of $T_1(F)$, which we define to be the group of elements of the form $\left(\begin{smallmatrix} a & \\ & 1 \end{smallmatrix}\right)$ with $a \in F^\times$. By Eq. (4.25), it follows that ϕ is locally constant. Also because π is smooth, ϕ is stabilized by $N(\mathfrak{p}^k)$ for some k, where $N(\mathfrak{p}^k)$ is the group of $\left(\begin{smallmatrix} 1 & x \\ & 1 \end{smallmatrix}\right)$ with $x \in \mathfrak{p}^k$. By Eq. (4.25), this means that $\phi(y) = \psi(xy)\, \phi(y)$ for all $x \in \mathfrak{p}^k$. If $|y|$ is sufficiently large, this means that $\phi(y) = 0$. Finally, we must prove the vanishing of elements of V_N for sufficiently small values of y. Because V_N is generated by elements of the form

$$\phi' = \pi \begin{pmatrix} 1 & x \\ & 1 \end{pmatrix} \phi - \phi,$$

it is sufficient to show that $\phi'(y) = 0$ when $|y|$ is small; but

$$\phi'(y) = \big(\psi(xy) - 1\big)\,\phi(y),$$

so this is clear. ■

Let $C_c^\infty(F^\times)$ be the "Schwartz space" consisting of all locally constant functions having compact support contained in $F^\times$. If $\phi \in C_c^\infty(F^\times)$, then $\phi(y) = 0$ when $|y|$ is sufficiently small, so it is natural to extend ϕ to F by putting $\phi(0) = 0$. Let $B_1(F)$ be the subgroup of $B(F)$ of elements of the form $\left(\begin{smallmatrix} a & b \\ & 1 \end{smallmatrix}\right)$.

Proposition 4.7.3 *Let $B_1(F)$ be allowed to act on $C_c^\infty(F^\times)$ by Eq. (4.25). Then $C_c^\infty(F^\times)$ is irreducible under this action.*

Proof Let U be a nonzero invariant subspace of $C_c^\infty(F^\times)$. We will show that if $a \in F^\times$, then U contains the characteristic function of any sufficiently small neighborhood of a. From this it is clear that $U = C_c^\infty(F^\times)$.

Let ϕ be a nonzero element of U. Then $\phi(b) \neq 0$ for some b, and replacing ϕ by $\pi\left(\begin{smallmatrix} b/a & \\ & 1 \end{smallmatrix}\right)\phi$, we may assume that $\phi(a) \neq 0$. Let $f \in C_c^\infty(F)$. We consider

$$\phi_1 = \int_F f(x)\,\pi\begin{pmatrix} 1 & x \\ & 1 \end{pmatrix}\phi\,dx.$$

Note that because π is smooth and $f(x)$ is compactly supported and π is smooth, this is actually a finite sum of elements of $C_c^\infty(F^\times)$ and hence lies in $C_c^\infty(F^\times)$. We have

$$\phi_1(y) = \int_F f(x)\,\psi(xy)\,\phi(y)\,dy = \hat{f}(y)\,\phi(y).$$

Because the Fourier transform $\hat{f}$ can be an arbitrary element of $C_c^\infty(F)$, we can choose f so that $\hat{f}$ is $f(a)^{-1}$ times the characteristic function of a small neighborhood W of a on which ϕ is constant. Then ϕ_1 is the characteristic function of W. ■

Theorem 4.7.1 *Let (π, V) be an irreducible representation of $GL(2, F)$. Assume that V is infinite dimensional, so that it has a Kirillov model; identify (π, V) with its Kirillov model. Then the kernel V_N of the projection of V onto its Jacquet module is precisely $C_c^\infty(F^\times)$.*

Proof By Proposition 4.7.2, V_N is contained in $C_c^\infty(F^\times)$. It is nonzero, because it is the kernel of a homomorphism from the infinite-dimensional space V to the finite-dimensional space $J(V)$. By Proposition 4.7.3, it must be all of V. ■

Proposition 4.7.4 *Let (π, V) be an irreducible representation of $GL(2, F)$. Assume that V is infinite dimensional, so that it has a Kirillov model; identify (π, V) with its Kirillov model. Let χ be a quasicharacter of $T(F)$, and let χ_1 and χ_2 be the characters of $F^\times$ associated with χ by Eq. (5.2). Assume that $\phi \in V$ is such that $\pi_N(t)\overline{\phi} = (\delta^{1/2}\chi)(t)\overline{\phi}$ for all $t \in T(F)$, where $\overline{\phi}$ is the image of ϕ in $J(V)$. Then there exist constants C and $\epsilon > 0$ such that*

$$\phi(t) = C\,|t|^{1/2}\,\chi_1(t) \tag{7.1}$$

if $|t| < \epsilon$.

Proof Let $t_0 \in \varpi\mathfrak{o}^\times$. We note that

$$\pi\begin{pmatrix} t_0 & \\ & 1 \end{pmatrix}\phi - (\delta^{1/2}\chi)\begin{pmatrix} t_0 & \\ & 1 \end{pmatrix}\phi$$

is in V_N, so by Theorem 4.7.1, it vanishes near zero. Thus there exists a constant $\epsilon(t_0) > 0$ such that

$$\phi(tu) - |t|^{1/2}\chi_1(t)\phi(u) = 0 \tag{7.2}$$

if $t = t_0$ and $|u| \leq \epsilon(t_0)$. Because π and χ are both smooth, Eq. (7.2) is also valid when t is near t_0 and $|u| \leq \epsilon(t_0)$. By the compactness of $\varpi\mathfrak{o}^\times$, there exists a constant ϵ such that Eq. (7.2) is true if $t \in \varpi\mathfrak{o}^\times$ and $|u| \leq \epsilon$. Multiplying u by t only decreases its absolute value, so we may factor any $0 \neq t \in \mathfrak{p}$ as a product of elements of $\varpi\mathfrak{o}^\times$ and apply Eq. (7.2) repeatedly to see that if $u \leq \epsilon$ and $0 \neq t \in \mathfrak{p}$, we have

$$\phi(tu) = |t|^{1/2}\,\chi_1(t)\,\phi(u).$$

This implies Eq. (7.1). ■

Theorem 4.7.2 *Let $(\pi, V) = \pi(\chi_1, \chi_2)$ be an irreducible principal series representation (so $\chi_1\chi_2^{-1}(t)$ does not identically equal $|t|$ or $|t|^{-1}$).*

(i) Assume that $\chi_1 \neq \chi_2$. Then the space of the Kirillov model of V consists of the functions ϕ on $F^\times$ that are locally constant, that vanish for large values of t and such that there exist constants C_1 and C_2 such that for $|t|$ small

$$\phi(t) = C_1\,t^{1/2}\,\chi_1(t) + C_2\,t^{1/2}\,\chi_2(t). \tag{7.3}$$

(ii) Assume that $\chi_1 = \chi_2$. Then the space of the Kirillov model of V consists of the functions ϕ on $F^\times$ that are locally constant, that vanish for large values of t and such that there exist constants C_1 and C_2 such that for $|t|$ small,

$$\phi(t) = C_1\,t^{1/2}\,\chi_1(t) + C_2\,v(t)\,t^{1/2}\,\chi_1(t). \tag{7.4}$$

Here $v : F^\times \to \mathbb{C}$ is the valuation map.

Proof First assume that $\chi_1 \neq \chi_2$. We identify V with the space of its Kirillov model. Let χ, χ' be as in Eq. (5.28). By Theorem 4.5.4, the Jacquet module $J(V)$ has a basis $\overline{\phi}_1, \overline{\phi}_2$ such that for $t \in T(F)$, we have

$$\pi_N(t)\,\overline{\phi}_1 = (\delta^{1/2}\chi)(t)\,\overline{\phi}_1, \qquad \pi_N(t)\,\overline{\phi}_2 = (\delta^{1/2}\chi')(t)\,\overline{\phi}_2.$$

Let ϕ_1 and $\phi_2 \in V$ have images $\overline{\phi}_1$ and $\overline{\phi}_2$ in $J(V)$. By Proposition 4.7.4, there exist constants c_1 and c_2 such that

$$\phi_1(t) = c_1\,|t|^{1/2}\,\chi_1(t), \qquad \phi_2(t) = c_2\,|t|^{1/2}\,\chi_2(t).$$

Let d_1 and d_2 be such that the image $\overline{\phi}$ of ϕ in $J(V)$ equals $d_1\overline{\phi}_1 + d_2\overline{\phi}_2$. Then by Theorem 4.7.1, $\phi - d_1\phi_1 - d_2\phi_2$ vanishes near zero, so we may take $C_1 = c_1 d_1$ and $C_2 = c_2 d_2$.

Now assume that $\chi_1 = \chi_2$. By Theorem 4.5.4, we may find ϕ_1 and $\phi_2 \in V$ such that if $\overline{\phi}_1$ and $\overline{\phi}_2$ denote their images in $J(V)$, we have

$$\pi_N(t)\,\overline{\phi}_1 = (\delta^{1/2}\chi)(t)\,\overline{\phi}_1$$

and

$$\pi_N(t)\,\overline{\phi}_2 = (\delta^{1/2}\chi)(t)\,\overline{\phi}_2 + v(t_1/t_2)\,(\delta^{1/2}\chi)(t)\,\overline{\phi}_1,$$

where if

$$t = \begin{pmatrix} t_1 & \\ & t_2 \end{pmatrix},$$

then $\chi(t) = \chi_1(t_1 t_2)$. By Proposition 4.7.4, there exists a constant C such that $\phi_1(u) = C\,|u|^{1/2}\,\chi_1(u)$ for u near zero, and because by assumption, ϕ_1 is not in V_N, by Theorem 4.7.1, C is nonzero; we may therefore assume without loss of generality that

$$\phi_1(u) = |u|^{1/2}\,\chi_1(u) \tag{7.5}$$

if $|u|$ is sufficiently small. Let $t_0 \in \varpi\mathfrak{o}^\times$. Then

$$\pi\begin{pmatrix} t_0 & \\ & 1 \end{pmatrix}\phi_2 - |t_0|^{1/2}\,\chi_1(t_0)[\phi_2 + \phi_1] \in V_N,$$

so there exists a constant $\epsilon(t_0) > 0$ such that (making use of Eq. (7.5))

$$\phi_2(tu) = |t|^{1/2}\chi_1(t)\,\phi_2(u) + |tu|^{1/2}\chi_1(tu) \tag{7.6}$$

when $t = t_0$ and $|u| \le \epsilon(t_0)$. By smoothness of χ_1 and π, Eq. (7.6) is true when t is near t_0, and by the compactness of $\varpi\mathfrak{o}^\times$, there exists an $\epsilon > 0$ such that Eq. (7.5) and Eq. (7.6) are both true if $t \in \varpi\mathfrak{o}^\times$ and $|u| \le \epsilon$. Now any nonzero element t of $\mathfrak{p}$ can be factored into a product of elements of $\varpi\mathfrak{o}^\times$, so it follows from an induction argument based on Eq. (7.6) that

$$\phi_2(tu) = |t|^{1/2}\,\chi_1(t)\,\phi_2(u) + v(t)\,|tu|^{1/2}\chi_1(tu)$$

when $0 \neq t \in \mathfrak{p}$ and $|u| < \epsilon$. This and Eq. (7.5) imply that the theorem is satisfied when $\phi = \phi_1$ or ϕ_2, and because an arbitrary element of V differs from a linear combination of these two by an element of V_N, which vanishes near zero, we are done. ■

Finally, there are the special representations.

Theorem 4.7.3 *Let $(\pi, V) = \sigma(\chi_1, \chi_2)$ be a special representation of $GL(2, F)$, where $(\chi_1\chi_2^{-1})(t) = |t|^{-1}$. Then the space of the Kirillov model of V consists of the functions ϕ on $F^\times$ that are locally constant and that vanish for large values of t and such that there exists a constant C such that for $|t|$ small*

$$\phi(t) = C\, t^{1/2}\, \chi_2(t). \tag{7.7}$$

Proof The proof is similar to Theorem 4.7.2(i), but makes use of Exercise 4.5.4. ■

Theorem 4.7.4 *Let (π, V) be an irreducible admissible representation of $GL(2, F)$, and let χ be a quasicharacter of $F^\times$. Then there are at most two values of s modulo $2\pi i/\log(q)$ such that the dimension of the space of linear functionals $L : V \to \mathbb{C}$ satisfying*

$$L\left(\pi\begin{pmatrix} y & \\ & 1 \end{pmatrix} v\right) = \chi(y)\,|y|^s\, L(v) \tag{7.8}$$

is greater than one dimensional.

We note that the quasicharacter $y \mapsto |y|^s$ depends only on the value of s modulo $2\pi i/\log(q)$.

Proof If V is one dimensional, this is trivial, so by Theorem 4.4.3, we may assume that V has a Whittaker model and hence also a Kirillov model; we identify V with the space of its Kirillov model. Let s be fixed, and let L_1 and L_2 be two functionals satisfying Eq. (7.8). First we consider the restriction of L_1 and L_2 to the kernel V_N of the projection map $V \to J(V)$. By Theorem 4.7.1, V_N is $C_c^\infty(F^\times)$ on which $T_1(F)$ acts by Eq. (4.25). It follows from Proposition 4.3.2 that L_1 and L_2 are linearly independent when restricted to V_N. Thus there exist constants c_1 and c_2, not both zero, such that $c_1L_1 + c_2L_2$ factors through $J(V)$; but because $J(V)$ is at most two dimensional, this implies that $c_1L_1 + c_2L_2 = 0$ for all but two possible choices of s in $\mathbb{C}$ modulo $2\pi i/\log(q)$. ■

If (π, V) is an irreducible admissible representation of $GL(2, F)$ such that $J(V) = 0$, then π is called *supercuspidal*.

If (π, V) is an irreducible admissible representation of $GL(2, F)$ that admits a Whittaker model, we define a *local L-function* $L(s, \pi)$ as follows. If π is supercuspidal, we define $L(s,\pi)=1$. If $(V,\pi)=\pi(\chi_1,\chi_2)$ is an irreducible principal series representation, we define

$$L(s,\pi)=(1-\alpha_1 q^{-s})^{-1}(1-\alpha_2 q^{-s})^{-1}, \tag{7.9}$$

where $\alpha_i = \chi_i(\varpi)$ if χ_i is nonramified and $\alpha_i = 0$ otherwise. Finally, if $(\pi, V)=\sigma(\chi_1,\chi_2)$, where $(\chi_1\chi_2^{-1})(y)=|y|^{-1}$, we define

$$L(s,\pi)=(1-\alpha_2 q^{-s})^{-1}, \tag{7.10}$$

with α_2 as above. It follows from Proposition 4.7.1 and the description of the irreducible subspaces of the $\mathcal{B}(\chi_1,\chi_2)$ in Section 4.5 that every irreducible admissible representation that has a Whittaker model is one of these types. More generally, if ξ is a quasicharacter of $F^\times$, we define $L(s,\pi,\xi)=L(s,\xi\otimes\pi)$, where $\xi\otimes\pi$ is the representation of $GL(2,F)$ on V given by $(\xi\otimes\pi)(g)=\xi\big(\det(g)\big)\,\pi(g)$. Thus, for example, $\xi\otimes\pi(\chi_1,\chi_2)=\pi(\xi\chi_1,\xi\chi_2)$.

Proposition 4.7.5 *Let (π,V) be an irreducible admissible representation of $GL(2,F)$ that admits a Whittaker model. If ϕ is an element of the space of the Kirillov model of π, consider the integral*

$$Z(s,\phi)=\int_{F^\times}\phi(y)\,|y|^{s-1/2}\,d^\times y, \tag{7.11}$$

where $d^\times y$ denotes the Haar measure on $F^\times$. This integral is convergent for re(s) *sufficiently large and has meromorphic continuation to all s. More precisely, $Z(s,\phi)=p_\phi(q^{-s})\,L(s,\pi)$, where p_ϕ is a polynomial. Moreover, ϕ can be chosen so that $p_\phi=1$.*

It is worth noting that the proposition implies that unless s is a pole of $L(s,\pi)$, the function $Z(s,\phi)$ is analytic at s. Thus we have the simultaneous analytic continuation of $Z(s,\phi)$ for *all* ϕ to almost all complex values of s, which is a stronger assertion than the separate meromorphic continuation of these functions individually.

Proof Proposition 4.7.5 is readily verified given the precise descriptions of the Kirillov spaces in Theorems 4.7.2, 4.7.3, and (for the supercuspidals) 4.7.1. We leave the verification to the reader. ■

More generally, we can define

$$Z(s,\phi,\xi)=\int_{F}\phi(y)\,\xi(y)\,|y|^{s-1/2}\,d^\times y, \tag{7.12}$$

and it follows by applying Proposition 4.7.5 to $\xi\otimes\pi$ that this integral has meromorphic continuation to all s and in fact equals $L(s,\pi,\xi)$ times a polynomial in q^{-s}.

If we are emphasizing the Whittaker model $\mathcal{W}$ of the representation π, and if ϕ is the element of the Kirillov model related to $W \in \mathcal{W}$ by $\phi(a) = \begin{pmatrix} a & \\ & 1 \end{pmatrix}$, then we will use the notations $Z(s, W)$ or $Z(s, W, \xi)$ rather than $Z(s, \phi)$ and $Z(s, \phi, \xi)$.

Theorem 4.7.5: Local functional equation *Let (π, V) be an irreducible admissible representation of $GL(2, F)$ with central quasicharacter ω that admits a Whittaker model, and let ξ be a quasicharacter of $F^\times$. Identify V with the space of its Kirillov model. There exists a meromorphic function $\gamma(s, \pi, \xi, \psi)$ such that for all $\phi \in V$, we have*

$$Z\left(1 - s, \pi(w_1)\phi, \omega^{-1}\xi^{-1}\right) = \gamma(s, \pi, \xi, \psi)\, Z(s, \phi, \xi),$$

$$w_1 = \begin{pmatrix} & 1 \\ -1 & \end{pmatrix}. \tag{7.13}$$

We include the additive character ψ in the list of parameters defining γ because the dependence of γ on ψ might easily be forgotten if we omitted it! This result is the analog of Proposition 3.1.5 in the $GL(1)$ theory.

Proof For fixed s, define two linear functionals L_1 and L_2 on V by

$$L_1(\phi) = Z(s, \phi, \xi), \qquad L_2(\phi) = Z\left(1 - s, \pi(w_1)\,\phi, \omega^{-1}\xi^{-1}\right).$$

We will prove that both L_1 and L_2 satisfy

$$L\left(\pi\begin{pmatrix} y & \\ & 1 \end{pmatrix}\phi\right) = \chi(y)^{-1}\,|y|^{-s+1/2}\,L(\phi). \tag{7.14}$$

To prove this for $L = L_1$, note that it is true by a simple change of variables within the region of absolute convergence; and for general s, it follows by analytic continuation. The identity for L_2 may be deduced from the identity for L_1 without difficulty.

Now by Theorem 4.7.4, the functionals L_1 and L_2, defined for almost all s, are proportional, so there exists a constant $\gamma(s, \pi, \xi, \psi)$ that makes Eq. (7.13) true. To see that it is meromorphic as a function of s, we simply pick a ϕ such that $L_1(\phi)$ and $L_2(\phi)$ not identically zero, and because they are meromorphic in s, so is their ratio $\gamma(s, \pi, \xi, \psi)$. ■

Next we show that the factors $\gamma(s, \pi, \xi, \psi)$ determine the representation π.

Proposition 4.7.6 *Let π_1 and π_2 be irreducible admissible representations of $GL(2, F)$. Suppose that π_1 and π_2 have the same central quasicharacter ω and that $\gamma(s, \pi_1, \xi, \psi) = \gamma(s, \pi_2, \xi, \psi)$ for all characters ξ of $F^\times$. Then $\pi_1 \cong \pi_2$.*

Proof We may take V_1 and V_2 to be the Kirillov models of the representations π_1 and π_2. Thus V_1 and V_2 are spaces of functions on $F^\times$ on which $B(F)$ acts by Eqs. (4.25) and (1.1). Let $V_0 = V_1 \cap V_2$. We will show that if $\phi \in V_0$, then $\pi_1(w_1)\,\phi = \pi_2(w_1)\,\phi$. Let us observe that this is sufficient. Indeed, it shows that V_0 is stable under $\pi_1(w_1)$ and $\pi_2(w_1)$, and it is clearly also stable under $B(F)$. It is nonzero because by Theorem 4.7.1 it contains $C_c^\infty(F^\times)$; because V_1 and V_2 are irreducible, it follows that $V_0 = V_1 = V_2$, and $B(F)$ and w_1 have the same effect on V_0. Because these generate $GL(2, F)$, this is enough.

Let $\phi_i = \pi_i(w_1)\,\phi$, $i = 1, 2$. We wish to show that these are equal, but making use of Eq. (4.25) it is sufficient to show that $\phi_1(1) = \phi_2(1)$; indeed, by Eq. (4.25) we have

$$\phi_i(a) = \left(\pi_i\begin{pmatrix} a & \\ & 1\end{pmatrix}\phi_i\right)(1) = \left(\pi_i(w_1)\,\phi'\right)(1),$$

$$\phi' = \pi_1\begin{pmatrix} 1 & \\ & a\end{pmatrix}\phi = \pi_2\begin{pmatrix} 1 & \\ & a\end{pmatrix}\phi.$$

Thus replacing ϕ by ϕ' will give us $\phi_1(a) = \phi_2(a)$, provided we can prove $\phi_1(1) = \phi_2(1)$.

If $n \in \mathbb{Z}$, let

$$F_\xi(n) = \int_{|y|=q^{-n}} \left(\phi_1(y) - \phi_2(y)\right)\xi(y)\,d^\times y.$$

Note that $F_\xi(1)$ depends only on the restriction of ξ to $\mathfrak{o}^\times$, and it is easy to see that $F_\xi(1) = 0$ for all but finitely many characters ξ of $\mathfrak{o}^\times$ and that

$$\phi_1(1) - \phi_2(1) = \sum_{\xi \in \hat{\mathfrak{o}}^\times} F_\xi(1). \tag{7.15}$$

Indeed, this is a special case of the Fourier inversion formula: If M is a compact Abelian group, written multiplicatively here, with the Haar measure normalized so the volume of M is 1, and if F is a continuous function on M, then

$$F(1) = \sum_{\chi \in \hat{M}} \int_M F(m)\,dm.$$

Equation (7.15) is the special case $M = \mathfrak{o}^\times$.

Our hypothesis implies that $Z(s, \phi_1, \xi) = Z(s, \phi_2, \xi)$ for all characters ξ of $F^\times$, because the right side of Eq. (7.13) is the same for $\pi = \pi_1$ or π_2, and therefore the left side is also. By Proposition 4.7.2, $F_\xi(n) = 0$ when n is sufficiently large, and denoting $x = q^{-s}$, we have

$$\sum_n F_\xi(n)\,x^n = Z(s, \phi_1, \xi) - Z(s, \phi_2, \xi) = 0$$

for x sufficiently small. Of course, this implies that $F_\xi(n) = 0$ for all n, and in particular, $F_\xi(1) = 0$. Thus by Eq. (7.15), $\phi_1(1) = \phi_2(1)$. ■

An important property of the gamma factors $\gamma(s,\pi,\xi,\psi)$ is their *compatibility with parabolic induction.* For $GL(n)$, this is formulated and proved in Jacquet, Piatetski-Shapiro and Shalika (1981b) and Jacquet and Shalika (1990a). For $GL(2)$, this means the following:

Proposition 4.7.7 *Suppose that χ_1 and χ_2 are quasicharacters of $F^\times$ such that $(\pi,V)=\mathcal{B}(\chi_1,\chi_2)$ is irreducible. Then*

$$\gamma(s,\pi,\xi,\psi)=\gamma(s,\xi\chi_1,\psi)\,\gamma(s,\xi\chi_2,\psi), \tag{7.16}$$

where the Tate gamma factors $\gamma(s,\chi_i\xi,\psi)$ are defined in Proposition 3.1.5. (We are suppressing the subscript v from the notation because in this chapter, F is a local field.)

We defer the proof of this fact to the next section where, following Jacquet and Langlands (1970), we will obtain it as an application of the Weil representation.

Exercises

Exercise 4.7.1 (a) The multiplicative group $F^\times$ of the non-Archimedean local field F acts on $C_c^\infty(F)$ by translation and hence also on distributions: If $\Phi\in C_c^\infty(F)$, $a\in F^\times$, let $\big(\rho(a)\Phi\big)(x)=\phi(ax)$, and if $D\in\mathfrak{D}(F)$, let $\big(\rho(a)D\big)(\Phi)=D\big(\rho(a^{-1})\Phi\big)$. If χ is a nontrivial quasicharacter of $F^\times$, show, using the methods of Section 4.3, that the dimension of the space of distributions on F that satisfy $\rho(a)\,D=\chi(a)\,D$ is at most one dimensional.
(b) Give a different proof of Proposition 2.1.3 as follows: Regard the numerator and denominator of Eq. (1.13) in Chapter 2 as distributions, and show using (a) that if s is in general position, these distributions are proportional.

Exercise 4.7.2 (a) Let F be a local field, either Archimedean or non-Archimedean. By a *finite function* on $F^\times$, we mean a function ϕ such that the functions $x\mapsto\phi(ax)$ with $a\in F$ span a finite-dimensional vector space. Show that a finite function is a finite linear combination of functions of the form $\chi(x)\log|x|^n$, where χ is a quasicharacter and $n\ge 0$. [HINT: Study Theorem 4.5.4.]
(b) Let $\mathcal{W}$ be a Whittaker model of an irreducible admissible representation of $GL(2,F)$. (If F is Archimedean, we mean a $(\mathfrak{g},K)$-module.) Let $W\in\mathcal{W}$, and let

$$\Omega(y)=W\begin{pmatrix} y & \\ & 1\end{pmatrix},\qquad y\in F^\times.$$

Show that there exist a finite number of finite functions $\xi_1,\cdots,\xi_h$ and Schwartz functions $\phi_1,\cdots,\phi_h\in S(F)$ such that $\Omega(y)=\xi_1(y)\,\phi_1(y)+\ldots+\xi_h(y)\,\phi_h(y)$. [HINT: If F is non-Archimedean, this follows easily from Theorem 4.7.2. If $F=\mathbb{R}$, this follows from the fact (proved in Section 2.8) that the function ϕ is a finite linear combination of confluent hypergeometric functions (because W is K-finite), which are of rapid decay as $y \to \infty$ and whose behavior at $y = 0$ is described in Whittaker and Watson (1927, 16.41 and 16.1). The complex case is treated in Jacquet and Shalika (1990b), in the more general context of $GL(n)$. See also Jacquet, Piatetski-Shapiro and Shalika (1979). The arguments in these two papers are well worth study.]

(c) Prove that if ξ is a finite function on $F^\times$ and ϕ is a Schwartz function on F, then

$$\int_{F^\times} \xi(y)\,\phi(y)\,|y|^s\,d^\times y$$

is convergent if $\mathrm{re}(s)$ is sufficiently large and has meromorphic continuation to all s.

[HINT: Split the integral into y large and y small. The integral over y large is entire. For the integral over y small, adapt the argument of Proposition 3.1.7.]

The following exercise gives the local functional equation for the local integrals arising from the Rankin–Selberg method for $GL(2) \times GL(2)$. See Section 3.8, Jacquet (1972) and Gelbart and Shahidi (1988, Appendix to Section 1.2) for further information on these local functional equations.

Exercise 4.7.3 Let F be a non-Archimedean local field, and let (π_1, V_1) and (π_2, V_2) be irreducible admissible representations of $GL(2, F)$ with central characters ω_1 and ω_2; let ξ_1 and ξ_2 be characters of $F^\times$, and let $\chi_1(y) = \xi_1(y)\,|y|^{s-1/2}$, $\chi_2(y) = \xi_2(y)\,|y|^{-s+1/2}$. Assume that $\omega_1\omega_2\xi_1\xi_2 = 1$, so that the product of the central characters of π_1, π_2 and $(\pi_0, V_0) = \mathcal{B}(\chi_1, \chi_2)$ is trivial.

(a) Prove that for all s outside a set of measure zero, there exists (up to constant multiple) at most one trilinear form $B : V_0 \times V_1 \times V_2 \to \mathbb{C}$ such that

$$B\big(\pi_0(g)\,v_0, \pi_1(g)\,v_1, \pi_2(g)\,v_2\big) = B(v_0, v_1, v_2) \tag{7.17}$$

for $g \in GL(2, F)$, $v_i \in V_i$.

[HINTS: the bilinear mapping B may be reinterpreted as a map $V_1 \times V_2 \to \hat{V}_0$, so (after replacing χ_i by χ_i^{-1} and hence $\hat{V}_0$ by V_0) it is enough to show that there exists at most one bilinear map $V_1 \times V_2 \to V_0$ such that $B\big(\pi_1(g)\,v_1, \pi_2(g)\,v_2\big) = \pi_0(g)\,B(v_1, v_2)$ or, equivalently, at most one intertwining operator $V_1 \otimes V_2 \to V_0$. Applying Frobenius reciprocity, this means that what we must prove is that for almost all s there exists at most one bilinear form $B_0 : V_1 \times V_2 \to \mathbb{C}$ such that for $b \in B(F)$

$$B_0\big(\pi_1(b)\,v_1, \pi_1(b)\,v_2\big) = \delta(b)^s\,\xi(b)\,B_0(v_1, v_2), \tag{7.18}$$

where ξ is a fixed character of $B(F)$ trivial on $N(F)$. Let ψ be a nontrivial additive character of F, and take V_1 and V_2 to be the Kirillov models, with respect to the characters ψ and ψ^{-1}, respectively, so that

$$\left(\pi_1\begin{pmatrix} 1 & x \\ & 1 \end{pmatrix}\phi\right)(a) = \psi(ax)\,\phi(a), \quad \left(\pi_2\begin{pmatrix} 1 & x \\ & 1 \end{pmatrix}\phi\right)(a) = \psi(ax)^{-1}\,\phi(a).$$

Let $V_{1,N}$ and $V_{2,N}$ be the kernels of the projections onto the Jacquet modules $J(V_1)$ and $J(V_2)$, so that by Theorem 4.7.1, $V_{i,N} = C_c^\infty(F^\times)$. Then B_0 satisfying Eq. (7.18) induces a bilinear map $B_{0,N} : V_{1,N} \times V_{2,N} \to \mathbb{C}$; prove as follows that $B_{0,N}$ is determined up to constant multiple. Begin with the identity

$$B_{0,N}\left(\pi\begin{pmatrix} 1 & x \\ & 1 \end{pmatrix}\phi_1, \phi_2\right) = B_{0,N}\left(\phi_1, \pi\begin{pmatrix} 1 & -x \\ & 1 \end{pmatrix}\phi_2\right).$$

Multiply this by $f(x)$, where $f \in C_c^\infty(F)$, and integrate to obtain the identity

$$B_{0,N}(\hat{f}\phi_1, \phi_2) = B_{0,N}(\phi_1, \hat{f}\phi_2),$$

where $\hat{f}$ is the Fourier tranform of f. Conclude that there exists a linear map $L : C_c^\infty(F) \to \mathbb{C}$ such that $B_{0,N}(\phi_1, \phi_2) = L(\phi_1\phi_2)$, and then invoke Proposition 4.3.2 to see that $B_{0,N}$ is determined up to constant multiple. Now it suffices to show that for almost all s, the vanishing of $B_{0,N}$ implies the vanishing of B_0. Show that if $B_{0,N} = 0$, then B_0 necessarily factors through the projection $V_1 \times V_2 \to J(V_1) \times J(V_2)$, and then make use of the finite dimensionality of the Jacquet modules.]
(b) Let $\mathcal{W}_1$ be the Whittaker model of π_1 with respect to the character ψ, and let $\mathcal{W}_2$ be the Whittaker model of π_2 with respect to the character ψ^{-1}. If $W_i \in \mathcal{W}_i$ and $f_s \in V_0$, define

$$Z(W_1, W_2, f_s) = \int_{Z(F)N(F)\backslash GL(2,F)} W_1(g)\,W_2(g)\,f_s(g)\,dg. \tag{7.19}$$

Show that this integral is convergent if $\mathrm{re}(s)$ is sufficiently large. Show that if $s \mapsto f_s$ is chosen to be a flat section, it has meromorphic continuation to all s, and show in fact that there exists a discrete subset Ω of $\mathbb{C}$ of measure zero such that if $s \notin \Omega$, then Eq. (7.19) has analytic continuation to all $s \notin \Omega$ (with simple poles in Ω) for all choices of W_1 and W_2 and all flat sections f_s. Show that for all $s \notin \Omega$, $Z(W_1, W_2, f_s)$ is a trilinear form satisfying Eq. (7.17). (We recall from the discussion of the intertwining integrals in Section 4.5 that a *flat section* is obtained by fixing a function f satisfying Eq. (5.22), and for every s, by taking f_s to be the unique extension of f to an element of V_0.)
(c) Let $(\pi_0', V_0') = \mathcal{B}(\chi_2, \chi_1)$, and let $M(s) : V_0 \to V_0'$ be the intertwining integral defined by the analytic continuation of Eq. (5.16) as in Section 4.5. Prove that

$$Z(W_1, W_2, f_s)/Z\big(W_1, W_2, M(s)\,f_s\big) \tag{7.20}$$

is a meromorphic function of s independent of the choice of data W_1, W_2, and f_s. (This is essentially Jacquet's (1972), Theorem 14.7.)

4.8 Supercuspidals and the Weil Representation

Let F be a non-Archimedean local field, let $\mathfrak{p}$, $\mathfrak{o}$, q, and ϖ be as in Section 4.2, and let ψ be a nontrivial additive character of F.

Let (π, V) be an irreducible admissible representation of $GL(2, F)$. We call π *supercuspidal* if $J(V) = 0$. The principal series representations that we have constructed are irreducible admissible representations that are *not* supercuspidal. We will construct many supercuspidals by means of the Weil representation. First, however, we will show by a different method that supercuspidal representations do exist – we will construct examples by induction from open subgroups. It is a theorem of Kutzko (1978) (generalized to $GL(n, F)$ by Bushnell and Kutzko (1993) and independently by Corwin (1993)) that all supercuspidal representations of $GL(2, F)$ can be constructed by induction from open subgroups; we will content ourselves with a special case by way of giving examples.

We require a partial converse to Schur's lemma, valid only for unitarizable representations. A representation (π, V) of a group G is *unitarizable* if there exists a positive definite invariant Hermitian inner product on V. In this case, we may complete V to a Hilbert space, and obtain a unitary representation.

Proposition 4.8.1 *Let (π, V) be a unitarizable admissible representation of a totally disconnected locally compact group G. Suppose that $\mathrm{Hom}_G(V, V)$ is one dimensional. Then π is irreducible.*

Proof Let U be a proper nonzero invariant subspace. The orthogonal projection onto U is a nonzero element of $\mathrm{Hom}_G(V, V)$. It is sufficient to show that this is not the identity map; to this end, we must show that there is an element of V that is orthogonal to U. If $\sigma \in \hat{K}$, where $K = GL(2, \mathfrak{o})$, let $U(\sigma) \subseteq V(\sigma)$ be the σ-isotypic component, so that (as in Proposition 4.2.5) U is the algebraic direct sum of the $U(\sigma)$, and V is the algebraic direct sum of the $V(\sigma)$. Because $U \neq V$, some $U(\sigma) \neq V(\sigma)$. As $V(\sigma)$ is finite dimensional, we may find a nonzero element v of $V(\sigma)$ that is perpendicular to $U(\sigma)$, and because the spaces $V(\sigma)$ are mutually orthogonal, v is orthogonal to all of U. ■

Let (π_0, V_0) be a cuspidal representation of $GL(2, \mathbb{F}_q)$, where $\mathbb{F}_q = \mathfrak{o}/\mathfrak{p}$ is the residue field. We recall that such cuspidal representations were constructed in Section 4.1. We lift π_0 to a representation (also denoted π_0) of K by means of the projection map $K \to GL(2, \mathbb{F}_q)$. The central quasicharacter of π_0 is lifted to a character ω_0 of $\mathfrak{o}^\times$, and we extend ω_0 to a unitary character of $F^\times$; then we extend π_0 to a representation of $KZ(F)$ such that $Z(F)$ acts by Eq. (1.1). Let (π, V) be the representation of $GL(2, F)$ obtained by compactly inducing (π_0, V_0) from $KZ(F)$ to $GL(2, F)$.

Theorem 4.8.1 *In this situation, (π, V) is an irreducible supercuspidal representation of $GL(2, F)$. It is unitarizable.*

Proof First let us show that (π, V) is admissible. It is sufficient to show that if (σ, U) is any finite-dimensional irreducible representation of K, then the dimension of $\mathrm{Hom}_{GL(2,F)}(V, U)$ is finite. By Exercise 4.5.5, this is isomorphic to the direct sum over a set of representatives γ for $K\backslash G/KZ(F)$ of the spaces $\mathrm{Hom}_{S^\gamma}(V_0, U^\gamma)$, where $S^\gamma = K \cap \gamma^{-1}K\gamma$, and U^γ is U endowed with the S^γ-action $g \mapsto \sigma(\gamma g \gamma^{-1})$. By the p-adic Cartan decomposition (Proposition 4.6.2), we may take γ to be of the form

$$\gamma_n = \begin{pmatrix} \varpi^n & \\ & 1 \end{pmatrix}, \tag{8.1}$$

where $0 \le n \in \mathbb{Z}$. The spaces $\mathrm{Hom}_{S^\gamma}(V_0, U^\gamma)$ are all finite dimensional, so what we must prove is that only finitely many such spaces are nonzero. If $b \in \mathfrak{o}$, then $\begin{pmatrix} 1 & b \\ & 1 \end{pmatrix} \in S^\gamma$, and there exists a constant N such that if $\gamma = \gamma_n$ with $n \ge N$, then $\sigma^\gamma \begin{pmatrix} 1 & b \\ & 1 \end{pmatrix} = 1$. Then if $\phi \in \mathrm{Hom}_{S^\gamma}(V_0, U^\gamma)$, $v \in V_0$, we have

$$\phi\left(\pi_0\begin{pmatrix} 1 & b \\ & 1 \end{pmatrix} v\right) = \phi(v) \tag{8.2}$$

for all $b \in \mathfrak{o}$. Because π_0 is cuspidal, this implies that $\phi = 0$. We have proved that π is admissible.

Now let us show that (π, V) is irreducible. We note that the finite-dimensional representation (π_0, V_0) of $KZ(F)$ is unitary, so by Exercise 4.5.6, (π, V) is unitarizable. By Proposition 4.8.1, it is therefore sufficient to show that $\mathrm{Hom}_{GL(2,F)}(V, V)$ is one dimensional. Let V' be the representation obtained from the representation (π_0, V_0) by ordinary induction. Then $V \subset V'$, so we have an injection $\mathrm{Hom}_{GL(2,F)}(V, V) \to \mathrm{Hom}_{GL(2,F)}(V, V')$, and it is sufficient to show that the latter is one dimensional. By Frobenius reciprocity (Proposition 4.5.1) and Exercise 4.5.5, it is sufficient to show that (with notation as above and $U = V_0$) all but one of the spaces $\mathrm{Hom}_{S^\gamma}(V_0, V_0^\gamma)$ are zero dimensional, and the single exception is one dimensional. Indeed, if $\gamma = \gamma_n$ with $n \ge 1$, then any $\phi \in \mathrm{Hom}_{S^\gamma}(V_0, V_0^\gamma)$ satisfies Eq. (8.2), so $\mathrm{Hom}_{S^\gamma}(V_0, V_0^\gamma) = 0$; whereas if $n = 0$, then $\mathrm{Hom}_K(V_0, V_0)$ is one dimensional because π_0 is irreducible. This completes the proof that (π, V) is irreducible.

Finally, we must show that π is supercuspidal. Let $L : V \to \mathbb{C}$ be a linear functional that satisfies

$$L\left(\pi\begin{pmatrix} 1 & x \\ & 1 \end{pmatrix} v\right) = L(v) \tag{8.3}$$

for all $v \in V$, $x \in F$. We will prove that $L = 0$. Such a linear functional is the same as a linear functional on $J(V)$, so this will show that $J(V) = 0$ and that V is supercuspidal. By Exercise 4.5.5, the space of linear functionals satisfying Eq. (8.3) is isomorphic to

$$\bigoplus_{\gamma \in N(F)\backslash GL(2,F)/KZ(F)} \mathrm{Hom}_{S^\gamma}(V_0, 1),$$

where $S^\gamma = KZ(F) \cap \gamma^{-1}N(F)\gamma$, and Hom_{S^γ} denotes the space of all linear functionals on V_0 that satisfy $L\big(\pi(g)\,v\big) = L(v)$ for all $s \in S^\gamma$. It may be easily deduced from the Iwasawa decomposition (Proposition 4.5.2) that we may take the double coset representatives for $N(F)\backslash GL(2,F)/KZ(F)$ to be the γ_n in Eq. (8.1), where now n may be any integer. We have $N(\mathfrak{o}) = \left\{ \left(\begin{smallmatrix} 1 & b \\ & 1 \end{smallmatrix}\right) | b \in \mathfrak{o} \right\} \subset S^\gamma$, and it follows that

$$\mathrm{Hom}_{S^\gamma}(V_0, 1) \subset \mathrm{Hom}_{N(\mathfrak{o})}(V_0, 1) = 0$$

by the cuspidality of π_0. Consequently, $L = 0$, and π is supercuspidal. ■

* * *

We turn now to the Weil representation. For the first part of the discussion, we will closely follow Weil (1964). Let G be a locally compact Abelian group, and let G^* be its dual. We will write G and G^* additively. Let $\mathbb{T} \subset \mathbb{C}^\times$ be the group of complex numbers of unit norm, and let $\langle\ ,\ \rangle$ denote the dual pairing $G \times G^* \to \mathbb{T}$. We introduce the group $A(G)$, which as a topological space is $G^* \times G \times \mathbb{T}$, with the group law

$$(v_1^*, v_1, t_1)(v_2^*, v_2, t_2) = \left(v_1^* + v_2^*, v_1 + v_2, t_1\, t_2 \left\langle v_1, v_2^* \right\rangle\right). \tag{8.4}$$

If $w = (v^*, v) \in G^* \times G$ and $t \in \mathbb{T}$, we may write (w, t) instead of (v^*, v, t) for the corresponding element of $A(G)$. If $w_1 = (v_1^*, v_1)$ and $w_2 = (v_2^*, v_2) \in G^* \times G$, we will write $[w_1, w_2] = \left\langle v_1, v_2^* \right\rangle$, so that the group law (Eq. 8.4) may be written

$$(w_1, t_1)\,(w_2, t_2) = (w_1 + w_2, t_1\, t_2\, [w_1, w_2]). \tag{8.5}$$

The group $A(G)$ will play the role of the Heisenberg group in the exercises in Section 4.1. It is a two-step nilpotent group.

We have a unitary representation ρ of $A(G)$ on $L^2(G)$, namely

$$\big(\rho(v^*, v, t)\Phi\big)(u) = t\, \left\langle u, v^* \right\rangle\, \Phi(u + v). \tag{8.6}$$

Our aim is to prove an analog of Exercise 4.1.10. Let $B(G)$ be the group of automorphisms of G, and let $B_0(G)$ be the subgroup of $B(G)$ consisting of elements that act trivially on the center $Z\big(A(G)\big) = \{(0, 0, t) | t \in \mathbb{T}\}$.

Theorem 4.8.2: Segal, Shale, and Weil *The representation ρ is irreducible. Let $\sigma \in B_0(G)$. Then there exists a unitary operator $\omega(\sigma)$ on $L^2(G)$, determined up to scalar multiple, such that (for $h \in A(G)$)*

$$\rho({}^\sigma h) = \omega(\sigma)\, \rho(h)\, \omega(\sigma)^{-1}. \tag{8.7}$$

Here *irreducible* means irreducible in the sense of unitary representations: There is no nontrivial invariant closed subspace of $L^2(G)$. There may well be invariant dense subspaces, such as the Schwartz space of G if G is the additive group of F^n, where F is an (Archimedean or non-Archimedean) local field. We follow Weil closely in the following proof.

Proof Let $\phi \in C_c(G^* \times G)$, the space of compactly supported continuous functions on $G^* \times G$. Let $\rho(\phi)$ be the endomorphism of $L^2(G)$ defined by

$$\big(\rho(\phi)\,\Phi\big)(u) = \int_{G^* \times G} \phi(w)\,\big(\rho(w, 1)\,\Phi\big)(u)\,dw. \tag{8.8}$$

We compute easily that for $u \in G$, $\Phi \in L^2(G)$

$$\big(\rho(\phi)\,\Phi\big)(u) = \int_G K_\phi(u, v)\,\Phi(v)\,dv, \tag{8.9}$$

where $K_\phi : G \times G \to \mathbb{C}$ is defined by

$$K_\phi(u, v) = \int_{G^*} \phi(v^*, v - u)\,\langle u, v^* \rangle\,dv^*, \tag{8.10}$$

or

$$K_\phi(u, u + v) = \int_{G^*} \phi(v^*, v)\,\langle u, v^* \rangle\,dv^*. \tag{8.11}$$

Thus $K_\phi(u, u + v)$ is the Fourier transform of $\phi(v^*, v)$ in the second variable. Because the Fourier transform is an L^2 isometry, we have

$$\int_{G \times G} |K(u, v)|^2\,du\,dv = \int_{G \times G} |K(u, u + v)|^2\,du\,dv = \int_{G^* \times G} |\phi(v^*, u)|^2\,du\,dv.$$

Thus $\phi \mapsto K_\phi$ extends to an L^2 isometry of $L^2(G^* \times G)$ into $L^2(G \times G)$; it is invertible because by the Fourier inversion formula and Eq. (8.11), we have

$$\phi(v^*, v) = \int_G K_\phi(u, u + v)\,\langle -u, v^* \rangle\,du. \tag{8.12}$$

We will denote by $\lambda : L^2(G \times G) \to L^2(G^* \times G)$ the inverse of this isometry, so that $\lambda(K_\phi) = \phi$ for $\phi \in L^2(G^* \times G)$.

Now let ϕ_1 and ϕ_2 be two elements of $C_c^\infty(G)$. We calculate easily that

$$\rho(\phi_1) \circ \rho(\phi_2) = \rho(\phi_1 * \phi_2), \tag{8.13}$$

where we define the convolution

$$(\phi_1 * \phi_2)(w) = \int_{G^* \times G} \phi_1(w_1)\,\phi_2(w - w_1)\,[w_1, w - w_1]\,dw_1. \tag{8.14}$$

Now Eqs. (8.9) and (8.13) imply that

$$K_{\phi_1 * \phi_2} = K_{\phi_1} \times K_{\phi_2}, \tag{8.15}$$

where the composition law in $L^2(G \times G)$ is defined by

$$(K_1 \times K_2)(u, v) = \int_G K_1(u, x)\, K_2(x, v)\, dx. \tag{8.16}$$

It is easily deduced from the Cauchy–Schwarz inequality that

$$||K_1 \times K_2||_2 \leq ||K_1||_2 \cdot ||K_2||_2,$$

so this composition law makes $L^2(G \times G)$ into a Banach algebra (without unit). Similarly, if P and $Q \in L^2(G)$, $K \in L^2(G \times G)$, we define

$$(K \times P)(u) = \int_G K(u, v)\, P(v)\, dv, \qquad (Q \times K)(v) = \int_G Q(u)\, K(u, v)\, du,$$

and the Cauchy–Schwarz inequality implies that

$$||K \times P||_2 \leq ||K||_2 \cdot ||P||_2, \qquad ||Q \times K||_2 \leq ||Q||_2 \cdot ||P||_2.$$

Thus $L^2(G)$ is both a left and right module for $L^2(G \times G)$.

Let $P, Q \in L^2(G)$. We define an element $P \otimes Q$ of $L^2(G \times G)$ by $(P \otimes Q)(u, v) = P(u) \otimes Q(v)$. By abuse of language, we will refer to elements of this type as *pure tensors*. If $K \in L^2(G \times G)$, we then have

$$(P \otimes Q) \times K = P \otimes (Q \times K), \qquad K \times (P \otimes Q) = (K \times P) \otimes Q, \tag{8.17}$$

and (with S and $T \in L^2(G)$)

$$(P \otimes Q) \times (\overline{S} \otimes T) = \left\{ \int_G Q(u)\, \overline{S(u)}\, du \right\} P \otimes T = (Q, S)_2\, P \otimes T, \tag{8.18}$$

where $(\,,\,)_2$ denotes the inner product in $L^2(G)$. ■

Lemma 4.8.1

(i) *An element K of $L^2(G \times G)$ has the form $P \otimes Q$ if and only if $K \times K' \times K$ is proportional to K for all K'.*

(ii) *Let $K_1 = P_1 \otimes Q_1$ and $K_2 = P_2 \otimes Q_2$. Then P_1 and P_2 are proportional if and only if $K_1 \times K$ and $K_2 \times K$ are proportional for all pure tensors $K \in L^2(G \times G)$, and similarly, Q_1 and Q_2 are proportional if and only if $K \times K_1$ and $K \times K_2$ are proportional for all pure tensors K.*

(iii) *Suppose that $s : L^2(G \times G) \to L^2(G \times G)$ is a unitary transformation respecting the composition law (Eq. 8.16). Then there exists a unitary transformation $s_0 : L^2(G) \to L^2(G)$ such that $s(P \otimes \overline{Q}) = s_0(P) \otimes \overline{s_0(Q)}$. If s is invertible, then so is s_0.*

Proof For (i), if K is nonzero, find P' and Q' in $L^2(G)$ such that both $P = K \times P'$ and $Q = Q' \times K$ are nonzero. Then if $K' = P' \otimes Q'$, we find by Eq. (8.17) that $K \times K' \times K = P \otimes Q$, and if K is proportional to this, we must have (after adjusting P and Q by a constant if necessary) $K = P \otimes Q$. Conversely, if $K = P \otimes Q$, by Eqs. (8.17) and (8.18) we have $K \times K' \times K = (Q, \overline{K' \times P})_2\, K$, which is proportional to K, so the condition of (i) is both necessary and sufficient.

For (ii), if $K = \overline{S} \otimes T$, we have by Eq. (8.18)

$$K_1 \times K = (Q_1, S)_2\, P_1 \otimes T, \qquad K_2 \times K = (Q_2, S)_2\, P_2 \otimes T.$$

These are proportional if P_1 and P_2 are, and because (if $Q_1 \neq 0$) S and T may be chosen so that (Q_1, S) and T are nonzero, the proportionality of $K_1 \otimes K$ and $K_2 \otimes K$ also implies the proportionality of P_1 and P_2.

For (iii), it is a consequence of the intrinsic characterization (i) of the pure tensors $P \otimes Q$ in $L^2(G \times G)$ that if K is a pure tensor so is $s(K)$. We choose P_0 of unit norm; because s is unitary, $s(P_0 \otimes \overline{P_0})$ has the form $P_0' \otimes \overline{Q_0'}$ for some P_0' and Q_0', both of norm one. Now if $P \in L^2(G)$, we may write $s(P \otimes \overline{P_0})$ in the form $P' \otimes \overline{Q'}$, and it follows from the intrinsic criterion (ii) that Q_0 and Q are proportional. Thus we may write $s(P \otimes \overline{P_0}) = \mu(P) \otimes \overline{Q_0'}$ for some uniquely determined $\mu(P) \in L^2(G)$. It is clear that μ defined this way is a unitary linear operator on $L^2(G)$; similarly, $s(P_0 \otimes \overline{Q}) = P_0' \otimes \overline{\nu(Q)}$ for a unitary linear operator ν. We have in particular $\mu(P_0) = P_0'$ and $\nu(P_0) = Q_0'$. By Eq. (8.18), we have $P \otimes \overline{Q} = (P \otimes \overline{P_0}) \times (P_0 \otimes \overline{Q})$. Therefore $s(P \otimes \overline{Q}) = c\mu(P) \times \nu(Q)$ where $c = (P_0', Q_0')_2$, and because s is unitary, applying this relation with $P = Q = P_0$ gives $c = 1$. We have

$$(P \otimes \overline{Q}) \times (P \otimes \overline{Q}) = (P, Q)_2\, (P \otimes \overline{Q}),$$

and applying s to this identity gives

$$\big(\mu(P), \nu(Q)\big)_2 \big(\mu(P) \otimes \overline{\nu(Q)}\big) = (P, Q)_2 \big(\mu(P) \otimes \overline{\nu(Q)}\big),$$

so $\big(\mu(P), \nu(Q)\big)_2 = (P, Q)_2$ for all P and Q. This (together with the unitaricity of μ and ν) implies that $\mu = \nu$, because expanding $\big(\mu(P) - \nu(P), \mu(P) - \nu(P)\big)_2$ gives zero. We may then take $s_0 = \mu = \nu$. If s is invertible, then we may apply the same reasoning to s^{-1} and construct an inverse to s_0. This concludes the proof of Lemma 4.8.1. □

Because $A(G)/Z\big(A(G)\big) \cong G^* \times G$, any $\sigma \in B_0(G)$ induces an automorphism s of $G^* \times G$, and so $\sigma(w, t) = \big(s(w), f(w)t\big)$, where $f : G^* \times G \to \mathbb{T}$ is some mapping. In order that σ be a homomorphism, we must have

$$f(w_1 + w_2) = f(w_1)\, f(w_2)\, [s(w_1), s(w_2)]\, [w_1, w_2]^{-1}. \tag{8.19}$$

We recall that if H is a locally compact Abelian group, and if $\alpha : H \to H$ is an automorphism, then α takes the Haar measure into a constant multiple of itself, and the *module* $|\alpha|$ of α is defined to be that constant. We need to know that the module of s is one. To prove this, we note that the commutator

$$(w_1, 1)\,(w_2, 1)\,(w_1, 1)^{-1}\,(w_2, 1)^{-1} = (0, \langle\!\langle w_1, w_2 \rangle\!\rangle)$$

lies in the center of $A(G)$, where we define $\langle\!\langle w_1, w_2 \rangle\!\rangle = [w_1, w_2]\,[w_2, w_1]^{-1}$. If $w_1 = (v_1^*, v_1)$ and $w_2 = (w_2^*, w_2)$, then

$$\langle\!\langle w_1, w_2 \rangle\!\rangle = \left\langle v_1, v_2^* \right\rangle \left\langle v_2, v_1^* \right\rangle^{-1}.$$

From this, it follows that the pairing $\langle\!\langle\, , \rangle\!\rangle$ of $G^* \times G$ with itself is skew-symmetric, bilinear, and nondegenerate. Because by definition of $B_0(G)$ it is assumed that σ acts trivially on the center of $A(G)$, it follows that s preserves the pairing $\langle\!\langle\, , \rangle\!\rangle$. Thus if we identify $G \times G$ with its dual by means of this pairing, $s = s^{*-1}$ in the notation of Exercise 4.8.1. Consequently, by that exercise, $|s| = |s|^{-1}$, and so $|s| = 1$.

We now define a unitary transformation Σ of $L^2(G^* \times G)$ by

$$(\Sigma\phi)(w) = f(w)^{-1}\,\phi\big(s(w)\big), \qquad \phi \in L^2(G^* \times G). \tag{8.20}$$

It is easily deduced from Eqs. (8.19) and (8.14) that

$$\Sigma(\phi_1 * \phi_2) = \Sigma\phi_1 * \Sigma\phi_2. \tag{8.21}$$

This is a straightforward verification, but it involves a change of variables, so it uses the fact that the module of s is one, and the unitaricity of Σ also uses the unimodularity of s. Now by means of the isometry $\lambda : L^2(G \times G) \to L^2(G^* \times G)$, we may transfer Σ to a unitary transformation of $L^2(G \times G)$, which we also denote Σ. In view of Eq. (8.15), Σ preserves the multiplication $\times$ on $L^2(G \times G)$, and by Lemma 4.8.1(iii) there exists a unitary map $\omega : L^2(G) \to L^2(G)$ (the inverse of s_0 in the lemma) such that

$$\Sigma(P \otimes \overline{Q}) = \omega^{-1}(P) \otimes \overline{\omega^{-1}(Q)}. \tag{8.22}$$

Next we prove that if $P, Q \in L^2(G)$, $w \in G^* \times G$, and $t \in \mathbb{T}$, then

$$\bar{t}\,\lambda(P \otimes \overline{Q})(w) = \big(P, \rho(w, t)\,Q\big)_2. \tag{8.23}$$

Indeed, by Eq. (8.12) with $\phi = \lambda(P \otimes \overline{Q})$, $K_\phi = P \otimes \overline{Q}$, and $w = (v^*, v)$, the left side of Eq. (8.23) equals

$$\int_G P(u)\,\overline{t\,Q(u+v)\,\langle u, v^* \rangle}\,du,$$

and Eq. (8.23) follows from Eq. (8.6). Now consider

$$\bar{t}\,\big(\Sigma\,\lambda(P \otimes \overline{Q})\big)(w). \tag{8.24}$$

On the one hand, by Eqs. (8.20) and (8.23), this equals

$$\overline{t\,f(w)}\,\lambda(P \otimes Q)\big(s(w)\big) = \big(P, \rho(s(w), f(t))\,Q\big)_2 = \big(P, \rho(\sigma(w, t))\big)_2.$$

On the other hand, using the fact that $\Sigma \circ \lambda = \lambda \circ \Sigma$ and Eq. (8.23) again, Eq. (8.24) equals

$$\bar{t}\,\lambda\big(\omega^{-1}(P) \otimes \omega^{-1}(Q)\big)(w) = \big(\omega^{-1} P, \rho(w, t)\, \omega^{-1} Q\big)_2 = \big(P, \omega\, \rho(w, t)\, \omega^{-1} Q\big)_2.$$

Because these are equal for all P and Q, we have $\rho\big(\sigma(w, t)\big) = \omega\, \rho(w, t)\, \omega^{-1}$. This proves the main assertion.

It remains to be shown that ρ is irreducible. This is equivalent to the assertion that any endomorphism T of $L^2(G)$ commuting with ρ is a scalar, because if ρ were reducible, orthogonal projection on a closed invariant subspace would contradict this assertion. Clearly T must also commute with the endomorphisms $\rho(\phi)$ defined by Eq. (8.8) with $\phi \in L^2(G^* \times G)$. We take $\phi = \lambda(P \otimes \overline{Q})$ with P and $Q \in L^2(G)$. Then for $\Phi \in L^2(G)$, it follows from Eq. (8.9) that $\rho(\phi)\Phi = (\Phi, Q)_2\, P$, and therefore $(T\Phi, Q)_2\, P = (\Phi, Q)_2\, TP$. We may choose Φ and Q such that $(\Phi, Q)_2 \neq 0$; then for all P, we have $TP = (\Phi, Q)_2^{-1}\, (T\Phi, Q)_2\, P$, showing that T is a scalar. This completes the proof of Theorem 4.8.2. ■

Now let F be a local field, which may be either Archimedean or non-Archimedean, but we will assume that the characteristic of F is not two. Let $\psi : F \to \mathbb{C}$ be a nontrivial additive character. Let V be a vector space over F of finite dimension d, and let $B : V \times V \to \mathbb{C}$ be a nondegenerate symmetric bilinear form. $O(V)$ will denote the orthogonal group of endomorphisms of V preserving B. As in the exercises to Section 4.1, let H be the Heisenberg group, which as a set is $V \times V \times F$, with group law

$$(v_1^*, v_1, x_1)\,(v_2^*, v_2, x_2) = \big(v_1^* + v_2^*, v_1 + v_2, x_1 + x_2 + B(v_1^*, v_2) - B(v_1, v_2^*)\big), \tag{8.25}$$

for $v_1, v_2, v_1^*, v_2^* \in V$, $x_1, x_2 \in F$. We may identify V with its dual by means of the pairing $(v, v^*) \mapsto \langle v, v^* \rangle = \psi\big(-2B(v, v^*)\big)$. Then the group $A(V)$ is as a set $V \times V \times \mathbb{T}$ with multiplication $(v_1^*, v_1, t_1)(v_2^*, v_2, t_2) = \big(v_1^* + v_2^*, v_1 + v_2, t_1\, t_2\, \psi(-2B(v_1, v_2^*))\big)$. We have a homomorphism $\tau : H \to A(V)$ by $\tau(v^*, v, x) = \big(v^*, v, \psi(x)\, \psi(-B(v, v^*))\big)$. As in the previous theorem, there exists a unitary representation ρ of $A(V)$ given by Eq. (8.6). Let $\pi = \rho \circ \tau$ be the corresponding representation of H on $L^2(V)$. Then it is easy to check that π is given by the same formulas (1.29) and (1.30) as in the case of a finite field.

We have actions of $SL(2, F)$ and $O(V)$ on H as follows: If $g = \left(\begin{smallmatrix} a & b \\ c & d \end{smallmatrix}\right) \in SL(2, F)$, we let ${}^g(v_1, v_2, x) = (av_1 + bv_2, cv_1 + dv_2, x)$, and if $k \in O(V)$, we let ${}^k(v_1, v_2, x) = \big(k(v_1), k(v_2), x\big)$. If $\Phi \in L^2(V)$, we define the Fourier transform

$$\hat{\Phi}(v) = \int_V \Phi(u)\, \psi\big(2\, B(u, v)\big)\, du, \tag{8.26}$$

where $\int_V du$ is the Haar measure on V that is self-dual with respect to the pairing $(v^*, v) \mapsto \psi\big(-2B(v, v^*)\big)$.

If $\mathfrak{H}$ is a (possibly infinite-dimensional) vector space over $\mathbb{C}$, let $GL(\mathfrak{H})$ be the group of endomorphisms of $\mathfrak{H}$, and let $PGL(\mathfrak{H})$ be $GL(\mathfrak{H})$ modulo the scalar transformations. If $\mathfrak{H}$ is a Hilbert space, let $U(\mathfrak{H})$ be the group of unitary transformations of $\mathfrak{H}$, and let $PU(\mathfrak{H})$ be $U(\mathfrak{H})$ modulo scalars; clearly, there is an injection of $PU(\mathfrak{H})$ into $PGL(\mathfrak{H})$. By a *projective representation* of a group G, we mean a homomorphism ω from G into $PGL(\mathfrak{H})$; we call it *unitary* if $\mathfrak{H}$ is a Hilbert space and ω may be lifted to a homomorphism $G \to PU(\mathfrak{H})$. The projective representation may be lifted to a mapping $\omega' : G \to GL(\mathfrak{H})$ or (if ω is unitary) $\omega' : G \to U(\mathfrak{H})$. As is discussed following Exercise 4.1.9, the lifting ω' is not uniquely determined, but it determines a cohomology class in $H^2(G, \mathbb{C}^\times)$ (or $H^2(G, \mathbb{T})$ if ω is unitary), and this cohomology class depends only on ω. As in the exercises to Section 4.1, this cohomology class gives rise to a central extension $\tilde{G}$ of G by $\mathbb{C}^\times$ (or $\mathbb{T}$), and ω may be reinterpreted as a true representation of $\tilde{G}$. By abuse of language, we will generally refer to the lifted representation ω' as a projective representation; in this language, a projective representation is a map $\omega' : G \to GL(V)$ such that $\omega'(g_1 g_2) = c(g_1, g_2)\, \omega'(g_1)\, \omega'(g_2)$, where $c(g_1, g_2) \in \mathbb{C}^\times$ (or $\mathbb{T}$ if ω' is unitary). Then c is a two-cocycle determining the above-mentioned cohomology class. If this cohomology class is trivial, then ω may be lifted to a true representation of G.

Theorem 4.8.3 *There exists a unitary projective representation ω_1 of $SL(2, F)$ on $L^2(V)$ such that for $g \in SL(2, F)$ and $h \in H$, we have $\omega_1(g)\, \pi(h)\, \omega_1(g)^{-1} = \pi({}^g h)$. There exists a (true) representation ω_2 of $O(V)$ on $L^2(V)$ such that $\omega_2(k)\, \pi(h)\, \omega_2(k)^{-1} = \pi({}^k h)$. The (projective) representations ω_1 and ω_2 commute with each other. The Schwartz space $S(V)$ is invariant under both these (projective) representations. We have*

$$\left(\omega_1 \begin{pmatrix} 1 & x \\ & 1 \end{pmatrix} \Phi\right)(v) = \psi\big(x\, B(v, v)\big)\Phi(v), \tag{8.27}$$

$$\left(\omega_1 \begin{pmatrix} a & \\ & a^{-1} \end{pmatrix} \Phi\right)(v) = |a|^{d/2}\, \Phi(av), \tag{8.28}$$

and

$$\omega_1(w_1)\, \Phi = \hat{\Phi}, \qquad w_1 = \begin{pmatrix} & 1 \\ -1 & \end{pmatrix} \tag{8.29}$$

and

$$\big(\omega_2(k)\Phi\big)(v) = \Phi(k^{-1}v). \tag{8.30}$$

Here, of course, $\omega_1(g)$ is only determined up to a complex number of absolute value one. Later, if V is even-dimensional, we will prescribe ω_1 more precisely so as to obtain a true representation. We recall that if F is Archimedean, the Schwartz space $S(V)$ consists of functions that (together with all derivatives) are of faster than polynomial decay, whereas if F is non-Archimedean, then $S(V) = C_c^\infty(V)$.

Proof There is an evident action of $SL(2, F)$ on $A(V)$ such that ${}^g\big(\tau(h)\big) = \tau({}^g h)$ for $g \in SL(2, F)$, $h \in H$. Thus g gives rise to an element of $B_0(V)$, and it follows from Theorem 4.8.2 that there exists a unitary transformation $\omega(g)$, determined up to a complex number of absolute value one such that

$$\omega(g)\,\pi(h)\,\omega(g)^{-1} = \pi({}^g h). \tag{8.31}$$

It follows from the uniqueness of $\omega(g)$ that $\omega(g_1 g_2)$ differs from $\omega(g_1)\,\omega(g_2)$ by a complex constant, which must have absolute value one because the operators $\omega(g)$ are unitary. Hence we have a projective representation. The formulas (8.27–29) are verified by checking Eq. (8.31) directly when g is one of the three types of matrices in question, and it is sufficient to do this when h is of the form $(v^*, 0, 0)$ or $(0, v, 0)$ with v^* or $v \in V$, because these generate H, and we leave this elementary check to the reader. Note that the factor $|a|^{d/2}$ in Eq. (8.28) is needed for unitaricity.

The group $O(V)$ also embeds in $B_0(V)$, and the existence of a projective representation follows similarly. However, it is unnecessary to make use of Theorem 4.8.2 here, because it is evident that ω_2 as defined by Eq. (8.30) is a unitary representation of $O(V)$ commuting with ω_1.

It is clear that the Schwartz space is preserved by the operations (Eqs. 8.27–8.30), and and so it is invariant under ω_1 and ω_2. ■

Our aim is to show that if the dimension d of V is even, then the cohomology class in $H^2\big(SL(2, F), \mathbb{T}\big)$ attached to the representation ω_1 is trivial. (If d is odd, it may be shown that unless $F = \mathbb{C}$, the cocycle is nontrivial, and the cocycle defines an important central extension of $SL(2, F)$, the *metaplectic group*.) We will make use of quaternion algebras and Hilbert symbols to prove this. A *quaternion algebra* over a field F is a two-dimensional central simple algebra. We will need only basic properties of quaternion algebras; what we need can be found in Weil (1967). Let us only point out that this topic is closely connected with a central issue in local class field theory, namely, the computation of the Brauer group of a local field. References for this theory are Weil (1967, Chapters 9 and 10), Serre (1979, Chapters 10, 12, and 13), O'Meara (1963, Sections 52 and 57), and Herstein (1968, Chapter 4).

A quaternion algebra over F is either isomorphic to $\mathrm{Mat}_2(F)$ or else it is a division algebra. It is known that if F is a local field (except $\mathbb{C}$) there exists exactly one isomorphism class of quaternion division algebras over F. We do not need this fact, however. Let $a, b \in F$. We define a quaternion algebra $\mathrm{Quat}(a, b)$ as follows: As a vector space, $\mathrm{Quat}(a, b)$ has a basis $\{1, i, j, k\}$, where i, j, and k are subject to the following laws: $i^2 = a$, $j^2 = b$, $k^2 = -ab$, $ij = -ji = k$, $jk = -kj = -bi$, and $ki = -ik = -aj$. This algebra has an antiautomorphism $\xi \mapsto \overline{\xi}$ defined by $\overline{(x+yi+zj+wk)} = x-yi-zj-wk$. The *reduced norm* and *reduced trace* are given by $N(\xi) = \xi\,\overline{\xi}$ and $\mathrm{tr}(\xi) = \xi+\overline{\xi}$. We define the *Hilbert symbol* $(a, b) = 1$ if $\mathrm{Quat}(a, b) \cong \mathrm{Mat}_2(F)$, and $(a, b) = -1$ if $\mathrm{Quat}(a, b)$ is a division ring.

We recall from local class-field theory that if F is a local field and E is a quadratic extension, then the multiplicative group of the norms from $E^\times$ is a subgroup of index two in $F^\times$. The unique nontrivial character of $F^\times$ that is trivial on this subgroup is called the *quadratic character of $F^\times$ attached to E.*

Proposition 4.8.2 *Let $a, b \in F^\times$. A necessary and sufficient condition for $(a, b) = 1$ is that $x^2 - a\,y^2 - b\,z^2 + ab\,w^2 = 0$ have a solution with x, y, z, and w not all zero, or equivalently, that a is a norm from $F(\sqrt{b})$. The Hilbert symbol has the following further properties:*

$$(aa', b) = (a, b)\,(a', b); \qquad (a, bb') = (a, b)(a, b'); \tag{8.32}$$

$$(a, b) = (b, a); \tag{8.33}$$

$$(a, 1-a) = (a, -a) = 1, \tag{8.34}$$

when $a \neq 0$ and (for $(a, 1-a)$) $a \neq 1$. The Hilbert symbol (a, b) only depends on the classes of a and b modulo squares.

Proof A necessary and sufficient condition that $(a, b) = 1$ is that $N(\xi) = 0$ for some nonzero $\xi \in \mathrm{Quat}(a, b)$. Indeed, $N(\xi) = 0$ if and only if ξ is a zero divisor, because if $N(\xi) \neq 0$, then $N(\xi)^{-1}\overline{\xi}$ is an inverse to ξ, which is thus a unit; so $N(\xi) = 0$ is solvable with nonzero ξ if and only if $\mathrm{Quat}(a, b)$ fails to be a division ring.

Thus, writing $\xi = x + yi + zj + wk$, a necessary and sufficient condition for $(a, b) = 1$ is that $\nu(\xi) = x^2 - ay^2 - bz^2 + ab\,w^2 = 0$ have a solution with x, y, z, and w not all zero. We may write this identity $a = N\big((x + z\sqrt{b})/(y + w\sqrt{b})\big)$, so this condition is also equivalent to a being a norm from $F(\sqrt{b})$. From the second necessary and sufficient condition, it is clear that the function $a \mapsto (a, b)$ is trivial if b is a square or is the quadratic character of $F^\times$ attached to the quadratic extension $F(\sqrt{b})/F$ if b is not a square. In either case, $a \mapsto (a, b)$ is a character, from whence comes Eq. (8.32).

It is clear from the first necessary and sufficient condition that (a, b) is symmetrical in a and b, which gives us Eq. (8.33), and if $b = -a$ or $b = 1 - a$, we have specific solutions $(x, y, z, w) = (0, 1, 1, 0)$ or $(1, 1, 1, 0)$ to $x^2 - ay^2 - bz^2 + ab\,w^2 = 0$, which gives us Eq. (8.34). Finally, we have $(ar^2, b) = (a, b)\,(r, b)^2 = (a, b)$ because $(r, b) = \pm 1$, so the Hilbert symbol only depends on square classes of a and b. ■

Now we turn to Weil's *analytic* interpretation of the Hilbert symbol. This is based on his observation that the function $F_B(v) = \psi\big(B(v, v)\big)$ on V has a *formal* Fourier transform. Because the function F_B has absolute value one for all v, it is of course not L^2, but it has a property that compensates for this defect, namely, it is rapidly oscillating. That is, if $|v|$ is large, F_B changes sign very rapidly, and so in some respects it behaves as if it were of rapid decay. We will show below that F_B has a Fourier transform in a natural sense.

Let $S(V)$ denote the Schwartz space of V. If Φ is an element of the Schwartz space $S(V)$, we will sometimes denote the Fourier transform $\hat{\Phi}$, defined by Eq. (8.26), by $\mathcal{F}\Phi$. If $\Phi_1 * \Phi_2$ denotes the convolution of two elements Φ_1 and Φ_2 of $S(V)$, defined by

$$(\Phi_1 * \Phi_2)(v) = \int_V \Phi_1(u)\, \Phi_2(v-u)\, du, \tag{8.35}$$

then it is well known and easy to check that

$$\mathcal{F}(\Phi_1 * \Phi_2) = \mathcal{F}\Phi_1 \cdot \mathcal{F}\Phi_2. \tag{8.36}$$

Proposition 4.8.3: Weil *If $\Phi \in S(V)$, then the convolution $\Phi * F_B$, defined by Eq. (8.35), lies in $S(V)$. There exists a complex number $\gamma(B)$ of absolute value one such that for all $\Phi \in S(V)$*

$$\mathcal{F}(\Phi * F_B) = \gamma(B)\, \mathcal{F}\Phi \cdot F_{-B}. \tag{8.37}$$

More generally, if $a \in F^\times$ and $d = \dim(V)$

$$\mathcal{F}(\Phi * F_{aB}) = |a|^{-d/2}\, \gamma(aB)\, \mathcal{F}\Phi \cdot F_{-a^{-1}B}. \tag{8.38}$$

In view of Eq. (8.36), this means that we should regard $|a|^{-1/2}\, \gamma(aB)\, F_{-a^{-1}B}$ as a Fourier transform of F_B, at least formally.

Proof It is easy to check that the convolution

$$(\Phi * F_B)(v) = \int_V \Phi(u)\, \psi\big(B(v-u, v-u)\big)\, du$$

equals $F_B(v)\, \mathcal{F}(\Phi \cdot F_B)(-v)$ and that $\Phi \cdot F_B$ is an element of $S(V)$; because the Schwartz space is stable under the Fourier transform, this proves the first assertion. We will deduce the second assertion from Theorem 4.8.3.

We make use of the matrix identity

$$w_1 \begin{pmatrix} 1 & 1 \\ & 1 \end{pmatrix} w_1 \begin{pmatrix} 1 & 1 \\ & 1 \end{pmatrix} \begin{pmatrix} -1 & \\ & -1 \end{pmatrix} = \begin{pmatrix} 1 & -1 \\ & 1 \end{pmatrix} w_1,$$

with w_1 as in Eq. (8.29). This is essentially Eq. (1.16) with $a = 1$. Because ω_1 is a unitary projective representation, we have

$$\omega_1(w_1)\, \omega_1 \begin{pmatrix} 1 & 1 \\ & 1 \end{pmatrix} \omega_1(w_1)\, \omega_1 \begin{pmatrix} 1 & 1 \\ & 1 \end{pmatrix} \omega_1 \begin{pmatrix} -1 & \\ & -1 \end{pmatrix}$$

$$= \gamma(B)\, \omega_1 \begin{pmatrix} 1 & -1 \\ & 1 \end{pmatrix} \omega_1(w_1)$$

for some constant $\gamma(B)$ of absolute value one. Now

$$\omega_1(w_1)\,\omega_1\begin{pmatrix}1 & 1\\ & 1\end{pmatrix}\omega_1(w_1)\,\omega_1\begin{pmatrix}1 & 1\\ & 1\end{pmatrix}\omega_1\begin{pmatrix}-1 & \\ & -1\end{pmatrix}\Phi = \hat{\Phi}_1,$$

where

$$\Phi_1 = \omega_1\begin{pmatrix}1 & 1\\ & 1\end{pmatrix}\omega_1(w_1)\,\omega_1\begin{pmatrix}1 & 1\\ & 1\end{pmatrix}\omega_1\begin{pmatrix}-1 & \\ & -1\end{pmatrix}\Phi.$$

It is easy to compute that

$$\Phi_1(v) = \int_V \Phi(u)\,\psi\big(B(v-u, v-u)\big)\,du,$$

so $\Phi_1 = \Phi * F_B$. On the other hand, we have

$$\left(\omega_1\begin{pmatrix}1 & -1\\ & 1\end{pmatrix}\omega_1(w_1)\,\Phi\right)(v) = \psi\big(-B(v,v)\big)\,\hat{\Phi}(v),$$

and comparing these two expressions, we obtain Eq. (8.37).

We recall that in Eq. (8.37), the Fourier transform is with respect to the Haar measure dx on V, which is self-dual with respect to the pairing $\psi\big(2B(u,v)\big)$ on V. The measure that is self-dual with respect to the pairing $\psi\big(2aB(u,v)\big)$ is $|a|^{d/2}\,dx$, and the Fourier transform of Φ with respect to this measure and this pairing is $\mathcal{F}'\Phi(v) = |a|^{d/2}\,\mathcal{F}\Phi(av)$. Also, multiplying the measure by $|a|^{1/2}$ multiplies the convolution on the left side of Eq. (8.37) by $|a|^{d/2}$. We have therefore $|a|^{d/2}\,\mathcal{F}'(\Phi * F_{aB}) = \gamma(aB)\,\mathcal{F}'\Phi\cdot F_{-aB}$. Evaluating this identity at $a^{-1}v$ gives Eq. (8.38) evaluated at v. ∎

If $\beta(v) = B(v,v)$ is the quadratic form associated with a symmetric bilinear form B, then we note that (because we are assuming the characteristic of F is not two) B can be reconstructed from β by

$$B(u,v) = \frac{1}{2}\big(\beta(u+v) - \beta(u) - \beta(v)\big). \tag{8.39}$$

A *quadratic space* is a pair (V,β) consisting of vector space V and a quadratic form β on V such that the associated symmetric bilinear form B is nondegenerate. If $X = (V,\beta)$ is a quadratic space, we will sometimes write $\gamma(\beta)$ or $\gamma(X)$ for $\gamma(B)$.

If (V_1,β_1) and (V_2,β_2) are quadratic spaces, then we can form the direct sum $(V_1\oplus V_2, \beta_1\oplus\beta_2)$, where the quadratic form $(\beta_1\oplus\beta_2)(v_1,v_2) = \beta_1(v_1) + \beta_2(v_2)$. If $a_1,\cdots,a_n \in F^\times$, we can form the quadratic space $\mathrm{QF}(a_1,\cdots,a_n) = (F^n,\beta)$, where $\beta(x_1,\cdots,x_n) = \sum_i a_i\,x_i^2$. Because the characteristic of F is not two, every quadratic form can be diagonalized, so every quadratic space has the form $\mathrm{QF}(a_1,\cdots,a_n)$ for some a_i (Lang (1993), Theorem VI.3.1, p. 575). We call a quadratic space *split* if it is isomorphic a direct sum of copies of $\mathrm{QF}(1,-1)$. Thus the dimension of a split quadratic space is even by definition.

The isomorphism classes of quadratic spaces form a commutative monoid with respect to direct sum, and the quotient by the submonoid of split quadratic spaces is a group, known as the *Witt group* (Lang (1993), Theorem XV.11.1, p. 594).

Proposition 4.8.4

(i) Let (V_1, β_1) and (V_2, β_2) be quadratic spaces. Then $\gamma(\beta_1 \oplus \beta_2) = \gamma(\beta_1)\,\gamma(\beta_2)$.
(ii) Let (V, β) be a quadratic space. Then $\gamma(-\beta) = \gamma(\beta)^{-1}$, where $-\beta$ is the quadratic form on V defined by $(-\beta)(v) = -\beta(v)$.
(iii) If (V, β) is a split quadratic space, then $\gamma(\beta) = 1$.
(iv) The function γ is a character of the Witt group.

Proof Part (i) is easily verified by taking a test function in $\Phi \in S(V_1 \oplus V_2)$ of the form $\Phi(v_1, v_2) = \Phi_1(v_1)\,\Phi_2(v_2)$ with $\Phi_i \in S(V_i)$. For (ii), one simply takes complex conjugation in the definition (Eq. 8.37) of γ, bearing in mind that the Fourier transform $\mathcal{F}$ is with respect to the nondegenerate pairing B on V associated with β. For (iii), it is necessary to show that $\gamma(\beta) = 1$ when $(V, \beta) = \mathrm{QF}(1, -1) = \mathrm{QF}(1) \oplus \mathrm{QF}(-1)$, and this follows immediately from (i) and (ii). Part (iv) follows from (i) and (iii). ■

If F is non-Archimedean and V is a finite-dimensional vector space over F, we recall that a *lattice* in V is a compact open subgroup.

Proposition 4.8.5 *Let (V, β) be a quadratic space over the local field F, and let B be the symmetric bilinear form associated with β. Let $\int_V dv$ denote the Haar measure on V that is self-dual with respect to the pairing $(u, v) \mapsto \psi\big(2B(u, v)\big)$.*
(i) If $\Phi \in S(V)$, then

$$\int_V (\Phi * F_B)(v)\,dv = \gamma(B) \int_V \Phi(v)\,dv. \tag{8.40}$$

(ii) If F is non-Archimedean, then

$$\gamma(B) = \int_L F_B(v)\,dv \tag{8.41}$$

for any sufficiently large lattice L.

Proof Equation (8.40) results from evaluating both sides of Eq. (8.37) at zero. As for (ii), let L' be the dual lattice to L, that is,

$$L' = \{u \in V \,|\, \psi\big(2B(u, v)\big) = 1\} \text{ for all } v \in L.$$

If L is sufficiently large, then L' is small enough that $F_B(u) = 1$ for all $u \in L'$. Now let Φ be the characteristic function of L'. Then

$$(\Phi * F_B)(v) = \int_{L'} F_B(v-u)\,du = \psi\big(B(v,v)\big) \int_{L'} \psi\big(2B(u,v)\big)\,du,$$

so $\Phi * F_B$ is $\mathrm{vol}(L')\, F_B$ times the characteristic function of L. Thus the left side of Eq. (8.41) is $\mathrm{vol}(L')$ times $\int_L F_B(v)\,dv$, and the right side is $\gamma(B)$ times $\mathrm{vol}(L')$. Comparing, we obtain Eq. (8.41). ■

If A is a quaternion algebra over F, and if $\nu : A \to F$ is the reduced norm, then (A, ν) is a quadratic space. Specifically, if $A = \mathrm{Quat}(a, b)$, then $(A, V) \cong \mathrm{QF}(1, -a, -b, ab)$, because $\nu(x + yi + zj + wk) = x^2 - ay^2 - bz^2 + abw^2$.

Theorem 4.8.4: Weil *Let A be a quaternion algebra over F, and let $\nu : A \to F$ be the reduced norm. Then $\gamma(\nu) = 1$ if A is a matrix algebra, and $\gamma(\nu) = -1$ if A is a division ring.*

Applying this to the quaternion algebra $\mathrm{Quat}(a, b) \cong \mathrm{QF}(1, -a, -b, ab)$, we obtain

$$(a, b) = \gamma\big(\mathrm{QF}(1, -a, -b, ab)\big). \tag{8.42}$$

Proof If A is a matrix ring, the reduced norm coincides with the determinant. Because $\det\left(\begin{smallmatrix} x & y \\ z & w \end{smallmatrix}\right) = xw - yz$, and because the quadratic forms $(x, w) \mapsto xw$ and $(y, z) \mapsto -yz$ are easily seen to be split (they are diagonalized by a simple change of variables), $\gamma(\nu) = 1$ by Proposition 4.8.4(iii).

For the remainder of the proof, we assume that A is a division ring. If F is Archimedean, the existence of a quaternion division algebra implies that F is real and A is the ring of Hamiltonian quaternions. Its norm form is $\mathrm{QF}(1, 1, 1, 1)$, and so $\gamma(\nu) = \gamma\big(\mathrm{QF}(1)\big)^4$. Because our main interest is in the non-Archimedean case, we will leave it to the reader to show that $\gamma\big(\mathrm{QF}(1)\big)$ is a primitive 8th root of unity. We only remark that to evaluate $\gamma(B)$, it is convenient to take $\Phi(x) = e^{-\pi x^2}$ in Eq. (8.37).

We assume for the remainder of the proof that F is non-Archimedean and that A is a quaternion division algebra. Because $|\gamma(\nu)| = 1$, it is sufficient to prove that the right side of Eq. (8.41) is real and negative for some conveniently chosen additive Haar measure. We will denote by dz an additive Haar measure on A. Then $d^\times z = \nu(z)^{-2}\,dz$ is a multiplicative Haar measure. We wish to evaluate the sign of

$$\int_L \psi\big(\nu(z)\big)\, |\nu(z)|^2\, d^\times z,$$

where L is a sufficiently large lattice. We recall that the module $z \mapsto |\nu(z)|^2$ is a non-Archimedean absolute value on A (Weil, 1967, Theorem I.6, p. 12).

Thus $L = \nu^{-1}(\varpi^{-N}\mathfrak{o})$ is a lattice, where $\mathfrak{o}$ is the ring of integers in F and N is a suitably large integer. Then we note that the integral factors though the homomorphism $\nu : A^\times \to F^\times$, and we are left to evaluate of the sign of

$$\int_{\varpi^{-N}\mathfrak{o}} \psi(x)\,|x|^2\,d^\times x = \int_{\varpi^{-N}\mathfrak{o}} \psi(x)\,|x|\,dx, \tag{8.43}$$

where for our Haar measures $d^\times x$ and dx on $F^\times$ and F, we take $d^\times x = |x|^{-1}\,dx$, with dx the measure with respect to which $\mathfrak{o}$ has volume one. Suppose that $\varpi^r\mathfrak{o}$ is the conductor of ψ, that is, the largest fractional ideal on which ψ is trivial. We note that

$$\int_{|x|=q^{-s}} \psi(x)\,dx = \begin{cases} q^{-s}(1-q^{-1}) & \text{if } s \geq r; \\ -q^{-r} & \text{if } s = r-1; \\ 0 & \text{if } s < r-1. \end{cases}$$

Thus, if N is large, the integral (Eq. 8.43) equals

$$-q^{1-2r} + \sum_{s=r}^{\infty} q^{-2s}(1-q^{-1}) = -q^{1-2r}(1-q^{-1})^{-1}.$$

This is negative, as required. ■

Proposition 4.8.6 *Suppose that (V, β) is a quadratic space of even dimension d. Write $(V, \beta) = QF(r_1, r_2, \cdots, r_d)$ with $r_i \in F^\times$, and let $\Delta = (-1)^{d/2} r_1 r_2 \cdots r_d$. Let $a \in F^\times$. Then $\gamma(a\beta) = (\Delta, a)\,\gamma(\beta)$.*

Proof Because $\mathrm{QF}(r_1, r_2, \cdots, r_d)$ decomposes into the direct sum of $d/2$ two-dimensional quadratic spaces, we reduce to the case $d = 2$ by Propositions 4.8.2 and 4.8.4. We begin by noting that by Eq. (8.42)

$$\gamma\big(\mathrm{QF}(r_1, r_2)\big) = (r_1, r_2)\,\gamma\big(\mathrm{QF}(1, r_1 r_2)\big). \tag{8.44}$$

This implies that with $(V, \beta) = \mathrm{QF}(r_1, r_2)$, we have

$$\gamma(a\beta)\,\gamma(\beta)^{-1} = (ar_1, ar_2)\,\gamma\big(\mathrm{QF}(1, a^2 r_1 r_2)\big)\,(r_1, r_2)^{-1}\,\gamma\big(\mathrm{QF}(1, r_1 r_2)\big)^{-1}.$$

The quadratic spaces $\mathrm{QF}(1, a^2 r_1 r_2)$ and $\mathrm{QF}(1, r_1 r_2)$ are equivalent, so this last expression equals simply $(ar_1, ar_2)\,(r_1, r_2)^{-1}$, which, using Proposition 4.8.2, simplifies to $(a, -r_1 r_2)$. ■

Theorem 4.8.5 *Suppose that (V, β) is a quadratic space of even dimension d. Then the cohomology class in $H^2\big(SL(2, F), \mathbb{T}\big)$ determined by the projective representation ω_1 in Theorem 4.8.3 is trivial. More specifically, let Δ be as in Proposition 4.8.6, and let $\chi : F^\times \to \{\pm 1\}$ be the quadratic character*

$\chi(a) = (\Delta, a)$. Then there exists a true representation ω_0 of $SL(2, F)$ on $L^2(V)$, having the Schwartz space $S(V)$ as an invariant subspace, such that

$$\left(\omega_0 \begin{pmatrix} 1 & x \\ & 1 \end{pmatrix} \Phi\right)(v) = \psi\big(x\,B(v,v)\big)\Phi(v), \tag{8.45}$$

$$\left(\omega_0 \begin{pmatrix} a & \\ & a^{-1} \end{pmatrix} \Phi\right)(v) = |a|^{d/2}\,\chi(a)\,\Phi(av), \tag{8.46}$$

and

$$\omega_0(w_1) = \gamma(B)\,\hat{\Phi}. \tag{8.47}$$

Proof The construction of the cohomology class associated with the projective representation is discussed after Exercise 4.1.9. It is clear that the cohomology class associated with a true representation is trivial. Hence it is sufficient to show that Eqs. (8.45–8.47) describe a true representation of $SL(2, F)$. We will deduce this from Lemma 4.1.2 by showing that the operators $\omega_0(g)$ defined by Eqs. (8.45–8.47) satisfy the relations (1.12), (1.13), and (1.14). The first two relations are simple to check, except that the second identity in Eq. (1.13) requires us to know that

$$\gamma(B)^2 = \chi(-1). \tag{8.48}$$

This reduces easily to the case where $d = 2$, so what we must prove is that $(-1, -r_1r_2) = \gamma\big(\mathrm{QF}(r_1, r_2)\big)^2$. By Eq. (8.42), we have

$$(-1, -r_1r_2) = \gamma\big(\mathrm{QF}(1, 1, r_1r_2, r_1r_2)\big) = \gamma\big(\mathrm{QF}(1, r_1r_2)\big)^2,$$

and squaring Eq. (8.44), this equals $\gamma\big(\mathrm{QF}(r_1, r_2)\big)^2$, as required.

We are left with the problem of verifying Eq. (1.14); given Eqs. (1.12) and (1.13), which are already verified, this is equivalent to showing that for $\Phi \in L^2(V)$

$$\omega_0(w_1)\,\omega_0\begin{pmatrix} a^{-1} & \\ & a \end{pmatrix}\omega_0\begin{pmatrix} 1 & -a \\ & 1 \end{pmatrix}\omega_0(w_1)\,\omega_0\begin{pmatrix} 1 & -a^{-1} \\ & 1 \end{pmatrix}\Phi$$

$$= \omega_0\begin{pmatrix} 1 & a \\ & 1 \end{pmatrix}\omega_0(w_1)\,\Phi. \tag{8.49}$$

The left side equals $\gamma(B)\,\mathcal{F}(\Phi_1)$, where

$$\Phi_1 = \omega_0\begin{pmatrix} a^{-1} & \\ & a \end{pmatrix}\omega_0\begin{pmatrix} 1 & -a \\ & 1 \end{pmatrix}\omega_0\begin{pmatrix} & 1 \\ -1 & \end{pmatrix}\omega_0\begin{pmatrix} 1 & -a^{-1} \\ & 1 \end{pmatrix}\Phi.$$

Unraveling the definitions, we find that $\Phi_1 = \gamma(B)\,|a|^{-d/2}\,\chi(-a^{-1})\,\Phi * F_{-a^{-1}B}$. Thus by Eq. (8.48), the left side of Eq. (8.49) equals $|a|^{-d/2}\,\chi(a)\,\mathcal{F}(\Phi * F_{-a^{-1}B})$, and by Eq. (8.38) and Proposition 4.8.6, this equals $\gamma(B)\,\mathcal{F}\Phi \cdot F_{aB}$. This equals the right side of Eq. (8.49), so Eq. (8.49) is proved. ■

For the remainder of the section, the field F will be non-Archimedean, and $\mathfrak{o}$, $\mathfrak{p}$, ϖ, and q will have their usual meanings. We are assuming that F does not have characteristic two but q can be even. Let (V, β) be a quadratic space, which we will assume to be even dimensional. (If V were odd dimensional, the group $SL(2, F)$ would be replaced in the sequel by its nontrivial central extension, known as the *metaplectic group* $\widetilde{SL}(2, F)$.) The representations ω_0 and ω_2 of $SL(2, F)$ and $O(V)$, described in Theorems 4.8.5 and 4.8.3 act on $L^2(V)$, which has an important invariant subspace, namely, $C_c^\infty(V)$. (Over an Archimedean field, this would be the Schwartz space.) Let ω be the representation $(g_1, g_2) \mapsto \omega_0(g_1)\,\omega_2(g_2)$ of $SL(2, F) \times O(V)$, and let ω_∞ be the restriction of ω to $C_c^\infty(V)$. (The notation is intended to suggest that $C_c^\infty(V)$ is the space of "smooth vectors" in ω.)

Let π_1 and π_2 be irreducible admissible representations of $SL(2, F)$ and $O(V)$. We say that π_1 and π_2 *correspond* if there exists a nonzero $SL(2, F) \times O(V)$ intertwining operator $\omega_\infty \to \pi_1 \otimes \pi_2$. Howe (1979) stated a general conjecture that in the present case amounts to the following assertion: Each π_1 can correspond with at most one π_2, and vice versa. After partial results towards Howe's conjecture were obtained by Howe and other authors, the conjecture was fully proved for local fields of odd residue characteristic by Waldspurger (1990). The correspondence between the representations π_1 and π_2 is known as the *theta correspondence*, and the conjecture of Howe proved by Waldspurger is known as *Howe duality*.

However, we are interested in $GL(2, F)$ rather than $SL(2, F)$. A modified correspondence relates representations of $GL(2, F)$ to representations of $GO(V)$, the group of orthogonal similitudes, that is, the group of automorphisms of V that take β to a constant multiple of itself.

If V is two dimensional, then (after replacing β by a constant multiple if necessary) we may identify the quadratic space (V, β) with the quadratic space (E, N), where E is a two-dimensional commutative semisimple algebra over F and $N : E \to F$ is the norm map. Specifically, if the quadratic form β is split, we may take $E = F \oplus F$ and $N(x, y) = xy$; if the quadratic form β is not split, then a constant multiple of β has the form $QF(1, -D)$, where $D \in F^\times$ is not a square, and then we may identify (V, β) with (E, N), where $E = F(\sqrt{D})$ and the norm $N(a + b\sqrt{D}) = a^2 - b^2 D$. There exists a homomorphism $\iota : E^\times \to GO(V)$, namely, $\iota(x)\,a = xa$ for $x \in E^\times, a \in V = E$. If we identify E with its image $\iota(E)$, then $E^\times$ is a subgroup of index two in the group $GO(V)$. Let $\sigma : E \to E$ be the nontrivial automorphism, which is a Galois conjugation if E is a field or $\sigma(u, v) = (v, u)$ if $E = F \oplus F$ and $u, v \in F$; we will also denote $\sigma(x) = \overline{x}$ for $x \in E$. Recalling that $V = E, \sigma$ is an element of $GO(V)$. Observe that $\sigma\,\iota(x)\,\sigma^{-1} = \iota(\overline{x})$ for $x \in E^\times$, so if we identify $E^\times$ with $\iota(E^\times)$, we arrive at the following description of $GO(V)$: $GO(V)$ is generated as a group by $E^\times$ and σ, subject to the relations $\sigma^2 = 1$ and $\sigma x \sigma^{-1} = \overline{x}$ for $x \in E^\times$.

We recall from the end of Section 2.5 that if G is a group and H is a subgroup of index two, there is a simple description of the irreducible representations of

G in terms of the irreducible representations of H – we ask the reader to review this description. Recall that we called a representation of H *Type II* if it cannot be extended to a representation of G; in this case, it is related to a unique representation of G, namely, its induced representation. With this language, a Type II representation of $E^\times$ is related to a unique representation of $GO(V)$, which in turn corresponds (by means of the theta correspondence) to a unique representation of $GL(2, F)$. The irreducible admissible representations of $E^\times$ are of course just quasicharacters, so we ask when a quasicharacter $\xi : E^\times \to \mathbb{C}$ is of Type II. However, ξ is *not* of Type II if it can be extended to a quasicharacter of $GO(V)$, that is, if its kernel contains the commutator subgroup of $GO(V)$. Now with the above description of $GO(V)$ as the group generated by $E^\times$ and σ, the commutator subgroup of $GO(V)$ is generated by elements of the form $x\sigma x^{-1}\sigma^{-1} = x\overline{x}^{-1}$ with $x \in E^\times$, and by Hilbert's Theorem 90 (Lang, 1993, p. 288), we see that the commutator subgroup of $GO(V)$ is exactly the kernel of the norm map $N : E^\times \to F^\times$. Thus a quasicharacter of $E^\times$ is of Type II if and only if it does not factor through this norm map. (Hilbert's Theorem 90 applies to the case where E is a field, but it is easy to see that this is also true if $E = F \oplus F$.)

Therefore Howe duality in this particular case gives rise to the following prediction: *If E is a two-dimensional commutative semisimple algebra over F and $\xi : E^\times \to \mathbb{C}$ is a quasicharacter that does not factor through the norm map $E^\times \to F^\times$, then ξ corresponds to a unique irreducible admissible representation of $GL(2, F)$ by means of the theta correspondence.* We will now prove this assertion under the assumption that E is a field.

The following result is extremely similar to the finite field result from Section 4.1; the main difference is that the norm map $E^\times \to F^\times$ is not surjective if E is local, so first we must construct a representation of a subgroup of index two of $GL(2, F)$, then we induce.

Theorem 4.8.6 *Let E/F be a quadratic extension of non-Archimedean local fields, and let ξ be a quasicharacter of $E^\times$ that does not factor through the norm map $N : E^\times \to F^\times$. Let $U_{\xi,\psi}$ be the space of functions $\Phi \in C_c^\infty(E)$ that satisfy*

$$\Phi(yv) = \xi(y)^{-1}\,\Phi(v) \text{ for all } y \in E^\times \text{ such that } N(y) = 1, \tag{8.50}$$

and let $\chi : F^\times \to \{\pm 1\}$ be the quadratic character attached to the extension E/F. Let $GL(2, F)_+$ denote the subgroup of $GL(2, F)$ consisting of elements whose determinants are norms from E. Then there exists an irreducible admissible representation $\omega_{\xi,\psi}$ of $GL(2, F)_+$ on $U_{\xi,\psi}$ such that

$$\left(\omega_{\xi,\psi}\begin{pmatrix} a & \\ & 1 \end{pmatrix}\Phi\right)(v) = |a|^{1/2}\,\xi(b)\,\Phi(bv) \tag{8.51}$$

if $b \in E^\times$, $Nb = a \in F^\times$,

$$\left(\omega_{\xi,\psi}\begin{pmatrix}1 & x\\ & 1\end{pmatrix}\Phi\right)(v) = \psi\big(x\,N(v)\big)\,\Phi(v), \tag{8.52}$$

$$\left(\omega_{\xi,\psi}\begin{pmatrix}a & \\ & a^{-1}\end{pmatrix}\Phi\right)(v) = |a|\,\chi(a)\,\Phi(av), \tag{8.53}$$

and

$$\omega_{\xi,\psi}(w_1)\,\Phi = \gamma(N)\,\hat{\Phi}, \tag{8.54}$$

where the Fourier transform

$$\hat{\Phi}(v) = \int_E \Phi(u)\mathrm{tr}(u\overline{v})\,du. \tag{8.55}$$

The representation of ω_ξ of $GL(2,F)$ induced from this representation of $GL(2,F)_+$ is irreducible and supercuspidal.

The supercuspidal representations constructed in this theorem are called *dihedral.*

Proof If (V,β) is the quadratic space (E,N), it is easy to verify that the character χ in Theorem 4.8.5 agrees with the quadratic character attached to the extension E/F if E is a field. Thus we obtain a representation of $SL(2,F)$ such that Eqs. (8.52–8.54) are valid, and in the case where E is a field, subspace $U_{\xi,\psi}$ is easily seen to be invariant. We may extend the action of $SL(2,F)$ to $GL(2,F)_+$ by Eq. (8.51).

First let us note that the resulting representation is smooth. To see this, it is sufficient to show that if Φ is given, then there is an ideal $\mathfrak{a} = \mathfrak{p}^N$ of $\mathfrak{o}$ such that Φ is stabilized by elements of the group $K(\mathfrak{a})$ consisting of elements of $GL(2,\mathfrak{o})$ that are congruent to the identity modulo ϖ^N. This group has an Iwahori factorization whose proof is similar to Eqs. (4.26) and (4.27): $K(\mathfrak{a}) = N_-(\mathfrak{a})\,T(\mathfrak{a})\,N(\mathfrak{a})$ in the notation of Section 4.4. Thus it is sufficient to show that Φ is fixed by $N_-(\mathfrak{a})$ and $T(\mathfrak{a})$ and $N(\mathfrak{a})$ if $\mathfrak{a}$ is sufficiently large; for $N_-(\mathfrak{a})$, this is equivalent to showing that $\omega_{\xi,\psi}(w_1)\,\Phi$ is fixed by $N(\mathfrak{a})$, where $w_1 = \begin{pmatrix} & 1\\ -1 & \end{pmatrix}$, so we may omit the verification for $N_-(\mathfrak{a})$. It follows easily from Eqs. (8.51–8.53) that Φ is fixed by $T(\mathfrak{a})$ and $N(\mathfrak{a})$ if $\mathfrak{a}$ is small, so $\omega_{\xi,\psi}$ is smooth.

To see that this representation is admissible, it is sufficient to show that if $\mathfrak{a}$ is a nonzero ideal of $\mathfrak{o}$, then the space of vectors fixed by both $N(\mathfrak{a})$ and $N_-(\mathfrak{a})$ is finite dimensional. It is clear from Eq. (8.52) that there exists an ideal $\mathfrak{a}'$ depending on $\mathfrak{a}$ and ψ such that if Φ is fixed by $N(\mathfrak{a})$, then the support of Φ is contained in $N^{-1}(\mathfrak{a}')$. If Φ is fixed by $N_-(\mathfrak{a})$, then $\omega_{\xi,\psi}(w_1)\,\Phi$ is fixed by $N(\mathfrak{a})$, so the Fourier transform of Φ has support contained in $N^{-1}(\mathfrak{a}')$, and this implies that Φ is constant on the cosets of some open subgroup of E. Thus for

Φ to be fixed by both $N(\mathfrak{a})$ and $N_-(\mathfrak{a})$, it must have bounded support and it must be constant on the cosets of some fixed open subgroup of E. The space of such functions is finite dimensional. This proves that ω_ξ is admissible.

To prove the irreducibility of $\omega_{\xi,\psi}$ and ω_ξ, we make use of Proposition 4.7.3. Let $B_1(F)$ be as in that proposition, and let $B_1(F)_+$ be the subgroup of index two in $B_1(F)$ consisting of elements of the form $\begin{pmatrix} a & b \\ & 1 \end{pmatrix}$ with $a \in N(E^\times)$. We will show that the restriction of $\omega_{\xi,\psi}$ to $B_1(F)_+$ is isomorphic to the space of all functions Φ on $N(E^\times) \subset F^\times$ that are compactly supported and locally constant, with group action (Eq. 4.25). Let $\mathcal{K}_+$ denote this representation of $B_1(F)_+$. We define a map $\Lambda : C_c^\infty(NE^\times) \to U_{\xi,\psi}$ by $(\Lambda\phi)(v) = \xi(v)^{-1}\,|Nv|^{-1/2}\,\phi(Nv)$. Because ξ is assumed not to be trivial on the elements of norm 1 in $E^\times$, any element Φ of $U_{\xi,\psi}$ vanishes at 0, and so Λ is bijective. It is easy to see that Λ is a $B_1(F)_+$-intertwining operator between $\mathcal{K}_+$ and $U_{\xi,\psi}$. Thus the restriction of $\omega_{\xi,\psi}$ to $B_1(F)_+$ is isomorphic to $\mathcal{K}_+$.

However, it is easy to see (compare Exercise 4.5.7) that $\mathcal{K}_+$ is isomorphic to the representation of $B_1(F)_+$ obtained by compact induction from the character $\psi_N\begin{pmatrix} 1 & x \\ & 1 \end{pmatrix} = \psi(x)$ of the group $N(F)$ of upper triangular unipotent matrices. Because ω_ξ is the representation of $GL(2, F)$ induced from the representation $\omega_{\xi,\psi}$ of $GL(2, F)_+$, and because $B_1(F)_+ = GL(2, F)_+ \cap B_1(F)$, it follows that the restriction of ω_ξ to $B_1(F)$ is isomorphic to the representation of $B_1(F)$ induced from the representation $\omega_{\xi,\psi}$ of $B_1(F)_+$, and as we have seen, the latter is itelf an induced representation. By transitivity of induction (Exercise 4.5.8), we may obtain the same result in one step: The restriction of ω_ξ to $B_1(F)$ is the representation obtained by compactly inducing ψ_N to $B_1(F)$, and, by Exercise 4.5.7 and Proposition 4.7.3, this is irreducible. *A fortiori*, ω_ξ is irreducible as a representation of $GL(2, F)$, and if $\omega_{\xi,\psi}$ were reducible, ω_ξ would be too. This completes the proof that $\omega_{\xi,\psi}$ and ω_ξ are irreducible admissible representations.

We have also proved that ω_ξ is supercuspidal, because we have shown that its restriction to $B_1(F)$ is isomorphic to the representation of Proposition 4.7.3, because by Theorem 4.7.1, this implies that the Jacquet module of ω_ξ is zero. ■

Instead of the field E, we may take $E = F \oplus F$ and try to construct principal series representations out of the Weil representation. A quasicharacter ξ of $E^\times$ has the form $\xi(x_1, x_2) = \xi_1(x_1)\,\xi_2(x_2)$ for quasicharacters ξ_1 and ξ_2 of $F^\times$, and we may therefore attempt to construct the principal series representation $\mathcal{B}(\xi_1, \xi_2)$ out of ξ_1 and ξ_2. We must proceed a bit differently from Section 4.1, however, because there are *no* nonzero compactly supported functions that satisfy Eq. (8.50) for all $y \in E^\times$ of norm 1. Thus we cannot construct $\mathcal{B}(\xi_1, \xi_2)$ as a subrepresentation of the Weil representation, but at least it is possible to construct it as a quotient.

There are two distinct $SL(2, F)$-module structures on $L^2(E)$. Firstly, there is the Weil representation ω_0 of Theorem 4.8.5. The bilinear form B in

Theorem 4.8.5 has the form

$$B\big((u_1, u_2), (v_1, v_2)\big) = \tfrac{1}{2}(u_1 v_2 + u_2 v_1).$$

Secondly, there is also the representation $\big(R(g)\Phi\big)(v) = \Phi(vg)$, where we regard an element v of $E = F \oplus F$ as a row vector and vg as the matrix product. Both these representations have the Schwartz space $S(E)$ as an invariant subspace.

Proposition 4.8.7 *Let $E = F \oplus F$. The two actions ω_0 and R of $SL(2, F)$ on $S(E)$ are equivalent. Specifically, if $\Phi \in S(E)$, let*

$$\mathfrak{f}\Phi(v_1, v_2) = \int_F \Phi(v_1, u)\, \psi(uv_2)\, du. \tag{8.56}$$

Then

$$\mathfrak{f} \circ \omega_0(g) = R(g) \circ \mathfrak{f} \tag{8.57}$$

for $g \in SL(2, F)$. The map $\mathfrak{f}$ extends to an isometry of $L^2(E)$ onto itself.

Proof It follows from the Fourier inversion formula that $\mathfrak{f}$ is a bijection of $S(E)$ onto itself. The fact that it extends to an isometry of $L^2(E)$ is simply the Plancherel theorem for the locally compact Abelian group F. One verifies Eq. (8.56) easily when

$$g = \begin{pmatrix} 1 & x \\ & 1 \end{pmatrix}, \quad \begin{pmatrix} a & \\ & a^{-1} \end{pmatrix}, \quad \text{or } w_1.$$

Because these generate $SL(2, F)$, the result follows. Note that the character χ in Theorem 4.8.5 is trivial, as is $\gamma(B)$, because the quadratic space (E, B) is split. ∎

It is not difficult to express the principal series representations as quotients of the representation R. Once we understand this, we may transfer our understanding to ω_0 by means of Proposition 4.8.7. Let χ_1 and χ_2 be quasicharacters of $F^\times$, and let χ be the character $\chi_1\chi_2^{-1}$. Let ξ_1 and ξ_2 be unitary characters and s_1 and s_2 be complex numbers such that $\chi_i(x) = \xi_i(x)\, |x|^{s_i}$, $i = 1, 2$. Then $\chi(y) = |y|^{s_1 - s_2}\, \xi(y)$, where $\xi = \xi_1\xi_2^{-1}$. We extend the action R of $SL(2, F)$ on $L^2(E)$ to an action of $GL(2, F)$ by

$$\big(R(g)\Phi\big)(v) = \sqrt{|\det(g)|}\, \chi_1\big(\det(g)\big)\, \Phi(vg). \tag{8.58}$$

Again, $S(E)$ is an invariant subspace. Let us show that with the group action extended in this way, the principal series representation $\mathcal{B}(\chi_1, \chi_2) =$

$(\pi_{\chi_1,\chi_2}, V_{\chi_1,\chi_2})$, defined in Section 4.5, is a quotient of R. Specifically, we have a map from $\tau_{\chi_1,\chi_2} : S(E) \to V_{\chi_1,\chi_2}$ defined by

$$\tau_{\chi_1,\chi_2}\Phi = \int_{F^\times} \Phi(0, v)\, \chi(v)\, |v|\, d^\times v. \tag{8.59}$$

This is convergent if $\mathrm{re}(s_1 - s_2 + 1) > 0$. To other values of s_1 and s_2, it has meromorphic continuation (regarding Φ and ξ as fixed and varying s_1 and s_2) with poles only when ξ is nonramified and $\xi(\varpi)\, q^{-s_1+s_2-1} = 1$. Indeed, this is a Tate integral of the form (Eq. 1.12) in Chapter 3, and the meromorphic continuation is proved in Proposition 3.1.5, the precise set of poles being determined in Proposition 3.1.8. If $s_1 - s_2 + 1$ is not at one of these poles, it is easy to see that

$$\tau_{\chi_1,\chi_2}\left(R\begin{pmatrix} y_1 & x \\ & y_2 \end{pmatrix}\Phi\right) = \chi_1(y_1)\,\chi_2(y_2)\,\sqrt{|y_1/y_2|}\,\tau_{\chi_1,\chi_2}\Phi, \tag{8.60}$$

and so we obtain an intertwining operator $T_{\chi_1,\chi_2} : S(E) \to V_{\chi_1,\chi_2}$ by

$$(T_{\chi_1,\chi_2}\Phi)(g) = \tau_{\chi_1,\chi_2}\big(R(g)\,\Phi\big). \tag{8.61}$$

We note that the possible poles of the integral τ_{χ_1,χ_2} correspond to values of χ_1 and χ_2 such that $\chi_1\chi_2^{-1}(x) = |x|^{-1}$, where the representation $\mathcal{B}(\chi_1, \chi_2)$ is reducible by Theorem 4.5.1. Hence we should be able to study at least the irreducible principal series representations $\mathcal{B}(\chi_1, \chi_2)$ by means of the Weil representation.

It follows from Proposition 4.8.7 that we can extend the action ω_0 of $SL(2, F)$ on $S(E)$ to an action of $GL(2, F)$ in such a way that $\mathfrak{f}$ remains an intertwining operator with the action R. It is clear from Eq. (8.58) that the correct definition must be

$$\left(\omega_0\begin{pmatrix} y & \\ & 1 \end{pmatrix}\Phi\right)(v_1, v_2) = \sqrt{|y|}\,\chi_1(y)\,\Phi(yv_1, v_2). \tag{8.62}$$

An advantage of realizing the principal series representation this way is that we can describe its Whittaker model explicitly. If $\Phi \in S(E)$, let

$$W_\Phi(g) = \int_{F^\times} \chi(t)\,\big(\omega_0(g)\,\Phi\big)(t, t^{-1})\,d^\times t. \tag{8.63}$$

It is easy to check that

$$W_\Phi\left(\begin{pmatrix} 1 & x \\ & 1 \end{pmatrix} g\right) = \psi(x)\,W_\Phi(g). \tag{8.64}$$

Our aim is to show that the space of functions W_Φ comprise the Whittaker model of $\mathcal{B}(\chi_1, \chi_2)$, when the latter representation is irreducible.

Proposition 4.8.8: Jacquet–Langlands *Assume that* $\mathrm{re}(s_1 - s_2 + 1) > 0$, *and let* $\Phi \in S(E)$, *where* $E = F \oplus F$. *Then*

$$\int_F W_\Phi \begin{pmatrix} y & \\ & 1 \end{pmatrix} \chi_2(y)^{-1} |y|^{-1/2} \psi(xy)\, dy = \left(T_{\chi_1,\chi_2} \mathfrak{f}\Phi\right) \begin{pmatrix} & -1 \\ 1 & x \end{pmatrix}, \tag{8.65}$$

the integral on the left being absolutely convergent for s_1 *and* s_2 *in this range.*

Note that we use the *additive* Haar measure in Eq. (5.65). The integrand is undefined at $y = 0$, but because the origin has measure zero, this doesn't matter.

Proof We replace t by t^{-1} in Eq. (8.63) to write

$$W_\Phi \begin{pmatrix} y & \\ & 1 \end{pmatrix} = \int_{F^\times} \chi(t^{-1}) \sqrt{|y|}\, \chi_1(y)\, \Phi(yt^{-1}, t)\, d^\times t. \tag{8.66}$$

Consequently, the left side of Eq. (8.65) equals

$$\int_{F^\times} \int_F \chi(yt^{-1})\, \Phi(yt^{-1}, t)\, dy\, d^\times t.$$

It is not hard to see that this double integral is absolutely convergent when $\mathrm{re}(s_1 - s_2 + 1) > 0$, which justifies the following manipulations. The right side of Eq. (8.65) equals

$$\begin{aligned} \tau_{\chi_1,\chi_2}\left(R\begin{pmatrix} & -1 \\ 1 & x \end{pmatrix} \mathfrak{f}\Phi\right) &= \int_{F^\times} \left(R\begin{pmatrix} & -1 \\ 1 & x \end{pmatrix} \mathfrak{f}\Phi\right)(0, y)\, \chi(y)\, |y|\, d^\times y \\ &= \int_{F^\times} \mathfrak{f}\Phi(y, yx)\, \chi(y)\, |y|\, d^\times y \\ &= \int_{F^\times} \int_F \Phi(y, t)\, \psi(xyt)\, \chi(y)\, |y|\, dt\, d^\times y. \end{aligned}$$

Now we note that the measures $|y|\, dt\, d^\times y$ and $|t|\, dy\, d^\times t$ are equal. Rewriting the measure this way, then interchanging the order of integration (justified by the absolute convergence of the integral under our assumption that $\mathrm{re}(s_1 - s_2 + 1) > 0$), and then making the variable change $y \to yt^{-1}$, this equals

$$\int_{F^\times} \int_F \chi(yt^{-1})\, \Phi(yt^{-1}, t)\, dy\, d^\times t,$$

which gives us Eq. (8.65). ∎

Proposition 4.8.9: Jacquet–Langlands *Assume that* $\mathrm{re}(s_1 - s_2 + 1) > 0$ *and that* $\mathcal{B}(\chi_1, \chi_2)$ *is irreducible. Then the space* $\mathcal{W}$ *of functions* W_Φ *comprises the Whittaker model of* $\mathcal{B}(\chi_1, \chi_2) = (\pi_{\chi_1,\chi_2}, V_{\chi_1,\chi_2})$. *If* ρ *denotes the action of*

$GL(2, F)$ on $\mathcal{W}$ by right translation, then for $g \in GL(2, F)$

$$\rho(g)\, W_\Phi = W_{\omega_0(g)\,\Phi} \tag{8.67}$$

Proof The formula (8.67) is an immediate consequence of the definition (Eq. 8.63) of W_Φ. Let us define $f_\Phi = T_{\chi_1,\chi_2}\, \mathfrak{f}\Phi$. Then $f_\Phi \in V_{\chi_1,\chi_2}$, and it follows from Eq. (8.57) and the fact that T_{χ_1,χ_2} is an intertwining operator from $S(E)$ (with the representation R) to V_{χ_1,χ_2} that

$$f_{\omega_0(g)\,\Phi} = \pi_{\chi_1,\chi_2}(g)\, f_\Phi. \tag{8.68}$$

We will prove that W_Φ is nonzero if and only if f_Φ is nonzero. Suppose for a moment that we have accomplished this. Then it follows from Eqs. (8.67) and (8.68) that $f_\Phi \mapsto W_\Phi$ is a $GL(2, F)$-equivariant bijection of V_{χ_1,χ_2} onto a space of functions satisfying Eq. (8.64). The theorem will then follow from the uniqueness of Whittaker models, Theorem 4.4.1.

Rewriting Eq. (8.65) as

$$\int_F W_\Phi\begin{pmatrix} y & \\ & 1\end{pmatrix} \chi_2(y)^{-1}\, |y|^{-1/2} \psi(xy)\, dy = f_\Phi\begin{pmatrix} & -1 \\ 1 & x\end{pmatrix}, \tag{8.69}$$

it is clear that if $W_\Phi = 0$, then f_Φ vanishes on the set of all elements of the form $g = \left(\begin{smallmatrix} & -1 \\ 1 & x\end{smallmatrix}\right)$. Because $f_\Phi \in V_{\chi_1,\chi_2}$, it then vanishes on all elements of the form bg, where $b \in B(F)$ and g is of this form. These elements are dense in $GL(2, F)$, and because F is locally constant, it is zero everywhere.

On the other hand, let us show that if $f_\Phi = 0$, then $W_\Phi = 0$. We have shown that there is a well-defined map $W_\Phi \mapsto f_\Phi$ from $\mathcal{W}$ to V_{χ_1,χ_2}, and we must show that this map is injective. By Eqs. (8.67) and (8.68), this map commutes with right translation, so to prove that $f_\Phi = 0$ implies $W_\Phi = 0$, it is sufficient to show that $W_\Phi(1) = 0$. We will deduce this from a form of the Fourier inversion formula.

Lemma 4.8.2 *Let $f \in L^1(F)$, and suppose that the restriction of f to $F^\times$ is locally constant and that the Fourier transform*

$$\hat{f}(y) = \int_F f(x)\, \psi(xy)\, dy$$

vanishes for all y. Then f vanishes on $F^\times$.

Proof Let N be a large integer. If $0 \neq a \in F$, consider

$$0 = \int_{\mathfrak{p}^{-N}} \hat{f}(y)\, \psi(-ay)\, dy = \int_{\mathfrak{p}^{-N}} \int_F f(x)\, \psi\big((x-a)y\big)\, dx\, dy$$

$$= \int_{\mathfrak{p}^{-N}} \int_F f(x+a)\, \psi(xy)\, dx\, dy.$$

Because $\mathfrak{p}^{-N}$ is compact and f is integrable, this integral is absolutely convergent and we are justified in interchanging the order of integration. Because $\int_{\mathfrak{p}^{-N}} \psi(xy)\,dy$ is a constant times the characteristic function of a fractional ideal, namely, $\mathfrak{p}^N$ times the conductor of $\mathfrak{o}$, and because f is constant near a, we see that $f(a) = 0$. ■

Applying this lemma, Eq. (8.69) implies that

$$W_\Phi\begin{pmatrix} y & \\ & 1 \end{pmatrix} \chi_2(y)^{-1}\,|y|^{-1/2} = 0$$

for $y \in F^\times$, and in particular, $W_\Phi(1) = 0$.

We may now give the following proof.

Proof of Proposition 4.7.7 Let $(\pi, V) = \mathcal{B}(\chi_1, \chi_2)$, which we assume to be irreducible. Because both sides of Eq. (7.16) remain unchanged if we interchange χ_1 and χ_2, there is no harm in assuming that $\mathrm{re}(s_1 - s_2 + 1) > 0$ so that we may use the concrete realization of Proposition 4.8.9 for the Whittaker model of π. We choose the Schwartz function Φ to be of the form $\Phi(u_1, u_2) = \Phi_1(u_1)\,\Phi_2(u_2)$, where $\Phi_i \in S(F)$. Let $W = W_\Phi$, which by Proposition 4.8.9 lies in the Whittaker model of π. By Eq. (8.66)

$$\begin{aligned} Z(s, W, \xi) &= \int_{F^\times} W_\Phi\begin{pmatrix} y & \\ & 1 \end{pmatrix} \xi(y)\,|y|^{s-1/2}\,d^\times y \\ &= \int_{F^\times}\int_{F^\times} \chi_1(t)^{-1}\,\chi_2(t)\,|y|^{1/2}\,\chi_1(y)\,\Phi\left(yt^{-1}, t\right) \\ &\qquad \times d^\times t\,\xi(y)\,|y|^{s-1/2}\,d^\times y. \end{aligned}$$

We now make a variable change $y \to yt$, and this integral splits up into two Tate local integrals of the type introduced in Eq. (1.12) (in Chapter 3). Specifically, we obtain

$$Z(s, W, \xi) = \zeta(s, \xi\chi_1, \Phi_1)\,\zeta(s, \xi\chi_2, \Phi_2), \tag{8.70}$$

where

$$\zeta(s, \xi\chi_1, \Phi_1) = \int_{F^\times} (\xi\chi_1)(y)\,\Phi_1(y)\,|y|^s\,d^\times y,$$

$$\zeta(s, \xi\chi_2, \Phi_2) = \int_{F^\times} (\xi\chi_2)(t)\,\Phi_2(t)\,|t|^s\,d^\times t.$$

Let $\omega = \chi_1\chi_2$ be the central quasicharacter of π. We note that the Fourier transform, defined by Eq. (8.26)

$$\hat{\Phi}(v_1, v_2) = \hat{\Phi}_1(v_2)\, \hat{\Phi}_2(v_1),$$

where the one-dimensional Fourier transforms

$$\hat{\Phi}_i(v) = \int_F \Phi_i(u)\, \psi(uv)\, dv.$$

Thus by Eqs. (8.70) and (8.47), with $\gamma(B) = 1$ by Proposition 4.8.4(iii)

$$\begin{aligned} Z\big(s, \rho(w_1)\, W, \xi^{-1}\omega^{-1}\xi^{-1}\big) &= \zeta\big(s, \xi^{-1}\,\omega^{-1}\,\chi_1, \hat{\Phi}_2\big)\, \zeta\big(s, \xi^{-1}\,\omega^{-1}\,\chi_2, \hat{\Phi}_1\big) \\ &= \zeta\big(s, \xi^{-1}\,\chi_2^{-1}, \hat{\Phi}_2\big)\, \zeta\big(s, \xi^{-1}\,\chi_2^{-1}, \hat{\Phi}_1\big). \quad (8.71) \end{aligned}$$

Now Eq. (7.16) follows from comparing Eqs. (7.13) and (1.12) (in Chapter 3), using Eqs. (8.70) and (8.71). ∎

Exercises

Exercise 4.8.1 Let H be a locally compact group and, H^* its dual group. Let $\alpha : H \to H$ be an automorphism. Recall that the *module* $|\alpha|$ of α is the factor by which α multiplies the Haar measure; more precisely, if $\phi \in C_c(H)$, then by definition

$$\int_H (\phi \circ \alpha)(h)\, dh = |\alpha|^{-1} \int_H \phi(h)\, dh.$$

Let $\alpha^* : H^* \to H^*$ be the automorphism induced by α. Prove that $|\alpha^*| = |\alpha|$.

[Hint: Let $\mathcal{F}\phi : H^* \to \mathbb{C}$ denote the Fourier transform of ϕ. Prove that

$$\mathcal{F}(\phi \circ \alpha) = |\alpha|^{-1}\, (\mathcal{F}\phi) \circ \alpha^{*-1},$$

and apply the Plancherel formula to this identity.]

4.9 The Local Langlands Correspondence

This section is not intended to be a complete and self-contained description of the local Langlands correspondence. It should be studied in conjunction with the references that we give below.

Let F be a non-Archimedean local field. If the residue characteristic of F is odd, then Tunnell proved in his dissertation that the dihedral supercuspidal representations constructed in Theorem 4.8.6 are the only supercuspidal representations of $GL(2, F)$. See Tunnell (1978) and (1979).

This means that for a non-Archimedean local field of an odd residue characteristic, as in the case of a finite field (Section 4.1), the irreducible admissible representations of $GL(2, F)$ are parametrized by characters of maximal tori. To be more precise, if T is a maximal torus in $GL(2, F)$, and if χ is a character of $T(F)$ that does not factor through the determinant map, then χ parametrizes an irreducible representation of $GL(2, F)$, and except for the special representations, which must be discussed separately, every irreducible admissible representation arises this way. If the torus T is split, the representations thus parametrized are the principal series representations. If T is nonsplit, then T is associated with a quadratic extension E/F, and the associated representations are the dihedral supercuspidals.

If the residue characteristic of F is even, however, this characterization breaks down, for there exist nondihedral supercuspidals. The *local Langlands conjecture* is a precise statement that accounts for these examples. The conjecture includes a hypothetical classification of the irreducible admissible representations of $GL(n, F)$, where F is a local field. It has been proved in many cases. Over an Archimedean field, the local Langlands conjecture (for arbitrary reducbive groups) is a theorem of Langlands. For $GL(2)$ over a non-Archimedean local field F, the conjecture was proved by Tunnell (1978) if the residue characteristic of F is odd, and by Kutzko (1980) in general. For $GL(3)$ over a non-Archimedean local field, the conjecture is a theorem of Henniart (1984). For $GL(n)$ over a local field of positive characteristic, it is a theorem of Laumon, Rapoport and Stuhler (1993). The survey paper of Kudla (1994) (with its companion piece by Knapp (1994)) is an important guide to the recent literature. The work by Tate and Borel (1979) are also fundamental and should be consulted.

The motivation for this local conjecture comes largely from global considerations. We review the theory of Artin L-functions from a more sophisticated standpoint than in Section 1.8. In that section, we considered the theory of Artin L-functions attached to Galois representations. There is another theory of L-functions, namely, the Hecke–Tate theory, which we considered in Section 3.1, and there is some overlap between the theories of these two classes. Let F be a global field and let A be its adele ring, and let $\rho : \mathrm{Gal}(\overline{F}/F) \to GL(1, \mathbb{C}) \cong \mathbb{C}^\times$ be a one-dimensional representation. Then ρ factors through $\mathrm{Gal}(F^{\mathrm{ab}}/F)$, where F^{ab} is the maximal Abelian extension, and the composition of ρ with the reciprocity law homomorphism $A^\times/F^\times \to \mathrm{Gal}(F^{\mathrm{ab}}/F)$ gives us a Hecke character whose L-function agrees with the Artin L-function of ρ.

Thus L-functions of some Hecke characters are also Artin L-functions. Not all are, however, because it follows from the "no small subgroups" argument (Exercise 3.1.1(a)) that the image of any continuous homomorphism $\rho : \mathrm{Gal}(\overline{F}/F) \to \mathbb{C}^\times$ is necessarily finite. Thus any Hecke character of infinite order, such as the ones we employed in Section 1.9, does not arise in this way. We see that although the two theories overlap, neither theory is contained in the other.

In order to obtain a theory that subsumes both these overlapping theories in a unified framework, Weil (1951) introduced a topological group W_F called the *(absolute) Weil group* of F, which is a substitute for the Galois group. Artin and Tate (1968), Tate (1979), and Deligne (1973b) are the basic references for the Weil group. Though not directly concerned with Weil groups, Cassels and Fröhlich (1967, Chapters VI and VII by Serre and Tate) and Serre (1979) contain proofs of the cohomological results of class field theory that underlie their construction. Readers of Jacquet and Langlands (1970) should consult Langlands (1970a), which is useful in place of his unpublished (1970b) since it describes Langlands' notations, used in Jacquet and Langlands (1970).

The Weil group is equipped with a homomorphism $W_F \to \mathrm{Gal}(\overline{F}/F)$, so every Galois representation induces a representation of W_F, but not every representation of W_F is of this form. With every continuous representation of W_F there is associated an Artin L-function (in Weil's generalized sense) that has a meromorphic continuation and functional equation. Thus enlarged, the class of Artin L-functions includes *all* Hecke L-functions, not just those of finite order, because the Abelianization of W_F is isomorphic to $A^\times/F^\times$, so the quasicharacters of W_F are the same as the quasicharacters of $A^\times/F^\times$.

Let F be a field that may be either local or global. If F is local, let $C_F = F^\times$, whereas if F is global and $A = A_F$ its adele ring, we will denote $C_F = A_F^\times/F^\times$. In either case, the *reciprocity law map* or *Artin map* is a homomorphism α_F : $C_F \to \mathrm{Gal}(F^{\mathrm{ab}}/F)$, where F^{ab} is the maximal Abelian extension of F.

Let E/F be a finite Galois extension. The *relative Weil group* is a topological group $W_{E/F}$ that fits into a short exact sequence

$$1 \to C_E \to W_{E/F} \to \mathrm{Gal}(E/F) \to 1. \tag{9.1}$$

Its construction is equivalent to the specification of an element of $H^2\big(\mathrm{Gal}(E/F), C_E\big)$. (The relationship between group extensions and cohomology classes was explained for *central* extensions in the exercises to Section 1. Because C_E is not a trivial Galois module, the Weil group is a *noncentral* extension of $\mathrm{Gal}(E/F)$ by C_E. The case of noncentral extensions is not much different from the special case considered in the exercises to Section 1. It is discussed in Artin and Tate (1968), or in any book on homological algebra.) It is shown in the class field theory that if $[E : F] = n$, then $H^2\big(\mathrm{Gal}(E/F), C_E\big)$ is cyclic of order n, and the cohomology class corresponding to the extension (Eq. 1.1) is known as the *fundamental class*.

If $K \supset E$ is a bigger Galois extension of F, then we have a map $W_{K/F} \to W_{E/F}$ that is compatible with the norm map $C_K \to C_E$. The *absolute Weil group* W_F is the inverse limit of the groups $W_{E/F}$ as E ranges through the Galois extensions of F. The relative Weil group $W_{E/F}$ can be recovered as follows. W_E may be naturally identified with a subgroup of W_F, and $W_{E/F}$ is isomorphic to W_F/W_E^c, where W_E^c is the closure of the commutator subgroup of W_E.

If F is a global field, then W_F is a somewhat mysterious object. If F is local, however, it admits a simple description. Suppose that F is a non-Archimedean local field. The Galois group $\mathrm{Gal}(\overline{F}/F)$ is a profinite group, which is the inverse limit of the Galois groups of the finite extensions of F. Let $\mathbb{F}_q$ be the residue field of F. Any automorphism of $\overline{F}/F$ induces an automorphism of $\overline{\mathbb{F}}_q/\mathbb{F}_q$, and the kernel I_F of the resulting homomorphism $\mathrm{Gal}(\overline{F}/F) \to \mathrm{Gal}(\overline{\mathbb{F}_q}/\mathbb{F}_q)$ is called the *inertia group*. Let us recall the structure of $\mathrm{Gal}(\overline{\mathbb{F}_q}/\mathbb{F}_q)$. If n is any integer, $\mathbb{F}_q$ has a unique extension of degree n, and the Galois group $\mathrm{Gal}(\mathbb{F}_{q^n}/\mathbb{F}_q)$ is cyclic of order n, with a canonical generator, namely, the Frobenius element $x \mapsto x^q$. Hence

$$\mathrm{Gal}(\overline{\mathbb{F}_q}/\mathbb{F}_q) = \varprojlim \mathrm{Gal}(\mathbb{F}_{q^n}/\mathbb{F}_q) \cong \varprojlim \mathbb{Z}/n\mathbb{Z}.$$

This is the profinite group often denoted $\hat{\mathbb{Z}}$ in the literature. It contains $\mathbb{Z}$ as a dense subgroup. The Weil group W_F may be identified with the group of elements of $\mathrm{Gal}(\overline{F}/F)$ whose image in $\mathrm{Gal}(\overline{\mathbb{F}_q}/\mathbb{F}_q)$ is a power of the Frobenius element. It therefore contains the inertia group, and we have exact sequences

$$\begin{array}{ccccccccc} 1 & \longrightarrow & I_F & \longrightarrow & W_F & \longrightarrow & \mathbb{Z} & \longrightarrow & 1 \\ & & \| & & \downarrow & & \downarrow & & \\ 1 & \longrightarrow & I_F & \longrightarrow & \mathrm{Gal}(\overline{F}/F) & \longrightarrow & \hat{\mathbb{Z}} & \longrightarrow & 1 \end{array} \tag{9.2}$$

Let us now go back and see how the exact sequence Eq. (9.1) works when F is a non-Archimedean local field. Let F^{ab} be the maximal Abelian extension of F, so $\mathrm{Gal}(F^{\mathrm{ab}}/F)$ is the Abelianization of $\mathrm{Gal}(\overline{F}/F)$, and let $F^{\mathrm{nr}} \subset F^{\mathrm{ab}}$ be the maximal nonramified extension. The Artin map $\alpha_F : F^\times \to \mathrm{Gal}(F^{\mathrm{ab}}/F)$ is not quite an isomorphism, but has a dense image. We have an isomorphism $\mathrm{Gal}(F^{\mathrm{nr}}/F) \cong \mathrm{Gal}(\overline{\mathbb{F}_q}/\mathbb{F}_q) \cong \hat{\mathbb{Z}}$, and so we have an exact sequence

$$1 \to \mathrm{Gal}(F^{\mathrm{ab}}/F^{\mathrm{nr}}) \to \mathrm{Gal}(F^{\mathrm{ab}}/F) \to \hat{\mathbb{Z}} \to 1.$$

An element of $\mathrm{Gal}(F^{\mathrm{ab}}/F)$ lies in $\alpha_F C_F$ if and only if it induces a power of the Frobenius element in $\mathrm{Gal}(\overline{\mathbb{F}_q}/\mathbb{F}_q)$, that is, if and only if its image in $\hat{\mathbb{Z}}$ lies in $\mathbb{Z}$. Now consider the exact sequence

$$1 \to \mathrm{Gal}(E^{\mathrm{ab}}/E) \to \mathrm{Gal}(E^{\mathrm{ab}}/F) \to \mathrm{Gal}(E/F) \to 1.$$

If we limit ourselves to elements of $\mathrm{Gal}(E^{\mathrm{ab}}/E)$ and $\mathrm{Gal}(E^{\mathrm{ab}}/F)$ that induce powers of the Frobenius element in $\mathrm{Gal}(\overline{\mathbb{F}_q}/\mathbb{F}_q)$, then we obtain an exact sequence in which the first term is the image of the Artin map $\alpha_E : C_E = E^\times \to \mathrm{Gal}(E^{\mathrm{ab}}/E)$, and the middle term is $W_F/W_E^c \cong W_{E/F}$. Hence we obtain the exact sequence Eq. (9.1).

Suppose that F is a global field. Weil (1951) showed that one could associate an L-function $L(s, \rho)$ having a functional equation with any continuous representation $\rho : W_F \to GL(n, \mathbb{C})$. The problem of showing that the epsilon factors for the functional equations satisfied by these L-functions can be defined purely locally was solved by Langlands (1970b), after some earlier work by Dwork. These notes were never published, and Tate (1979) comments that the "experience of Dwork and Langlands indicates that the local proof ... is too involved to publish completely." A proof of Langlands' local result using global methods was published by Deligne (1973b), so the fact that Langlands (1970b) was never published is not a gap in the literature.

The result of Langlands is stated precisely in Tate (1979) Theorem 3.4.1, Deligne (1973b) Theorem 4.1, or Langlands (1970a) Theorem 1. If F_v is a local field, and if $\rho_v : W_{F_v} \to GL(n, \mathbb{C})$ is a representation, then there are defined local L- and ϵ-factors $L(s, \rho_v)$ and $\epsilon(s, \rho_v, \psi_v)$, the latter depending on the choice of an additive character ψ_v. These must satisfy certain axioms, the most important of which is a compatibility with induction. If ρ_v is one dimensional, then in view of the isomorphism $W_{F_v}^{\text{ab}} \cong F_v^\times$, ρ_v is essentially a quasicharacter of $F_v^\times$, and $L(s, \rho_v)$ and $\epsilon(s, \rho_v, \psi_v)$ agree with the Tate factors that were defined in Section 3.1. The consistency of the conditions that $\epsilon(s, \rho_v, \psi_v)$ must satisfy is by no means obvious because some representations might be expressible as linear combinations of representations induced from one-dimensional characters in more than one way, and when this occurs, an identity between Gauss sums must be satisfied. This consistency is the content of the result of Langlands.

If F is a global field, v is a place of F, and if $\rho : W_F \to GL(n, \mathbb{C})$ is a representation, let $\rho_v : W_{F_v} \to GL(n, \mathbb{C})$ be the composition of ρ with the canonical homomorphism $W_{F_v} \to W_F$. Then we have an L-function $L(s, \rho) = \prod_v L(s, \rho_v)$ and an epsilon factor $\epsilon(s, \rho) = \prod_v \epsilon_v(s, \rho_v, \psi_v)$. Langlands proved a functional equation

$$L(s, \rho) = \epsilon(s, \rho)\, L(1 - s, \hat{\rho}),$$

where $\hat{\rho}$ is the contragredient representation to ρ. If $n = 1$, then χ is essentially a character of $A^\times / F^\times \cong W_F^{\text{ab}}$, and this functional equation is the same as that in Theorem 3.1.10.

Langlands' functoriality conjecture, as set forth in Langlands (1970c, 1980) and Borel (1979), implies that there exists an automorphic representation $\pi(\rho)$ whose L-function and ϵ-factors agree with that of ρ.

In order to discuss the cases where this has been proved, we review the classification, due to Felix Klein (1913), of the finite subgroups of $PGL(2, \mathbb{C})$. Consider the adjoint square homomorphism $GL(2, \mathbb{C}) \to GL(3, \mathbb{C})$. The kernel of this representation consists of the scalar matrices, and its consists of the orthogonal rotations; so we have an isomorphism $PGL(2, \mathbb{C}) \cong SO(3, \mathbb{C})$. We consider the image of an arbitrary finite subgroup of $PGL(2, \mathbb{C})$ in $SO(3, \mathbb{C})$. After conjugating, we may assume that its image is contained in $SO(3)$, the real rotation group. This image is, up to conjugacy, one of the groups described in the following Proposition.

Proposition 4.9.1: Klein *Every finite subgroup of $SO(3)$ is conjugate to one of the following:*

(1) *the cyclic group of order n generated by a rotation in an angle $2\pi/n$;*
(2) *the dihedral group of order $2n$ obtained by adjoining a reflection to the cyclic group of order n;*
(3) *the group A_4 of rotations of the tetrahedron;*

(4) the group S_4 of rotations of the octahedron; or
(5) the group A_5 of rotations of the icosahedron.

Proof See Dornhoff (1971, Part A, Chapter 26). ■

Let F be either a local or global field. If $\rho : W_F \to GL(n, \mathbb{C})$ is any representation, then the kernel of ρ contains an open subgroup by the "no small subgroups" argument, so ρ factors through $W_{E/F}$ for some finite extension E/F. A representation $\rho : W_F \to GL(n, \mathbb{C})$ is called *primitive* if it is not induced from a proper subgroup. Imprimitive Weil group representations are important, but their L-functions are the same as the L-functions of the Weil group representations from which they are induced, so we may concentrate on the imprimitive representations. It is a useful fact that in a primitive irreducible representation, a normal Abelian subgroup must act by scalars – this is a consequence of Clifford's theorem (see Dornhoff (1971, Part A, Chapter 14) for Clifford's theorem in the case of finite groups)–so (consulting the exact sequence Eq. (9.1)) the normal Abelian subgroup C_E lies in the kernel of the composition $W_{E/F} \to GL(n, \mathbb{C}) \to PGL(n, \mathbb{C})$. Therefore this composition factors through $\mathrm{Gal}(E/F)$. In particular, the image of $W_{E/F}$ in $PGL(n, \mathbb{C})$ is a finite group.

Suppose that $\rho : W_F \to GL(2, \mathbb{C})$ is a primitive representation. Because, as we see its image in $PGL(2, \mathbb{C})$ is finite, it falls into the classification of Proposition 4.9.1. If the type of the representation is cyclic, dihedral, tetrahedral, or octahedral, then the existence of $\pi(\rho)$ was proved by Langlands (1980) or by Tunnell (1981). If the type of the representation is icosahedral, the existence of $\pi(\rho)$ is not known in general, but Buhler (1978) gave an example of an icosahedral Galois representation for which the existence of $\pi(\rho)$ could be established, and recently, a few other examples have been described in Frey (in press).

The *local Langlands conjecture* asserts that if F is a local field and $\rho : W_F \to GL(n, \mathbb{C})$ is an irreducible representation, then there exists a supercuspidal representation π of $GL(n, F)$ whose L- and ϵ-factors agree with those of ρ. (If $n = 2$, we defined these L- and ϵ-factors in Section 3.5 and Section 4.7.) It is assumed that if χ is a quasicharacter of $F^\times$, then $L(s, \rho) = L(s, \chi)$ and $\epsilon(s, \chi \otimes \rho, \psi) = \epsilon(s, \pi, \chi, \psi)$. (We are suppressing the subscripts v from the notation, of course.) It is a consequence of Proposition 4.7.6 that at most one representation π can have this characterization. This representation is denoted $\pi(\rho)$.

The local Langlands conjecture thus asserts that the supercuspidal representations of $GL(n, F)$, where F is a non-Archimedean local field, are in bijection with the irreducible n-dimensional representations of W_F. We can expand the scope of the conjecture to include all irreducible admissible representations of $GL(n, F)$, which are in rough bijection with all semisimple representations of W_F, including the reducible ones – for example, if $n = 2$, the principal series representations are parametrized by the pairs of quasicharacters of $F^\times = C_F$, which correspond to representations of W_F that are the direct sum of a pair of quasicharacters. One must be somewhat careful because this scheme does not account for the special representations. To obtain a precise bijection, one employs not the Weil group, but a slightly larger group, the *Weil–Deligne group* (Deligne (1973b), Tate (1979) and Borel (1979)). We avoid this issue by considering only supercuspidal representations.

Let $s \in \mathbb{C}$. We define a quasicharacter $\omega_s : W_F \to \mathbb{C}^\times$ by $\omega_s(g) = q^{-ks}$ if g induces the kth power of the Frobenius element in $\mathrm{Gal}(\overline{\mathbb{F}_q}/\mathbb{F}_q)$. If $\rho : W_F \to GL(n, \mathbb{C})$ is a representation (always assumed continuous), we may then consider $\omega_s \otimes \rho$.

Proposition 4.9.2 *Let F be a non-Archimedean local field, and let $\rho : W_F \to GL(n, \mathbb{C})$ be an irreducible representation. There exists $s \in \mathbb{C}$ such that $\omega_s \otimes \rho$ factors through the homomorphism $W_F \to \mathrm{Gal}(\overline{F}/F)$. The image of $\omega_s \otimes \rho$ is a finite subgroup of $GL(n, \mathbb{C})$.*

Proof Let $I_F = \mathrm{Gal}(\overline{F}/F^{\mathrm{ab}})$ be the inertia group. Then I_F is compact and so by the "no small subgroups" argument, ρ is trivial on a subgroup of finite index in I_F. Let $\sigma \in W_F$ be an element whose image induces the Frobenius element in $\mathrm{Gal}(\overline{\mathbb{F}_q}/\mathbb{F}_q)$. Then W_F is generated by I_F and σ. Conjugation by σ induces an automorphism of the finite group $I_F/\big(I_F \cap \ker(\rho)\big)$, and so there exists a positive integer n such that conjugation by σ^n induces the identity on $I_F/\big(I_F \cap \ker(\rho)\big)$. Now we claim that $\rho(\sigma^n)$ must be a scalar. Indeed, $\rho(\sigma^n)$ commutes with $\rho(\sigma)$ and with $\rho(g)$ for $g \in I_k$, and so it commutes with the entire image of ρ in $GL(n, \mathbb{C})$; hence if λ is an eigenvector, the λ-eigenspace of $\rho(\sigma^n)$ is an invariant subspace. Because ρ is irreducible, it follows that $\rho(\sigma^n)$ is just the scalar λ.

Now we can choose s so that $(\omega_s \otimes \rho)(\sigma^n) = 1$. Then the image of $\omega_s \otimes \rho$ is finite, because its kernel contains $I_F \cap \ker(\rho)$ and σ^n, and together these generate a subgroup of W_F of finite index. The closure of this group in G_F is a subgroup of finite index, and $\omega_s \otimes \rho$ extends to this group by continuity. ■

Let F be either a local or global field, and let E/F be a quadratic extension. Then W_E is a subgroup of index two in W_F. We say that a continuous representation $\rho : W_E \to GL(2, \mathbb{C})$ is *dihedral* if there exists a quadratic extension E/F such that ρ is induced from a quasicharacter of W_E.

The Abelianization of W_E is isomorphic to C_E. It follows that the quasicharacters of W_E are essentially the same as the quasicharacters of C_E.

Proposition 4.9.3 *Let F be a non-Archimedean local field of odd residue characteristic, and let $\rho : W_F \to GL(2, \mathbb{C})$ be an irreducible continuous representation. Then ρ is dihedral.*

It is possible to say much more: If the residue characteristic of F is prime to n, then an irreducible representation $\rho : W_F \to GL(n, \mathbb{C})$ is induced from a one-dimensional representation of W_E where E is an extension field of degree n. See Tate's equation (1979) (2.2.5.3). This result relies on a classification by Koch (1977) of the primitive (i.e., noninduced) Galois representations of local fields. Related results are contained in Buhler (1978). Weil (1974) considers the delicate question of two-dimensional Galois representations of a local field of even characteristic, and his results are generalized to $GL(n)$ by Koch (1977).

Proof The class of dihedral representations is preserved under twisting by characters ω_s, so there is no loss of generality in invoking Proposition 4.9.2 to assume that ρ extends to a homomorphism $\mathrm{Gal}(\overline{F}/F) \to \mathbb{C}$. Every such homomorphism has an open kernel and factors through a finite quotient $\mathrm{Gal}(K/F)$, where K is a finite extension of F. We will show that if the residue characteristic of F is odd, then every representation of $\mathrm{Gal}(K/F)$ is induced from a subgroup of index two. The proposition follows, with E being the fixed field of this subgroup of index two.

We will show that the image of $\mathrm{Gal}(K/F)$ in $GL(2, \mathbb{C})$ must be dihedral in Klein's classification. The dihedral group D_n has a cyclic subgroup C_n of index two, and the two-dimensional representation of D_n is induced from a character of C_n. Thus every dihedral representation is induced from a character of a subgroup of index two.

Because $\rho : \mathrm{Gal}(K/F) \to GL(2, \mathbb{C})$ is irreducible, its image is not a cyclic group. We must show that it cannot be A_4, S_4, or A_5. We recall some facts about the structure of $G = \mathrm{Gal}(K/F)$, whose proofs may be found in Serre (1979, Chapter IV). Let G_0 be the inertia group, and let G_1 be the first ramification group. Let p be the residue characteristic; by assumption, $p \neq 2$. Then G/G_0 is cyclic, G_0/G_1 is cyclic of order prime to p, and G_1 is a (possibly non-Abelian) p-group. Clearly, G cannot admit A_5 as a homomorphic image, because G is solvable. To eliminate the possibilities of A_4 and S_4, we must make use of the fact that $p \neq 2$. Considering the images G', G'_0, and G'_1 of G, G_0, and G_1 under a hypothetical surjective homomorphism $G \to A_4$ or S_4, we see that $G' = A_4$ or S_4 would admit a filtration $G' \supseteq G'_0 \supseteq G'_1$ such that G'_1 is a p-group and is normal in G'_0, G'_0/G'_1 is cyclic of order prime to p, G'_0 is normal in G', and G'/G'_0 is cyclic. If $p \neq 2$, neither A_4 nor S_4 admits such a filtration. ■

Theorem 4.9.1: Jacquet–Langlands *Let F be a non-Archimedean local field, let E/F be a quadratic field extension, let χ be a quasicharacter of W_E, and let $\rho : W_F \to GL(2, \mathbb{C})$ be the induced representation of W_F. Then $\pi(\rho)$ exists.*

Proof (sketch) Because the Abelianization of W_E is isomorphic to $C_E = E^\times$ we may regard χ as a character of $E^\times$ and construct a supercuspidal representation of $GL(2, F)$ by Theorem 4.8.6. To show that this representation is $\pi(\rho)$, it is necessary to compute $\epsilon(s, \pi, \chi, \psi)$ when χ is a character of $F^\times$. This computation is based on the same idea as the proof of Proposition 4.7.7, which was given in Section 4.8: The Whittaker models of representations that are constructed by means of the Weil representation are explicitly known, with the group element w_0 acting by Fourier transform, so it is possible to compute the ϵ-factors explicitly. For details of this computation, see Jacquet and Langlands (1970, Theorem 4.7(iii)). The notation $\lambda(E/F, \psi_F)$ that occurs in that proposition and in Jacquet and Langlands (1970, Section 1), is explained in Langlands (1970a). ■

The content of Proposition 4.9.3 is that if F is a non-Archimedean local field of odd residue characteristic, then every irreducible two-dimensional Weil group representation is dihedral, that is, induced from W_E, where E/F is quadratic. If the local Langlands conjecture is true, the same should be true of supercuspidals. In any case, Theorem 4.9.1 shows that if $\rho : W_F \to GL(2, \mathbb{C})$ is dihedral, then $\pi(\rho)$ exists.

Bibliography

Arthur, J. and L. Clozel, *Simple Algebras, Base Change, and the Advanced Theory of the Trace Formula*, Princeton University Press, Princeton (1989).

Arthur, J. and S. Gelbart, Lectures on automorphic L-functions, in *L-functions and Arithmetic, Proceedings of the Durham Symposium, July 1989*, J. Coates and M. Taylor, ed., Cambridge University Press, Cambridge (1991).

Artin, E., Zur Theorie der L-Reihen mit allgemeineren Gruppencharakteren, *Abh. Math. Sem. Univ. Hamburg* **8** (1930), 292–306.

Artin, E., and J. Tate, *Class Field Theory*, Benjamin, New York (1968). Reissued by Addison-Wesley, 1974.

Asai, T., On certain Dirichlet series associated with Hilbert modular forms and Rankin's method, *Math. Ann.* **226** (1977), 81–94.

Atkin, O., and J. Lehner, Hecke operators on $\Gamma_0(m)$, *Math. Ann.* **185** (1970), 134–160.

Baily, W., Satake's compactification of V_n. *Amer. J. Math.* **80** (1958), 348–364.

Baily, W., *Introductory Lectures on Automorphic Forms,* Iwanami Shoten, Tokyo, and Princeton University Press, Princeton (1973).

Baily, W. and A. Borel, On the compactification of arithmetic quotients of bounded symmetric domains, *Annals of Math.* **84** (1966), 442–528.

Banks, W., *Exceptional representations on the metaplectic group*, Dissertation, Stanford University, Stanford (1994).

Bernstein, J., Letter to Piatetski-Shapiro (1985). To appear in Cogdell and Piatetski-Shapiro (book in preparation).

Bernstein, J. and A. Zelevinsky, Representations of the group $GL(n, F)$ where F is a local nonarchimedean field, *Russian Mathematical Surveys* **3** (1976), 1–68.

Bernstein, J. and A. Zelevinsky, Induced representations of reductive p-adic groups I, *Ann. Sci. Ecole Norm. Sup.* 4^{e} serie **10** (1977), 441–472.

Bers, L. and M. Schechter, Elliptic Equations, Part II of L. Bers, F. John and M. Schechter, *Partial Differential Equations*, Wiley-Interscience, New York (1964).

Borel, A., Automorphic L-functions, 27–62 in part 2 of Borel and Casselman (1979).

Borel, A., *Linear Algebraic Groups*, second edition, Springer Verlag, New York (1991).

Borel, A., *Introduction to Automorphic Forms in One Variable*, lecture notes at the Academia Sinica of Beijing, Cambridge University Press (to appear).

Borel, A. and W. Casselman, ed., *Automorphic Forms, Representations, and L-functions*, AMS Proceedings of Symposia in Pure Mathematics **33** (1979), two parts.

Borel, A. and H. Jacquet, Automorphic forms and automorphic representations, 189–202 in part 1 of Borel and Casselman (1979).
Borel, A. and G. Mostow, ed., *Algebraic Groups and Discontinuous Subgroups*, AMS Proceedings of Symposia in Pure Mathematics **9** (1966).
Borel, A. and J. Tits, Groupes Réductifs, *Publ. Math. IHES* **27** (1965), 55–150.
Borevich, Z. and I. Shafarevich, *Number Theory*, Newcomb Greenleaf trans., Academic Press, Boston (1966).
Bourbaki, N., *Éléments de Mathematique, Livre II, Algébra, Chapitre 8: Modules et Anneaux Semi-simples*, Hermann, Paris (1958).
Bourbaki, N., *Integration*, Actualites scientifiques et industrielles **1343**, Nouvelle Edition, Hermann, Paris (1969).
Brauer, R., On Artin L-series with generalized group characters, *Ann. of Math.* **48**, (1947), 502–514.
Bruhat, F., Sur les représentations induites des groupes de Lie, *Bull. Soc. Math. France* **84** (1956), 97–205.
Bruhat, F. Distributions sur un groupe localement compact et applications à l'edude des representations des groupes p-adiques. Bull. Soc. Matn. France **89** (1961), 43–75.
Bruhat, F., and J. Tits, Groupes réductifs sur un corps locaux I, *Publ. Math. IHES* **41** (1972), 1–276.
Buhler, J., *Icosahedral Galois Representations*, Springer Lecture Notes in Mathematics **654** (1978).
Bump, D., The Rankin-Selberg Method: A Survey, in *Number Theory, Trace Formulas and Discrete Groups*, Symposium in Honor of Atle Selberg, K. Aubert, E. Bombieri and D. Goldfeld ed., Academic Press, Boston (1989).
Bump, D., W. Duke, J. Hoffstein and H. Iwaniec, An estimate for the Hecke eigenvalues of Maass forms, *International Research Notices of the Duke Mathematical Journal* **4** (1992), 75–81.
Bump, D. and S. Friedberg, The "exterior square" automorphic L-functions on $GL(n)$, 47–65 in part 2 of Gelbart, Howe and Sarnak (1990).
Bump, D. and D. Ginzburg, Symmetric square L-functions on GL(r), *Annals of Math.* **136** (1992), 137-205.
Bump, D. and J. Huntley, Unramified Whittaker functions for $GL(3, \mathbb{R})$, *Journal D'analyse Mathématique* **65** (1995), 19–44.
Bushnell, C. and P. Kutzko, *The admissible dual of $GL(N)$ via compact open subgroups*, Princeton University Press, Princeton (1993).
Carter, R., *Simple Groups of Lie Type*, Wiley-Interscience, New York (1972).
Carter, R., *Finite Groups of Lie Type: Conjugacy Classes and Complex Characters*, Wiley-Interscience, New York (1985).
Cartier, P., Representations of p-adic groups: a survey, 111-156 in part 1 of Borel and Casselman (1979).
Casselman, W., *Introduction to the theory of admissible representations of reductive p-adic groups* (unpublished book).
Casselman, W., *On some results of Atkin and Lehner*, Math. Ann. **201** (1973), 301–314.
Casselman, W., The unramified principal series of p-adic groups I: the spherical function, *Compositio Math.* **40** (1980), 387–406.
Casselman, W. and J. Shalika, The unramified principal series of p-adic groups II: the Whittaker function, *Compositio Math.* **40** (1980), 207–231.
Cassels, J. and A. Fröhlich, *Algebraic Number Theory*, Academic Press, Boston (1967).
Cogdell, J. and I. Piatetski-Shapiro, Converse Theorems for GL_n, *Pub. Math. IHES* **79** (1994).
Cogdell, J. and I. Piatetski-Shapiro, *L-functions for GL_n* (book in preparation).

Cohn, H. On the shape of the fundamental domain of the Hilbert modular group, in *Theory of Numbers, AMS Proceedings of Symposia in Pure Mathematics* **8** (1965).

Corwin, L., A construction of the supercuspidal representations of $GL_n(F)$, F p-adic, *Trans. AMS* **337** (1993), 1–58.

Deligne, P., Formes modulaires et représentations l-adiques, in *Séminaire Bourbaki, 1968/69 exposés*, Springer Lecture Notes in Mathematics **179** (1971), 347-363.

Deligne, P., Les constants des équations fonctionelles des fonctions L, in *Modular functions of one variable II*, Springer Lecture Notes in Mathematics **349** (1973a), 501–595.

Deligne, P., Formes modulaires et représentations de $GL(2)$ in *Modular functions of one variable II*, Springer Lecture Notes in Math. **349** (1973b), 55–105.

Deligne, P., and G. Lusztig, Representations of reductive groups over finite fields, *Ann. of Math.* **103** (1976), 103–161.

Deligne, P., and J.-P. Serre, Formes modulaires de poids 1, *Ann. Sci. Ec. Norm. Sup.* **7** (1974), 507–530. In vol. 3 of Serre's *Collected Papers*.

Deshouillers, J.-M. and H. Iwaniec, The nonvanishing of Rankin-Selberg zeta-functions at special points, in *The Selberg trace formula and related topics (Brunswick, Maine, 1984)*, Contemp. Math. **53**, The American Mathematical Society, Providence, R.I., (1986), 51–95.

Deshouillers, J.-M., H. Iwaniec, R. Phillips, R and P. Sarnak, Maass cusp forms, *Proc. Nat. Acad. Sci. U.S.A.* **82** (1985), 3533–3534.

Doi, K., and H. Naganuma, On the functional equation of certain Dirichlet series, *Invent. Math.* **9**, 1–14 (1969).

Dornhoff, L., *Group Representation Theory*, Marcel Dekker, New York (1971).

Dunford, N. and J. Schwartz, *Linear Operators*, 3 volumes. Wiley-Interscience, New York (1958), (1963) and (1971).

Edwards, H., *Riemann's Zeta Function*, Academic Press, Boston (1974).

Flath, D., Decomposition of representations into tensor products, 179–184 in part 1 of Borel and Casselman (1979).

Freitag, E., *Hilbert Modular Forms*, Springer Verlag, New York (1990).

Frey, G., ed., *Proceedings of the Workshop On Artin's Conjecture for Odd 2-Dimensional Representations*, Springer Lecture Notes in Mathematics **1593** (1994).

Friedman, A., *Partial Differential Equations*, Holt, Reinhart and Winston, New York (1969).

Fulton, W. and J. Harris, *Representation Theory*, Springer Verlag, New York (1991).

Garrett, P., Decomposition of Eisenstein series: Rankin triple products, *Ann. Math.* **125** (1987), 209–237.

Garrett, P., *Holomorphic Hilbert Modular Forms*, Wadsworth and Brooks/Cole, Belmont, Calif. (1990).

Garrett, P. and M. Harris, Special values of triple product L-functions, *Amer. J. Math.* **115** (1993), 161–240.

Gelbart, S., *Automorphic Forms on Adele Groups*, Princeton University Press, Princeton, (1975).

Gelbart, S. Automorphic forms and Artin's conjecture, in *Modular Functions of One Variable IV*, Springer Lecture Notes in Mathematics **627** (1977), 241–276.

Gelbart, S., An elementary introduction to the Langlands program, *Bull. Math. Soc. AMS* **10** (1984), 177–219.

Gelbart, S., R. Howe and P. Sarnak, ed., *Festschrift in honor of I. I. Piatetski-Shapiro on the occasion of his 60-th birthday*, two volumes, The Weizmann Science Press of Israel, Jerusalem (1990).

Gelbart, S. and H. Jacquet, A relation between automorphic representations of $GL(2)$ and $GL(3)$, *Ann. Sci. Ecole Normale Sup. 4e série*, **11** (1978), 471–552.

Gelbart, S. and I. Piatetski-Shapiro, L-functions for $G \times GL(n)$, in Springer Lecture Notes in Mathematics **1254** (1985).

Gelbart, S. and F. Shahidi, *Analytic properties of automorphic L-functions*, Academic Press, Boston (1988).

Gelfand, I. and S. Fomin, Geodesic flows on manifolds of constant negative curvature, *Usp. Mat. Nauk.* **7** (1) (1952), 118–137, *Transl., II. Ser., Am. Math. Soc.* (1955), 49–65. Reprinted in Gelfand's *Collected Papers*, vol. 2, Springer Verlag (1988).

Gelfand, I. and M. Graev, Construction of irreducible representations of simple algebraic groups over a finite field, *Dokl. Akad. Nauk SSSR* **147** (1962), 529–532; translated in *Sov. Math., Dokl.* **3** (1962), 1646–1649.

Gelfand, I., M. Graev and I. Piatetski-Shapiro, *Representation Theory and Automorphic Functions*, Saunders, Philadelphia (1968). Reissued by Academic Press, 1990.

Gelfand, I. and D. Kazhdan, Representations of the group $GL(n, K)$ where K is a local field, in *Lie Groups and their Representations*, John Wiley and Sons, New York (1975).

Ginzburg, D. L-functions for $SO_n \times GL_k$, *J. reine. angew. Math.* **405** (1990), 156–180.

Ginzburg, D., I. Piatetski-Shapiro and S. Rallis, L-functions for the orthogonal group, Memoirs of the AMS (to appear).

Godement, R., Analyse spectrale de fonctions modulaires, *Séminaire Bourbaki* **278** (1965).

Godement, R., The decomposition of $L^2(G/\Gamma)$ for $\Gamma = SL_2(\mathbb{Z})$, 211–224 in Borel and Mostow (1966a).

Godement, R., The spectral decomposition of cusp forms, 225–234 in Borel and Mostow (1966b).

Godement, R., *Notes on Jacquet-Langlands Theory*, available from the Institute for Advanced Study, Princeton (1970).

Godement, R. and H. Jacquet, *Zeta Functions of simple algebras*, Springer Lecture Notes in Mathematics **260** (1972).

Green, J., The characters of the finite general linear groups, *Trans. AMS* **80** (1955), 402–447.

Gross, B., Some applications of Gelfand pairs to number theory, *Bull. Amer. Math. Soc.* **124** (1991), 277–302.

Gross, B. and S. Kudla, Heights and the central critical values of triple product L-functions, *Compositio Math.* **81** (1992), 143–209.

Gunning, R. *Lectures on Riemann Surfaces*, Princeton University Press, Princeton (1966).

Halmos, P., *Measure Theory*, Van Nostrand, New York (1950). Reissued by Springer Verlag, 1974.

Harder, G., R. Langlands and M. Rapoport, Algebraische Zykeln auf Hilbert-Blumenthal Flächen, *J. Reine Angew. Math.* **366** (1986), 53–120.

Hardy, G., A note on Ramanujan's function $\tau(n)$, *Proc. Cambridge Philos. Soc.* **23** (1927), 675–680.

Harish-Chandra, Invariant eigendistributions on a semisimple Lie group, *Trans. Amer. Math. Soc.* **119** (1965), 457–508. In vol. 3 of his *Collected Works*.

Harish-Chandra, Discrete series for semisimple Lie groups, II. Explicit determination of the characters. *Acta. Math.* **116** (1966), 1-111. In vol. 3 of his *Collected Works*.

Harish-Chandra, *Automorphic forms on semisimple Lie groups*, Lecture Notes by J. G. M. Mars, Springer Lecture Notes in Mathematics **62** (1968).

Harish-Chandra, Eisenstein series over finite fields, in *Functional Analysis and Related Fields*, Springer Verlag (1970), 76–88.

Harris, M. and S. Kudla, The central critical value of a triple product L-function, *Annals of Math.* **133** (1991), 605–672.

Hartshorne, R. *Algebraic Geometry*, Springer Verlag, New York (1977).

Hecke, E., Eine neue Art von Zetafunktionen und ihre Beziehung zur Verteilung der Primzahlen, *Math. Zeitschrift* **1** (1918), 357–376 and **6** (1920), 11-51. Reprinted in Hecke (1983), 215–234 and 240–289.

Hecke, E., Zur Theorie der elliptischen Modulfunktionen, Math. Ann. **97** (1926), 210–242. Reprinted in Hecke (1983), 428–460.

Hecke, E., Über das verhalten von $\sum_{m,n} e^{\pi i \tau \frac{|m^2-2n^2|}{8}}$ und ähnlichen Funktionen bei Modulsubstitutionen, *J. reine u. angew. Math.* **157**, 159–170 (1927). Reprinted in Hecke (1983), 487–504.

Hecke, E., Über Modulfunktionen und die Dirichletschen Reihen mit Eulerscher Produktenwicklungen, I and II, *Math. Ann.* **114** (1937), 1–28 and 316–351. Reprinted in Hecke (1983), 644–707.

Hecke, E., *Mathematische Werke,* third edition, Vanderhoeck and Ruprecht, Göttingen (1983).

Heilbronn, H., Zeta-functions and L-functions, 204–230 in Cassels and Fröhlich (1967).

Hejhal, D., *The Selberg Trace Formula for* $PSL(2, \mathbb{R})$, Springer Lecture Notes in Mathematics **548** and **1001**, (1976) and (1983).

Henniart, G., La conjecture de Langlands locale pour $GL(n)$ sur un corp local, *Mem. Soc. Math. France* **11–12** (1984), 1–186.

Herstein, I., *Noncommutative Rings*, Carus Mathematical Monographs **15**, Mathematical Association of America and John Wiley, New York (1968).

Hewitt, E. and K. Ross, *Abstract Harmonic Analysis*, second edition, Springer Verlag, New York (1979).

Hilton, P. and S. Wylie, *Homology Theory*, Cambridge University Press, Cambridge (1960).

Hirzebruch, F., Hilbert modular surfaces, *L'Enseignement Math.*, **71** (1973), 183–281.

Howe, R., Theta series and invariant theory, 275–286 in part 1 of Borel and Casselman (1979).

Howe, R. with the collaboration of A. Moy, *Harish-Chandra Homomorphisms for* $\mathfrak{p}$*-adic groups*, CBMS Regional Conference Series **59**, AMS, Providence, R. I. (1985).

Humphreys, J., *Linear Algebraic Groups*, Springer Verlag, New York (1975).

Humphreys, J., *Arithmetic Groups*, Springer Lecture Notes in Mathematics **789** (1980).

Ihara, Y. Hecke Polynomials as congruence ζ functions in elliptic modular case, *Ann. of Math.* **85** (1967), 267–295.

Ikeda, T., On the location of poles of the triple L-functions, *Composition Math.* **83** (1992), 187–237.

Iwaniec, H., Non-holomorphic modular forms and their applications, in *Modular Forms*, R. Rankin, ed., Ellis Horwood Limited, Chichester (1984).

Iwaniec, H., *Introduction to the Spectral Theory of Automorphic Forms*, Revista Matemática Iberoamericano, Madrid (1995).

Jacquet, H., *Automorphic Forms on* $GL(2)$*, Part II*, Springer Lecture Notes in Mathematics **278** (1972).

Jacquet, H., Principal L-functions of the linear group, 63–86 in part 2 of Borel and Casselman (1979).

Jacquet, H. and R. Langlands, *Automorphic Forms on* $GL(2)$, Springer Lecture Notes in Mathematics **114** (1970).

Jacquet, H., I. Piatetski-Shapiro and J. Shalika, Automorphic forms on GL(3), I and II, Ann. Math. **109**, (1979).

Jacquet, H., I. Piatetski-Shapiro and J. Shalika, Relèvement cubique non normal, *C. R. Paris Sér. I Math.*, **193** (1981a), 13–18.

Jacquet, H., I. Piatetski-Shapiro and J. Shalika, Rankin-Selberg Convolutions, *Am. J. Math.*, **105** (1981b), 367–464.

Jacquet, H., and J. Shalika, A nonvanishing theorem for zeta functions of GL_n, *Invent. Math.* **38** (1976), 1–16.

Jacquet, H. and J. Shalika, On Euler products and the classification of automorphic representations, *Am. J. Math.* **103** (1981), 499-588 and 777-815.

Jacquet, H. and J. Shalika, Rankin-Selberg Convolutions: Archimedean Theory, 125–208 in part 1 of Gelbart, Howe and Sarnak (1990a).

Jacquet, H. and J. Shalika, Exterior square L-functions, in *Automorphic Forms, Shimura Varieties and L-functions II*, L. Clozel and J. Milne, ed., Academic Press, Boston (1990b).

Janssen, U., S. Kleiman and J.-P. Serre, *Proceedings of the Summer Research Conference on Motives, held at the University of Washington, Seattle, Washington, July 20–August 2, 1991*, AMS Proceedings of Symposia in Pure Mathematics **55** (1994).

Kato, S., On an explicit formula for class-1 Whittaker functions on split reductive groups over p-adic fields, preprint, University of Tokyo (1978).

Katznelson, Y., *An Introduction to Harmonic Analysis*, John Wiley, New York (1968). Second corrected edition published by Dover (1976).

Kazhdan, D., On lifting, in *Lie Group Representations II*, R. Herb, S. Kudla, R. Lipsman and J. Rosenberg, eds., Springer Lecture Notes in Mathematics **1041** (1983).

Kazhdan, D. and S. Patterson, *Metaplectic Forms*, Publ. Math. IHES **59** (1984).

Kelley, J., *General Topology*, Van Nostrand, Toronto (1955). Reissued by Springer Verlag, 1975.

Klein, F., *Lectures on the Icosahedron and the Solution of Equations of the Fifth Degree*, Kegan Paul, Trench, Trubner and Co., London (1913). Reprinted by Dover Publications, 1956.

Knapp, A., *Representation theory of semisimple groups, an overview based on examples*, Princeton University Press, Princeton (1986).

Knapp, A., Local Langlands correspondence: the Archimedean case, 393–510 in Janssen, Kleiman and Serre (1994).

Knapp, A. and D. Vogan, *Cohomological Induction and Unitary Representations*, Princeton University Press, Princeton (1995).

Kneser, M., Strong approximation, 187–196 in Borel and Mostow (1966).

Koch, H., Classification of the primitive representation of the Galois groups of local fields, *Invent. Math.* **40** (1977), 195–216.

Kudla, S., The local Langlands correspondence: the non-Archimedean case, 365–391 in Janssen, Kleiman and Serre (1994).

Kutzko, P., On the supercuspidal representations of $GL(2)$, *Amer. J. Math.* **100** 43–60 and 705–716 (1978)

Kutzko, P., The local Langlands conjecture for $GL(2)$ of a local field, *Ann. Math.* **112** (1980), 381–412.

Lang, S., *Algebraic Number Theory*, Addison Wesley, Reading, Mass. (1970). Reissued by Springer Verlag, 1986.

Lang, S., $SL(2, \mathbb{R})$, Addison Wesley (1975). Reissued by Springer Verlag, 1985.

Lang, S., *Modular Forms*, Springer Verlag (1976).

Lang, S., *Introduction to Algebraic and Abelian Functions*, second edition, Springer Verlag (1982).

Lang, S., *Algebra*, third edition, Addison Wesley, Reading, Mass. (1993).

Langlands, R., On Artin's L-functions, in *Complex Analysis, 1969*, H. Resnikoff and R. Wells, ed., Rice University Studies, **56** No. 2 (1970a).

Langlands, R., On the functional equations of Artin's L-functions, unpublished notes, (1970b).
Langlands, R., Problems in the theory of automorphic forms, in Springer Lecture Notes in Mathematics **170** (1970c).
Langlands, R., *Euler Products*, Yale Mathematical Monographs **1** (1971).
Langlands, R., *On the Functional Equations satisfied by Eisenstein Series*, Springer Lecture Notes in Mathematics **544** (1976).
Langlands, R., *Base change for $GL(2)$*, Princeton University Press, Princeton (1980).
Langlands, R., On the classification of irreducible representations of real algebraic groups, in *Representation theory and harmonic analysis on semisimple Lie groups*, P. Sally and D. Vogan, ed., AMS Mathematical Surveys and Monographs **31** (1989).
Laumon, G., M. Rapoport and U. Stuhler, $\mathcal{D}$-elliptic sheaves and the Langlands correspondence, *Invent. Math.* **113** (1993), 217–338.
Li, W., Newforms and functional equations, *Math. Ann.* **212** (1975), 285–315.
Luo, W., On the nonvanishing of Rankin-Selberg L-functions, *Duke Math. J.* **69** (1993), 441–425.
Luo, W., Z. Rudnick and P. Sarnak, On Selberg's eigenvalue conjecture, *Geometric and Functional Analysis* **5** (1994), 387–401.
Maass, H., Über eine neue Art von nichtanalytischen automorphen Funktionen und die Bestimmung Dirichletscher Reihen durch Funktionalgleichungen, *Math. Ann.* **121** (1949), 141–183.
Maass, H., Die Differentialgleichungen in der Theorie der elliptische Modulfunktionen, *Math. Ann.* **125** (1953), 235–263.
Macdonald, I., *Spherical functions on a group of p-adic type*, the Ramanujan Institute, Madras (1971).
Macdonald, I., *Symmetric functions and Hall polynomials*, Oxford University Press, Oxford (1979); second edition, 1995.
Mac Lane, S., *Categories for the Working Mathematician*, Springer Verlag, New York (1971).
Miyake, T., On automorphic forms on GL_2 and Hecke operators, *Ann. Math.* **94** (1971), 174–189.
Miyake, T., *Modular Forms*, Springer Verlag, New York (1989).
Moeglin, C. and J.-L. Waldspurger, Spectral decomposition and Eisenstein series, a paraphrase of the scripture, Cambridge University Press, Cambridge (1995).
Moeglin, C. and J.-L. Waldspurger, le spectre residuel de $GL(n)$, *Ann. Sci. Ecole Norm. Sup.* (4) **22** (1989), 605–674.
Montgomery, D. and L. Zippin, *Topological Transformation Groups*, Wiley-Interscience, New York (1955).
Mordell, L., On Mr. Ramanujan's empirical expansions of modular functions, *Proc. Cambridge Phil. Soc.* **19** (1917), 117–124.
Naganuma, H., On the coincidence of two Dirichlet series associated with cusp forms of Hecke's Nebentypus and Hilbert modular forms over a real quadratic field, *J. Math. Soc. Japan* **25** (1973), 547–555.
O'Meara, T., *Introduction to Quadratic Forms*, Springer Verlag, New York (1963). Corrected Version, (1973).
Patterson, S. and I. Piatetski-Shapiro, The symmetric square L-function attached to a cuspidal automorphic representation of $GL(3)$, *Math. Ann.* **283** (1989), 551–572.
Phillips, R. and P. Sarnak, On cusp forms for co-finite subgroups of $PSL(2, \mathbf{R})$, *Invent. Math.* **80** (1985), 339–364.
Piatetski-Shapiro, I., Euler subgroups, in *Lie Groups and their Representations*, John Wiley and Sons, New York (1975).

Piatetski-Shapiro, I., Multiplicity One Theorems, 209–212 in part 1 of Borel and Casselman (1979).

Piatetski-Shapiro, I., *Complex Representations of $GL(2, K)$ for Finite Fields K*, Contemporary Mathematics **16**, The American Mathematical Society, Providence, R. I. (1983).

Piatetski-Shapiro, I. and S. Rallis, Rankin triple L-functions, *Compositio Math.* **64** (1987), 31–115.

Ramanujan, S., On certain arithmetical functions, *Trans. Cambridge Phil. Soc.* **22** (1916), 159–184. In Ramanujan's *Collected Works*, Cambridge University Press (1927) and Chelsea (1967).

Randol, B., Small eigenvalues of the Laplacian operator on compact Riemann surfaces, *Bull. Amer. Math. Soc.* **80** (1974), 996–1000.

Rankin, R., Contributions to the theory of Ramanujan's function $\tau(n)$ and similar arithmetical functions, I and II, *Proc. Cambridge Phil. Soc.* **35** (1939), 351–356.

Riemann, G., Über die Anzahl der Primzahlen unter einer gegebenen Grösse, *Monatsberichte der Berliner Akademischie* (1859). In Riemann's *Gesammelte Mathematische Werke*, H. Weber (ed.), Teubner, Leipzig (1892), Reissued by Dover, 1953. A translation of Riemann's paper may be found in Edwards (1974).

Riesz, F. and B. Sz.-Nagy, *Functional Analysis*, trans. by Leo Boron from the French. F. Ungar, New York (1960). Reprinted by Dover (1990)

Roelcke, W., Über die Wellengleichung bei Grenzkreisgruppen erster Art, *S.-B. Heidelberger Akad. Wiss. Math.-Nat. Kl. 1953/1955* (1956), 159–267.

Roelcke, W., Automorphe Formen in der hyperbolische Ebene, I, *Math. Ann.* **167** (1966), 292–337.

Rudin, W., *Fourier Analysis on Groups*, Wiley-Interscience (1962).

Rudin, W., *Real and Complex Analysis*, second edition, McGraw-Hill, New York (1974).

Rudin, W., *Functional Analysis*, second edition, McGraw-Hill, New York (1991).

Saito, H., *Automorphic forms and algebraic extensions of number fields,* Lectures in Math., Kyoto Univ. Published by the Kiyokuniya Book Store, Tokyo (1975).

Satake, I., On the compactification of the Siegel space, *J. Indian Math. Soc.* **20** (1956), 259–281.

Satake, I., Theory of spherical functions on reductive algebraic groups over $\mathfrak{p}$-adic fields, *Publ. Math. IHES* **18** (1963), 1–69.

Satake, I., *Classification Theory of Semi-Simple Algebraic Groups*, Marcel Dekker, New York (1971).

Selberg, A., Bemerkungen über eine Dirichletsche Reihe, die mit der Theorie der Modulformer nahe verbunden ist, *Arch. Math. Naturvid.* **43** (1940), 47–50.

Selberg, A., Harmonic analysis and discontinuous groups in weakly symmetric Riemannian spaces, *J. Indian Math. Soc.* **20** (1956).

Selberg, A., On the estimation of Fourier coefficients of modular forms, in *Number Theory*, AMS Proceedings of Symposia in Pure Mathematics **8** (1965).

Serre, J.-P., *Cohomologie Galoisienne*, Springer Lecture Notes in Mathematics **5** (1965).

Serre, J.-P., Modular forms of weight one and Galois representations, in *Algebraic Number Fields*, A. Fröhlich ed., Academic Press, Boston (1977), 193–268. In vol. 3 of Serre's *Collected Papers.*

Serre, J.-P., *Local Fields*, translated by Martin Greenberg, Springer Verlag, New York (1979).

Shahidi, F., *On certain L-functions*, Amer. J. Math. **103** (1981), 297–355.

Shahidi, F., On the Ramanujan conjecture and the finiteness of poles for certain L-functions, *Annals of Math.* **127** (1988), 547–584.

Shahidi, F., Best estimates for Fourier coefficients of Maass forms, in *Automorphic Forms and Analytic Number Theory*, M. Ram Murty, ed., CRM (1990a), 135–141.

Shahidi, F., Automorphic L-functions: a survey, in Clozel and Milne, ed., *Automorphic Forms, Shimura Varieties and L-functions*, vol. 1, Academic Press, Boston (1990b), 415–437.

Shahidi, F., Symmetric power L-functions for $GL(2)$, in *Elliptic Curves and Related topics*, H. Kisilevsky and R. Murty, CRM Procedings and Lecture notes **4**, The American Mathematical Society, Providence, R. I. (1994), 159-182.

Shalika, J., The Multiplicity One Theorem on GL(n), *Ann. Math.* **100** (1974), 171–193.

Shalika, J. and S. Tanaka, On an explicit construction of a certain class of automorphic forms, *Amer. J. Math.* **91** (1969), 1049–1076.

Shimura, G., *Introduction to the Arithmetic Theory of Automorphic Functions*, Iwanami Shoten, Tokyo and Princeton University Press, Princeton (1971).

Shimura, G., On the holomorphy of certain Dirichlet series, *Proc. London Math. Soc.* (3) **31** (1975), 79–98.

Shintani, T., On an explicit formula for class-1 "Whittaker Functions" on $GL(n)$ over P-adic fields, *Proc. Japan Acad.* **52** (1976), 180–182.

Shintani, T., On liftings of holomorphic cusp forms, 97–110 in part 2 of Borel and Casselman (1979).

Siegel, C. L., *Topics in Complex Function Theory*, 3 volumes, Wiley-Interscience, New York (1969), (1971) and (1973). Reissued in 1989.

Siegel, C. L., *Lectures on Advanced Analytic Number Theory*, Notes by S. Raghavan, the Tata Institute, Bombay (1961).

Silverman, J., *The Arithmetic of Elliptic Curves*, Springer Verlag, New York (1986).

Soudry, D., A uniqueness theorem for representations of $GSO(6)$ and the strong multiplicity one theorem for generic representations of $GSp(4)$, *Israel J. Math.* **58** (1987), 257–287.

Soudry, D., Rankin-Selberg convolutions for $SO_{2l+1} \times GL_n$: local theory, *Memoirs of the AMS* **500** (1993).

Spanier, E., *Algebraic Topology*, McGraw Hill (1966). Reissued by Springer Verlag, 1981.

Springer, T., *Linear Algebraic Groups*, Birkhäuser (1981).

Stark, H., Values of L-functions at $s = 1$, Part I: L-functions for quadratic forms, *Advances in Math.* **7** (1971), 301–343; Part II: Artin L-functions with rational characters, *Advances in Math.* **17** (1975), 60–92; Part III: Totally real fields and Hilbert's seventh problem, *Advances in Math.* **22** (1976), 64–84, and Part IV: First derivatives at $s = 0$, *Advances in Math.* **35** (1980), 197–235.

Tate, J., Fourier analysis in number fields and Hecke's zeta-functions, Harvard Dissertation (1950). 305–347 in Cassels and Fröhlich, 1967.

Tate, J., *Number Theoretic Background*, 3–26 in part 2 of Borel and Casselman (1979).

Terras, A., *Harmonic Analysis on Symmetric Spaces and Applications, I and II*, Springer Verlag, New York (1985) and (1987).

Tits, J., Classification of algebraic semi-simple groups, 33–62 in Borel and Mostow (1966).

Tunnell, J., On the local Langlands conjecture for $GL(2)$, *Invent. Math.* **46** (1978), 179–200.

Tunnell, J., Report on the local Langlands conjecture for $GL(2)$, 135–138 in part 2 of Borel and Casselman (1979).

Tunnell, J., Artin's conjecture for representations of octahedral type, *Bull. Amer. Math. Soc.* **5** (1981), 173–175.

Van Der Geer, G., *Hilbert Modular Surfaces*, Springer Verlag (1987).

Vaserstein, L., The group SL_2 over Dedekind rings of arithmetic type, Mat. USSR Sbornik **18** (1972), 321–332.

Venkov, A., *Spectral Theory of Automorphic Functions*, Kluwer Academic Publishers Group, Dordrecht (1990).

Waldspurger, J.-L., Correspondance de Shimura, *J. Math. pures et appl.* **59** (1980), 1–133.

Waldspurger, J.-L., Démonstration d'une conjecture de Dualité de Howe dans le cas p-adic, $p \neq 2$, 267–327 in part 1 of Gelbart, Howe and Sarnak (1990).

Wallach, N., Asymptotic expansions of generalized matrix coefficients of representations of real reductive groups, in Springer Lecture Notes in Mathematics **1024** (1983).

Wallach, N., *Real Reductive Groups*, 2 volumes. Academic Press, Boston (1988).

Warner, G., *Harmonic Analysis on Semisimple Lie Groups*, 2 volumes. Springer Verlag (1972).

Washington, L., *Introduction to Cyclotomic Fields*, Springer Verlag (1982).

Weil, A., Sur les fonctions algébriques à corps de constants fini, *Comptes Rendus* **210** (1940). In Weil, Collected Works, Vol. 1, Springer Verlag, 592–594.

Weil, A., Numbers of solutions of equations in finite fields, *Bull. Am. Math. Soc.* **55** (1949). In Weil, Collected Works, Vol. 1, Springer Verlag, 497–508.

Weil, A., Sur la théorie de corps de classes, *J. Math. Soc. Japan* **3** (1951). In Weil, Collected Works, Vol. 1, Springer Verlag, 1–53.

Weil, A., Sur certains groupes d'opérateurs unitaires, *Acta Math.* **111** (1964). In Weil, Collected Works, Vol. 3, Springer Verlag, 143–211.

Weil, A., Über die Bestimmung Dirichletscher Reihen durch Funktionalgleichungen, *Math. Ann.* **168** (1967). In Weil Collected Works, vol. 3, Springer Verlag, 149–156.

Weil, A., *Basic Number Theory*, Springer Verlag, New York (1967). Second and third editions 1973 and 1974.

Weil, A., Exercices Dyadiques, *Invent. Math.* **27** (1974). In Weil, Collected Works, Vol. 3, Springer Verlag, 1–22.

Whittaker, E. and G. Watson, *A Course of Modern Analysis*, fourth edition, Cambridge University Press, Cambridge (1927).

Wiles, A., Modular Elliptic Curves and Fermat's Last Theorem, *Annals of Mathematics* **141** (1994), 443–551.

Zelevinsky, A., *Representations of the Finite Classical Groups*, Springer Lecture Notes in Mathematics **869** (1981).

Index